WIND TURBINE TECHNOLOGY

FUNDAMENTAL CONCEPTS OF WIND TURBINE ENGINEERING
SECOND EDITION

Frontispiece. The historic Brush wind turbine generator, which operated from 1888 to 1908 in Cleveland, Ohio, was the world's first wind power plant. It powered the residence and laboratory of Charles F. Brush, an inventor and manufacturer of electrical equipment. Dr. Brush successfully integrated the highly-developed technology of 19th-century windmills with the emerging technologies of electric power and feedback controls. (Specifications: rotor diameter = 17 m; rated power = 12 kW DC; tower height = 18.3 m; tail vane area = $112\,m^2$; yaw control by tower rotation; side vane for automatic horizontal furling in high winds.)

WIND TURBINE TECHNOLOGY

FUNDAMENTAL CONCEPTS OF WIND TURBINE ENGINEERING
SECOND EDITION

Editor

DAVID A. SPERA, PH.D.

© 2009 by ASME, Three Park Avenue, New York, NY 10016, USA (www.asme.org)

ISBN: 978-0-7918-0260-1

The Library of Congress has cataloged the first edition as follows:

Library of Congress Cataloging-in-Publication Number 94-11137

"We are Newtonians, fervent and devout, when we speak of forces and masses, of action and reaction; ....and when we stretch out an arm and feel the force of gravity all around, pulling earthward. Newton's laws are our laws. What Newton learned remains the essence of what we know, as if by our own intuition."

James Gleick, 2003, *Issac Newton*, Pantheon Books, New York

Preface to the First Edition

Wind Turbine Technology is designed to be both a text book and a reference volume for all those interested in the field of wind energy and in modern wind turbines. This book is written for educators and students as well as for practicing engineers, designers, meteorologists, researchers, utility project managers and planners, wind power plant developers, and manufacturers of wind turbine equipment. Its objective is to document the significant technical advances in wind power engineering that have taken place during the past 20 years, and to do so in a format that emphasizes fundamental principles.

Expert authors in the fields of aerodynamics, structural dynamics and fatigue, wind characteristics, acoustic and electromagnetic emissions, commercial wind power applications, and utility power systems have contributed to this book. They have drawn from many published papers in their specialities and from their own research and experience to inform readers of the lessons they have learned about wind turbines. Even the historical chapters in this book, which deal in depth with the origin and development of windmills and the evolution of modern wind turbines, emphasize the common theme of the book: What technical choices have been made by inventors, designers, and builders of wind machines, and what lessons can we learn from those choices?

Where possible, mathematical models are developed from basic principles, and many numerical examples and case studies are described. Some chapters are application-oriented, with numerous sketches and descriptions of past and present wind turbine configurations, discussions of economic and environmental considerations, and the integration of wind power plants into electric utility systems. References and bibliographies guide the reader to additional information in the literature.

Wind Turbine Technology follows in the tradition of Putnam's *Power From The Wind* and Golding's *The Generation of Electricity by Wind Power* in its treatment of modern wind turbines as complete engineering systems. It is an authoritative survey of the wind energy field, including design concepts and philosophies, research and demonstration projects, technical analysis and mathematical models, and application economics and integration. An important aspect of this book is that the various chapters have been unified by common terminology, nomenclature, and graphic styles as well as significant amounts of cross-referencing. This is done so that the reader may have as integrated a view as possible of wind power engineering, while benefitting from the personal experiences of a variety of experts in the field.

David A. Spera, Ph.D.
Editor
1994

First Edition Acknowledgments

There were many contributors to the different phases of this book, and the editor would like to acknowledge and thank all those listed below. First are the sponsors of the work and the project board, without whom this book would not have been possible. The backbone of a project such as this is, of course, its writers (who are listed on page iii) as well as the technical reviewers who kept a quality check on the timeliness and accuracy of the information included. The editor would also like to acknowledge the contribution of the graphics designers for their diligence and technical know-how.

Wind Turbine Technology Primary Sponsor
U.S. Department of Energy, Office of Renewable Technologies

Co-Sponsors
National Aeronautics and Space Administration, Lewis Research Center
National Renewable Energy Laboratory, Midwest Research Institute
American Society of Mechanical Engineers

Project Board
Daniel F. Ancona, Chairman, *U.S. Department of Energy*
Darrell F. Baldwin and Marvin H. Hirschberg, *NASA Lewis Research Center*
Linda Brown, *National Renewable Energy Laboratory*

Editorial Review Committee
A. Craig Hansen, Ph.D., *University of Utah*, Principal Technical Reviewer
Amir Mikhail, Ph.D., *Zond Systems, Inc.*, Principal Technical Reviewer
Dale E. Berg, Ph.D., *Sandia National Laboratories*
Darrell M. Dodge, *National Renewable Energy Laboratory*
William R. Johnson, *NASA Lewis Research Center*
Erik Nelson, *National Renewable Energy Laboratory*
Paula Pitchford, *National Renewable Energy Laboratory*
James L. Tangler, Ph.D., *National Renewable Energy Laboratory*
R. J. Templin, *National Research Council of Canada*
Larry Wendell, Ph.D., *Battelle Pacific Northwest Laboratory*

Graphics Designers
Caren J. McMillen, Managing Artist
Evie Frenchik
Mary Jane Barlak
Sverdrup Technologies, Inc.

Preface to the Second Edition

The wind energy industry has experienced rapid growth since the first edition of *Wind Turbine Technology* was published in 1994. According to the American Wind Energy Association (AWEA), wind power generating capacity in the U.S. is projected to be approximately 25,000 megawatts (MW) in 2008, an 11-fold increase over the nation's 2,000 MW capacity in 1995. Wind power projects accounted for over 30 percent of all new power generating capacity added in the U.S. in 2007.

During this period of exponential growth, *Wind Turbine Technology* has served as a reliable text book, reference volume, and guide to the ever-increasing numbers of persons in the U.S. and abroad who are contributing to the design and development of modern wind turbines and wind power stations.

Many wind turbine technology projects have also been completed successfully in the past decade, in parallel with the growth of the wind power industry. New technology has emerged from these, based on lessons learned in the field.

During the past few years it became clear to the American Society of Mechanical Engineers and ASME Press, publishers of *Wind Turbine Technology,* that the release of an updated edition of the book at this time was very desirable. Original authors and potential new authors were contacted, and many agreed to participate in the project. The result of their work is *Wind Turbine Technology, Second Edition.*

Because the format of the first edition emphasized fundamental principles, it was possible to preserve almost all of the first edition and add the updated material to it. As a result, none of the basic theoretical contributions by expert authors in the first edition has been lost. New material has been integrated into the existing chapters, preserving the common terminology and nomenclature that are important characteristics of the first edition.

David A. Spera, Ph.D.
Editor
2009

Second Edition Acknowledgements

Updating of the first edition of *Wind Turbine Technology* is possible only with the timely contributions from expert authors and the professional management by the staff at ASME Press. The editor would like to give special thanks to Ms. Tara Smith at ASME Press, who managed this book project so well; to Dr. Robert W. Thresher, recently Director of the Wind Technology Center at the National Renewable Energy Laboratory, who organized extensive research and writing efforts; to Dr. Larry A. Viterna at the NASA-Glenn Research Center, for his many contacts with wind turbine manufacturers that helped make this book relevant to today's wind industry; and to Ms. Tami Sandberg, NREL Librarian, who researched and compiled the long bibliographies so valuable to our readers.

Finally, many thanks are extended to all the wind turbine manufacturers who supplied photographs for our new gallery of color pictures that convey not only the technical sophistication, strength, and power of their products, but also the grace and beauty of modern wind turbines.

Contents

1

Historical Development of the Windmill

Dennis G. Shepherd
Professor of Mechanical and Aerospace Engineering
Cornell University
Ithaca, New York

Introduction

The wind turbine has had a singular history among prime movers. Its genesis is lost in antiquity, but its existence as a provider of useful mechanical power for the last thousand years has been authoritatively established. Although there are a few earlier mentions in the literature, these are generally not acceptable for recognition as historical fact by most professional historians of technology. The windmill, which once flourished along with the water wheel as one of the two prime movers based on the kinetic energy of natural sources, reached its apogee of utility in the seventeenth and eighteenth centuries. Its use then began to decline, as prime movers based on thermal energy from the combustion of fuel took precedence. Steam engines, steam turbines, and oil and gas engines provided more powerful and more compact machines, adaptable to a multitude of uses other than just the grinding of grain and the pumping of water. These new heat engines also were continuously available rather than subject to the vagaries of nature, and they could be located at the job site rather than requiring that the job be brought to them.

Nevertheless, the windmill persisted through the industrial revolution; it even continued to supply essential service into the twentieth century in sparsely populated areas where relatively small amounts of scattered power were required, and constant availability was not essential. This was particularly true in the United States and in other countries having vast land masses, such as the former USSR, Australia, and the Argentine. In the industrialized regions of the world, where greater population densities and increasing manufacturing enterprises were the rule, the windmill was becoming moribund by the end of the first third of the twentieth century.

Wind machines, however, were revived by the emergence and proliferation of two major technologies: the rapid spread of electricity as a versatile transducer of energy between the prime mover and the job, and the burgeoning of the engineering science of aerodynamics, which was occasioned by the development of the airplane. At the end of the last century, in the years from 1888 to 1900, experiments began in which windmills were used to generate electricity, both in the United States and in Denmark, which has no fossil fuels of its own.

The history of wind power in Denmark, provides a salutary example of how socio-economic conditions can both inhibit a technology's gradual but continuous development and rescue it from an apparently endless decline, when politics or economics intervene. From 1900 to 1910, many Danish wind-driven electricity plants were in use, particularly for agriculture, but diesel engines were beginning to give them some stiff competition by 1910 to 1914 because of their convenience and economy [Juul 1956, 1964]. However, during World War I, oil supplies to the country were virtually shut off and wind power was resurrected; many 20- to 35-kW plants were built at that time. After the war, electrification took place throughout the country, and once more the windmill languished. In 1939, World War II caused another cut-off in fuel supplies, and wind power was once more called upon to the fullest extent possible. After the hostilities ended, further rapid, extensive electrification took place, but this time the utility of wind power was not discounted; instead, a research program was begun to consider it as a supplement to the large, central plants.

These compelling reminders of the perils of relying on transported energy sources, together with a degree of awareness of the eventual dissipation of these sources, prompted some of the development of windmills into wind turbines. In the period after 1945, there were developments not only in Denmark, but in France, Germany, and Britain as well. These developments might have been stimulated by the 1.25-MW *Smith-Putnam turbine* in the United States, which operated intermittently from 1941 to 1945. All these ventures enjoyed some degree of technical success, but that was not sustained, because they were not considered to be cost-effective. However, they did provide a useful starting point for the renaissance of wind power in the early 1970s, which was provoked by the international oil crisis of 1972.

In contrast to the uneven history of the windmill, the water mill has enjoyed 2,000 years of slow but continuous development up to this very day. Built in multi-megawatt sizes, it achieves the highest efficiencies of any prime mover. The water mill has changed, in its general physical characteristics, considerably over the centuries; the windmill, but little (in detail, yes, but not in form). In fact, if a European miller of the thirteenth century were given the opportunity to observe the operation floor of a modern hydroelectric plant, he would be utterly confounded. On being taken to see a wind power station in California, however, he might immediately express his delight in seeing so many windmills in use, although he might also inquire as to why so many of them had lost one or two sails.

This brief review might serve to place the evolution of the windmill in its general context. Now, however, we shall go back in time to its origins and its early development, exploring its mechanical design and some problems that were solved by early engineers and some that were not. But first, it is desirable to establish an important feature of nomenclature, to avoid confusion in discussing the literature of the past and of the present.

Notice that we have used both the terms *windmill* and *wind turbine* in the preceding overview. Modern technology has firmly and rightly established the *wind turbine* as the prime mover of a wind machine capable of being harnessed for a number of different applications, none of which are concerned with the milling of grain or other substances (at least in industrialized countries), together with the various other pieces of apparatus necessary for a complete power plant: *mechanical transmission, nacelle, tower, load (e.g. generator or pump), control gear*, and so forth. The wind turbine is also described as a *wind energy conversion system (WECS)* or, if used to produce electric power, as a *wind turbine generator (WTG)*.

However, up to recent times the term *windmill* was used for the whole system, whatever its duty, be it generating electricity, pumping water, sawing wood, or, as we shall see a little later, pumping air for an organ. It seems that in nearly all previous writings, *windmill* is used, whether it is directed toward technological, historical, or simply antiquarian aspects (or combinations of them). Since this chapter is concerned with past events, it is convenient and has a certain logic in it to retain this term in its historic sense, using such words as *wheel, rotor*, or *blades* to refer to what we now call the *turbine* itself.

The adjectives *horizontal* or *vertical* attached to the two major classes of wind machine are a potential source of misunderstanding. In modern terminology, they refer to the geometrical aspect of the driving shaft on which the rotor (turbine wheel) is mounted. For example, an oldtime wooden machine with four sails (a type typified by the term *Dutch windmill*) is now called a *horizontal-axis wind turbine (HAWT)*. Past usage terms it a *vertical windmill*, because the path of a point on a moving blade lies in a vertical plane. A machine built like a carousel with a central vertical axis and a number of straight or bent vanes arranged in a direction more or less parallel to the shaft is today called a *vertical-axis wind turbine (VAWT)* but in past times a *horizontal windmill*. This typology is emphasized here because the new and old usages are diametrically opposed. To avoid confusion, the modern terms *horizontal-axis windmill* and *vertical-axis windmill* will be used here, although molinologists continue to study vertical and horizontal windmills.

The International Molinological Society (TIMS) was formed following a meeting in Portugal in 1965 of savants and others interested in windmills. The word *molinology* was introduced and officially adopted at that time to define the field of studies concerned with windmills, water mills, and animal-powered mills. The term is derived from the root *molino* found in differing national forms in many countries of the Western Hemisphere; it can be combined with such Greek suffixes as *-ology, -graph, -phile*, etc. It was proposed to the philological authorities of every country for adoption into their vocabularies and dictionaries. Since the first gathering in 1965, meetings of TIMS have been held in 1969 (Denmark), 1973 (The Netherlands), 1977 (England), 1982 (France), 1985 (Belgium), 1989 (W. Germany), 1993 (Wales), 1997 (Hungary), 2000 (USA), 2004 (Portugal), and 2007 (Netherlands). The published transactions of these symposia are a most interesting and valuable source of current matters connected with the history, cataloging, and preservation of windmills.

This chapter is aimed toward engineers and technology-minded readers and not toward molinological scholars. Hence, the author has tried to steer a middle course with respect to quoted sources by endeavoring to give proper credit for specific material without overloading the text with references of interest only to the history specialist. In addition, he has leaned heavily on many of these sources for the general information as well as the particular material contained in them, and he acknowledges this debt with gratitude.

Ancient Times

The earliest mentions of the use of wind power come from the East: India, Tibet, Afghanistan, Persia. Ancient manuscripts, however, have often suffered from mistranslations, revisions, and interpolations by other hands over the centuries. In some, even diagrams were changed to suit the whims of revisionists, and there are instances of forgeries. Drachmann [1961], Needham [1965], Vowles [1930], and White [1962] all cite examples of these aberrations. Marie Boas provides a good illustration of the treatment a manuscript can undergo in her detailed monograph, "Hero's Pneumatica - A Study of its Transmission and Influence" [1949].

Mentioning the Boas monograph is apposite here because of the well-known ascription of the invention of the windmill to Heron (a variant of *Hero*) of Alexandria, by virtue of his account of it as one of the many devices in his *Pneumatica* of 2,000 years ago. This ascription is now discounted by most authorities in varying degrees, ranging from outright rejection, through wistful reluctance to relinquish the idea, to acceptance as only a toy. There is difficulty with respect to the provenance of a sketch in the *Pneumatica* and some disagreement as to the exact meaning of certain key words. This story is reviewed here because it is a classic example of the difficulty of making a positive attribution from an ancient manuscript. Was Heron really the inventor of the windmill as a practical prime mover, and was his invention the inspiration for those that followed, even though centuries elapsed between the birth of the idea and its fulfillment?

The exact dates of Heron's birth and death are not known. Surmises lie between the second century B.C. and the third century A.D.; some time in the first century A.D. is perhaps the most probable. His *Pneumatica* (best known to many as the source of the reaction steam turbine) consists of descriptions of various ingenious apparatuses that operate on the basis of air or water; some of them are what we would now call toys or even "magic" devices. He himself said that he added some of his own inventions, but he did not say which ones they were. The book was known and referred to in medieval times, but many transcripts and translations into Latin or Greek have been lost either in whole or in part. An English translation was provided by Bennett Woodcraft [1851] and a German one by Wilhelm Schmidt [1899]. The latter contains the original Greek wording side by side with the German, and it is generally accepted as a standard text.

The opening sentence in the relevant chapter in the *Pneumatica* is given by Woodcroft as "the construction of an organ from which when the wind blows the sound of a flute shall be produced." Schmidt provides essentially the same translation in his German version. Both contain diagrams [Figures 1-1(a) and (b)] showing a shaft with blades at one end and four pegs at the other, the pegs intermittently striking a lever rod which then lifts a piston contained in a cylinder. Between lifts, the piston falls in the cylinder of its own weight, resulting in air being pumped to a musical organ. Although both drawings are based on the description in the text, each suits the translator's own imagination: Woodcroft presents a horizontal-axis rotor having four sails, a type unknown until the twelfth century, and Schmidt shows a water-mill type of rotor, again from a much later era. But Schmidt does discuss in his introduction a much cruder version of the rotor illustrated in Figure 1-1(c). According to Drachmann [1961], who has made a detailed reassessment in recent years, this is as close to the original sketch as we are likely to get. Vowles also discusses the *Pneumatica* puzzle [1930] and shows four examples of transmogrified images from various later manuscripts that help to confound the confusion.

In addition to the difficulty we have with the drawings of Heron's device, the exact meaning of some of the words leaves us in doubt. Two prime examples are the word *anemurion*, meaning a windmill or only a weathervane, and whether Heron uses the word

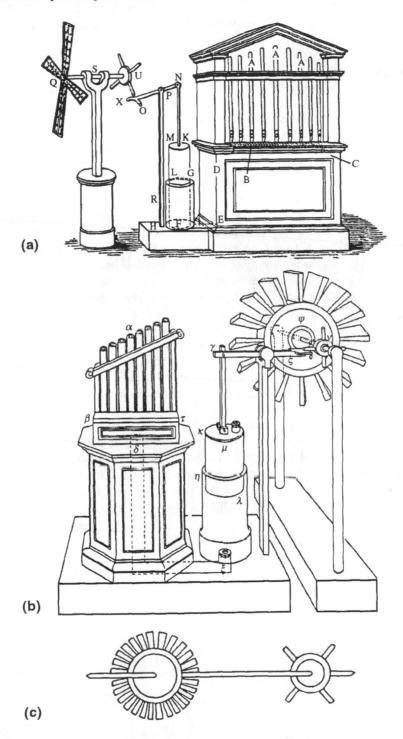

(a)

(b)

(c)

Figure 1-1. Some concepts of the windmill-like device, or organ, described by Heron of Alexandria in his *Pneumatica*, possibly in the first century A.D. (a) A Dutch-style rotor driving a disk with four pegs that repeatedly raise the piston in an air pump [Woodcroft 1851]. (b) The device with a water-mill type of rotor [Schmidt 1899]. (c) Heron's sketch of the drive, according to Schmidt [1899]. Note the needle bearing ends of the shaft.

organon to mean a musical organ or just as a generic term, as we might speak of an organ of the body. There are design problems as well, such as turning the device into the wind (stated as being possible in the text) and the peg-driven tripping and return motion (in lieu of a crank) requiring a very rapid oscillating movement of the piston. Neither of these operations seems to be possible except in a very small model that could be moved by hand, and one having a very light piston. So perhaps the device was meant to be a toy and not an invention to be taken seriously as a useful working machine.

If the inventor of the windmill was not Heron of Alexandria, then who, and where? This we cannot answer, nor can we think it reasonable that we ever will. Perhaps there was no one person who ever left such a clear record as Heron's, however suspicious we are of his particular account. Instead, perhaps a number of small steps, taken by trial and error in a number of different places over the years, eventually were diffused to yield a working machine with no specific birth to record. Vowles points out at some length that, despite the lack of any direct evidence, the windmill might have been known in Graeco-Roman times by other than "the not too clear example from Heron's work" [1930]. Thus, we are not justified in rejecting the possibility of its invention in that era.

The first part of the thousand-year medieval period, from the end of the Roman Empire to the Renaissance, is called the Dark Ages, although perhaps they were not quite so dark as they were once thought to be. The Greek culture had spread to many places beyond the mainland (including Alexandria, of course), and schools of learning were established throughout the Near and Middle East, Palestine, Syria, Anatolia, Mesopotamia, and Persia, to call them by their older names. This region became part of Islam by the Mohammedan conquest in the seventh century, and so the Greek heritage became available to the whole of Islam. Many Greek texts were translated into Arabic and thus had a much wider, and less academic, audience. However, although these are known to have included the *Pneumatica*, the time of its appearance is problematical and may have been as early as the ninth century [Vowles 1930]. Thus Heron's work might have stimulated the use of wind power in the Islamic world, but there is no hard evidence to substantiate that. Nearly all the stories and the records we have from between the first and the twelfth centuries come from the Near East and Central Asia, and so those regions of the world are generally considered to be the birthplace of the windmill.

The Vertical-Axis Persian Windmill

There have been suggestions that the Tibetan prayer wheel was the inspiration for the windmill, but although this is a possibility, the provenance of the prayer wheel itself is very doubtful. The next chronological reference to the windmill that we have refers to the seventh century, but the text itself appeared at least 200 years later. The story was well known and was repeated by several writers in more than one version. The first mention is by Alī al-Tabarī (A.D. 834-927).[1] According to Needham [1965], the Caliph Omar was murdered in the Islamic capital of Medina in A.D. 644 by a captured Persian technician/slave, Abū Lu'lu'a, who was bitter about the high taxes; he also claimed to be able to build mills driven by the wind. This story was repeated by the geographer Alī al-Mas'ūdī (c. A.D. 956), apparently in a slightly variant form in which the Caliph asked the Craftsman if he had boasted about being able to build a mill driven by the wind, to which he received the reply, "By God, I will build this mill of which the World will talk" [Wulff 1966]. Because it was orally transmitted over many years, and there is no record of it by any writer around the time of Omar's death, the incident can be used only to lend some plausibility to the possible existence of the windmill in the Islamic world in the middle of the seventh century.

First Record of a Windmill

We now come to the first accepted establishment of the use of windmills; this was in the tenth century in Persia. It is documented by reliable writers and reinforced by ancient drawings that correspond to the remains of old mills and to modern ones still in use. The region is Sīstān in eastern Persia (in present-day Iran), which bordered on Afghanistan; al-Mas'ūdī describes it as

"...the land of winds and sand. There the wind drives mills and raises water from the streams, whereby gardens are irrigated. There is in the world, and God alone knows it, no place where more frequent use is made of the winds."

Al-Istakhrī, also of about A.D. 950, provides a similar description as well as an interesting account of how the inhabitants coped with a shifting dune. They enclosed it in a high fence with a door in a lower part that allowed the wind to blow the sand away from the fence, which adds credibility to their skill in mastering the windmill. The wind of Sīstān is famous, or rather notorious, as it is said to reach 45 m/s (100 mph) and to blow at gale force with little respite for four months in late spring and early summer [Bellew 1874; Hedin 1910].

Some three hundred years later we have confirmation from the geographer al-Qazwīnī (A.D. 1203-1283), who wrote

"There the wind is never still, so in reliance on it mills are erected; they do all their corn grinding with these mills, it is a hot land and has mills which depend on the utilization of the wind" [Wulff 1966].

---

[1] Authors differ frequently as to spelling and diacritical marks in Arabian names. Here, Needham's usage has been followed for consistency, and because he introduces all such names given in this text.

About A.D. 1300, we have the report of al-Dimashqī (A.D. 1256-1326), a Syrian cosmographer who provided a detailed description and a drawing, shown as Figure 1-2. The two-storied, walled structure had millstones at the top and a rotor at the bottom, the latter consisting of a spoked reel with 6 to 12 upright ribs, each covered with cloth to form separate sails like longitudinal fins on a heat-exchanger tube. The sketch shows the bellying of the cloth coverings as they catch the wind and push the reel around. Each wall had an offset opening the height of the rotor, with its perimeter beveled to decrease the free area through the thickness of the wall and hence accelerate the wind from any direction.

The Windmills of Neh

Windmills of this basic type were still in use in Sīstān in 1963, when Wulff saw 50 of them operating in the town of Neh; they might still be used today. They throw much light on what al-Dimashqī saw, so we will skip a few centuries in this generally chronological presentation and condense Wulff's description of the mills in Neh.

The construction is shown in Figure l-3(a). The rotor is about 5.5 m high and 4.3 m in diameter, enclosed in side walls about 6.5 m high and a half-wall that leaves a 2.2-m-wide opening facing north, the main wind direction. The central wooden shaft is about 43 cm in diameter. Its bottom end, which extends downward into the mill room, contains a

Figure 1-2. A Persian vertical-axis windmill in Sīstān, according to al-Dimashqī, c. A.D. 1300 [Wulff 1966]. The earliest windmill design on record. Grinding stones are above the rotor with its bellying cloth sails. The walls have openings to let the wind in and out. *(Reprinted by permission of MIT Press; ©1966, MIT Press)*

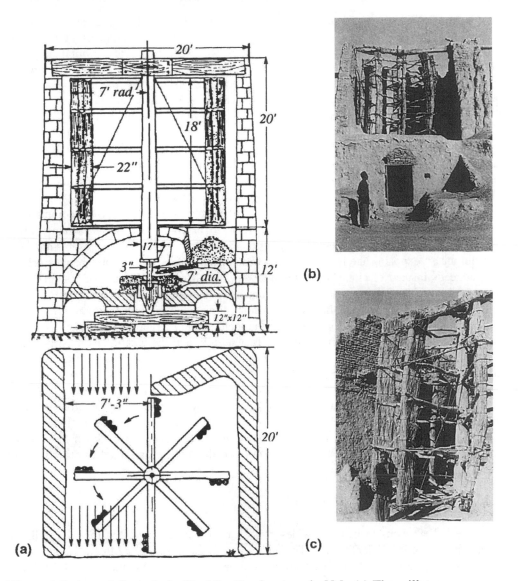

Figure 1-3. An existing windmill of the Persian type in Neh. (a) The millstones are now below the rotor and the sails are bundles of reeds [Wulff 1966]. (b) a general view of the downwind (south) wall of the mill. (c) Close-up view of the reed sails. *(Reprinted by permission of MIT Press; ©1966, MIT Press)*

steel thrust pin that rests on a wooden thrust block in a cavity lined with a tallow-soaked lubrication pad of many layers of cloth. The upper millstone is attached to this steel pin, and the lower stone rests on a brick structure. These stones are about 2 m in diameter.

The rotor shaft supports five tiers of spokes, with eight spokes in each tier. Each sail is composed of several long bundles of reeds pressed against the end of the spokes by tie bars and is about 56 cm wide and 15 cm thick. The tie bars are secured to the spokes by pegs and ropes, and there are several sets of stay ropes to tie the whole rotor assembly together. The system is certainly elastic, as we see in Figures 1-3(b) and (c), and it seems as if it might need frequent adjustment so the millstones can operate with proper clearance. The thrust block with its "tallow-soaked lubrication pad" also does not seem to make a precise bearing.

Now Wulff gives some figures relating to the performance of these mills. When he visited Neh in 1963, 50 mills were still operating. Each milled an average of a ton of grain in 24 hours, so in a 120-day windy season, they were said to produce a total output of 6,000 tons. Based on a wind speed of about 30 m/s, an effective exposure at any time of 1.5 blades, and a mill efficiency of 50 percent, Wulff estimates a power output of about 75 hp per mill. It is difficult to accept this high figure. A power calculation based on aerodynamic drag in accordance with Equation (5-2) would yield a maximum 22 hp in a 30 m/s wind for the same assumptions.

The basic design of these primitive vertical-axis mills has lasted at least 1,000 years, although a major change has come about in that the machine has been inverted, placing the sails above the millstones, as we see in comparing Figures 1-2 and 1-3. Wulff suggests that the earlier concept may simply have been taken from the ancient Norse or Greek water mill, in which the mill itself had to be placed over the water wheel [1966]. No comment on when the change might have been made seems to be available, but it certainly made operation much more convenient. Elevating the driver to a more open exposure improved the output by exposing the rotor to higher wind speeds. Other noticeable changes are the use of reeds instead of cloth to provide the working surface and the use of a single entry port in place of the four described by al-Dimashqī, although this may have been limited to the region around Neh, where the summer wind is almost constantly from the north.

It would seem, then, that we can take the tenth century as the earliest known date for acceptable documentation of the vertical-axis windmill, and the location as most probably West or Central Asia. Forbes apparently goes as far as to take the birthplace to be Sīstān and to place the invention in early Muslim or even pre-Muslim times. If we take the founding of Islam to be in the first part of the seventh century, this is a good deal earlier than most science historians would consider to be proven. Forbes also asserts that, after having been first confined to Persia and Afghanistan, the invention subsequently spread in the twelfth century throughout Islam and beyond to the Far East. On the other hand, Lynn White states that there is no evidence that mills of this type ever spread to other parts of the Islamic world [1962].

Chinese Vertical-Axis Windmills

The belief seems to be quite widespread that the Chinese invented the windmill and have been using it for 2,000 years. This might well be so, considering that they developed so many engineering artifacts, but there is little or no evidence that the windmill was one of them. The eminent scholar Joseph Needham, whose monumental work in many volumes, *Science and Civilization in China* [1965], is the recognized classic text in the field, states that the earliest really important reference dates back to 1219. There is a report of a visit in that year to Samarkand by a celebrated Chinese statesman and patron of astronomy and engineering, Yehlü Chhu-Tshai, who in a poem wrote that stored wheat was milled by the rushing wind and that the inhabitants used windmills just as the people of the south used water mills. Later Chinese references to the windmill again all point to its transmission from lands adjacent to western China as being the most likely supposition, and that it was carried there by sailors or merchants from Central and Southwest Asia. Needham points out further that the references suggest that the introduction of the windmill took place no earlier, because it never before received a specific character or specific wording; it might have been confused with the rotary winnowing fan, however, which is much older.

The first European to report windmills in China was Jan Nieuhoff in 1656. These mills had a distinctive form, with eight junk slat-sails mounted on masts around a vertical axis and disposed so that they could be positioned automatically. Figure 1-4(a) shows the arrangement of the masts (for clarity, only one sail is rigged), supported from the rotor shaft by radial cross-arms. The power take-off is through a right-angle drive composed of gears with pin teeth. Figure 1-4(b) illustrates the action of the sails as the rotor turns. Each sail is mounted asymmetrically on its mast and held against the wind by a rope (positions *G, H, A,* and *B*) until it reaches a point in its rotation where it *jibes*, reversing its orientation and swinging outward (*C* and *D*), coming thence into the eye of the wind or *luffing*, which gives little resistance to shaft rotation (*E* and *F*). Thus, it does not require shielding walls like the Sīstān models and can utilize wind from any quarter. This type of mill is still used in eastern China.

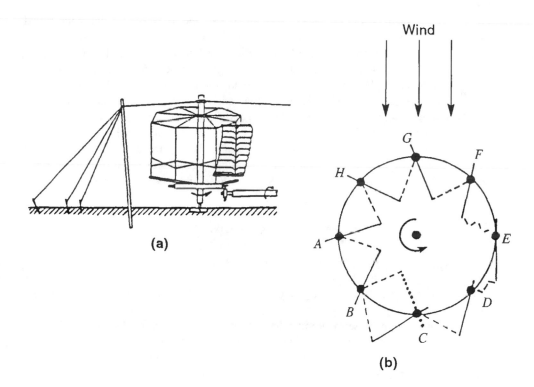

Figure 1-4. The vertical-axis Chinese windmill, with eight junk slat-sails [Needham 1965]. (a) General arrangement of sails, masts, main shaft, and power take-off. (b) Automatic positioning of sails during rotation. At *E* and *F*, sails are luffing, presenting little resistance to the wind. (*Reprinted by permission of Cambridge University Press; from Chhen Li, Khao Hsung Thung Pao, Vol. 2, No. 3, 1951, p. 266*)

The real mystery, however, lies in the fact that the vertical-axis Persian windmill never came into use in Northwest Europe. At the end of the twelfth century, there was an efflorescence of a completely different type, the *horizontal-axis windmill*. This development presents a second enigma in the technical development of the wind turbine that occurred some thousand years after the enigma left by Heron's device. But this is a story in itself, and it requires a separate presentation.

The Horizontal-Axis European Windmill

Northwest Europe, particularly France, Germany, Great Britain, Iberia, and the Low Countries, eventually became the region that developed the most effective type of windmill, one in which the shaft carrying the sails was oriented horizontally rather than vertically as in the Persian mill. In a relatively short time, tens of thousands of what we will call *horizontal-axis European windmills* were in use for a variety of duties. The familiar cruciform pattern of their sails prevailed for almost 800 years, from the twelfth to the twentieth century.

The Domesday Book registers between 5,000 and 6,000 mills in England by A.D. 1086, but without distinguishing among kinds of mills, whether hand, animal, water, or wind, so the last of these cannot be assumed to have been in use. For many years, the first windmill in the West was believed to have been located near Bury St. Edmunds in Suffolk, England, because of its unimpeachable provenance from the famous *Chronicle of Jocelin of Brakelond*, which dealt with the affairs of the abbey, and because of the widespread knowledge of that text from Thomas Carlyle's "Past and Present," which was based on the *Chronicle*. The story told there is that of Dean Herbert, who in 1191 built a windmill on glebe land. *Glebe* signifies land attached to an ecclesiastical benefice controlled by the lord of the manor, in this case the Abbot Samson who made the Dean tear the mill down.

Diligent and patient scholarship, however, has uncovered the unqualified existence of other mills of about the same period, *c.* 1180 to 1190. There are some half-dozen of these: two in France, three in Britain, and one in Syria. There is some disagreement or doubt as the exact year of some; hence, the hesitancy about the dating. Including the Syrian mill may appear to be irregular in an account of windmill development in Northwest Europe, but the following verse will explain this. White [1962] gives us an eyewitness account (from 1190) of the Third Crusade, which states

> "The German soldiers used their skill
> to build the very first windmill
> that Syria had ever known"

Apart from its early date, this verse is important for another reason, which we will discuss shortly.

The provenances of these windmills have been the few acceptable ones until quite recently, when two monographs on medieval mills in England have appeared. That of Kealey [1987] explores the social context of a technological revolution, and his information comes mainly from the financial records of landowners of that time. Technical details are almost non-existent for the period, but he does register mill locations, first datable appearances, sponsors, owners, and lessees. He asserts evidence of some 56 horizontal-axis windmills, all before the year 1200, including those he dates earlier than 1185 and one as far back as 1137. A second monograph by Holt [1988], who has made a very considerable contribution pertaining to medieval windmills in England, criticizes many of Kealey's evidences as unsound. Holt says there are only 23 English windmills firmly dated before the year 1200, nine of them newly identified by Kealey. Of these 23, three are firmly dated in the 1180's (none earlier than about 1185), 15 in the 1190's, and the remaining five before 1200. With this number of mills, it seems reasonable to regard England as the origin of the horizontal-axis windmill, rather than the Middle or Far East as is thought by some.

At the end of the following century, windmills were becoming common in Northwest Europe, but little or no penetration is known elsewhere; in the fourteenth century, they were a major source of power. In spite of the vagaries of the wind, the many appropriate sites for the mills allowed much more flexibility of application than did water mills. Numbers in use are not known for these early times: only very approximate figures can be given for later eras. From the figures of Jannis Notebaart we can obtain order-of-magnitude estimates of possible maximum numbers at any time in the eight centuries from their introduction to the heyday of the multibladed windmill in the United States: 3,000 in Belgium, 10,000 in England, 650 in Anjou in France, and 9,000 in Holland [1972].

Changing the Axis

So we have what seems to have been the sudden eruption and very rapid spread of these windmills, which requires some explanation. But there is also an even more important fact to contend with: the complete change from the vertical-axis mill to the horizontal-axis mill. The European windmill's four sails, possible flat boards in the earliest instances, were mounted on a horizontal shaft, with each sail set at a small angle with respect to the plane of rotation of the whole wheel. This presented several engineering problems. Three major ones were (1) transmission of power from a horizontal rotor shaft to a vertical shaft, on which the grindstones were set; (2) turning the mill into the wind; and (3) stopping the rotor when necessary, because the wind could not be diverted or blocked.

The first problem was solved by adopting the *cog-and-ring gear* shown in Figure 1-5, designed long before by Vitruvius for his horizontal-axis water wheel. To solve the second problem, the bold step was taken of rotating the whole system on a central spindle composed of a stout post supported by heavy beams. This is suggested in Figure 1-6, which is the earliest know representation in a book of this type of windmill and appears as part of an illuminated letter in an English psalter of 1270 [Wailes 1956]. The third problem, stopping the mill, could be solved by turning it out of the wind and applying a frictional braking action at the outer edge of the large gear wheel shown in Figure 1-5.

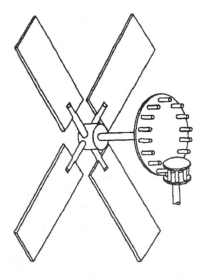

Figure 1-5. Vitruvius' cog-and-ring gear mechanism. This gear made it possible for early horizontal-axis windmills to drive grindtones.

Figure 1-6. The earliest known illustration in a book of horizontal-axis windmill, from the "Windmill Psalter" of 1270. The mill is supported on a post and hence is called a *post mill* [Wailes 1956]. *(Reprinted by permission of Oxford University Press)*

The horizontal-axis windmill was a considerably more complex mechanism than the Persian vertical-axis type, but its adoption is readily explained by the fact that it was so much more efficient. The puzzle lies in the origin of its invention and its sudden appearance in a region that seems to have had no knowledge of vertical-axis windmills. Was it a completely separate invention, or was it a question of diffusion of knowledge about wind power from the East, perhaps with a prototype to copy of which no record remains? Or was it a natural development from the horizontal-axis water wheel? There is no firm consensus, but the opinion among authorities seems to be that this new mill developed naturally from the water wheel. In his discussion of Vowles' paper, "An Inquiry into Origins of the Windmill" [1930], Wailes declares that

> "he considered the step from the Roman watermill with its horizontal wheelshaft to the ordinary windmill to be easier than from the horizontal windmill with a vertical shaft to the ordinary windmill with a horizontal shaft. All the essentials of a windmill were ready to hand except the sails which replace the blades and rim on the water wheel."

Even more flatly, Wailes declares in his book *The English Windmill* that

> "the mechanism of the early postmill is like that of an early watermill turned upside-down, the watermill drive being from below upwards and the windmill drive from above downwards; and it is my opinion, in spite of absence of proof, that the origin of the (horizontal-axis) windmill considered from the mechanical standpoint is just that and no more" [1954].

The diffusion theory of origin from the East has some support from those who cannot accept the notion of a single flash of invention. It takes two different forms; one is by virtue of trade routes from East to West, and the other is by virtue of returning Crusaders. Both have credence, but there is no positive evidence for either. Vowles details the spread of the Greek inheritance and the establishment of schools of learning throughout the Near East and eventually the whole Islamic empire [1930]. Many of the centers of learning were distributed along a great trade route, the old Royal Road, so that a large volume of ideas was spread far and wide, along with the merchandise, from the Baltic to the Far East. He believes that the European windmill was first established in Holland, and that because of the omnidirectional wind, as opposed to the Sīstān environment, "a new form would therefore be evolved to suit the conditions." Arguments against this notion hold that Holland's first attested mill came 70 years after those in France and England, and that the Sīstān mills were actually far from a major trade route.

The other diffusion theory holds that between the First Crusade in 1096 and the first known appearance of the horizontal-axis windmill some 50 years later, a large number of people who recognized the potential of the windmill moved to and fro between Northwest Europe and the Middle East, and that many of them then pursued the mill's potential in their homelands. This is a feasible theory, but it too lacks positive evidence; the verse quoted earlier about the German soldiers also indicates that the horizontal-axis mill was unknown in the Levant. So it seems that the first solid records we have of European windmills are those from the time of the peevish Abbot Samson of Bury St. Edmunds in 1191, with perhaps the possibility of some of those analyzed by Kealey being of earlier standing if additional evidence could be found.

Lift Replaces Drag

Both arguments, however — that for the diffusion theory and that for straightforward development by upending the driven end of the ordinary water mill — require consideration of another, very innovative step that warrants somewhat more attention than it has received. Although the right-angle gear mechanism allowed the rotor axis to be transposed from vertical to horizontal, the action of the sails also had to be turned through 90 deg. This was revolutionary, because it meant that the simple, straightforward push of the wind on the face of the sail was replaced by the action of the wind in flowing smoothly around the sail, providing a force normal to the direction of the wind. As a concept, it is indeed a sophisticated one that was not fully developed until the advent of the airplane at the end of the nineteenth century and the engineering science of aerodynamics.

In fact, although they were not aware of it, the first builders of the horizontal-axis windmill had discovered *aerodynamic lift* and had used it to achieve a greatly improved design over that of *drag*, which is the force that powered the Persian windmills. That the concept of lift is difficult to comprehend, except in precise terms of flow over an airfoil or an edgewise flat plate with an *angle of attack*, is illustrated by a modern example quoted in a small book of short essays accompanied by ink-wash renderings of some European windmills by Sir Frank Brangwyn, RA, and Hayter Preston [1923]. In an introduction, there is a discussion of the comparative power of the horizontal- and vertical-axis types, and another knight of the realm, Sir David Brewster (1781-1868), is quoted thus:

> "The planes of these sails of the horizontal-axis windmill are placed obliquely to the plane of the revolution; so that, when the wind blows in the direction of the axle, it impinges on the surface obliquely, and thus the effort of the sail to recede from the wind causes it to turn upon its axle."

This picturesque, if somewhat humanized, description is based on drag and not lift, but apparently Brewster's nineteenth-century explanation was acceptable to the twentieth-century authors of the book.

Because understanding the concept of lift and its application to the horizontal-axis windmill does not seem to have been possible in the twelfth century, it is argued that the form came about empirically. However, historians disagree as to the steps that were taken along this empirical path. There seems still to be a logical gap in understanding the transition from the drag to the lift concept; hence, we have no convincing explanation the sudden ubiquity of the horizontal-axis windmill in Northwest Europe to the exclusion of the vertical-axis type until much later on.

The upshot is that we don't know the answer. Shall we ever do so? The odds might be against it, but Forbes has said (with respect to water mills) that many of the ancient manuscripts have not yet been translated or studied. Likewise, Wailes, in connection with the many earlier references that had proved false before those of c. 1180, states that "there is always the hope that some new and fully authenticated ones may be found." One wonders also if all of the manuscripts studied have come under the eye of a scholar who is aware of the importance of references to molinology, and whether heretofore they might have been passed over unremarked. We might remember, too, that it took some 2,000 years for the Dead Sea Scrolls to come to light.

Windmills spread rapidly throughout Europe in the thirteenth century, more or less from west to east, although in view of the relative paucity of authenticated records the appearances we know about are certainly not necessarily the earliest uses. Notebaart provides worldwide coverage, with information on the type of windmill and other details when they are available. It must be emphasized that attested dates of use must have been preceded by a period of development of unknown length, when all designs were guided only by trial and error, and there must have been many early failures (a state of affairs not unknown in the present day).

The people of a particular region, which might be as small as a town or as large as a country, are likely to be at least somewhat chauvinistic about their ancient artifacts. This is certainly true of windmills, with respect to who had the earliest, the most, the largest, the best built, the most aesthetically pleasing, and so on. The palm would seem to go to Holland for the greatest number per unit of land area and for the widest utilization. The average density of windmills in Holland at their peak was perhaps three times that of England. But then, what country could compete with between 800 and 1,000 windmills in a few square miles, as at Zaandam, a suburb of Amsterdam? From the seventeenth century onward, it was a highly industrialized town that used windmills for power as we use electric motors. Or what country had the equal of the 19 large drainage windmills along a mile or so of the Kinderdijk in South Holland?

The windmill is recognized everywhere as a logo of Holland, along with the tulip and the young lady in the traditional costume and white cap. But there the unanimity of opinion ends, and there is much spirited competition to identify the earliest record of the European horizontal-axis windmill and the "best developed" mill. The former is subject to the degree of proof required by the exponent or arbiter. As to the latter, we have no clear winner, and chauvinism could reach its height in attempts to find one.

The Post-Mill Design

The psalter picture of 1270 (Figure 1-6), referred to earlier as the first know illustration in a book of a European windmill, shows a mill with a long handle to turn it into the wind; the whole body is mounted on a central post supported by offset struts to the ground. This is known as a *post mill*, and it is the simplest type of horizontal-axis windmill. There are a number of drawings and sketches from the fourteenth and early fifteenth centuries, all showing crude representations, particularly with respect to perspective. Figure 1-7, depicting a mill of 1430, is given here because it attempts to show a cross section of the interior and the exterior [Usher 1954]. It is very simple and not to scale, but it does show the cog-and-ring gear, the feed of grain to the stones, and the pole for turning the whole structure on the post, which itself is not shown clearly. Note that the pole seems to be drawn as an afterthought, with a peculiar placement; it was normally on the back of the mill for safety and to help in balancing the weight of the sails.

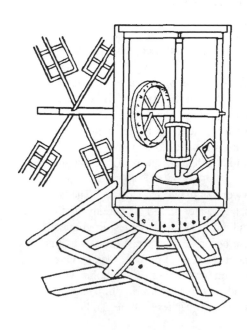

Figure 1-7. A crude sketch of the components of a post mill, c. 1430. The pole at the lower left is presumably for winding the mill [Usher 1954]. *(Reprinted by permission of Harvard University Press)*

We can only speculate on the development of post mills for some four centuries from their known inception in the middle of the twelfth century, because descriptions and drawings having any significant detail have not come to light from that period. It would seem that the first we know of are those of Ramelli in 1588, and excellent drawings they are. It is reasonable to suppose that the first horizontal-axis mills were small, with a fixed orientation, having their foundation members sunk in the ground, because the thrust of the wind on the sails gave rise to large overturning moments that presented a problem. However, the directional variability of the European winds would severely limit the energy output of a fixed-direction mill, so once a sufficiently strong structural design that resisted the large moments had been achieved, a method of turning the mill into the wind must have been sought at an early stage of development. The mounting of the sails, transmission, and grinding stones on a frame that could rotate on a heavy, central post securely fastened to a fixed bottom structure was a bold step; one could expect that many mills were damaged or destroyed by the wind before a suitable design was found.

Before looking at the complete engineering design of a post mill, which is quite complex in its full development, let us look at a diagrammatical sketch of a very common type of post with supporting members, which is the heart of the structural design and gave the mill its name. This is shown in Figure 1-8. The *main post*, usually a great block of oak perhaps 80 cm square, is supported by *quarterbars* mortised in at the top end and anchored to the *crosstrees* at the bottom with *bird's mouth joints*. Secured by the *iron straps*, these joints take the longitudinal outward thrust of the quarterbars. The ends of the crosstrees are supported on *masonry piers* but are not fastened to them. Sometimes the bottom of the main post is cut out to provide *side tongues* fitting loosely over the crosstrees as a steady. It is interesting to note that the best oak for windmills was considered to be from trees about 100 years old!

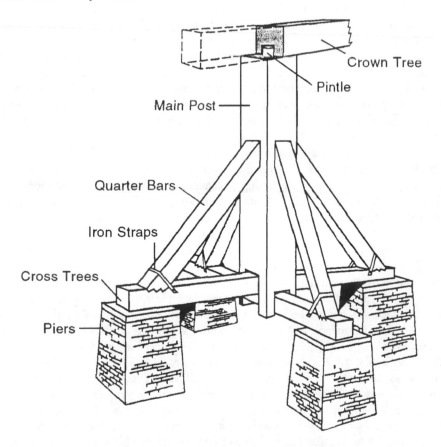

Figure 1-8. Structural components in the base of a post mill. The great main post was usually oak and might be 70 to 80 cm square and cut from a 100-year old tree.

The top of the main post is shaped to make a stepped bearing or pintle, perhaps 20 cm in diameter and depth, fitting into the great *crowntree*, which takes the entire weight of the rotatable assembly, including the sails, drive shaft, gearing, and millstones, all contained in the housing called the *buck*. A *steady* (not shown in the figure) is provided to brace the crowntree against the main post just above the quarterbars to prevent excessive swaying in high winds. This was formed by transverse and longitudinal joists, making a frame that fits easily around a circular cross section of the main post at this point. In later models, the wooden steady could be replaced by *centering wheels*.

In spite of the weight and the absence of antifriction bearings (until later years), the turning maneuver known as *winding the mill* was apparently carried out without great trouble. The rotation is made slowly and sometimes infrequently, so that the wear in the crowntree bearing is relatively small. As confidence grew in the structural design, the post mill would have developed in size and longer sails would have been used, so that the height of the working floor increased. Eventually the ladder required to climb into the mill grew to a considerable length, and in the full development of the post mill it could contain some three dozen or more steps and weigh half a ton.

Let us turn now from the basic structure of the foundation and the rotatable body to that of the moving parts that the structure carried, namely, the sails, the transmission, and the auxiliaries such as the brakes and hoists. These represent a more straightforward design exercise, although each had its problems and probably negligible analytic theory behind it for guidance.

Figure 1-9 is again a simplified diagram of the essentials. The sails, which are not shown in detail because this will come later, are carried on the *windshaft*, which is supported at its forward or *breast* end by the large *breast beam* or *weather beam* and its rear or *tail* end by the *tail beam*. The weather beam supports the *neck bearing* and thus the weight of the sail assembly, which is considerable. The *tail bearing* takes the axial thrust which again is considerable, because in this kind of wind turbine, operating on the lift principle, the axial component of force generated by the wind is much greater than the tangential component of force that provides the torque. The weather beam and tail beam transmit their loads by a variety of methods to the buck and the ends of the crowntree, and thence to the main post.

Each pair of opposing sails had a single longitudinal spar or *stock*, which was mortised through the breast end of the windshaft. In later years, the stocks were fitted into a cast-iron *poll end* or *canister*, thus preventing a weakening of the windshaft by mortising and by exposure to the weather. The neck of the shaft was clad with iron, and the neck bearing was of stone or hardwood, which was later replaced with brass or bronze. Judging by the crude sketches of the first post mills, the windshaft was initially placed horizontally, but quite early on it was inclined some 10° or so upward from tail to breast. The reasons put forward for this are that it shifted some of the weight of the sail assembly to the tail end and so improved stability, that it allowed longer sails to be used that could still clear the base structure or possibly permit a larger buck, and the sails "could catch more wind." Based on modern experience, inclining the rotor axis presents design difficulties without any improvement in performance, and so it is done only to provide clearance between the blades and the tower. Thus, the second explanation above is the most reasonable one.

The power take-off from the windshaft is made by a large *brake wheel*, so called because it also carries the brake on its rim. It was fitted originally with hardwood pegs that transmitted the torque to the *wallower*, or *lantern pinion*, the vertical shaft of which either directly or via intermediate gears powered the millstones or other devices. As time went on, the wooden pegs or staves become shaped cogs. Iron parts replaced some of the wood, and eventually the brake wheel and wallower developed into iron *cross-helical gears*. The wooden pegs were lubricated and lasted a surprisingly long time; some are still in use.

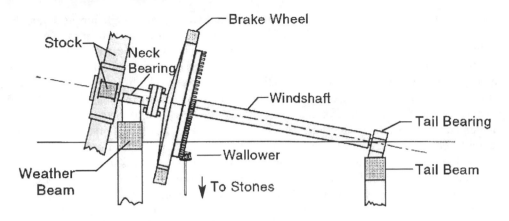

Figure 1-9. The drive train of a typical post mill.

As illustrated in Figure 1-10, the brake is simply a friction band around the circumference of the brake wheel, made of a number of curved wooden *brake blocks* banded together, with one end of the band anchored to a timber of the buck, and the other to a *brake lever*, itself pivoted at a fixed point on the structure. The active end of the brake lever could be pulled up or down by a rope. The brake lever had an iron pin that could engage with a notch in a *catch plate*, free to swing from a pin in its head. The brake lever was very heavy, and when it was unsupported by the rope or the catch plate, it pulled the brake blocks sufficiently hard against the rim of the brake wheel to hold the windshaft at

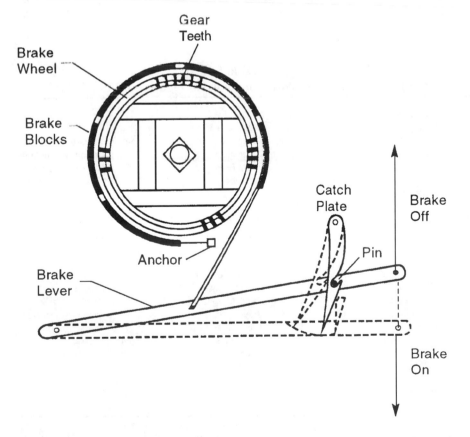

Figure 1-10. Diagram of a post-mill break.

rest. Note that the brake wheel has a large diameter, so that a high torque could be produced by the friction on its rim band. The braking device had an element of "fail-safe" in that it had to be lifted to the "off" position.

It was advantageous to have the brake be capable of operation at a distance, with the miller on the lower working floor. Application of the brake in a high wind or with full sail, either by design or accident, could start a fire from sparks of metal or ignition of wood because of the heavy friction effect. The miller must have had to keep a sharp eye for sudden storms that might catch him with his sails up, so to speak. There was on auxiliary action he could take to control over-speeding, and that was to absorb as much energy in the grinding process as he could — for as long as the supply of grain held out.

The post mill was a triumph of mechanical engineering and the most complex power device of medieval times, even up to the beginning of the Industrial Revolution, which was initiated by the steam engine of Newcomen at the beginning of the eighteenth century. A cutaway view of a complete post mill [Freese 1957] is given in Figure 1-11, and Table 1-1 lists its major parts. This is a view of the mill at Brill, Bucks, England, which was built in 1686; it was largely restored by knowledgeable hands, so it is quite representative of the English post mill. The drawing shows many devices and mechanisms, such as those for grinding the grain, dressing the millstones, and hoisting the sacks. The cramped operating quarters will no doubt be noted, and this brings us to the other major type of horizontal-axis European windmill, the *tower mill*, which was larger than the post mill.

This description gives the basic construction of the post mill, but designs change in detail from mill to mill, region to region, and country to country. This is necessarily true with respect to materials available locally. It is very difficult to date mills, because even when a date is carved into the wood, who knows when it was actually carved? Some documentary evidence is required before the date can be authenticated. Of course, all the

Table 1-1.
Parts List of the Brill Post Mill Shown in Figure 1-11 [Freese 1957]

1.	Brick piers	24.	Sail cleat	47.	Hopper
2.	Main post	25.	Brake wheel	48.	Damsel
3.	Crosstrees	26.	Brake	49.	Feed shoe
4.	Quarterbars	27.	Brake chain	50.	Spring stick or "rabbet"
5.	Retaining straps	28.	Brake lever	51.	Feed adjustment cord
6.	Heel, or tongue of main post	29.	Wallower	52.	Feed adjustment screw
7.	Centering wheels	30.	Upright shaft	53.	Meal spout
8.	Crowntree	31.	Glut box on spindle beam	54.	Bell alarm
9.	Side girt	32.	Bridge beam	55.	Sack-gear "take-off"
10.	Diagonal brace	33.	Great spur wheel	56.	Sack bollard
11.	Cap ribs	34.	Stone nut	57.	Sack control lever
12.	Steps or ladder	35.	Crossbar or bridgepiece	58.	Sack chain
13.	Weather beam	36.	Upper or runner stone	59.	Sack trap
14.	Windshaft	37.	Rhynd or mace	60.	Dresser case
15.	Tail beam and bearing	38.	Bridgetree	61.	Auxiliary "take-off"
16.	Sail stock	39.	Brayer	62.	Auxiliary gear frame
17.	Poll head or canister	40.	Tentering rod	63.	Dresser gears
18.	Sailshaft or whip	41.	Tentering screw	64.	Dresser
19.	Sail bars	42.	Steelyard	65.	Wire brushes
20.	Uplongs	43.	Governors	66.	Dresser spout
21.	Hemlath	44.	Grain bin	67.	Tail pole
22.	Windboard	45.	Grain spout		
23.	Curtain rail	46.	"Horse"		

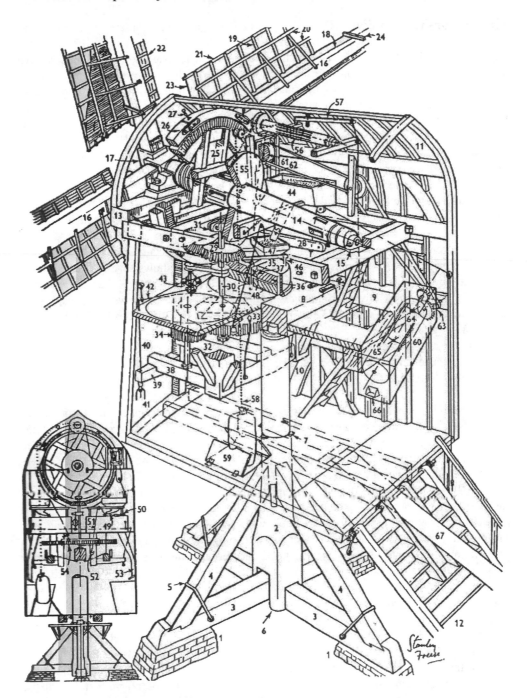

Figure 1-11. Details of the complex construction of a fully-developed English post mill at Brill. The names of the parts are listed in Table 1-1 [Freese 1957]. *(Reprinted by permission of Cambridge University Press)*

mills had a limited life, even when what we now call *retrofits* were made. The oldest post mill still standing in the United Kingdom is certified to be from at least 1636, by documentary evidence. This is not the date of its building but of its known existence at the time, so the mill may be older yet.

The Tower-Mill Design

In order to be able to make larger mills, builders had to take another major inventive step: changing the design from one in which the whole body of the mill had to move to face the wind to one in which only the sails, windshaft, and brake wheel had to move. This was accomplished by mounting the windshaft assembly in the cap of the mill, which turned in a curb or track mounted on the top of a fixed tower. The fact that the mill tower was fixed allowed it to be larger in cross section and higher than the post mill, it could now be made of brick with a circular cross section or of wood in an octagonal shape. The mills made of timber were covered with clapboarding in England and often painted white, so that they came to be called *smock mills*, from their supposed likeness to the rural smock or frock. Many Dutch tower mills had a brick base and a rush-thatched body. There was little difference in the machinery and sails of either type, except for those engaged in specific applications, such as sawmills, which did require some special design considerations.

The tower mill seems to have been introduced in the fourteenth century. The earliest representation is given as 1390; a traveler's sketch dated 1420 shows several located in the Byzantine town of Gallipoli. Many such tower mills are still extant in Holland. Some now inside the modern towns were originally on walls bounding the old towns and were more easily defended than post mills. Many observers might agree with Wailes [1957] that

> "the best-constructed tower- and smock-mills are to be found in The Nether-lands; these cannot be rivalled elsewhere, and the Dutch always led in the design of cloth-spread mills"

Figure 1-12 shows an elevation section through a large Dutch tower mill [Stokhuyzen 1965]. Its essential difference from the post mill is the *cap*, which contains only the windshaft and the brake wheel. Its size was kept small, and its external design was varied according to the degree that the effect of its shape on the wind flow behind the sails was recognized, and perhaps according to the aesthetic sense of the miller or the builder.

The top of the tower had to be of stout construction and have two essential features. The first was the provision of a fixed *curb* or *rail* on which the cap could turn with a minimum of friction between the horizontal surfaces through which the gravity load was transmitted. The second feature as a means of keeping the cap truly centered, again with a minimum of friction between vertical fixed and moving surfaces through which thrust loads were carried. The horizontal bearing was initially wood blocks sliding on a curb, well greased, or with iron plates fixed below the cap frame. Later, iron trolley wheels were mounted on the cap ring, and finally iron rollers were placed between special iron tracks attached to both tower curb and cap ring, so that a *roller-bearing* was effectively formed.

Figure 1-12 shows the greater room available with twin drives to a pair of stones, an economical way of increasing output without increasing the tooth loads on the *great spur wheel* or the size of the stones. Many tower mills were built with a *staging* or *deck* around the tower at the level of the sail tips, to reduce the amount of effort that had to be expended to climb up and down the steep ladders to make changes in the sails. There was also room for living quarters for the miller and his family, which made the mill quite as commodious as the small, "two-up, two-down' cottages of the workers. A tower constructed of bricks could be very sturdy and resistant to weather, but it was not easy to repair if splits appeared as a result of a shift in the foundation or because of the constant vibrations. Thus, the usual practice of placing windows in a linearly symmetrical pattern (as in Figure 1-12) was sometimes changed to a spiral pattern to avoid lines of structural weakness. Wooden smock-mill towers, on the other hand, were subject to joints opening and subsequent rotting from water seepage, Their multi-sided design included walls with slanted corner posts and beams with beveled ends, all of which required expert craftsmanship and constant maintenance to make them secure and leak-resistant.

Brown states that the largest windmill ever built in Europe was a tower mill 37 m high to the top of the cap and 12 m in diameter at the base [1976]. Its rotor diameter would have been at least 30 m, based on the proportions in Figure 1-12. It was built in East Anglia (U.K.) in 1812 and demolished in 1905, after being severely damaged in a storm. What may have been the largest-diameter tower mills ever built were a pair erected in San Francisco's Golden Gate Park to pump water to a hilltop lake [Torrey 1976]. Built in 1902 and 1905, they had rotor diameters of 34.7 m and towers 24 m tall.

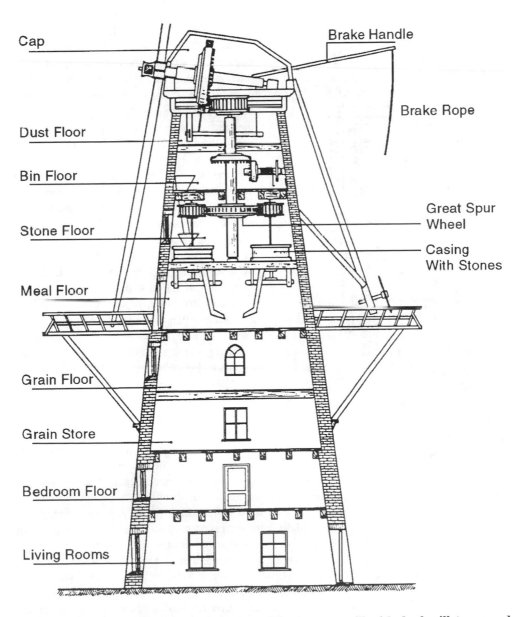

Figure 1-12. Cross-sectional view of a large Dutch tower mill with dual millstones and living quarters [Stokhuyzen 1965]. *(Courtesy of DeHaan/Unieboek)*

Applications of European Windmills

Thus far we have taken the European mill to be a grain mill, but it was also used in a host of other ways. One very considerable application was for pumping water; one naturally thinks of The Netherlands in this regard, because the Dutch were always engaged in either keeping out the sea or turning it back when it got in. Although they were not the first to adapt the windmill for drainage, the Dutch quickly developed the *wipmolen* or *hollow post mill*, one of the designs developed specifically for pumping water. This "pumping" was done by a *scoop wheel*, as illustrated in Figure 1-13, with a lift limited to 1.5 m at the most. The plane of rotation of the scoop wheel had to remain stationary when

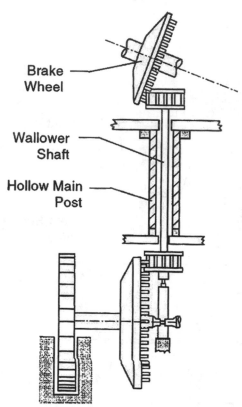

Brake Wheel

Wallower Shaft

Hollow Main Post

the mill structure *yawed* with the wind. To accomplish this, the solid main post of the original post mill was replaced with a hollow one that allowed the vertical shaft from the wallower to extend down the inside of it. At the bottom end, there was a right-angle drive to the scoop wheel formed by a second cog-and-ring mechanism. Note the provision for the shaft of a tandem scoop wheel. In later years, the tower mill was also adapted as a drainage mill; these were often grouped together to drain the Dutch polders and keep them drained. A most striking example is the Kinderdijk group mentioned earlier.

Another special windmill application, which again was developed and used widely in The Netherlands from 1600 onwards, was as a sawmill. It, too, was originally an adaptation of the post mill, called a *paltrok,* in which the brake wheel drove a vertical gear wheel that turned a *horizontal crank-shaft.* The end of a long, *vertical connecting rod* from the crank was fixed to a saw assembly, which cut the timbers longitudinally into planks by an up-and down motion. Tower mills were also used later as saw-

Figure 1-13. Diagram of the power train of a Dutch *wipmolen*, a post mill adapted for pumping water. The wallower shaft passes through a hollow main post.

mills, because their greater size and power allowed somewhat larger logs to be sawn or more saws to be used in parallel, leaving the smaller work to the paltroks.

There was a multiplicity of industrial applications of windmills in the seventeenth century and onward, particularly in The Netherlands. The countless waterways from the sea allowed timber, spices, cocoa, snuff, mustard, dyes, chalk, paint pigments, etc., from all over the world to be processed straight from the cargo ship and transported throughout Europe and beyond. The Zaan district was the industrial center, and windmills were the electric motors of their day. Some odd applications have been suggested over the centuries, and one must admire the ingenuity of two of these, both for military use. One was as the motive power for what might be called an armored chariot (a suggestion which preceded that of the tank by about six centuries), and the other for hurling hives of bees over the walls of fortifications (perhaps a precursor of shrapnel or anti-personnel bombs).

Windmill Technology Development

Throughout history, windmill technology represented the highest levels of development in those technical fields we now refer to as mechanical engineering, civil engineering, and aerodynamics. The best technical minds of their day were constantly seeking to improve the design and operation of windmills. A continuous series of modest changes introduced and tested by builders and millers must have occurred that finally resulted in the refinement and advancement of windmill technology. We will now consider some important technical developments in three areas: sails, the heart of the windmill; control devices, particularly those making the miller's work easier and safer; and technical analysis, which determines the factors that influence performance and efficiency.

Development of Windmill Sails

Windmill sails and their development are topics about which we would like more information, since most of the early designs are known only from a few dated pictures of a general nature. There is no adequate record of sail designs for the earliest years of the horizontal-axis windmill. Through the fifteenth century, we have to rely on crude illustrations and carvings that were quite small and showed little detail. Perhaps sails were at first flat boards, but these were soon replaced by a cloth-covered wooden lattice consisting of transverse and longitudinal battens. This lattice was fixed symmetrically on a central spar, forming two "ladders" through which the sail cloth was laced over and under alternate transverse battens. The sails of the large post mills and tower mills developed chiefly in the countries of Northwest Europe; most of the discussion that follows is concerned with the type called the *common sail*. Although it was replaced in many instances by improved designs, it was still in use at a very late date.

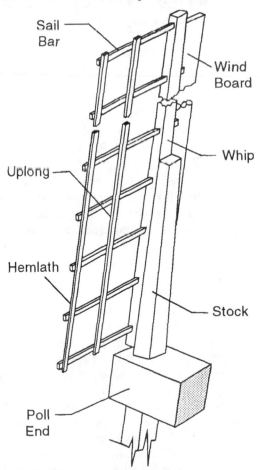

Figure 1-14 shows the structure of the common sail diagrammatically. The main structural element was the *stock*, which either was mortised right through the windshaft or was fitted into the iron *canister* or *poll end*. It could be as long as 27 m; although anything longer might have led to structural problems, it was most likely limited to this length by the availability of the right kind of lumber. Along the length of the stock were fastened narrower timbers called *whips*, through which transverse *sail bars* were mortised at intervals. To the sail bars were nailed longitudinal laths, the outside ones called *hemlaths* and the intermediate ones (one or two) called *uplongs*. In this way, a lattice was formed on which

Figure 1-14. Structural components of the common sail.

cloth could be attached. The sail bars were initially placed symmetrically on either side of the stock, but in later times the common sail had the stock positioned as shown in Figure 1-14. We would now call this a *fractional chord position* with respect to the sail bars, at 1/3- or 1/4-chord, for example. This chordwise location of the stock is discussed in more detail later.

The forward ends of the sail bars supported a leading-edge *wind board* that directed the wind onto the sail and helped to hold the cloth firmly against the frame. At the poll end there was a transverse iron bar onto which the end of the sail was attached by rings and eyelets in the fashion of a present-day curtain. Ropes were attached along both lengthwise edges of the sail so that it could be drawn radially outward and fastened at the tip. Note that the tip had to be within reach of the miller, standing on or near the ground, or on the tower stage, for those mills that had one.

Furling of the common sail acted to control power and rotor speed much as the variable pitch control does for the modern wind turbine. *Pointing lines*, additional cords near the tip, allowed the outer part of the sail to be furled back to the whip. When the mill was not operating, the sail was unfastened at the tip, twisted into a roll, and cleated to the whip. This was a considerable improvement over the earlier sail design, in which the spar was at mid-chord and separate cloths on each side of it would over and under alternate sail bars, making furling a very awkward task. However, the design of the common sail still required that the mill be brought to rest when the sail area had to be rearranged, so the miller had to be able to draw the brake very tightly to ensure that the sails did not sweep him away in a gust of wind.

The earliest sails were inclined at a constant angle to the plane of rotation of about 20 deg, whereas the common sail was given a *twist* from root to tip to vary the inclination continuously along its length. This was called *weathering* the sail, and it was done by mortising the sail bars through the whip at different angles which, according to Wailes, might vary from 22.5 deg at the root to less than zero at the tip [1954]. This was undoubtedly an empirical discovery, because it is unlikely that the millwrights were aware of the concepts of *relative velocity* and *angle of attack*. Perhaps weathering was prompted by observations of the behavior of the stretched cloth along its length "catching the wind" or "filling the sail." What seems somewhat surprising is that it was carried out as far as placing tip sail bars at a negative angle to the plane of rotation. Although negative blade pitch at the tip is now recognized as being theoretically correct in some instances, in those days it must have looked wrong when the mill was at rest.

Jan Drees put forth some observations on the design and performance of windmill sails based on his studies of sixteenth- and seventeenth-century paintings, etchings, and engravings of rotors, comparing these with the features of modern rotary wings, i.e. of helicopters and the like [1977]. As a rotary-wing engineer, he was amazed to find that such modern design features as *nonlinear blade twist, leading-edge camber* ("droop snoot"), and fractional-chord position of the main spar (stock) could all be found on the sails of large windmills of the seventeenth century. Although many of the examples he quotes have been recognized for some time, Drees' contribution is to emphasize a pattern of continuous historical development of windmill sail technology in relation to modern concepts.

Drees developed a diagram of what he considers took place in windmill sail design between the thirteenth century and today, and this is shown in Figure 1-15. Some of his statements are conjectural, and he acknowledges that a great deal more research is necessary to establish the validity of his contentions. According to Drees, a 1550 engraving by Pieter Breughel the Elder shows for the first time a windmill sail with the stock moved forward from the 1/2-chord to the 1/3-chord position. This approaches the optimum position of 1/4-chord for the aerodynamic center and the center of gravity of modern airfoils, which minimizes twisting moments. This eventually allowed a threefold increase in rotor diameter

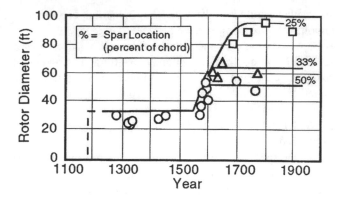

Figure 1-15. Historical development of windmill sail geometry, showing trends in rotor size and the location of the spar or stock [Drees 1977].

and a corresponding tenfold increase in power over about the next 150 years. There is still no leading-edge camber or twist shown in the Breughel engraving.

Drees considers the 1550 date very close to that of the initiation of the revised stock position, because pictures by Stradanus and Ramelli from the latter half of the sixteenth century and a patent granted to Cornelius Muys in 1589 (perhaps the first patent of a windmill) still show the 1/2-chord stock position. Another example of a windmill sail with the stock near the 1/4-chord position is given in a 1589 sketch by Simon Stevin, in his specifications for a drainage mill in Kralinger [Forbes 1966].

By 1650 several pictures by Dutch artists show the stock moved forward, including one by Ruisdael in which the stock is at the 1/4-chord position. There is also one by J. van de Velde in 1617 that shows nonlinear twist, apparently for the first time. In 1664 Jan van Goyen showed a stock all the way forward. Not long after these examples appeared, the wind board shown in Figure 1-14 (corresponding to modern leading-edge camber when inclined) was introduced, utilizing the 25 percent of chord between the leading edge and the whip to direct the wind efficiently around the sail itself. Thus, windmill sails were provided with four characteristics of today's rotary-wing blades: leading-edge camber, 1/4-chord spar position, 1/4-chord center of gravity, and nonlinear twist.

There were, of course, some variations from the common sail. The Mediterranean countries developed triangular *lateen* or *jib-type* sails, allied to those of local sailboats, with lighter spars braced to each other by guy ropes. Although this design is still in use, the date of its first appearance is not known.

Windmill Control Devices

We see, then, that from sometime in the latter half of the twelfth century, the sail developed from a flat board to an efficient cloth sail with twist, a cambered wind board, and a near-optimum stock position. It took some 500 years to accomplish this, and it was not until another century had elapsed that any other major design improvements appeared. When they did so, they were mostly in the form of operational innovations to ease the heavy work of the miller and to make that work more effective by automating some of the necessary but time-consuming and unproductive procedures. Three of the major improvements were made to (1) mechanically change the sail setting, (2) improve braking, and (3) automatically wind the mill. These were all specific inventions with known sponsors, although several variations in design appeared some time later. Most of these improvements took place concurrently with the Industrial Revolution, a time rich in mechanical invention.

The first device we consider, for the setting of the sail, was the *spring sail* invented by Andrew Meikle in 1772, in which a number of hinged wooden shutters mounted transversely across the stock replaced the sail cloth. These shutters operated like venetian blinds tied together by a longitudinal bar and iron cranks that opened and closed them simultaneously. At full power, the shutters were closed and presented a flat surface to the wind. Excessive wind speeds caused the shutters to open and spill the wind. The movement was controlled by a spring; the initial setting was made with the mill at rest, and once it was adjusted, the speed of the mill could be kept reasonably constant. Spring sails were somewhat less efficient than the common sail because the shutters could not be weathered satisfactorily. Thus, sometimes a mill would have two common sails and two spring sails, as a compromise between output and steady running.

The second device was that of Stephen Hooper, who by 1798 had replaced each of the stiff slats in the spring sail with a flexible *roller-reefing sail.* His main contribution to the state of the art was a rotor control device in which furling was activated by a long *striking rod* that passed right through a hole bored centrally down the entire length of the windshaft. A cross or *spider* was affixed to the poll end of the striker. Each of the four "legs" of the spider was connected by a lever to a *bell crank,,*which operated a longitudinal rod along the stock of a sail. In turn, this rod operated bell cranks fixed to each roller or shutter. In this manner, a linear motion of the striking rod caused simultaneous opening or closing of all the individual segments of the sail.

At the tail end of the windshaft was a *rack-and-pinion* mechanism attached to the striking rod together with an endless chain that reached to the ground or a convenient lower level. Pulling the chain turned the pinion, thus moving the rack and hence the striking rod. As Wailes notes, the hole for the striking rod was drilled from the tail end of the windshaft to the poll with the mill turning, the stationary drill being lengthened 0.5 m at a time. This must have required excellent craftsmanship as well as courage on the part of the millwrights. There were no laser-guided drills in those days!

Apparently, the roller-reefing sail itself was not successful, but the spider control mechanism was retained by William Cubitt, who amalgamated it with the shutters of the spring sail to form his 1804 *patent sail.* A further improvement was to attach weights to the endless chain to keep the sail shut until the wind speed increased and exerted sufficient pressure to open the shutters. Thus, the patent sail effected automatic sail setting and relieved the millers of much arduous labor.

In France, Berton introduced a different orientation of the shutter around 1840. Instead of being in the transverse position, the shutters were fitted longitudinally for the whole length of the stock and were pivoted so that they could close to make an almost flat surface. When open, they were positioned one behind the other, as seen from the leading edge of the sail. They had to be operated manually from inside the mill and were used extensively.

In addition to their use as sail elements, shutters were also utilized as aerodynamic brakes. There were several varieties, with one, two, or three shutters fitted longitudinally at the outer end of the sail in remarkably similar fashion to the aileron control surfaces of some modern experimental wind turbine blades. In his 1860 design, Catchpole used two such shutters placed at the leading edge of the outboard third of the patent sails. These longitudinal shutters were operated by the same mechanism as the transverse shutters. When closed, they provided additional sail area. When opened, they were at an angle to the plane of rotation of the sail assembly and hence, being outboard, could rapidly produce a strong braking action. They were called *Catchpole's skyscrapers.* Another aerodynamic brake design consisted of two or three shutters again arranged longitudinally near the tip but located on the trailing side of the whip in place of some four or five transverse shutters.

The third major advance was the invention of the *fantail*, a device that automatically carried out the operation of winding or yawing the mill. Before this time, post mills and the caps of tower mills were turned manually to follow the wind. As the size of windmills increased, the effort to lift up the tail pole and turn the structure became considerable, so a winch was added to do the job. The large handwheel of the winch on the Dutch tower mill shown in Figure 1-12 is seen on the right just above the stage. The invention of the fantail in 1745 by Edmund Lee allowed the wind itself to do the work.

Lee's device consisted of a small windmill with some half-dozen vanes turning on a horizontal axis set parallel to the plane of rotation of the main sail assembly. With the wind blowing squarely on the latter, the fantail was edge on and did no work. When the wind direction changed and blew at an angle to the main sails, it did the same to the fantail, which then developed power. Through gearing, the small fantail turned the mill bodily until the sails were again athwart the wind the fantail action stopped. On the post mill, the fan was mounted on a frame fixed to the ladder near the ground, but on the tower mill it was up high on the cap, with a short ladder for access from the cap.

Thus, by the middle of the nineteenth century, these three devices — patent sails, aerodynamic brakes, and fantail — led to a substantial increase in productivity, as we would now say. Note too that the risk of fire and accident was correspondingly reduced, because the sail shutters and brakes permitted faster shutdown and greatly lessened the danger of fire from the friction brake.

It seems remarkable that these inventions did not spread from England to the rest of Europe for many decades, and even then they were very sparsely adopted. This seems particularly true for The Netherlands, which had led so much development in the previous centuries. The Dutch people have a high reputation in the world of commerce and industry; their resistance to these control devices might possibly have been based on a judgment that they were not cost-effective. On the other hand, the Dutch have also produced much great art and architecture, and they continue to do so. So perhaps their aesthetic sense was affronted by the replacement of the cloth sail by the wooden shutter and the addition of an odd-looking contraption perched on the cap, which would have spoiled the clean, sparse lines of so many of the Dutch tower mills. Although the main emphasis in Stokhuyzen's excellent small volume on the Dutch windmill [1965] is on history and technology, there is more than a hint of the aesthetic quality of such mills.

Technical Analysis of Windmills: Stevin, Leeghwater, and Smeaton

Up to this point, this story of the windmill has been one of relating the development from records that consist largely of pictures, drawings, descriptions, and, from later years, remaining structures in whole or in part. That is to say, this has been an account based on empiricism in design and on practice in working. Beginning at the end of the sixteenth century, however, Simon Stevin and Jan Adriaanzoon Leeghwater[2] made some attempts to analyze the performance and construction of windmills. These were largely given to the calculation of loads on the internal components and to the hydraulics of the pumping process, particularly of the scoop wheel, so important for drainage uses. Around the middle of the eighteenth century (that is, 150 years later), John Smeaton made a major con-

---

[2] His name is handled in different ways. Some authors simply use Leeghwater; others, Jan Adriaanzoon. L. E. Harris tells us that Leeghwater was an adopted addition in later life [1957]. Drees states that the name *Leeghwater* was given to Adriaanzoon because he emptied 27 lakes with windmills in his lifetime, and that the name literally means "empty water" [1977, 1984].

tribution in formulating the rules governing the output of a mill. This he did by ingenious experimentation with rotor models; he then reduced his test data to simple mathematical expressions.

Stevin (1548-1620), born in Bruges, was well-known indeed for a range of writings on mathematics, mechanics, hydraulics, military devices, and so on, including music and precepts for citizens! Along with his writings, he had many inventions patented and devised many improved components for a variety of engineering artifacts. In his work on windmills, he concentrated on calculating the relationships between dimensions, speeds, and number of gear teeth in order to estimate output. Mechanics was not a very exact science in his day. The concepts of energy, power, and efficiency were unformulated, and there were often no names for them. However, he was the leader in applying what was available in mechanics and mathematics to practical attempts to calculate the output of a mill in terms of the water raised by a scoop-wheel and to use such calculations to prove his suggested improvements. In a published collection of this notes, "Van De Molens" [see Forbes 1966, p. 327], he gives calculations for some 20 mills, either existing, improved, or projected.

Leeghwater (1575-1650), born in North Holland, lived a generation after Stevin. In his early life, he was a carpenter and millwright who had a natural talent as an inventor and engineer on a large scale. He used windmills in large numbers for his drainage schemes and made many improvements over the years. None of these resulted in patents, but his skills were widely recognized at home and abroad. He made his reputation by draining a large lake called the Beemster (which had an average depth of 3 m) in one year by using 26 windmills. In another scheme to reclaim a polder (swamp), he used 51 drainage mills pumping out water at the rate of 1,000 m^3/min. His vision was such as to encompass the draining of the Haarlemmermeer with the aid of 160 mills. His study of this was published in 1641 and passed through 17 reprints; the last edition was published in 1838. In 1848, the drainage was finally accomplished after plans were revised several times, but the power was supplied by steam engines.

Smeaton (1724-1792), who was born in the north of England, came much later than the other two engineers, but he contributed the first sound, basic rules of windmill performance. He was a remarkable civil engineer in several branches of the field, primarily in structures such as bridges, lighthouses (the Eddystone among them), windmills, and water wheels, but also in land drainage, canals, harbors, steam engines, and materials (such as cement and cast iron). Smeaton was above all a very careful experimenter; he taught himself and, by example, many others the application of systematic experiment to technological improvement.

Smeaton's chief contribution to molinology was his paper entitled "On the Construction and Effects of Windmill Sails," which was given to the Royal Society [1759; see also Tredgold 1836], The paper makes very interesting reading, and although it was given some 230 years ago, it is still available in many technical libraries. This paper describes what might have been the first use of scale models for obtaining the algebraic relationships governing full-size machines. Smeaton had no wind tunnel to use and so he invoked the principle of *Galilean invariance* and mounted his model on a whirling arm in still air, as shown in Figure 1-16. He still had few accepted laws of energy or standard units with which to obtain numerical values of power, but at least he could run tests at constant speeds and measure, by the raising of weights, the work done by the rotor.

Figure 1-16 follows the original drawing of Smeaton's apparatus. The rotation of the rotor support arm *(FG)* was accomplished by the mysterious hand *(Z)* at the left pulling the cord that turned the barrel on the shaft *(DE)*. Speed was adjusted so that the support arm made one turn in the time the pendulum *(VX)* made two vibrations. This whirling-arm apparatus was not a new idea, but in his customary manner Smeaton improved on others'

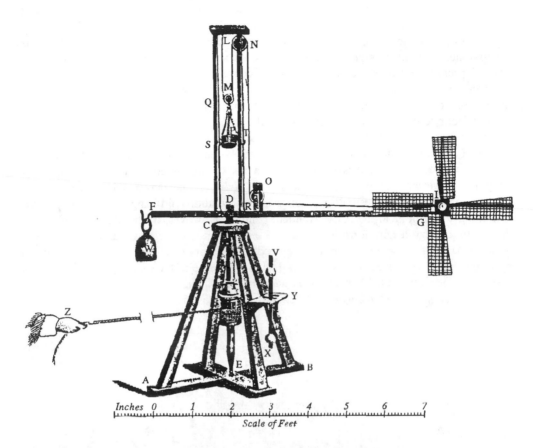

Figure 1-16. **Smeaton's laboratory apparatus for testing the performance of model windmill rotors** [Smeaton 1759].

work in its construction and then performed a series of considered, methodical tests that would not otherwise have been possible.

Smeaton's model rotor had a sail-tip radius of 53 cm, a sail length of 46 cm, and a sail breadth of 14 cm. The maximum "wind speed" developed appears to be about 2.7 m/s; hence, the Reynolds number for these tests was very low, about 25,000. This may have affected his conclusions quantitatively if not qualitatively. His results can be summarized as follows:

(a)   For flat, untwisted sails, the optimum angle to the plane of rotation should be 15 deg to 18 deg.

(b)   Weathered (twisted) sails should have a twist that gives a concave surface to the wind, and a total twist of 15 deg with a tip angle of 7.5 deg.

(c)   A 25 percent increase in the sail area toward the tip (*i.e.* the model in Figure 1-16) together with an increase in blade pitch of 2.5 deg yields 28 percent more power, achieving optimum power per sail.

(d)   Increasing the sail area beyond a certain point decreases the output.

(e) The maximum load that can be lifted is nearly proportional to the square of the wind speed.

(f) The tip speed of the sails in lifting the maximum load is nearly equal to the wind speed (*i.e.* the optimum *tip-speed ratio* is unity). Therefore, the rate of lifting the maximum load is nearly proportional to the wind speed.

(g) Hence, as a consequence of (e) and (f), the maximum power of the sails is nearly proportional to the cube of the wind speed.

Smeaton's work, however imprecise in measurement, gives basic insights into the process of wind energy conversion, and much of it is still valid today.

In spite of all Smeaton's good work in analyzing performance, the builders who continued with their established designs and practices seem to have taken little notice of it. Of course, some of the labor-saving improvements we discussed earlier were made, but otherwise, there were only minor changes. The steam engine put a brake on the use of wind power, although the established design of the windmill did hang on for a long time, continuing to be built in the early twentieth century. However, in the second half of the nineteenth century a new form emerged, the *multivane* or *annular windmill*, which was usually small but capable of satisfying the needs of a farm or ranch with respect to pumping water. In the United States, every farm had one or more of the multivane mills, and the type has come to be called the *American windmill*. This is a useful characterization in the same sense as the term *Dutch windmill*, and thus it is so designated here.

Figure 1-17. Some of the many designs of the American windmill, which was used for pumping water [Baker 1985]. *(Reprinted by permission of University of Oklahoma Press; ©1985, the University of Oklahoma Press)*

The American Windmill

For this history of the American windmill, we lean heavily on *A Field Guide to American Windmills* by T. Lindsay Baker [1985]. This is not a pocket handbook like the well-known field guides on birds, flowers, and so on, but a remarkable, large-page book of more than 500 pages. It is certainly the most comprehensive catalog of American windmills and their makers, culled largely from manufacturers' literature and from periodicals of the farm implement industry, in considerable detail where available. The "field guide" part is preceded by some 100 pages of history, development, and general information about the industry and its technology. It is concerned only with the period from the introduction of a prototype unit by Halladay in 1854 to the essential end of production of these machines in the 1930s.

Figure 1-17, from Baker, shows many of the different designs of the American windmill. Dozens of manufacturers erected windmills at the World's Columbian Exposition in Chicago in 1893, with every conceivable configuration competing for recognition and scrambling to get a piece of the market; an excellent photograph of this event is reproduced in Torrey [1976].

First, let us give consideration to the origin of the design and function of the American windmill. Daniel Halladay is credited with the invention of the first commercially successful, self-governing windmill in 1854; although there were some previous ventures, the *Halladay Standard* came to be the archetypal American model. It was intended as a small unit (2 to 5 m in diameter) for pumping, as were most other American mills. The output of a mill 5 m in diameter in a 7-m/s wind was about a horsepower or so, quite adequate for a "water pumper" on farms all over the country. In later years, some much larger ones (up to about 18 m in diameter) were built for industrial and railroad water-supply systems.

Sectional Wheels: The Halladay Standard

The Halladay mill initially had four rotatable, flat wooden blades joined to an iron shaft with a crank at its opposite end. A solid wooden *tail vane* attached to the shaft and set perpendicularly to the plane of the wheel provided the winding, and a centrifugal governor changed the *blade-setting angle* according to the wind speed. Within a few years, the four paddle-type blades were replaced with a large number of thin, wooden blades nailed to wood rims. These blades were grouped in perhaps a dozen sections of a half-dozen blades each; each section was mounted on a hinged casting. This method allowed the groups of blades to be pivoted backward and forward like an umbrella, controlled by wind pressure or centrifugal effect, with weights or springs for reverse action. Rotors of this sort were called *sectional wheels*, and constituted one of the two main styles of American windmills in use throughout the nineteenth century. One variant of the sectional mill did not have a tail vane, in which case the wheel would become a free-swinging, downwind type, called *vaneless sectional* (the far-left machine in Figure 1-17). It had a speed control akin to that of the vaned, upwind model.

Solid Wheels: The Eclipse

The second major style of the American windmill was that of the *solid wheel*, as opposed to the sectional wheel. Blades were mounted together in a single rigid section, and control was effected by moving the whole wheel to some angle to the wind commensurate with the wind speed. The restraint or return motion was effected in a variety of ways. The prototype of the solid wheel was introduced in 1866 by a Reverend Leonard R. Wheeler in Wisconsin and was named the *Eclipse* (second from left in Figure 1-17). A hinged *tail vane* on the rear framework of the system could turn from being perpendicular to the wheel to a parallel position for shutting it down. A much smaller fixed *side vane* was mounted on an arm parallel to the wheel, with its tip projecting just beyond the circumference. The drag force on this small vane was such that it tended to turn the wheel out of the wind, thus reducing the effectiveness of the blades in high winds and hence controlling the speed. The restoring force was supplied by a weighted lever or spring. This form of control remained the most common throughout the era of the American windmill to the present day, but there was considerable variety in the form and use of the side vane. Instead of a side vane, the wheel was mounted somewhat off the tower pivot centerline in a number of designs, its thrust providing the spoiling effect as the wind speed increased.

The number of manufacturers of windmills multiplied rapidly after these beginnings. Steel blades were introduced in the late 1870s, but there was considerable skepticism about them at first. There is a reasonable explanation for this, in that wood construction usually means more rapid repairs in case of breakage and does not require much machinery, and straightforward carpentry can be applied. Metalworking is more difficult, and repairs sometimes require that replacement parts be ordered, with a consequent increase in machine downtime.

All-Steel Windmills: The Aermotor

A third major step in the development of the American windmill occurred in 1888 when two men, LaVerne Noyes, and inventor and manufacturer, and Thomas Perry, an engineer with a scientific bent, joined in organizing The Aermotor Company in Chicago. By 1900, the *Aermotor windmill* had captured more than half the market; by the middle of this century, the company claimed to have 800,000 mills in service, more than half of them operating for more than 40 years [Baker 1985]. The Noyes and Perry combination appeared to be a symbiosis of entrepreneur and engineer akin to that of Matthew Boulton and James Watt (albeit in a minor key), achieving both economic and technical success. Perry might also be likened to Smeaton. Not only did he test an enormous number of existing technical artifacts to instruct himself, and then design and test his own constructions, he also devised a steam-powered whirling arm for testing model wheels, together with instrumentation to measure speeds, temperatures, and pressures.

The Aermotor did not have any major new operating features, but it did have thin, curved, sheet-metal blades, properly angled and supported by steel members offering minimum drag resistance. This resulted in a much lighter wheel with improved aerodynamic performance, capable of useful work at both lower and higher wind speeds than wooden wheels. However, its rotational speed was too fast for a reciprocating pump. This necessitated a reduction gear, effected by means of a small *pinion* on the wheelshaft and a larger gear on a separate crankshaft. This was called a *back gear*, and it had several advantages, including a high *starting torque*, a longer *pump stroke*, and a division of the bearing load between the wheel and the pump. The Aermotor was not the first all-steel mill, but it performed very effectively in terms of efficiency, structural design, and economy of manufacture.

Automatic Lubrication

One notable technical problem must be related, however — a problem that surfaces in all engineering designs that include dynamic action — namely, bearings and their lubrication. Two mainstays were the *poured babbitt bearing* and (perhaps strangely) wood, particularly maple. The latter was used especially to support the *pitman*, or connecting rod, which converts rotary to reciprocating motion. Nominally "oilless" bronze and graphite came to be used in later years, followed by ball and roller bearings. Lubrication does not sound like a difficult problem, except when the bearings are on top of a tower 6 to 15 m high, quite open to the weather. Perhaps a weekly chore of climbing the ladder and hanging on while using the oil can or replenishing a container does not appear too risky or arduous for a reasonably active man, but it certainly could be so during months of high winds and icy steps in many parts of the country. Hence, there was a continuous effort to find better ways of reducing the need for attention to lubrication.

One of the most useful automatic lubrication devices was a canister of oil above the bearings, normally closed at the bottom by a spring-loaded valve. A wire from the valve down to the ground could then be pulled to release enough oil to last for a number of days, as judged by the operator. An alternative method was to supply the tower with a hinged center and then tilt it to permit lubrication at ground level. Raising and lowering the top part of the tower was accomplished by using the bottom half as a derrick. A number of hinged towers were used in the last years of the nineteenth century, but they lost favor when they were found to be too susceptible to wind damage.

The real solution was introduced by the *Elgin Wind Power and Pump Company*, which enclosed bearings and other parts requiring lubrication in a casing in the manner of an automobile crankcase. The gears carried oil up to the shaft and pinion from the case reservoir, with feeds to the other elements of the transmission, and excess oil flowed back to the reservoir. This served all the necessary purposes, keeping dust and water out and eliminating splashing and loss of oil. A self-oiling mill also allowed more sophisticated methods of transmission, such as the use of cams or worm wheels.

The Decline of the American Windmill

The production of multivane windmills was at its peak at about the time of World War I. Although sales were still brisk in the 1920s, there were signs that their heyday might be over. Baker attributes the decline of windmill production and sales that ensued in the 1930s to the Great Depression, and describes it as a blow from which the industry never recovered [1985]. The economic decline occurred throughout most of the world, and so the very considerable export sales of U.S. companies were likewise reduced. Another concurrent reason was that the growing demand for electricity brought power lines into the heartland of the country. So, although times were hard, such funds as were available could be used for electric pumps. Whatever contentment a farmer felt while listening to his windmill working for free was apparently replaced by the ease of pressing a button to turn on the electricity. *Sic transit gloria molini Americani.*

From Windmills to Wind Turbines: 1888 to World War II

The First Wind-Powered Generation of Electricity

At the end of the nineteenth century, interest developed in using wind power for electrical generation, particularly to service scattered habitations. According to Wolff [1885], using a windmill to drive a generator to charge storage batteries ("accumulators") was first suggested by Sir William Thomson in 1881. In his address on energy sources to the mathematical and physical science section (of which he was President) of the British Association for the Advancement of Science, Sir William observed

"Even now, it is not utterly chimerical to think of wind superseding coal in some places for a very important part of its present duty – that of giving light. Indeed, now that we have dynamos and Faure's accumulator, the little want to let the thing be done is cheap windmills." [Thomson 1881].

So, from the very beginning the cost of the wind machine was a major consideration. On the proposed electric lighting by windmill-driven generators, Wolff predicted, "The application of the windmill to this purpose will soon come actively into play when storage batteries have been developed to a greater success than is attained at the present time."

At that time, it had to be direct-current (DC) power not only for charging batteries but also because of the varying speed on the windmill, which of course could also cause continual variations in the power delivered (if it was available at all). But the demand was there, and experimental work was carried out in several countries around the turn of the century. Here, we describe two very different pioneering windmill-generators.

The first, shown in the Frontispiece and in Figure 1-18, is the *Brush windmill*, so-named for its inventor and builder, Charles F. Brush, a Cleveland, Ohio, industrialist in the electrical field [*Scientific American* 1890; Spera 1977; Righter 1991]. In 1888, Brush erected a windmill to supply 12 kW of DC power for charging storage batteries on his own large estate, mostly for 350 incandescent lights. The configuration he used was the post mill. The wheel, with its 144 blades, was 17 m in diameter on a tower 18 m high, the latter supported by a central iron post 36 cm in diameter extending 2.4 m into a masonry foundation. To provide a steady to relieve strain on the main post in extremely heavy winds, the tower had arms at the four corners carrying casters at their bottom ends which had a small clearance with a concentric rail let into the foundation.

The upwind rotor was a solid-wheel type, with an 18-m by 6-m tail vane and a side vane to turn it out of the wind, like the Wheeler *Eclipse.* The whole system operated automatically, and maintenance was said to be minimal. It ran for 20 years until the rotor was removed in 1908.

The Brush windmill was a landmark in the history of the multivane type. In the first place, it was among the largest built, placing it in the same category as the 18-m-diameter machines used for flour milling and railroad water-pumping. Second, it introduced the high *step-up ratio* (50:1) to windmill transmissions, in this case by two *belt-and-pulley sets* in tandem, to yield a full-load dynamo speed of 500 rpm. Third, it was the first (and most ambitious) attempt to combine

Figure 1-18. The Brush windmill built in 1888 in Cleveland, Ohio. The first use of a windmill to generate electricity. [Scientific American 1890; Spera 1977; Righter 1991]

the best-developed structural and aerodynamic windmill technology with newly developing electrical technology. At the same time that it was doing this, it demonstrated that the production of electrical power was unlikely to be a future application of low-speed, multi-blade rotors. It could be said that this was a very successful operation, but the patient died.

The next important step in the transition from windmills to wind turbine generators was taken by professor Poul LaCour in Denmark [Juul 1956], again at the turn of the century. LaCour was a scientist who conducted wind turbine research from 1891, the year of his appointment to an experimental station at Askov, until his death in 1907. He put the principles of the new engineering science of aerodynamics into use in the *LaCour windmill*, and he was one of the first in the world to use a wind tunnel. Figure 1-19(a) shows LaCour's wind tunnel, with a model rotor in position for testing in the free stream just forward of the outlet of the tunnel. The four-bladed rotor at the tunnel inlet is the fan that drove the air flow.

As we see in Figure 1-19(b), LaCour's rotors still followed the four-bladed, twisted, rectangular pattern of the conventional European windmill, but he appreciated the advantages of low *solidity* (ratio of sail area to *swept area*), leading-edge camber, and low drag. After several years of experimentation, LaCour laid down a set of rules for obtaining optimum rotor performance and succeeded in developing practical wind machines for producing electricity. He designed wind power plants generating 5 to 25 kW for agricultural and village use, and by 1910 several hundred of these were operating in Denmark. Then along came the diesel, and the beginning of the oscillating fortunes of the windmill, whether for generating electricity or pumping water.

In summary, the transition from windmills supplying mechanical power to wind turbines producing electrical energy took place during the last dozen years of the nineteenth century. Brush's high-solidity post-mill design and LaCour's more practical, low-solidity tower-mill configurations pioneered the stand-alone units that generate DC electricity for charging storage batteries. As we see in the next section, this was practically the only application of wind-generated electricity until the beginning of World War II.

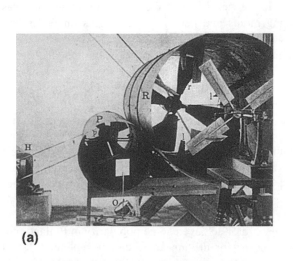

(a) **(b)**

Figure 1-19. LaCour's adaptation of a windmill for generating DC electricity at Askov in Denmark around the turn of the century. (a) LaCour's wind tunnel, c. 1895, one of the first in the world. (Juul 1956) (b) A four-bladed LaCour windmill. *(©1955, E.W. Golding; 1976, E.&F.N. Spon Ltd.; reprinted by Halsted Press, John Wiley & Sons, Inc.)*

The Development of Small Wind Turbine Generators

The advent and development of the airplane in the first decades of the twentieth century gave rise to intense analysis and design studies of the propeller that could immediately be applied to the wind turbine, as we may now properly call it. Professor Albert Betz evolved in clear form the magic value of 16/27 as the ideal maximum utilization factor (or the *Betz limit*, as it is sometimes called) pertaining to the flux of energy available from the wind [1926]. He was also interested in several other detailed aspects of aerodynamic performance, but the times were not auspicious for any considerable undertakings because of the worldwide depression of the 1930s. One notable feature, however, was the introduction in several countries of a new type of fast-running wind turbine having two or three blades with an aerodynamic profile more suitable for generating electricity.

In the United States, the design of Marcellus and Joseph Jacobs could be called the archetypal *battery-charging wind generator*; it was immediately popular for a quarter of a century [Jacobs 1973]. Figure 1-20 shows the *Jacobs Wind Electric power plant*. Its development started in 1925, commercial installation began in 1931, and production ended in 1957. Jacobs' wind generators had a reputation for high performance, minimal maintenance, and excellent structural integrity. A 32-V DC model delivered up to 2,500 W, and a 110-V DC unit produced up to 3,000 W. An interesting feature of the development period was the selection of a three-bladed rotor. A serious vibrational condition was observed in a two-bladed model during changes in wind direction. As it yawed, the rotor experienced a series of jerks produced by changes in *gyroscopic inertia* about the tower axis each time the blades passed from a vertical to a horizontal position. Three blades, however, provided a constant inertia and solved the vibration problem. (This difficulty with two-bladed rotors appears to have been overcome by the use of a *teetered hub* in these later years.) Note that nearly all Danish wind turbines are still three-bladed, along with most other European types, and these turbines appear to be among the leaders in *on-line availability* in wind power stations worldwide.

A successful two-bladed, *propeller-type turbine* for providing direct current was the *Wincharger*, originally available in sizes from 200 to 1,200 W. The 200-W model had a rotor only 1.8 m in diameter, was much less expensive than the Jacobs machine, and was sold in very large numbers for powering radios and perhaps a few lights on farms and ranches.

An innovative type of wind turbine rotor, the *Savonius* rotor, was named after its inventor, Finnish engineer S. J. Savonius, and first tested in 1925 [Savonius 1931; Bach 1931]. The inventor's interest had been aroused by the *Flettner rotor ship* with its large, rotating cylindrical "sails." Wind passing over these cylinders created lift by the *Magnus effect*, which propelled the ship forward. He was intrigued by the possibility of substituting wind power for the external motor power used to rotate these cylinders on the Flettner ship. His experiments resulted in a rotor with an S-shaped cross section which, in its simplest form, could be constructed by cutting a circular cylinder in half longitudinally and rejoining opposite edges along an axle. Tests did indeed show that his *S-rotor* could spin a Flettner cylinder with enough power to propel a small boat at speeds up to 15 knots, sailing at all angles to the wind.

According to the inventor, the Savonius rotor achieved some popularity in Finland, but it has not prospered commercially as a means for driving an electrical generator. It did have a vogue in the 1960s and '70s, largely because its design and construction were very simple. It could be mode from an oil drum and a piece of pipe, which indeed it was by enthusiastic amateurs in many places around the world. Its other advantages were high starting torque and the ability to accept wind from any direction; its drawbacks were low

Figure 1-20. A typical stand-alone, battery-charging wind turbine of the 1930s: the Jacobs Wind Electric power plant, shown here with Marcellus Jacobs, co-inventor and manufacturer, aboard. *(Courtesy of the Jacobs Wind Electric Co., Inc.)*

speed and heavy weight. The Savonius rotor has appeared in more sophisticated forms of cross section, sometimes with three blades to even out the torque variation during each rotor revolution. Its power coefficient is in some doubt, but it would seem that the most reliable maximum value is between 0.18 and 0.23.

Another innovative rotor design introduced in the early 1930s was a type of vertical-axis turbine invented by F. M. Darrieus [1931]. The *Darrieus rotor* has two or three curved blades attached top and bottom to a central column, accepting the wind from all directions without yawing. This column rotates in upper and lower bearings and transmits torque from the blades to the power train, which is located below the rotor, where maintenance is easier and weight are not quite so important. The upper bearing is supported by a set of guy cables. The curved shape of the Darrieus turbine blades approximates that which a perfectly flexible member would assume under the action of centripetal forces, with little or no bending stresses along its length. This theoretical curve is called a *troposkien*, from the Greek for "turning rope." After its initial appearance, the Darrieus turbine was generally neglected until it was re-invented in Canada in the early 1970s and developed further in that country and in the United States.

Thus, the typical wind turbine generator of the 1930s had evolved from the pioneering machines of Brush and LaCour into a two- or three-bladed HAWT with the rotor upwind of the tower and low solidity, using the same type of tail vane as a water pumper for directional control. Its electrical system usually operated at 12 or 32 V DC and utilized *lead-acid batteries* for energy storage. Being a stand-alone, direct-current-producing unit, our typical 1930s wind turbine generator usually operated at variable speed with its blades at a fixed pitch angle. Some designs, however, incorporated variable pitch as a means of overspeed control. The Jacobs Wind Electric rotor, for example, allowed centrifugal loads on a *flyball governor* to pitch its blades. This provided a simple but effective passive pitch control system that feathered the blades with increasing rotor speed.

By the late 1930s, these wind turbines had developed into generally reliable and long-lived machines, given reasonable maintenance. They did not, however, have the performance or cost-effectiveness (let alone the capability of producing AC power) to compete with central stations as a source of electricity. This mattered little in outlying areas until the Rural Electrification Act of 1937, enacted as part of the Roosevelt administration's program to bring the nation out of the Great Depression. The REA authorized the construction of new central power plants, many of them hydroelectric, and subsidized the installation of vast electrical distribution systems from these plants with low-interest loans. Unable to compete with central-station power available almost everywhere, the stand-alone wind turbine, generating low-voltage direct current, became extinct in the United States by the 1940s, in all but the most remote locations.

The former USSR, like the United States, has tremendous areas under cultivation that need electrical power. Thus, the Russians, too, developed multiblade farm windmills of a simple type with high-torque, low-speed rotors suitable for lifting water with piston pumps. We have some translations [*e.g.,* Fateev (also spelled *Fateyev*) 1959] that allow a general picture to emerge which in many ways is similar to that of the West. According to Fateev, wind turbine research had been under government sponsorship in the USSR since the end of World War I and had covered wind tunnel experiments and the establishment of trial grounds for testing complete systems for operational efficiency and mechanical endurance. Modern, high-speed turbines with two or three propeller-type blades have been made in several sizes, up to 20 m in diameter, generating up to 35 kW. Fateev points out that long-term, successful utilization is possible only with good maintenance, and that this was not always available in Russia (or elsewhere, we might add).

The First Large Wind Power Plants

Almost parallel with the demise of the small wind turbine was the beginning of interest in larger wind power plants for incorporation into electric utilities. This interest was fostered by an increasing concern about worldwide energy shortages. As early as 1924, L. Constantin, a French engineer, gave definition to feelings of insecurity with respect to oil supplies; 50 years later, this became a widespread alarm that fossil fuels would soon run out. An excerpt from one of his writings [Constantin 1924] can be translated freely as

"The earth's reserves of fuels, solid and liquid, are rapidly being exhausted, and whatever be the hopes, splendid but distant, that give rise to the study of radioactivity of matter, this suspended threat to our economic life merits the attention of every thinking person"

One of the first steps in the development of large-scale wind power plants for electric utility applications was taken in Russia in 1931, with the construction of a 100 kW 30-m-diameter *Balaclava wind turbine*, on the Black Sea (Figure 1-21). Although its blades had rough surfaces and some of its gears were made of wood, it ran for two years or more and generated 200,000 kWh in that time [Sektorov 1934].

Figure 1-21. The 100-kW, 30-m diameter Balaclava wind turbine in 1931. It was the first wind turbine interconnected with an AC utility system. [Sektorov 1934]

The last wind turbine discussed in this review is the product of a landmark achievement in wind power. This work was conceived and led by Palmer C. Putnam, a brilliant, forceful American engineer who for many years contended that large wind turbines should be utilized to supplement central power plants. In the late 1930s, Putnam interested the S. Morgan Smith Company of York, Pennsylvania, in his plan to build a prototype of a *megawatt-scale* wind turbine generator using the latest technology. The Smith Company, experienced in the construction of hydroelectric turbines and electrical power equipment, agreed to provide financing and manage all phases of the engineering, construction, and operation. The result of this collaboration was the construction in 1941 of the largest wind turbine up to that time and for almost 40 years afterward.

The *Smith-Putnam wind turbine*, shown in Figure 1-22 [Voaden 1943; Putnam 1948; Koeppl 1982], had a two-bladed rotor that swept an area 53.3 m in diameter. It had a truss-type tower and a rotor axis 33.5 m above grade. The rotor powered a 1.25-MW *synchronous generator* through a geared step-up transmission. This configuration was chosen to meet Putnam's goal of producing energy at the lowest possible unit cost and in large enough quantities to be useful to electric utilities. The preliminary design effort was led by Putnam, who recruited a nationwide team of consultants for assistance, including aerodynamicists such as Theodore von Karman at the California Institute of Technology, engineering faculty members at the Massachusetts Institute of Technology, and staff engineers at the General Electric Company.

Technological innovations in the Smith-Putnam wind turbine included *full-span active control* of blade pitch; the use of individual *flapping hinges* on the blades to reduce gyroscopic loads on the shaft (a generic problem with two-bladed rotors on rigid hubs that the Jacobs brothers had experienced earlier); and *active yaw control* by means of a *servomotor* turning a pinion meshing with a large *bull gear* between the machinery house or *nacelle* and the tower.

The Smith Company organized an experienced industrial team for the final design work, which began in 1939, and for the fabrication of parts, which began in 1940. Blades were built by the Budd Company of Philadelphia with stainless steel skins to resist corrosion, applying a technology they developed for railway passenger cars. After shop assembly and system tests by the Wellman Engineering Company in Cleveland, the machinery and blades were shipped to the site — a hill in Vermont called "Grandpa's Knob." On October 10, 1941, electricity flowed for the first time from "the windmill on Grandpa's Knob" into the network of the Central Vermont Public Service Corporation.

Several hundred hours of testing proved that the system could operate satisfactorily as a utility power plant. However, a bearing failure in 1943 caused operations to be delayed for two years, because of wartime supply difficulties. It was also discovered that the blade spars were under-designed at the root, and so reinforcing doublar plates were welded in place. The rotor was then locked in place for the duration of the war, to endure winter storms with little or no maintenance. When operations recommenced in 1945, cracks were discovered in a blade root at a repair weld. Unable to obtain funding for a replacement rotor, the project managers took the risk that the test program could be completed successfully before the cracks propagated to failure. The turbine was run continuously for several weeks with excellent power production, but the risk was too high. In the early hours of March 26, 1945, the faulty blade spar separated at the repair weld. Unfortunately, the S. Morgan Smith company could not afford to continue the work.

As we close this review of the technological development of windmills and early wind turbines — an evolution that stretches back to ancient Persia — we should note that the Smith-Putnam team made two other significant contributions to the technology: They were pioneers in organizing a wind turbine research and development project along the lines of a modern industry/university partnership, and they served as an inspiration for the eventual

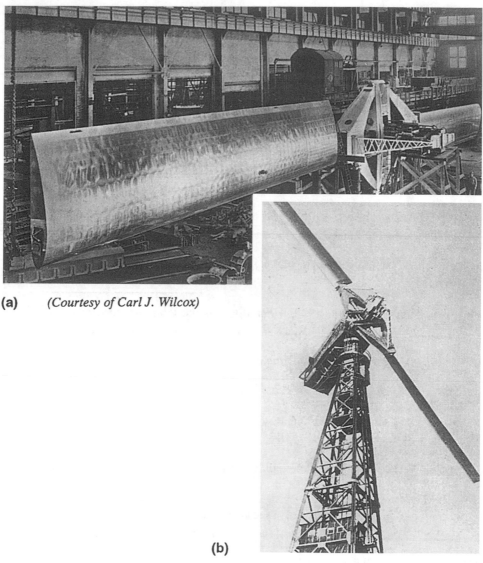

(a) *(Courtesy of Carl J. Wilcox)*

(b)

(Reprinted by permission of
Van Nostrand Reinhold Company)

Figure 1-22. The Smith-Putnam wind turbine, c. 1941 — the world's first megawatt-scale wind power plant. (a) Shop assembly of the 53-m rotor. The individually-hinged blades have stainless steel skins. (b) Operating atop Grandpa's Knob, near Rutland, Vermont, this power plant supplied up to 1.25 MW of AC power to the utility system. [Voaden 1943; Putnam 1948; Koeppl 1982]

rebirth of wind power. When the so-called "fuel crisis" of the 1970s caused worldwide shocks that demanded that concrete attention be paid to alternative energy sources, wind power was in the forefront because its technology was already relatively advanced. But we must now end this history of the windmill and move forward to the evolution of modern wind turbines.

References

Bach, G., June 1931, "Untersuchungen über Savonius-Rotoren und verwandte Strömungs-maschinen," *Forsch. Geb. Ingr.*, 2. Bd./Heft 6, Technische Hochschule, Berlin, pp. 218-231; English translation by C. A. Henkel, 1974, SAND74-6018, Albuquerque, NM: Sandia National Laboratories.

Baker, T. L., 1985, *A Field Guide to American Windmills*, Norman, Oklahoma: University of Oklahoma Press.

Bellew, H. W., 1874, *From the Indus to the Tigris*, London: Trubner and Co.; reprinted in 1977, Royal Book Co., Karachi-3, Pakistan, pp. 239-240.

Betz, A., 1926, *Windenergie und Ihre Ausnutzung durch Windmühlen*, Göttingen, Germany: Vandenhoeck und Ruprecht.

Boas, M., 1949, "Hero's Pneumatica — A Study of its Transmission and Influence," *Isis*, 40, pp. 38-48.

Brangwyn, F., and H. Preston, 1923, *Windmills*, London: John Lane the Bodley Head; republished in 1975 by Gale Research Co., Detroit, Michigan, p. 25.

Brown, R.J., 1976, *Windmills of England*, London: R. Hale, p. 18.

Constantin, L., 1924, "Le Vent," *La Nature*, 52, Pt. 1, pp. 395-400.

Darrieus, F. M., Dec. 1931, "Turbine Having its Rotating Shaft Transverse to the Flow of Current," U.S. Patent No. 1,834,018.

Drachmann A. G., 1961, "Heron's Windmill," *Centaurus*, 7, pp. 145-151.

Drees, J. M., Spring 1977, "Blade Twist, Droop Snoot, and Forward Spars," *Wind Technology Journal*, I:1, pp. 10-16; see also "Speculations about the Origin of Sails for Horizontal Axis Turbines," *Wind Technology Journal*, 1984, 2:1/2, pp. 13-31.

Fateev, E. M., 1959, *Wind Power Installations, Present Condition and Possible Lines of Development*, Moscow; translated by Krammer Associates, 1975, published as NASA TT-F-16204, Washington, DC: National Aeronautics and Space Administration, 73 pp.

Forbes, R. J., ed., 1966, *Principal Works of Simon Stevin, Vol. 5, The Works on Engineering of Simon Stevin*, E. Crone *et al.*, eds., Amsterdam: Swets and Zeitlinger, Plate I.

Forbes, R. J. 1956, "Power," *A History of Technology*, Vol. II, C. Singer *et al*, eds., London: Oxford University Press, p. 617.

Freese, S., 1957, *Windmills and Millwrighting*, London: Cambridge University Press, Fig. 3.

Golding, E. W., 1955, *The Generation of Electricity by Wind Power*. London: E. & F.N. Spon Ltd.; reprinted with an additional chapter by R. I. Harris, 1976, New York: Halsted Press, Division of John Wiley & Sons, Inc.

Harris, L. E., 1957, "Land Drainage and Reclamation," *A History of Technology*, Vol. III, C. Singer *et al.*, eds., London: Oxford University Press, p. 305.

Hedin, S., 1910, *Overland to India*, Vol. II, London: Macmillan, pp. 135, 292.

Holt, R., 1988, *The Mills of Medieval England*, Oxford, England: Basil Blackwell, Ltd., Appendix 1.

Jacobs, M. L., 1973, "Experience with Jacobs Wind-Driven Electric Generating Plant, *Proceedings, First Wind Energy Conversion Systems Conference*, NSF/RANN-73-106, Washington, DC: National Science Foundation, pp. 155-158.

Juul, J., 1956, "Wind Machines," *Wind and Solar Energy - Proceedings, New Delhi Symposium*, Paris: UNESCO, pp. 56-73.

Juul, J., 1964, "Design of Wind Power Plants in Denmark," *Wind Power - Proceedings, United Nations Conference on New Sources of Energy*, Vol. 7, New York: The United Nations, pp. 229-240.

Kealey, E. J., 1987, *Harvesting the Air*, Berkeley: University of California Press.

Koeppl, G. W., 1982, *Putnam's Power from the Wind*, 2nd ed., New York: Van Nostrand Reinhold Co.

Needham, J., 1965, *Science and Civilization in China, Vol. 4, Physics and Physical Technology, Pt. II: Mechanical Engineering*, London: Cambridge University Press, pp. 556-560.

Notebaart, J., 1965, *Windmühlen*, Den Haag: Mouton Verlag; see the summary in English, pp. 357 ff.

Putnam, G. C, 1948, *Power from the Wind*, New York: Van Nostrand Reinhold Co.

Righter, R. W., 1991, "A Wind-Fueled Electric Power Plant in the Backyard, 1888, *Invention and Technology*, pp. 28-31.

Savonius, S. J., 1931, "The S-Rotor and Its Applications," *Mechanical Engineering*, 53(5), pp. 333-338.

Schmidt, W., ed., 1899, *Herons von Alexandria, Vol. I, Pneumatica et Automata*, Leipzig, Germany: Teubner, pp. xxxix-xl, 203-207.

Scientific American, Dec. 20, 1890, "Mr. Brush's Windmill Dynamo." Vol. LXIII, No. 25, cover and p. 389.

Sektorov, V. R., 1934, "The First Aerodynamic Three-Phase Electric Power Plant in Balaclava, *L'Elettrotecnica*, 21(23-24), pp. 538-542; Translated by Scientific Translation Service, NASA TT-F-14933, Washington, DC: National Aeronautics and Space Administration, 13 pp.

Smeaton, J., 1759, "On the Construction and Effects of Windmill Sails," *An Experimental Study Concerning the Natural Powers of Water and Wind*, Philosophical Transactions of the Royal Society of London, 51, Pt. 1, pp. 138-174; reprinted in Tredgold, T., 1836, *Tracts in Hydraulics*, 2nd ed., pp. 47-78.

Spera, D. A., 1977, "The Brush Wind Turbine Generator as Described in *Scientific American* of Dec. 20, 1890," *Proceedings, Workshop on Wind Turbine Structural Dynamics*, NASA CP-2034, DOE CONF-771148, Cleveland, Ohio: NASA Lewis Research Center, pp. 275-283.

Stokhuyzen, F., 1965, *The Dutch Windmill*, Bussum, The Netherlands: van Dishoeck, p. 47.

TIMS, 1971-1986, *Transactions of The International Molinological Society: 1965 Meeting,* 1977; *1969 Meeting,* 1971; *1973 Meeting,* 1977; *1975 Meeting,* 1979; *1979 Meeting,* 1984; *1984 Meeting,* 1986.

Thomson, W. 1881, "On the Sources of Energy in Nature available to Man for the Production of Mechanical Effect," *Report of the British Association for the Advancement of Science*, Transactions of the Sections, Section A. – Mathematical and Physical Science, pp. 513-518.

Torrey, V., 1976, *Wind-Catchers*, Brattleboro, VT: Stephen Greene Press, pp. 90-91, 148.

Usher, A. P., 1954, *A History of Mechanical Inventions*, Cambridge, Massachusetts: Harvard University Press, p. 174.

Voaden, G. H., 1943, "The Smith-Putnam Wind Turbine — A Step Forward in Aero-Electric Power Research," *Turbine Topics*, 1(3); reprinted 1981 in NASA CP-2230, DOE CONF-810752, Cleveland, Ohio: NASA Lewis Research Center, pp. 35-42.

Vowles, H. P., 1930-1931, "An Inquiry into Origins of the Windmill," *Transactions of the Newcomen Society of London,* XI, pp. 1-14.

Wailes, R., 1954, *The English Windmill*, London: Routledge and Kegan Paul, pp. 92, 150.

Wailes, R., 1956, "A Note on Windmills," *A History of Technology*, Vol. II, C. Singer *et al.,* eds., London: Oxford University Press, p. 623.

Wailes, R., 1957, "Windmills," *A History of Technology*, Vol. III, C. Singer *et al.,* eds., London: Oxford University Press, p. 107.

White, L., Jr., 1962, *Medieval Technology and Social Change*, London: Oxford University Press, pp. 86-87.

Wolff, A.R., 1885, *The Windmill as a Prime Mover*, New York, John Wiley & Sons, p. 4.

Woodcroft, B., 1851, *The Pneumatics of Hero of Alexandria*, London: Taylor, Walton, and Maberley, pp. 108-109.

Wulff, H. E., 1966, *The Traditional Crafts of Persia, Their Development, Technology, and Influence on Eastern and Western Civilization*, Cambridge, Massachusetts: M.I.T. Press, pp. 284-289.

2

Introduction to Modern Wind Turbines

David A. Spera, Ph.D.
Formerly Chief Engineer of Wind Energy Projects
NASA Lewis (now Glenn) Research Center
Cleveland, Ohio

and

Wind Energy Consultant
DASCON Engineering, LLC
Bonita Springs, Florida

Introduction

The preceding chapter traces the development of wind power technology from ancient windmills through the historic *Smith-Putnam* project which ended in 1945. During the next 40-odd years, wind energy enthusiasts proposed turbine designs covering every conceivable concept, shape, and size. Many of these resulted in complete designs, and a large number were built and tested. In the last decade of the 20th century typical configurations of modern horizontal-axis and vertical-axis wind turbines (HAWTs and VAWTs) began to emerge, and these will be described in this chapter. The evolutionary process that led to modern wind turbine configurations is described in detail in Chapter 3.

Wind turbine nomenclature has also evolved during the past several decades, and definitions of common terms will be presented here. In addition, parameters which wind turbine engineers use to analyze power and energy output will be introduced, in preparation for the more-detailed discussions of the wind characteristics and wind turbine performance that appear later in this book. A bibliography of publications recommended for additional introductory information on modern wind turbines is given in Appendix A.

Two early large-scale prototype wind turbines will be described here to illustrate the antecedents of modern horizontal-axis and vertical-axis configurations. These noteworthy machines are the *DOE/NASA Mod-5B HAWT* (Fig. 2-1) and the *DOE/SNL 34-meter VAWT* (Fig. 2-2). At the time of their construction, each represented an advanced stage in the design evolution of its configuration. Many of the features of these two prototype machines are now common to modern wind turbines of various sizes.

The Mod-5B, with its 97.5-m rotor, 3.2-MW rating, and variable-speed generating system, was at one time the largest wind turbine in the world. Designed and built by the *Boeing*

Figure 2-1. The configuration of the 3.2-MW DOE/NASA Mod-5B prototype wind turbine contains many of the features typical of modern HAWTs. With its swept area of 7,470 m², it was the largest wind turbine in the world when constructed in 1988. (*Courtesy of NASA Glenn Research Center; photograph by R. Ensign*)

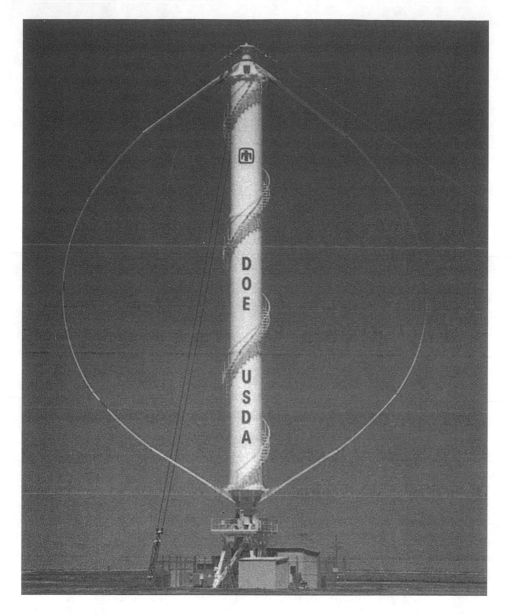

Figure 2-2. The 625-kW DOE/SNL 34-m VAWT prototype was an advanced test bed dedicated to the development of technology and components for modern vertical-axis turbines. (*Courtesy of Sandia National Laboratories*)

Aerospace Company for the U.S. Department of Energy and the National Aeronautics and Space Administration, it was installed on the Hawaiian Island of Oahu and later owned by the *Makani Uwila Power Corporation*, which operated it for many years as a commercial power plant [Spera and Miller 1991].

The 34-meter VAWT, designed and built by the Sandia National Laboratories (SNL), had a variable-speed generating system rated at 625 kW. Its high-performance Darrieus rotor, 34 m in diameter and 42 m in height, had variable-chord blades with airfoils designed specifically for VAWT service. [Dodd *et al.* 1990].

Modern Large-Scale Wind Turbines

Often described as *utility-scale* because of their application in wind power stations feeding power to electrical utility grids, modern large-scale wind turbines are typically 3-bladed horizontal-axis machines. Figure 2-3 shows typical utility-scale wind turbines, *General Electric (GE) 1.5-MW Model 1.5s HAWTs*, with 70.5-m diameter rotors and 64.7 m hub heights. The turbines pictured here are part of a cluster of 136 units installed at the New Mexico Wind Energy Center, producing a rated station power of 204 MW.

Figure 2-3. These GE Model-1.5s wind turbines, each rated at 1.5 MW, are typical of modern utility-scale machines, with their 3-bladed 70.5-m diameter rotors mounted upwind of tubular towers at a hub-height of 64.7 m. (*Courtesy of GE Energy, Inc.*)

A relatively new but growing application of large-scale wind turbines is as units in *off-shore wind power stations*. Figure 2-4 shows the first wind turbines in a planned 300-MW offshore power station 30 km off the Belgium coast. These *REpower Model 5M* wind turbines, each of which has a rated power of 5.0 MW and a rotor diameter of 126 m, are installed in water 25 m deep.

Figure 2-4. An example of off-shore wind power stations in Europe is this installation of REpower Model 5M wind turbines off the Belgium coast. Specifically designed for the marine environment, these 5.0-MW turbines have 3-bladed rotors 126 m in diameter mounted upwind of tubular towers. (*Courtesy of REpower Systems*)

General Configurations of Horizontal-Axis Wind Turbines

Figures 2-3 and 2-5 illustrate two general configurations of modern HAWTs. The principal subsystems which make up the total wind energy conversion system are (1) the rotor, (2) the power train, (3) the nacelle structure, (4) the tower, (5) the foundation, and (6) the ground equipment station.

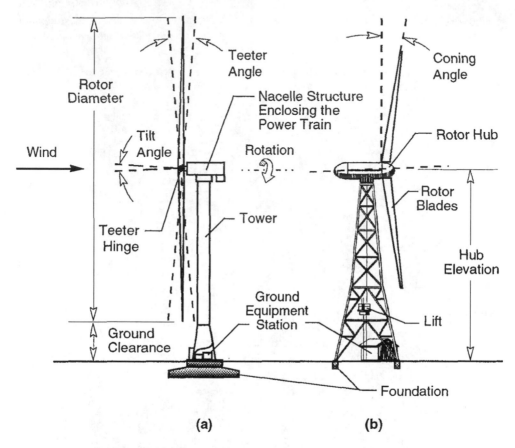

(a)   **(b)**

Figure 2-5. Principal subsystems of a HAWT. (a) Two-bladed teetered-hub upwind rotor (b) Two-bladed rigid-hub downwind rotor. Many current HAWTs have three-bladed rigid-hub upwind rotors

The Turbine Rotor Subsystem

HAWT rotors are often described as "propeller-type", indicating correctly that many of the aerodynamic and structural-dynamic principles that are basic to aircraft propeller technology are also applicable to HAWT rotors. Of course, HAWT rotors decelerate rather than accelerate the air, and their tip speeds (typically less than Mach 0.4) are much lower than those of aircraft propellers. As shown in Figure 2-5, the main components of the HAWT rotor are its *blades* fastened to a central *hub*. HAWT rotors usually contain either two or three blades. One-bladed rotors with counterweights are technically feasible but rare. As the rotor turns, its blades generate an imaginary surface whose projection on a vertical plane is called the *swept area*. For purposes of calculating the swept area, blades are assumed to be undeformed by applied loads.

The terms *downwind rotor* and *upwind rotor* denote the location of the rotor with respect to the tower. An *unconed rotor* is one in which the spanwise axes of all of the blades lie in the same plane. Blade axes in a *coned rotor* are tilted downwind from a plane normal to the rotor axis, at a small *coning angle*. This helps to balance the downwind bending of the blade caused by aerodynamic loading with upwind bending by radial centrifugal forces. *Tower clearance* (the minimum distance between a blade tip and the tower) is influenced by blade coning, rotor teetering, and elastic deformation of the blades under load. Elastic deformation can be significant for blades fabricated from composite materials, such as fiberglass. Often an *axis-tilt angle* is required to obtain sufficient clearance. Axis tilt is kept to a minimum because of potential negative side effects, such as reduced swept area and a vertical component to the rotor torque that can cause a yaw moment on the nacelle.

The two general types of rotor hubs are *rigid* and *teetered*. In a typical rigid hub, each blade is bolted to the hub and the hub is rigidly attached to the *turbine shaft*. The blades are, in effect, cantilevered from the shaft and therefore transmit all of their dynamic loads directly to it. To reduce this loading on the shaft, a two-bladed HAWT rotor usually has a teetered hub, which is connected to the turbine shaft through a pivot called a *teeter bearing* or a *teeter hinge*, as shown in Figure 2-5(a). This bearing permits cyclic, rigid-body motion of the rotor perpendicular to the plane of rotation through small *teeter angles* (less than ±10 deg) at a frequency equal to the rotor speed (*i.e.* one cycle per rotor revolution or *1P*). Teeter bearings may contain rolling, sliding, or elastically-deforming elements such as *elastomeric* bearings composed of alternate rubber and steel sheets.

Teeter motion is a passive means for balancing air loads on the two blades, by cyclically increasing the lift force on one while decreasing it on the other. Teetering also reduces the cyclic loads imposed by a two-bladed rotor on the turbine shaft to levels well below those caused by two blades on a rigid hub. One-bladed rotors have been attempted, and these are almost always downwind of the tower and teetered, because their inherently-large aerodynamic imbalance causes large teeter angles.

A three-bladed rotor usually has a rigid hub. In this case, cyclic loads on the turbine shaft are much smaller than those produced by a two-bladed rotor with a rigid hub, because three or more blades form a *dynamically-symmetrical rotor*: One with the same mass moment of inertia about any axis in the plane of the rotor and passing through the hub. This is fortunate, since practical *gimballing* (teetering about two orthogonal axes) is difficult to achieve.

Medium- and large-scale HAWT rotors usually contain a mechanism for adjusting *blade pitch*, which is the angle between the blade chordline (at a specified reference radius) and the plane of rotation. This *pitch-change mechanism*, which may control the angle of only the outboard section of each blade (*partial-span pitch control*, Fig. 2-1) or of the entire blade (*full-span pitch control*, Figs. 2-3, 2-4, and 2-5(b)) provides a means of controlling starting torque, peak power, and stopping torque. Pitch-change mechanisms are high-quality structural/mechanical devices with strong actuators (usually hydraulic) and computerized controls. Some small- and medium-scale HAWTs have fixed-pitch *stall-controlled* blades, avoiding the cost and maintenance of pitch-change mechanisms by relying on aerodynamic stall to limit peak power.

A simplified form of aerodynamic control mechanism is a *tip brake* or a *tip vane*, in which a short outboard section of each blade is turned at right angles to the direction of motion, stopping the rotor by aerodynamic drag or at least limiting its speed.

A wide variety of materials have been used successfully for HAWT rotor blades, including glass-fiber composites (both laid-up in molds and filament-wound over mandrels), laminated wood composites, steel spars with non-structural composite fairings, and welded steel airfoils. The choice of blade materials is a system engineering decision involving considerations of size, strength, stiffness, weight design and manufacturing expertise, maintenance, and cost. Whatever the blade material, HAWT rotor hubs are almost always fabricated from steel forgings, castings, or weldments.

Figures 2-6 and 2-7 illustrate the basic shape of a modern wind turbine blade with full-blade pitch. A typical *planform* shape [Griffin 2001] is shown in Figure 2-6, in which the radial distance from the center of rotation, *r,* and the width or *chord* of the blade, *c,* are normalized by the rotor tip radius, *R*. The airfoil section of the blade typically tapers from a narrow chord at the tip to the maximum chord width at approximately 25 percent of the rotor radius. Thus, the aerodynamic sections of the turbine blades cover about 94 percent of the rotor swept area.

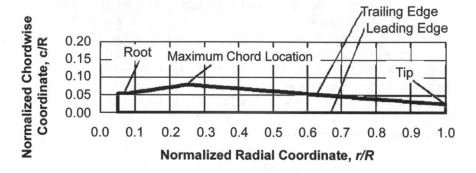

Figure 2-6. Typical normalized planform shape of a wind turbine blade. [Griffin 2001]

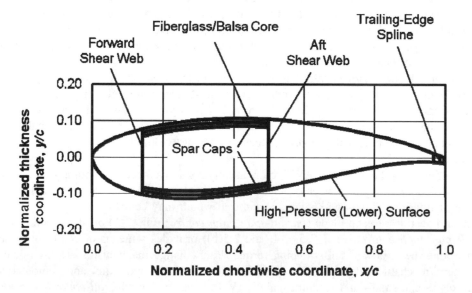

Figure 2-7. Schematic view of the cross-section of a composite wind turbine blade, showing typical internal structural elements. [Griffin 2001]

A circular blade *root section* usually extends from a flange at approximately 5 percent of the rotor radius where it is attached to the pitch-change mechanism in the hub, to approximately 7 percent of the radius where it transitions to the airfoil cross-section at the maximum chord width. While this basic planform has straight leading and trailing edges and a squared-off tip, advanced blade designs may add curvature to both leading and trailing edges to improve performance. In addition, the blade tip is often shaped to reduce noise by rounding leading and trailing edge corners.

Figure 2-7 shows a schematic diagram of the internal structure of a typical composite wind turbine blade. The standard *x-y coordinate system* for describing the complex curved surfaces of an airfoil has its origin at the leading edge, with the positive *x* or *chordwise* dimension extending aft to the trailing edge, and the positive *y* or *thickness* dimension extending toward the *low-pressure* or *upper surface* of the airfoil. Negative *y* dimensions then point to the *high-pressure* or *lower surface*. The upper- and lower-surface designations have their basis in airplane wing descriptions. Both the *x* and *y* dimensions are typically normalized by the chord width *c*.

The structural arrangement in Figure 2-7 is representative of current commercial blade designs [Griffin 2001]. The primary structural member is a *box-spar*, with *shear webs* located approximately 15 percent and 50 percent of the chord width measured from the leading edge, with a substantial build-up of *spar cap* material between these two webs. Exterior skins and internal shear webs are both of sandwich construction, in which fiberglass laminate skins are separated by balsa cores.

The Power-Train Subsystem

The *power train* of a wind turbine consists of the series of mechanical and electrical components required to convert the mechanical power received from the rotor hub to electrical power. In a HAWT, this equipment is atop the tower, so low maintenance is an important design requirement. Examples of small-, medium-, and large-scale power trains are illustrated in Figure 2-8. A typical HAWT power train consists of a *turbine shaft assembly* (also called a low-speed or primary shaft), a *speed-increasing gearbox*, a *generator drive shaft* (also called a high-speed or secondary shaft), a *rotor brake*, and an *electrical generator*, plus auxiliary equipment for control, lubrication, and cooling functions. As illustrated in Figure 2-8(a), some small-scale HAWTs and a few medium- and large-scale HAWTs have a *direct-drive* from the turbine to the generator, with no gearbox.

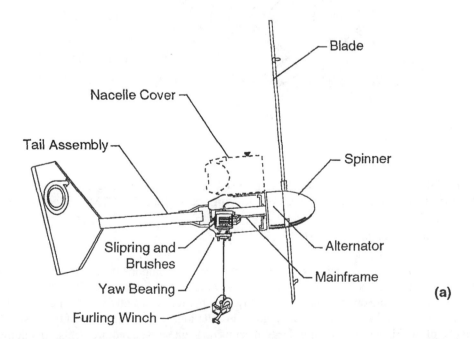

Figures 2-8. Typical HAWT power trains. (a) Small-scale: Bergey BWC-1500/1.5 kW (*Courtesy of Bergey Windpower Company, Inc.*)

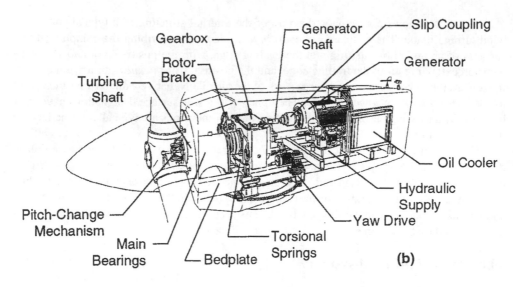

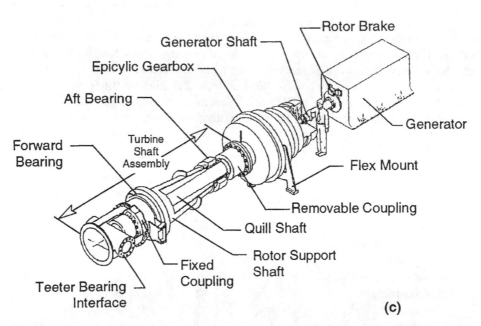

Figure 2-8 (Concluded). Typical HAWT power trains. (b) Medium-scale: Vestas V27/225 kW (*Courtesy of Vestas Wind Systems A/S*) (c) Large-scale: Mod-5B/3.2 MW [Boeing 1988]

Turbine Shaft Assembly

The *turbine shaft assembly* is one of the most critical components in a HAWT, because of its dual structural/mechanical function. Rotor weight, thrust, torque, and lateral forces all cause fatigue loading on this component whose design lifetime usually equals or exceeds that of the total system. The design, manufacture, installation, and maintenance of this assembly must be of the highest quality in order to meet its difficult service requirements. In addition to the primary shaft with its bearings, lubrication, and couplings, turbine shaft assemblies may include rotor control and safety devices (such as sensors and a rotor brake), rotary hydraulic couplings, sliprings (for power and data transfer), and attached electrical and mechanical equipment with their necessary cables and piping.

The proper amount of *torsional flexibility* in the turbine shaft can attenuate rotor torque oscillations before they reach the gearbox. However, the turbine shaft also needs to be strong in bending, in order to support the rotor weight. In order to meet both of these requirements, the assembly may be composed of two concentric shafts connected at the hub end and separated at the gearbox end (Fig. 2-8(c)). The flexible inner shaft, sometimes called a *quill shaft*, transmits only torque to the gearbox.

A HAWT *speed-increasing gearbox* has a step-up ratio (equal to the generator shaft speed divided by the turbine shaft speed) that is as high as 100 in a large-scale HAWT, depending on blade tip speed, rotor diameter, and generator design. Parallel-shaft, epicyclic (planetary), and hybrid designs are used. Parallel-shaft gearboxes are more readily available than epicyclic units, cost less, and provide easier access to the rotor for cables and piping. To minimize hub and gear case deflections that can cause premature failure of bearings and gears, nacelle structural supports must be designed to provide adequate stiffness [Rahlf *et al.* 1998]. Strength alone is not sufficient. This is particularly true for parallel-shaft gearboxes.

The *generator drive shaft* is a conventional machine element with bolted flanges on both ends. If the HAWT has a rotor brake, its disk may be mounted on this shaft rather than on the turbine shaft for multiplication of the braking power (equal to the square of the step-up ratio). If there is a pitch control mechanism for stopping the rotor, the rotor brake is usually used only for emergencies, parking, and maintenance. Other components that may be on the generator drive shaft are a rotor-positioning device, a maintenance lock, and a torque-limiting device (with friction or break-away action).

Generators

All types of *electrical generators* are used in HAWTs. Small-scale turbine rotors may drive variable-speed *alternators* and *DC generators*, minimizing the amount of rotor speed control required. Medium- and large-scale HAWTs use *AC generators*. The requirements of size, weight, efficiency, and durability for HAWT applications can usually be met by modifying commercially-available generators. The requirements that most often govern the design of HAWT generating systems are those of *cost, power quality* (harmonic distortion), *power factor* (reactive power), and *torsional damping* (to attenuate cyclic torques).

Induction AC generators are often selected because of the torsional damping provided by their inherent *slip* (shaft speed in excess of synchronous speed by an amount proportional to torque) and because their cost is relatively low. A high-slip induction generator can provide a modest amount of "softness" to the power train, although efficiency is reduced in the process. *Synchronous AC generators* produce higher power quality and higher efficiency than induction generators, but they require external voltage regulators and are unable to provide significant softness or damping to the power train.

A *variable-speed constant-frequency* (VSCF) generating system can provide more damping than a high-slip induction generator, electrical efficiency and power quality approaching that of a synchronous machine, and the ability to control turbine speed (within limits) for increased aerodynamic efficiency. VSCF systems require additional *power electronic equipment* which is usually located on the ground. Two general types of VSCF systems are used in HAWTs. The first type employs a *doubly-wound generator* (windings in both rotor and stator), in which most of the power is generated in the stator at line frequency. The fraction of power generated in the rotor is equal to the ratio of the slip frequency to the line frequency. Rotor power is generated electronically from the slip frequency to the line frequency using a *cycloconverter*. In the second type of VSCF, the generator is a conventional synchronous machine, and the frequency of all of the power generated is restored to line frequency through *AC-DC-AC conversion.*

Direct-Drive vs. Speed-Increasing Gearboxes for Utility-Scale HAWTs

Large-scale HAWT rotors turn at relatively low shaft speeds, to maintain an optimum *tip-speed ratio* (tip forward speed divided by free-stream wind speed) and partially because of the necessity of limiting blade tip speeds in order to control noise. For example, the GE Model 1.5s rotor shown in Figure 2-3 operates at speeds from 10 to 22.2 rpm. Thus, the design of a direct-drive generator for a utility-scale HAWT is, by necessity, unconventional. The issue of direct-drive *vs.* gear-driven generators was addressed by Bywater *et al.* [2005]. These authors observed that direct drives for large-scale wind turbines have attracted increased commercial support because of their simplicity, quiet operation, a small efficiency advantage, and (most importantly) avoidance of costly gear failures.

Because of its slow shaft speed, a direct-drive generator of a utility-scale HAWT is a large, heavy machine. To be competitive with relatively light gearbox-driven generators, designers of direct-drive power trains must employ innovative measures to reduce the size, weight, and cost of the generator. Generally, the increased size and strength of the structure of a direct-drive generator can be used to good advantage in supporting the turbine rotor.

Since 1994, the *Enercon Corporation* has been the principal innovator of large-scale direct-drive HAWTs, producing machines with rated powers from 500 kW to 6.0 MW. Several other manufacturers have recently introduced direct-drive generators with ratings from 750 kW to 2.0 MW. However, speed-increasing gearboxes remain the dominant design approach for large-scale HAWTs. A recent survey [Bywater *et al.* 2005] found that an estimated 85 percent of the worldwide wind power capacity utilizes gearbox-driven generators.

The Nacelle Structure Subsystem

Referring to Figure 2-9, the HAWT *nacelle structure* is the primary load path from the turbine shaft to the tower. Figure 2-9(a) illustrates a *bed plate assembly* that provides a stiff floor on which to mount the turbine shaft bearings, the power train components, and the yaw drive mechanism. Nacelle structures are usually a combination of welded and bolted steel sections which form trusses or box beams, with an enclosure to protect equipment and maintenance personnel (Figure 2-9(b)). Stiffness and static strength, rather than fatigue strength, are the usual design drivers of nacelle structure.

HAWTs require a *yaw drive mechanism* so that the nacelle can turn to keep the rotor shaft properly aligned with the wind. As shown in Figure 2-9(c), this mechanism includes a large bearing that connects the bed plate to the tower and serves as a primary load path. An *active yaw drive* (one which turns the nacelle to a specified azimuth) contains one or two motors (electric or hydraulic), each of which drives a *pinion gear* against a *bull gear*, and an automatic yaw control system with its *wind direction sensor* mounted on the nacelle. A *passive yaw drive* permits wind forces to orient the nacelle.

A two-bladed rotor produces cyclic loads on the yaw drive that are much larger than those from a three-bladed rotor. Teetering removes some but not all of these additional yaw loads. The yaw drive mechanism of a two-bladed HAWT, therefore, is a robust component designed to resist both fatigue and wear. Even then, *yaw brakes* may be required to hold the nacelle in position.

HAWTs also require a *yaw slip ring* or *cable-wrap device* to transfer electrical power, control signals, and data from the moving nacelle to stationary cables in the tower. Other

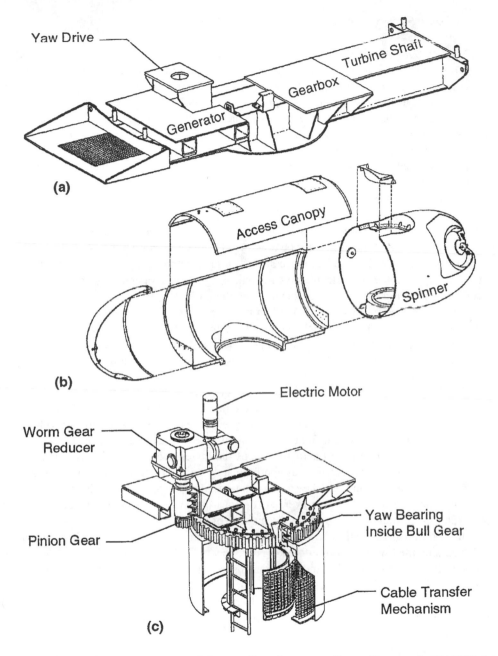

Figure 2-9. Typical components of the nacelle subsystem of a medium-scale HAWT. (a) Bedplate (b) Protective shroud (c) Yaw drive and power transfer (*Courtesy of Westinghouse Electric Corporation*)

auxiliary equipment in the nacelle may include a control computer, hydraulic and lubrication pumps and reservoirs, safety equipment (*e.g.*, for fire suppression), and service power for lights and maintenance tools.

The Tower Subsystem

A HAWT *tower* raises the rotor and power train to the specified *hub elevation*, the distance from the ground to the center of the swept area. Minimum tower height is determined by the required *ground clearance* (distance between the lowest point of the swept area and the ground). An increase in tower height above this minimum depends on the trade-off between the marginal increases in energy capture (because average wind speeds generally increase with increasing elevation) and the marginal increase in system cost, including construction and maintenance costs.

The principal component in the HAWT tower subsystem is the tower structure itself, which can be a steel or reinforced-concrete shell, or a steel truss. Cylindrical shell towers such as those shown in Figures 2-3 and 2-4 are now used almost exclusively to support large-scale wind turbines. A type of lightweight tower construction is shown in Figure 2-10 in which tension cables are used for truss diagonals in place of structural sections. This reduces wake-induced or *tower shadow* loads on downwind rotors with rigid hubs. A typical HAWT tower subsystem also includes a device such as a ladder and/or a powered lift for maintenance, as well as cables for carrying power, control signals, and operational data between the nacelle *yaw slip ring* (or *cable-wrap* device) and the ground.

HAWT towers are usually supported on massive *spread foundations* of reinforced concrete, although local site conditions may make a smaller foundation tied down by *rock anchors* the most economical design. Anchor bolts securing the tower to the foundation usually extend down to the bottom of the concrete. Resistance to overturning and allowable soil pressures are the key design requirements for HAWT foundations.

The dimensions of the tower structure (*e.g.* height, diameter, wall thicknesses, and base shape) can be adjusted within limits to obtain the desired *fundamental system frequency*, for purposes of minimizing any structural-dynamic responses of the HAWT to unsteady rotor loads. The fundamental system frequency is slightly lower than the frequency of a flexural pendulum with the same length and stiffness as the tower and a lumped mass equal to the masses of the rotor and the nacelle plus approximately one-third of the tower mass. The ratio of the fundamental system frequency to the rotor speed is an indication of the *relative rigidity* of the support provided by the tower. A *stiff tower* is one in which the frequency-to-rotor speed ratio is substantially larger than the number of blades. In other words, a stiff tower's natural vibration modes all have frequencies higher than the repetitive forcing caused by blades passing the tower. A *soft-soft tower* is one in which the fundamental frequency ratio is substantially less than one. This indicates that a soft-soft tower can attenuate vibratory loads with frequencies of 1P and higher, at least to some degree. An intermediate ratio characterizes a *soft tower*. Modern HAWTs usually have soft or soft-soft towers. Stiff towers are generally not cost effective because of their relatively high weight, nor do they bend enough to significantly attenuate transient thrust loads from wind gusts.

The Ground Equipment Station

Located in the *ground equipment station* are those components which are necessary for properly interfacing the HAWT with the electric utility or other distribution system. Power conditioning and control equipment (*e.g.* transformers, circuit breakers, and the electronic components of variable-speed generating systems), a ground control unit, and data recording devices are typical components in the ground equipment station. Some or all of these may be located within the base section of the tower.

Figure 2-10. Cables in a truss tower reduce both its weight and the magnitude of wake-induced loads on a downwind rotor. (*Courtesy of Kenetech/U.S. Windpower Corporation*)

General Configuration of a Vertical-Axis Wind Turbine

There are several types of vertical-axis wind turbines, each with its own distinctive features. Figure 2-11 presents the general configuration of a modern VAWT of the *Darrieus* type, named after the inventor of this distinctive curved-blade rotor. Its principal subsystems are (1) the rotor, (2) the power train, (3) the support structure, (4) the foundations, and (5) the ground equipment station. While some of the subsystem and component names are common to both VAWTs and HAWTs, their configurations are often quite different. Symmetry about its vertical axis allows a VAWT to accept winds from any direction, so no yaw drive mechanism is needed. This is one of its primary advantages.

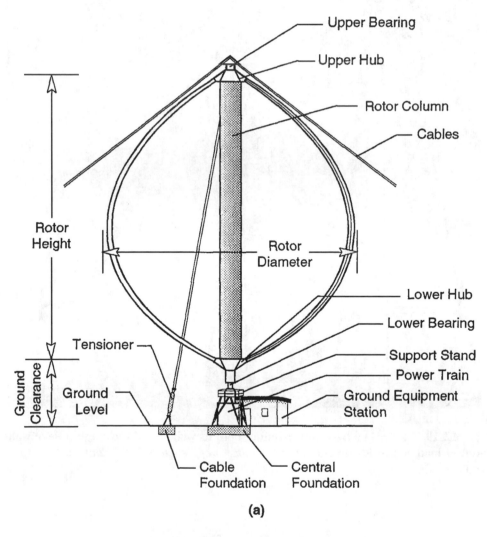

(a)

Figure 2-11. Schematic views of the principal components of a modern VAWT. (a) General configuration

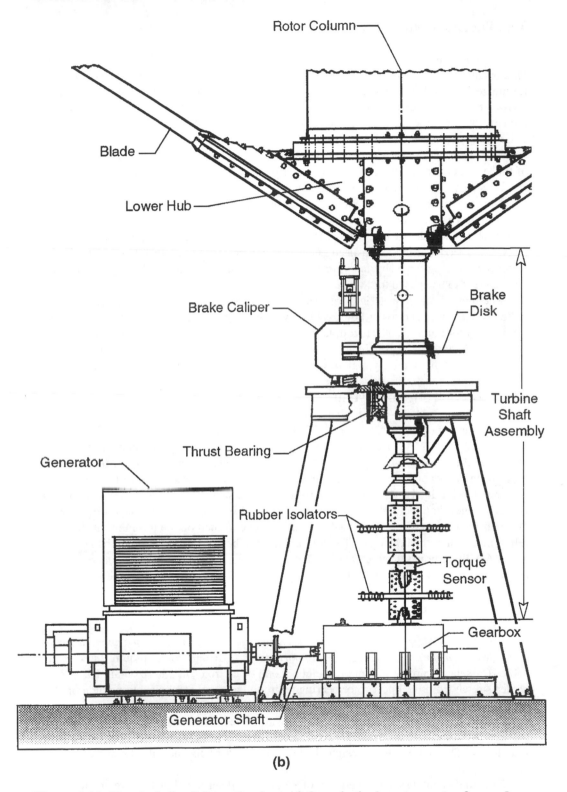

(b)

Figure 2-11 (Concluded). Schematic views of the principal components of a modern VAWT. (b) Power train (*Courtesy of Sandia National Laboratories*)

The Turbine Rotor Subsystem

As shown in Figure 2-11, the main components of a Darrieus VAWT rotor are its *curved blades* with ends fastened to rigid *upper and lower hubs* separated by the *rotor column*. To minimize internal bending stresses during rotation, blades are shaped to approximate a *troposkien* (from the Greek for "turning rope"), a shape with zero bending stress. VAWT rotors contain two or three fixed-pitch blades, usually symmetrical in cross-section and without twist or taper. As with a HAWT, the *swept area* of a VAWT is defined by the projection on a vertical plane of the surface generated by the moving blades. *Rotor diameter* is the width of the swept area at its equator. *Rotor height* is the distance between upper and lower hubs and is usually 15 percent to 30 percent larger than the diameter.

Darrieus rotors are stall-controlled, because pitch-change mechanisms have not been found to be cost-effective. Motoring of the generator is the usual method for starting Darrieus rotors, since the blades develop lift and torque only through a superposition of rotational (forward) speed and the wind speed and, therefore, are not normally self-starting. VAWT rotors are usually stopped by applying a *rotor brake* in the power train, although trailing-edge flaps have also been used for this purpose.

The most common material for Darrieus blades is extruded aluminum alloy. Blades are bolted to the upper and lower hubs, each of which is rigidly connected to the rotor column. Thus, the rotor column collects torque from the two hubs and transmits it to the power train. Buckling strength is the principal structural requirement on the rotor column, since this column must react the relatively high downward loads produced by the supporting cables.

The Power Train Subsystem

Comparison of Figures 2-11(b) and 2-8 shows that there are three major differences between HAWT and VAWT power trains. First, VAWT power-train components are located at or near the ground, which provides for easier maintenance and requires a relatively low support stand. Second, the VAWT turbine shaft assembly carries axial and torque loads only, with no bending loads like those on a HAWT turbine shaft. Third, the VAWT rotor brake is much larger than the parking brake typical of a HAWT, because it must be able to stop a Darrieus rotor operating at top speed. It may even be located on the turbine shaft for added reliability, so that the braking torque does not have to be transmitted through the gearbox.

VAWT gearboxes, generator-drive shafts, and generators have the same general configurations and functions as described previously for HAWT power-train components.

The Support Structure Subsystem

The VAWT support structure consists of *upper and lower rotor bearings, structural cables with tensioning devices*, and a *support stand*. Darrieus rotors require three or four cables (or sets of cables) to support the upper end of the rotor in a horizontal plane. These cables stretch from the upper rotor bearing to ground anchors at an elevation angle of about 30 to 40 degrees. Cable tension causes a downward thrust load on the upper rotor bearing equal to one-half or more of the tensile loads in all cables. This thrust load passes downward through the rotor column, to the lower rotor bearing, the support stand, and finally to the foundation. Depending on the design, the upper and lower hubs may also be in the compressive load path.

The fundamental system frequency of a VAWT is determined by the size and tension of the cables, because they are the elastic springs which restrain motions of the center of mass of the VAWT rotor. Like a HAWT tower, cables can provide stiff, soft, or soft-soft support to the rotor. Cables are usually sized and tensioned to a soft condition, producing a fundamental system frequency greater than the rotor speed but less than the blade passing frequency (*i.e.*, the number of blades times the rotor speed).

The height of the support stand is equal to the ground clearance. The minimum ground clearance required for safety is usually much less for a VAWT than for a HAWT, because the speed of a VAWT blade near the ground is relatively low. Any increase in support-stand height above the minimum ground clearance again depends on a trade-off between the marginal increase in energy capture and the marginal increase in system cost. The latter is relatively high because many cable changes must be made in order to maintain the desired fundamental system frequency as the upper bearing is raised. These include larger cables, more cables, and higher cable tension, which in turn require more capacity in rotor bearings, more buckling strength in the rotor column and support stand, and more weight in the cable foundations.

All these cost factors combine to keep the elevation of the center of a typical Darrieus VAWT rotor lower than that of a HAWT with the same swept area. Because average wind speed generally increases with elevation, the annual average *wind power density* of a VAWT rotor (in watts of wind power per square meter of swept area) is usually lower than that of a comparable HAWT rotor at the same site.

The Foundation Subsystem

VAWT foundations include a *central foundation* under the support stand and a *cable foundation* at the lower end of each set of support cables. Because the central foundation is not subject to uplift or overturning loads, its weight is usually less than the combined weights of the cable foundations, and it is not as wide or as heavily reinforced as a HAWT tower foundation. VAWT cable foundations contain steel cable anchors, are heavily reinforced to resist tensile stresses, and are sized to prevent uplift or shifting which would result in loss of cable tension.

The Ground Equipment Station

The equipment on the ground that interfaces a VAWT with the electric utility system or other user is essentially the same as that required for a HAWT of the same power rating. This equipment may all be housed in an enclosure separate from the VAWT, or part of it may be located on the central foundation.

Wind Turbine System Performance

The principal measure of the performance of a wind turbine system is *annual energy output*, which is the *electrical energy* delivered to the customer during a complete year [ASME 1989]. *System power output* is often used as an intermediate measure of performance and is defined by electrical power output as a function of steady wind speed (graphed as the *power-versus-wind speed curve* or, simply, the *power curve*). Obviously, the net electrical energy produced by a wind turbine system will depend on the energy of the wind passing through its swept area, as well as on the efficiencies of its components.

A measure of the energy-conversion efficiency of a wind turbine system is its *coefficient of energy* or *energy recovery factor*, defined as the ratio of its electrical energy output to its wind energy input over the course of a year. Thus

$$C_E = \frac{AEO_G}{E_W} = \frac{\displaystyle\int_{year} P_O \, dt}{\displaystyle\int_A \left(\int_{year} p_W \, dt \right) dA} \tag{2-1}$$

where

$$
\begin{aligned}
C_E &= \text{coefficient of energy or energy recovery factor} \\
AEO_G &= \text{gross annual energy output (kWh/y)} \\
E_W &= \text{annual wind energy input (kWh/y)} \\
P_O &= \text{system output power (W)} \\
t &= \text{elapsed time (h)} \\
A &= \text{swept area of the turbine rotor; projection on a vertical plane (m}^2\text{)} \\
p_W &= \text{wind power density (W/m}^2\text{)}
\end{aligned}
$$

The term *gross annual energy output, AEO_G*, refers to the calculated electrical energy delivered by a turbine to a specified system output point in one year under the following conditions:

--- Wind speeds and durations are equal to a specified annual wind regime.
--- Wind speeds and durations are not affected by other turbines.
--- Turbine operates whenever wind speeds are in the specified operating range
--- HAWT rotor axis is always aligned with the wind direction.
--- Turbine performance is not degraded by age, wear, dirt on blades, *etc.*
--- Shutdowns for maintenance occur only when wind speeds are below the operating range.

Thus, gross annual energy output is a theoretical upper bound on the turbine energy output.

It is often convenient during the design process to calculate the coefficient of energy on the basis of a hypothetical *reference wind regime* that is representative of the wind speed distributions in time and space expected to be present during operations. The principal factors involved in determining reference values of annual energy production and annual wind energy will be introduced here and discussed in more detail in later chapters.

Annual Wind Energy Input

Consider a horizontal *streamtube* of wind, which is air flowing in an imaginary pipe with all particles passing a given cross-section at the same speed. The *wind power density* at a point in this streamtube is the fluid-dynamic power per unit of cross-sectional area, given by the following equation:

$$p_W = 0.5\rho U^3 \tag{2-2}$$

where

ρ = air density = 1.2250 kg/m^3 at *sea-level standard* (SLS) conditions
U = horizontal component of the steady free-stream wind speed (m/s)

Thus, wind power density is directly proportional to the cube of the wind speed, and this fact is fundamental to both wind turbine design and site selection.

For purposes of calculating wind power density, wind speeds are usually averaged for about 0.1 hour to obtain the *steady wind speed*. This averaging process eliminates higher-frequency *turbulence* (instantaneous deviations from the average wind speed and direction) whose effects would be too rapid or too local to influence long-term energy conversion. The term "steady" is a relative one and relates only to a selected averaging period and elevation. The steady wind itself will vary over longer periods of time and with changes in elevation, even at a specific geographic location. Because of these variations, a meaningful measure of the wind as a power source is the *annual wind energy density*, or

$$e_W(z) = \int_{year} p_W(z)\,dt = 0.5\,\rho \int_0^\infty U^3 f_W(z)\,dU \tag{2-3}$$

where

e_W = annual wind energy density at elevation z (Wh/m^2/y)
z = elevation above ground level, *AGL* (m)
f_W = frequency distribution function of U at elevation z [(h/y)/(m/s)]

Models of the Steady Wind

Two simple models are commonly used together to calculate the frequency distribution function f_W in terms of both wind speed and elevation. These are a *Weibull model* for the frequency distribution function of wind speed at a specified *reference elevation*, and a *power-law model* for the variation of wind speed with elevation. The power-law equation for the vertical profile of the steady wind speed or *wind shear profile* is

$$U(z) = U_R(z/z_R)^\alpha \tag{2-4}$$

where

U_R = steady wind speed at the reference elevation, at the same time as U (m/s)
z_R = reference elevation (m)
α = empirical wind shear exponent

The exponent α is not a constant, but varies with the roughness of the terrain, temperature gradients in the atmosphere over the site, and steady wind speed [Justus and Mikhail 1976, Spera and Richards 1979]. Equations (2-3) and (2-4) can be combined to give

$$e_W = 0.5\rho \int_0^\infty U_R^3 (z/z_R)^{3\alpha} f_W (z_R) \, dU \tag{2-5}$$

Placing the wind shear exponent α inside the integration permits it to vary with wind speed U. The Weibull model for f_w at the reference elevation z_r is as follows:

$$f_W = (8,760/C_R) k_R (U/C_R)^{k_R - 1} \exp\left[-(U/C_R)^{k_R}\right] \tag{2-6}$$

where

C_R = empirical Weibull *scale factor* for winds at the reference elevation (m/s)
k_R = empirical Weibull *shape factor* for winds at the reference elevation
exp [ ] = exponential function of [ ]

The *annual average wind speed* at a selected elevation can be expressed in terms of the Weibull factors as

$$U_A(z) = C(z)\, \Gamma[1 + k(z)] \approx (0.90 \pm 0.01)\, C(z) \tag{2-7}$$

where

$U_A(z)$ = annual average wind speed at elevation z (m/s)
$C(z), k(z)$ = Weibull scale (m/s) and shape factors at elevation z, respectively
Γ [ ] = gamma function of [ ]

DOE/NASA Reference Design Site and Reference Wind Regime

The characteristics of a *reference design site* were specified for the U. S. Federal Wind Energy Program (described in Chapter 3, Part A) in the 1970's in order to provide a uniform basis for research and development projects. This design site was one with level terrain and an annual average wind speed of 14 mph (6.24 m/s) at a reference elevation of 30 ft (9.1 m), and a wind shear exponent of 1/7. Early wind resource studies indicated that this site description was representative of large areas in the U. S. and around the world suitable for installation of wind turbines. In the early 1980's, during the development of megawatt-scale HAWTs for the U. S. Department of Energy, NASA engineers further defined the frequency distribution and vertical profile of a *reference wind regime* in the following terms:

$$\begin{aligned}
z_R &= 10.0 \, m \\
\alpha_R &= 0.351 - 0.192 \log(U_R) \\
C_R &= 7.17 \, m/s \\
k_R &= 2.29 \\
\rho_R &= 1.225 \, kg/m^2
\end{aligned} \tag{2-8}$$

Figure 2-12 shows the reference frequency distributions of steady wind speed, f_w, and wind power density, e_w, at an elevation of 10 m. The area under the latter curve equals an annual wind energy density of 2,324 kWh/y/m². The wind speed at which the energy frequency is a maximum lies in the middle of the most energetic wind range at the design site. This speed is designated here as the *design wind speed, U_D,* because annual energy production is usually a maximum if a wind turbine is designed for maximum aerodynamic efficiency at $U \approx U_D$. With the model specified by NASA for the vertical profile of the wind speed [see Figure 8-13; Spera and Richards 1979], frequency distributions can be calculated for any given elevation. The parameters C, k, U_A, U_D, and e_w for the reference wind regime are given in Table 2-1 *vs.* elevation. It can be seen that the reference wind energy density increases significantly with elevation.

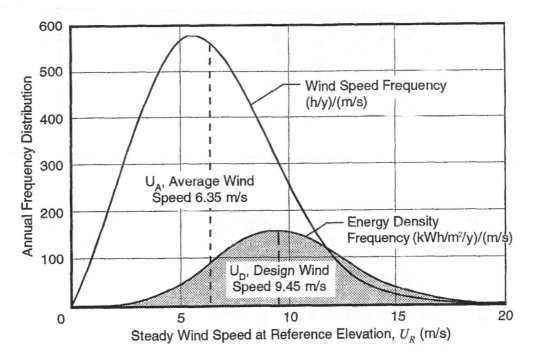

Figure 2-12. Frequency distribution functions for wind speed and wind energy density at the DOE/NASA Design Reference Site. Elevation = 10 m; air density = 1.225 kg/m³.

Referring to Equations (2-1) and (2-3), the reference annual wind energy becomes

$$E_W = \int_A e_W(z)\, dA = \int_H e_W(z) w(z)\, dz \qquad (2\text{-}9)$$

where

E_W = reference annual wind energy input to the rotor swept area (W/y)
H = vertical height of the swept area; includes *tilt*, if any (m)
w = width of the swept area at elevation z (m)

Table 2-1. Parameters of the DOE/NASA Reference Wind Regime *vs.* Elevation

Elevation z (m)	Scale C (m/s)	Shape k	Average Speed U_A (m/s)	Design Speed U_D (m/s)	Energy Density e_w (kWh/m²/y)
5	6.30	2.16	5.58	8.55	1,656
10	7.17	2.29	6.35	9.45	2,324
15	7.73	2.37	6.85	10.05	2,836
20	8.16	2.43	7.24	10.50	3,276
25	8.51	2.48	7.54	10.85	3,664
30	8.80	2.52	7.80	11.15	4,009
35	9.06	2.56	8.04	11.40	4,331
40	9.28	2.59	8.24	11.65	4,621
45	9.49	2.62	8.44	11.85	4,907
50	9.68	2.64	8.60	12.00	5,183
55	9.85	2.67	8.76	12.20	5,425
60	10.01	2.69	8.90	12.35	5,669
65	10.16	2.71	9.03	12.50	5,902
70	10.30	2.73	9.16	12.65	6,123
75	10.44	2.75	9.29	12.77	6,350
80	10.56	2.77	9.40	12.90	6,545
85	10.68	2.79	9.51	13.00	6,744
90	10.80	2.80	9.61	13.10	6,958
95	10.91	2.82	9.72	13.22	7,145
100	11.01	2.83	9.81	13.30	7,327

For example, taking e_w from Table 2-1 and performing the integrations numerically, the reference annual wind energy inputs to the Mod-5B HAWT and the 34-m VAWT are

Mod-5B HAWT:

$$(12.2\ m \leq z \leq 109.7\ m,\ A = 7,470\ m^2):\ E_W = 41,710\ MWh/y \qquad (2\text{-}10a)$$

34-m VAWT:

$$(7.1\ m \leq z \leq 49.0\ m,\ A = 955\ m^2):\ E_W = 3,640\ MWh/y \qquad (2\text{-}10b)$$

Estimating Rotor Shaft Power

Rotor shaft power is the resultant of the aerodynamic *lift and drag forces* acting on the turbine blades and transmitted by them to the rotor hub. Lift forces act at right angles to the wind vector at the blade section, and drag forces act in the same direction as the local wind vector. Computer models for estimating rotor power are usually based on *strip theory*, in which each incremental length or "strip" of each rotor blade is assumed to produce aerodynamic forces independently of all other strips and blades. This assumption has been found to be satisfactory for analyzing the performance of modern wind turbine rotors with relatively low *solidity,* which is the ratio of the total *planform* (maximum projected) areas of the blades to the swept area of the rotor.

Aerodynamic Forces

Figure 2-13 illustrates the aerodynamic wind and force vectors acting on a section of a rotor blade of a horizontal-axis wind turbine. In this simplified example, the wind speed is steady, the wind direction is parallel with the rotor shaft, the blade *pitch* (longitudinal) axis is perpendicular to the shaft, and the blades section is perpendicular to the pitch axis. The blade section moves tangentially forward in the plane of rotation and has no axial motion. In a more general case, the blade sections may experience cyclic axial motions as a result of elastic deformation and/or hinging. These cyclic velocities must then be added to or subtracted from the axial airspeed (which itself may have cyclic components) when calculating *aeroelastic* fatigue loads.

The vectors in Figure 2-13 that relate directly to the estimation of rotor shaft power are the *relative airspeed*, the *tangential force* (also called the *torque force*), and the *axial force* (also called the *thrust force*). The magnitudes of these vectors are calculated at discrete sections or *stations* along the length of the blade axis, usually at 5 percent to 10 percent intervals, as follows:

$$V_R = \sqrt{V_X^2 + V_Y^2} \tag{2-11a}$$

$$V_X = (1 - a)U \tag{2-11b}$$

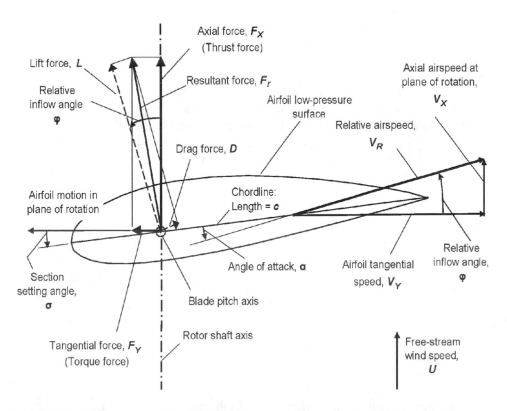

Figure 2-13. Diagram of wind and force vectors acting on a section of an airfoil.

$$V_Y = r\Omega \tag{2-11c}$$

$$F_X = L\cos\phi + D\sin\phi \tag{2-11d}$$

$$F_Y = L\sin\phi - D\cos\phi \tag{2-11e}$$

$$L = 0.5\rho V_R^2 c\, C_L(\alpha) \tag{2-11f}$$

$$D = 0.5\rho V_R^2 c\, C_D(\alpha) \tag{2-11g}$$

where
- V_R = relative airspeed (m/s)
- V_X = steady axial airspeed; wind speed at plane of rotation (m/s)
- V_Y = airfoil tangential speed (m/s)
- a = wind speed axial induction (retardation) factor
- U = free-stream wind speed (m/s)
- r = radial distance from rotor axis to station (m)
- Ω = rotor shaft speed (rad/s)
- F_X = axial airload intensity (N/m)
- F_Y = tangential airload intensity (N/m)
- L = lift force intensity (N/m)
- D = drag force intensity (N/m)
- ϕ = relative wind inflow angle; measured clockwise from plane of rotation to local relative airspeed vector (deg)
- ρ = air density (kg/m^3)
- c = chordline length (m)
- C_L = airfoil lift coefficient; function of *angle of attack*
- C_D = airfoil drag coefficient; function of *angle of attack*
- α = angle of attack; angle from relative airspeed vector to chordline (deg)

The term *intensity* designates forces and loads per unit length along the blade axis. *Rotor shaft thrust and torque* are calculated by integrating the effects of the relevant airload intensities over the lengths of all blades, as follows:

$$T = B \int_{R_0}^{R} F_X\, dr \tag{2-12a}$$

$$Q = B \int_{R_0}^{R} F_Y\, r\, dr \tag{2-12b}$$

where
- T = rotor shaft thrust (N)
- Q = rotor shaft torque (m-N)
- B = number of rotor blades
- R = rotor tip radius (m)
- R_0 = rotor hub radius; inner end of active airfoils (m)

The integrations in Equations (2-12) are normally performed numerically, applying Simpson's Rule. The magnitude of the *rotor shaft power* can then be calculated from shaft torque and speed, as follows:

$$P_R = Q\,\Omega \tag{2-13}$$

where
$\quad P_R$ = rotor shaft power (N-m/s)

Rotor Power Coefficient

The conventional measure of the aerodynamic performance of a wind turbine rotor (regardless of configuration) is its *rotor power coefficient*, which is the ratio of the *rotor power density* (mechanical power at the turbine shaft per unit of swept area) to the *wind power density*, or

$$C_{P,R} = \frac{P_R/A}{p_W} = \frac{P_R}{0.5\rho U^3 A} \tag{2-14}$$

where
$\quad C_{P,R}$ = rotor power coefficient
$\quad P_R$ = mechanical power at the turbine rotor shaft (W)

Axial Induction (Retardation) of the Wind

The axial induction factor, a, in Equation (2-11b) is the fractional decrease in wind speed between the upwind free-stream and the plane of the rotor, as power is extracted at the rotor from the wind. According to the *axial momentum theory* [Rankine 1865, W. Froude 1878, R. Froude 1889], the wind speed downstream of the rotor is reduced further by the same amount as upstream. On this basis, the total retardation of the free-stream wind, from upstream of the rotor to the *far wake* downstream, is $2a$. This would indicate that the maximum value of a for a wind turbine is 0.5, when the wind speed in the far wake is zero. However, momentum theory is considered to be invalid for induction factors larger than about 0.4 [Wilson 1994].

By applying the Rankine-Froude theory to the calculation of rotor power, it has been shown that

$$C_P = 4a(1 - a)^2, \quad a \leq 0.4 \tag{2-15a}$$

from which

$$C_{P,max} = 0.593 \quad \text{at } a = 1/3 \tag{2-15b}$$

Therefore, an axial induction factor of approximately 1/3 is a common goal for a wind turbine operating at its *design wind speed* (see Figure 2-12).

As discussed in detail by Eggleston and Stoddard [1987], the axial induction factor can be determined from the rotor thrust force, normalized as follows:

$$C_T = \frac{T}{0.5\rho U^3 A} \tag{2-16}$$

where

C_T = rotor thrust coefficient

A combination of the momentum theory and test data is required in order to derive an equation relating a to C_T, which is valid for values of a larger than about 0.4. Figure 2-14 [data from Wilson and Lissaman 1974, Stoddard 1977] shows the momentum theory relationship at lower induction factors combined with rotor test data at higher induction factors. A simple empirical equation has been fitted here to cover both ranges of a [Spera 2008], as follows:

$$a = 0.27 \, C_T + 0.10 \, C_T^3 \tag{2-17}$$

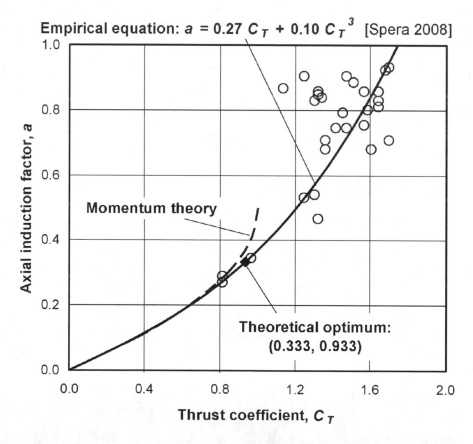

Figure 2-14. Empirical equation derived for calculating an axial induction factor from the rotor thrust coefficient. Equation is based on the wind turbine and helicopter test data shown in the figure. [Data: Lock *et al.* 1926, Wilson *et al.* 1976, and Eggleston and Stoddard 1987].

Applying Equation (2-16) to condition (2-14b), a thrust coefficient of 0.933 produces an induction factor of 1/3 and the theoretical maximum rotor power coefficient. This "theoretical optimum" point is indicated in Figure 2-14.

One of two approaches can now be taken to produce a *rotor power curve* for the turbine, which defines estimated rotor power output, P_R, versus free-stream wind speed, U. The first approach requires iteration and the second does not, but iteration can be done rapidly in spreadsheet software.

(1) Specify a free-stream wind speed, U; assume a trial value of $a = a'$, calculate C_T by Equation (2-15), calculate a by Equation (2-16), iterate until $a = a'$, and finally, calculate P_R; or

(2) Specify an axial airspeed at the rotor, V_X, calculate P_R and C_T, calculate a from Equation (2-16), and finally, calculate U from Equation (2-11b).

Airfoil Angle of Attack and Twist

As illustrated in Figure 2-13, the angle of attack is measured counter-clockwise from the geometric chordline of the airfoil section to the relative airspeed vector and is calculated as follows:

$$\alpha = \phi - \sigma \tag{2-18a}$$

$$\phi = \arctan(V_X / V_Y) \tag{2-18b}$$

$$\sigma = \beta + \theta \tag{2-18c}$$

where

σ = section setting angle; measured counter-clockwise from the plane of rotation to the local chordline (deg)

β = section pre-twist angle; built-in rotation of local chordline; measured counter-clockwise from an arbitrary datum (deg)

θ = blade pitch angle; rotation of blade pitch axis and all chordlines; measured counter-clockwise from pre-twist datum; applied during blade installation and/or by active pitch control system (deg)

As discussed further in Chapter 5 (see Table 5-3), the blade pre-twist angle (or simply twist angle), β, is smallest at the tip and largest at the root, in order to keep the angle of attack, α, in an optimum range for maximum energy capture. Near the root, the forward speed of an airfoil section, V_Y, becomes small, so the relative inflow angle, ϕ, becomes large in accordance with Equation (2-18b). As ϕ increases from tip to root, β must increase accordingly if the angle of attack is to remain in the preferred design range.

Models of Airfoil Lift and Drag Coefficients

Mathematical models of the lift and drag coefficients in Equation (2-11f) and (2-11g), as functions of the local angle of attack, are required before rotor power and turbine performance can be estimated. As discussed in more detail in Chapter 6 and illustrated in Figure 6-2(a), airfoil behavior can be characterized by three flow regimes: the *attached regime*, for angles of attack from about -15 to +15 deg; the *high-lift/stall-development* regime, for angles of attack between about +15 and +30 deg; and the *deep-stall regime*, with angles of attack from +30 to more than +90 deg. In the attached regime, lift and drag coefficients are strongly airfoil-dependent, while in the deep-stall regime most airfoils exhibit similar lift and drag behavior.

In addition to its angle of attack, an airfoil's lift and drag behavior depend strongly on its *aspect ratio*, which is a measure of its slenderness. For a wind turbine rotor, blade aspect ratio is defined as the ratio of twice the active span length of a blade to its *mean chord*. For example, consider the typical blade *planform* shape shown in Figure 2-15 [Griffin 2001].

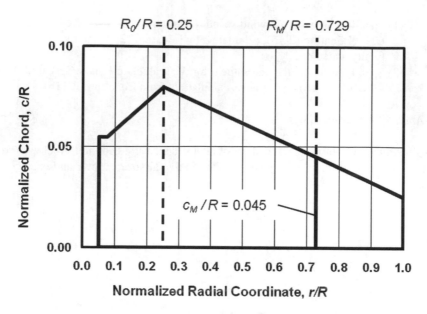

Figure 2-15. Typical planform of a HAWT rotor blade. The vertical scale is magnified for clarity. [Griffin 2001].

The aerodynamically active length of this blade is assumed to extend from 25 percent of span to the tip. The *mean radius* is defined as the radius that divides the active flow area into two equal areas. With these assumptions in mind, the aspect ratio of this sample blade is calculated to be 33.3, as follows:

$$R_M/R = \sqrt{\frac{1 + (R_0/R)^2}{2}} = 0.729 \tag{2-19a}$$

$$c_M/R = \frac{c(R_0/R = 0.729)}{R} = 0.045 \quad \text{by interpolation} \tag{2-19b}$$

$$\mu = \frac{2(1 - R_0/R)}{c_M/R} = \frac{1.500}{0.045} = 33.3 \tag{2-19c}$$

Aerodynamic force data are usually measured in a wind tunnel using sub-scale models that extend from wall to wall, eliminating any airflow around the ends of the model. These models are said to have an *infinite aspect ratio*, and the force coefficients obtained are referred to as *two-dimensional*. Mathematical models were developed for modifying two-dimensional coefficients to represent the aerodynamic behavior of finite-length airfoils, not only in the attached regime [Jacobs and Abbott 1932] but also in the far-stall regime [Viterna and Corrigan 1981]. These early model equations are given in Chapter 6.

Previous models of airfoil lift and drag coefficients have recently been updated and extended to include effects of *airfoil thickness* in addition to the effects of angle of attack, aspect ratio, and stall [Spera 2008]. Figures 2-16(a) and (b) illustrate these updated models. Lift and drag behavior in the attached and stall-development regimes, where forces are airfoil-contour dependent, are represented by one set of equations in the reference, labeled *pre-stall*. A second set of equations are labeled *post-stall* and represents behavior in the deep-stall

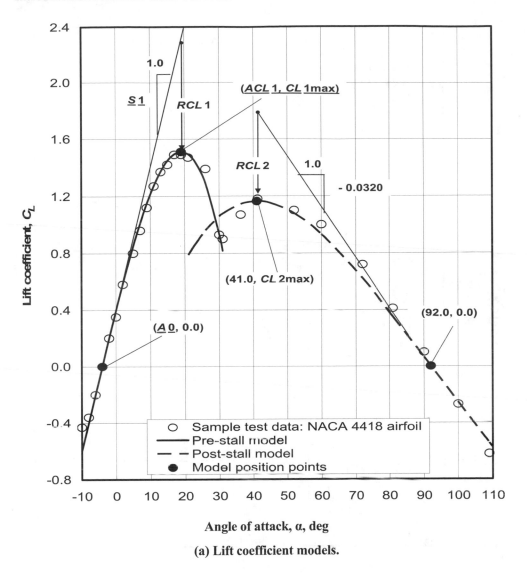

(a) Lift coefficient models.

Figure 2-16. Pre- and post-stall models of the variation of airfoil lift and drag coefficients with angle of attack. Input data are underlined. [Spera 2008]

regime where it is assumed that airfoil contour has only a secondary effect on aerodynamic forces. All the governing equations are available in the reference.

The peak power of wind turbine rotors equipped with modern airfoils designed specifically for *fixed-pitch* wind turbines [Tangler and Somers 1985, 1986; Tangler 1987] is limited by aerodynamic stall in higher winds. Therefore power calculations for these rotors require the joint application of both the pre- and post-stall equations.

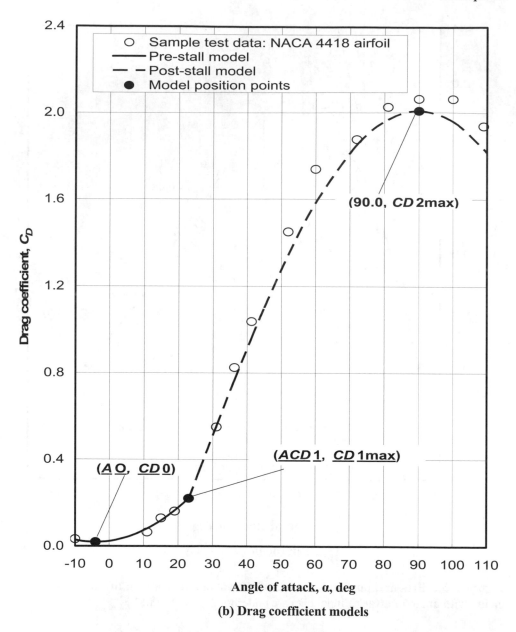

(b) Drag coefficient models

Figure 2-16 (Concluded). Pre- and post-stall models of the variation of airfoil lift and drag coefficients with angle of attack. Input data are underlined. [Spera 2008]

A convenient scaling parameter which integrates the principal aerodynamic effects of wind speed, rotor speed, and rotor size on the rotor power coefficient is the *tip-speed ratio*:

$$\lambda = \Omega R / U \qquad\qquad (2\text{-}20)$$

where

λ = tip-speed ratio
Ω = rotor speed (rad/s)
R = rotor radius, from axis to tip (m)

Two basic physical processes limit the *maximum rotor power coefficient* of an unducted wind turbine. First, a rotor increases the upwind static pressure, reducing the mass flow rate through its swept area and the wind energy available for conversion. Second, a rotor converts some of the linear kinetic energy of the wind to rotational kinetic energy in its wake, which is no longer available for conversion to mechanical energy. Figure 2-17 is a typical graph of rotor power coefficient *vs.* tip-speed ratio and illustrates the effects of these two limiting processes. The first or *retardation* process limits the rotor power coefficient at all tip-speed ratios to 0.593 (16/27), which is referred to as the *Betz* or, more-accurately, the *Lanchester-Betz limit* [Bergey 1980]. The second or *wake rotation* process reduces the maximum rotor power coefficient further, but this is important only if the tip-speed ratio is less than about 3.

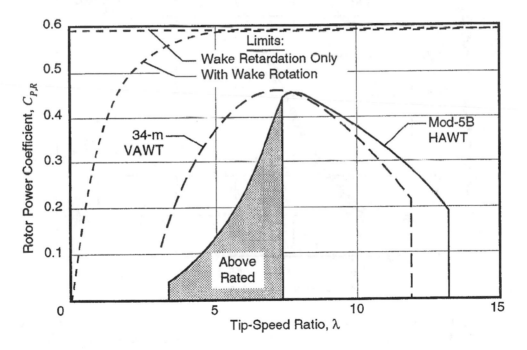

Figure 2-17. Sample variations of HAWT and VAWT rotor power coefficients with tip-speed ratio. Power coefficients are limited by retardation and wake rotation effects.

The design rotor power coefficients of the Mod-5B HAWT [Boeing 1988] and the 34-m VAWT [Dodd 1990] are also shown in Figure 2-17. These are typical of modern wind turbine rotors which have a small number of slender blades designed to operate at higher tip-speed ratios. As such, their power coefficients are not significantly affected by wake rotation losses.

For a HAWT with blade pitch control, when the wind speed exceeds the *rated wind speed* power is limited to the maximum permitted through the power train. In Figure 2-17, *above-rated operation* is indicated on the Mod-5B curve by the shaded region in which power coefficients are purposely reduced at lower tip-speed ratios. Rotor power coefficients have little significance in the above-rated regime. The power limit at above-rated wind speeds is determined by an economic trade-off study in which the value of the power lost is balanced by the reduced cost and maintenance expense of a smaller-capacity power train.

Figure 2-18 shows sample HAWT and VAWT *power output density curves* (power output per unit of swept area *vs.* wind speed) calculated from rotor power coefficients, specified air density, schedules of rotor speed *vs.* wind speed (for a variable-speed turbine), mechanical and electrical losses in the power trains, and power lost in transmission to the specified system output point. The calculation procedure is illustrated in Table 2-2.

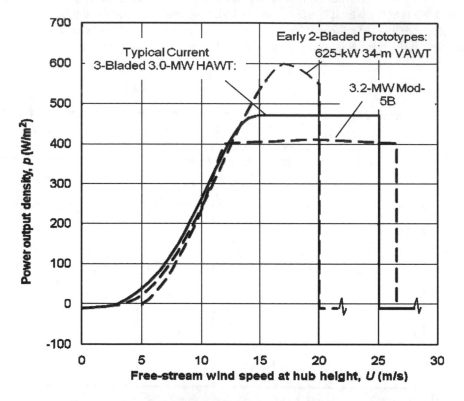

Figure 2-18. Sample HAWT and VAWT power density curves. The HAWT power is limited by blade pitch control, while the VAWT power is limited by aerodynamic stall.

Before calculating gross annual energy output, it is necessary to determine the design frequency distribution of free-stream winds at the *mid-elevation* of the area swept by the rotor. Weibull factors for this elevation are obtained by interpolation within Table 2-1. For the example wind turbines, these factors are

Mod-5B HAWT:

$$z_m = 61.0\,m$$
$$C_m = 10.04\,m/s \tag{2-21a}$$
$$k_m = 2.70$$

Table 2-2.
Power Output Density and Gross Annual Energy Output of the Mod-5B HAWT

Wind speed bin	Rotor speed	Tip-speed ratio	Rotor power coef.	Rotor power density	Losses (W/m²)			Power output density	Time in bin
					Gearbox & main bearings	Generator & power electronics	Auxiliary electrical (c)		
U (m/s)	Ω (rad/s)	λ	$C_{P,r}$ (a)	p_r (b) (W/m²)				p_O (W/m²)	Δt (h/y)
< 4.8	0.00	0.00	0.000	0	0	0	10	-10	1,088
5.0	1.35	13.19	0.192	14	3	8	5	-2	309
5.5	1.35	11.99	0.273	27	3	8	6	10	348
6.0	1.35	10.99	0.332	43	3	9	6	25	382
6.5	1.35	10.15	0.376	62	3	10	6	43	413
7.0	1.35	9.42	0.407	84	3	10	7	63	437
7.5	1.35	8.79	0.429	108	4	11	8	86	455
8.0	1.35	8.24	0.450	137	4	12	8	114	465
8.5	1.43	8.20	0.455	167	4	13	9	141	469
9.0	1.50	8.15	0.455	199	5	14	10	170	464
9.5	1.58	8.12	0.455	233	5	16	10	202	453
10.0	1.66	8.08	0.455	272	6	17	11	238	435
10.5	1.73	8.05	0.455	315	6	18	12	279	411
11.0	1.81	8.02	0.455	363	7	20	13	323	382
11.5	1.81	7.67	0.455	415	7	22	15	371	350
12.0	1.81	7.35	0.432	446	8	23	15	401	316
12.5	1.81	7.06	0.383	448	8	23	15	402	281
13.0	1.81	6.54	0.341	448	8	23	15	402	245
13.5	1.81	6.30	0.305	449	8	23	15	403	211
14.0	1.81	6.09	0.274	450	8	23	15	404	178
14.5	1.81	5.88	0.247	451	8	23	15	404	148
15.0	1.81	5.69	0.223	452	8	23	15	405	121
15.5	1.81	5.52	0.203	452	8	23	16	406	98
16.0	1.81	5.52	0.185	453	8	23	16	407	77
16.5	1.81	5.35	0.169	454	8	23	16	407	60
17.0	1.81	5.19	0.155	455	8	23	16	408	46
17.5	1.81	5.04	0.142	455	8	23	16	409	34
18.0	1.81	4.90	0.131	456	8	23	16	410	25
18.5	1.81	4.77	0.121	457	8	23	16	411	18
19.0	1.81	4.65	0.111	458	8	23	16	412	13
19.5	1.81	4.53	0.103	459	8	23	16	412	9
20.0	1.81	4.41	0.096	459	8	23	16	412	6
20.5	1.81	4.31	0.089	459	8	23	16	411	4
21.0	1.81	4.20	0.083	458	8	23	16	411	3
21.5	1.81	4.10	0.077	457	8	23	16	410	2
22.0	1.81	4.01	0.071	456	8	23	16	409	1
22.5	1.81	3.92	0.067	455	8	23	16	408	0
26.5	1.81	3.33	0.040	448	8	23	16	402	0

Gross Annual Energy Output: $AEO_G = A \times \Sigma\,(p_O\,\Delta t) = 12{,}040$ MWh/y   Total: 8,760

(a) Includes effects of wind shear and yaw heading error
(b) Air density = 1.225 kg/m³
(c) Includes transformer losses and power input for standby and starting operations

34-m VAWT:

$$z_m = 28.0\ m$$

$$C_m = 8.68\ m/s \tag{2-21b}$$

$$k_m = 2.50$$

The subscript m denotes a parameter at the mid-elevation of the rotor. The gross annual energy output in Equation (2-1) can now be expressed as

$$AEO_G = \int_{year} P_O(t)\,dt = A \int_0^\infty p_O(U)\,f_W(z_m)\,dU \tag{2-22}$$

where

AEO_G = gross annual energy output (W/y)

p_O = system output power density (W/m²)

The integration in Equation (2-14) is usually performed numerically, by summing the products of the last two columns in Table 2-2. The product $f_W\,dU$ has been replaced by a *histogram* in which the duration in a wind speed interval or *bin* of width ΔU is calculated from the Weibull model as

$$\Delta t(U) = 8,760 \left\{ \exp\left[-\left(\frac{U - \Delta U/2}{C_m} \right)^{k_m} \right] - \exp\left[-\left(\frac{U + \Delta U/2}{C_m} \right)^{k_m} \right] \right\} \tag{2-23}$$

where

Δt = duration of $(U - \Delta U/2) \le U \le (U + \Delta U/2)$ (h/y)

Using the reference annual wind energy inputs from Equations (2-10) and the gross annual energy outputs calculated in accordance with Table 2-2, the gross coefficients of energy for the example wind turbines are

$$\text{Mod-5B:} \quad C_E = \frac{12,040\ MWh/y}{41,710\ MWh/y} = 0.29 \tag{2-24a}$$

$$\text{34-m VAWT:} \quad C_E = \frac{1,240\ MWh/y}{3,640\ MWh/y} = 0.34 \tag{2-24b}$$

Coefficients of energy tend to increase somewhat at lower hub-level wind speeds, as illustrated in Figure 2-19 for representative current commercial and early prototype wind turbines. The aerodynamic designs of early 2-bladed prototype rotors established a level of performance that designers of modern wind turbines now seek to achieve and exceed. The design energy production of large-scale 3-bladed wind turbines is now roughly 45 percent to 70 percent of the theoretical *Lanchester-Betz* limit of 0.593.

Wind tunnel testing of a scale-model rotor can be used to verify design power coefficients if the tip-speed ratio, λ, of the model is approximately equal to that of the full-scale turbine. This leads to the following scaling requirements:

$$U_M \approx U; \quad \Omega_M \approx \Omega(R/R_M) \tag{2-25}$$

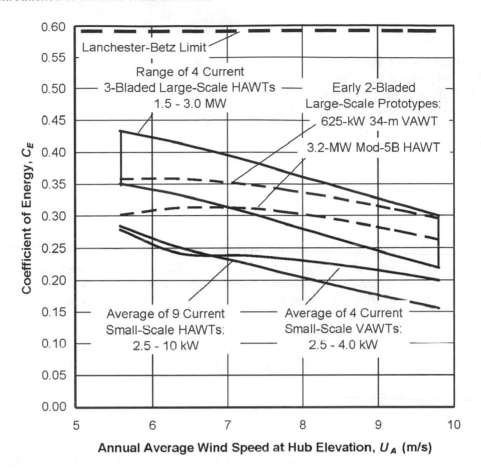

Figure 2-19. Calculated variation of coefficients of energy with wind speed for current representative large- and small-scale wind turbines, compared with coefficients of two early prototype wind turbines.

where

M = subscript denoting model parameters

Unfortunately, the *Reynolds Number* (aerodynamic parameter proportional to the product of the resultant wind speed and a characteristic length such as the rotor radius) of the model will be significantly smaller than that of the full-scale rotor, by the ratio R_M/R. This may affect the lift and drag characteristics of the model blades.

Wind Turbine Economics

The economic *figure-of-merit* most often used as the basis for choosing the configuration of a wind power station -- including the type and size of turbines, number and location of units, electrical collection and distribution system, and operation and maintenance strategies -- is the *unit cost of energy*, or *COE*, delivered to the customer. The COE of a wind turbine system, like that of a conventional power plant, contains the following three general elements: (1) Capital cost, (2) operating and maintenance costs, and (3) energy output. The following simplified formula can be used for estimating COE:

$$COE = \frac{ICC \times FCR + AOM}{AF \times AEO_G} \tag{2-26}$$

where

COE = unit cost of energy delivered to the customer ($/kWh)

ICC = total initial capital cost of the wind turbine system; includes complete cost exposure for land, equipment, and start-up operations ($)

FCR = levelized fixed charge rate; includes return on capital, income and property taxes, and insurance (1/y)

AOM = annual cost of operations and maintenance of the system ($/y)

AF = availability factor accounting for system downtime, partial or total

AEO_G = gross annual energy output of the system; without downtime (kWh/y)

Initial Capital Cost

Items usually included in the initial capital cost of a wind power station are listed in Table 2-3, together with an early estimate of their relative contributions to the total cost. The hypothetical station in this study was composed of 60 large-scale HAWTs [Boeing 1988]. Land costs are not included because they are highly variable. In this early estimate, made when only prototype utility-scale wind turbines were available, the cost of the turbine itself is about 51 percent of the total installed cost.

Table 2-3.
Early Estimate of Relative Contributions to Initial Capital Cost (ICC) of Components in a Wind Power Station with 60 Large-Scale HAWTs [Boeing 1988]

Initial Cost Item	Contribution	
Rotor Assemblies	23.2 %	
Nacelle Structures and Auxiliary Equipment	15.1 %	
Power Train Equipment	13.0 %	**Turbine: 51.3%**
Site Preparation and Roads	10.3 %	
Towers and Foundations	8.9 %	
Profit	8.7 %	
Ground Equipment Stations	8.2 %	
Maintenance Equipment and Initial Spares	5.0 %	
Electrical Interconnections	3.8 %	
Transportation	2.5 %	
Other Non-recurring Costs	1.3 %	
Initial Capital Cost of Wind Power Station	**100.0 %**	

The recent development of cost models based on many wind turbines with three-bladed, pitch-controlled, variable-speed rotors [Fingersh *et al.* 2006] permits estimates of component cost contributions in more detail now than previously possible. For example, Table 2-4 is an extended list of items to be considered when estimating the cost of a wind power station. In this example, the station has a rated capacity of approximately 51 MW and is composed of 34 mature-design turbines rated at 1.5 MW each. Component costs have again been normal-

Table 2-4.
Recent Estimate of Relative Costs of Turbine and Site Components in a
51-MW Wind Power Station

Rotor Assemblies ....**16.9 %**

      Blades... 10.8 %
      Hub...3.1 %
      Pitch Mechanism & Bearings....................2.7 %
      Spinner...0.3 %

Nacelles and Power Trains .... **46.4 % Turbine: 63.3 %**

      Gearbox... 10.8 %
      Variable-Speed Electronics ....8.4 %
      Generator ....7.0 %
      Main Structure ....6.6 %
      Electrical Connections...........................4.3 %
      Control, Safety, & Condition Monitoring.......2.5 %
      Low-Speed Shaft.................................1.5 %
      Nacelle Cover ....1.5 %
      Yaw Drive and Bearing ....1.4 %
      Hydraulic & Cooling Systems ....1.3 %
      Bearings ....0.9 %
      Mechanical Brakes, High-Speed Shaft, *etc* .....0.2 %

Towers .... **10.5 %**

Balance of Station .... **26.2 %**

      Electrical Interface & Connections..............8.7 %
      Roads, Civil Work, & Staging...................5.6 %
      Transportation..................................3.6 %
      Foundations ....3.3 %
      Assembly & Installation.........................2.7 %
      Permits, Engineering, & Site Assessment ......2.3 %

Initial Capital Cost of Wind Power Station ....**100.0 %**

ized by the total installed capital cost. In this current estimate, the cost share of the rotor has decreased while the nacelle and power train cost share has increased significantly, compared with the 1988 estimate. This results in a net turbine share increase of about 12 percent, with a comparable reduction in the balance of station share.

Fixed Charge Rate

The fixed charge rate in Equation (2-26) may be a composite of rates for different items in Tables 2-3 and 2-4. It is sensitive to the cost of capital, method of capitalization, tax rates, tax incentives, and the lifetime of the wind turbine system. Wind turbine economic studies have been made with fixed charge rates from 0.10 to 0.20. In their baseline cost model, Fingersh *et al.* [2006] set the fixed charge rate at 0.116.

Annual Operations and Maintenance Cost

Operating and maintenance budgets for a wind power station are sensitive to such factors as the maturity and durability of the wind turbine equipment, the size and number of units in the system, ease of maintenance and availability of spare parts, and weather conditions.

Availability Factor

On-line availability is closely related to operations and maintenance costs and is also sensitive to the maturity of the wind turbine equipment. Availability factors from 0.90 to 0.97 are usually assumed when estimating the cost of energy from a wind power station after it has been in operation for several years.

Annual Energy Output

Both the long-term energy content of the local winds and the overall efficiency of the turbines determine the annual energy output of the wind power station in the absence of downtime. However, as illustrated by the coefficients of energy in Figure 2-19, modern utility-scale wind turbines are designed to extract about 33 percent of the incident wind energy, on average, regardless of their configuration. Since the energy content of the wind is proportional to the cube of the wind speed, it has been found that the annual average wind speed at the mean elevation of the turbine rotors is a dominant parameter in determining the cost of energy and the economic viability of a wind power station.

Figure 2-20 illustrates the sensitivity of the cost of energy to the annual average wind speed, using data from Table 2-1. For this example, the mean elevation of the rotors is

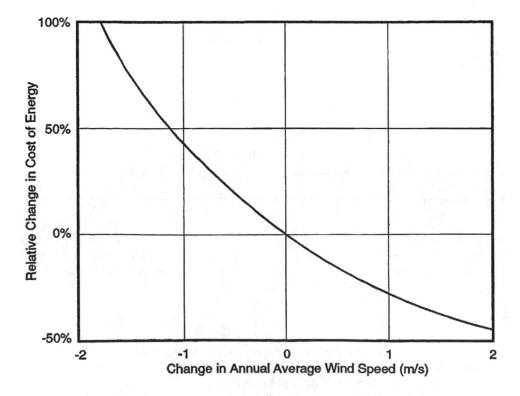

Figure 2-20. Sensitivity of the cost of energy to annual average wind speed. The reference annual average wind speed is assumed to be 7.8 m/s at mid-rotor elevation.

assumed to be 30 m, at which the reference annual average wind speed U_A is 7.80 m/s. Assuming the Weibull factors C and k are related to average wind speed as given in this table, cost of energy is inversely proportional to the annual energy density e_w. As shown in Figure 2-20, a reduction of 1.0 m/s in the annual average wind speed can increase the cost of energy by about 40 percent. Similarly, an increase of 1.0 m/s can decrease COE by almost 30 percent.

The annual average wind speed at an individual wind turbine site may decrease as the result of the operation of turbines upwind of it. This reduction in wind speed within the wind power station itself is often referred to as a *wake effect* or *array effect* on performance and is discussed in more detail in Chapter 6.

Offshore *vs.* Onshore Wind Power Station Costs

Many utility-scale wind turbines are now being installed in offshore wind power plants in Europe, as illustrated in Figure 2-4 and listed in Table 3-9. According to the European Wind Energy Association (*EWEA*) if certain institutional barriers are removed, up to 40 GW (*gigawatts*) of offshore generating capacity could be operating in the European Union by 2020, producing 4 percent of Europe's electricity supply [EWEA 2007]. In the Eastern U.S., the amount of space available for off-shore wind turbines is many times larger than for on-shore wind power stations [Musial and Butterfield 2004]. Two economic issues affecting the cost of energy from offshore wind turbines will be addressed here, namely (1) estimated comparative costs of offshore and onshore components, and (2) estimated benefits of potentially higher energy production as a result of lower wind shears over water.

Comparative Cost Estimates

Early estimates of the additional costs that builders of offshore wind turbines might face were made by Madsen and Svenson [1995 and 1997] who concluded that an offshore system would cost approximately 40 percent more than an equivalent system onshore. The added costs were primarily for the foundation (six times as expensive), the grid connection (three times as expensive), and assembly and transportation (each two times as expensive).

These early estimates have now been updated to reflect current experience, costs, and additional items required for offshore applications in relatively shallow water [Fingersh *et al.* 2006]. Results from this study are summarized in Table 2-5. Cost data in this study were calculated for a 51 MW wind power station, composed of 17 turbines rated at 3.0 MW each. Overall, offshore application costs are now estimated to be approximately 80 percent higher than equivalent onshore costs. Costs of the turbines themselves are expected to be roughly equal. The major increase is in the offshore *balance of station* costs, which are estimated to total about 2.2 times those of onshore wind power stations. As listed in Table 2-6, offshore foundation costs are estimated to be approximately 17 times those for onshore foundations, even for installations in relatively shallow water. Economical deep-water foundations are a subject of current research [Butterfield *et al.* 2007].

The additional items specific to offshore applications listed in Table 2-5 add about 28 percent to the system cost. These items include a *warranty premium* (10 percent) and *turbine marinization* (8 percent), which refers to the application of special materials and coatings to turbine components to resist deterioration from moisture and salt in the marine environment.

Potential for Increased Energy Output of Offshore Wind Turbines

In the long term, offshore winds are expected to be more advantageous for wind energy conversion than onshore winds. Winds at sea tend to blow faster and more uniformly than over land [Musial and Butterfield 2004]. Winds over open water are characterized by significantly *lower wind shears* because of the lower surface roughness of large expanses of water (see Table 8-3). In order to measure annual wind speed durations and wind shears undisturbed by land roughness and temperature variations, special installations of high-elevation anemometers are required miles from shore. Such wind monitoring sites are rare.

Figure 2-21 shows the installation by helicopter in 2005 of one such anemometer tower on the *water intake crib* located in Lake Erie 3.5 miles north-northwest (NNW) of Cleveland, Ohio. The Cleveland Water Crib is a large offshore structure 50 ft in diameter, containing the primary inlet to the municipal water supply. Cup anemometers, wind direction vanes, and thermometers are now mounted on this tower at elevations of 30, 40 and 50 m above water

Table 2-5.
Estimated Costs for an Offshore 3.0-MW Wind Turbine in a 51-MW Wind Power
Station, Compared to Costs for an Equivalent Onshore System
[Data from Fingersh *et al.* 2006]

Item	Offshore Turbine Cost (2005 $1,000)	Onshore Turbine Cost (a) (2005 $1,000)	Ratio of Offshore Costs to Onshore Costs
TURBINE CAPITAL COST	**2,377**	**2,136**	**1.1**
Nacelle and Drive Train	1,425	1,276	
Rotor	477	498	
Tower	415	324	
Control, Safety, and Condition Monitoring	60	38	
BALANCE OF STATION COSTS	**2,855**	**872**	**3.3**
Foundations	1,114	66	17.0
Electrical Interface & Connections	926	241	3.8
Assembly & Installation	371	71	5.2
Transportation	281	272	
Engineering, Permits, & Site Assessment	119	75	
Roads, Civil Work, Port, & Staging	74	147	
OFFSHORE ADDITIONAL COSTS	**1,169**	**0**	**NA**
Off-Shore Warranty Premium	405	0	
Turbine Marinization	321	0	
Scour Protection	204	0	
Surety Bond for Decommissioning	176	0	
Special Personnel Access Equipment	64	0	
INITIAL CAPITAL COST (2005 $1,000)	**6,431**	**3,008**	**2.1**

(a) Scaled for inflation from 2002$ (reference data) to 2005$, by factor of 1.074

Table 2-6.
Gross Annual Energy Outputs for Mod-5B 2.5-MW HAWTs at Representative
Offshore and Onshore Sites with the Same Wind Resource at 100 m Elevation (a)

Parameter	Offshore Site: Cleveland Water Intake Crib	Onshore Site: DOE/ NASA Reference
Surface roughness length, z_0	0.000115 m	0.0445 m
Annual average wind speed at 10m, $U_{A,R}$	6.71 m/s	4.66 m/s
Annual average wind speed at 100m, $U_A(100)$	7.63 m/s	7.63 m/s
Wind shear power-law exponent, α	0.0561	0.2141
Gross Annual Energy Output, AEO_G	7,840 MWh/y	6,420 MWh/y
Ratio: Offshore output to onshore output	**1.22**	

(a) Rotor diameter = 97.5 m; rotor mid-elevation = 60.9 m

(a) General view of wind measurement station. Skyline of
Cleveland, Ohio, is in background

**Figure 2-21. Installation of an offshore wind measurement station in Lake Erie atop
the city's water intake structure known as the 5-mile Crib** (*Courtesy of The Renaissance
Group, Kirtland, Ohio*)

level, with duplicate instruments at each level. A propeller-style anemometer at an elevation
of 18 m provides lower-level wind data.

This unique offshore monitoring facility was sponsored by two local non-governmental
organizations: *The Renaissance Group*, designer/installer, an appropriate technology consulting and implementation company; and *Green Energy Ohio* (GEO), a non-profit organization

(b) Upper tower with booms supporting anemometers, direction sensors, and thermometers being lowered into place by helicopter. Instruments are at 18-, 30-, 40-, and 50-m levels above the water.

Figure 2-21 (Concluded). Installation of an offshore wind measurement station in Lake Erie atop the city's water intake structure known as the 5-mile Crib (*Courtesy of The Renaissance Group, Kirtland, Ohio*)

performing land-based wind resource assessments in Ohio since 1998. Analysis of two years of wind speed data from the water crib tower indicated that this offshore wind resource has a significantly higher annual average wind speed than any of the ten onshore locations in northern Ohio where measurements were made previously by GEO [Dykes *et al.* 2008]. Another significant finding documented in that comprehensive report is that the measured wind shear was very low, with monthly power-law exponents in the range of -0.09 to + 0.09.

Figure 2-22 shows the vertical profile of the annual average wind speeds measured at the Cleveland Water Crib monitoring site from October 2005 to September 2007. A power-law

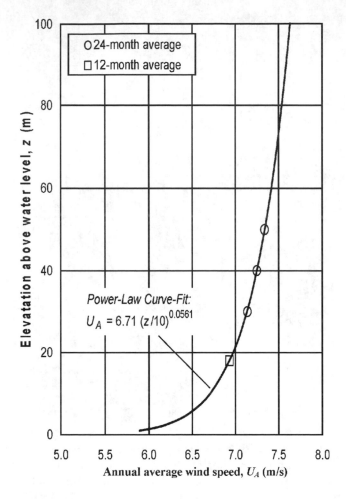

Figure 2-22. Vertical profile of annual average wind speeds measured at the Cleveland Water Intake Crib, illustrating the low wind shears characteristic of offshore sites. [Data: Dykes *et al.* 2008].

curve-fit of these data points gives a reference annual wind speed, $U_{A,R}$, of 6.71 m/s and a wind shear exponent, α, equal to 0.0561. The following equations and Figure 8–13 can now be used to calculate a surface roughness length for this site [Spera and Richards 1979]:

$$\alpha_0 = \alpha / [1 - 0.55 \log(U_R)] \tag{2-27a}$$

$$z_0 = 10 \alpha_0^5 \tag{2-27b}$$

where

α_0 = wind shear power law exponent at 1 m/s reference wind speed
z_0 = empirical surface roughness length (m)

Applying Equations (2-27), the surface roughness length associated with the Cleveland Water Intake site is 0.000115 m, which falls within the typical range in Table 8-3 for large expanses of water: 0.0001 m (calm open sea) to 0.001 m (coastal areas with off-sea winds).

Accurate estimates of the capacity of the wind resource for utility-scale HAWTs depend on wind data taken at higher elevations, and this is particularly true for offshore sites. Assume, for example, that wind speeds at elevations near 100 m are approximately the same for an onshore and an offshore site. If that is the case, the comparative wind power densities for these two sites could be as illustrated in Figure 2-23. Because of the lower wind shear at

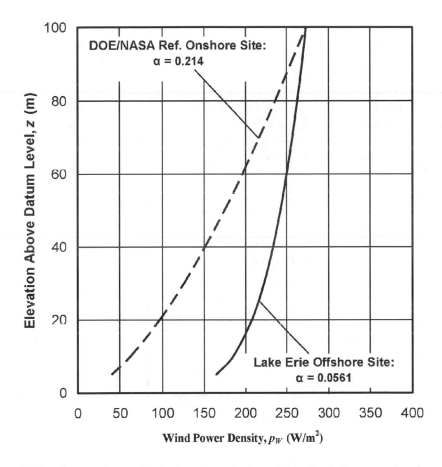

Figure 2-23. Comparison of calculated vertical profiles of wind power density at on-shore and offshore sites when higher-level winds are the same at both locations, indicating a potential advantage in energy production at the offshore site. In this example, the wind resource is the same at an elevation of 100 m.

the offshore site, wind power densities there at elevations below 100 m may be significantly larger than at onshore sites. This illustrates the potential for increased energy production from an offshore turbine that may offset to some degree its higher initial capital cost.

Table 2-6 lists the results of calculations of comparative gross annual energy outputs of utility-scale wind turbines represented by the DOE/NASA Mod-5B 3.2-MW HAWT, following the procedure described in Table 2-2. The potential energy output advantage for the offshore location is 22 percent in this example.

Rotationally-Sampled Wind Turbulence

The static and dynamic loads acting on a wind turbine strongly affect its initial capital cost, its operations and maintenance costs, and its reliability. The structural dynamic behavior of wind turbines subject to aerodynamic, inertial, and gravity loads is discussed in detail in Chapters 10, 11, and 12. An important category of aerodynamic loading is that resulting from non-uniform wind speeds within the swept area of the rotor. If the wind field is envisioned as a mixture of local regions of higher and lower wind speeds as illustrated in Figure 8-16, a section of a rotor blade following a circular path through these regions will experience a repetitively time-varying wind velocity. The rotary motion of the turbine blade converts steady winds that are non-uniform in space into a wind speed at a given blade section that varies with time. This time variation is referred to as *rotationally-sampled turbulence* [Veholek 1978, Connell 1981, Kristensen and Brandsen 1982, Connell and George 1983a and 1983b, George 1984, and Connell 1995].

The basic method for measuring rotationally-sampled (R-S) turbulence is illustrated in Figure 2-24 [Verholek 1978]. Mounted on a line of instrument towers perpendicular to the prevailing wind, a set of anemometers is arranged in a circular pattern in a vertical plane, with center elevation H and radius R (in this case, 24.4 m and 12.2 m, respectively). This set of towers and anemometers is called a *vertical plane array* or VPA, and represents the path

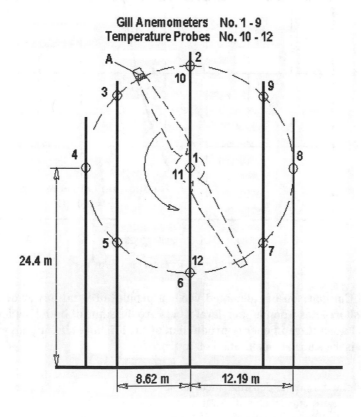

Figure 2-24. Schematic diagram of a vertical plane array (VPA) of anemometers for measuring wind turbulence experienced by a rotating HAWT blade. Sampling of signals from anemometers 2 to 9 is synchronized with the passage of the hypothetical blade section A. [Verholek 1978]

followed by a blade section of an imaginary HAWT rotor. The wind speed at each anemometer is continuously recorded. Samples from each record are then taken sequentially from consecutive anemometers around the circle, with the sampling interval time determined by the rotational speed and the number of anemometers. Interpolations are made between consecutive samples in order to synthesize a continuous record which represents the wind speeds that a blade section would experience in the rotating frame of reference.

VPA measurements have shown that the time-varying wind speed acting at a section of a rotating wind turbine blade is *harmonic* in nature, with frequencies equal to multiples of the rotor speed. These wind speed harmonics can produce a *forced harmonic response* (cyclic motions at the harmonic frequencies) in the flapwise direction (*i.e.* out of the plane of revolution), thereby increasing blade fatigue stresses.

Model of R-S Turbulence for Predicting Blade Fatigue Loads

Equations have been developed with which to estimate the sizes of the wind harmonic speeds as input into HAWT structural-dynamics computer codes [Spera 1995]. These equations are based on R-S turbulence test data and verified against loads measured on the 2.5 MW Mod-2 HAWT [Boeing 1982]. Additional information on turbulence defined in a rotational frame of reference is given in Chapter 8.

Rotationally-Sampled Turbulence Data

The R-S turbulence data used in this model were measured by researchers from Battelle's Pacific Northwest Laboratory (PNL) using a VPA located near Clayton, New Mexico (Connell and George 1983a and 1983b, and George and Connell 1984). The dimensions of this array were a center elevation $H_C = 30.5$ m and a sampling radius $R_C = 19.0$ m, and the rotational frequency was 0.667 Hz. *Power spectral densities* (PSDs) of 8.5-min segments of the synthesized wind speed were created using a Fast Fourier Transform (FFT) technique. Integration of a PSD over a selected frequency band then gave the *variance* of wind speeds within this band. R-S turbulence in the frequency band is the square-root of this variance. Dividing the turbulence by the steady wind speed at the center of the VPA (Anemometer 1 in Fig. 2-24) then gives the R-S turbulence intensity for the selected frequency band.

Table 2-8 presents typical data reported by PNL researchers for one data segment. A total of 17 data segments form the basis of the following R-S turbulence model. Of these, 10 were for atmospheric stability conditions ranging from neutral to unstable, while 7 were for stable conditions. Stable atmospheres typically result in larger vertical gradients in wind speed and smaller mixing between winds at different elevations. For a discussion of the influence of atmospheric stability on wind shear, see Chapter 8 [Frost and Aspliden]. Only the longitudinal component of turbulence was measured during these VPA tests. Therefore, lateral and vertical turbulence components are not included explicitly in the R-S turbulence model presented here.

Dependence on Wind Shear

The turbulence intensities for each harmonic frequency of the 17 data segments like that in Table 2-7 were examined to determine their dependence, if any, on the vertical wind shear across the VPA. It was found that

- Only the first harmonic turbulence intensity varies significantly with normalized wind shear (wind shear per unit of wind speed at hub elevation), and its variation is linear.

Table 2-7. Typical Clayton VPA Wind Segment Data
[George and Connell, 1984]

Data Segment Number = C2c
Date; Starting Time = 06/30/82; 1740
VPA Hub Height, H_c = 30.5 m
VPA Sampling Radius, R_c = 19.0 m
Test Segment Length = 8.5 min
Rotational Sampling Speed, P = 0.67 Hz
Center Mean Wind Speed, U_0 = 10.71 m/s
Center Stationary Turbulence, σ_0 = 0.856 m/s
Wind Shear Across Disk, ΔU = 1.506 m/s

Frequency band f (Hz)	Mid-band harmonic frequency f/P	Variance μ (m/s²)	Turbulence σ (m/s)	Turbulence intensity σ/U_0
< 0.33		0.498	0.706	0.066
0.33 – 1.00	1	0.431	0.656	0.061
1.00 – 1.67	2	0.146	0.382	0.036
1.67 – 2.33	3	0.070	0.264	0.025
2.33 – 3.00	4	0.037	0.192	0.018
3.00 – 3.67	5	0.032	0.179	0.017

- Turbulence intensities of harmonics above the first decrease with increasing harmonic number, following a roughly exponential decay.
- Atmospheric stability has little, if any, direct effect on harmonic turbulence intensities. An indirect effect of a stable atmosphere on the first harmonic intensity is through the increased wind shear often present in a stable atmosphere.

These observations lead to the following model for representing the Clayton VPA data:

$$(\sigma_1/U_0)_C = 0.0311 + 0.297 \, \Delta U/U_0 \qquad (2\text{-}28a)$$

$$(\sigma_n/U_0)_C = 0.059 \, n^{-0.75} \quad n > 1 \qquad (2\text{-}28b)$$

where

σ_n = rotationally-sampled turbulence in the frequency range from $(n - \frac{1}{2})P$ to $(n + \frac{1}{2})P$ (m/s)
P = rotational sampling frequency (rad/s)
C = subscript denoting Clayton VPA parameters
H = hub elevation above ground level (m)
R = radius of circular sampling path (m)
U_0 = steady free-stream wind speed at hub elevation (m/s)
ΔU = total steady wind shear from top to bottom of circular sampling path (m/s)

We can determine the steady wind shear across the circular path, ΔU, from anemometer test data, or we can estimate it using one of several models, including the well-known *power law model* (Eqns. (2-4) and (8-11), and Fig. 8-13) of the vertical gradient in the steady wind speed. With this model the normalized wind shear can be calculated, as follows:

$$\Delta U / U_0 = [(H + R)/z_R]^{\alpha} - [(H - R)/z_R]^{\alpha} \tag{2-29}$$

where

$H =$ hub elevation above ground level (m)
$R =$ tip radius of rotor (m)
$z_R =$ reference elevation above ground level = 10 m

Scaling Clayton VPA Data to Other Radii and Elevations

To generalize Equations (2-28), we must now make additional assumptions about the size effects of R and H on R-S turbulence intensity. First, the longer sampling paths and sampling periods of larger rotors are assumed to result in larger wind speed variations around the perimeter of the sampled circle. For simplicity, this relationship between increasing sampling radius and increasing R-S turbulence intensity is assumed to be linear.

Second, the effects on R-S turbulence of center elevation, H, and surface roughness length, z_0, are assumed to be approximately the same as these effects on longitudinal turbulence measured at a fixed point. In accordance with Eqn. (8-19a) [Frost and Aspliden], the following model is used:

$$\frac{\sigma_{0,x}}{U_0} = \frac{0.52}{\ln(H/z_0)(0.177 + 0.00139\,H)^{0.4}} \tag{2-30}$$

where

$\sigma_{0,x} =$ longitudinal turbulence at a fixed point (m/s)
$z_0 =$ surface roughness length; see Table 8-3 (m)

Combining a linear effect of sampling radius with the effect of center elevation given in Equation (2-30), we obtain the following equations for scaling the Clayton VPA data to a circular path of a different radius and hub elevation:

$$\frac{\sigma_n / U_0}{(\sigma_n / U_0)_C} = \frac{R}{R_C} \frac{\left[\ln(H/z_0)(0.177 + 0.00139\,H)^{0.4}\right]_C}{\ln(H/z_0)(0.177 + 0.00139\,H)^{0.4}} \tag{2-31}$$

Substituting the Clayton VPA magnitudes of R and H into Equation (2-31), the scaling equations for the R-S turbulence intensities along the tip path of a general HAWT rotor are as follows:

$$\sigma_n / U_0 = S(\sigma_n / U_0)_C \tag{2-32a}$$

$$S = \frac{0.204\,R}{\ln(H/z_0)(0.177 + 0.00138\,H)^{0.4}} \tag{2-32b}$$

where

$S =$ scaling factor at radius R relative to Clayton VPA harmonic amplitudes in Equations (2-28)

Spatial Distribution Around the Sampling Circle

R-S turbulence is assumed to be quasi-static, which means that a HAWT rotor makes several revolutions before any significant changes occur in the spatial distribution of the wind within the rotor's swept area. Dynamic responses of the rotor blades are assumed to approach a steady state before the wind turbulence changes significantly. With these quasi-static assumptions, the time-varying unsteady wind can be converted to a spatially-varying steady wind in a harmonic format, as

$$U_{RS} = U_0 + \Sigma(A_n \cos n\psi) \quad n = 1, 2, \ldots \tag{2-33}$$

where

U_{RS} = rotationally-sampled free-stream horizontal wind speed, quasi-steady in time (m/s)
A_n = amplitude of nth harmonic of wind speed (m/s)
ψ = azimuthal position in rotor swept area; 0 = down (deg)

Wind turbulence is equal to the standard deviation of the wind speed from its steady value. For each cosine wave in Equation (6), standard deviation and amplitude are related as follows:

$$|A_n| = \sqrt{2}\,\sigma_n \tag{2-34}$$

While the absolute values of the harmonic amplitudes can be determined from R-S turbulence data, the signs of these amplitudes cannot. Various patterns of positive and negative signs (equivalent to in-phase and out-of-phase harmonics in Eq. (2-33)) have been used. For example, Zimmerman *et al.* [1995] assumed negative amplitudes for odd-numbered harmonics and positive amplitudes for even-numbered harmonics. Further examination of the shapes of various vertical wind profiles defined by different combinations of positive and negative harmonic amplitudes indicates that a reasonable vertical profile is obtained with negative harmonics except for the third and fourth.

Spatial Distribution Along a HAWT Blade

In order to apply R-S turbulence wind speeds in a structural-dynamics code for calculating blade fatigue loads it is necessary to define the wind speed distribution from hub to tip. Earlier, the assumption was made that the size of each R-S turbulence harmonic varies linearly with the sampling radius. However, this assumption alone is not sufficient to define the simultaneous wind speed distribution along a turbine blade. The spanwise distribution of turbulence also depends strongly on the *transverse coherence* of the wind. As explained in Chapter 8, coherence is a dimensionless quantity between zero and unity that represents the degree to which two unsteady events, separated in space, are alike in their time histories. If the two time histories are identical their coherence is unity, and if they are completely unrelated their coherence is zero.

Coherences of individual harmonics of R-S turbulence were measured by Zimmerman *et al.* [1995] using wind speed sensors mounted at two locations 69.2 ft apart on a 150-ft Mod-2 HAWT blade. The coherence between the first harmonic at the inboard location ($r = 30.8$ ft) and the first harmonic at the outboard station was found to be high, but equivalent coherences of the higher harmonics were all low for this separation distance. To generalize these observations, it is assumed that the first harmonic of the R-S turbulence has a coherence of unity along the entire blade length. Thus, the first harmonic turbulence acts simultaneously along the blade from hub to tip. Higher harmonics are assumed to have a coherence of unity in the outer half of the swept area and a coherence of zero in the inner half. In the inner half of the swept area the higher harmonic amplitudes are assumed to be zero and have no affect on fatigue loads. The radius separating the two half-areas is the mean radius, R_m, defined by Equation (2-18a) as equal to 0.73 R, where R is the tip radius.

Summary of Equations in the R-S Turbulence Model

$$U_{RS} = U_0 + \frac{r}{R}\Sigma(A_n \cos n\psi) \tag{2-35a}$$

$$U_0 = U_R(H/z_R)^{\alpha} \tag{2-35b}$$

$$\Delta U = U_R[(H+R)/z_R]^{\alpha} - [H-R]/z_R]^{\alpha} \tag{2-35c}$$

$$S = \frac{0.204 / R}{\ln(H/z_0)(0.177 + 0.00138 H)^{0.4}} \tag{2-35d}$$

$$A_1 = -0.0440\, S\, U_0 + 0.4120\, \Delta U_0 \tag{2-35e}$$

If r $\geq 0.73\ R$:

$$A_2 = -0.0496\, S\, U_0 \tag{2-35f}$$

$$A_3 = +0.0366\, S\, U_0 \tag{2-35g}$$

$$A_4 = +0.0295\, S\, U_0 \tag{2-35h}$$

$$A_5 = -0.0250\, S\, U_0 \tag{2-35i}$$

$$A_6 = -0.0218\, S\, U_0 \tag{2-35j}$$

If r $< 0.73\ R$:

$$A_n = 0 \quad n = 2, 3, \ldots \tag{2-35k}$$

where
U_{RS} = rotationally-sampled free-stream horizontal wind speed; quasi-static (m/s)
U_0 = steady free-stream wind speed at hub elevation (m/s)
U_R = steady free-stream wind speed at reference elevation (m/s)
r = radial distance from rotor axis (m)
R = tip radius of rotor (m)
H = elevation of hub above ground level (m)
S = scale factor; referenced to Clayton VPA
z_0 = surface roughness length (m)
z_R = reference elevation above ground level = 10 m
n = harmonic number
α = exponent in wind shear power law
ψ = azimuthal position in rotor swept area; 0 = down (deg)
A_n = amplitude of nth harmonic of wind speed distribution around circle of radius
    R (m/s)

Correlation of Model Load Calculations with Test Data

A spectrum of flapwise cyclic bending loads measured at two blade stations on the DOE/ NASA Mod-2 2.5 MW HAWT shown in Fig. 3-37 [Boeing 1982] was used for preliminary verification of this R-S turbulence model. Equation (2-35d) was used to scale the Clayton VPA data to the Mod-2 hub elevation of 61.0 m and tip radius of 45.7 m. The two blade stations at which load measurements were made were located at 21 percent and 65 percent of the tip radius. Good correlation was found between calculated blade loads and the 50th percentile measured loads in the spectrum at both stations [Spera 1995]. The conclusion drawn is that an R-S harmonic wind model like that in Equations (2-35) is a useful tool for predicting average blade fatigue loads in the flapwise direction, as the wind input in a structural-dynamic computer code. The 99.9th percentile cyclic loads at the two measurement stations on the Mod-2 blade were found to be approximately 3 times this average.

Sample Application of R-S Turbulence Model

To illustrate the application of R-S turbulence modeling of the wind input, consider the hypothetical onshore and offshore vertical wind profiles whose wind power densities are shown in Figure 2-23. Steady wind speeds in both of these two profiles at an elevation of 100 m are assumed to be 7.63 m/s. However, the offshore profile has a much smaller wind shear, which should lead to lower R-S wind turbulence and lower flapwise blade loads. We can use Equations (2-35) to compare the expected harmonic amplitudes at the onshore and offshore locations, as a basis for the future calculation of rotor fatigue loads. For this example, the turbine rotor dimensions are taken to be those of a Mod-2 2.5 MW HAWT, with the site data listed in Table 2-6.

Table 2-8 lists calculated parameters for the wind fields at the offshore and onshore sites, followed by a comparison of the first six harmonic turbulence amplitudes at the tip circumfer-

Table 2-8.
Comparison of Estimated Amplitudes of Wind Turbulence Harmonics at Offshore and Onshore Sites with the Same High-Level Winds. The sample rotor swept area at both sites is that of the 2.5 MW Mod-2 HAWT

Parameter	Wind Vertical Profile		Ratio: Offshore to Onshore
	Onshore Site	Offshore Site	
Hub elevation, H	61.0 m		Same
Rotor tip radius, R	45.7 m		Same
Annual average wind speed at 100 m	7.63 m/s		Same
Wind shear exponent, α	0.2141	0.0561	0.0026
Annual average wind speed at reference elevation (10 m), U_R	4.66 m/s	6.71 m/s	1.44
Annual average wind speed at hub elevation, U_0	6.86 m/s	7.43 m/s	1.08
Vertical wind shear across disk, ΔU	2.63 m/s	0.79 m/s	0.30
Scale factor, S	2.21	1.21	0.55
First harmonic amplitude, A_1	-1.448	-0.630	0.44
Second harmonic amplitude, A_2	-0.752	-0.446	0.59
Third harmonic amplitude, A_3	0.555	0.329	"
Fourth harmonic amplitude, A_5	-0.447	-0.265	"
Fifth harmonic amplitude, A_5	-0.379	-0.225	"
Sixth harmonic amplitude, A_6	-0.330	-0.196	"

ence of the swept area. Figure 2-25 shows the variable wind speeds around the two swept areas, as experienced by the blade tips. Based on this R-S turbulence model and the relative wind shears assumed for the two sites, a large-scale HAWT located offshore would experience cyclic wind loads that are only 44 percent to 59 percent as large as those acting on the onshore machine. In addition, the annual average wind speed at hub height would be about 8 percent higher offshore compared to onshore. This type of information can be very useful during the preliminary design phase of a wind power station.

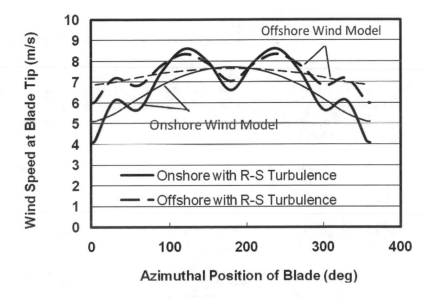

Figure 2-25. Sample comparison of onshore and offshore variable wind speeds acting on the rotating blade tips of representative Mod-2 HAWTs, as calculated using the harmonic amplitudes in Table 2-8. Rotor diameter is 45.7 m and hub height is 61.0 m. The one-per-rev sinusoidal lines represent wind shear without additional turbulence.

References

ASME, 1989, *Performance Test Codes - Wind Turbines*, ASME/ANSI PTC 42-1988, New York: American Society of Mechanical Engineers.

Bergey, K. H., 1980, "The Lanchester-Betz Limit," *Journal of Energy*, Vol. 3, No. 6, pp. 382-384.

Boeing Aerospace Company, 1988, *Mod-5B Wind Turbine System Development Final Report - Vol. II, Detailed Report*, NASA-CR-180896, DOE/NASA/0200-3, Cleveland, Ohio: NASA Lewis Research Center.

Butterfield, C. P., W. Musial, and J. Jonkman, 2007, *Overview of Offshore Wind Technology*, NREL/CP-500-42242, Golden, Colorado: National Renewable Energy Laboratory.

Dodd, H. M., 1990, "Performance Predictions for an Intermediate-Sized VAWT Based on Performance of the 34-m VAWT Test Bed," *Proceedings, Ninth ASME Wind Energy Symposium*, D. E. Berg, ed., SED-Vol. 9, New York: American Society of Mechanical Engineers, pp. 131-136.

Dykes, K., F. Miller, R. Weinberg, A. Godwin, and E. Sautter, 2008, *A Wind Resource Assessment for Near-Shore Lake Erie*, Report prepared by Green Energy Ohio: Cleveland, Ohio.

EWEA, 2007, "Developing Offshore Wind Power in Europe," Policy Publication, *Renewable Energy World*: European Wind Energy Association.

Fingersh, L., M. Hand, and A. Laxson, 2006, *Wind Turbine Design Cost and Scaling Model*, NREL-TP-500-40566, Golden, Colorado: National Renewable Energy Laboratory.

Griffin, D. A., 2001, *WindPACT Turbine Design Scaling Studies Technical Area 1-Composite Blades for 80- to 120-Meter Rotor*, Global Energy Concepts, NREL/SR-500-29492, Golden, Colorado: National Renewable Energy Laboratory.

Justus, C. G., and A. S. Mikhail, 1976, "Height Variation of Wind Speed and Wind Distribution Statistics," *Geophysical Research Letters,* 3: pp. 261-264.

Musial, W., and S. Butterfield, 2004, *Future for Offshore Wind Energy in the United States*, NREL-TP-500-36313, Golden, Colorado: National Renewable Energy Laboratory.

Rahlf, U., R. Osthorst, W. Gobel, 1998, "Bearing Overload Caused by Imprecise Definition of Connective Stiffness," *Proceedings, DEWEK 1998 Conference*, Wilhelmshaven, Germany: Aerodyn Corporation.

Spera, D. A., 1995, "A Model of Rotationally-Sampled Wind Turbulence for Predicting Fatigue Loads in Wind Turbines," *Collected Papers on Wind Turbine Technology*, D. A. Spera, ed., NASA CR-195432, Cleveland, Ohio; NASA Lewis Research Center, pp. 17-26.

Spera, D. A., 2008, *Models of Lift and Drag Coefficients of Stalled and Unstalled Airfoils in Wind Turbines and Wind Tunnels*, NASA-CR-215434, Cleveland, Ohio: NASA Glenn Research Center.

Spera, D. A., and T. R. Richards, 1979, *Modified Power Law Equations for Vertical Wind Profiles*, NASA-TM-79275, DOE/NASA/1059-79/4, Cleveland, Ohio: NASA Lewis Research Center.

3

Evolution of
Modern Wind Turbines
Part A: 1940 to 1994

Louis V. Divone
Director, Solar Energy Technologies
U.S. Department of Energy
Washington, DC

Introduction

The modern electricity-generating horizontal-axis wind turbine (now often labeled with the acronym HAWT) is an obvious descendant of the historic European windmill and the small, DC-generating wind turbines of the 1930s. The resemblance is somewhat deceptive, however, since the HAWT and its less-familiar vertical-axis cousin, the VAWT, have evolved as sophisticated products of current technology. Their high performance and reliability are the result of steady improvement in methods of aerodynamic and structural design, in new materials, and in mechanical and electrical engineering. The evolution of wind turbine technology in the United States and elsewhere since World War II is described in this chapter, along with some of the problems that have arisen and how these problems have been faced.

The world's largest wind power plant prior to the 1970s was the *Smith-Putnam* wind turbine (Fig. 1-22) erected in 1939 on Grandpa's Knob near Rutland, Vermont [Putnam 1948]. With a rotor diameter of 53.3 m and a power rating of 1.25 MW, this pioneering HAWT was a major work of mechanical engineering. The Smith-Putnam project was a milestone between the decline of fully-developed, stand-alone wind turbines generating DC power only for local use and the new growth of wind power plants connected to utility lines and producing AC power for distribution throughout the system.

Wind Turbine Development from 1945 to 1970

During the twenty-five years from 1945 to 1970, new growth in wind turbine technology took place principally in western Europe and at a very modest pace. Some of the research and development activities during this period are described in the following sections, according to the country in which they took place.

Denmark

Pre-World War II wind turbine development in Europe took place principally in Denmark under the direction of Poul LaCour (called by many "the Danish Edison") and his protege, Johannes Juul [Juul 1964]. Denmark, a country lacking in indigenous energy sources, utilized wind power to a significant degree during both World Wars when oil supplies were curtailed. With the onset of the World War II occupation the *F. L. Smidth Company* (F.L.S.) developed a series of wind turbines in the 45-kW range. Wind power eventually produced 4 million kilowatt-hours annually during this period.

The principal product line of F.L.S. was concrete manufacturing equipment; hence the use of concrete towers on the F.L.S. machines and the continued propensity for many years for the towers of larger Danish wind turbines to be built of concrete. Initially, rotors were of the two-bladed configuration, but F.L.S. (like the Jacobs brothers in the U.S. in the '20s) soon switched to three blades to alleviate tower vibration problems. Generally, DC generators were installed, since portions of outlying areas in Denmark were still supplied by small DC grids at that time.

The relative success of the small-scale F.L.S. wind power plants led to further experiments with larger machines in the years immediately following the end of World War II. With the help of Marshall Plan funding and the design experience of Juul, a 200-kW, 24-meter diameter wind turbine was installed during 1956-57 on the island of Gedser in the far southeast of Denmark (Fig. 3-1). Like its smaller predecessors, the *Gedser wind turbine* had a three-bladed rotor located upwind of a concrete tower. It supplied AC power to the local utility, Sydøstsjaellands Elektricitets Aktieselskab (SEAS), from 1958 until 1967. *Capacity factors* (*i.e.*, ratios of annual energy output to rated power times 8760 hr) of 20 percent were achieved in some years.

Research on medium- and large-scale wind energy development was discontinued in Denmark in the mid-'60s, but in the mid-1970s the simplicity, ruggedness, and reliability of the Gedser wind turbine provided valuable lessons to Danish engineers who responded to new demands for alternative energy generation. In 1977 the machine was refurbished, equipped with modern instrumentation, and operated intermittently for research purposes [Merriam 1977, Lundsager *et al.* 1980]. Tests of aerodynamic performance and structural loads were successfully conducted. Modern, commercially-successful Danish wind turbines owe much to the pioneering work of F.L.S.

France

During the period from 1958 to 1964, three large-scale HAWTs were built and tested in France by *Electricité de France* (EDF), in collaboration with two companies: *BEST* and *Neyrpic* [Bonnefille 1974]. The first turbine was called the *Type Best-Romani* and was erected at Nogent-le-Roi near Paris. Its three-bladed rotor had a diameter of 30 m, and the system rating was 800 kW at a wind speed of 16 m/s. It operated for five years, from 1958 to 1963, connected to the EDF network. There were some difficulties with gear lubrication,

Figure 3-1. The rugged and reliable 200-kW 34-m diameter Gedser HAWT in Denmark, after its refurbishment in 1977. (*Courtesy of Risø National Laboratory Station for Wind Turbines*)

drive-train clutching, and mechanical braking, but electrical braking was satisfactory. Most importantly, the connection of the wind turbine's generator to the AC grid functioned well.

The second French machine, of a design called *Type Neyrpic*, had a smaller diameter of 21 m. Its rated power was 132 kW at a wind speed of 13.5 m/s. Erected near the English Channel at Saint-Remy-des-Landes, it operated successfully for three years and accumulated only 60 days of outage for various technical reasons.

A larger Type Neyrpic turbine (Fig. 3-2) was built at the same site and operated for seven months in 1963 and 1964. Its three-bladed rotor had a diameter of 35 m and its maximum power was 1,085 kW. During November 1963 it produced 200,000 kWh of electricity. Its total energy output during a period of seven months was about 28 percent of the wind energy available, which is a performance level seldom achieved even by modern turbines. The tests ended in June 1964, when the turbine shaft broke. Although these three prototype turbines clearly demonstrated the feasibility of grid-coupled operation, the French decided in 1964 to discontinue further wind energy research.

Figure 3-2. France's 1.1-MW 35-m Type Neyrpic turbine. It was the largest of three French prototypes tested during the 1958-64 period. [Bonnefille 1974]

Figure 3-3. The 100-kW John Brown HAWT in the Orkney Islands. Its 18-m rotor was later reduced to 15 m. [Stodhart 1974]

United Kingdom

A variety of electricity-generating wind turbines was developed and tested in the U.K. from 1948 to the early 1960's [Stodhart 1974]. The three largest of these were 100-kW HAWTs of entirely different designs, each developed as a prototype for a wind power plant connected to a utility grid. The first of the prototypes (Fig. 3-3), designed and built by *John Brown & Co.*, was installed in the Orkney Islands in the early 1950's. It had a downwind rotor with three wood blades that were similar in design to helicopter blades. The

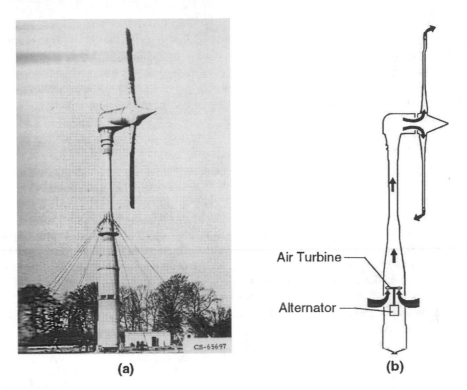

(a) **(b)**

Figure 3-4. The 100-kW 25-m Enfield-Andreau turbine in the early '50s. Hollow rotor blades with tip vents drew air through a turbogenerator in the tower. (a) General view (©*1955, E.W. Golding; 1976, E.&F.N. Spon Ltd.; reprinted by Halsted Press, John Wiley & Sons, Inc.*) (b) Diagram of the flow path.

rotor diameter was initially 18 m, but this was reduced to 15 m after an accident in which one of the blades struck the tower in a high wind. A series of modifications was required to solve structural problems in the hub and resonant vibrations in the tower. Operation of the *John Brown* wind turbine ceased in 1956.

Also in the early 1950's, a 100-kW HAWT 25-m in diameter was built by *Enfield Cables* and installed initially at St. Albans in the U.K. (Fig. 3-4). It was of the *Andreau* design, a unique concept in which mechanical coupling between the turbine and the generator is eliminated by driving the generator pneumatically. The turbine rotor has hollow blades with open tips and acts as a centrifugal air pump. As illustrated in Figure 3-4(b), air is drawn in through side vents in the tower shell, passing upward to drive an enclosed high-speed air turbine coupled directly to the generator. After flowing through the rotor hub into the hollow turbine blades, it is finally expelled from the blade tips. While the *Enfield-Andreau turbine* operated successfully, it had a low overall efficiency. High drag losses in the internal flow paths were suspected to be the cause. The turbine was later moved to Algeria, where it is said to have operated intermittently for about 180 hours. It was shut down permanently after suffering bearing failures at the blade roots.

A third 100-kW wind turbine, built by *Smith (Horley) Ltd.*, was installed on the Isle of Man in the late 1950's and operated until 1963. The *Isle of Man* wind turbine was relatively low in cost ($20,000 installed), and it pioneered two rotor design features that have been used successfully in more recent times: control of peak power through aerodynamic stall of fixed-pitch blades, and blades made inexpensively from extruded aluminum. Operation ended in 1963 after damage to the blades in a severe storm.

Considerable research on wind flow patterns and wind characteristics was also accomplished during this time by E. W. Golding and Arthur Stodhart. Golding describes this work in one of the first modern texts on wind power [Golding 1955, 1976], a book which continues to be a valuable reference for wind turbine engineers. Many aspects of wind power generation are discussed by Golding, including wind characteristics, design configurations, field testing, and economics. An example of the economic projections of the time is shown in Figure 3-5.

United States

One person who had been impressed by the work of Putnam and the Grandpa's Knob machine was Percy Thomas of the then U.S. Federal Power Commission. Thomas firmly believed in the future of windpower and the need for it in this country. He wrote a series of monographs on the subject from 1945 through 1954 [*e.g.* Thomas 1946, 1949], stressing the economics and requirements of size from a utility perspective. He also advocated *multiple rotors* on a single tower (Fig. 3-6) as a method for obtaining multi-megawatt capability within the constraints of current rotor blade technology. In the United States Thomas was a lone voice "in the wind", for he received no funds, even though there was a Congressional hearing on the subject in 1951. No actual design work (much less experimental work) was undertaken.

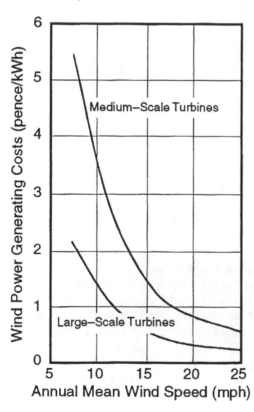

Figure 3-5. Early estimate of the relative costs of electricity from large- and medium-scale wind turbines, as a function of the site's annual mean wind speed. [Golding 1955, 1976]

Figure 3-6. Percy Thomas' dual-rotor concept for a multi-megawatt wind turbine. [Thomas 1949]

Germany

Wind turbine activities in Germany resumed after World War II, based on earlier tests at Weimar of *Ventimotor GMBH* wind turbines (8 m and 18 m in diameter), and continued through the '50s and '60s under the guidance of Professor Ulrich Hütter [Hütter 1973a]. During the 1950s Hütter developed and tested a 10-kW, *10-m Hütter-Allgaier* HAWT. In the early '60s this work culminated in the 100-kW *34-m Hütter-Allgaier* wind turbine, shown in Figure 3-7 [Hütter 1973b, 1974], the most technologically advanced system of its time and for decades to follow. Hütter concentrated on pushing the state of rotor technology toward lower solidity, higher tip speed, and flexibility. The 34-m rotor had two blades with full-span pitch control and very low solidity. These slender, flexible blades were constructed of fiberglass and mounted on a *teetered hub*. The rotor was downwind of the tower and incorporated 7 deg of coning. Because of very limited funding, experiments on this turbine proceeded slowly into the 1960s, hampered by flutter problems in the long, thin rotor blades.

(b)

(a)

Figure 3-7. The technologically-advanced 100-kW 34-m Hütter-Allgaier wind turbine. (a) General view of the turbine mounted on its 22.3-m guyed shell tower. (b) View of the fiberglass blade roots, teetered hub, and in-line power train. [Hütter 1973b, 1974]

The Beginnings of Modern Developments
1970 to 1974

Professor Hütter's wind turbine program in Germany ended for the same reasons as those that halted experiments in other countries: Energy from fossil-fuel and nuclear power plants was inexpensive, and little emphasis was placed on research on alternative sources of energy. At the end of the 1960's there was, unfortunately, little useful documentation and almost no experimental data from these several decades of activities around the world. For all the large advances in the field of wind energy since the end of the nineteenth century, prospective wind turbine designers in the '70s had little firm information upon which to build.

Revival of Interest in Wind Power

By 1970, there was little or no activity world-wide for producing electricity by wind power. Some water-pumping windmills were still being produced, principally for use in the developing world. In the U. S., a few enthusiasts were rebuilding *Jacobs Wind Electric* (Fig. 1-20) and other small DC wind generators from the 1930s for use in remote rural applications. *Dunlite* (Australia), *Elektro* (Switzerland), and *Aerowatt* (France) were essentially the only active manufacturers. Their systems were imported in very small quantities by the *Solar Wind Company* in Maine. A number of companies in the U.S., staffed by young enthusiasts, were beginning to attempt the design of machines in the 1-kW to 10-kW range. In the academic community, Hughes at Oklahoma State University and Heronomous at the University of Massachusetts and their students were studying small-scale and large-scale wind energy concepts, respectively.

At the *National Research Council* of Canada (NRC), the curve-bladed vertical-axis rotor patented by G. J. M. Darrieus in France in 1925 and in the U.S. in 1931 was re-invented by Peter South and Raj Rangi in the late 1960s. The *4.3-m NRC Darrieus* VAWT which they constructed was tested both in a wind tunnel (Fig. 3-8) and outdoors, producing some of the first wind turbine performance data obtained under controlled testing conditions.

In the U.S. in 1972, engineers from the *Lewis Research Center* of the *National Aeronautics and Space Administration* (NASA) were involved in measuring winds in Puerto Rico (for other purposes) and encountered some local interest in wind power. This was the start of wind energy research that continued at that laboratory for over 20 years. Aerodynamicists at NASA's *Langley Research Center* also began theoretical and experimental research on the Darrieus rotor.

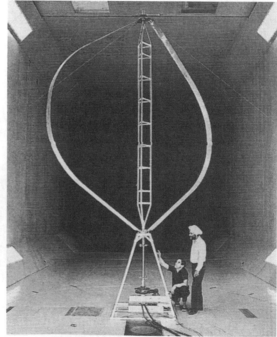

Figure 3-8. Testing the 4.3-m NRC Darrieus VAWT in 1972, observed by developers Peter South and Raj Rangi. (*Courtesy of the National Research Council of Canada*)

In the U.S. federal sector, the *National Science Foundation* (NSF), under their new "Research Applied to National Needs" (RANN) program, had been examining the overall long-term issues of energy supply, and had concluded (along with others) that renewable energy sources could have a major role in the future. However, individual views of that future varied enormously. The NSF was initially given the responsibility and a small budget for a federal research program on renewable energy. That program included solar energy, of which wind energy was considered to be a constituent part. The NSF, without any laboratories of its own, turned to the NASA Lewis Research Center (now the Glenn Research Center) in Cleveland, Ohio, for technical and management assistance [Thomas 1982].

This was the start of the NASA/DOE wind turbine research program that continued for over 20 years. This program resulted in the design, construction, and testing of 12 medium- and large-scale horizontal-axis wind turbines, supported by an extensive series of research and technology projects. NASA-related wind energy publications eventually totaled approximately 620 citations by over 500 authors and co-authors [Spera 1995].

The first step undertaken was the sponsorship of a Wind Energy Workshop by NSF and NASA-Lewis in 1973 [Savino 1973], to which were invited all those who had any prior or current interest in wind power. Pioneers from the 1930s -- such as Marcellus Jacobs, Palmer Putnam, and Beauchamp Smith from this country; and Ulrich Hütter and Arthur Stodhart from Europe -- and a younger generation of wind power developers presented papers and recommended research needs. In 1974 the Swedish government hosted a second international wind energy workshop [Ljungström 1974]. Similar conferences and workshops, held in different countries, became annual events. The *Wind Workshop Proceedings* from these annual conferences, supplemented by those from more specialized meetings, form a detailed record of the technical development of wind power from the mid-1970s until today [see Appendix B].

Just prior to the 1974 Stockholm workshop, it was discovered that not all of the experimental wind turbines of the 1940-1960 era had been dismantled or destroyed. To everyone's surprise, the Danish *Gedser wind turbine* was found to still exist. With a modest expenditure under a U.S./Danish bi-lateral agreement, the Gedser machine was later refurbished and retested. Data were used to validate new computer codes in the U.S. for predicting aerodynamic and structural dynamic performance. At the same time, the Gedser tests stimulated renewed interest in wind power in Denmark.

The U.S. Federal Wind Energy Program

In 1974, following recommendations from that first workshop, NSF and NASA drew up an initial wind energy research plan, although with little optimism that significant funding would be forthcoming. The shock of the Arab oil embargo a few months later, however, ensured rapid growth in research funds not only in the U. S. but worldwide. In 1975 the NSF program was absorbed into the newly-formed *Energy Research and Development Administration* (ERDA). The core of ERDA was the Atomic Energy Commission, and government-owned/contractor-operated national laboratories were selected to operate elements of the wind program. In 1977, ERDA was combined with other Federal organizations to form the U.S. *Department of Energy* (DOE).

The U. S. *Department of Agriculture* (USDA) was also asked to add its expertise in farm machinery applications which led to the continuing involvement of the USDA Agricultural Research Service at Bushland, Texas. Thus, while research was undertaken, supported, and reported in the literature by several agencies and laboratories, that work was all part of a single integrated program.

The initial federal plan envisioned research and technology projects closely coupled with the design and testing of experimental wind turbines that would incorporate increasingly advanced developments as they became available. The plan also assumed that three cycles or "generations" of experimental turbines would be required. First-generation turbines would be necessary merely to develop an understanding of design issues and to obtain basic data. The second generation was needed to put new developments into practice. Finally, a third generation of wind turbines would be required to reach a level of performance and reliability that could be cost effective on a broad scale. This series of wind turbines was designed to prove the technology and to reduce technical risk to the point where significant private capital could be attracted for continued development and commercial production.

Since the role of turbine size in the economics of the wind machine market was not understood at the time (and, to some extent, is still not clearly defined) a second major feature of the federal plan was that it supported the parallel development of prototypes in three sizes: Small-scale turbines (1 kW to 99 kW) for rural and remote use; medium-scale turbines (100 kW to 999 kW) for a remote community or industrial market; and large-scale systems (1 MW to 5 MW), primarily for the electric utility market.

NASA/DOE Mod-0 100-kW Experimental HAWT: 1975 to 1987

One of the first activities under the Federal Wind Energy Program was the design and construction of an experimental, medium-scale HAWT to serve as a test bed. This size was clearly needed in order to reach reasonable risk levels before proceeding to large-scale turbines. Conversely, many of the test results from a medium-scale turbine could well be applied to small-scale systems. This new research wind turbine was designated the *Mod-0* to emphasize its role as a test bed. It was designed and built for NSF by an engineering and fabrication team at the NASA Lewis Research Center [Puthoff and Sirocky 1974]. Installed in 1975 at NASA's Plum Brook Test Station near Sandusky, Ohio, it became a mainstay of experimental work on HAWTs in the U.S. for the next dozen years.

Original Mod-0 Configuration

The diameter of the Mod-0 rotor was selected to be 38.1 m, and a very low rated power of 100 kW (at a rated wind speed of 8.0 m/s at hub elevation) was chosen. This low rating was determined to be suitable for such a large rotor because of the modest wind speeds in the Sandusky area. Available running time for experimental work was a much higher priority than cost optimization at that time.

As shown in Figure 3-9, a two-bladed rotor located downwind of the tower was selected, following the examples of the Smith-Putnam and Hütter turbines and in accordance with economic studies that indicated a third blade was not cost-effective in large-scale systems. The Mod-0 rotor and power train were located in a streamlined *nacelle* atop a stiff, four-legged truss tower, with the rotor axis at an elevation of 30.5 m. Its original set of blades were of aluminum rib/spar/skin construction, following airplane wing design. While quite expensive, they were very light in weight (9,000 N each), which was considered a necessity because of the many unknowns in the structural dynamic behavior of the system. Details of the Mod-0 power train and yaw drive subsystems are illustrated in Figure 3-10. The rotor drove the *turbine shaft* at 40 rpm which, through a *parallel-shaft step-*

Figure 3-9. Final assembly of the 100-kW Mod-0 HAWT test bed in 1975. It was located at the NASA-Lewis Plum Brook Test Station near Sandusky, Ohio. (*Courtesy of NASA Lewis Research Center*)

up gearbox was increased to the 1,800-rpm speed of the 100-kW *synchronous generator*. At winds above rated, power was held constant at 100 kW by *full-blade pitch* under computer control, with the blades positioned by hydraulic actuators mounted on the *rigid hub*. Wind direction was sensed by a *wind vane* on top of the nacelle and monitored by the automatic yaw control system. When a change in the *nacelle azimuth* was needed, a pair of electric motors operated through a *worm-gear reduction drive* and a *pinion gear* to drive a *bull gear* attached to the *bedplate*. Yawing speed was 1/6 rpm.

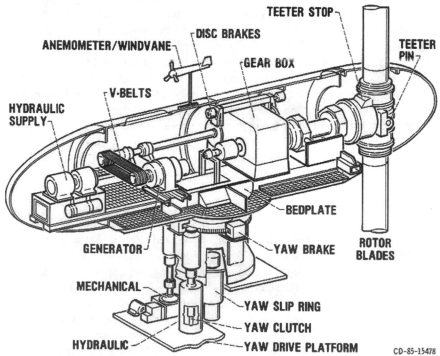

Figure 3-10. Power train and yaw drive equipment in the Mod-0 nacelle. The pulleys and belts at the generator permitted changes in rotor speed, to study tip-speed effects and avoid structural resonances. (*Courtesy of NASA Lewis Research Center*)

Mod-0 Research Tests and Configuration Changes

Over the next decade, more testing to investigate new ideas and new configurations (Fig. 3-11) was accomplished with the Mod-0 HAWT than with probably any other wind turbine before or since. Initially designed as a downwind, two-bladed configuration with full-blade pitch, it was later tested with one (counterbalanced) blade and with two blades in both upwind and downwind configurations. New materials were developed for wind turbine blades and first tested on the Mod-0. These included *laminated wood-epoxy* (Spera *et al.* 1990], a material now used on many small and intermediate wind turbines, and *fiberglass composite blades* fabricated with *transverse-filament-tape* [Weingart 1981].

One detrimental result from the original truss tower with its central stairway was high impulsive pressure loading on the blades, from the excessive wind speed reduction in the wake behind this high-drag tower. The staircase was soon removed, which reduced cyclic loads to a tolerable level. However, *tower wake-induced fatigue loads* had been identified as a major design driver for downwind rotors. First-generation machines used rigid hubs in an attempt to overcome the effects of dynamic loads by a simple brute force technique. This approach was never wholly satisfactory, so the Mod-0 test configuration expanded to include a *teetered hub* (Fig. 3-12).

As structural dynamic knowledge increased, the Mod-0 truss tower was placed on a new base composed of flexural steel beams (Fig. 3-11(a)). This allowed the natural vibration frequency of the turbine to be lowered and "tuned" in order to simulate "soft" tower structural concepts. Such concepts would have the potential for lower tower weight and cost, but structural dynamic loads could be lower or much higher, depending on resonances and instabilities that were not well-understood with the available analysis tools. Later, the Mod-0's truss tower was replaced by a slender shell tower (Fig. 3-11(b)) to prove out those tools and the effectiveness of soft structural systems.

Another area which received considerable emphasis was *variable-speed constant-frequency (VSCF)* operation. Prior to the need for interconnection with the utility grid, many small machines operated at variable rotor speed depending on electric load and wind speed, but this had rarely been attempted on any turbine much over 15 m in diameter. The desire to operate at variable rpm is engendered by the potential for higher energy capture (always operating near peak rotor efficiency) and the potential for gust load alleviation.

Solutions to the electrical problems associated with a variable generator speed could now be envisioned that were not available in a practical way two decades earlier. The structural dynamics issues associated with the need to preclude harmonic vibrations over a range of rotor speeds were not viewed sanguinely. It was enough of a problem in the 1970s to accomplish this at one speed. In spite of this, the Mod-0 was operated as a variable-speed machine, testing several generator and power-conditioning components. More importantly, the Mod-0 tests probed the structural dynamics "envelope" on a relatively large, flexible system and thereby provided the data to develop and validate complex computer models needed to predict natural frequencies and loads under such conditions.

Also tested were rotors with *partial-span pitch control*, *flaps* (often less-correctly termed *ailerons*), *fixed-pitch stall regulation*, and *free-yaw* response to changing wind directions. Tests at Plum Brook determined the effects of precipitation, various airfoils, blade root and tip design innovations, and auxiliary aerodynamic devices such as *vortex generators*. The validation of computer models and control algorithms, though less visible than hardware changes, was probably one of the most valuable contributions of the test bed program. After a useful life of over a dozen years, the Mod-0 experimental HAWT was dismantled in February 1987, leaving as its legacy an extensive set of documentation that forms a principal basis of modern wind turbine technology.

(a)

(b)

(c)

Figure 3-11.   Different Mod-0 configurations during a decade of research.
(a) **1979:** Upwind rotor, steel spar blades, partial-span pitch, teetered hub, and a spring base; simulated the "soft" second-generation Mod-2 HAWT. (b) **1982:** Shell tower, flap control, inboard blade sections of laminated wood. (c) **1985:** One-bladed, teetered rotor with tip control. (*Courtesy of NASA Lewis Research Center*)

Figure 3-12. Mod-0 teetered hub. (*Courtesy of NASA Lewis Research Center*)

Development of Modern Small-Scale Wind Turbines:
1976 to 1981

During the early 1970s a number of small companies, recognizing the potential of wind power and the limitations and costs of the few Dunlite, Aerowatt, and Elecktro wind turbines being imported, began to develop small wind turbines in the 1 kW to 10 kW range. Several supporting activities were initiated by the U.S. government, as an outgrowth of early studies on the use of wind power for small rural or remote applications.

Major Testing Facilities

For all the long history of wind power, there was actually very little detailed, measured performance or other engineering data available. Few of the limited design tools and analytical models had been validated to any significant degree, and most information was anecdotal in nature. Hence, testing of available machines became an early priority, both in the U.S. and abroad.

The Rocky Flats Test Station

The Rocky Flats Plant near Golden, Colorado, operated by *Rockwell International Inc.* for ERDA (later DOE), was given the responsibility for the development and testing of small wind turbines. The Test Station consisted of an array of 32 test pads, each capable of supporting, servicing, and collecting data from a small-scale turbine (Fig. 3-13). During the years 1976 through 1981, when the bulk of the testing was performed, 54 different wind turbines in the 1-kW to 50-kW range were tested at Rocky Flats. A permanent building housed both personnel and extensive test equipment. Shops with dynamometers allowed rapid check-out testing of components prior to outdoor systems tests as well as rapid turnaround during modifications. The site's low mean wind speed (6 m/s) yielded only modest annual energy, which was of some concern to the manufacturers in the early days, but the relatively frequent high winds and turbulence were excellent for determining design loads, system ruggedness, and reliability within reasonably short test periods.

Machines developed by private industry could be tested at Rocky Flats under two different procedures: First, a wind turbine could be tested at government expense, the results then being public information. However, the degree of testing, instrumentation, and analysis could in many cases be larger than the manufacturer could otherwise perform. Secondly, the manufacturer could pay for testing, in which case the results would remain proprietary.

An important outgrowth of the testing at Rocky Flats (as well as at Sandia National Laboratories and NASA Lewis Research Center), was the development of standardized testing methods for determining *power curves* (power output *vs* wind speed). This most basic performance characteristic is one of the most difficult to measure because of the variability of the wind. Prior to 1979, most power performance curves for commercial wind turbines were generally not reproducible nor comparable within acceptable bounds of uncertainty. Standardization of terms and methodology eventually brought major improvements to the comparability of advertised power curves and *annual energy output*. Data analysis techniques developed at Rocky Flats and Sandia (such as the *method of bins*) were later included in voluntary standards adopted by the International Energy Agency, the American Wind Energy Association, and the American Society of Mechanical Engineers [*e.g.* Frandsen 1982, AWEA 1985, ASME 1989].

Figure 3-13. The Rocky Flats Test Station, near Boulder, Colorado. The performance and reliability of a wide variety of small-scale commercial and experimental turbines were tested here. (*Courtesy of the U.S. Department of Energy*)

Difficulties in determining a power curve in the rapidly-varying natural wind and the need to validate analytical models led to supplementary performance testing at the nearby *High-Speed Rail Test Facility* operated by the Department of Transportation. Full-scale small turbines were placed on a tower mounted on a flatcar and pushed by a diesel locomotive at speeds covering the turbines's operating wind speed range. Control of the relative velocity between the turbine and the air allowed rapid validation of power curves and the analytical techniques being developed for normal on-site testing.

USDA Bushland Test Station

The *U.S. Department of Agriculture* (USDA) has also tested a variety of small-scale wind turbines in actual agricultural and rural applications, including irrigation, crop drying (both electrically and by churning a viscous fluid), diesel interconnection, and general rural electrical use [USDA 1979, Clark 1983]. Testing is conducted principally at the USDA Agricultural Research Service facility near Bushland, Texas, with some on private farms. Prototypes of turbines developed privately as well as with federal funds are evaluated against particular agricultural requirements. Darrieus VAWTs are also tested at the USDA test station by engineers from the *Sandia National Laboratories* in Albuquerque, New Mexico, because of the superior wind regime at Bushland.

Risø Test Station

Several small wind turbine test stations were set up in Europe during the late 1970s. Most notable was the *Risø National Laboratory Station for Wind Turbines* in northern

Denmark. In addition to research and development testing, the Risø station supports the evaluation and certification of commercial wind turbines. With certification manufacturers are eligible for various Danish tax credits and incentives, as well as for those from certain other European countries. While the merits of government certification can be argued relative to the merits of industry self-certification coupled with the workings of the market place, the Danish program was effective in precluding undeveloped or poorly-developed turbines from reaching the European market in any quantity. This led to the reputation for reliability and performance enjoyed by Danish turbines during the highly-competitive upsurge in the development of wind power stations in the U.S. in the mid-1980s.

U.S. National Wind Technology Center

In the early 1980s reductions in the U.S. government funding of wind research brought an end to major testing activities at Rocky Flats. Other responsibilities of the test station were turned over to the nearby *Solar Energy Research Institute* (SERI), which has recently become the *National Renewable Energy Laboratory* (NREL) of the U.S. Department of Energy. However, with the resurgence of interest in wind energy in the 1990s, the facility at Rocky Flats was reconditioned as the *National Wind Technology Center* and operated by NREL. This upgraded facility provides both government and private industry with the laboratory and field test capabilities required to support the development of advanced small-, medium-, and large-scale wind turbines.

Small-Scale Wind Turbine Development Program

Shortly after the start of tests at Rocky Flats on existing wind turbines, a series of competitive solicitations were issued for the development of new small-scale systems with government support [Healy and Dodge 1981]. A number of these solicitations provided "set-asides" for small business, to stimulate that segment of the industry. Table 3-1 lists the small-scale turbines developed under this program, with summary design data.

The first series of development projects was initiated in 1978 and included three levels of rated power: 1 kW to 2 kW, for extremely reliable systems for remote use; 8 kW, for rural residential and other uses; and 40 kW, for irrigation and other agricultural applications.

High-Reliability Turbines

Three high-reliability prototype machines in the 1-2 kW range were completed in 1980 by *Enertech Corporation, North Wind Power Company*, and *Aerospace Systems, Inc./Pinson Engineering*. The first two were HAWTs with upwind rotors and two and three blades, respectively. The last was a 3-bladed giromill VAWT configuration. Several copies of the Enertech turbine were built for development purposes. The North Wind turbine evolved into commercial products, the 2-kW *HR-2* and 3-kW *HR-3* HAWTs, of which a significant number have been built that operate successfully in very remote locations. Their reliability stems from having only a few moving parts. Power and speed are controlled by *vertical furling* (*i.e.* tipping up the rotor) in high winds.

8-kW Turbines

Three manufacturers who developed 8-kW prototypes were *Windworks, Inc., Grumman Energy Systems,* and *United Technologies Research Center* (UTRC). The first two designed 3-bladed rotors, while the 2-bladed rotor of the UTRC machine incorporated a unique "self-twisting" concept for speed and power control. Of fiberglass construction, the blade spars

Table 3-1. Small-Scale Wind Turbines Developed with U.S. Government Support
[Healy and Dodge 1981]

Manufacturer	Rating (kW)	Rotor Configuration	Annual Energy[1] (kWh/y)	Power Trian	Weight on Tower (Tower Weight) (kg)
ASI/Pinson	2.0	4.6-m giromill VAWT 3 aluminum blades 2.4-m high Cyclical pitch	1,830	Gearbox Alternator	272 (680)
North Wind	2.0	5.0-m upwind HAWT 3 wood blades Vertical furling	6,870	Direct-drive Alternator	361 (499)
Enertech	2.0	5.0-m downwind HAWT 2 wood blades Centrifugal control	8,400	Gearbox Alternator	345 (556)
Structural Composites Industries	4.0	9.5-m downwind HAWT 3 fiberglass/foam blades Pitch control	22,000	2-Stage Spur GB Induction generator	726 (576)
North Wind	6.0	9.5-m downwind HAWT 2 wood blades Passive pitch control	22,100	Direct-drive Lundel synch. alternator	477 (1,776)
Tumuc	4.0	11.9-m Darrieus VAWT 3 fiberglass blades Mech. & aero. brakes Cantilevered tower	22,300	Planetary GD MSCF induction generator	2,365 (946)
UTRC	8.0	9.45-m downwind HAWT 2 GRP blades Flex-beam pitch control	25,000	Gearbox Induction generator	839 (1,120)
Windworks	8.0	10-m downwind HAWT 3 aluminum blades Hydraulic pitch control	30,000	Direct-drive Alternator	735 (844)
Grumman	11.0	10.1-m downwind HAWT 3 aluminum blades Electric pitch control	32,000	Gearbox Induction generator	1,174 (1,189)
UTRC	15.0	14.6-m downwind HAWT 2 fiberglass/foam blades Self-twist pitch control	50,000	Gearbox Induction generator	703 (544)
Enertech	15.0	13.4-m downwind HAWT 3 wood blades Stall control Tip brakes	51,500	Gearbox Induction generator	431
McDonnell	40	17.7-m giromill VAWT 3 steel/aluminum blades 12.8 m high Cam pitch control	128,400	Gearbox Induction	7,128 (4,730)
Kaman	40	19.5-m downwind HAWT 2 fiberglass blades Pitch control	134,000	Gearbox Induction generator	2,766 (1,814)

[1] Predicted for a site with a 5.3-m/s annual-average wind speed

were laid-up with fiber orientations that gave them specific torsional properties. Changes in aerodynamic and centrifugal forces caused the spars to twist and untwist, enabling a totally passive blade pitch control with varying wind and rotational speed.

40-kW Turbines

The larger 40-kW prototype wind turbines were developed by the *Kaman Aerospace Corporation* and the *McDonnell Aircraft Company*. The Kaman design was a HAWT with a 2-bladed, downwind rotor (Fig. 3-14), while the McDonnell design was a 3-bladed giromill VAWT (Fig. 3-15). Both prototypes were delivered to Rocky Flats in 1980 and tested in 1981.

17-m Alcoa VAWT

Alcoa Corporation, under contract to Sandia National Laboratories, developed a pre-commercial prototype of the *Sandia/DOE 17-m VAWT* (Fig. 3-20) and built four units. One of these was installed at Rocky Flats for comparative testing, and to verify predictions of performance and loads. Using the data obtained, two manufacturers developed the design further for commercial use (Fig. 3-24).

Additional Small-Scale Wind Turbine Development Projects

A second series of competitions followed in 1980 with contract awards to *Tumac Industries, Structural Composites Industries* (SCI), and *North Wind Power Company* for 4-kW prototypes. The application for this size wind turbine were seen as a power source for remote residences, either standing alone or inter-connected with a utility line. SCI designed a HAWT with a 3-bladed, downwind rotor, while Tumac developed a small Darrieus VAWT with a cantilevered tower and no supporting cables. The last of the competitions for government support of small-scale wind turbine development was for a light agricultural application, on farms and ranches. It resulted in awards to the Enertech and UTRC companies, both of which selected HAWTs with downwind rotors.

The wide spectrum of design concepts listed in Table 3-1 is indicative of the early uncertainty amongst wind turbine designers as to what constitutes an optimum configuration for a specific application. While some blind alleys were followed for a time, a number of these configurations were eventually selected for commercial machines. The inclusion of wind energy in the solar energy tax-credit legislation in the early 1980s encouraged numerous other companies to enter the wind turbine development and manufacturing field. Future development of wind turbine systems in the U.S. was undertaken without direct government support, while the federal program concentrated on supporting research to expand the technology base for all sizes and applications of wind turbines.

Supporting Technology for Small-Scale Wind Systems

A series of analytical and experimental projects conducted under the Federal Wind Energy Program provided data for both a technology base and decision-making related to the further development of small-scale turbines. Pacific Northwest Laboratories tailored the techniques of wind resource assessment for the use of manufacturers, planners, distributors, and agricultural extension agents. This work led to the development of "stand-alone" courses on the siting of turbines, using slides, audio tapes, and workbooks. Economic and market analyses were made by organizations such as A. D. Little, JBF Scientific, and General Motors on subjects crucial to the developing wind turbine industry, such as

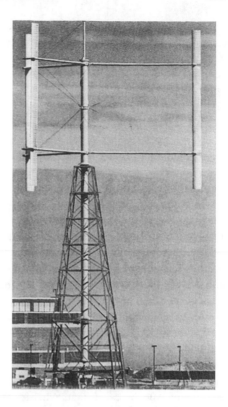

Figure 3-14. The prototype Kaman 40-kW HAWT. (*Courtesy of the U.S. Department of Energy*)

Figure 3-15. The prototype McDonnell 40-kW giromill VAWT. (*Courtesy of the U.S. Department of Energy*)

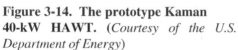

- -- market penetration barriers;
- -- market characterization and demographics;
- -- markets with near-term high potential;
- -- decision analysis for the small-system user;
- -- economics of small systems owned by the end user;
- -- product liability insurance;
- -- industry consensus standards;
- -- producibility of small-scale turbines;
- -- effects of production rates on the economics of small systems.

The electrical stability and safety implications of many wind turbines distributed over a utility distribution grid were studied, as well as the potential electrical and aerodynamic interactions between multiple units at a given location.

Similar studies relating specifically to the farm and rural use of small-scale wind turbines were performed or sponsored by the USDA. Topics included economic analyses of wind-powered

- -- irrigation pumping;
- -- space heating systems for farm houses and other buildings;
- -- grain drying;
- -- refrigeration;
- -- water heating.

Experimental research at the Rocky Flats plant included dynamometer, vibration, and electrical testing of small-scale turbines, and the validation of computer models for the prediction of power performance and loads. At the USDA, research on both shallow- and deep-well pumping and the matching of wind turbines and pumps was conducted to optimize the performance of the total system.

Small Turbine Deployment and Applications

In the 1970s many institutional problems had to be faced by owners of wind turbines interconnected with utility power lines, and by developers of small wind turbines. These included safety, power quality, standby and reserve power criteria, buying and selling rates for electricity, zoning requirements, and even the determination of the authority and office to which the turbine owner should apply. Since most of these problems were at the state and local levels and outside federal responsibility, another type of testing was undertaken to solve them. In 1979 the "Field Evaluation Program" (FEP) was initiated, in which federal funds were provided for the installation of 100 or more commercial, small-scale wind turbines throughout each of the United States and its territories, and to provide advice and assistance. In return, utilities and state and local authorities would facilitate the interconnection of private turbines and develop local permitting and approval processes.

Many of the early turbines available commercially at the time had not, unfortunately, been adequately tested, and the difficulty of achieving adequate reliability had been underestimated by many of the new entries into the field. Failure rates were excessively high, giving a negative image to wind energy in some quarters. On the other hand, there were significant successes. Much was learned about correcting operation and installation problems that may have been missed in the design phase. Utility companies and state and local governments became familiar with both the potential benefits and issues associated with wind power systems. About 40 of the planned 100 turbines were installed before the FEP was discontinued in 1983.

The New England Wind Project, a program similar to the FEP, was implemented in 1980 to evaluate some of the DOE-sponsored small-scale turbines under field conditions. Fourteen sites were selected, and a number of prototypes were installed. These included an 8-kW UTRC HAWT on Moon Island in Boston harbor.

There is no doubt that efforts under the Field Evaluation Program and the New England Wind Project helped mature the small-scale wind turbine industry and added impetus to the development of new designs. Many institutional issues were resolved in key states, aiding the introduction of reliable commercial machines in large numbers in the mid-1980s.

Innovative and Unconventional Wind Turbine Concepts

Before describing *unconventional* wind turbines, a definition of a *conventional* turbine is in order. For the purpose of this discussion, a conventional configuration is that which was established as practical by the DC wind power plants of the 1930s and by the Smith-Putnam project. Thus, a conventional wind turbine system is a HAWT with a low-solidity rotor powered by aerodynamic lift driving an electrical generator, with all rotating components mounted on a tower. There have been, throughout history, a myriad of attempts to use other techniques for harnessing wind power. In principle any movable device subject to wind forces can be made to rotate (on some axis), oscillate, or translate. It can then be coupled to some sort of converter and be made to produce mechanical and/or electrical power. Whether an innovative device can generate energy at a lower cost than a conventional wind turbine is quite a different question, however.

Early Unconventional Systems

Panemones

The earliest practical windmills (see Chapter 1) were of the vertical-axis type driven by drag forces and known generically as *panemones*. Panemones are multi-bladed machines often designed with some type of articulating mechanism to fold or *feather* the blades that are moving upwind, while deploying the working blades moving downwind. Panemone devices are re-invented at frequent intervals. Their relatively poor efficiency, low blade speed (a tip-speed ratio necessarily much less than unity), and large surface area have led panemones into a continual dead end. While simple to manufacture in small sizes, they have not proven to be cost-effective in other than primitive installations, owing to the need for large amounts of material and the problem of withstanding high wind loads.

The *Savonius* rotor (Fig. 3-16) is another innovative vertical-axis device and one which has somewhat better performance. With its "S-shaped" cross section, the Savonius rotor, (named for its Finnish inventor) is really more of a semi-lift or low lift-to-drag device. Since the blades are simple to form from sheet metal, the Savonius finds an occasional niche in developing countries for tasks such as water pumping. For the same reasons as the panemone, the Savonius turbine has not been successful for general use.

By the 1950s a number of unconventional wind turbine configurations had been explored in some detail. One of the more unusual of these was the *Enfield-Andreau* HAWT described earlier (Fig. 3-4).

HAWTs with Multiple Rotors

Honnef, in Germany, was an advocate of multiple rotors in the same plane on a single tower as a means of achieving high power levels with rotors of intermediate size [Honnef 1932, Hütter 1974]. In the early 1940s he tested systems with two counter-rotating rotors in the 8-m and 10-m diameter range, but his large-scale designs never got further than artist's drawings. Nevertheless, the idea of multiple rotors continues to attract followers. Studies in the early 1970s concluded that it was more cost-effective to use multiple turbines or larger turbines than to pay for the complex structure needed to support and yaw a multiple-rotor system. One company briefly went into limited production of a six-rotor, single-tower HAWT in 1987.

Multiple rotors in the same plane increase net swept area and should not be confused with multiple rotors on the same axis. In some cases the latter are counter-rotating. Advocates of such systems usually misunderstand the physics behind wind energy conversion. The effects of induced velocity limit the theoretical maximum power coefficient of two rotors on the same axis to little more than that of a single rotor. Counter-rotation does decrease the rotational energy lost in the wake, but this benefit is trivial compared to the costs of the second rotor and associated gearing. These systems have rarely been successful, much less cost-effective. The *Noah* HAWT, privately developed in northern Germany in the mid-1970s, is the most recent example known.

Translating Units

Translating wind power systems have been conceived as a means of achieving power levels far beyond that of an individual wind turbine. These concepts generally involve a large circular or oval track upon which a "train" of vehicles moves, each vehicle supporting a sail or vertical wing of some type. The wheels of the vehicles turn generators, and power is conducted away through a third rail, in the reverse of a conventional electric railway

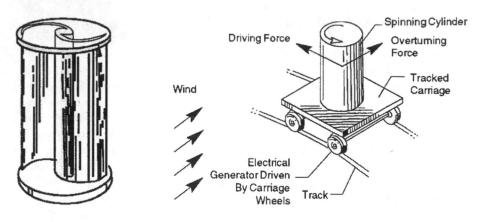

Figure 3-16. Savonius rotor.   **Figure 3-17. The Madaras concept for generating electricity using the Magnus effect.**

system. By increasing the lengths of the track and the train, more power could be obtained, presumably without limit.

The only known large-scale test conducted to develop the technology needed for a translating wind power plant was the 1933 *Madaras* experiment. Madaras conceived a plan for a 40-MW plant in which the railway cars were propelled around a track by "sails" in the form of large, vertical *rotating cylinders*, as illustrated in Figure 3-17. Wind flowing across a spinning cylinder or sphere can create a high lift force at right angles to the flow. This is the *Magnus effect*, known since the 1850s. In a preliminary test of the Madaras design, one full-scale cylinder -- 27.4 m high and 8.5 m in diameter -- was built and mounted on a stationary platform in order to measure the Magnus-effect forces. The test was performed in conjunction with the Public Service Company of New Jersey, in Burlington. Test results were mixed, and high cost estimates combined with the economic climate of the Great Depression led to abandonment of the project.

The basic idea of the translating system re-emerged in the early 1970s with some enthusiastic sponsors. Several small test articles were constructed privately with various types of sails or wings. Translation was usually horizontal, but at least one translated vertically like a conveyor belt. None of these was successful because of mechanical complexity, low "tip speed" with an associated low aerodynamic efficiency, high track loads and overturning moments, the need to reverse the blade angle (or direction of cylinder rotation) at each end of the oval track, and lower wind speeds near the ground.

Searching for Modern Unconventional Concepts

To ensure that promising ideas were examined, the Federal Wind Energy Program funded an "Innovative Concepts Research" activity [Vas and Mitchell 1979], inviting proposals on any concept that could extract energy from the atmosphere. Many proposals were received, and a number were supported through the exploratory research phase. In addition, privately-funded unconventional concepts for wind energy conversion appeared regularly. Some examples from both the public and private sectors are described here.

Vertical Axis Systems

Several vertical-axis concepts have been proposed for wind turbines that are driven by lift forces, rather than the drag forces that turn a panemone. These include the *Darrieus rotor* which has been the most successful, and this VAWT configuration is discussed in the following section. The *giromill*, another lift-driven vertical-axis machine, has been developed by several manufacturers. Giromills use straight vertical blades whose pitch angles vary cyclically during rotation and are independently controlled. The 40-kW prototype built and tested by the *McDonnell Aircraft Company* was discussed previously (Fig. 3-15). The mechanical complexity of the blade support and pitch-change systems has usually prevented giromill designs from being cost-effective.

An innovative VAWT rotor, invented by Peter Musgrove at Reading University in England, is the *Arrow* configuration shown in Figure 3-18. This prototype at Carmarthen Bay, South Wales, is 25 m in diameter and rated at 130 kW [Musgrove and Clare 1987]. Its vertical turbine shaft and a horizontal crossarm form a rotating "T". Two blades are attached to the ends of the crossarm by struts, and each is hinged at its mid-length. Below the rated wind speed the blades are vertical, but at wind speeds above rated they are furled or "reefed" to an arrow-head shape by hydraulic actuators in the crossarm. This reduces their swept area and limits both the output power and blade loads.

(a)

Figure 3-18. The 125-kW, 25-m diameter Musgrove Arrow VAWT at Carmarthen Bay, South Wales. (a) Blades vertical for maximum swept area (b) Blades furled. (Courtesy of *Vertical Axis Wind Turbines Ltd.*)

(b)

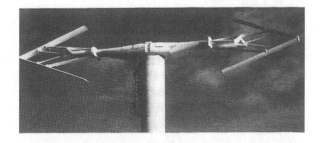

Wind Flow Augmentation

Another category of innovative concepts encompasses methods of augmenting (accelerating) the wind stream. For example, a *diverging exit cone* or *diffuser* extending downwind from the rotor plane can augment the mass flow into and through the turbine. To prevent separated flow the angle of divergence in a conventional diffuser must usually be rather shallow, which would make the exit cone very heavy and expensive. Research at *Grumman Aerospace Corporation* [Foreman and Gilbert 1979, Loeffler 1981] showed that placing annular slots in the cone allowed external air to energize the boundary layer and delay separation. This could lead to a shorter exit cone with an *augmentation factor* (*i.e.*, maximum augmented power divided by the maximum power of an unducted rotor of the same size) of two to three. Even with these favorable aerodynamic results, the cost of the exit cone and its supporting structure has been estimated to exceed that of using a larger conventional rotor to obtain the additional power.

An attractive alternative, conceived at the University of Delft (Netherlands), is the use of *tip vane dynamic inducers* to provide the same flow effect as a diffuser. Considerable research on dynamic inducers was also undertaken in the U.S. by *AeroVironment, Inc.* [Lissaman *et al.* 1980]. Performance was found to be extremely sensitive to tip-vane operating parameters, and this has prevented significant augmentation over a practical operating range, at least to-date.

Atmospheric Energy Conversion

Several laboratory-scale research efforts examined methods of extracting the *latent heat of vaporization* existing in the humidity in the atmosphere (the same energy source that causes thunderstorms) and the possibility of tapping the energy of *tornado-like vortices*. It turned out that these would require structures several thousands of meters high to achieve reasonable efficiency. Cost and institutional issues precluded further consideration.

Another "tower-type" energy concept that is re-discovered at intervals utilizes atmospheric convection currents, either natural or augmented. The most-notable attempt in recent times was the construction of a large tower in Spain by a German-Spanish team. Several acres of black Mylar plastic sheet were suspended a meter above the ground around the tower, forming a solar "oven" to heat the air below. Convection forces then drew the heated air to the base of the tower and through an internal turbine, exhausting upward through the tower. The predictable result of such experiments is a finding that the system has a low efficiency and a high capital cost.

A more esoteric scheme utilized *electrostatic generators* incorporating charged screens which could extract kinetic energy from the wind by electric fields operating on minute charged particles in the air. In theory, that could lead to a "no-moving-parts" wind power device. In experiments at the University of Dayton and at *Marks Polarized, Inc.* power was actually extracted, but not enough to justify further research and development.

Unconventional HAWT Subsystems

An interesting innovative concept developed with private funds was the 50.3-m diameter *Schachle-Bendix* HAWT shown in Figure 3-19, sponsored by the *Southern California Edison Company* and tested at its Devers site near Palm Springs [Rybak 1981]. One of its innovative features was a tripod truss tower rotating on a track, there being no yaw drive at the nacelle. This approach was similar to that of some early French HAWTs and the pioneering Russian *Balaclava* turbine of the 1930s (Fig. 1-21). A second unusual feature was the use of a *variable-speed hydraulic drive* in the power train. Fourteen fixed-displacement hydraulic pumps in the nacelle, driven by the turbine shaft, supplied high-pressure fluid to 18 variable-displacement hydraulic motors on the ground, and these in turn drove a synchronous generator. This allowed flexibility in the rotor speed and in the control system. A third unconventional feature was the use of new, non-aircraft airfoils designed by the manufacturer for very high lift. The system was found to have large power losses in the hydraulic drive and an aerodynamic performance which did not meet expectations. Although it did attain a power output of 1.1 MW, a very high wind speed would have been required to achieve its 3-MW rating.

Figure 3-19.   The Schachle-Bendix 3-MW HAWT. (*Courtesy of Southern California Edison Co.*)

Sandia/DOE Experimental VAWTs

The Darrieus Wind Turbine Rotor

An elegant rotor design had been invented in France in the 1920s by G. J. M. Darrieus [Darrieus 1931]. It utilized a vertical axis around which rotated curved airfoil blades in the shape of a hoop, somewhat resembling an egg-beater. This curved-blade rotor was re-invented in the late 1960s by engineers at the National Research Council in Canada (Fig. 3-8), where it was the subject of extensive study and development for two decades [*e.g.* South and Rangi 1972].

The *Darrieus rotor* has long been recognized as containing the seeds of a highly efficient and intrinsically simple VAWT. It has its gearbox and generator located at the base of the rotor for ease of maintenance. It will accept the wind from any direction without a yaw mechanism. Although Darrieus blades are about three times as long as the blades of a HAWT with the same swept area, they are generally so slender that the actual rotor solidities (*i.e.*, planform area per unit of swept area) are similar. However, a cost-effective Darrieus VAWT required considerable development to overcome several disadvantages. Guy cables required to support the top rotor bearing make it difficult to mount the rotor very far above ground level to take advantage of higher wind speeds. Also, a Darrieus rotor produces higher gearbox torques, both steady and cyclic, than a HAWT of the same power and requires more material.

Sandia/DOE 17-m Experimental VAWT

Darrieus research was undertaken briefly at the NASA Langley Research Center in the early '70s [Muraca *et al.* 1975, Muraca and Guillote 1976]. However, the Sandia National Laboratories in Albuquerque became the center of VAWT research and development in the U.S., building on technology from the NRC in Canada. After initial analysis, laboratory research, and small-scale field tests, a *17-m VAWT* was designed and built at Sandia (Fig. 3-20). The main purpose of this turbine was to determine what design and manufacturing improvements would be required to make Darrieus VAWTs competitive with HAWTs.

Economic studies supporting the early research tests on the 17-m VAWT suggested several improvements: First, two blades would be more cost-effective than the original three on this turbine. Second, the long struts used to strengthen the rotor should be eliminated. These struts were adding drag that consumed significant power, and their cost was of the same order of magnitude as that of the working blades. Third, the blade airfoil shape should be changed to one specifically tailored for the VAWT application. It is worth noting that all three of these findings also applied to HAWTs. Strut drag losses, for example, were found to be significant on the Gedser HAWT (Fig. 3-1) during tests in the late 1970s, and few (if any) modern HAWT rotors have struts.

VAWT Blade Airfoils

Each time the blade of a VAWT makes a 360-deg rotation around its central column, its angle of attack changes rapidly and over a wide range. This presents VAWT designers with a complicated aerodynamic problem not present in aircraft and only to a small degree in HAWT rotors. Airfoils tailored for VAWT use must have not only enhanced lift at low angles of attack, both plus and minus, but must also be shaped to control stall behavior during normal operation. In high winds, stall moderates lift so that constant power is

Figure 3-20. The Sandia/DOE 17-m experimental Darrieus VAWT under test near Albuquerque, New Mexico. (Courtesy of Sandia National Laboratories)

achieved, and regulating power in high winds without movable aerodynamic surfaces helps to control system costs. Sandia researchers patented an early attempt to tailor stall behavior by *pumped spoiling*, which involves drilling lines of holes along the blade and forcing small quantities of air into the boundary layer. While technically feasible, initial cost and maintenance expenses were found to be a major disadvantage of pumped spoiling. Redesigning the blade shape proved to be much more practical.

Engineers at Sandia and at various universities have developed airfoils that can regulate power through stall. In 1980 researchers at Ohio State University originated the concept of *natural laminar flow airfoils*, which met the two requirements of enhanced lift at low angles of attack and regulated stall at high angles. These specially-tailored components were first tested on a 5-m research VAWT at Sandia, and then their improved performance was verified on the 17-m VAWT [Klimas and Worstell 1986]. As a result of the NRC and Sandia airfoil development programs, peak power coefficients of today's Darrieus turbines have reached the same level as those of modern HAWTs.

Blade Manufacturing Improvements

The first VAWT blades were expensive because they were made of aluminum, fiberglass, and honeycomb materials, all of which had to be carefully fitted together by hand to form precise compound curves. In the mid-1970s *Alcoa Industries* became interested in aluminum VAWT blades and systems as potential product lines and investigated methods for reducing blade manufacturing costs. Working with Sandia engineers, the company developed an extrusion process for Darrieus blades in which partially molten bars of aluminum are forced through an airfoil-shaped die, after which the shaped structure cools and solidifies. This process dramatically reduced the cost to manufacture VAWT blades, produced long sections of uniformly high quality material, and controlled the airfoil shape to very close tolerances.

Extruded aluminum blades were first flown on the 17-m VAWT, and extrusion continues to be used today to manufacture VAWT blades. Figure 3-21 shows a steel extrusion die and extruded aluminum sections for a large VAWT airfoil. The leading and trailing-edge sections in Figure 3-21(b) are joined together before the complete airfoil is given its curved shape.

Figure 3-21. Extrusion of aluminum VAWT blades.
(a) Steel die used to form the leading-edge section of an airfoil integrally with its interior spars. (b) Leading- and trailing-edge extrusions prior to being joined and bent to their final, curved shape. (*Courtesy of Sandia National Laboratories*)

(b)   **(a)**

Structural Dynamics

Because of its relatively long and slender blades, a detailed knowledge of the structural-dynamic behavior of a Darrieus VAWT is critical to achieving an acceptable fatigue lifetime. Engineers must be able to accurately predict a VAWT's many modes and frequencies of vibration, as well as static and dynamic stresses caused by gravity, centrifugal forces, and the wind. Predicting natural frequencies requires a computer model of the complete VAWT system, including the supporting cables, that must be validated by *modal testing* of the actual turbine [Lauffer *et al.* 1987]. To conduct a modal test, accelerometers are mounted at strategic points on the structure; a cable is attached to the VAWT, tensioned, and then quickly released. The resulting dynamic signals from the accelerometers are analyzed by computer for their frequency content, which identifies the vibration modes. The modal testing procedures developed at Sandia using the 17-m VAWT as a test bed are now applied throughout the world on both VAWTs and HAWTs.

Sandia/DOE 34 m VAWT

Between 1984 and 1987 a *34-m Darrieus VAWT* test bed for developing advanced concepts was designed and fabricated in modules at Sandia and then installed at the U.S. Department of Agriculture Test Station at Bushland, Texas [Ashwill *et al.* 1987]. This highly-instrumented 625-kW turbine (Fig. 2-2) has *variable-chord blades* for aerodynamic and structural optimization, *tailored airfoils* designed specifically for VAWTs, and a *variable-speed constant-frequency (VSCF)* generating system. The test program began in 1988, in cooperation with the USDA personnel, and by early 1989 full rated power was achieved.

Figure 3-22 shows the *operating map* of the control system. The test bed uses a variable-speed system in which the generator's variable-frequency AC output is converted to DC and then converted back to AC at the utility line frequency. *Elastomeric couplings* are incorporated in the drive train to attenuate the *torque ripples* associated with VAWTs (caused by the blades traveling in alternating upwind and downwind paths).

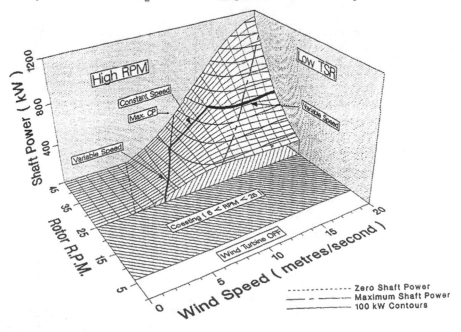

Figure 3-22. Operating map of the 34-m VAWT test bed. [Ashwill *et al.* 1987]

The 34-m diameter rotor has two *step-tapered* blades, each with five sections of extruded 6063-T6 aluminum. Airfoil cross-sections along the curved blade are illustrated in Figure 3-23. The use of step-tapered blades, with a longer chord near the hubs, a smaller chord at the equator, and a transition section in between, maintains a more uniform *Reynolds Number* over most of the blade length. The high-lift, *natural laminar flow* airfoils used in the equatorial sections are part of a series developed specifically for use on VAWTS. Root sections use conventional *NACA airfoils*, which offer superior aerodynamic properties at lower Reynolds numbers and higher structural strength. The chord widths of the various blade sections are too large for extrusion in one piece, so each section is made up of either two or three extrusions with bolted lap joints in the spanwise direction, as shown in Figure 3-23.

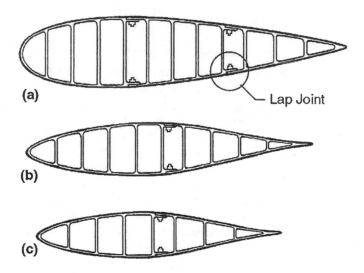

Figure 3-23. Airfoil cross-sections in the 34-m VAWT blades. (a) Root sections, near the hubs (b) Transition sections (c) Center (equatorial) section [Berg *et al.* 1990]

Continued research work, centered on the 34-m VAWT test bed, includes validating aerodynamic and structural-dynamic computer models, testing airfoil designs and new blade materials, and developing various control strategies. The purpose of this work is to assist industry to improve the VAWTs in commercial operation, as well as to develop technology that can be used in the next generation of VAWTs.

Developing a Commercial VAWT System

Testing and modification of the 17-m VAWT at SNL in the late 1970s rapidly improved its performance and reliability. A low-cost, pre-commercial design was then derived from it in 1979 under a Sandia development contract won by *Alcoa Industries*. Specifications required a rotor diameter of 17 m and a power output of 100 kW, and that the system be designed for manufacturing at low cost. By 1981 three units were installed on sites representing specific applications: Bushland, Texas, at the U.S. Department of Agriculture Test Station, to demonstrate an agricultural application; Rocky Flats, Colorado,

Figure 3-24. A typical commercial VAWT in a California wind power station whose design was derived from the Sandia/DOE 17-m experimental VAWT.

at the Wind Turbine Test Station, to confirm structural and performance predictions; and Martha's Vineyard, Massachusetts, to demonstrate the VAWTs applicability to a small utility system.

The successful operation of the pre-commercial prototypes, including more than 10,000 hours for the Bushland machine, convinced two other manufacturers to further develop the Sandia/DOE 17-m VAWT into commercial products. The *VAWTPOWER* and *Flowind* companies each manufactured VAWTs for use in California wind power stations, with two-bladed configurations like that of the turbine shown in Figure 3-24. By the late 1980s, over 600 Darrieus VAWTs were installed in California, with a total rated capacity of over 90 MW.

Supporting Research and Technology

While the construction and testing of particular wind turbines engendered the most interest and publicity, these systems evolved due to the advances in the basic technologies that underlie the design and siting of the turbines themselves. Research and technology development were financed principally by government grants and contracts because of the limited financial resources of the wind industry and the uncertain markets during the 1970s and 1980s. The three technical needs underlying the future success of wind power were (1) the need to *improve performance*, particularly annual energy output, (2) the need to *extend structural lifetime and integrity*, while decreasing material costs and design complexity, and (3) the need to be able to rapidly *locate and evaluate productive wind turbine sites*, with reasonable accuracy. The disciplines of aerodynamics, structural dynamics, and the atmospheric sciences are interrelated in all three of these areas.

Wind Characteristics

In wind energy conversion systems, the wind "fuel" is the source of both the energy and the principal structural loads. Hence, the wind's characteristics directly affect both energy production and system cost. Moreover, the wind speed varies with time on scales from seconds to years, affecting mean and transient loads, control of peak power, and utility dispatch and planning. The wind also varies with location, affecting siting, turbine spacing, and profitability. Furthermore, the wind's power can range from zero to an order of magnitude above the mean. This inconstant behavior of the wind leads to unique design problems that require an understanding of wind fundamentals for solution. Lastly, wind measurement and research in the past have been aimed at weather prediction, with the bulk of the existing data being collected for aviation purposes and, more recently, environmental studies. This data base, therefore, contains major biases toward airports and urban sites and away from potential wind turbine locations.

In 1974, the Battelle Pacific Northwest Laboratory (PNL) undertook the task to develop both the data base and the analytical modeling tools for site evaluation and description of wind characteristics. On the siting front, an initial *U.S. Wind Atlas* was developed [Barchet *et al*. 1980-81] from National Atmospheric and Oceanographic Administration data (principally from airports), utilizing early interpolation tools. This Atlas presented estimates of the annual and quarterly mean wind speeds for every 25 square kilometers of the U.S., and the level of uncertainty in those wind speeds. It formed the basis for the first estimates of the national wind resource and served as a guide for the early "wind prospecting" by entrepreneurs endeavoring to develop and commercialize wind power. The California Energy Commission, recognizing the state's potential for wind power, extended the California atlas to a finer scale [Miller and Simon 1980].

PNL developed the first *world wind resource map* (on a coarser scale) for the World Meteorological Organization [WMO 1981]. At the United Nations Conference on New and Renewable Sources of Energy in Nairobi in 1981, this work was a stimulus to the investigation of the worldwide potential of wind power. Later, more advanced interpolation tools and analytical techniques, utilizing climatological and upper air data as well as additional wind data from numerous sources, led to the creation of an advanced *Wind Energy Resource Atlas of the United States* in 1987 [Elliott *et al*. 1987].

While wind atlases are useful for estimating large-scale resources and identifying potentially high wind areas, terrain and climate cause variations too localized to be identified on any reasonable map scale. Thus, models were developed to estimate flow patterns and wind velocities across relatively small geographical areas. Such models and other techniques allow estimation of wind speeds over a site less than 10 square kilometers in area, given a few local measuring locations. These computer models are used today to lay out wind power stations, although their accuracy varies depending on the complexity of the terrain (PNL 1980, 1981].

On an even finer scale, the character of both the wind *inflow* into a turbine rotor and the *wake* behind it are critical. Knowledge about the wake is needed to determine effects on downwind turbines in wind power stations, such as energy losses and increases in structural loads, as a function of wind turbine spacing. The turbine wake structure is also important in assisting analysis of the flow pattern through the rotor. Improved mathematical models of turbine wakes have resulted from wind tunnel test results and field measurements. The latter can be extremely demanding.

Knowledge of the characteristics of the wind inflow is also critical to understanding the performance, dynamics, and structural loads imposed on the wind turbine. Early analyses assumed a steady wind across the turbine rotor disk, allowing for *wind shear* (variation with elevation) and *tower shadow* (a sector downwind of the tower with reduced speed). Such inputs to structural analyses predicted mean loads fairly well, but badly underestimated cyclic loads caused by *wind turbulence*. This in turn led to underestimation of fatigue damage. Extensive research was undertaken in the 1980s to understand the expected *turbulence spectra* under various climatic, terrain, and wind velocity conditions.

Vertical and horizontal *planar arrays of anemometers* (Fig. 3-25), in which rings of anemometers are located upwind or downwind of the rotor swept area, have provided experimental data from which the turbulence experienced by a blade element can be synthesized.

Whirling arm tests, utilizing instrumented booms rotating as if they were a turbine blade, directly measured local wind velocities as an airfoil section would actually see them. Kites, smoke tests, and sonic techniques have also been used. By 1986, *empirical turbulence models* became available which significantly improved the ability to analyze rotor fatigue loads through knowledge of wind inflow details smaller than the rotor itself.

Figure 3-25. A vertical plane array of anemometers near the 25-kW research HAWT at the NREL test station. By sequentially sampling the data around this ring of anemometers (a process called "rotational sampling"), inflow turbulence experienced by a moving rotor blade can be measured. (*Courtesy of the National Renewable Energy Laboratory*)

Validation of these models continues, using turbulence data from sites where turbines have experienced wind damage. Data from U.S. sites were collected at the Pacific Northwest Laboratory. Turbine wake turbulence data are currently shared under an International Energy Agency (IEA) agreement amongst eight countries [Milborrow and Ainslie 1992]. A better understanding of small-scale turbulence is considered a major key to improved fatigue life and performance in future wind turbines.

Aerodynamics

While the fundamental aerodynamics of wind turbines was well understood over half a century ago, adequate analytical tools were not readily available to assist designers to rapidly and accurately predict power performance. In addition, the characteristics of hundreds of different airfoils for airplane wings, propeller blades, and helicopter rotors, well-documented in wind-tunnel test reports, were less useful to wind turbine blade designers than had been expected. Trends in aviation led to relatively thin airfoils, while wind turbines (with less concern for weight and more for extremely long fatigue life) require moderate-to-thick airfoils that are proportionately stronger while retaining good performance. In addition, the *lift and drag* characteristics of most available airfoils had been measured thoroughly for a relatively narrow range of *angles of attack*, only slightly beyond the angles of *stall* (+/-18 to 20 degrees or so). Yet, portions of wind turbine blades often operate at angles of attack well beyond stall.

The fundamentals of wind turbine aerodynamics, started by Prandtl and Betz, were re-derived and extended by the initial work of Wilson and Lissaman at Oregon State University and AeroVironment, Inc. [Wilson and Lissaman 1974, 1976]. A compendium of aerodynamics design methods (as well as methods for other aspects of wind turbine design) was developed at the Massachusetts Institute of Technology [Miller *et al.* 1978]. From this base, persistent in-house and contract research at the NASA Lewis Research Center, the Solar Energy Research Institute, and Sandia Laboratories led to continual improvements in aerodynamic modeling tools and understanding of rotor aerodynamic issues. Early models based on *single stream tube* analysis and simple *blade element* theory were replaced by computer codes using *multiple stream tubes* and *lifting line* models. These, in turn, have been giving way to improved multiple stream tube models, *lifting surface* models, and models incorporating three-dimensional flow and turbulence effects.

At first, both aerodynamic analyses and wind tunnel tests consistently under-predicted the power output and blade loads encountered in the field, for all but the smallest wind turbines. Causes of these discrepancies included *three-dimensional flow*, *inflow turbulence*, and *dynamic stall* during rapid changes in angles of attack.

Airfoils specifically tailored to the needs of wind turbine designers were developed in the 1980s, including *natural laminar flow* airfoils, new moderate-to-thick airfoils for increased strength and stiffness, and airfoils with *lower sensitivity to roughness*. Accumulated dirt and insects have caused significant energy losses in many wind power stations. Development of advanced airfoils continues today, aimed particularly at increased energy capture in low-to-moderate winds and limited output power and loads in higher winds.

Structural Dynamics and Fatigue Life Design

The limitations of analytical models for wind characteristics and aerodynamic behavior were aggravated in the mid-1970s by serious limitations in the state-of-the-art for predicting structural dynamic loads and fatigue life. Widely-used structural analysis codes such as *NASTRAN* were useful for predicting *natural frequencies* and static loads, but were totally inadequate for calculating *aeroelastic* dynamic loads, in which wind forces are coupled to structural motions. These limitations became more evident as designs of advanced turbines incorporated flexible towers, teetered hubs, and variable-speed power trains.

The effects of unattended, all-weather operations on loads, fatigue life, and damage mechanisms became better understood as first-generation turbines (heavily instrumented with anemometers, *strain gages*, and *accelerometers*) provided the first valid field data (Fig. 3-26). Various computer models (with a plethora of acronyms) were soon developed which, in turn, led to more accurate load predictions and better insight into design and operating issues [Spera 1977, Thresher 1981]. Codes such as *REXOR, MOSTAB, FLAP* [Wright *et al.* 1988], and *DYLOSAT* [Finger 1985] incorporated more and more capabilities, although limitations were still significant.

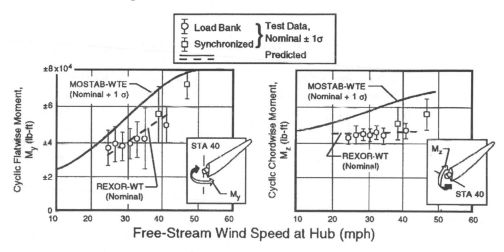

Figure 3-26. Typical comparison between measured and calculated dynamic loads on the blades of first-generation HAWTs. Early Mod-0 loads are shown here. [Spera 1977]

Today, codes named *HAWTDYN, VAWTDYN* (for the Darrieus), *NASTRAN/FFEVD*, and *DUWECS* [Bongers *et al.* 1993], and models using the general-purpose *ADAMS* soft-ware[1] [Wright *et al.* 1993] enable the structural analyst to include *statistical techniques*, non-linear *damage accumulation* models, integrated *rotor/tower/controls coupling*, and complex wind inflow conditions. A wealth of field test data has led to significant validation of structural modeling techniques and the identification of remaining limitations.

Public Acceptance

Research on the public's acceptance of the *aesthetics* of wind turbines has not produced any definitive conclusions, a not wholly unexpected result. Early research in the vicinity of isolated prototype turbines and subsequent experience with multiple installations have shown that the presence of a few turbines normally does not create more than isolated complaints. The same is true for wind turbines in larger numbers if they are located in relatively remote areas, are not exposed to much public visibility, and do not occupy particularly scenic ground. With only a few exceptions, most medium- and large-scale turbines in small numbers were usually greeted as beneficial, as potential tourist or advertising attractions, or as representative of the pioneering spirit of the community. The apparently leisurely turning of the larger rotors with their slender blades has generally led to a positive aesthetic reaction, in most surveys undertaken. Conversely, some *wind power*

---

[1] *ADAMS* is a registered trademark of Mechanical Dynamics, Inc.

stations with hundreds of smaller turbines in highly-visible locations have stimulated considerable public opposition.

Two additional factors clearly improve the degree of public acceptance. The first factor is the perceived energy benefit to the community, including lower cost and higher availability. The second factor is the degree of careful planning and open communication with the community and its leaders well before construction starts. However, one can recall that European and American windmills, which society now views with nostalgic fondness as romantic artifacts from our past, were work-a-day machines in their time, no different from our tractors and factory smokestacks. Only history will determine whether the same will be said in the future about today's modern wind turbines.

System Configuration Tradeoffs

A wind turbine, like an aircraft, is a complex system which is the result of many tradeoff decisions made in the search for optimum overall performance and economy. In the evolution of modern wind turbines, there are a number of major configuration variables whose trends over time (and the reasons for those trends) are fairly evident. In other cases, there is still doubt as to the "best" or optimum approach. In some cases the "optimum" approach changes with time as technological developments in one subsystem affect the overall design.

Turbine Size

The question of optimum size of wind turbines for large-scale electrical generation is probably the most controversial one in the field of wind power, and there is yet no clear answer. From a first-order standpoint, there are two countervailing forces: Rotor *swept area* and power increase as the square of the rotor diameter, but weight and cost would be expected to increase roughly as the cube, for geometrically similar structures. This *square-cube relationship* should favor small systems and limit the maximum turbine size.

However, a converse effect occurs for several reasons. Most sites have a *positive wind shear*, so wind speed increases with elevation. Input wind energy increases as the cube of the wind speed, so a larger wind turbine, being taller, should capture more energy per unit of swept area. This additional energy capture also implies reduced *land requirements* for a wind power station of a given rating composed of larger turbines, when the turbine separation is a fixed number of diameters. In addition, the size and cost of many components (*e.g.* the control system) do not increase at a cubic rate with increasing rotor diameter, and sometimes increase very little. Clearly then, there are some *economies of scale* to offset the square-cube relationship. Thus, there is some optimum size at which these two countervailing forces are economically balanced.

Several studies in the 1970s attempted to determine optimum size by examining families of hypothetical wind turbines, estimating scaling laws for individual subsystems, and calculating *cost-of-energy* (COE, in cents per kilowatt-hour) *vs* turbine size. Many in the field concluded that there were, in fact, two minimum energy cost points. Very small, simple turbines would be expected to be relatively expensive because of the threshold cost of many components and services. A site maintenance call, for example, could cost the equivalent of a month's electrical output. As this small turbine is made larger, then, energy costs would be expected to decrease. However, this downward trend would rapidly reverse and energy costs would increase again, because simple "brute force" components become very heavy as dynamic loads increase with size (*e.g.*, a tail fin for yaw control of a HAWT). This reversal establishes the *first minimum of energy cost*.

Changing the system configuration to one utilizing advanced structural concepts and active aerodynamic, electronic, and electrical controls reduces loads and increases energy capture, causing the cost of energy to decrease again. This assumes that the system is large enough to afford such controls. Finally, as rotor size continues to increase, the square-cube law eventually dominates, and further increases in rotor size lead to increases in the cost of each unit of energy produced. This establishes the *second minimum of energy cost*.

In practice, of course, many other factors come into play. *Technical risk*, at today's stage of technology, is higher in larger machines. The higher *capital investment* required for manufacturing large-scale turbines represents a major commitment that is unlikely to be made unless a very firm market and market advantage are discerned. For a given capital investment, more small units can be fabricated, and the advantages of *production tooling* and improvements based on *production experience* can be realized. Small- and medium-scale turbines may also be able to use standard *off-the-shelf components*, such as gearboxes or low-cost automotive brakes. A number of large-scale prototype turbines have been successful in both European and U.S. government programs, and many private ventures have occurred in this size range. The size of new commercial wind turbines has, been steadily increasing over time.

Tip Speed Ratio and Solidity

Tip-speed ratio (tip speed divided by wind speed) is a key configuration variable that has increased over several centuries and continues to do so today, albeit at a slower rate. The reasons are numerous. A higher tip-speed ratio reduces *rotational wake losses*, and hence increases the theoretical peak power coefficient. More importantly, a higher tip-speed ratio means a higher *turbine shaft speed* (for a given rotor diameter) and thus a lower *torque* for a particular power output. This in turn means smaller gearbox shafts, cases, bearings, and gears. Higher turbine shaft speed may permit fewer *step-up stages* in the gearbox to reach an efficient generator speed or a smaller generator, if a direct-drive concept is used. This can be significant in anything other than the smaller wind turbines.

For highest aerodynamic efficiency, tip-speed ratio and *rotor solidity* (total blade *planform area* divided by swept area) are inversely related, so higher tip-speed ratios lead to lower blade area and hence less blade material. This is particularly important at the larger sizes in which the blades form a larger percentage of total system cost.

Factors that make higher tip speed ratios difficult to achieve include the added *starting* difficulties associated with a low-torque rotor, particularly in moderate or low wind conditions and the potential for *acoustic noise* from the tips. The starting problem can be solved by using a generator which can motor the rotor up to operating speed. This involves a small amount of energy consumption and added equipment cost, but *motor-starting* is an attractive solution. The Darrieus VAWT, for example, does not produce torque (of any significance) when stopped, and it is essentially always motored up to a rotational speed where aerodynamic forces can take over.

Number of Blades

At very low solidities, blade dimensions become sufficiently small that it becomes difficult to design them with adequate structural strength and stiffness. For a given solidity, dividing the total planform area amongst fewer blades maximizes the cross-sectional size and strength of each blade. Thus, modern wind turbine rotors have almost universally either two or three blades. Some experiments have been undertaken on one-bladed (counterbalanced) machines as the logical limit of this trend. One-bladed rotors appear to have peak power coefficients only 5 percent to 10 percent below those with two or three blades. The structural dynamic behavior of a HAWT becomes increasingly more difficult to analyze and to accommodate as the number of blades is reduced from three to two to one.

Research on one-bladed rotors has been conducted in Germany (since the 1930s), Italy, and in the U.S. on the Mod-0 experimental HAWT. Structural dynamics questions associated with asymmetric loading are a worrisome issue for the one-bladed concept. A one-bladed, 24-m diameter 370-kW *Monopteros* wind turbine erected near Bremerhaven, Germany, was tested with sufficient success that three 50-m, 640-kW commercial units were erected at nearby Wihelmshaven. An Italian-German team produced several sizes of one-bladed *Riva Calzoni* wind turbines, including a 350-kW unit 33 m in diameter.

During the early stages of design evolution, large-scale wind turbines around the world, both government and privately funded, tended more to use two blades, rather than one or three. The benefits of higher tip-speed ratios were expected to be more important to larger turbines with their intrinsically lower shaft speeds, higher step-up ratio requirements, very large torques, and higher ratios of rotor cost to total installed cost. Moreover, large-scale system manufacturers generally have proportionately more engineering and analytical facilities and budgets available to solve structural dynamics problems. Exceptions to this trend, retaining the three-bladed configuration, included the two *630-kW Nibe turbines* in Denmark and larger commercial units (up to 500 kW) built with Danish utility and European Economic Community (EEC) funds. For the converse set of reasons, smaller commercial turbines have generally used three blades.

Rotor Blade Materials

Early electricity-producing wind turbines used wood blades, as did many of their water-pumping predecessors. Early experiments into larger systems (such as the Mod-0 HAWT) used *aluminum rib/spar/skin construction* typical of airplane wings. This design, however, is labor intensive and expensive, and the many rivets and bolts are subject to fatigue failure. Other early machines used everything from *riveted and bolted steel* (the Smith-Putnam turbine) to *laid-up fiberglass*, cured in molds (the Hütter turbines). Fiberglass, manufactured in this manner, remains one of the dominant materials for blades of all sizes.

The *Hamilton Standard-KKRV WTS-3 and -4 wind turbines* successfully utilized filament-wound fiberglass blades each 24 m long. Other large turbines, such as the *Mod-2* and the *3.2-MW Mod-5B* have utilized *welded steel rotors*, and one would expect that to remain the case for rotor diameters larger than about 60 m, particularly if the blades contain outboard mechanisms for tip pitch control. However, an increasing market may allow further investment in automated filament-winding machinery, which could cause the trend to shift toward fiberglass and away from welded steel. Composite materials more exotic than conventional fiberglass have been used, but rarely. The German *Growian* and *Aeolus II* HAWTs, for example, had some *carbon fiber composite* in their blades, but in most cases the economics of wind turbines preclude the large-scale use of such expensive materials.

In a blade technology program at the NASA Lewis Research Center, numerous blade designs and materials were examined and tested [Linscott *et al.* 1984]. The chronology of this extensive program, with field installations of the various materials developed, is illustrated in Figure 3-27. Some techniques failed, but several fiberglass variants were among those that were successful. Both *filament- and tape-wound fiberglass* were shown to be amenable to fabrication over mandrels on automated machinery [*e.g.* Weingart 1981]. Reducing hand labor should lead to lower costs and improved quality control, always an issue in highly-stressed composites.

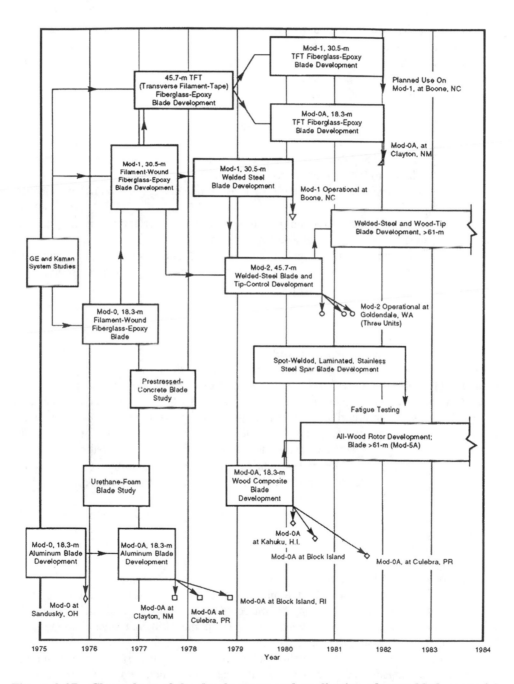

Figure 3-27. Chronology of the development and application of rotor blade materials for medium- and large-scale HAWTs, at the NASA-Lewis Research Center. [Linscott *et al.* 1984]

One of the most successful families of blades were (surprisingly) made of wood. They evolved from the *WEST* laminated wood-epoxy technique developed by *Gougeon Brothers, Inc.* (GBI) for racing yachts. Under contract to NASA, GBI adapted the process to the building of wind turbine blades of their own design. Carried further by private development, laminated wood blades have been utilized on a significant number of turbines up to 43 m in diameter. GBI became a major supplier of wood-epoxy blades which competed successfully with fiberglass blades in the U.S.

A comprehensive review and assessment of the status of materials for wind turbine rotor blades is contained in a report from a special committee of the National Research Council [Dieter *et al.* 1991]. This committee reviewed the three related subjects of structural loading characteristics, materials properties and life prediction, and wind turbine rotor design, drawing conclusions regarding the following issues:

-- adequacy of existing models to predict dynamic stress patterns;
-- properties of wind turbine materials in dynamic and fatigue failure;
-- understanding of the performance of joints, fasteners, and critical sections in relation to failure modes and fracture;
-- adequacy of analytical tools such as computer design models and material databases;
-- need for special laboratory facilities and turbine prototypes to improve the design and operation of wind energy systems;
-- opportunities for new materials, better manufacturing processes, and advanced control techniques to improve wind turbine performance and durability.

Aerodynamic Control

Some form of aerodynamic control is generally required for speed and power regulation, normal startup and shutdown, overspeed protection, and emergency shut-down situations. In particular cases, some of these functions can be performed mechanically, electrically, or even with passive aerodynamic techniques. Mechanical brakes (other than for parking) may be prohibitive in size and cost in the largest turbines because of the amount of rotational energy contained in the rotor and power-train equipment. Part of the tradeoff in selecting the method of aerodynamic control depends on whether a fixed- or variable-speed system is contemplated. In any event, the synchronous speed of the electrical generator provides the basic speed regulation, but separate overspeed protection is also normally required in the event of electrical system failure.

Traditionally, many of the smaller systems have been able to use the *fixed-pitch stall-regulated* approach, albeit with some penalty in energy capture. Medium- and large-scale systems have utilized *full-span variable-pitch* rotors, much like conventional constant-speed aircraft propellers. This does, however, involve relatively complex and heavy *pitch change mechanisms* at the blade root where structural loads are high. A number of rotors have been built with *tip controls*, wherein the majority of the blade is at a fixed pitch while the pitch of the remaining tip section may be varied. Examples of this approach include the *2.5-MW Mod-2* and the *3.2-MW Mod-5B HAWTs* in the U.S. and several commercial Danish, Dutch, and British turbines.

Research has been performed using outboard *flap control surfaces* which show a number of mechanical and structural advantages, but which have only been used experimentally in the U.S. and Japan. The *multiple hinge points* of a flap control surface have intrinsically better structural and safety features (because of *redundancy*) than the single spindle shaft which typically anchors a tip control surface.

Several approaches have been undertaken to develop *passive pitch control* techniques that automatically adjust the blade pitch angle without a need for hydraulic or electro-mechanical actuators and their power supplies and controls. The design of the *Jacobs Wind Electric* turbines of the 1940s allowed the rotor blades to slide in and out of the hub for a short distance, balancing centrifugal force against springs. During the sliding action, cams changed the blade pitch angle, and a simple passive pitch control was obtained that responded to rotor speed. A number of modern technology equivalents have been developed in the 1980s. One concept is the *self-twisting blade*, in which the *blade spar* in the vicinity of the hub is fabricated from a composite material carefully tailored so that increasing thrust and centrifugal loads cause the blade to twist toward the feathered position. Utilized in several small wind turbines, this concept may find broader application as experience is gained.

A unique method of power regulation for a modern large-scale wind turbine is embodied in the Italian *Gamma 60* 1.5-MW machine (Fig. 10-1) which employs an active yaw control system for the purpose. The blades of this turbine are fixed in pitch, and peak power is controlled by yawing the rotor out of the wind.

Considerable research and testing have been undertaken to develop *aerodynamic yaw control* systems for HAWTs that would face the rotor into the wind without active control, but these have not yet been wholly successful. This has been partly due to erratic, unstable performance and also to low cost savings. Often, *high-wind safety* and "parking" under storm conditions can be more easily satisfied by an active yaw control system. The old techniques of *tail vanes*, or auxiliary *side rotors* and *fantails* have given way to *active yaw control* using wind-direction sensors and electric or hydraulic drive motors.

Power Train Configuration

As illustrated in Figure 3-10, a typical first-generation wind turbine utilized a heavy *steel bedplate* on which to mount mechanical and electrical equipment, with the *turbine shaft* supported by separate bearings and attached to a conventional *parallel-shaft gearbox*. Medium- and large-scale systems rapidly moved to lighter-weight *planetary gearboxes*. Some also introduced a *duplex turbine shaft*, composed of two concentric shafts: An inner flexible *quill shaft* transmits only rotor torque, while a stiff outer shaft supports the weight and thrust of the rotor. Various forms of gearbox *shock mountings* have been used to reduce and dampen dynamic torsional loads entering the power train, for both structural and electrical reasons. Increased understanding of power-train dynamics and variable-speed generators have reduced the dependence on such devices in recent years.

Two other design features have appeared on wind turbines and have the possibility of future development. One is the use of the *gearbox case as primary structure*, thus reducing the need for a bedplate, or even the *nacelle* itself. The difficulty with this approach is the need for extra strength and stiffness in the gearbox, which would now have to be a custom-designed structure.

A second innovation is to omit the gearbox entirely and use a direct-drive to a very low speed, multiple-pole generator. This could result in a weight increase (such generators are relatively large), but this may not be critical for a VAWT generator on the ground. Recently, the large-scale Canadian *Eolé 64-m VAWT* was constructed with a direct drive between its rotor and a hydroelectric-type generator, with power electronics that permit operation at sub-synchronous speed. Several very small HAWTs use direct-drive generators or alternators, and a German manufacturer, *Enercon*, developed a 500-kW HAWT with a direct drive to its ring-type generator.

Type of Electrical Generation

Early wind power plants produced DC electricity, as do many small systems today which are designed for remote stand-alone applications. Batteries are used for energy storage, and generator speed is allowed to vary or is controlled only within modest limits. Interfacing with conventional utility grids or diesel-electric systems, however, requires an AC output and more stringent controls on *power quality* and *synchronization*. Until recently, most medium- and large-scale wind turbines utilized *synchronous generators*. Larger turbines can afford the controls necessary for the more-difficult synchronization requirements. Utilities generally favor synchronous generators, because they provide their own *reactive power* or *VARS* and can usually deliver excess VARS to the line when needed.

Most designers of wind turbines in the 1- kW to 100-kW range have selected *induction generators*, because they are relatively inexpensive and easy to synchronize with the grid. They also provide some valuable *power-train damping*. Their principal disadvantage is that they consume considerable reactive power. When induction generators were used in limited numbers, this was not significant and was usually ignored by the utility. The problem can be solved through the use of *capacitors* in the system at a small additional cost. However, capacitors may introduce a potential safety hazard, which is the possibility that *self-excitation* after a *fault* opens the utility line could cause the turbine to keep generating. Appropriate additional controls can eliminate this type of hazard.

The efficiency of an electrical generator usually falls off rapidly below its rated output. Since the power in the wind fluctuates widely, this becomes a major consideration in the selection of rated wind speed and rated power. Some small- and medium-scale systems have utilized two generators of different sizes. The smaller generator operates near its rated power at low wind speed, switching over to the larger generator during higher winds. The costs of the additional smaller generator and controls must be balanced against the losses associated with operating the larger generator at a low percentage of its rated power.

Of necessity, both induction and synchronous generators must be operated at constant speed in order to maintain the *grid frequency*. It has long been recognized, however, that variable speed operation has two major advantages over synchronous operation. First, the aerodynamic efficiency of the rotor is improved at low wind speeds if the rotor speed is also reduced. Second, system dynamic loads are attenuated by the "flywheel" action of the rotor, as it speeds up and slows down in response to wind gusts. There are several methods to produce constant frequency power from a variable speed generator, at some cost. However, the primary deterrent to the incorporation of variable speed in all but the smallest wind turbines has been the difficulty in predicting and preventing harmonic resonances.

By the mid-1980s, advances in structural dynamic analysis, *variable-speed constant-frequency (VSCF)* generators, and *power electronics* combined to make variable speed operation practical in larger sizes. The Mod-5B HAWT, the Sandia 34-meter and the Eolé VAWTs, and most new systems under development in Europe have VSCF generators.

HAWT Tower Stiffness

The towers for most early HAWTs were made of heavy *steel trusses* designed more for stiffness than strength. High stiffness was needed to keep the *fundamental (lowest) natural frequency* of the system higher than the blade passage frequency, in order to minimize the possibilities of resonant vibrations and associated structural dynamic problems. Some smaller wind turbines also used *guy cables* for stiffening. The development of better analytical design tools allowed a change to *steel shells* for towers, which is now the predominant configuration. These are so-called *soft tower* designs, in which the fundamental system frequency is less than the blade passage frequency. Care must be taken that the rotor speed passes rapidly through resonance with tower bending modes (and others) during starting and stopping. As shown in Figure 3-28, low-frequency cylindrical steel designs require much less material than stiff towers of the same height, and therefore cost considerably less. Safe, soft-tower design technology also permits taller towers, to take advantage of positive wind shears and increase the average wind speed at the rotor.

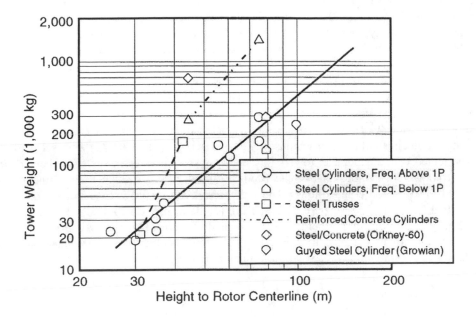

Figure 3-28. Trend of HAWT tower weight with height, stiffness, and type of construction. [Frederick and Savino 1986]

NASA/DOE Mod-0A Experimental HAWTs

During the 1970's, progress continued on several developmental cycles of medium- and large-scale machines. Soon after the initial testing of the 38.1-m diameter Mod-0 HAWT test bed, the U.S. Department of Energy decided to install a pair of upgraded replicas (later increased to four through Congressional actions) into actual utility operation. These machines were designated the *Mod-0A* wind turbines (Fig. 3-29), and each had a rated power of 200 kW, twice that of the Mod-0. While their power was still quite small as viewed by a utility, these would be the largest wind turbines integrated into a utility since the Smith-Putnam turbine of 1939. In fact, at the time there was almost no experience in the U.S. in operating a wind turbine of any size in an electric utility environment.

The purpose of the Mod-0A program was to identify and resolve technical and operational *utility interconnection issues*. These included questions of power quality, transient effects, safety, re-closure, and startup/synchronization/shutdown procedures. In addition, the Mod-0A HAWTs would form a visible validation of such operations. Through a competition aimed at the utility industry, 17 sites (later expanded to 35) were selected and instrumented with anemometer towers for detailed site wind assessments. This became the base from which the locations of the Mod-0As (and later NASA/DOE machines) were selected in follow-on competitions. The four Mod-0A HAWTs were installed from 1977 to 1979 at Clayton, New Mexico; Block Island, Rhode Island; Culebra Island, Puerto Rico; and near Kahuku on the northern tip of the island of Oahu, Hawaii. Mod-0A sites were selected at relatively small utilities or isolated locations, so that some understanding of the problems of *significant penetration* of wind power into a grid could be investigated. Some

Figure 3-29. The 200-kW 38.1-m diameter Mod-0A wind turbine on Block Island, Rhode Island. It was one of four turbines of the same design installed on small grids in the U.S. to identify and resolve utility interconnection and operation issues. (*Courtesy of NASA Lewis Research Center*)

issues could have remained unidentified if the wind turbine rating was an extremely small percentage of the local generating capacity.

The first unit was fabricated at NASA Lewis, while a parallel contract was awarded to *Westinghouse Electric Corporation* to construct the remainder. While ostensibly identical, each machine received detailed improvements based on the experiences of the prior installation. The *Lockheed Corporation*, which had built the original fabricated-aluminum blades for the Mod-0, provided the initial sets of Mod-0A blades to the same general design but with thicker skin panels. However, the downwind configuration and rigid hub introduced high and (at that time) uncertain dynamic loads that caused fatigue cracks in the aluminum skins and ribs near the blade roots. Eventually all four rotors were fitted with laminated wood-epoxy or fiberglass blades and operated successfully for extended periods.

Probably the most severe operational test of the Mod-0A came in the installation at Block Island, which was one of the reasons for the selection of that site. The Block Island grid is powered by several diesel-electric generators and is not interconnected with any other utility. Block Island has many summer vacationers and only a very small year-round population. Thus, summer peak loads reach over 1,800 kW, while during night hours in winter (which is also the high-wind season) the total load can go down to only a few hundred kilowatts. Occasionally the Mod-0A at 200 kW was producing over 50 percent of the power for the island. This large penetration introduced several problems in terms of both *voltage and frequency stability* and diesel operating problems caused by excessive throttling. The Block Island Mod-0A was therefore derated to 150 kW during winter operations, unless under special test.

When the Mod-0A project was completed in June 1982, the four machines had accumulated over 38,000 hours of operating time and had fed some 3.6 million kWh into their host utility grids [Shaltens and Birchenough 1983]. At a *Hawaiian Electric Company* site near Kahuku, the fourth and most reliable Mod-0A (Fig. 3-30) achieved a *capacity factor* of 0.48 during its last months of operation and was a principal cause of the developing interest in wind power in the Hawaiian Islands. The highly successful operation of the Kahuku turbine also led its builder, the Westinghouse Corporation, to privately develop a 600-kW HAWT and Hawaiian Electric Industries (the parent corporation of the utility)

to participate in the later Mod-5B program and encourage private wind power developers. The most important contribution of the four Mod-0A HAWTs was that they produced the first visible evidence that wind turbines, while not yet cost-effective, could be successfully integrated into a utility's normal operations and could produce high-quality AC power of value to that utility. They also provided a technology base that paved the way for the growth in size of privately-developed wind turbines, from the 10- to 15-m diameter and 10- to 25-kW sizes of the early 1970s to the 100- to 300-kW and 20- to 30-m diameter turbines that were developed and installed in the late 1980s.

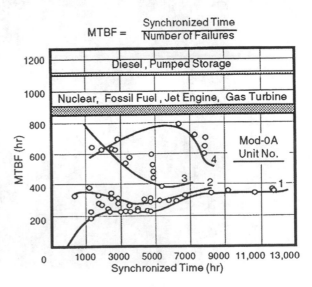

Figure 3-30. Mod-0A mean times between failure. Unit No. 4, on Oahu in the Hawaiian Islands, was the most durable.

[Shaltens and Birchenough 1983]

NASA/DOE Mod-1 HAWT and Environmental Issues

The Wind Turbine System

Development of the *Mod-1 Experimental HAWT* (Fig. 3-31), the first megawatt-scale wind turbine on a utility grid since the 1939 Smith-Putnam turbine, began in parallel with the installation of the Mod-0As. Rated at 2.0 MW and with a rotor 61 m in diameter, the Mod-1 HAWT used the same general design configuration as the Mod-0: a two-bladed, rigid hub rotor with full-blade pitch control, mounted downwind of a stiff, truss tower. The experimental system was designed and built by the *General Electric Company*, with welded-steel blades fabricated by the *Boeing Aerospace Company* (Fig. 3-32). It was installed on a small mountain called Howard's Knob near Boone, North Carolina, and was dedicated on July 11, 1979. The local utility, the *Blue Ridge Electric Membership Cooperative*, operated the Mod-1 for two years, proving that megawatt-scale wind turbines could be successfully interfaced with a large, conventional utility power system [Collins *et al.* 1982].

Figure 3-31. The 2.0-MW 61-m diameter Mod-1 experimental HAWT in 1979 on Howard's Knob overlooking Boone, North Carolina. It was the first megawatt-scale wind turbine since the Smith-Putnam HAWT. (*Courtesy of NASA Lewis Research Center*)

Figure 3-32. A 30-m long Mod-1 blade during final assembly at the Boeing Aerospace plant in Seattle, Washington. The structural spar was fabricated from welded steel plates, to which trailing edge sections of foam-filled fiberglass were bonded. [Linscott *et al.* 1984]

The Mod-1 design, however, was not completely successful. While it achieved its rated power and operated safely in an unattended, automatic mode, the wind velocity decrements behind the stubby truss tower applied high, impulsive loads to the rigidly-mounted blades. While no major blade problems actually occurred, it was clear that this design configuration would not have the 20- to 30-year life necessary in a commercial turbine. Improperly-torqued bolts attaching the turbine shaft to the hub did fail near the completion of the test program, but no major damage ensued because the rotor was supported by a large bearing mounted directly to the nacelle structure. However, it was deemed wisest to dismantle the turbine. The 60-m welded steel rotor, at that time the largest that had ever been built, was placed on display at the Science Museum in Raleigh.

Environmental Problems

The Mod-1 encountered two environmental problems: interference with television signals and acoustic noise. In parallel with the technology portion of the wind energy development program at the NSF, a major study was initiated with *Battelle Memorial Institute* in Columbus, Ohio, to identify any conceivable environmental effects that could be caused by either an individual wind turbine or by the large-scale use of wind power [Rogers *et al.* 1977]. These ranged from the possibility of affecting the micro-climate to striking birds. The latter was, in fact, a major initial worry regarding the large scale use of wind power, regardless of the size of the individual turbines. Extensive tests on and around the Mod-0 HAWT showed that there were no significant ecological effects. However, while the locale is rich in herbivorous and migratory birds, it has few local raptors. Some later wind power stations in California did encounter bird strikes with raptors, for which potential ameliorating approaches were taken.

Three potential problem areas were identified: the possibilities of *acoustic noise*, *electromagnetic interference* with local microwave radio or TV reception, and uncertain public acceptance of the *aesthetics* of wind turbines on the landscape. Following the initial Battelle study, specialized research projects concentrating on these potential issues were undertaken [Balombin 1980, Senior *et al.* 1977, Ferber 1977].

Electromagnetic Interference

The first environmental issue to be actually encountered at the Mod-1 site was the electromagnetic interference (EMI) problem. Results of the research showed that, while there would not be a significant effect across most frequencies (unless the turbine was close to, and literally in the path of, a microwave or other antenna), the *upper VHF* and *lower UHF television bands* were the most vulnerable and could be affected. Analytical tools began to be developed to predict the possibility of EMI in any given installation. Early analysis showed that, of the 17 sites under consideration by DOE, the Block Island site would likely experience TV interference, given the already marginal television reception in that area. Because it was ideal in all other respects, Block Island was selected for the third Mod-0A site partially to allow measurement of actual TV interference under complex real-world situations. A television cable system was first installed in cooperation with the nearby town in order to mitigate any effects on the public.

Like Block Island, the area around Boone had TV signals of marginal strength, and some EMI was encountered there also. EMI measurements around the large-scale Mod-1 turbine with its steel blades, coupled with the Block Island measurements and laboratory tests at the University of Michigan, led to the development of accurate tools for the evaluation of future sites. The EMI potential was found to be a predictable function of blade size and material, rotor speed, and the local transmitter/receiver/turbine geometry (see Chapter 9).

Acoustic Noise

A second environmental issue at the Mod-1 site was, for a time, an intractable noise problem. Prior wind turbines were generally relatively quiet, and the Mod-1 itself was not noticeably noisy close-up. Under certain conditions, however, it emitted low frequency pressure pulsations. At seemingly random intervals, this would produce unacceptable noise at various locations a considerable distance from the site, even though at other locations or times, no noise was detectable. The source was determined to be coupling between the blade passage and wakes from the heavy tubular legs of the truss tower. Atmospheric conditions, particularly inversions, combined with the complex mountain terrain could then focus the noise at some locations distant from the site. Research conducted in the area eventually led to the development of refined and verified methods for analyzing and predicting wind turbine noise (see Chapter 7). The Mod-1 noise problem was eventually solved by reducing the rotor speed, an operation that required replacing the generator.

Noise generated from blade tips or protrusions (for tip control mechanisms or aerodynamic brakes) and some gearboxes and hydraulic motors have occasionally led to some "noisy machines," regardless of turbine power or rotor size. Careful tip design, fairings, upwind rotor location, and component selection have led to generally quiet machines.

Foreign Medium- and Large-Scale Wind Turbines

In the late 1970s a resurgence of interest in wind power also reached a high level of momentum in Europe. Denmark, Germany, and the Netherlands developed broad programs which included basic technology efforts, the direct and indirect support of the private development of smaller wind turbines, and government-funded development of medium-scale or larger systems. Several countries commissioned testing centers: Denmark at Risø, Germany at Pellworm Island, and the Netherlands at Petten. These centers allowed for the testing of both experimental and commercial machines as well as setting in place *certification programs* as a requirement for tax or subsidy benefits, which effectively precluded turbines in Europe from entering the market prematurely.

Since the mid-1970s, some degree of international information exchange has been accomplished through the *International Energy Agency* (IEA). The IEA, headquartered in Paris, was modelled after the International Atomic Energy Agency, and was conceived as a way for western nations to coordinate and cooperate in energy policy, research, and development after the shock of the 1973 oil embargo. Two IEA agreements were implemented covering wind energy. One, a general research and development agreement, initially involved 12 countries (later increased to 16). A second agreement involved the exchange of information between those countries developing megawatt-class turbines. The *IEA Wind Energy Annual Reports* are an excellent source of information on national wind energy development programs. Table 3-2 is an example of the detailed data in these reports, listing wind turbines in Europe and the U.S. in 1993 with ratings larger than 500 kW, together with specification and performance data [IEA 1993].

IEA cooperative projects have included research activities on comparison of siting models, wake flow, wind flow over terrain features, and analysis of system test results. Development of comparative testing methodology and resource assessment studies are also undertaken. Of particular interest to the European countries are studies of the potential for off-shore wind power. With their higher population densities and fewer open areas, this possibility is of importance even in view of the expected higher cost associated with shallow water foundations and marinization requirements.

Denmark

For their first entries into the field of larger wind turbines, Denmark constructed a pair of 40-m diameter, 630 kW turbines, placed side by side, at *Nibe*, near Alborg in northern Jutland. Conservative in concept, they were a direct outgrowth of the old Gedser machine. Both had concrete towers, as had been traditional with the larger Danish wind turbines. Each had an induction generator, a three-bladed rotor upwind of the tower, and steel blade spars. One of the rotors was tip-controlled with cables for external bracing, while the other had full-span pitch-controlled cantilevered blades. These were the first of the new genera-tion of wind turbines in Europe to reach the testing stage. Operating for over 15 years with various types of blades, they provided some of the first information on operation and maintenance costs, reliability, and flow interactions between machines, supporting the expansion of the Danish wind energy industry [Godtfredsen *et al.* 1993].

Netherlands

The Netherlands government installed a 300 kW, variable-speed experimental machine at Petten in 1980. It was designed for maximum test flexibility. It still represents one of the most versatile of test machines and could be rapidly re-configured and operated in various modes, with fast turn-around on data reduction.

Table 3-2. Wind Energy Systems Larger Than 500 kW [IEA 1993]

| Country | Machine | | Rotor Specifications | | | | | | |
Member	Manufacturer	Model	Axis	B	L	M	D (m)	H (m)	A (m²)
Belgium	Windmaster-VUB		H	2	U	GRP	46.0	63.0	1,662
Canada	Shawinigan	Eolé	V	2		S	64.0	56.0	4,000
Canada	Indal Technol	6400	V	2		A	24.4	21.3	595
Denmark	Several	Nibe-A	H	3	U	Wood	40.0	45.0	1,256
Denmark	Several	Nibe-B	H	3	U	Wood	40.0	45.0	1,256
Denmark	Several	Tvind	H	3	D	GRP	54.0	53.0	2,290
Denmark	DWT	Windane 40	H	3	U	GRP/W	40.0	45.0	1,257
Denmark		Tjæreborg 2MW	H	3	U	GRP	60.0	60.0	2,827
Germany	M.A.N.	WKA 60	H	3	U	GRP	60.0	50.0	2,827
Germany	MBB	Monopt. 50	H	1	D	CRP	50.0	60.0	1,963
Germany	M.A.N.	WKA 60 Land	H	3	U	GRP	60.0	60.0	2,827
Germany	MBB	Aeolus II	H	2	U	CRP	80.0	77.0	5,077
Germany	Husumer Sch.	HSW 750	H	3	U	GRP/W	40.0	45.0	1,257
Germany	Tacke	TW 500	H	3	U	GRP	36.0	35.0	1,018
Italy	Aeritalia	Gamma 60	H	2	U	GRP	66.0	60.0	2,827
Netherlands	Holec	Holec 500	H	3	U		35.0		962
Netherlands	Stork-FDO	NEWECS-45	H	2	U	GRP	45.0	60.0	1,590
Netherlands	Windmaster NL	500	H	2	U	GRP	33.0		855
Netherlands	Windmaster	Windmast750	H	2	U	WE	40.0	50.0	1,257
Netherlands	Newinco	Newinco 500	H	2	U	S	34.0		908
Spain	Asinel, M.A.N.	AWEC 60	H	3	U	GRP	60.0	46.0	2,827
Spain	MADE								
Sweden	KMW AB	WTS-75	H	2	U	S/GRP	75.0	77.0	4,418
Sweden	KKRV	WTS-3	H	2	D	GRP	78.0	80.0	4,778
Sweden	Kvaerner-MBB	Nasudden I	H	2	U	S	75.0		4,418
Sweden	Kvaerner-MBB	Nasudden II	H	2	U	CRP	80.0	77.0	5,027
U.K.	WEG	LS-1	H	2	U	S/GRP	60.0	45.0	2,827
U.K.	WEG	LS-2							
U.K.	Howden	750 kW	H	3	U	WE	45.0	35.0	1,590
U.K.	Howden	1MW	H	3	U	WE	55.0	45.0	2,376
U.K.		HSW	H	3	U	GRP/W	40.0	45.0	1,257
U.K.	VAWT Ltd	500kW	V	2		GRP			
USA	Boeing	Mod-5B	H	2	U	S	97.5	61.0	7,472
USA	Ham. Std.	WTS-4	H	2	D	GRP	78.0	80.4	4,778
USA	Westinghouse	WWG-0600	H	2	U	WE	43.0	31.0	1,452

Axis:
  H = Horizontal
  V = Vertical
A = Swept Area
B = Number of Blades:
D = Rotor Diameter
H = Elevation of Center of
     Swept Area
L = Location of Rotor:
  U = Upwind of Tower
  D = Downwind

M = Blade Material:
   A = Aluminum
 CRP = Carbon Fiber Reinforced Plastic
 GRP = Glass Fiber Reinforced Plastic
   W = Wood
 WE = Wood Epoxy
   S = Steel

Table 3-2 (Continued). Wind Energy Systems Larger Than 500 kW [IEA 1993]

Weight		Rating		Generator		Performance					
W_{OT}	W_T	U_R	P_R	Type	No. of	Run	Output	From		To	Source
(1000 kg)		(m/s)	(kW)		Units	Hours	(MWh)				
		16.0	1,000	I							
344.0 (Rotor)		23.0	4,000	A	1	10,395	6,730	07/87	-	11/90	EMR
20.2 (Rotor)		18.2	522	I	2	5,873	557	05/87	-	07/90	EMR
80.0	47.0	13.0	630	I	1	6,146	1,313	09/79	-	04/92	DEFU
80.0	47.0	13.0	630	I	1	25,699		08/80	-	12/91	
			2,000	S							
60.5	47.0	15.0	750	I	5	87,500		11/86	-	12/91	DEFU
156.0	500.0	15.0	2,000	I	1	6,779		12/87	-	12/91	DEFU
	470.0	12.2	1,200	S	1	1,800	950	09/89	-	11/92	IEA
		11.0	640	S	3			09/89	-	11/92	IEA
		17.2	1,400	S	1			04/92	-	11/92	IEA
			3,000		1			09/92	-	11/92	IEA
20.0	47.0	14.0	750	A				1993			IEA
25.1	35.0	14.5	500	A				1993			IEA
95.0	145.0	13.5	1,500	A	1			9/91	-		IEA
			500		1						
31.7	59.0	13.9	1,000	A	1			06/86	-	10/89	IEA
			500	I	1			05/89	-	10/89	IEA
41.0		14.0	750	I							MFG
			500	I	1			11/89	-		IEA
186.0	92.0	12.2	1,200	I	1	2,400	559	11/89	-	02/92	IEA
			500								
205.0	1,500.0	12.5	2,000	I	1	11,350	12,600	02/83	-	10/89	IEA
191.0	291.0	14.0	3,000	S	1	23,079	29,847	07/82	-	02/92	NUTEK
			2,000		1	11,400	13,000			1988	IEA
	1,500.0		3,000	S	1			11/91	-		IEA
197.0	654.0	17.0	3,000	S	1	4,758	6,375	10/87	-	03/92	NSHEB
								10/87			IEA
			750	S	1	2,480	1,206	06/88	-	11/91	NSHEB
			1,000		1	2,150	798	02/90	-	04/91	
20.0	47.0	14.0	750	A							
			500	I							
265.3	160.5	13.0	3,200	C	1	20,561	26,776	07/87	-	09/92	HERS
198.2	192.8	15.0	4,000	S	1	4,100	8,000	01/82	-	08/87	MFG
36.2	34.8	13.0	600	S	15		36,384	12/85	-	09/92	HERS

W_{OT} = Weight on Top of the Tower
W_T = Weight of the Tower
U_R = Rated Wind Speed at Mid-Elevation of Rotor
P_R = Rated Power

Generator Type:

A = AC/DC/AC
C = Cycloconverter
I = Induction
S = Synchronous

Sweden

Sweden proceeded rapidly into a large-scale turbine research program, after first experimenting with the *SAAB-Scania 100-kW HAWT*, which was tested near Uppsala. A Swedish consortium named *KaMeWa* developed a 2.5-MW, 75-m diameter turbine with two blades on a rigid hub upwind of the tower, and installed it at Nasudden on the island of Gotland. The *KaMeWa HAWT* contained two unusual features: As shown in Figure 3-33, the last stage of the gearbox utilized bevel gears to drive its generator. This component was mounted vertically in the tower just below the nacelle, thus eliminating the need for power sliprings. A second unusual feature was a carriage assembly mounted on vertical rails on the side of the concrete cylindrical tower, for raising or lowering all major components (including the nacelle with rotor blades mounted). This eliminated the need for a large crane during construction or maintenance.

The second Swedish turbine was designed as a joint venture between Karlskronavarvet (KKRV) in Sweden and *Hamilton Standard* in the U.S. Called the *WTS-3*, it was built at Maglarp, near Malmo in southern Sweden (Fig. 3-34). Although more conventional than the KaMeWA design, the 3-MW WTS-3 was nonetheless technologically advanced. It had a tall, "soft-soft" tower and a two-bladed, teetered, downwind rotor 78-m in diameter. Its gearbox was mounted on springs to absorb dynamic torques. The fiberglass blades of the WTS-3 were designed by engineers at Hamilton-Standard and fabricated on a specially-built and automated filament-winding machine.

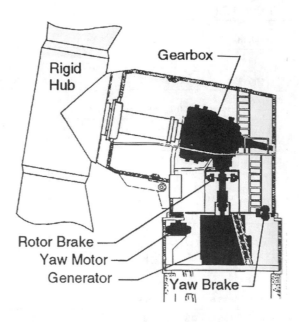

Figure 3-33. The unique drive train in the 2.5-MW KaMeWa HAWT eliminated sliprings to carry power from the generator. (*Courtesy of the National Energy Administration, Sweden*)

Figure 3-34. The 3-MW WTS-3 HAWT near Maglarp, Sweden. (*Courtesy of Hamilton Standard*)

While both turbines encountered various early problems (the KaMeWa turbine was once nearly destroyed due to a sheared yaw drive shaft), they both operated successfully for an extended period of time.

An up-rated version of the WTS-3, the *Hamilton Standard/KKRV WTS-4* was purchased by the U.S. *Bureau of Reclamation* (USBR) and installed at a site near Medicine Bow, Wyoming (Fig. 4-24), under a project managed by the NASA Lewis Research Center. With its rating of 4 MW, it was the most powerful wind turbine ever built. The Bureau wanted to examine the possible large-scale use of wind power in connection with hydroelectric systems. Percy Thomas' ideas were finally being realistically investigated!

Germany

The development of the German *3-MW Growian HAWT* (Grosse Windenergie Anlage) in 1982 (Fig. 3-35) represented the greatest technological leap of the times, as well as the highest technological risks [Windheim 1983]. It encompassed just about every advanced feature yet considered. At 100 m in diameter with a 100-m tall tower, it was the largest wind turbine ever built at that time. The rotor used two, full-span pitch-controlled, carbon-filament blades with a high degree of *coning*. It utilized the downwind rotor configuration

Figure 3-35. The 3-MW Growian HAWT near Bremerhaven, Germany. It was one of the largest wind turbines ever built, with a 100-m diameter rotor and a 100-m tall tower. *(Courtesy of MAN-Neue Technologie)*

combined with a very slender, flexible tubular steel tower stabilized by many guy cables. The majority of the turbine was assembled at the site near Bremerhaven, with less factory assembly than most other machines.

The Growian HAWT was the only other large-scale turbine besides the KaMeWa to be erected without a crane in this time period. The nacelle, with a central opening through which the tower passed, was winched up the tower with rotor attached. The guy cables were tightened, and the generator was then slid forward into place over the tower hatch, completing the installation. The Growian was also the first large-scale wind turbine to attempt variable-speed operation. Unfortunately, the technology at the time could not support the number and magnitude of the innovations undertaken. The Growian project encountered an inordinate number of problems, including fatigue cracking of major components in the hub. While it made significant contributions to the understanding of large wind turbines, it never operated satisfactorily and was dismantled after only limited testing time.

A more successful turbine was the 370 kW, 48-m diameter *Monopteros* HAWT [Stahl and Windheim 1987]. The Monopteros, constructed in 1981 near Bremerhaven, was the first large experiment in achieving very low solidities by utilizing a one-bladed rotor.

United Kingdom

Another European country investigating wind power on a large scale was the United Kingdom. The Orkney Islands, with the interest of the North of Scotland Hydro Board, became the major test site. The UK took a much slower and more deliberate approach than most of the other countries and installed a 250-kW scale model, designated as the *MS-1*, of its proposed large-scale system in 1982. The scale model was of rather rigid design with a two-bladed downwind rotor of 20-m diameter. A privately developed prototype, the *300-kW Howden* wind turbine, 22 m in diameter and of much more flexible tower construction, was installed nearby. Thus two machines -- one of stiff design and one of "soft", flexible construction -- could be compared side-by-side. The megawatt-class prototype, the 3-MW *LS-1*, was developed at a deliberate pace. It evolved through several design configurations, ending as a two-bladed upwind turbine. Built by the *Wind Energy Group* (a consortium of *Taylor-Woodrow Construction, British Aerospace*, and *GEC*), it was installed in the Orkneys, and testing commenced in 1987 [Page and Bedford 1987].

Canada

In Canada the technological route was different from other countries. Based on early research there on Darrieus VAWTs, Canadian planners elected to concentrate future projects in that direction. In addition to smaller systems for use in remote areas, a 230-kW experimental Darrieus turbine of 24.4-m diameter was installed in 1977 at a Hydro-Quebec utility facility on the Magdalen Islands in the Gulf of St. Lawrence. Basic testing of the turbine proceeded satisfactorily, until a maintenance error caused the generator to be disconnected without locking the rotor. While a Darrieus rotor has low starting torque, the unsecured rotor started without load, oversped, and destroyed the turbine. This occurred, however, after sufficient positive results had been obtained to justify continuing the program. The *Magdalen Island* VAWT was rebuilt, and the development of two prototypes of the same size, but up-rated to 500 kW, was initiated with *Indal Technologies, Inc.*

Installation of the *Eolé* VAWT (Fig. 3-36) was completed in early 1987 at Cap-Chat, Quebec, near the banks of the St. Lawrence River [Richards 1987]. This giant machine was 64 m in diameter and was originally rated at 4 MW. It thus represented the first *megawatt-class Darrieus* turbine. Like a hydroelectric turbine, it had no gearbox. Instead, the rotor drove directly a large-diameter 162-pole alternator at ground level, which was then connected through an *AC-DC-AC* link to the Hydro-Quebec grid, to provide variable-speed constant-frequency operation. The system was generally operated at reduced speed to ensure longevity, and power was limited to 2.5 MW. Hydraulically-deployed *aerobrakes* at the rotor equator were a back-up to the primary dual-disk mechanical brake above the generator.

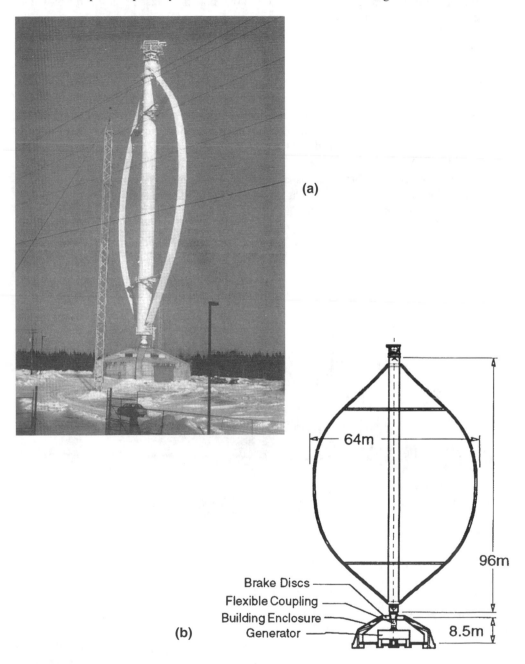

Figure 3-36. The 4-MW Eolé VAWT. (a) General view of the turbine at Cap-Chat, Quebec (*Courtesy of the Hydro-Quebec Company*) (b) Sketch of major components, including the direct-drive generator [Richards 1987]

NASA/DOE Second- and Third-Generation HAWTs

The Mod-2 Program

Even before the first-generation Mod-1 HAWT began its test program, development was started at NASA Lewis on a second-generation large-scale machine, to be named the *Mod-2 HAWT*, which would represent a major advance from earlier systems. *Boeing Aerospace Corporation* received the development contract, and a site was selected at Goodnoe Hills, near the Columbia River Gorge and the town of Goldendale, Washington. The *Bonneville Power Administration* was the cooperating utility, with the power to be fed into the grid of the Klickitat County Public Utility District.

The state of knowledge of structural dynamics of wind turbines had then reached the stage where a "soft" structural design could be attempted on a large turbine. An upwind rotor configuration was selected to reduce the possibility of noise and cyclic loads from tower shadow. As shown in Figure 3-37, the two-bladed rotor concept of the Mod-0 and Mod-1 was retained. Instead of full-span pitch control, however, partial-span tip control was selected, which allowed a teetered hub that was less expensive and structurally superior.

Economic optimization from the utility viewpoint, coupled with the federal role of pursuing technology levels that were beyond the risk limits acceptable to a private company, led to the selection of a power rating of 2.5 MW at a rated wind speed of 12.3 m/s (at the hub) and a rotor diameter of 91.4 m. The 900-kN (100-ton) rotor was fabricated from welded steel plates in five separate sections: a hub, two mid-blades, and two pitchable tips. These were bolted together at the site and lifted into place in one piece [Boeing 1982].

Figure 3-37. The 2.5-MW Mod-2 HAWT, the second-generation turbine in the large-scale segment of the Federal Wind Energy Program. Its welded-steel, teetered rotor was 91.4 m from tip to tip, with pitchable tips. (*Courtesy of NASA Lewis Research Center*)

Figure 3-38. Arrangement of the three Mod-2 HAWTs on the Goodnoe Hills site near Goldendale, Washington. The triangular pattern permitted research on wake interference effects at different downwind distances. (*Courtesy of Boeing Aerospace Corporation*)

Three Mod-2 turbines were built and tested at the Goodnoe Hills site in order to investigate initial problems associated with large-scale wind power stations and interactions and interferences between multiple large wind turbines. The three turbines were placed in a triangular pattern 5, 7, and 10 diameters apart such that wake interactions could be examined under different spacings as changing wind directions placed the various units behind one another (Fig. 3-38). Extensive use of anemometers as well as kites, smoke, balloons, and other techniques were used to characterize the local wind flow over the site in detail.

As in any high-risk technological advance, a number of problems were encountered, particularly during the first two years of operation. These included fatigue cracks in the turbine shafts originating at component mounting holes, contaminated hydraulic oil, and leaking grease seals causing premature failure of blade tip support bearings. While these difficulties caused significant delays and costs because of the size of the components and the necessity to special-order or fabricate each part, the turbines themselves operated according to design. Investigations continued into operating strategies and control algorithms, to increase energy capture and reduce structural loads.

Other modifications were made to improve the Mod-2 system, based on experience gained during the program. For example, lines of small vortex generators were installed on the low-pressure side of the blade, which delayed flow separation, enhanced control stability, and significantly increased energy capture. As a result, the turbine produced more power than its design predictions, in spite of the tips operating at higher angles of attack than expected.

One problem that was directly associated with the design of the wind turbine was that rotor cyclic loads were higher than calculated. While no structural failures occurred, the Mod-2 rotor as built would probably have had an approximately 10-year fatigue life, rather than the 30-year design value. It was felt that the rotor could be redesigned for a 30-year life, based on continued research which has shown the foundations of the problem. The structural loading estimates of that time used overly-simplified models of the wind. The effects of atmospheric turbulence on unsteady local lift forces, then on loads and stresses, and finally upon fatigue life were not adequately represented in design models. Based on the Mod-2 test results, major improvements in these models have been made. However, achieving long fatigue life in the presence of small-scale turbulence remains a principal technical challenge to the development of advanced wind turbines that are still reliable and cost-effective.

After modifications and repairs associated with initial mechanical problems, the Mod-2 test program and the turbines themselves went on to accomplish major test and performance objectives. The three Mod-2 HAWTs at Goodnoe Hills together accumulated over 16,000 hours of operating time and supplied over 10 million kWh to the local grid, over 60 percent of that amount in the final year of testing. A major test objective that was achieved was proving conclusively that groups of modern wind turbines could operate in a totally automatic, unattended mode. Another was establishing a data base on rotor wake structures and wake effects on downwind turbines that has led to the development of improved wake models being applied now to wind power stations (see Chapter 6).

Two additional Mod-2 turbines were built by Boeing for utility companies who wished to examine the performance of large-scale wind turbines directly from their perspective. One, purchased by the Pacific Gas and Electric Company, was installed on a test site in Solano County, northeast of San Francisco. A second, for the Bureau of Reclamation, was installed near the WTS-4 HAWT at Medicine Bow, Wyoming, for comparative testing. While all five of these machines achieved significant (and sometimes spectacular) test and operating results, large wind turbines at that time were not yet at the stage where they could compete successfully for utility company operating funds, nor was there an industry willing to take the financial risks associated with their commercial development. All Mod-2 turbines were later dismantled, while the technology continued to be developed and commercial use awaited a change in energy economics.

The Mod-5 Program

Plans had already been laid in the late 1970s for the development of third-generation wind turbines. First, conceptual studies of advanced large- and medium-scale turbines (designated as Mod-3 and Mod-4) were conducted, but these were never carried beyond the preparatory stage. Later, consideration was given to two hardware-development programs: A large-scale *Mod-5 HAWT* program and a medium-scale *Mod-6 HAWT and VAWT* program. Two contractors were chosen to design and develop what became known as the *General Electric Mod-5A* and the *Boeing Mod-5B* HAWTs. At that point energy and fuel costs peaked and then turned downward, and energy was no longer a major national priority. As a result, the Mod-6 program (whose contractors had not yet been selected) was canceled, and General Electric chose not to proceed past the design stage. Thus, only one third-generation turbine, the Mod-5B, was completed under the Federal Wind Energy Program. NASA-Lewis engineers managed the project and *Hawaiian Electric Industries* became Boeing's utility partner.

The Mod-5B (Fig. 2-1 and 4-25), installed at the Kahuku wind power station on Oahu, has an overall configuration similar to that of the Mod-2, with a two-bladed, partial-span controlled, teetered, and upwind rotor atop a steel shell tower [Boeing 1988]. The pitchable tip sections were extended 3 m longer than those on the Mod-2 rotor, leading to a rotor diameter of 97.5 m. This made the Mod-5B HAWT the largest wind turbine in the world of its day. Its rated power was 3.2 MW, up 28 percent from the Mod-2 rating.

The major advancement in technology achieved by the Mod-5B is that it was the first large-scale wind turbine to operate successfully at *variable speed*. The speed varied from 13 to 17.3 rpm depending on wind speed, thus improving energy capture as well as reducing structural loads. Maintaining constant 60-cycle power was accomplished through the use of a *doubly-wound generator*, associated *cycloconverter* power electronic equipment, and advanced control algorithms. *Vortex generators* (shown being installed in Fig. 5-24) and *trailing-edge tabs* improved the aerodynamic performance of the rotor. Most importantly, the Mod-5B was designed using advanced structural-dynamic computer codes like *Dylosat* [Finger 1985] and incorporated experience from the Mod-2 test program [Bovarnick and Engle 1985]. Thus it appeared to be the first large-scale turbine with a reasonable expectation of a 30-year structural lifetime.

Final assembly commenced in January, 1987, at the site on the north shore of the island of Oahu, not far from where the fourth Mod-0A 200-kW wind turbine had been located. The turbine nacelle and rotor had been shipped by barge to Oahu from the mainland in subassemblies, together with the very large *ringer crane* needed for final assembly. The hub and mid-blade subassemblies of the 1.4-MN rotor, also of all-steel construction, were welded together on-site, rather than having bolted joints, as was the case with the Mod-2.

First wind-powered rotation of the Mod-5B HAWT occurred on July 1, 1987. The test program began in earnest in August, 1987, and consisted of two 500-hour phases. The first phase accomplished checkout of the turbine through its operating envelope and adjustment of controls and the variable-speed generating system. The second phase consisted of testing under a utility acceptance test scenario, the purchase price to the utility being a function of both *performance* and *availability* during a test period of at least 500 hours. The turbine achieved an energy capture performance of 106 percent of the basic contract requirement, producing 988 MWh during 660 hours of testing. It also achieved an availability of 95 percent, 5 percent over the basic requirement, an unparalleled level for that early a stage of testing of a new, advanced, and very large wind turbine.

In January of 1988, the Mod-5B was sold to the project's utility partner to operate as an integral part of the power generation mix on Oahu. During its first 55 months of service as a commercial power plant, the Mod-5B operated for 18,920 hours and produced 24,533 MWh of electricity, for an average power output of 1.32 MW [Spera and Miller 1992].

Eventually, the Kahuku wind power station consisted of the Mod-5B and 15 *Westinghouse WWG-0600* turbines rated at 600 kW (Fig. 3-39), owned and operated by the Makani Uwila Power Corporation. The design of these Westinghouse machines combined successful technology from the Mod-0A program (wood/epoxy blades, nacelle structure, yaw drive by hydraulic actuators) with that from the Mod-2 (upwind/teetered rotor, shell tower, and dynamic flexibility).

Figure 3-39. A 600-kW Westinghouse WWG-0600 HAWT, 43.3-m in diameter. This design applied technology from both the Mod-0A and Mod-2 wind turbines. (*Courtesy of Hawaiian Electric Industries*)

Advanced Wind Turbine Development

The development of technologically-advanced, higher-efficiency wind turbines continues to be a high priority of the wind industry worldwide. In the U.S., the Department of Energy sponsored a range of programs with the goal of developing wind power plants that can compete with conventional electric generation, producing energy at a cost of $0.04/kWh by the year 2000 at sites with average wind speeds of 5.8 m/s (at a 10-m elevation). In the near term, an "Advanced Wind Turbine Program" [Laxson *et al.* 1992] is assisting U.S. industry to apply the latest technology to (1) improve existing wind turbine configurations, designs, and manufacturing methods, and (2) to initiate conceptual design studies of advanced wind power systems. Some of the concepts being explored for HAWTs under this program are

-- increased rotor sizes to take advantage of new airfoils designed to limit maximum lift coefficients (see Chapter 6);
-- "flow-through" rotor structures without bolted joints between blades and hub;
-- aileron control surfaces;
-- integrated gearbox and mainframe structures in the nacelle;
-- aerodynamically-shaped, wood/epoxy rotating towers;
-- totally integrated power trains, mounted directly on tower-top castings;
-- taller towers.

Many of the recent advances in technology and analytical capability have yet to be applied to commercial wind turbines. This is expected to occur over the next several years, particularly with the resurgence of interest in alternate sources of energy in the early 2000s. Passage of the Energy Policy Act of 1992 re-introduced tax credits in the U.S., in the form of production credits. This is believed by many to be more effective than the earlier tax credits based on capital investment. Increased emphasis on reducing power-plant emissions and on global climate changes are also providing a spur to non-polluting energy sources.

Concluding Remarks

Whether commercial machines grow in size to that of the Mod-5B HAWT in the future, only time, the international marketplace, and the vagaries of energy cost and availability will determine. Meanwhile, several other advanced medium-scale and large-scale systems with rated powers from 500 kW to 2 MW and rotor diameters from 30 m to 60 m are under development in Europe. Commercial development of small- and medium-scale wind turbines continues, albeit this development was slowed somewhat by the slackening of the energy market and changes in energy incentives and tax policy. Most of all, research and advanced technology development continue, in both the private and public sectors, in a quest for higher-performance coupled with more reliability.

Stepping back from individual details, one can assess the significant changes in the technology of wind turbines over the two decades by examining their overall performance. *Specific annual energy production* (kilowatt-hours per square meter of swept area) has increased approximately 40 percent from the mid-1970s to the mid-1980s because of improved aerodynamic performance. *On-line availability* has improved from the 60 percent range to over 95 percent (for the better systems) during the same period. In California in 1992, these two factors plus improvements in operating and maintenance strategies significantly increased specific annual energy production to an average of almost 900 kWh/m² for machines rated above 200 kW. The average rated power of commercial wind turbines installed in the U. S. has increased from less than 50 kW in 1981 to well over 100 kW in one decade. *Installed costs* have declined from over $3,000 per square meter in the mid-1970s to the $500 level in the same period of time.

While significant progress has been made, additional improvements are both necessary and possible. Cost-effective wind turbines are required for the much-more prevalent sites with medium wind speeds, in order to expand the geographic distribution of wind power stations. This still remains the principal challenge to the developing technology of wind power. In the short term, improvements in the structural lifetime of components, particularly of rotor blades, is required.

Predicting the longer term future is more difficult. One can anticipate a continuing increase in performance and energy capture from improved airfoils, control systems (including increased adoption of variable-speed designs), operating strategies, and siting techniques. One also anticipates decreased structural weight and complexity from improved understanding of the interaction between wind turbulence, air loads, and structural response.

While there remain many areas in which research and development will continue, the evolution of modern wind turbines since World War II (and particularly during the past two decades) have brought us to a stage where the large-scale use of wind power with minimal technical risk is a probability. The success of commercial wind power stations, after overcoming early difficulties, has established the technical and economic potential of wind power. More energy is being produced today by wind turbines than has ever been produced in the history of wind power.

References

ASME, 1989, *Performance Test Codes - Wind Turbines*, ASME/ANSI PTC 42-1988, New York, New York: American Society of Mechanical Engineers.

Balombin, J. R., 1980, *An Exploratory Survey of Noise Levels Associated with a 100 kW Wind Turbine*, NASA TM-81486, Cleveland, Ohio: NASA Lewis Research Center.

Barchet, W. R., A. Bass, S. R. Andersen, *et al*, 1980-81, *Wind Energy Resource Atlas, Vols. 1 - 12*, PNL-3195-WERA-1 to -12, Richland, Washington: Battelle Pacific Northwest Laboratories.

Berg, D. E., P. C. Klimas, and W. A. Stephenson, 1990, "Aerodynamic Design and Initial Performance Measurements for the Sandia 34-m Diameter Vertical-Axis Wind Turbine," *Proceedings, Ninth ASME Wind Energy Symposium*, D. E. Berg, ed., New York: American Society of Mechanical Engineers, pp. 93-99.

Boeing Engineering and Construction Company, 1982, *Mod-2 Wind Turbine System Development Final Report - Vol. I, Executive Summary*, NASA CR-168006, DOE/NASA/0002-1, Cleveland, Ohio: NASA Lewis Research Center.

Boeing Aerospace Company, 1988, *Mod-5B Wind Turbine System Development Final Report - Vol. I, Executive Summary*, NASA CR-180896, DOE/NASA/0200-3, Cleveland, Ohio: NASA Lewis Research Center.

Bongers, P. M. M., G. E. van Baars, and S. Dijkstra, 1993, "DUWECS: A Wind Turbine Design Tool," *Proceedings, Windpower '93 Conference*, Washington, DC: American Wind Energy Association, pp. 305-312.

Bonnefille, R., 1974, "French Contribution to Wind Power Development - by EDF 1958 - 1966," *Proceedings, Advanced Wind Energy Systems,* Vol. 1, (published 1976), O. Ljungström, ed., Stockholm: Swedish Board for Technical Development and Swedish State Power Board, pp. 1-17 to 1-22.

Bovarnick, M. L., and W. W. Engle, 1985, "The Evolution of the Mod-2 and Mod-5B Wind Turbine Systems, " *Proceedings, Windpower '85*, SERI/CP-217-2902, Washington, DC: American Wind Energy Association, pp. 158-163.

Clark, R. N., 1983, "Agricultural Wind Energy Research -- An Overview," *Proceedings, Sixth Biennial Wind Energy Conference and Workshop*, B. H. Glenn, ed., Boulder, Colorado: American Solar Energy Society, pp. 31-34.

Collins, J. L., R. K. Shaltens, R. H. Poor, and R. S. Barton, April 1982, *Experience and Assessment of the DOE-NASA Mod-1 2000-kW Wind Turbine Generator at Boone, North Carolina*, NASA TM-82721, DOE/NASA/23066-2, Cleveland, Ohio: NASA Lewis Research Center and The General Electric Company.

Dieter, G. E., J. Chapman, H. T. Hahn, D. H. Hodges, C. W. Rogers, L. Valavani, and M. D. Zuteck, 1981, *Assessment of Research Needs for Wind Turbine Rotor Materials Technology*, National Research Council Report NAT S-324, Washington, DC: National Academy Press.

Elliott, D. L., C. G. Holloday, W. R. Barchet, H. P. Foote, and W. F. Sandusky, 1987, *Wind Energy Resource Atlas of the United States*, DOE/CH-10094-4, Golden, Colorado: National Renewable Energy Laboratory; also available from American Wind Energy Association, Washington, DC.

Ferber, R., 1977, "Public Reactions to Wind Energy and Windmill Designs," *Proceedings, Third Biennial Conference and Workshop on Wind Energy Conversion Systems*, CONF 770921, Washington, DC: JBF Scientific Corporation, pp. 413-419.

Foreman, K. M., and G. L. Gilbert, 1979, "Technical Development of the Diffuser Augmented Wind Turbine (DAWT) Concept," *Wind Engineering*, Vol. 3, No. 4: pp. 153-166.

Frandsen, S. (ed.), 1990, *Recommended Practices for Wind Turbine Testing and Evaluation - Vol. 1: Power Performance Testing*, Second Edition, Lyngby, Denmark: Department of Fluid Mechanics, Technical University of Denmark.

Frederick, G. R., and J. M. Savino, 1986, *Summary of Tower Designs for Large Horizontal Axis Wind Turbines*, NASA TM-87166, DOE/NASA/20320-68, Cleveland, Ohio: NASA Lewis Research Center.

Golding, E. W., 1955, *The Generation of Electricity by Wind Power*, London: E. & F.N. Spon Ltd.; reprinted with an additional chapter by R. I. Harris, 1976, New York: Halsted Press, Division of John Wiley & Sons, Inc.

Godtfredsen, F., P. H. Jensen, and P. E. Morthorst, 1993, "Wind Energy in Denmark: Development in Wind Turbine Technology and Economics Since 1980," *Proceedings, Windpower '93 Conference*, Washington, DC: American Wind Energy Association, pp. 110-117.

Honnef, H., 1932, *Windkraftwerke*, Braunschweig: F. Vieweg & Sohn Verlag.

Hütter, U., 1973a, *The Development of Wind Power Installations for Electrical Power Generation in Germany*, NASA Technical Translation TT F-15,050 (L. Kanner Associates), Washington, D.C.: National Aeronautics and Space Administration.

Hütter, U., 1973b, *A Wind Turbine with a 34-m Rotor Diameter*, NASA Technical Translation TT F-14,879 (L. Kanner Associates), Washington, D.C.: National Aeronautics and Space Administration.

Hütter, U., 1974, "Review of Development in West-Germany," *Proceedings, Workshop on Advanced Wind Energy Systems*, Vol. 1, 1974 (published 1976), O. Ljungström, ed., Stockholm: Swedish Board for Technical Development and Swedish State Power Board, pp. 1-51 to 1-72.

IEA, 1993, *IEA Wind Energy Annual Report 1992*, B. Pershagen, ed., Stockholm: NUTEK.

Juul, J., 1964, "Design of Wind Power Plants in Denmark," *Wind Power, Proceedings of United Nations Conference on New Sources of Energy*, Vol. 7, New York: The United Nations, pp. 229-240.

Klimas, P. C., and M. H. Worstell, 1986, "Performance Testing of the Sandia National Laboratories 17-m Diameter Research Turbine with Natural Laminar Flow Blade Elements," *Proceedings, Fifth ASME Wind Energy Symposium*, New York: American Society of Mechanical Engineers.

Lauffer, J. P., T. G. Carne, and T. D. Ashwill, 1987, "Modal Testing in the Design Evaluation of Wind Turbines," *Proceedings, Windpower '87 Conference*, SERI/CP-217-3315, Golden, Colorado: National Renewable Energy Laboratory, pp. 79-87.

Laxson, A. S., S. M. Hock, W. D. Musial, and P. R. Goldman, 1992, "An Overview of DOE's Wind Turbine Development Program," *Proceedings, Windpower '92 Conference*, Washington, DC: American Wind Energy Association, pp. 426-430.

Loeffler, A. L., Jr., 1981, "Flow Field Analysis and Performance of Wind Turbines Employing Slotted Diffusers," *Transactions ASME, Journal of Solar Engineering*, Vol. 103: pp. 17-22.

Linscott, B. S., P. Perkins, and J. T. Dennett, 1984, *Large, Horizontal-Axis Wind Turbines*, NASA TM-83546, DOE/NASA/20320-58, Cleveland, Ohio: NASA Lewis Research Center.

Lissaman, P. B. S., A. D. Zalay, and B. Hibbs, 1980, "Development of the Dynamic Inducer Wind Turbine," *Proceedings, Second Wind Energy Innovative Systems Conference*, Vol. II, SERI/CP-635-1061, Golden, Colorado: National Renewable Energy Laboratory, pp. 113-132.

Lundsager, P., S. Frandsen, and C. J. Christensen, 1980, *Analysis of Data from the Gedser Wind Turbine, 1977-1979*, Risø-M-2242, Roskilde, Denmark: Risø National Laboratory Station for Wind Turbines.

Milborrow, D. J., and J. F. Ainslie, 1992, *Intensified Study of Wake Effects Behind Single Turbines and in Wind Turbine Parks*, Leatherhead, U.K.: National Wind Power.

Miller, A., and R. L. Simon, September 1980, *Wind Resource Potential in California*, California Energy Commission Report P-500-82-052, San Jose, California: San Jose State University.

Miller, R. H., *et al.*, 1978, *Wind Energy Conversion - Vol. 2: Aerodynamics of Horizontal Axis Wind Turbines*, C00-4131-T1, Cambridge, Massachusetts: Massachusetts Institute of Technology.

Muraca, R. J., Stephen, S., V. Maria, and R. J. Dagenhart, 1975, *Theoretical Performance of Vertical Axis Windmills*, NASA TMX-72662, Hampton, VA: NASA Langley Research Center.

Musgrove, P. J., and R. Clare, 1987, "Development of the U.K. Vertical Axis Wind Turbine," *Proceedings, Windpower '87 Conference*, SERI/CP-217-3315, Washington, DC: American Wind Energy Association, pp. 28-34.

Page, D. I., and L. A. W. Bedford, 1987, "A Progress Report on the UK Department of Energy's Wind Energy Programme," *Proceedings, Windpower '87 Conference*, SERI/CP-217-3315, Washington, DC: American Wind Energy Association, pp. 246-252.

PNL, 1980, *Siting Handbook for Small Wind Energy Conversion Systems*, PNL-2521, Richland, Washington: Battelle Pacific Northwest Laboratories.

PNL, 1981, *Meteorological Aspects of Siting Large Wind Turbines*, PNL-2522, Richland, Washington: Battelle Pacific Northwest Laboratories.

Puthoff, R. L., and P. J. Sirocky, 1974, *Preliminary Design of a 100-kW Wind Turbine Generator*, NASA TMX-71585, ERDA/NASA/1004-77/6, Cleveland, Ohio: NASA Lewis Research Center.

Putnam, P. C., 1948, *Power from the Wind*, New York: Van Nostrand Reinhold Company.

Richards, B., 1987, "Initial Operation of Project Eolé 4 MW Vertical Axis Wind Turbine Generator," *Proceedings, Windpower '87 Conference*, Washington, DC: American Wind Energy Association, pp. 22-27.

Rogers, S. E., B. W. Cornaby, P. R. Sticksel, and D. A. Tolle, 1977, *Environmental Studies Related to the Operation of Wind Energy Systems*, COO-0092-77/2, BCL-0092, Columbus, Ohio: Battelle Memorial Institute.

Rybak, S. C., 1981, "Description of the 3 MW SWT-3 Wind Turbine at San Gorgonio Pass California," *Proceedings, Fifth Biennial Wind Energy Conference and Workshop*, Vol. I, I. E. Vas, ed., Washington, DC: American Wind Energy Association, pp. 193-206.

Senior, T. B. A., D. L. Sengupta, and J. E. Ferris, 1977, *TV and FM Interference by Windmills*, DOE/TIC-11348, Ann Arbor, Michigan: University of Michigan.

Shaltens, R. K., and A. G. Birchenough, 1983, *Operational Results for the Experimental DOE/NASA Mod-0A Wind Turbine Project*, NASA TM-83517,DOE/NASA/20320-55, Cleveland, Ohio: NASA Lewis Research Center.

South, P., and R. S. Rangi, 1972, *A Wind Tunnel Investigation of a 14-ft Diameter Vertical Axis Windmill*, NRC Laboratory Technical Report LTR-LA-105, Toronto: National Research Council of Canada.

Spera, D. A., 1977, *Comparison of Computer Codes for Calculating Dynamic Loads in Wind Turbines*, NASA TM-73773, DOE/NASA/1028-78/16, Cleveland, Ohio: NASA Lewis Research Center.

Spera, D. A., J. B. Esgar, M. Gougeon, and M. D. Zutcck, 1990, *Structural Properties of Laminated Douglas Fir/Epoxy Composite Material*, NASA Reference Publication 1236, DOE/NASA/20320-76, Cleveland, Ohio: NASA Lewis Research Center.

Spera, D. A., and M. W. Miller, 1992, "Performance of the 3.2-MW Mod-5B Horizontal-Axis Wind Turbine During 55 Months of Commercial Operation in Hawaii," *Proceedings, Windpower '92 Conference*, Washington, DC: American Wind Energy Association, pp. 231-238.

Spera, D. A., 1995, "Bibliography of NASA-Related Publications on Wind Turbine Technology 1973-1995," DOE/NASA/5776-3, NASA CR-195462, Cleveland, Ohio: National Aeronautics and Space Administration.

Stahl, W., and R. Windheim, 1987, "Wind Energy Utilization: Status of Research and Development in the Federal Republic of Germany," *Proceedings, Windpower '87*, SERI/CP-217-3315, Washington, DC: American Wind Energy Association, pp. 240-245.

Stodhart, A. H., 1974, "Review of the UK Wind Power Programme 1948-1960," *Proceedings, Workshop on Advanced Wind Energy Systems*, Vol. 1, (published 1976), O. Ljungström, ed., Stockholm: Swedish Board for Technical Development and Swedish State Power Board, pp. 1-23 to 1-34.

Thomas, P. H., 1946, *Electric Power from the Wind*, Federal Power Commission Report.

Thomas, P. H., 1949, "Harnessing the Wind for Electric Power," *Proceedings, United Nations Scientific Conference on the Conservation and Utilization of Resources*, Vol. III, New York: The United Nations, pp. 310.

Thomas, R. L., 1982, *DOE/NASA Lewis Large Wind Turbine Program*, NASA TM-82991, Cleveland, Ohio: NASA Lewis Research Center.

USDA, 1979, *Wind Energy Applications in Agriculture*, Report No. 1109-20401/79/2, Washington, D.C.: U. S. Department of Agriculture.

Vas, I. E., and R. L. Mitchell, 1979, "A Review of the Wind Energy Innovative Systems Program," *Proceedings, Fourth Biennial Conference and Workshop on Wind Energy Conversion Systems*, CONF-791097, Washington, DC: JBF Scientific Corporation, pp. 349-418.

Weingart, O., 1981, *Design, Evaluation, and Fabrication of Low-Cost Composite Blades for Intermediate-Size Wind Turbines*, NASA CR-165342, DOE/NASA/0100-1, Cleveland, Ohio: NASA Lewis Research Center.

Wilson, R. E., and P. B. S. Lissaman, 1974, *Applied Aerodynamics of Wind Power Machines*, Corvallis, OR: Oregon State University.

Wilson, R. E., P. B. S. Lissaman, and S. N. Walker, 1976, *Aerodynamic Performance of Wind Turbines*, ERDA/NSF/04014-76/1, Corvallis, OR: Oregon State University.

Windheim, R., 1983, "The Wind Energy Program of the Federal Republic of Germany and State of Wind Energy Projects," *Proceedings, Sixth Biennial Wind Energy Conference and Workshop*, B. H. Glenn, ed., Boulder, Colorado: American Solar Energy Society, pp. 75-85.

Wright, A. D., M. L. Buhl, and R. W. Thresher, 1988, *FLAP Code Development and Validation*, SERI/TR-217-3125, Golden, Colorado: National Renewable Energy Laboratory.

Wright, A. D., M. L. Buhl, and A. S. Elliott, 1993, "Development of a Wind Turbine Systems Dynamics Model for a Two-Bladed Teetering Hub Turbine Using the Automatic Dynamic Analysis of Mechanical Systems (ADAMS) Software," *Proceedings, Windpower '93 Conference*, Washington, DC: American Wind Energy Association, pp. 299-304.

WMO, 1981, *Meteorological Aspects of the Utilization of Wind as an Energy Source*, WMO Report No. 575, Geneva, Switzerland: World Meteorological Organization.

3

Evolution of
Modern Wind Turbines
Part B: 1988 to 2008

Robert W. Thresher, Ph.D., Michael Robinson, Ph.D., and Walter Musial

National Renewable Energy Laboratory
Golden, Colorado

and

Paul S. Veers, Ph.D.
Sandia National Laboratories
Albuquerque, New Mexico

Introduction

Evolving from the birth of modern electricity-generating wind turbines in the 1970s to now, wind energy technology has dramatically improved. Guided by the operational experience gained in the California wind power plants during the 1980s, initial capital costs have plummeted, reliability has improved, and energy capture has increased. Well-capitalized multinational corporations now mass-produce wind turbines for delivery around the world. Modern wind turbines are routinely installed and operated by wind power plant developers, generating energy at a cost approaching – and in some cases below – that of fossil fuel generating plants. Wind-powered generating plants of 500 megawatts (MW) and larger are now being integrated into electrical grids in accordance with exacting utility specifications.

Wind power generating capacity worldwide has grown at a rate of 20 percent to 30 percent per year over the past decade [AWEA 2008, BTM Consult ApS 2005]. By the end of 2008, the total installed wind power capacity in the U.S. had grown to 25,200 MW, enough to power more than 6.7 million homes. Table 3-3 lists the capacities in individual states at the

Table 3-4
**Distribution by States of Wind Power Generating Capacity in the
U.S. at the End of 2008**

Rank	State	Wind Power Capacity (MW)
1	Texas	7,356
2	Iowa	2,584
3	California	2,538
4	Minnesota	1,554
5	Washington	1,367
6	Colorado	1,068
7	New York	1,036
8	Oregon	991
9	Kansas	921
10	Illinois	916
11	Oklahoma	708
12	Wyoming	676
13	North Dakota	518
14	New Mexico	497
15	Wisconsin	449
16	Pennsylvania	361
17	Montana	272
18	West Virginia	230
19	South Dakota	187
20	Missouri	163
21	Nebraska	152
22	Michigan	144
23	Idaho	95
24	Hawaii	63
25	Maine	42
26	Tennessee	29
27	New Hampshire	25
28	Utah	20
29	New Jersey	8
30	Ohio	7
31	Vermont	6
32	Massachussetts	6
33	Rhode Island	1

end of 2008. Despite this rapid growth, however, wind power currently provides only about 1 percent of the total electricity consumed in the United States.

Worldwide, 94,000 MW of wind power was installed by the end of 2007. According to analysts at the National Renewable Energy Laboratory (NREL) and BTM Consult ApS, by the end of 2011 wind power capacities are expected to grow to 40,000 MW in the U.S. (the majority added in the West and Midwest) and 237,000 MW worldwide [BTM 2008, Zervos and Kjaer 2008].

Today's Modern Commercial Wind Turbines

General Configuration

The typical modern wind turbine deployed throughout the world today is shown in Figure 3-40. It is a horizontal-axis machine with a three-bladed rotor about 70 m to 80 m in diameter mounted upwind on a 60- to 80-m tower. Its power rating is about 1.5 MW, and it is installed in groups with a combined peak power output of 50 MW or more. A typical layout of the power train components in the nacelle is illustrated in Figure 3-41. Newer and larger turbine designs of 2 MW to 3 MW are beginning to be deployed in large quantities. These larger machines have rotor diameters of 90 m to 100 m and often have hub heights at 80- to 100-m elevations.

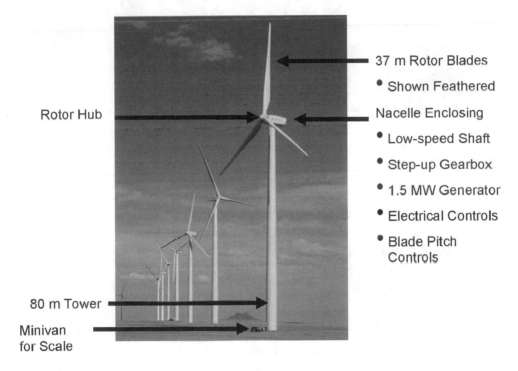

Figure 3-40. General view of a typical modern wind turbine installed in a wind power plant. [Photographed by M. Rumsey, Sandia National Laboratories]

The power output of today's wind turbines is regulated by *blade pitch control*, in which the blades are rotated about their long axis to adjust their aerodynamic *angle of attack* with respect to the relative wind. The rotor is pointed into the wind by rotating the nacelle about the tower axis, which is referred to as *yaw control*. Wind direction sensors on the nacelle tell the yaw controller where to point the rotor. Torque and speed sensors on the generator and drivetrain enable the blade-pitch controller to regulate the power output and rotor speed to prevent overloading structural components. Modern turbines generally start operating in winds at hub height of about 4 m/s (9 mph) and reach their rated (usually maximum) power output at about 13 m/s (29 mph). At about 25 m/s (55 mph) the control system will shut down the turbine by *feathering the blades* (pitching them to stop rotation).

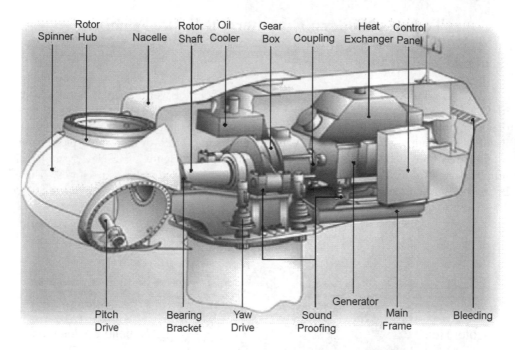

Figure 3-41. Schematic view of the component layout within the nacelle of a typical modern wind turbine.

Recent Developments in Wind Power Plant Operation

Turbine Performance

The power performance of modern commercial wind turbines has improved dramatically over the past 20 years. Rotor systems can now capture about 80 percent of the theoretically extractable energy in the flow stream. This has been made possible partly through the design of custom airfoils for wind turbines. In fact, it is now commonplace for turbine manufacturers to have special airfoil designs for each individual turbine design. These special airfoils attempt to optimize aerodynamic efficiency at low wind speeds and limit aerodynamic loads in high winds. They also attempt to minimize sensitivity to *blade fouling*, which is caused by dirt and bugs that accumulate on blade leading edges and can greatly reduce aerodynamic efficiency. Although rotor design methods have improved significantly, there is still much room for improvement [Tangler 2001].

As the power performance of commercial turbines has increased with time during the past decade, resulting *capacity factors* have also slowly increased. Capacity factor is defined as the ratio of the annual average power output divided by the rated power of the generator. Table 3-5 presents capacity factors from the Lawrence Berkeley National Laboratory database versus the commercial operation start date [Wiser and Bollinger 2008]. These data show that turbines that began commercial operation prior to 1998 have an average capacity factor of about 22 percent, whereas the turbines that began commercial operation after that show an increasing capacity factor trend, reaching close to 35 percent in 2004–05. Much of this increase can be attributed to increases in turbine size and power, plus installation on taller towers reaching into more-energetic winds. This increasing trend in capacity factors is expected to continue over time.

Cost of Energy

At first, the cost of wind-generated electricity, COE, dropped dramatically after 1980, when the first commercial wind plants began operation in California. However, after 2003 costs of energy began to increase. Table 3-5 lists cost data taken from public records for some more recent wind energy projects. These figures represent the price of electricity as sold by a wind-plant operator to the host utility. The price includes the benefit of the federal production tax credit, any state incentives, and revenue from the sale of any renewable energy or green energy credits. Thus, the true cost of the delivered electricity would be higher than the costs in Table 3-5 by approximately 2 cents per kWh, which is the value of the federal tax credit.

Table 3-5. Evolution of Capacity Factors for Wind Power Plants
[Wiser and Bolinger 2008]

Years	No. of Wind Power Plants	Total Rated Power (MW)	Capacity Factor	
			Weighted Average	Range
Pre-1998	20	936	22%	12% - 29%
1998-99	21	892	30%	16% - 37%
2000-01	26	1,743	30%	8% - 36%
2002-03	25	1,911	34%	21% - 41%
2004-05	26	2,669	35%	20% - 44%
2006	17	2,225	33%	28% - 46%

Table 3-6. Evolution of Sales Cost of Energy for Wind Turbine Projects
[Wiser and Bollinger 2008]

Years	No. of Projects	Total Rated Power (MW)	Cost of Energy (¢/kWh)	
			Weighted Average	Range
1998-99	14	624	3.8	2.6 – 5.6
2000-01	16	860	3.5	2.2 – 6.1
2002-03	24	1,781	3.1	2.1 – 5.6
2004-05	21	1,681	3.7	2.4 – 6.8
2006	10	732	4.9	3.0 – 6.5
2007	20	2,412	4.5	3.5 – 6.8

As can be seen from the cost data in Table 3-6, prices for wind-generated electricity increased significantly in 2006 and 2007. In 2007, the price paid for electricity generated in large wind power plants was between 3.5 and 6.8 cents per kilowatt-hour (kWh) with an average near 4.5 cents per kWh. Accounting for the tax credit then indicates that the unsubsidized cost for wind-generated electricity for projects completed in 2007 ranges from about 5.5 to 8.8 cents per kWh.

The reasons generally offered for recent increases in the price of wind-generated electricity after the long downward price trend of the past 25 years include the following:

- Weakening of the U.S. dollar relative to the euro, because many major turbine components are imported from Europe. There are relatively few wind turbine component manufacturers in the United States at this time, although the number of new wind turbine manufacturing facilities in the United States is rapidly increasing [Sterzinger and Svrcek 2004].

- Significant increases in the cost of materials such as steel, copper, and fuel, caused by increased world demand for these commodities.
- Shortages of turbine components caused by the recent dramatic growth of the wind industry in the United States and Europe
- Hindrance of investment in new turbine production facilities by uncertainty during on-again/off-again cycles of the *wind energy production tax credit*.
- Hurried and expensive production, transportation, and installation of projects encouraged when tax credits are available.

To reestablish a decreasing trend in the cost of wind-generated energy and to continue the evolution of the technology, industry and government will need to continue their research and development efforts.

Wind Turbine Technology Development: 1988 to 2008

The latter half of the 20th century saw spectacular changes in the technology of wind turbines. Blades that had once been made of sail or sheet metal progressed through wood to advanced fiberglass composites. DC alternators gave way to induction generators that were synchronized to the transmission grid. From mechanical cams and linkages to feather or furl rotor blades, designs moved to high-speed digital controls. Custom wind turbine airfoils designed for insensitivity to surface roughness and dirt replaced airplane wing airfoils. Knowledge of *aeroelastic loads* and the ability to incorporate this knowledge into *finite element models* and *structural dynamics codes* have made the machines of today more robust and yet much less expensive than those of decades ago.

Trends in Turbine Size

The tower heights, rotor diameters, and rated powers of wind turbines have increased during recent years in order to capture the more energetic winds that occur at higher elevations and to produce more energy per turbine installation. For land-based turbines, however, size is not expected to grow as dramatically in the future as it has in the past. With each new and larger wind turbine design, there have been predictions that 3 MW to 5 MW machines are as large as they will ever be. Larger sizes are physically possible, but the cost and logistical constraints on transporting very large components over highways and lifting them into place are potential barriers.

As illustrated in Figure 3-42, over the past 20 years average wind turbine size and capacity rating have grown at an increasing rate. The majority of utility-scale wind turbines installed in the U.S. from 2005 to 2007 have a rating of 1.5 MW. A growing number of new designs rated at 2 MW to 2.5 MW with rotor diameters of 90 m to 100 m are now being sold globally by multinational manufacturers. Even larger commercial wind turbines, with rated powers of 3 MW to 5 MW or more are being developed, the largest designed for off-shore deployment.

Although a reduction in *life-cycle cost of energy* (cost calculation including all maintenance and operational costs during the entire life of the wind turbine) has been achieved with each increase in size, the primary argument for a size limit on wind turbines is based on the *square-cube rule*. This rule states that as a wind turbine rotor increases in size, its

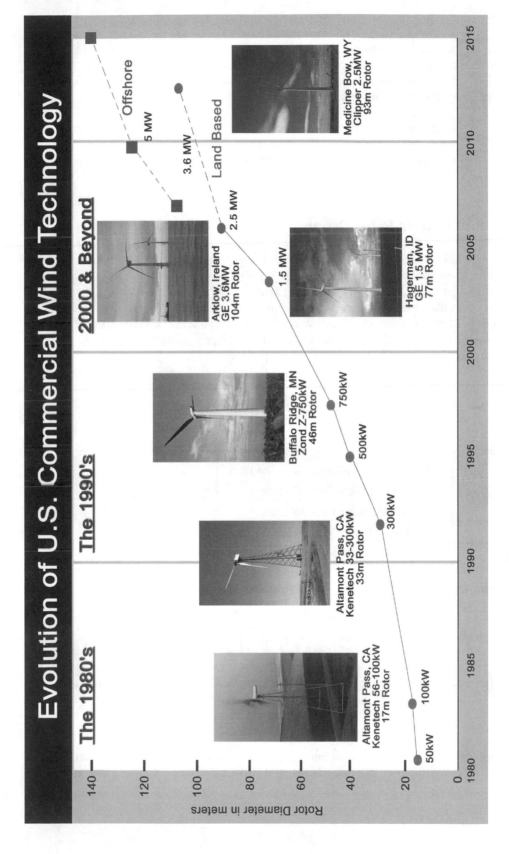

Figure 3-42. The Development Path and Size Growth of Modern Wind Turbines.

energy output (proportional to its swept area) increases as the *square* of the diameter, while the volume of material (and therefore its mass and cost) increases as the *cube* of the diameter. According to this simplified rule, costs for a larger turbine will grow faster than the energy revenue, making further scale-up a losing economic game.

For years, engineers have successfully skirted the square-cube rule by making design changes that alter the square-cube ratio of energy capture and cost, as follows:

- Taller towers lift the rotor into more-energetic winds, so energy capture increases faster than the square of the rotor diameter.
- More efficient use of material trims weight, so masses and costs increase slower than the cube of the rotor diameter.

Advancements in Rotor Design

As wind turbines grow in size, so do their blades—from about 8 m in 1980 to more than 40 m for many land-based commercial systems today. Improved blade designs have enabled the weight growth with increasing size to be kept to a much lower rate than simple geometric scaling. As illustrated in Figure 3-43 [Griffin 2001], blade mass has recently been scaling with rotor diameter at roughly an exponent of 2.3 (dashed line) *versus* the expected cubic exponent of 3 (dotted line). Also shown in this figure are preliminary estimates of even lower weights possible for blades on a 104 m diameter rotor.

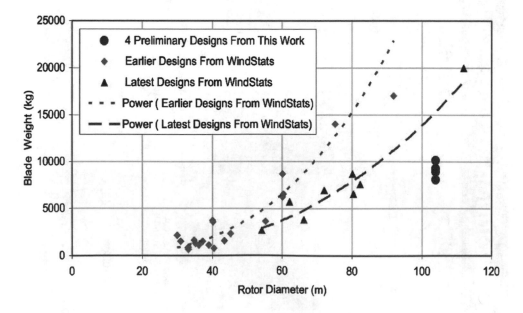

Figure 3-43. Results of a study indicating that the rate of increase in wind turbine blade weight with rotor diameter is being significantly reduced by the introduction of new technology. The dotted trend line indicates a 3.0-power weight growth according to an elementary cubic rule, while the dashed line fits the latest design weights with a 2.3-power trend. [Griffin 2001]

As turbines grow larger and larger, rotors must improve their ability to handle large dynamic loads with increased structural efficiency to avoid the costly cubic-rule of weight growth described previously. Today's blade designs are subjected to rigorous evaluation using the latest computer analysis tools that remove excess weight. Designers are also applying lighter and stronger carbon fiber in highly stressed locations to stiffen the blade and improve fatigue resistance while reducing blade weight. However, carbon fiber must be used judiciously, because its cost is about 10 times the cost of fiberglass.

Another approach to reducing the relative cost of larger rotor blades involves developing new blade airfoil shapes that are much thicker at the root where the blade needs the most strength. In general, thin streamlined structures like airfoils are very inefficient at carrying structural loads applied to their upper and lower surfaces. The design challenge is to make a thick, structurally efficient airfoil cross-section that doesn't give up much in aerodynamic performance. Figure 3-44 illustrates such an airfoil shape, called a *flat-back thick airfoil*, which is used near the root of the blade where aerodynamic performance is less important [Berg and Zayas 2008]. An additional benefit is that blades with this type of shorter root chord are more easily transported over the highway, where width and height restrictions apply.

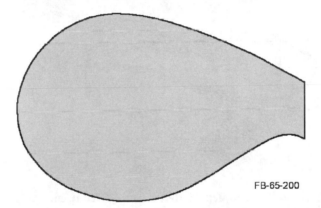

FB-65-200

Figure 3-44. A flat-back thick airfoil shape with high structural efficiency and moderate aerodynamic performance for use in the root sections of large rotor blades. The relatively short chord of this cross-section also allows for easier highway transport. [Berg and Zayas 2008]

Reducing Fatigue Loads

Another approach to increasing blade length while restraining the weight and cost growth is to reduce the fatigue loading on the blade. There can be a big payoff in this approach because the approximate rule of thumb for fiberglass blades is that a 10 percent reduction in cyclic stress can provide about an order of magnitude increase in fatigue life. Blade fatigue loads can be reduced by controlling the blade's aerodynamic response to turbulent wind inputs. This is done by using the turbine control system to continuously adjust the blade pitch in response to fluctuating winds. This approach is now being explored using modern *state-space control strategies* so that future wind turbines can take advantage of this innovation [Wright and Fingersh 2008].

An elegant concept for increasing blade fatigue life is to build a passive means of reducing cyclic loads directly into the blade structure. By carefully tailoring the structural properties of the blade using the unique attributes of composite materials, the blade can be built in a way that couples the bending deformation of the blade to a twisting (pitching) deformation. This is referred to as *flap-pitch* or *bend-twist coupling* and allows the outer portion of the blade to twist as it bends as illustrated in Figure 3-45. Flap-pitch is accomplished by designing the internal structure of the blade (*e.g.* by orienting the fiberglass and carbon plies within the composite layups) in such a way as to make the blade twist as it bends. This twisting changes the angle of attack over much of the blade. If properly designed, this change in angle of attack will reduce the lift forces caused by wind gusts and therefore passively reduce cyclic fatigue loads.

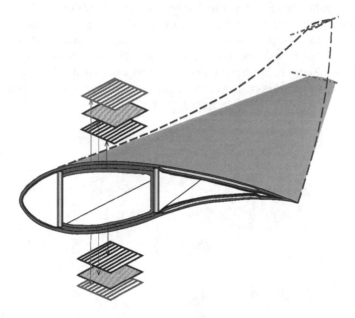

Figure 3-45. A twist–flap coupled blade fabricated from composite material with a layup pattern designed to alleviate fatigue loads. Note that the bending deflection of the blade causes the pitch angle at the tip to change, thereby reducing cyclic lift loads from wind gusts. [Lobitz *et al.* 2001]

Another approach to achieve flap-pitch coupling is to build the blade *planform* in a curved shape so that aerodynamic load fluctuations apply twisting movements to the blade that will vary local angles of attack [Lobitz *et al.* 2001, Paquette *et al.* 2007]. These new blade design concepts are complex and must be developed, tested, and optimized so as not to adversely impact energy production or result in unstable vibrations.

Concepts such as on-site manufacturing and segmented blades are also being explored to help reduce overland transportation costs. It may be possible to segment molds and move them into temporary buildings close to the site of a major wind installation so that over-the-road transportation is not necessary.

Limiting Factors on Onshore Turbine Size

As mentioned previously, land transportation constraints can pose limiting factors on the future sizes of wind turbines installed in inland power stations. Cost-effective transportation can only be achieved by remaining within standard, over-the-road trailer dimensions of 4.1 m high by 2.6 m wide. Rail transportation is even more dimensionally limited, although some rail lines now have special cars designed for transporting large wind turbine components.

Crane requirements also limit the practical size of wind turbines, because of the combination of larger nacelle masses, higher lifts, and larger boom extensions. As these three factors increase, the number of available cranes decreases. Other limiting factors are that cranes with very large lifting capacities are difficult to transport overland, require large crews, and therefore have high mobilization, operation, and de-mobilization costs.

Offshore Turbine Size

It should be noted that the previous discussion of turbine size *vs.* cost is largely applicable only to land-based turbines, because the economic considerations driving the cost tradeoffs for offshore turbines are much different. For offshore wind power plants, the cost of the underwater support structure is a very significant portion of the total cost, as well as the cost of working and installing a wind turbine at sea. Additionally, there is a much larger cost advantage for high reliability in the operation and maintenance (O&M) of an offshore turbine relative to an onshore turbine of the same size. This is due to the difficulty of servicing the turbine during high wind periods, when access to offshore turbines may be prohibited for periods of several days for safety reasons. As a result, minor faults causing turbine outages must be eliminated by high-reliability design, even if the resulting costs are higher.

Variable Speed Controls

Today's commercial wind turbine controllers integrate the signals from dozens of sensors to control rotor speed, blade pitch angle, generator torque, and power conversion voltage and phase. The controller is also responsible for critical safety decisions, such as shutting down the turbine when extreme conditions are realized. Today, most turbines operate at variable speed for peak aerodynamic efficiency. The control system regulates the rotor speed to keep the ratio of the blade tip speed to the wind speed approximately constant at its optimum value as the wind speed varies. Continuously updating the rotor speed and generator loading maximizes output power. The controller also regulates the electrical torque of the generator to reduce drivetrain transient torque loads.

Operations at variable speed require the use of power converters, which also enable turbines to deliver *fault ride-through protection*, voltage control, and dynamic reactive power support to the grid. Because there are now large quantities of wind turbines supplying power to the electrical system, utilities are requiring that wind power plants remain on line and capable of providing power in spite of electrical faults that last as long as several electrical cycles.

Evolution of Power Trains

Background

Wind generation of electricity places an unusual set of requirements on electrical drive systems. Most applications for electrical drives are aimed at using electricity to produce torque, rather than using torque to produce electricity. Moreover, other electricity generation sources that produce electricity from torque usually operate at a constant power, or load. Wind turbines, on the other hand, must generate electricity at all power levels and spend a substantial amount of time at low power levels. Unlike most electrical machines, wind generators must operate at the highest possible aerodynamic and electrical efficiencies in the low-power/low-wind speed region to squeeze every kilowatt-hour out of the available wind energy.

Traditional electrical machines and power electronics are not very efficient in the low-power generating regime, because in most conventional generating systems there is power to spare, and consequently efficiency is less important in this low-power region. In addition, conventional generators do not operate in this low-power region for long periods of time. On the other hand, for wind systems, it is not critical for the generation system to be efficient in high-wind, high-power conditions, because the rotor is spilling energy to keep the power at the rated level. Therefore, wind systems can afford inefficiencies at high power while they require maximum efficiency at low power—just the opposite of most other electrical generating systems.

Generators

Current commercial generator designs are either *squirrel-cage induction* or *wound-rotor induction*, with some newer machines using the *doubly-fed induction* design for variable speed. In the latter design, the turbine rotor's variable frequency electrical output is fed into the collection system through a solid-state power converter. However, full power conversion and permanent-magnet, low-speed synchronous machines are being introduced in some newer turbines, because of their fault-ride-through capability, higher efficiency, and other attributes.

Several unique designs are under development to reduce drivetrain weight and cost while improving reliability. These have been explored in design studies under the NREL's Wind-PACT project by Poore and Lettenmaier [2003] and Bywaters *et al.* [2004]. One approach for improving reliability is to build a *direct-drive generator* that eliminates the complexity of the gearbox, which is the component where most of the current reliability problems are occurring. The tradeoff is that the slowly rotating generator must have a high *pole count* and is therefore large in diameter. Depending on the design, a direct-drive generator can be in the range of 4 m to 10 m in diameter and quite heavy.

The decrease in cost and increase in availability of rare-earth permanent magnets is expected to significantly affect the size and cost of future permanent-magnet generator designs. Permanent-magnet designs tend to be quite compact and lightweight, with reduced electrical losses in the windings. *An Enercon* 1.5-MW direct-drive generator using rare-earth permanent magnets has been studied and a prototype has been built. This design uses 56 poles and is only 4 m in diameter, versus the 10 m for a wound rotor design [Poore and Lettenmaier 2003]. This direct-drive machine has undergone testing at NREL's National Wind Technology Center.

Gearboxes

Converting rotor torque to electrical power in a wind turbine has historically been achieved using a speed-increasing gearbox and an induction generator. Many current megawatt-scale turbines use a three-stage gearbox consisting of varying arrangements of planetary gears and parallel shafts. Fleet-wide gearbox maintenance issues and related failures with some past designs have occurred. For this reason extensive dynamometer testing of new drivetrain configurations has become standard practice, to verify their durability and reliability prior to serial production. The life-cycle reliability of the current generation of megawatt-scale drivetrains has not yet been fully verified with long-term, real world operating experience. There seems to be a broad consensus that wind turbine drivetrain technology can and will evolve significantly in the next several years in order to meet future reliability and cost requirements.

A hybrid of the direct-drive approach that offers promise for future large-scale designs is the single-stage gearbox driving a low- or medium-speed generator. This allows the use of a generator that is significantly smaller than a comparable direct-drive design. The WindPACT drivetrain project has developed a prototype for such a drivetrain that was brought to practice by *Multibrid Gmbh*. This design uses a single-stage planetary drive operating at a gearbox ratio of 9.16:1. This gearbox drives a 190 RPM, 72-pole, permanent-magnet generator. This design, which reduces the diameter of a 1.5-MW generator to 2 m [Bywaters *et al.* 2004], was fabricated and then tested on the dynamometer at NREL's National Wind Technology Center.

Distributed Drivetrains

Another approach that offers promise for reduced size, weight, and cost is the distributed drivetrain. This concept is based on splitting the drive path from the rotor to power several parallel generators. Studies have shown that by distributing the rotor torque on the *bull gear* (the main gear attached to the rotor shaft) over a number of parallel secondary *pinions*, significant size and weight reductions are achieved. Figure 3-46 shows a 2.5-MW prototype distributed power train [Clipper Windpower 2006] that is currently installed in Clipper Liberty wind turbines. The development of new technology of this type and incorporation of that technology into production wind turbines requires significant R&D resources and a number of years to ensure a reliable production product.

Figure 3-46. An example of a distributed power train, in which the turbine rotor torque (center shaft) is split among four generators. [*Courtesy of Clipper Wind Power, Inc.*]

Towers

For the large turbines deployed since 2000, the tower configuration used almost exclusively has been a steel monopole tower on a reinforced-concrete foundation that is custom designed according to local site conditions. The major tower variables are its height and base diameter. Depending on the site's wind characteristics, tower height is selected to optimize energy capture with respect to cost. Generally, utility-scale turbines are placed on a tower 60 m to 80 m tall, but 100-m towers are now being used more frequently. Bridge clearances limit the base diameters of land-based wind turbines.

There are ongoing efforts to develop advanced tower configurations that are less costly and more easily transported and installed. The cost impact of extremely large cranes and the transport premiums for large tower sections and blades are driving the exploration of novel tower design approaches. Several concepts are under development or being proposed that would eliminate the need for cranes for very high, heavy lifts [LaNier 2005, Global Energy Concepts 2001 and 2002]. One concept is the telescoping or self-erecting tower. Other self-erecting designs include lifting dollies or tower-climbing cranes that use tower-mounted tracks to lift the nacelle and rotor to the top of the tower.

Operations, Reliability, and Availability

The costs of operations and maintenance (O&M) of wind power plants have also dropped significantly since the 1980s because of improved component designs and increased quality of manufacture. Wiser and Bollinger [2007] present data showing that current O&M expenses are a significant portion of total system cost of energy. O&M costs are reported to be as high as 3 to 5 cents/kWh for wind power plants constructed with 1980s technology, whereas the latest generation of turbines has reported O&M costs below 1 cent/kWh. Availability, defined as the percentage of time during which the equipment is ready to operate on an annual basis, is now more than 95 percent and is often reported to exceed 98 percent.

Future Advances in Wind Energy Technology

The U.S. Department of Energy (DOE) in conjunction with the American Wind Energy Association (AWEA), the National Renewable Energy Laboratory (NREL), and the consulting firm of Black & Veatch undertook a study to explore the possibility of producing 20 percent of the nation's electricity by 2030 using wind energy. Their comprehensive report, titled "20 percent Wind Energy by 2030: Increasing Wind Energy's Contribution to the U.S. Electricity Supply" [U.S. Department of Energy 2008], describes in detail the many important developments needed in the future to achieve this 20 percent goal, including advances in the following areas:

- – wind turbine technology development
- – manufacturing, materials and resources
- – transmission lines and integration into the U.S. electric system
- – siting and resolution of environmental issues
- – markets for wind-generated power.

The Wind Energy Deployment System model developed at NREL [Short *et al.* 2003] was used to estimate many of the important outcomes associated with producing 20 percent of the nation's electricity from wind power by 2030. This model of expanding generation capacity uses data from a wide range of electricity generation technologies, including pulverized coal plants, combined-cycle natural gas plants, combustion turbine natural gas plants, nuclear plants, and wind power stations. Estimates of technology development costs, performance projections, and transmission operation and expansion costs are outputs of this model.

In Europe, the 2008 report *The European Wind Energy Technology Platform for Wind Energy* envisions that in 2030, wind energy will be a major modern energy source, reliable and cost competitive in terms of cost per kWh. This study also predicts that wind energy will contribute 21 percent to 28 percent of the electricity demand in the European Union (EU), and describes a long series of research and development improvements that will be necessary to make wind cost competitive by 2030. The conclusions of the European study are similar to those described previously for the United States.

No major technology breakthroughs are envisioned for wind turbine technology either in the U.S. or European studies. Instead, many evolutionary steps executed over the next two decades through incremental technology advances are seen to cumulatively bring about a 30 to 40 percent improvement in the cost effectiveness of wind power, as has been achieved over the past two decades.

Summary of Potential Future Turbine Technology Improvements

Several studies of advanced wind turbine concepts have been reported under the Wind-PACT Project [1999] that identified a number of areas where technology advances would result in changes to the capital cost, annual energy production, reliability, and O&M. Many of these potential improvements from a 2002 WindPACT study are summarized in Table 3-7. Also included in this table are estimates of the effects of the *manufacturing learning-curve*, generated by several doublings of the turbine manufacturing output over the coming

Table 3-7. Areas of Potential Technology Improvement

Technical Area	Potential Advances	Increments from Baseline (Best/Expected/Least)	
		Annual Energy Production (%)	Turbine Capital Cost (%)
Advanced Tower Concepts	* Taller towers in difficult locations * New materials and/or processes * Advanced structures/foundations * Self-erecting, initial or for service	+11/+11/+11	+8/+12/+20
Advanced (Enlarged) Rotors	* Advanced materials * Improved structural-aero design * Active controls * Passive controls * Higher tip speed/lower acoustics	+35/+25/+10	-6/-3/+3
Reduced Energy Losses and Improved Availability	* Reduced blade soiling losses * Damage tolerant sensors * Robust control systems * Prognostic maintenance	+7/+5/0	0/0/0
Drivetrain (Gearboxes and Generators and Power Electronics)	* Fewer gear stages or direct drive * Medium/low-speed generators * Distributed gearbox topologies * Permanent-magnet generators * Medium-voltage equipment * Advanced gear tooth profiles * New circuit topologies * New semiconductor devices * New materials (GaAs, SiC)	+8/+4/0	-11/-6/+1
Manufacturing and Learning Curve	* Sustained, incremental design and process improvements * Large-scale manufacturing * Reduced design loads	0/0/0	-27/-13/-3
Totals		+61/+45/+21	-36/-10/+21

years. The learning-curve effect on capital cost reduction is assumed to range from zero in a worst case scenario to the historic level in a best-case scenario, with the most likely outcome halfway in between. The most likely scenario is a sizeable increase in capacity factor with a modest drop in capital cost, compared to the 2002 levels of each.

Status of Wind Turbine Design Standards

The development of a suite of international standards for wind turbines has been a major contributor to the evolution of wind turbine technology over the past 15 years. The International Electrotechnical Commission (IEC) is a global organization that prepares and publishes international standards for electrical, electronic, and related technologies. These standards serve as the basis for national standards and as a reference when drafting international tenders and contracts. The IEC standards for wind turbines address certification procedures, design requirements, engineering integrity, measurement techniques, and test procedures. Their intended purpose is to provide a uniform basis for design, quality assurance and certification of wind turbines and wind projects. These standards cover wind turbine systems and subsystems, such as mechanical and internal electrical systems, support structures, and control and protection systems.

In the U.S. there are no laws enforcing the IEC standards. However, many lending institutions financing wind projects require that the turbines meet the IEC standards as a condition of the loan agreement.

The IEC certification document explains how the IEC suite of turbine standards should be used to ensure a consistent design process. This design process requires an internally consistent set of design conditions, design quality system, validation through prototype testing and manufacturing quality system. The flow chart of Figure 3-47 illustrates how this suite of complementary standards should be used to design, manufacture, and install a *type certified* wind turbine. The starting point in this process is the Wind Turbine Design Standard IEC 61400-1, which specifies minimum requirements for the design of land-based wind turbines. The standard is intended to ensure the integrity of wind turbines and provide an appropriate level of protection against damage from all hazards during the turbines planned lifetime. This standard applies to wind turbines of all sizes, but there is an alternate standard, IEC 61400-2, which may be used for wind turbines with power ratings less than 100 kW. Offshore turbine designs are covered by a separate standard, IEC 61400-3.

Wind Turbine Design Classes

When turbines are in the design stage, designers rarely know the specific sites where they may be installed, but a set of site design conditions is still needed. To help solve this problem, the IEC standards committees set up a consistent set of wind conditions or *design classes* that are each representative of a type of potential site. Each class is defined by specified wind conditions, including a reference site wind speed, a reference turbulence intensity level, extreme wind speeds, *etc.* Table 3-8 summarizes the wind turbine design standard classes, with specified wind conditions at hub elevation. The design standard specifies that the cumulative probability distribution of the 10-minute average wind speed at hub height shall be taken as a Rayleigh distribution (see Chs. 2 and 8).

The reference wind speeds in Table 3-8 serve to establish the severity of the wind conditions used in the load cases used to design the turbine components. From the table, it is clear that a Class IA wind turbine is designed for deployment at sites that are expected to see higher average wind speeds, higher gust wind speeds, and higher turbulence levels. This method of

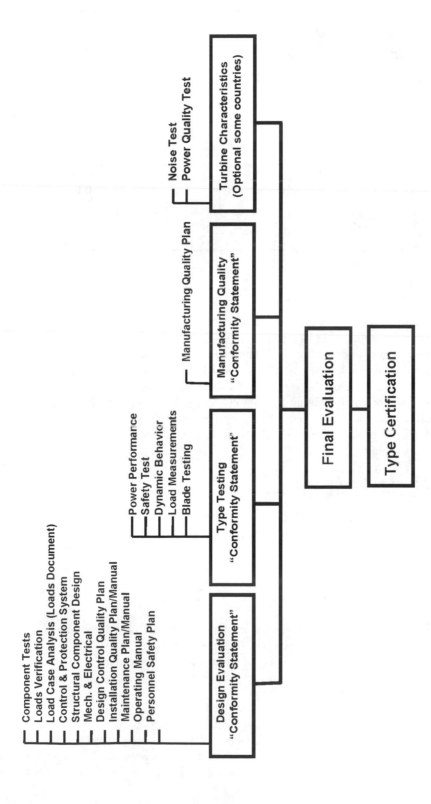

Figure 3-47. Flow chart illustrating the use of International Electrotechnical Commission (IEC) standards for Type Certification.

Table 3-8. Basic wind parameters defining wind turbine design classes
[IEC 61400-1]
Wind conditions apply at rotor hub elevation

Wind Turbine Design Class	Reference Wind Speed over 10 min, V_{Ref} (m/s)	Reference Turbulence Intensity at $V = 15$ m/s, I_{Ref}
IA	50	0.16
IB		0.14
IC		0.12
IIA	42.5	0.16
IIB		0.14
IIC		0.12
IIIA	37.5	0.16
IIIB		0.14
IIIC		0.12
S		Specified by Designer

classifying turbines seeks to insure that a wind turbine selected for installation at a specific site will perform reliability over its intended lifetime.

It is difficult to over-emphasize how important the IEC standards have been in establishing a basis for comparing the cost effectiveness and long-term performance of different turbine designs. These standards have facilitated the development of large markets for commercial wind generation by defining a common set of transparent design requirements and common design processes. Type certification is generally a required by the lender and banks as a requirement for project finance, which shows the confidence placed on the standards by the financial community.

Standard Design Load Cases

A wind turbine must be designed to a set of relevant load cases that represent the fatigue and extreme loading situations the turbine will see over its lifetime. These load cases are representative of the operational modes combined with the appropriate wind conditions from Table 3-8. Other situations must be addressed in the design process, such as assembly, erection, maintenance, and external conditions. The design load cases used to verify the structural integrity of a wind turbine shall be calculated by combining:

- normal design situations and appropriate normal or extreme conditions, including wind loading combined with other events.
- fault design situations and appropriate external conditions.
- transportation, installation, and maintenance design situations, combined with appropriate external conditions.

It is also required that when a correlation exists between an extreme external event and a fault event, the combination of these events must be considered as a design load case.

A brief summary and description of the IEC load cases is presented in Table 3-9. The IEC 61400-1 wind turbine design requirements document specifies a matrix of eight basic

Table 3-9. Summary of Design Load Cases for IEC 614000-1 Type Certification

Design Situation	Wind and Other Conditions for Load Case Analysis
1. Power production	Analyses are to be performed for: • Overall operating wind speeds using the normal and extreme turbulence models for ultimate and fatigue loads cases, • Extreme coherent deterministic gust combined with direction change for the ultimate loads case, • Extreme wind shear for the ultimate loads case.
2. Power production plus fault	Ultimate strength analyses are to be performed for: • Overall operating wind speeds using the normal turbulence model with a control system fault or loss of the electrical grid, • Overall operating wind speeds using the normal turbulence model with a protection system or preceding an internal electrical including loss of the electrical grid, • Extreme operating gust model with an external or internal electrical fault including loss of the electrical grid. Fatigue analyses are to be performed for: • Overall operating wind speeds using the normal turbulence model with a control system fault or loss of the electrical grid.
3. Startup	Fatigue analyses are to be performed for: • Startups under the normal wind shear profile. Ultimate strength analyses are to be performed: • Using the extreme operating gust model, • Extreme gust with direction change.
4. Normal shut-down	Fatigue analyses are to be performed for: • Shut-downs under the normal wind shear profile. Ultimate strength analyses are to be performed: • Extreme operating gust model.
5. Emergency shut- down	Ultimate strength analyses are to be performed using the normal turbulence model at rated wind speed, and plus and minus 2 m/s around the rated wind speed.
6. Parked – standing still and idling	Ultimate strength analyses are to be performed using the extreme wind speed model for a 50-year recurrence, including loss of the grid. Fatigue analyses are to be performed using the normal turbulence model for a wind speed of $0.7V_{Ref}$.
7. Parked with fault conditions	Ultimate strength analysis performed using the extreme wind speed model for a one-year recurrence period.
8. Transport, assembly, maintenance, and repair	Ultimate strength analyses are to be performed: • Using the normal turbulence model at the highest maintenance wind speed allowed by the manufacturer, • Using the extreme wind model with one year recurrence period.

situational load cases that include numerous external wind and operational scenarios and events. Some of these are *deterministic*, discrete events to be simulated. Others are *probabilistic* in nature and require multiple *stochastic time domain simulations* from which the extreme loading events must be extrapolated. All final load cases to be used in type certification are expected to be simulated using a dynamically coupled aeroelastic model of the full wind turbine. This model must have been validated using prototype test data.

The standard document provides detailed computational descriptions of the deterministic wind models, such as the extreme coherent gust with direction change, the extreme wind direction change model, the extreme wind speed model, the extreme wind shear model, and the normal wind profile model. The standard also provides mathematical models for normal and extreme turbulence. The application of these models for calculating the resulting loads and stresses on turbine components requires a full-field simulation of the three dimensional turbulent flow fields in time and space across the swept area of the rotor. The turbulent variations are defined to be the random variations in the wind field from a 10-minute mean flow. The standard provides information on two models that may be used for turbulence calculations.

The standard also provides information on how to account for wake turbulence effects from neighboring turbines in a wind power station when performing fatigue calculations.

Computer Simulation

The computer simulation of the operational environment of a modern wind turbine has markedly advanced over the past 30 years. Techniques for performing the complex analyses required by the IEC standards have been developed by the U.S. national laboratories in collaboration with laboratories in Europe. Documentation of the theory and practical application of these analysis techniques with description of the related computer methods is available from the National Renewable Energy Laboratory in Golden, Colorado, and Sandia National Laboratories in Albuquerque, New Mexico.

Measurement and Testing Standards

A second suite of standards has been developed by the IEC for the measurement and testing of wind turbines, as follows:

1) IEC 61400-1, Wind turbines – Part 1: Design requirements
2) IEC 61400-2, Wind turbines – Part 2: Design requirements for small wind turbines
3) IEC 61400-3, Wind turbines – Part 3: Design requirements for offshore wind turbines
4) IEC 61400-11 Wind turbine generator systems – Part 11: Acoustic noise measurement techniques
5) IEC 61400-12, Wind turbines – Part 12: Power performance measurements of electricity producing wind turbines
6) IEC/TS 61400-13, Wind turbine generator systems – Part 13: Measurement of mechanical loads
7) IEC/TS 61400-14, Wind turbines – Part 14: Declaration of apparent sound power and tonality values
8) IEC 64100-21, Wind turbines – Part 21: Measurement and assessment of power quality characteristics of grid connected wind turbines
9) IEC/TS 61400-23, Wind turbine generator systems – Part 23: Full scale structural testing of rotor blades

10) IEC/TR 61400-24, Wind turbine generator systems – Part 24: Lightning protection
11) IEC 61400-25-1, Wind turbines – Part 25-1: Communication for monitoring and control of wind power plants – Overall description of principles and models
12) IEC 61400-25-2 ,Wind turbines – Part 25-2: Communication for monitoring and control of wind power plants – Information models
13) IEC 61400-25-3, Wind turbines – Part 25-3: Communication for monitoring and control of wind power plants – Information exchange models
14) IEC 61400-25-4, Wind turbines – Part 25-4: Communication for monitoring and control of wind power plants – Mapping to communication profile
15) IEC 61400-25-5, Wind turbines – Part 25-5: Communication for monitoring and control of wind power plants – Conformance testing
16) IEC 614000-SER, Wind Turbine generator systems – ALL PARTS
17) IEC 61400-22, IEC System for Conformity Testing and Certification of Wind Turbines – Rules and procedures
18) ISO 61400-4, Wind turbines – Part 4: Design and specification of gearboxes

Evolution of Offshore Wind Energy

Background

The concept for offshore wind power generation can be traced back to Hermann Honnef, a German engineer in the 1930s [Dörner 2007]. In the United States, William Heronemus [1972], a naval architect and professor of Ocean Engineering at the University of Massachusetts, was the first to introduce offshore wind energy as an engineered solution for large-scale energy production. In the 1970s, before the latest wind renaissance, Heronemus wrote about his vision for large-scale power production from deep water, multi-rotor, floating, offshore wind turbines that produced hydrogen instead of electricity in a series of papers.

The first European studies, published in the late 1970s, laid the groundwork for the first round of offshore projects [Elkintin *et al.* 2008]. These studies focused on assessing the offshore resource and investigating the feasibility of large offshore wind plants. No turbines were actually built until 1990, when a single demonstration wind turbine was erected in Nogersund, Sweden. Shortly thereafter, the Vindeby wind plant was erected in Denmark in 1991.

Current Status of Offshore Wind Power Stations

Offshore wind energy got a big boost in 2001 when Demark installed two large 160-MW offshore wind power plants: Horns Rev and Nysted. According to the European Wind Energy Association (EWEA), by the end of 2007, the worldwide offshore installed capacity was 1,079 MW with 21 different projects in five countries, as illustrated in Figure 3-48 [EWEA 2008]. Table 3-10 lists the offshore wind power plants installed world-wide by 2008. Recently, EWEA announced targets to expand offshore wind in the European Union to 40,000 MW installed by 2020 and as much as 150,000 MW by 2030 [Elkintin *et al.* 2008]. In the United States, at the end of 2007 there were several project proposals for offshore wind power stations in both state and federal waters totaling to more than 1,700 MW, but no projects have been installed as yet.

Offshore wind power is both a globally dispersed and abundant resource that has the potential to make a large impact in meeting future energy needs. In the United States, 28 of the states in the contiguous lower 48 states have a coastal boundary, and these states use 78 percent of the nation's electricity. Of these 28 states, only six have a substantial land-based wind energy resource. However, 26 of the 28 states have enough wind resources to meet 20 percent to 100 percent of their electricity needs if *shallow water offshore potential* (waters less than 30 m in depth) is added into their wind resource mix. If deeper water wind resources are included, the offshore wind power potential often overwhelms other local options [Musial and Butterfield 2004, Musial 2007].

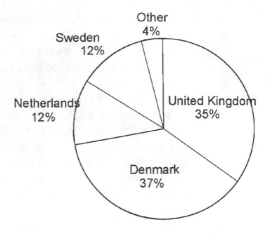

For many coastal states, offshore wind resources are the only renewable energy source capable of making a significant en-ergy contribution. In many congested and

Figure 3-48. Distribution of offshore wind power generation capacity in Europe at the end of 2007. [EWEA 2008]

energy- constrained regions, offshore wind-generated energy may be necessary to supple-ment dwindling fossil fuel supplies. NREL analysts estimate that the potential U.S. offshore wind power resource (excluding Hawaii and Alaska) is greater than 2,500 GW, assuming (1) all of the windy regions (*Class 5* or above) from the shoreline out to 50 nautical miles (nm) are included, and (2) turbines can be installed in deep waters, up to 900 m deep on floating platforms. This estimate assumes that one 5 MW wind turbine is placed on every square kilometer of water with Class 5 winds. If Class 4 wind regions are included, the estimated wind resource for the same assumptions grows to 3,570 GW of capacity.

Baseline Technology Description of Offshore Wind Power Systems

Wind Turbine Configurations

Today's typical shallow-water offshore wind turbine is essentially an enhanced version of the standard land-based turbine shown in Figure 3-40, with some system modifications to account for ocean conditions. These modifications include the following:

- Structural enhancements to the tower to address the added loading from waves
- Pressurized nacelles and environmental controls to prevent corrosive sea air from degrading critical drivetrain and electrical components
- Personnel access platforms to facilitate maintenance and provide emergency shelter
- Corrosion protection systems at the sea interface and high-grade marine coat-ings on most exterior components
- Warning lights, vivid markers on tower bases, and fog signals for marine navi-gational safety.

To minimize expensive servicing, offshore turbines may be equipped with the following systems that exceed the standard for land-based designs:

- Enhanced condition monitoring systems
- Automatic bearing lubrication systems
- On-board service cranes
- Oil temperature regulation systems

Table 3-10.
Worldwide Offshore Wind Power Stations in 2008:
Operating, Under Construction, and Planned. [Wikipedia 2008]

(a) Operating offshore stations

Name	Installed capacity (MW)	Country	Turbine Model	Commissioned
Vindeby	5	Denmark	Siemens 450 x 11	1991
Lely	2	Denmark	NEG Micon x 4	1994
Tuno Knob	5	Denmark	Vestas 500kW x 10	1995
Dronten Isselmeer	16.8	Netherlands	NEG Micon 600kW x 28	1996
Bockstigen	2.75	Sweden	NEG Micon 550kW x 5	1997
Utgrunden	10.5	Sweden	GE 1.5 x 5	2000
Blyth	4	UK	Vestas V66-2MW x 2	2000
Middelgrunden	40	Denmark	Bonus 2MW x 20	2001
Yettre Stengrund	10	Sweden	NEG Micon 2MW x 5	2001
Horns Rev	160	Denmark	Vestas V80-2MW x 80	2002
Paludans Flak	23	Denmark	Siemens 2.3 x 10	2003
Nysted Wind Farm	166	Denmark	Siemens 3.6 x 72	2003
Arklow Bank	25	Ireland	7 x 3.6 GE	2003
North Hoyle	60	UK	Vestas V80-2MW x 30	2003
Scroby Sands	60	UK	Vestas V80-2MW x 30	2004
Kentish Flats	90	UK	Vestas V90-3MW x 90	2005
Barrow Offshore Wind	90	UK	Vestas V90-3MW x 30	2006
Egmond aan Zee	108	Netherlands	Vestas V90-3MW x 36	2006
Burbo Bank Offshore Wind Farm	90	UK	Siemens 3.6-107 x 25	2007
Lillgrund Wind Farm	110	Sweden	Siemens 2.3 x 48	2007
Beatrice	10	UK	RePower 5M x 2	2007
Princess Amalia Wind Farm	120	Netherlands	Vestas V80-2MW x 60	2008

(b) Offshore stations under construction

Name	Planned capacity (MW)	Country	Turbine Model	Completion expected
Rhyl Flats	90	UK	Siemens 3.6-107 x 25	2009
Lynn and Inner Dowsing	194	UK	Siemens 3.6-107 x 54	2009
Robin Rigg	180	UK	Vestas V90-3MW x 60	2009
Thornton Bank	300	Belgium	RePower 5M	2009
Gunfleet Sands	108	UK	Siemens 3.6-107 x 30	TBD

(c) Planned offshore station

Delaware Offshore Wind Farm, Delaware Bay, Atlantic Ocean; 450 MW; Vestas V90-3.0 MWx170; Approved 2008; Construction Start 2009; Online 2011, Completion 2015

Lightning protection is mandatory for both land-based and offshore systems. The major portion of the turbine nacelle covers and towers are painted light blue or gray to minimize their visual impact, especially at long distances.

The rated power of a typical offshore wind turbine ranges from 2 MW to 5 MW, and its three-bladed horizontal-axis upwind rotors is typically 80 m to 130 m in diameter. Offshore machines are generally larger than land-based wind turbines because there are fewer constraints on transportation of components and erection equipment, which limit the size of land-based machines. Blade tip speeds of offshore wind turbines (80 m/s or greater) are also typically higher than those of land-based turbines because aerodynamic noise is less of an issue.

The basic drivetrain topology of an offshore wind turbine differs very little from that illustrated in Figure 3-41 for land-based systems. Typical offshore drivetrains consist of a modular, three-stage, hybrid planetary-helical gearbox that steps up the rotor speed to generator speeds of 1,000 to 1,800 revolutions per minute and generally runs with variable-speed torque control. However, direct-drive generators may prove to be a viable alternative. Offshore towers are shorter than land-based towers because wind shear profiles are more gradual, diminishing the gains in energy capture with increased height (see Figure 2-22).

Offshore Foundations

Offshore foundation and substructure systems differ most substantially from those of land-based turbines. The most common substructure is the *monopole*, which is a large steel tube ranging in length from 35 to 50 m, with a diameter up to 6 m and a wall thickness up to 60 mm. A monopole substructure extends from the foundation to above the water surface where a transition piece is attached, which contains a flange to fasten the monopole to the tower.

Foundation embedment depths vary with the soil type of the sea bottom, but typical North Sea installations require pile embedments of 25 to 30 m below the mud line. This type of monopile foundation requires special installation equipment for driving piles into the seabed and lifting the monopole into place. In several projects, gravity-based foundations (large concrete slabs) have been deployed as a viable alternative, which avoids the need for large pile driving equipment. Gravity-based systems require a significant amount of bottom preparation prior to installation and are only compatible with firm soil substrates in relatively shallow waters. Newer *multi-pile foundations* (*i.e.* multiple-legged substructures) may provide yet another alternative to the conventional monopiles.

The costs of infrastructure installation and logistical support are significant portions of the total cost for a large offshore wind power station. As with land-based stations, offshore wind turbines are arranged in arrays that take advantage of the measured prevailing wind conditions at the site, and turbine spacing is chosen to minimize aggregate power plant power losses, interior plant turbulence, and the cost of cabling between turbines [Elkinton *et al.* 2008].

Connections to the Electrical Grid

In a typical offshore plant, the output from each turbine at a voltage of 480 V to 690 V is stepped up with individual turbine transformers (which can be dry air cooled) to a distribution voltage within the station of about 34 kilovolts (kV). An *electrical service platform* provides a common electrical interconnection in the distribution system and serves as a substation where the collection cables are combined and brought into common phase. For smaller arrays of wind turbines that are closer to shore, this function can be done entirely onshore. For larger projects, the voltage would be stepped up at the offshore substation to about 138 kV for transmission to a land-based substation connected to the onshore grid.

The offshore electrical service platform also provides a central service facility for the wind power station and may include the following facilities:

- helicopter landing pad
- wind plant control room, staff and service facilities, and temporary living quarters
- supervisory control and data acquisition monitoring systems
- crane, rescue boat, and firefighting equipment
- communication station
- emergency diesel backup generators

Power is transmitted from the electrical service platform to shore through a number of buried high-voltage sub-sea cables, where a shore-based interconnection point sends the power to the grid. The voltage may need to be increased again onshore to (nominally) 345 kV for offshore power plants larger than 500 MW [Green *et al.* 2007].

Cost Trade-Offs

Currently, the *installed capital cost* (ICC) of an offshore wind power project is significantly higher than that of one with land-based wind turbines. Factors that contribute to this additional cost are the costs of *marinizing* turbine components for service at sea, additional costs for higher reliability components, and higher balance of station costs. In addition, operation and maintenance (O&M) costs are two to three times higher than for land-based systems. This is partly because accessibility to turbines is more difficult, external conditions are more severe and more difficult to measure and characterize, and extreme wind and wave loads add substantial uncertainty to the design methods used on the turbines and substructures.

Higher offshore costs can be offset by higher offshore wind speeds, higher pricing for electric energy near offshore wind plants, proximity to existing transmission grids, and avoided costs of building new transmission lines into heavily congested areas. Because offshore wind technology is less mature than land-based wind energy, it has a greater potential for cost reduction through technology development and innovation, and through learning-curve reductions that take advantage of more efficient manufacturing and deployment processes.

Technology Development Pathways

The first generation of offshore wind turbines was deployed using extrapolations of proven land-based technology. However, a departure from land-based systems seems inevitable as the technology evolves. Offshore systems must be more reliable, accessible from marine vessels, designed to withstand a load regime consisting of dual dominate wind/wave spectra, and larger in rated power and size to take advantage of the available offshore infrastructure. In addition, as water depths increase with increasing distances from shore, floating wind turbine foundations may need to be developed. These foundations will demand far-reaching wind turbine innovations that incorporate lighter components and the ability to tolerate increased tower deflections and nacelle motions.

Figure 3-49 illustrates the pathway for offshore wind energy technology development, beginning on land and progressing into shallow water (less than 30 m deep), where most of the projects are located as of 2008. The one exception to this is a single demonstration project deployed by *Talisman Energy* in 2006 in 45 m water depth near Aberdeen, Scotland. As the industry gains experience through deployment of wind turbines in the shallower and more sheltered sites, more projects will be sited farther from shore in deeper and deeper waters to allow full exploitation of the wind resource. This progression to deeper water, possibly as deep as 900 m, will require increasing levels of technical complexity and risk. The potential

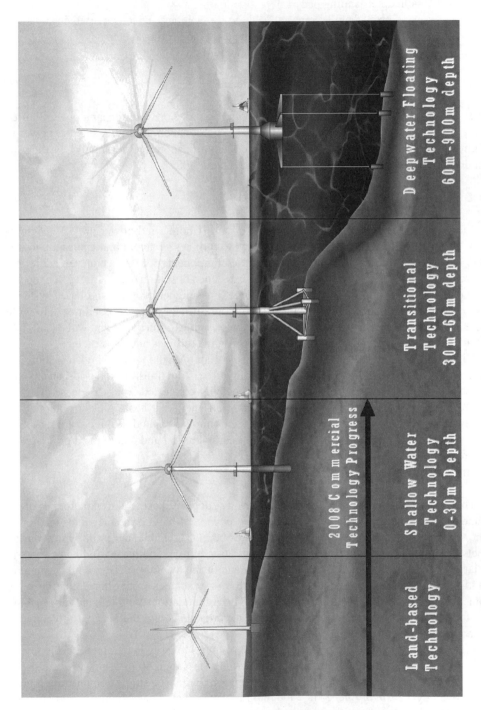

Figure 3-49. Offshore technology development pathway, from land-based to deep-water floating wind turbines.

benefits will be access to a far greater number of potential sites with fewer conflicts with human activity and wildlife.

First, the industry must gain experience deploying wind turbines in water depths up to 30 m. Basic technical issues must be understood before moving into deeper waters, issues such as the following:

- structural loading from wind and wave combinations
- environmental impacts
- wind turbine operation and maintenance at sea
- regulatory issues

Logically, the next step is to deploy systems at transitional depths (30 m to 60 m) to expand the number of available sites. Transitional technology deals mostly with substructures that can be adapted from existing offshore oil and gas practices. Finally, technology for deep water deployment (60 m to 900 m) will be developed. This technology may require floating systems [Musial *et al.* 2004], which will necessitate a more advanced course of research and development to optimize turbines that are lightweight and can survive additional tower-top motion induced by buoyant platforms. No floating offshore wind turbines have yet been deployed, but several companies are coming close to demonstrating the technology at sea.

Technologically, the design of deep-water foundations and substructures will be the most challenging step, but a step that also holds the most promise for the long term. The development of deep-water technology would open up major areas of the outer continental shelf to wind energy development where the turbines would not be visible from shore and where competition with other human activities would be minimal. Deep-water platforms may in fact be easier to mass produce with greater possibility for onshore assembly and full system transport, introducing a major new opportunity for cost reduction [Butterfield *et al.* 2005].

The pathway to deep water deployment should not be considered as a mutually-exclusive alternative to development of shallow water offshore wind systems. There is a high degree of interdependence among the technology requirements for shallow-water, transitional-water, and deep-water wind systems. To understand the technical and commercial complexities of deep-water floating wind turbines, it will be necessary to approach their design from experience and knowledge about shallow-water offshore wind power stations.

Research Needs

The advancement of offshore wind power stations is hindered by many technical, regulatory, socioeconomic, and political barriers that can only be mitigated through targeted R&D efforts. Short-term research can address impediments that prevent the early commercial projects from proceeding. This research is required to build the needed experience for long-term optimization and commercial deployment of advanced systems. Short-term research topics include the development of standards for offshore deployment, remote sensing tools for monitoring the offshore wind, characterization of the offshore ocean environment, and the development of advanced design tools that will allow accurate modeling of new concepts.

Long-term research will involve the design and testing of new offshore systems developed for the specific requirements of the marine environment. The design of larger turbines, from 5 MW to 10 MW in size, for offshore deployment will require enabling technologies that overcome current scaling limits in blade length, tower height, and rotor/nacelle weight. New machines will need to incorporate higher reliability components that must evolve through rigorous analysis and testing. Holistic turbine designs will integrate the logistics of installation, operations, and decommissioning with the turbine design, resulting in lower life-cycle costs. Lightweight materials and turbine weight reduction strategies must be developed to optimize offshore systems, especially floating designs.

Summary of Wind Turbine Technology Developments
from 1988 to 2008

Energy production from wind power has evolved very rapidly over the past two decades. Capital costs have plummeted, reliability has improved, and efficiency has dramatically increased, resulting in a robust commercial product that is competitive with fossil-fuel power generation systems. Investments in research and development projects as well as the development of robust standard design criteria have helped to mitigate risk and attract market capital for large wind power stations. High-quality products built to international standards are now provided by every major turbine manufacturer. Complete wind power stations are now being engineered to seamlessly interconnect with the electrical grid.

The importance of the role played by the IEC standards in the rapid development and deployment of wind turbine technology cannot be overestimated. The IEC design standards allow buyers of wind turbines to match the turbine design class to their specific site wind environment. Testing standards provide protocols for performance, loads, noise, and acoustic output to ensure that test results are comparable under a set of standard conditions. The IEC 61400-22 wind turbine certification standard provides the buyer with a full system certification process.

The COE metric remains the principal technology indicator, incorporating the key elements of capital cost, efficiency, reliability, and durability. The unsubsidized cost of wind-generated electricity ranges from about 5.5 to 8.8 cents/kWh for projects completed in 2006.

The technology of wind turbines is now highly developed. Major technical breakthroughs in land-based technology are no longer needed for a broad geographic penetration of wind power on the electric grid. Instead, incremental technology advances at research laboratories such as the U.S. National Wind Technology Center shown in Figure 3-50 are

Figure 3-50. Aerial view of the National Wind Technology Center near Golden, Colorado, where a wide variety of wind turbine technology research is conducted.

required, in parallel with a systems development and integration approach on the part of wind power station developers. The NWTC is part of the National Renewable Energy Laboratory, sponsored by the U.S. Department of Energy. Technical breakthroughs are not needed now because of the high level of engineering already incorporated into modern machines.

Capacity factors can be increased over time using larger rotors on taller towers. Market incentives will remain necessary to sustain the industry growth in the near term, but in the longer term, subsidies can probably be eliminated. The development of cost-effective offshore wind energy systems is the next big step in the evolution of wind turbines. The offshore wind energy resource is very large and in close proximity to the load centers where energy is needed. The technology challenges that the industry faces to harvest this wind resource are no greater that those already achieved for land-based wind power during the past two decades. No new technological breakthroughs are needed to develop wind turbines for these offshore stations. However, developing the innovations necessary to adapt current wind turbine technology for long-term operation in a marine environment and reaching deeper, more remote sites will take the coordinated efforts of government agencies, researchers, project developers, and investors.

References

AWEA, 2008, "Another Record Year for New Wind Installations," *AWEA's Wind Energy Fact Sheets*, Washington, DC: American Wind Energy Association.

Berg, D.E., and J. R. Zayas, 2008, "Aerodynamic and Aeroacoustic Properties of Flatback Airfoil," *Proceedings, 46th AIAA Aerospace Sciences Meeting and Exhibit, 27th ASME Wind Energy Symposium*, Albuquerque, New Mexico: Sandia National Laboratories. Available at www.sandia.gov/wind/.

BTM Consult ApS, 2005, *Ten Year Review of the International Wind Power Industry 1995 – 2004*, Ringkobing, Denmark.

BTM Consult ApS, 2008, *BTM World Market Update 2007: Forecast 2007 - 2012*, Ringkobing, Denmark.

Butterfield, C.P; W. D. Musial, J. Jonkman, P. Sclavounos, and L. Wayman, 2005, *Engineering Challenges for Floating Offshore Wind Turbines*, Proceedings, Copenhagen Offshore Conference 2005, Copenhagen, Denmark.

BWEA, 2000, *Prospects for Offshore Wind Energy*, London, England: British Wind Energy Association. Available at www.bwea.org.

Bywaters, G., V. John, J. Lynch, P. Mattila, G. Norton, J. Stowell, M. Salata, O. Labath, A. Chertok, and D. Hablanian, 2004, *Northern Power Systems WindPACT Drivetrain Alternative Design Study Report; April 12, 2001 to January 31, 2005*, NREL/SR-500-35524, Golden, Colorado: National Renewable Energy Laboratory. Available at www.nrel.gov/publications/.

Dörner, H.H., 2007, *Wind Energy History*, Stuttgart, Germany: University of Stuttgart.

Elkinton, C., J. Manwell, and J. McGowan, 2008, "Optimizing the Layout of Offshore Wind Energy Systems," *Marine Technology Society Journal, Fall Issue,* Columbia, Maryland: The Marine Technology Society.

European Wind Energy Technology Platform for Wind Energy, 2008, *Strategic Research Agenda – Market Deployment Strategy from 2008 to 2030, Synopsis – Preliminary Discussion Document*, Brussels, Belgium: TPWind Secretariat.

EWEA, 2008, *Wind Power Installed in Europe by 2007, On Shore and Offshore* European Wind Energy Association, Brussels, Belgium: European Wind Energy Association.

Global Energy Concepts, 2001, *WindPACT Turbine Design Scaling Studies Technical Area 3.Self-Erecting Tower and Nacelle Feasibility*, NREL/SR-500-29493, Golden, Colorado: National Renewable Energy Laboratory.

Global Energy Concepts, 2002, *Addendum to WindPACT Turbine Design Scaling Studies Technical Area 3.Self-Erecting Tower and Nacelle Feasibility*, NREL/SR-500-29493a, Golden, Colorado: National Renewable Energy Laboratory.

Green, J., A. Bowen, L. J. Fingersh, and Y. Wan, 2007, "Electrical Collection and Transmission Systems for Offshore Wind Power," *Proceedings, 2007 Offshore Technology Conference*, OTC-19090-PP, Houston, Texas.

Griffin, D. A., 2001, *WindPACT Turbine Design Scaling Studies Technical Area 1 -- Composite Blades for 80- to 120-Meter Rotor; 21 March 2000 - 15 March 2001*, NREL/SR-500-29492, Golden, Colorado: National Renewable Energy Laboratory. Available at www.nrel.gov/publications.

Heronemus, W.H., 1972, "Pollution-Free Energy from Offshore Winds," *Proc. 8th Annual Conference and Exposition*, Washington, D.C.: Marine Technology Society.

International Electrotechnical Commission (IEC). Various Electrical Standards. Descriptive information and standards available at http://www.iec.ch/.

LaNier, M. W., 2005, *LWST Phase I Project Conceptual Design Study: Evaluation of Design and Construction Approaches for Economical Hybrid Steel/Concrete Wind Turbine Towers, June 28, 2002–July 31, 2004*, NREL/SR-500-36777, Golden, Colorado: National Renewable Energy Laboratory. Available at www.nrel.gov/publications/.

Lobitz, D., P. Veers, G. Eisler, D. Laino, P. Migliore, and G. Bir, 2001, *The Use of Twist-Coupled Blades to Enhance Performance of Horizontal Axis Wind Turbine*, SAND 2001-1303, Albuquerque, New Mexico: Sandia National Laboratories. Available at www.sandia.gov/wind/.

Musial, W.D. C. P. Butterfield, and A. Boone, 2004, *Feasibility of Floating Platform Systems for Wind Turbines*, NREL/CP-500-34874, Golden, Colorado: National Renewable Energy Laboratory.

Musial, W.D., and C.P. Butterfield, 2004, *Future for Offshore Wind Energy in the United States*, NREL/CP-500-36313, Golden, Colorado: National Renewable Energy Laboratory.

Musial, W., 2007, "Offshore Wind Electricity: A Viable Energy Option for the Coastal United States," *Marine Technology Society Journal*. Vol. 41(3), Fall Issue, pp. 32-43, NREL JA-500-41338, Golden, Colorado: National Renewable Energy Laboratory.

Paquette, J., J. van Dam, and S. Hughes, 2007, "Structural Testing of 9 m Carbon Fiber Wind Turbine Research Blades," *Proceedings AIAA 2007 Wind Energy Symposium*, NREL/CP-500-40985, Golden, Colorado: National Renewable Energy Laboratory. Available at www.nrel.gov/publications/.

Poore, R., and T. Lettenmaier, 2003, *Alternative Design Study Report: WindPACT Advanced Wind Turbine Drivetrain Designs Study; November 1, 2000–February 28, 2002*, NREL/SR-500-33196, Golden, Colorado: National Renewable Energy Laboratory. Available at www.nrel.gov/publications/.

NREL National Wind Technology Center Website, 2008, *Description of Dynamic Computer Models for Wind Turbine Systems and Components*, Golden, Colorado. Reports available at http://www.nrel.gov/wind/pubs_research.html#computer.

Sandia National Laboratories Wind Energy Technology Program, 2008, *Reports and Publication on the Analysis of Wind Technology Systems and Components*, Albuquerque, New Mexico. Publications available at http://www.sandia.gov/wind/topical.htm.

Short, W., N. Blair, D. Heimiller, and V. Singh, 2003, *Modeling the Long-Term Market Penetration of Wind in the United States*, NREL/CP-620-34469, Golden, Colorado: National Renewable Energy Laboratory. Available at www.nrel.gov/publications/.

Sterzinger, G., and M. Svrcek, 2004, *Wind Turbine Development: Location of Manufacturing Activity*, Renewable Energy Policy Project Technical Report, Washington, DC: U.S. Department of Energy.

Tangler, J. L., 2001, *Evolution of Rotor Blade Design*, NREL/CP-500-28410, Golden, Colorado: National Renewable Energy Laboratory. Available at www.nrel.gov/publications/.

U.S. Department of Energy, 2008, *20 percent Wind Energy by 2030: Increasing Wind Energy's Contribution to the U.S. Electricity Supply*, DOE/GO-102008-2567, Golden, Colorado: National Renewable Energy Laboratory. Available at www.nrel.gov/publications/.

WindPACT Project, 1999, *Wind Partnerships for Advanced Component Technology, Various Projects on Advanced Wind Technology*, various NREL reports, Golden, Colorado: National Renewable Energy Laboratory. Descriptive information and study reports available at http://www.nrel.gov/wind/advanced_technology.html.

Wiser, R. and M. Bolinger, 2008, *Annual Report on U.S. Wind Power Installation, Cost, and Performance Trends: 2007*, Washington, DC: U.S. Department Office of Scientific and Technical Information. Available at www.osti.gov/bridge.

Wright, A.D., and L. J. Fingersh, 2008, *Advanced Control Design for Wind Turbines: Part I, Control Design, Implementation, and Initial Tests'* NREL /TP-500-42437, Golden, Colorado: National Renewable Energy Laboratory. Available at www.nrel.gov/publications/.

Zervos, A., and C. Kjaer, 2008, *Pure Power: Wind Energy Scenarios Up to 2030.*

4

Commercial Wind Turbine Systems and Applications[1]

Larry A. Viterna, Ph.D.
National Aeronautics and Space Administration Cleveland, Ohio

and

Daniel F. Ancona III
Vice-President for Renewable Energy
Princeton Energy Resources International
Rockville, Maryland

Introduction

This chapter describes commercial wind turbine systems in the U.S. and other countries, including their application for generating electrical energy as well water pumping. Commercial wind systems, as defined here, include production turbines as well as pre-production turbines that could be manufactured commercially within the next five years. Although the history of wind system development was discussed in the previous chapters, some historical details will be introduced here as well, when they help describe the development of commercial markets and products. This discussion of commercial systems and applications considers

---

[1] Update of Chapter 4, First Edition, by Robert Lynette, *R. Lynette and Associates, Inc., Redmond, Washington;* and Paul Gipe, *Paul Gipe and Associates, Tehachapi, California.* [Lynette and Gipe 1994]

large-, medium-, and *small-scale* horizontal-axis wind turbines (HAWTs), and also reviews vertical-axis wind turbines (VAWTs).

Wind power applications described include *wind power stations* delivering electricity on utility grids, *distributed (dispersed) turbines* on utility grids, turbines on *isolated and/or small electrical grid systems*, and *stand-alone units* for mechanical and electrical power. Trends in cost of energy (COE) from wind systems in various applications are also discussed. *Operation and maintenance* (O&M) requirements for current systems are included that reflect the commercial experience, including typical maintenance scenarios, associated costs, and trends within the industry.

The development of current wind systems did not occur in a vacuum. This chapter concludes with material that describes the *social, business, and regulatory environment* that led to the development of the current wind power industry and provides the past, current, and future *cost goals* for commercial wind turbines. These nontechnical factors strongly affect the present wind power industry and the development and deployment of future wind turbines and applications.

Wind turbines today have an established reputation as practical and reliable systems with widespread use throughout the world. Examples of wind turbines in commercial service are pictured in Appendix E. The vast majority of wind turbines have been installed on land. In the past few years some wind turbine systems have been installed in waters offshore to capture higher wind speeds and to locate near energy needs such as population centers.

For remote, unattended locations, *battery chargers* employing highly reliable rotor control methods are selected. If power is required to supplement the needs of homes or small businesses, *small alternating current (AC) units* with outputs up to 10 kW are the machines of choice.

For farms and agricultural applications where grid interruptions are common, multi-unit applications with turbines up to abut 25 kW, can provide standby AC power provided there are batteries and a *direct current (DC)/AC inverter* are available. An engine-driven generator can also be used to set line voltage and frequency. Operating in this mode is mandatory to disconnect from the grid for safety reasons and to prevent frequency synchronization issues when power is restored. In Europe it is common for farmers, individually or in small groups, to purchase several *medium-scale* systems ranging from 100 up to 1,000 kW that are installed in clusters that are producing farmstead power but are also sending electricity to the grid.

Clusters of small- to medium-scale turbines are attractive to be used as fuel-savers in *mini-grids* in isolated communities. In these applications isolated communities have *continuous duty* diesel engine-driven generators providing power day and night serving electrical loads of 10 to 50 MW. But fuel cost and delivery can make wind plants desirable for such applications.

Finally, the largest market is for large-scale multi-unit wind power stations containing *multi-megawatt* machines with rotor diameters up to 100 m and larger, providing bulk electric power to utility grids. Wind power stations are mainly deployed in North America and Europe, but applications in developing countries will become critically important as their appetite for electricity increases and indigenous wind resources are available.

The HAWT configuration continues to dominate wind power production, as it has for most of the modern era. Since the interconnection of wind turbines to utilities became their principal application during the 1980s and continues to be so today, the average size of wind turbines has grown. In the United States, many small machine designs have simply disappeared as manufacturers have scaled them up into the medium-size range. However, several small turbines are being mass produced with more than 9,000 units sold in 2007 alone, adding 9.7 MW of new capacity and bringing the total small-turbine capacity to more than 55 MW.

Scale Classifications

For purposes of this chapter, wind turbines are classified as shown in Table 4-1, according to their diameters and/or their rated powers. This table follows the terminology used in the

Table 4-1. Scale Classifications of Wind Turbines

Scale	Rotor Diameter	Power Rating
Small	Less than 12 m	Less than 40 kW
Medium	12 to 45 m	40 to 999 kW
Large	46 m and larger	1.0 MW and larger

early U.S. Federal Wind Energy Program (described in Chapter 3), but with some variation in numbers.

Significant expansion of the wind turbine market and engineering experience has occurred since the 1980s, but the early wind turbines of this period were instrumental in providing data to refine the technology required for the following generations of HAWTs. Over the next 15 years the commercial wind energy industry began scaling up small-scale machines to the 100- to 200-kW power range and more government programs began emphasizing long-term research applicable to all machine sizes. By the early 1990s the commercial sector produced machines with ratings of 300 kW and more. The average capacity rating of wind turbines installed in California wind power stations almost quadrupled from 49 kW in 1981 to 191 kW in 1992. Over the same period, the maximum rating increased from 55 to 600 kW. The trend to produce larger machines is a result of at least four factors:

- Wind turbine costs per rated kilowatt generally decrease as machine size increases.
- *Balance-of-station costs* per kilowatt (the price of the land and infrastructure) decrease as the turbine size increases (*e.g.* less wiring and land are required).
- Many good wind sites comprise ridges that enable the production of more energy per unit of land using larger wind turbines, because ridges frequently have only enough space for a single line of wind turbines.
- A segment of the O&M costs increases with the number of turbines, and therefore decreases with increasing turbine size for a given wind power station rating.

Although wind turbines continue to be produced in each of the small, medium, and large size categories, the distribution of turbine has shifted over time to the large sizes, as listed in Table 4-2. It should be noted, however, that the total number of small machines sold is very large and increasing.

Table 4-2.

Size Distribution of Turbines Deployed in the U.S from 1998 to 2006 [AWEA 2006]

Rated power range	1998-99 1,013 MW 1,418 turbines	2000-01 1,758 MW 1,987 turbines	2002-03 2,125 MW 1,784 turbines	2004-05 2,782 MW 1,937 turbines	2006 2,454 MW 1,523 turbines
0.00 to 0.5 MW	1.3%	0.4%	0.5%	1.9%	0.7%
0.51 to 1.0 MW	98.4%	73.9%	44.2%	17.6%	10.7%
1.01 to 1.5 MW	0.0%	25.4%	42.8%	56.6%	54.2%
1.51 to 2.0 MW	0.3%	0.4%	12.3%	23.9%	17.6%
2.01 to 2.5 MW	0.0%	0.0%	0.0%	0.1%	16.3%
Over 3.0 MW	0.0%	0.0%	0.1%	0.0%	0.5%

Large-Scale Horizontal-Axis Wind Turbines

As the dominant size turbine category today, large-scale wind turbines offer advantages that include

-   the ability to extract more wind energy per unit of land area
-   improved aerodynamic performance, because of higher *Reynolds numbers* associated with larger blade chord dimensions
-   lower sensitivity of larger blades to dirt, rain, and insects, because of larger blade thickness dimensions
-   potential economies of scale for maintenance and some components, such as control system cost per unit of installed power

While all HAWTs share many similar design considerations, some of these may be more critical for multi-megawatt machines and must be carefully addressed in the design process. These include

-   handling, storing, and transporting components in the shop and to the site
-   availability of cranes and limitations on lift weights and heights
-   dependence on aerodynamic control of speed and power
-   provisions for personnel access and safety, including access inside the rotor
-   good exterior condition with minimum maintenance
-   site selection to avoid electromagnetic interference
-   structural design for weather extremes and long-term (30- year) life
-   grid compatibility and power quality
-   quality control to high standards for steel weldments, castings, and forgings

The configuration of the *General Electric* 1.5-MW wind turbine shown in Figure 4-1 is typical of the modern commercial wind turbines. Table 4-3 contains a summary of characteristics of representative commercial large-scale wind turbines. The majority of current large-scale HAWTs have a rotor with three blades attached to a rigid hub, with integral full-span blade pitch control. Other typical features include a rotor that is located upwind of the support tower. These design features, characterized by the early Danish HAWT designs, have demonstrated good reliability and performance over time. *Soft* tubular steel towers (*i.e.*, with lowest tower bending natural frequency less than the number of blades per rotor revolution) and variable speed drive trains follow the design approach of early U.S. prototype wind turbines.

Two-bladed wind turbines are also commercially available and promote the cost advantage of elimination of one blade and lower weight throughout the system for a given energy output. *Teetered hubs* (rotor hubs containing a hinge permitting upwind/downwind motion of the blades) are used on two-bladed rotors to reduce or eliminate cyclic loads into the drive train and tower. The Nordic Windpower turbine shown in Figure 4-2 is an example of this configuration.

Figure 4-1. Typical configuration of current commercial wind turbines, with three-bladed rotors upwind of tubular towers. (*Courtesy of NREL*)

Trends in Large-Scale HAWT Designs

With the market maturing for land-based turbines, overall system designs are tending to stabilize. This is due in part to the significant investment by the customer and required confidence that the turbines will perform reliably throughout their operating lifetimes. However, economic incentives to reduce operating costs continue to drive designs toward even larger rotor diameters. Increased diameters in turn affect subsystems and components resulting in the introduction of new materials to meet reliability requirements. Limits on transport and assembly of very large components and subsystems can also drive the development of designs such as segmented blades, new high-torque drives, and poured concrete towers.

Table 4-3. Configurations of Representative Large-Scale Commercial HAWTs

Manufacturer	Vestas	GE	Gamesa	Enercon	Suzlon	Siemens	Clipper	Nordic	DeWind
Model	V90-3.0	1.5 sle	G87-2.0	E-126	S88-2.1	SWT-3.6-107	C96	N1000	D8.2
Rotor diam.	90 m	77 m	87 m	126 m	88 m	107 m	96 m	59 m	80 m
Rated power	3.0 MW	1.5 MW	2.0 MW	7.0 MW	2.1 MW	3.6 MW	2.5 MW	1.0 MW	2.0 MW
Rotor location	Upwind	Upwind	Upwind	Upwind	Upwind	Upwind	Upwind	Upwind	Upwind
No. of blades	3	3	3	3	3	3	3	2 - teetered	3
Blade material	Glass/carbon	Fiber glass	Glass/carbon	Fiber glass	Fiber glass	Fiber glass	Fiber glass	Glass/Carbon	Glass/Carbon
Power control	Full blade pitch	Full blade pitch	Full blade pitch	Full blade pitch	Full blade pitch	Full blade pitch	Full blade pitch	Passive stall (fixed pitch)	Full blade pitch
Braking	Mechanical disk	Hydraulic parking	Mechanical disk	Mechanical disk	Hydraulic disk	Mechanical	Hydraulic disk	Hydraulic disk	Hydraulic disk
Overspeed control	Full blade pitch	Full blade pitch	Full blade pitch	Full blade pitch	Full blade pitch	Full blade pitch	Full blade pitch	Blade tip pitching	Full blade pitch
Gearbox (no. of stages)	Planetary, helical (3)	Planetary (3)	Planetary, parallel (3)	None (0)	Planetary, helical (3)	Planetary, helical (3)	Planetary (2) multiple drives	Planetary, helical (3)	Planetary, variable hydrodynamic
Generator	Asynchronous	Asynchronous	Asynchronous	Direct drive synchronous	Asynchronous	Asynchronous	4 Permanent magnet, synchronous	Asynchronous induction	Synchronous
Voltage	1,000	690	690	N/A	690	690	690	690	N/A
Yaw system	Active	Active, electric	Active, hydraulic	Active	Active, electric	Active, electric	Active, electric	Passive, hydraulic	Active, electric
Tower type	Tubular steel	Tubular steel	Tubular steel	Pre-cast concrete	Tubular steel	Tubular steel	Tubular steel	Tubular steel	Tubular steel

N/A Not Available

Figure 4-2. Example of a two-bladed 1-MW wind turbine, with its teetered rotor up-wind of a tubular tower. (*Courtesy of Nordic Windpower*)

Medium-Scale Horizontal-Axis Wind Turbines

By the guidelines in Table 4-1, medium-scale wind turbines are those with rotor diameters between 12 and 45 m and/or power ratings between 40 and 999 kW. There are, of course, many similarities between large-scale and medium-scale machines, since the dividing line separating them is somewhat arbitrary. Also, they share a common technology base. In general, more design variations in the overall system, as well as subsystems and components, are found in medium-scale turbines. These include the number of blades, construction materials, types of power control, and drive train design. Table 4-4 contains a summary of the characteristics of representative commercial medium-scale wind turbines.

Table 4-4. Configurations of Representative Medium-Scale HAWTs

Manufacturer	EWT	Goldwind	Vergnet	Windflow
Model	Direct Wind 750	750	GEV MP 250	500
Rotor diam.	51.5 m	49 m	32 m	33.2 m
Rated power	750 kW	750 kW	275 kW	520 kW
Rotor location	Upwind	Upwind	Downwind	Upwind
No. of blades	3	3	2-teetered	2 - teetered
Blade material	Fiberglass	Fiberglass	Fiberglass	Wood/epoxy
Power control	Full blade pitch	Passive stall (fixed pitch)	Full blade pitch	Full blade pitch
Braking	Mechanical	2 Mechanical disks	Mechanical disk	Mechanical disk
Overspeed	Aerodynamic pitching	Aerodynamic tip pitching	Aerodynamic pitching	Aerodynamic pitching
Gearbox (no. of stages)	None 0	Planetary, spur 2	Planetary 2	Planetary, parallel 4
Generator	Direct drive synchronous	Asynchronous	2 speed asynchronous	Synchronous
Voltage, volts	690	690	690	415
Yawing system	Active, electric	Active, electric	Passive	Active, hydraulic
Tower type	Tubular steel	Tubular steel	Guided/tilting tubular steel	Tubular steel

The majority of medium-scale wind turbines presently feature a rotor located upwind of the support tower. Two or, more often, three blades are used with full-span blade pitch control, although stall-control, fixed-pitch systems also exist. Generator types include synchronous, asynchronous, two-speed, and direct drive. Tubular steel towers dominate and most have active yaw systems. The *EWT Direct Wind 750*, shown in Figure 4-3 does not have a gearbox but instead uses a direct-drive synchronous generator. The *Vergnet 250* turbine shown in Figure 4-4 emphasizes reliability and survivability in hurricane zones. It uses a guided tubular steel tower that can be lowered when the power station is threatened by severe weather conditions.

Trends in Design of Medium-Scale HAWTs

Compared to multi-megawatt turbines, medium-scale machines offer potential economies associated with lower rotor costs per unit of swept area as well as easier transportation and installation. The overriding advantage of medium-sized machines is lower development

Figure 4-3. Representative of the variety in the designs of medium-scale HAWTs, these EWT Direct Wind 750-kW turbines have no gearboxes. (*Courtesy of EWT, BV*)

Figure 4-4. Vergnet GEV MP-250 275-kW wind turbines in hurricane-prone locations can be lowered for safety in very high winds. (*Courtesy of Vergnet SA*)

risk and availability of many mechanical and electrical components from other industries. This advantage facilitates continued innovation through this medium scale size into the broad wind turbine market. It is expected that new types of drive train components, particularly gearboxes and generators, will be developed and integrated into medium-scaled commercial wind turbines first before being adapted for use in large-scale HAWTs. Because of the larger growth rates of the large-scale turbine market, many medium-scale designs will be scaled up into the multi-megawatt sizes.

Small-Scale Horizontal-Axis Wind Turbines

From the earliest days of electricity generation, small wind turbines have been used to drive generators. Until the late 1970s, small wind generators were designed for stand-alone *battery-charging* applications. Since then, wind turbines less than 12 m in diameter also have been designed for generating *utility-grade electricity* suitable for interconnection with the local network. In many respects, small-scale wind turbines resemble their larger counterparts that were described earlier. Similar to the medium-scale turbines, the small-scale turbines exhibit more variation in system configuration, subsystems, and components.

Table 4-5 contains a summary of the characteristics of representative commercial small-scale wind turbines. Upwind, three-bladed rotors are used on a majority of small-scale wind

Table 4-5. Configurations of Representative Small-Scale HAWTs

Manufacturer	Aerostar	Bergey	Kestrel	Southwest
Model	6 Meter	Excel	e400	Skystream
Diameter, m	6.7	6.7	4	3.7
Rated power, kW	10	7.5	3	1.9
Rotor location	Downwind	Upwind	Upwind	Downwind
No. of blades	2	3	3	3
Blade material	Fiberglass	Fiberglass	Fiberglass	Fiberglass
Power control	Passive stall (fixed pitch)	Passive stall (fixed pitch)	Full blade pitch	Passive stall (fixed pitch)
Braking	Mechanical disk	Electric brake	N/A	Electronic
Overspeed	Aerodynamic tip brake	Aerodynamic furling	Aerodynamic pitching	Electronic controlled stall
Gearbox (no. of stages)	Planetary 2	None 0	None 0	None 0
Generator	Asynchronous	Permanent magnet, synchronous	Permanent magnet, synchronous	Permanent magnet, synchronous
Voltage, volts	240 AC	24-240 DC	300 DC	120-240 inverted
Yawing system	Passive	Passive tail	Passive tail	Passive
Tower type	Guided/tilting tubular steel	Guided lattice	Tripod	Steel tubular

turbines, but two blades and downwind configurations are also quite prominent. One aspect of small HAWTs that makes them different from larger ones is their frequent reliance on tail vanes for orienting the rotor into the wind, although downwind turbines offer passive orienting without any additional devices. A variety of overspeed control methods are found, including aerodynamic tips, furling of tail vanes, blade pitching, and the use of generator load control. Direct-drive permanent-magnet generators are most commonly used. A variety of tower systems continue to be available for these small-scale turbines.

Figures 4-5 and 4-6 show representative small-scale commercial wind turbines.

Figure 4-5. Bergey Excel 10-kW wind turbine utilizes a tail vane for yaw control. (*Courtesy of the National Renewable Energy Laboratory*)

Figure 4-6. The rotor of this two-bladed Aerostar 6-Meter 10-kW wind turbine is located downwind of the tower, providing passive yaw control. (*Courtesy of Aerostar, Inc.*)

Trends in the Design of Small-Scale HAWTs

Compared to large- and medium-scale machines, small-scale wind turbines offer much lower development and manufacturing costs. However, because the market size is much smaller, the small-scale systems are often tailored to niche markets such as remote sites and severe climate applications. In addition, market channels are more diverse for the small-scale turbines in that they are sold to a wide variety of customers such as research laboratories, home and boat owners, and offshore oil rigs. It is expected that a variety of design approaches will continue with support from these niche markets. A common design driver will be the requirement for very high reliability to compensate for the fact that these small-scale systems often operate remotely with minimal maintenance and support systems and quiet operation due to proximity to residences.

Subsystems for Horizontal Axis Wind Turbines

As discussed in Chapter 2, wind turbine costs consist of major subsystems and cost elements that are not only convenient for analysis but also represent significant aspects of the commercial turbine *manufacturing supply chain*. For the purposes of this chapter, we define five major subsystems and cost elements as follows:

-- rotor blades and hub
-- drive train, nacelle, and yaw drive
-- tower and foundation
-- generator, electrical system, and controls
-- transport and assembly

Figure 4-7 illustrates a typical breakdown of the relative costs of these major subsystems and cost elements. A more detailed discussion of each of these subsystems follows.

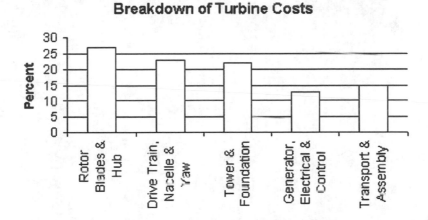

Figure 4-7. Typical breakdown of wind turbine system costs, by major subsystem.

Rotor Blades and Hub

The rotor blades and the hub of a wind turbine are the most critical elements in determining the amount and efficiency of energy capture, as well as the magnitude of static and dynamic loads transferred to the turbine. They also represent the highest cost subsystem. In addition, control of turbine power output is often through rugged and precise mechanisms in the hub for changing blade pitch angle or adjustable aerodynamic devices on the rotor blades.

Most of the major manufacturers of wind turbines also manufacture their own blades. However, *LM Glasfiber* is the single largest independent supplier to those manufacturers that purchase blades [Efiong and Crispin 2007]. The supply and cost of blades has stayed fairly stable in recent years.

Blade lengths for commercial turbines have steadily grown from approximately 20 m in 1990 to over 60 m by 2008. Although a simple analysis suggests that blade mass and costs should scale as a near-cubic of the rotor diameter, actual commercial blade designs have maintained a scaling exponent closer to 2.4 [Griffin 2001, 2002]. This has been a remarkable achievement resulting from significant advances in structural analysis (see Chapters 10, 11, and 12) as well as in manufacturing. However, without significant and yet-to-be-developed changes in materials or manufacturing processes, further cost increases with rotor diameter may follow closer to a scaling exponent of 3.0 [Williams 2008].

Blades are commonly made of composite materials, predominantly epoxy and fiberglass, and remain the most labor intense components of wind turbines. As shown in Figure 4-8, blade fabrication typically involves hand layup, open-mold wet processing. Critical parameters that must be closely controlled for strength and stiffness are fiber direction, resin content, and void content. The use of carbon fiber and other high-strength materials is increasing but remains limited to small, local applications in the structure. *Vestas* and *Gamesa* are among the manufacturers of large-scale wind turbines that are using carbon fiber composites in some blades to reduce weight.

Figure 4-8. Typical hand layup processing in the manufacture of a large-scale wind turbine blade.

Number of Blades

The great majority of today's large-scale wind turbines have three blades, an evolution of successful early commercial designs in Denmark. Two blades are becoming more common in small- and medium-scale turbines, such as the Vergnet turbine shown in Figure 4-4. Vergnet is also one of the new entries into the large-scale class with a 1-MW two-bladed machine.

One-blade rotors have been developed for wind turbines but none are commercially available. The addition of a second blade increases the rotor's aerodynamic efficiency by about 7 percent. Further increases in the number of blades yield minimal improvements in rotor aerodynamic efficiency.

Additional factors in the determination of the number of blades involve design considerations for other components such as the hub, as well as visual aesthetics as viewed by the public. Each blade is not only a weight and cost contributor itself, but blade weight increases component weight and cost throughout the rest of the system. Factors that increase noise emissions from the blades must be considered, such as the speed of the blades due to rotation and the interaction of downwind-rotors with the wake of the tower. These noise sources are characteristic of all wind turbine rotors, although two-bladed rotors are often designed to operate at higher rotational speeds than three-bladed rotors, for increased aerodynamic efficiency.

Rotor Hub

The rotor hub connecting the blades together is typically made of ductile cast steel, as illustrated in Figure 4-9. Although large casting and machining facilities are required, current procurement of hubs does not typically experience design or supply chain issues. Corresponding to the great majority of three-blade turbines, most rotor hubs rigidly connect the blades to one another and to the drive train. This rigid design affects the structural dynamic loads both within the hub as well as the loads transferred to the drive train. For example, during alignment of the wind turbine to changes in wind direction (yawing), each blade experiences a cyclic load at its root end that varies with blade position. This is true of one, two, three blades, or more. For a three-bladed rotor, these cyclic loads when combined together at the hub are nearly symmetric and balanced, thus transferring reduced cyclic loading into the drive train and the rest of the system during yawing.

Figure 4-9. Typical cast-steel hubs for large-scale wind turbines, showing bolt circles for rigid attachment of the blades. (*Courtesy of the National Renewable Energy Laboratory*)

As illustrated in Figure 4-10, turbines with two blades typically use a *teetered hub* with a pivot bearing. Teetering motion of the connected blades substantially reduces unbalanced aerodynamic and yawing-induced cyclic loads. Transfer of these cyclic loads into the drive train is essentially eliminated. The pivot system is most often an *elastomeric bearing* with teeter stops to limit deflections. These components have been found to be reliable, but they are additional cost components that also affect the design of the blade aerodynamic controls.

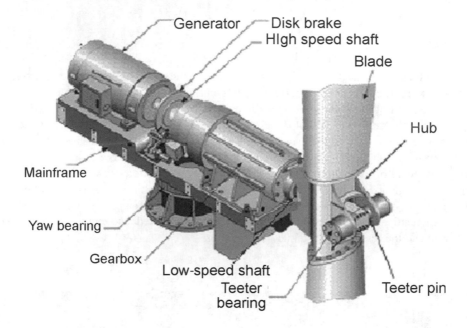

Figure 4-10. Schematic view of the teetered hub and drive train of a two-bladed wind turbine. Teeter motion greatly reduces cyclic loads transmitted from the hub to the drive train, nacelle, and tower. [Cotrell 2002]

Blade Aerodynamic Control

Most wind turbines today have some type of aerodynamic control on the rotor for power and/or speed control. One of the most popular means for limiting rotor power is changing the *pitch angle* of the blades. *Full-span pitch* (turning the entire blade about its longitudinal axis) of the blades is used on most large- and medium-scale wind turbines today. Hydraulic cylinders are typically used in the pitch actuator, although electromechanical gear motors are now also widely used. Full-span pitch facilitates rapid control of power output and, perhaps more importantly, effective speed control following loss of electrical load. This does come at the expense of a somewhat more complex and expensive rotor hub.

Nearly all small turbines using pitch control employ a *governor* activated by centrifugal forces. The classic *fly-ball governor* on the *Jacobs wind generator* (Figure 1-20) used weights that mechanically changed blade pitch towards feather. A later version used the weight of the blade itself to activate the governor. All mechanical governors require periodic service, and most have demonstrated only modest reliability. As a successful alternative, one of the most popular product lines of small-scale wind turbines, from Bergey Windpower Company (Figure 4-5) employs a patented *Powerflex* rotor system that uses weights near the rotor blade tips. Aerodynamic and centrifugal forces act together to twist the blades toward their optimum running position, once the turbine is operating. This form of pitch control does not limit power, but aids in starting the rotor in low winds.

Another technique used by several small-scale turbine manufacturers for speed and power control is *furling*, in which the rotor is turned partially out of the wind by yawing to one side (called *horizontal furling*) or pitching upward (*vertical furling*). It is possible to accomplish this through the downwind force of aerodynamic thrust balanced by spring or gravity forces.

Stall regulation of power is commonly used on rotors that do not feature full-span blade pitch. For a set rotational speed and blade pitch angle, airfoil angle of attack and lift increase as the wind speed increases, and the rotor produces more power. This continues until the angle of attack reaches a point at which aerodynamic stall occurs, and power passively levels off. Although stall regulation offers great control simplicity and reliability, it requires significant aerodynamic modeling, design, and testing. Techniques developed in the early 1980s [Viterna 1981] allowed wind turbine designers to utilize stall regulation more effectively. By 1993, 60 percent of the operating wind turbines used stall regulation for power control [Hansen 1993]. It continues to be used today, predominantly in small- and medium-scale wind turbines.

For overspeed control, *tip brakes, buckets, and pitchable blade tips* are sometimes used. Two types of tip devices for stopping a wind turbine with fixed-pitch blades are illustrated in Figure 4-11. *Tip brakes*, Figure 4-11(a), are flat metal or fiberglass plates attached to the

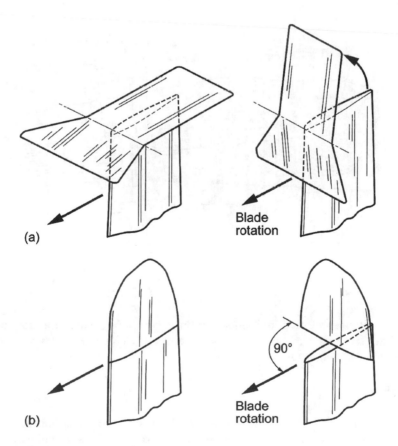

Figure 4-11. Tip devices for overspeed control in the undeployed (left) and deployed (right) positions. (a) Tip brakes (b) Pitchable tips.

outer end of each blade, normally oriented parallel to the blade's path. During an emergency they operate centrifugally to deploy perpendicular to the blade's path, slowing the rotor by drag. They are simple and effective, and have saved many small fixed-pitch rotors from destruction. However, tip brakes have two significant drawbacks. They are not aerodynamically part of the blade, and when not deployed they add significant drag, lessening the efficiency of the rotor. When deployed, they do not reduce the blade's lift, which is the principal force causing the overspeed. *Buckets* are semicircular plates which, like tip brakes, are centrifugally deployed to create drag and slow the rotor.

Pitchable tips, Figure 4-11(b), differ from tip brakes in that they are an integral part of the blade airfoil. When deployed, they rotate from within the plane of the blade to an out-of-plane position. This not only creates drag near the tip but also takes about 10 percent of each blade out of operation. Since the outer third of the blade creates most of the lift, the deployed tips dramatically reduce rotor torque. The Nordic Windpower N1000 (Fig. 4-2) and the Aerostar 6 Meter wind turbine (Fig. 4-6) are examples of commercial turbines today that use stall regulation with pitchable tips along with two-bladed and teetered hubs.

Drive Train, Nacelle, and Yaw Drive

Figure 4-12 illustrates the common mechanical components within the drive trains of wind turbines. These are a *turbine shaft* (often called the *low-speed* shaft), a *gearbox* (speed increaser), a *brake* (located on either side of the gearbox), a *generator shaft* (often called the *high-speed shaft*), and a *yaw drive*.

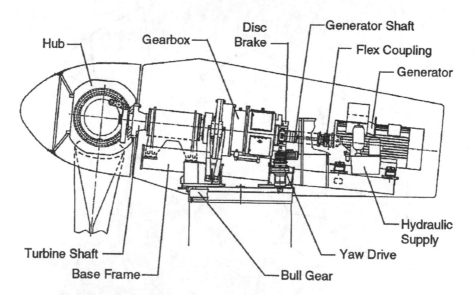

Figure 4-12. Representative components in the drive trains of medium- and large-scale wind turbines. The drive train shown is that of the Vestas V39/500 wind turbine. (*Courtesy of the Vestas Wind Systems A/S*)

Turbine Shaft and Gearbox

The gearbox increases the turbine shaft rotational speed to the speed required for the generator to produce AC power at frequencies of 50 or 60 Hz. The vast majority of all commercial wind turbines employ two- or three-stage *gearboxes* using *planetary* gears in combination with *parallel, helical,* or *spur* gears. Some variations are evolving to handle the large torques for large turbines, such as the *Clipper Liberty* gearbox that drives multiple generators directly from the planetary gears (see Fig. 3-46).

Even more variations in gearboxes are found in the design of smaller wind turbines. *Enertech* uses a *cycloidal-drive* with no gears on its 5-kW model. Many other small-scale turbines that operate at higher rotor speeds do not use a gearbox at all, but instead drive the generator directly with the turbine shaft. A very few medium- and large-scale turbines also use direct-drive generators, including the EWT Directwind and the *Enercon E-126* 7-MW wind turbine. Direct-drive generators will be discussed later in this chapter.

An alternative to the conventional mechanical gearbox is the use of *hydraulic transmission* systems. *Voith Turbo's WinDrive* system couples a hydraulic torque converter to a planetary gearbox to provide variable speed ratios and reduced torque dynamics. This drive system is used in the *DeWind* 2-MW wind turbine.

Because the generator is electrically locked to the utility grid (*i.e.*, constrained to stay at the same frequency as the grid) under normal operation, the drive train must also serve to partially dampen *torque fluctuations* caused by wind gusts. This is generally accomplished by the use of *shock/spring mounts* that allow the gearbox to rotate by a few degrees (see Fig. 10-3). Other approaches to dampen torque fluctuations are (1) the use of one or more *flexible couplings* located on either side of the gearbox or (2) the use of a *quill shaft*, a torsionally flexible tube between the hub and the gearbox, which is located inside a hollow turbine shaft.

Most manufacturers of wind turbines have experienced failures of the gears, gear surfaces, or bearings in their gearboxes. Gearbox problems have generally been the result of (1) lack of understanding of the *cyclic and random loads* imposed by the rotor, resulting in premature fatigue failures; (2) inadequate alignment of the high-speed shafts, resulting in seal degradation; (3) inadequate servicing resulting in premature failures from friction and wear; (4) poor assembly quality control resulting, for example, in excessive bearing endplay and misalignment; and (5) inadequate control of impact loads imposed by brakes on the high-speed shaft, resulting in fracture of critical components, and (6) lack of lubrication and impact loads during rotor parked (stopped) conditions.

As large-scale commercial turbines grow in physical size, power increases while rotor speed decreases. This trend simultaneously increases both rotor torques into the gearbox and gearbox speed-up ratios. With turbine sizes continuing to increase in the foreseeable future, the low-speed end of the drive train will continue to be a significant design and maintenance challenge.

Brakes

Mechanical brakes are typically employed for normal stopping of the rotor or holding the rotor in place after it has stopped. Mechanical brakes are activated electromechanically, hydraulically, or pneumatically. The most common brake systems are conventional *disc and caliper* types. If the drive train has a gearbox, brakes are usually placed on the generator drive shaft (gearbox high-speed shaft) or aft on the generator shaft itself, because of the torque-multiplying effect of the gearbox. In addition to mechanical brakes, several manufacturers have incorporated circuitry that allows the use of the generator as an *electrical brake* or, as it is commonly termed, a *dynamic brake*.

There is general agreement within the wind turbine industry that if mechanical brakes are the primary means of stopping the rotor, the turbine must either provide aerodynamic braking to augment them or provide fail-safe protection (such as redundant brakes) to stop the rotor should the primary brakes fail.

Mechanical brake wear has been a problem at some sites. Wear is accelerated by excessive *start-stop cycles* which can occur in areas with high, gusty winds (causing frequent high-wind shut downs) and in areas where the utility grid is frequently disabled (causing repeated losses of electrical load followed by emergency stops).

Yaw drive system

All HAWTs are designed to track the wind by orienting the nacelle in *azimuth* (compass direction) so that the rotor plane is normal to the wind. Most HAWTs with upwind rotors use an active yaw system for wind alignment. Almost all downwind machines use *weather vaning* or *passive yaw* for tracking the wind. Many small wind turbine rotors are located upwind of the tower and track the wind through the action of wind forces acting on a tail vane. Passive yaw systems reduce cost, weight, and maintenance by eliminating yaw drive components. However, if not properly designed, passive yaw can result in higher fatigue loads caused by operations at relatively large *yaw errors*

The majority of yaw drive systems for medium- and large-scale upwind machines are electromechanical, with the remainder being predominantly hydraulic. Figure 4-13(a) illustrates a common yaw drive system in which a motor turns a small *pinion gear* through a *worm-gear reducer*. These components are mounted on the movable bedplate of the nacelle. The pinion gear engages a large, stationary *slewing-ring* or *bull gear* mounted rigidly on the tower, as illustrated in Figure 4-13(b). The tower structure provides the stiff vertical support required for the yaw bearing. As turbine sizes have increased, so have tower heights and diameters, yielding larger diameters and costs of yaw bearings. These very large diameters are pushing the capability of machining yaw bearings, putting pressure on the manufacturing supply chain for large-scale wind turbines.

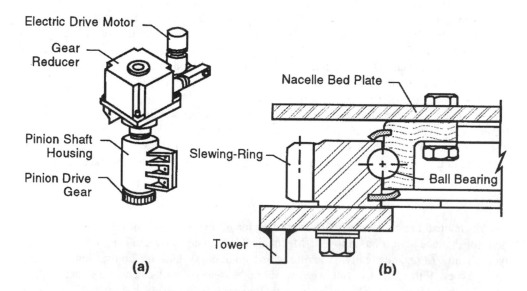

Figure 4-13. Schematic diagram of a motor-driven yaw system. (a) Electric or hydraulic motor drive and pinion gear (b) Bull gear and yaw bearing assembly.

Upwind machines have experienced fatigue-related problems with yaw slew rings and the pinion drives. Most of these result from unbalanced rotors and underestimating wind loads from turbulence and wind shear, both horizontal and vertical. *Yaw damping*, either by the use of sliding *yaw brakes* or flexible drive mechanisms, has proven beneficial in reducing yaw slewing rates and cyclic loads in all sizes of HAWTs. The problems that have been encountered most frequently with downwind machines are associated with *hunting*, which causes premature wear of yaw drives and bearings. Hunting refers to the tendency of some machines to continuously and slowly oscillate plus or minus 10 to 15 deg around the optimum rotor orientation.

Tower and Foundation

Figure 4-14 shows some of the many configurations of wind turbine towers that have been tried in the past, such as tapered tubular shells, stepped tubular shells, trusses, and guyed shells. Today, the vast majority of towers for large- and medium-scale commercial wind turbines use cylindrical and tapered tubular shells made of rolled steel plates and configured as free-standing *monopoles* (see Figs. 4-1 to 4-4). A concrete foundation is typically custom designed for these towers based on the soil conditions at the site. Tower heights normally range from 40 m up to more than 160 m. Towers are often manufactured as close to the site as possible to reduce transportation and logistics costs.

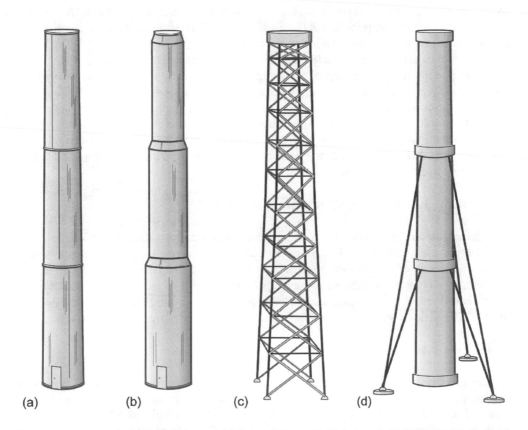

(a)   (b)   (c)   (d)

Figure 4-14. Typical HAWT tower configurations. (a) Tubular shell (b) Stepped shell (c) Truss (or lattice) (d) Guyed shell (or, similarly, guyed truss)

The expertise for manufacturing steel shell towers is widely available, although the thickness of the steel plates and the diameter of the sections for large-scale towers sometimes exceed the limits of local rolling facilities. Concrete is being used more frequently in the lower sections of larger towers, to mitigate high steel prices and transportation limits. *Enercon*, builder of some of the largest wind turbines in the world, is using pre-cast concrete towers.

Towers for small wind turbines are generally taller in relation to their rotor diameters than those of large turbines, in order to place the rotor at heights where the winds are more energetic. Unlike the predominance of free-standing towers with larger commercial turbines, *guyed towers* have proven popular for small turbines (see Fig. 2-10). Small turbines are often mounted on tubular towers made from a variety of materials including steel and fiberglass as well as solid wood poles. For turbines rated from 1 to 5 kW, wood poles offer an inexpensive tower option, particularly where concrete and fiberglass towers have limited availability such as in remote areas of developing countries. Wood poles are graded for different loads and can be sized to the loads produced by the turbine.

Truss or *lattice* towers are less costly and can sometimes be easier to install than free-standing tubular or pole towers of the same height. Installation of a lattice tower usually requires only simple hand tools. However, use of truss towers is declining because of problems associated with the fasteners (which can become loose) and with fatigue of one or more of the tower legs. Fatigue problems have been caused by improperly transferring the loads from the tower top plates to the tower legs. Tower-to-foundation interface problems have generally been associated with poor foundation installations. Tower problems are accentuated by excessive yaw motion and rotor vibration in highly turbulent wind regimes

Although some machines employ guyed towers to achieve additional height and therefore energy capture, the majority of wind turbine towers are not guyed. Guyed systems require additional maintenance, primarily because thermal cycling causes periodic loosening of guys and turnbuckles at most sites. Some manufacturers employ *tilt-down* towers, which generally permit the use of a winch rather than a crane to lower and raise the tower. There is no consensus within the industry regarding the "best" approach, though tilt-down towers have clear advantages in remote or hurricane-prone areas, or for stand-alone applications where crane availability is a problem.

Generator, Electrical System, and Controls

Generators and associated electrical control systems convert the mechanical power of the rotor into electrical power. The generator type is chosen on the basis of the turbine's rated power and the use of the electrical energy. The generator choice is also highly dependent on the method of controlling rotor aerodynamic power and speed, as well as on the choice of the speed increaser in the drive train. Today, both *synchronous* and *asynchronous* generators are used in all sizes of wind turbines, but the majority of generators are asynchronous.

Asynchronous AC Generators

Single-speed asynchronous generators, typically *induction generators*, have been very common in the wind turbine industry around the world and comprise much of the installed base. They are often used with fixed-pitch stall-regulated wind turbines. In large- and medium-scale turbines, most induction generators are *three-phase* with 690 VAC output. In small-scale turbines, *single-phase* 120/240 or 400 VAC outputs are most common. Induction generators are mechanically and electrically identical to induction motors. They produce power when their rotational speed exceeds the synchronous speed of equivalent induction motors.

Key advantages of induction generators result from their simplicity and include characteristics of good reliability, low cost, and reduced electrical components and control. Induction generators, however, require that the magnetizing flux be provided by the utility grid or in the case of a stand-alone system, an electrical energy storage device. Induction generators also contribute to inductive reactance on the power line, often to the detriment of the utility system operations. The *Nordic Windpower N1000* (Fig. 4-2) is an example of a wind turbine with an induction generator. Some commercial wind turbines have induction generators that include multiple sets of windings, allowing for multiple rotational operating speeds. The *Vergnet GEV MP 250* (Fig. 4-4) is an example of a wind turbine with a two-speed induction generator.

Doubly-fed induction generators, pioneered by the large-scale prototype Mod-5B and Growian wind turbines in the 1980s (Figs. 2-1 and 3-35), are now some of the most widely used generators for commercial large-scale wind turbines because of their ability to operate at variable speed [Herrera *et al.* 1985]. This feature increases energy capture by allowing the rotor speed to increase and decrease as wind speed changes, maintaining a high aerodynamic efficiency. Doubly-fed induction generators are asynchronous induction generators with a multiphase wound rotor. A multiphase slipring assembly contains brushes to contact the windings on the rotor. Using an electronic converter, both the rotor and grid currents are controlled.

In addition to being able to vary the wind turbine rotor speed by ±30 percent, it is also possible to control reactive power to the grid with a doubly-fed induction generator. Overall generator efficiency remains high as well. Initial costs and maintenance expenses are higher than those for a conventional induction generator, because of the required power-electronics converter and the slip ring assembly. Examples of manufacturers of large-scale commercial wind turbines with doubly-fed induction generators are *Suzlon, GE,* and *Vestas.*

Synchronous AC Generators

Synchronous generators are also widely used on all sizes of wind turbines. When connected to a utility grid, a synchronous generator rotates at a speed that corresponds directly to the line frequency. For synchronous generators connected to a power inverter, the generator rotor frequency is equal to the inverter frequency. A stationary magnetic field is provided by the generator's stator. Turbine rotors connected to synchronous generators often operate at constant rotational speed. Key advantages of a synchronous generator are its ability to provide power without line excitation and its ability to control reactive power by adjustments of the rotor field current. An example of a commercial wind turbine using a synchronous generator is the *Windflow 500.*

Permanent magnet synchronous generators are becoming more prevalent in all sizes of wind turbines. An advantage of this type of generator, which uses permanent magnets to generate a stationary magnetic field, is that it can either be connected to the utility grid or used as a stand-alone power source. Back-to-back power converters can be used to enable variable speed operation for increased energy capture. Furthermore, with increases in the physical diameter of the generator, it is possible to design a *direct drive* power train with permanent-magnet generator, which eliminates the need for a gearbox. The *Enercon E-126* is an example of a large-scale commercial wind turbine that uses a direct drive synchronous generator. Examples of medium- and small-scale turbines with directly driven generators include the *EWT Direct Wind 750* and the *Southwest Windpower Skystream.*

DC Generators and AC Alternators

DC generators have been used for years in small-scale commercial wind turbines, particularly at remote sites where battery charging is needed. During the 1930s, wind turbines charged batteries using shunt-wound DC generators, which were relatively massive. These pass current directly from the generator rotor through brush contacts, requiring regular brush replacement. Today, only a few very small turbines (ratings from 200 to 500 W) use DC generators, remaining essentially unchanged since they were first designed 75 years ago.

Many small-scale wind turbines use *AC alternators* which utilize materials more effectively than DC generators and draw current off the stator. As a result, the slip rings in a typical alternator carry only enough current to magnetize the field. There are no brushes that need replacement. Like the DC generators before them, today's alternators are driven at variable speeds. Variable-frequency AC output is then rectified to DC for charging batteries. In other applications, the variable-frequency, variable-voltage AC is first *rectified* to DC and then *inverted* back to constant-frequency AC, producing utility-grade electricity. This is a so-called *AC-DC-AC* electrical system. Most wind turbine alternators use *electromagnets* in the spinning field, but *permanent magnets* also play an important role because they do not require slip rings. The Bergey Excel and the Kestrel e400 are examples of small-scale wind turbines using permanent-magnet direct-drive AC alternators.

Inverters

As mentioned above, some form of *inverter* is needed to produce constant-frequency, constant-voltage power from a variable-speed alternator. Inverters often use solid-state switches to approximate the *AC wave form* of a true synchronous generator. Because inverters can also convert DC electricity to AC, some power systems use a wind turbine to charge batteries, and then use the battery-supplied DC to drive a motor-generator set (called a *rotary inverter*) to generate utility-grade AC. However, rotary inverters are only 60 percent efficient, while solid-state inverters can be up to 90 percent efficient.

In interconnected applications, a *synchronous* or *line-commutated inverter* is used to convert the alternator's output to line-quality AC. The synchronous inverter uses the utility-line wave form as a signal to fire or switch a *thyristor bridge*. Thyristors act as gates that pass current at the proper voltage as necessary to produce an AC wave form shaped like that of the electric utility. Synchronous inverters are susceptible to line transients and lightning surges and are costly to repair. Nevertheless, they are working reliably in hundreds of installations across the United States.

Electrical Connections

Figure 4-15 is a *one-line electrical diagram* for a typical wind turbine connected to a utility grid. Standard electrical *protection devices* are employed on commercial wind turbines to detect the following fault characteristics:

- over- and under-voltage
- over- and under-frequency
- over-current (current detectors for each phase)
- line out-of-phase
- generator over-temperature

Lightning protection and capacitors for *power factor correction* are generally used with commercial wind turbines. Electrical energy from the turbine generator is transferred from the movable nacelle to the ground by use of *slip rings* or a *droop cable*. Droop cables are fixed at both ends and can be wound up to the point where they are destroyed unless some provision is made to periodically unwind them. Many wind turbines use an active yaw system to unwind the cable. Other approaches employ a *pop-out connector* at the base of the tower to prevent damage to the droop cable from excessive twisting. If there is no active yaw system on the wind turbine, the cable must be untwisted periodically by shutting the turbine down, which has proven to be only a minor inconvenience. Because droop cables lower initial costs and maintenance requirements by eliminating the need for slip rings, they are now employed in most commercial wind turbines.

Most of the problems that have occurred with wind turbine electrical systems have been traced to the use of components manufactured with marginal or substandard quality. The use

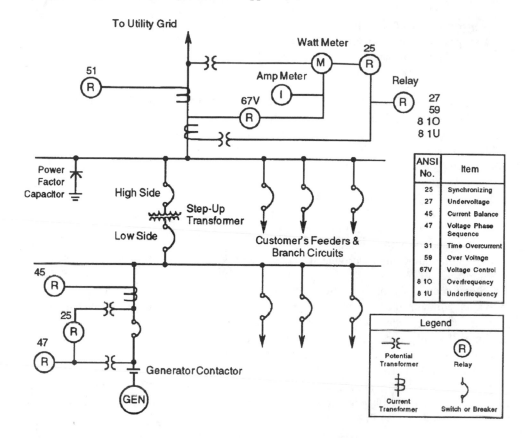

Figure 4-15. Typical electrical one-line diagram for a grid-connected wind turbine.

of improper electrical connectors has caused significant machine downtime. Many commercial developers have used step-up transformers with marginal ratings, and failures of these components have been a problem at some wind power stations. Generator failures, such as short circuits and overheating, have sometimes occurred and are usually caused by moisture in the windings. Today's generator designs generally employ heaters and temperature sensors in the windings to prevent these problems.

Control and Monitoring Systems

Nearly all commercial wind turbines employ a *system control and data acquisition system* (SCADA) with a *microprocessor* to control wind turbine operation and monitor performance. Most new wind power stations are equipped with central monitoring and control, in which an *on-board microprocessor* in each wind turbine sends performance and machine status information to a *central computer* equipped to constantly monitor and record these data. Automatic alarms are widely used to notify an operator of a wind turbine or utility line problem. Most of the problems encountered with microprocessor controls have been related to the associated software programs. The onboard microprocessor generally controls the following functions:

- *cut-in* and *cut-out* wind speeds (defining the operating range)
- connection of the generator output lines to the grid
- nacelle orientation (active yaw systems)
- blade pitch angle (active pitch systems)

- rotor speed (variable speed systems)
- power factor (capacitor or generator control systems)
- wind turbine normal and emergency shutdowns
- diagnostics for fault detection

Solid-state or conventional mechanical *relays* are used to activate the associated mechanisms. *Time-delay* devices or the control software are used to prevent frequent wind turbine starts and stops when the wind speeds loiter around the turbine's cut-in and cut-out speeds. In addition to its operational duties, the central microprocessor generally monitors and records the following performance parameters:

- wind turbine status (on-line or disabled)
- cumulative wind turbine availability
- wind speed and direction
- generator output
- cumulative energy produced

Some central computers compare the actual power and energy performances of each individual wind turbine to its expected performance (*i.e.*, expected output *vs.* wind speed) and provide a *figure of merit* for each machine. This information allows the operator to identify machines that may need adjustment or maintenance. Another feature that has been incorporated into central processors is the ability to calculate *cumulative revenues* for each wind turbine and for the entire wind power station.

In addition to electrical protection, wind turbines employ various other sensors to protect the machinery from damage and to assist in the control of the machine. The following items have been monitored for these purposes:

- wind speed
- wind direction
- nacelle orientation
- rotor speed
- blade pitch angle (for variable pitch wind turbines)
- equipment vibration levels
- gearbox oil level, temperature, and pressure
- generator speed
- brake pad thickness
- hydraulic fluid level, temperature, and pressure
- bearing temperatures
- structural integrity (*e.g.*, by strain gauges, crack detectors and bolt damage sensors)

Transport and Assembly

Transport of wind turbine components from manufacturing plants and system assembly at the site remain significant cost factors. Table 4-6 [Griffin 2001] summarizes the significant transportation limitations on the major components of a large-scale wind turbine. While a number of the components of a wind turbine add to transport and assembly costs, the blades and the tower are the most significant.

Table 4-6. Dimensional Breakpoints in Transportation Costs. Exceeding these breakpoints results in major logistic problems and excessive costs. [Griffin 2001]

Component	Height	Width	Length	Mass
Blades	4.4 m (14.5 ft)	7.6 m (25 ft)	45.7-48.8 m (150-160 ft)	Not Problematic
Hubs (w/o permits)	3.7 m (12 ft)	Not Problematic	Not Problematic	17,200-19,100 kg (38,000-42,000 lbs)
Nacelles	3.7 m (12 ft)	Not Problematic	Not Problematic	79,400-83,900 kg (175,000-185,000 lbs)
Towers (w/o pemits)	3.7 m (12 ft)	Not Problematic	1.2 m (5.3 ft)	17,200-19,100 kg (38,000-42,000 lbs)
Towers (w/ permits)	4.5 m (14.5 ft)	Not Problematic	Not Problematic	79,400-83,900 kg (175,000-185,000 lbs)

Blade Transportation

Blades are commonly transported by ship, rail, and truck. Blade length is the most difficult transportation challenge for large-scale wind turbines. Some early large-scale wind turbines (*e.g.*, the Mod-2 2.5 MW HAWT and the Mod-5B 3.2 MW HAWT) used blades made of welded steel plates that were manufactured in several pieces to satisfy transportation constraints. These sections were then bolted and welded together on-site. However, today's wind turbine blades are made of composite materials and are manufactured in one piece. This is partly because of the difficulty in designing and fabricating joints between composite material sections. Using metal fasteners at a potential blade joint can result in significant internal stresses caused by the difference in the moduli of elasticity of the two materials. Figure 4-16 shows the type of special truck and trailer equipment that is required to transport very large wind turbine blades.

Figure 4-16. Specialized wheeled fixtures are required for transporting very long wind turbine blades. The blade shown here is 61.5 m long. (*Courtesy of LM Glasfiber*)

Tower Transportation

Towers are primarily made of rolled steel with the primary restriction for transport over land being tower diameter. In most countries, tower diameters are limited to 4 to 4.5 m. In countries that are mountainous, the restrictions can be less than 4 m.

The *Repower R5M 5 MW* turbine in Germany (see Fig. 2-4), with a 6 m diameter at the tower base, demonstrates the criticality of the transportation issue facing the large wind turbine industry. Large-scale towers are typically manufactured in 20 to 30 m sections because of limits on both size and mass. Because of these constraints, it does not pay to transport towers more than 1,000 km, so they are often manufactured locally for large installations and regional markets. Assembly costs also rise rapidly with tower height, due in part to limitations on the availability of very large cranes and other construction equipment. For this reason, the rental of the crane is the largest cost of assembly and erection. A 300-ft crane with necessary 200 ton capacity can cost as much as $80,000 per day. Weather-related or other delays can often add 10 percent per day to these costs.

As discussed earlier, medium- and small-scale wind turbines often use towers that can be assembled from relatively small pieces. Some are assembled while horizontal on the ground and then tilted up for installation back to horizontal for maintenance and repair. Offshore installations can be done with jack-up barge mounted cranes (barges with retractable legs). Assembly in ship yards and transport can be on barges, virtually eliminating weight and height restrictions.

Water-Pumping Wind Turbines

Uses of water-pumping wind turbines include land *irrigation*, human or livestock *water supply*, and *drainage*. Pumping water with wind power requires the rotor to deliver higher torque than that required by an electrical generator of the same rated power. Prior to the 19th century development of *back gearing* (speed reduction), torque requirements were particularly high when one revolution of the rotor completed one pump cycle.

Designs of water-pumping windmills and wind turbines reflect their application, depending on the *flow required*, the type of *well*, and the *head* (the distance from the well water to the outlet). In the United States, most of these applications are *low flow* (approximately 1 m^3/h) and *medium head* (from 15 to 45 m). In the developing world, the flow desired may be similar, but the head is often less. The windmill may be driving a *piston*, *air-lift*, or *diaphragm pump*.

American Windmills

Because the demand for water is most critical during the dry or summer season, water-pumping windmills must function when the winds are light. Most builders of water-pumping windmills began to solve this problem by using rotors with *multiple blades*, consisting of many flat panels or paddles placed around either a vertical or horizontal axis (see Fig. 1-17). The result was the appearance of the machine known as the classic *American windmill*, a much-copied design that is now found commercially throughout the world. The abundance of non-U.S. manufacturers of water-pumping wind machines, especially in Australia, has led some outside the United States to describe the design today as a *classical windmill*.

For pumping small amounts of water in flat terrain, no substitute has been found for the classical, multi-bladed American windmill driving a mechanical pump. However, when the turbine must be located away from the water source, modern *electrical water pumpers* are a better choice. These machines drive a piston-type pump in a well that normally ranges from 1 m to more than 150 m in depth. For remote, unattended locations, *battery chargers* employing highly reliable rotor control methods are selected. If power is required to supplement the needs of homes or small businesses, small AC units with outputs up to 25 kW are the machines of choice.

The major components of water-pumping windmills are the rotor, the *crank* mechanism, the *tracking* mechanism, the *pump*, and the *tower*. Attempts have been made to improve upon these multi-blade turbines by using articulating paddles or blades in vertical-axis designs, but with little success.

Windmill Rotors

The "mathematical" windmill, as the first American rotors were called, substituted curved metal sheets for the wooden slats previously used for blades. This design, which nearly doubled the system performance to about 15 percent efficiency, dominated the market until the 1980s. Most of the rotors have *high solidity* (ratio of total blade area to rotor swept area) and contain 10 to 20 radial blades. These rotors operate at slower speeds and have higher starting torques than two- or three-bladed, low-solidity wind turbine rotors. A high starting torque is desirable for those wind machines that pump water from deep wells. The deeper the well, the longer and heavier the pump rod and the higher the water has to be lifted, so more starting force is required. Rotor diameters range from about 2 to 7 m. Wind tracking is generally achieved by the use of a *tail vane*.

Crank Mechanism

The crank mechanism illustrated in Figure 4-17 converts rotational motion into *vertical reciprocating* motion. A slight *gear reduction* (about 1:3) is often used between the turbine shaft and the crank shaft. This reduction decreases crankshaft speed and increases available torque.

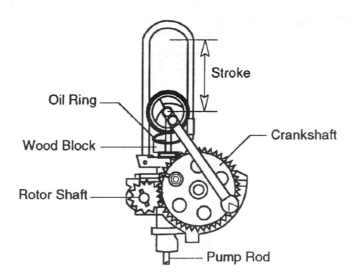

Figure 4-17. Schematic view of the power train of an American water-pumping windmill. Shown are the gear reduction between the rotor shaft and the crankshaft and the crack mechanism for converting rotary motion to reciprocating motion.

Pump and Tower

The most common type of pump being used with an American windmill is the *single-action piston pump*, where water is lifted on the upward stroke. Cylinder diameter sizes range from about 25 mm for deep wells to about 250 mm for shallow wells. Most towers are of a truss design.

Recent Technical Developments in Water-Pumping Windmills

During the late 20th century, there were few fundamental changes in the design of the American windmill or its water pump. One of the objectives in recent improved designs is to reduce the required *starting torque*. This improvement would allow the machine to start pumping water at lower wind speeds, as well as increase the flow rate at higher wind speeds. The most direct approach is to balance most of the weight of the pump rod and the water column using *counterweights* and/or *springs*. Another design approach is a *quick-return cam mechanism* that directs more of the available energy during each cycle into the upward or lifting stroke and less energy into the downward stroke.

The most promising new design is called an *automatic stroke controller* that can be attached to a conventional windmill pump. This device varies the pump stroke proportionally to the wind speed. At lower wind speeds the stroke is shorter, which results in a mechanical advantage that decreases the rotor torque required to start pumping water. As the wind speed increases, the stroke automatically increases at a predetermined rate. This design produces a better match between the pump load and the available wind power. The highest known windmill/pump system efficiency was obtained using this mechanism. With an automatic stroke controller it should be possible to use high-speed, low-solidity rotors to achieve higher efficiencies in pumping and lower rotor weights.

Improvements in the water pump designs include the use of *double-acting* rather than *single-acting pumps*, *long-life cylinders*, *larger flapper valves*, and *air lift pumps*. Many different power transmission schemes are being tried. These include mechanical (direct or indirect), electrical, pneumatic, and hydraulic.

Modern Wind Turbines for Water Pumping

Another approach for irrigation is to use conventional HAWTs or VAWTs to power an electric deep-well pump [Clark *et al.* 1981, Gaudiosi and Pirazzi 1987]. This method allows flexibility when the turbine is at a different location than the pump. In 1987, the U.S. Department of Energy sponsored a test of a wind-electric system utilizing a *Bergey Excel* 10-kW HAWT. In a trade-off study [Clark and Mulh 1992] the calculated annual output of a modern windmill (a 2.44-m-diam. *Aermotor*) and its mechanical pump was compared to that of a small-scale wind turbine (a *Bergey 1500* 1.5 kW HAWT) powering a submersible electric motor and pump. The basis for these calculations was short-term performance test data and long-term wind data. For approximately the same installed cost, the wind-electric system was predicted to pump about 68 percent more water per year for the same lift. However, a consideration of the total system costs (*i.e.*, cost per cubic meter of water delivered with the same lift) may show more balance between the two approaches [Moroz 1993].

Most electrical configurations for water pumping will employ an *induction generator* connected to the grid for *excitation*. In an early study, however, USDA researchers demonstrated that stand-alone wind turbines with alternators producing *variable-frequency* and *variable-voltage* electricity can also be used to drive well motors [Clark and Pinkerton 1988]. This is an important consideration when designing pumping systems for remote applications where utility service is nonexistent. In a variable-speed system it is necessary to match the performance characteristics of the turbine, motor, and pump components over the speed range expected.

Another system for pumping water solely with wind power is illustrated in Figure 4-18, in which two different wind power sources are used in tandem. The well pump is driven mechanically by a multi-bladed windmill, lifting water first to a ground-level storage tank. An electrical *booster pump* powered by a small-scale wind turbine is then used to distribute the water to higher elevations. Shown here is a typical commercial system that uses this approach, combining a 6-m-diameter windmill, which drives a reciprocating well pump, with a 1.5-m-diameter, 600-W HAWT for powering the booster pump.

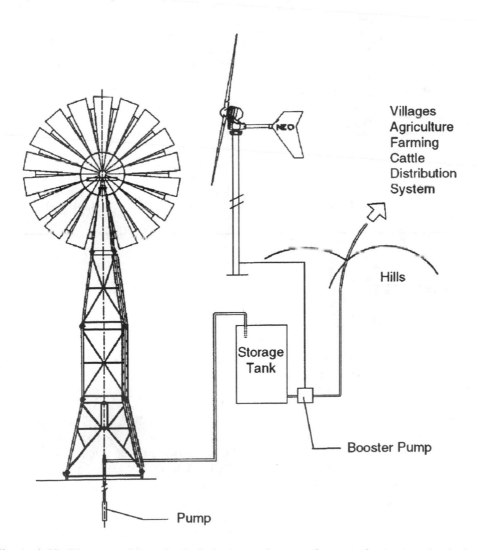

Figure 4-18. Diagram of a tandem wind-powered system for pumping water. A windmill drives the well pump, while a wind turbine powers an electric booster pump for the distribution system. (*Courtesy of Wind Baron Corporation*)

Vertical-Axis Wind Turbines

Vertical-axis wind turbines (VAWTs) had limited success in the early commercial markets of the 1980s. They were nearly nonexistent in the 1990s but are now re-entering the small-scale wind turbine market particularly for in applications on or near buildings. VAWTs are often classified into two types, *lift-based* and *drag-based,* depending on the dominant aerodynamic forces on their blades. Table 4-7 lists the characteristics for some representative VAWTs.

Table 4-7. Representative Commercial VAWTs

Manufacturer	Quietrevolution	Ropatec	PacWind
Model	QR5	Big Star	Seahawk
Rotor size, HxD	5.0 m x 3.1 m	8.5 m x 4.3 m	1.2 m x 0.8 m
Rated power	6 kW	20 kW	1 kW
Rotor type	Darrieus helical	Darrieus giromill	Savonius
No. of blades	3	5	6
Blade material	Carbon fiber	Fiberglass	PVC
Power control	Speed controlled stall	Speed controlled stall	None
Braking	Mechanical disk	None	None
Overspeed control	Mechanical disk	None required	None required
Gearbox	None	None	None
Generator	Permanent magnet	Permanent magnet	Permanent magnet
Voltage	415 VAC inverted	380 VAC inverted	12, 24, or 48 VDC
Yawing system	None req'd	None req'd	None req'd
Tower type	Steel tubular	Steel truss	Steel tubular

Darrieus VAWTs

Among the lift-based vertical-axis turbines, only the *Darrieus* has been developed commercially. In the 1980s extensive research and development programs on Darrieus VAWTs were conducted at government laboratories and in industry. Significant advances were made in developing design tools, components, and systems. Most of the early commercial Darrieus rotors emulated the designs developed at the Sandia National Laboratories and at the National Research Council of Canada (see Figs. 2-2, 3-20, 3-24, and 3-36).

The installation and operation of the 625-kW *34-m Sandia/DOE VAWT* testbed in 1987 (Fig. 2-2) represented a significant step in the development of larger and more efficient commercial machines. Several manufacturers, including *FloWind Corporation* and *Vawtpower* in the United States and *DAF-Indal Technologies* in Canada, deployed approximately 650 commercial VAWTs of the Darrieus configuration in the United States. All of these companies, however, discontinued production when the renewable energy market contracted in the 1990s. Early operational problems with Darrieus VAWTs were associated with their rotor blades and supporting struts. Fatigue-related failures have been encountered at the connections of the blade to the upper and lower hubs, at the blade splices, and at the blade-to-strut connections. Additional problems have occurred because of loosening or fatigue of the guy wires and turnbuckles.

Today, Darrieus wind turbines are again commercially available, especially for use in urban areas or even on top of buildings. Most designs have peak powers less than 20 kW and emphasize characteristics of good performance in wind conditions that are variable in direction and low in speed. Figure 4-19 shows an example of a commercially available Darrieus wind turbine with helical blades.

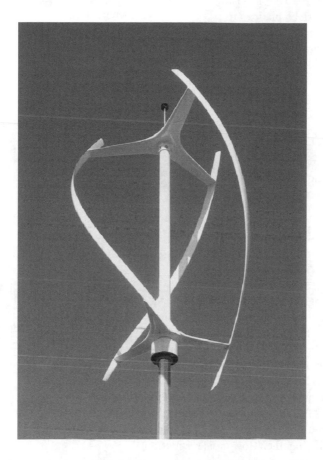

Figure 4-19. Quietrevolution QR5 6-kW Darrieus wind turbine.

Design Improvements in Darrieus VAWTs

The most promising design improvements for Darrieus machines appear to be in the areas of increasing aerodynamic performance and reducing manufacturing costs. VAWT analytical codes are available now that yield satisfactory agreement between design calculations and test data for mean (average) aerodynamic and structural characteristics. One approach to improving aerodynamic performance is the use of a combination airfoil that has tapered (or stepped) sections to optimize the local airfoil shape from the upper and lower hubs to the equator. This change alone may increase energy capture by 10 to 20 percent. Another approach to increasing energy output which is receiving consideration in the industry is the use of either a *multiple constant-speed* or a *continuously-variable speed* power train. Either method permits lower cut-in wind speeds (higher cut-in speeds are a problem with Darrieus VAWTs) and higher energy capture over the wind speed range at a particular site.

The relative manufacturing costs for a Darrieus VAWT may be divided among subsystems approximately as shown in Table 4-8. There may be some margin for cost reduction in each of these areas. Some Darrieus blades are *extruded aluminum*. The cost of this metal is relatively high, and extrusion does not offer the opportunity to vary the shape of the airfoil over the length of the blade. Blades may be fabricated from fiber-reinforced plastics or other composites, and may involve a box-spar construction or another fabrication technique but yield reduced material costs.

Table 4-8. Typical relative costs of Darrieus VAWT Subsystems

Subsystem	Cost share
Blade support system: (Struts, root connections, central column, upper bearing, guys)	25%
Power Train	20%
Control System	20%
Base Structure	20%
Blades	15%

Savonius VAWTs

A few drag-based vertical-axis turbines, mostly based on the Savonius design, are commercially available. Figure 4-20 shows an example of a modern commercial Savonius wind turbine. Such systems are often designed for sites that have significant variability in both wind speed and direction. This includes installations near buildings and on roof tops. In addition to their variable wind capabilities, manufacturers promote environmental benefits including their low noise emissions and their cross sections being more visible to approaching birds and bats. These benefits can overcome the lower efficiencies of drag-based systems for some sites.

Most commercial Savonius VAWTs generate DC electricity which can be stored in batteries or inverted to AC power. The slower rotational speed and higher torque of Savonius turbines may be appropriate for commercial applications in water pumping and grain grinding.

Figure 4-20. The PacWind Sea-Hawk 1-kW VAWT is an example of a commercial Savonius wind turbine.

Early Commercial Wind Turbine Installations

The birth of the modern wind turbine market began in the mid-1980s and included a wide variety of small-, medium-, and large-scale turbines that operated commercially during this early period. Sketches of some of the wind turbines installed in major U.S. wind power stations in the 1980s are shown in Appendix C, together with pertinent technical data. Wind power stations were installed on a wide variety of terrains, as shown by the general views in Figures 4-21 through 4-25. In California, these include the rolling inland terrain of Altamont Pass (Fig. 4-21), the rugged terrain of the Tehachapi Mountains (Fig. 4-22), and the flat desert areas in San Gorgonio Pass (Fig. 4-23). Two other types of terrain with wind turbines are represented by the flat high-plains area near Medicine Bow, Wyoming (Fig. 4-24), and the coastal hills on Oahu in the Hawaiian Islands (Fig. 4-25).

Figure 4-21. Rolling terrain in the Altamont Pass region of California. Lines of *USW 56-100* HAWTs are normal to the prevailing wind, with spacings larger in the windwise and smaller in the crosswind directions. (*Courtesy of Kenetech/U.S. Windpower, Inc.*)

While many of the wind turbine models described in Appendix C are no longer being manufactured, their designs define the scope of commercial wind power development in the mid-1980s. Plant operators have modified many of the medium-scale machines to upgrade their durability and performance. Retrofitting of rotors, power trains, and yaw drives has been common, and some power ratings have had to be lowered. Since the mid-1980s, many small-scale turbines have been retired and, in most cases, replaced by larger units. The large-scale prototype Mod-5B and WTS-4 HAWTs [Johnson and Young 1985], built for technology development under U.S. Government sponsorship, operated for a number of years as commercial power plants. No further production of these two machines occurred, however, because the companies that built them withdrew from the market when oil prices declined in the late 1980s. Many of the design features such as the tubular steel towers and variable speed generators pioneered by these early multi-megawatt turbines are in wide-spread use today. Operating experience of the Mod-5B machine over a period of 55 months, including production and maintenance data, has been reported by Spera and Miller [1992].

Figure 4-22. Rugged terrain in the Tehachapi Mountains of California. These 120-kW *Bonus 120/20* HAWTs are typical of the three-bladed, upwind-rotor turbines manufactured in Denmark. (*Courtesy of Arbutus Energy Company*)

Figure 4-23. Flat desert terrain in the San Gorgonio Pass near Palm Springs, California. The two-bladed, downwind-rotor turbines shown are 25-kW *Carter 25* HAWTs. (*Courtesy of NASA Langley Research Center*)

Figure 4-24. Flat high-plains terrain in Wyoming. The 4.0-MW *WTS-4* HAWT shown here is a former research prototype of the U.S. Bureau of Reclamation, put into commercial service. (*Courtesy of the Medicine Bow Energy Company*)

Figure 4-25. Coastal hills on Oahu in the Hawaiian Islands. In the foreground is the 3.2-MW *Mod-5B* HAWT, and 600-kW *WWG-0600* machines are in the background. (*Courtesy of Hawaiian Electric Industries*)

Comprehensive documentation of the early commercial development of wind turbines worldwide in the late 1980s is given by Jaras [1987]. In addition to extensive statistics on the world market for wind turbines (*e.g.*, by region, rotor diameter, and type of application), descriptions are included of 160 manufacturers of small- and medium-scale machines, with sketches, specifications, and estimated numbers of the wind turbine models produced.

Most of the first wind power stations used relatively small wind turbines because that was the only size available in large quantities. Research on the small-scale systems produced at that time by industry (with ratings from 1 to 40 kW) was begun in 1976. A series of prototype turbines, 5 to 25 m in diameter, was developed under the U.S. Federal Wind Energy Program from 1977 to 1981, and several were used in the first wind power stations. These systems included the *Enertech 44/40* and several derivatives of the experimental flex-hub rotor system developed at the *United Technologies Research Center.*

Commercial Wind Power Applications

Commercial generation of energy by wind turbines nearly died out during the era of abundant low-cost fossil fuels, but the resurgence of wind power in the 21st century is expected to continue at an accelerating rate. Wind-generated power is predicted to soon surpass hydropower as the largest renewable energy source. Markets for wind energy continue to evolve as new and improved equipment is developed and various deployment incentives are implemented by governments.

Wind energy applications can be divided into four major categories, as follows:

- wind power stations – for the business of generating bulk electric power for delivery via utility grid systems
- distributed wind systems – individual turbines or small clusters connected to utility grid systems
- isolated grid systems – multiple generators supplying power to isolated communities or villages
- remote off-grid systems – turbines interconnected with energy storage and hybrid power generation possibly including solar, fuel cells, or engine-driven generators

Each market segment has its own characteristics, driving forces, issues, and incentives.

Wind Power Stations on Utility Grids

Commercial wind power plants in the United States are typically built with clusters of turbines connected by a common transmission line to a utility grid. These installations are called *wind power stations* or *wind power plants* and have also been referred to as *wind parks*, and *wind farms*, primarily in European terminology. From 1982 to 1992, about 16,000 commercial wind turbines were built, mostly in California wind power stations with a total installed capacity of 1,584 MW. Since then the market has grown dramatically, to the point where in 2007 some individual wind power stations are that large. About 5,330 MW of wind power was added in the United States and more than 20,000 MW was added worldwide during 2007 alone [EIA 2008]. Wind power contributed 35 percent of all new generating capacity added in the United States during 2007, increasing from 19 percent added in 2006 and 12 percent added in 2005 [Wiser 2008].

Turbines in a wind power station are generally sited in regions that have growing demand for electricity, high COE (power purchase price), high *annual average wind speed* (6 m/s or more), a reasonable match of higher wind speeds (diurnal and annual) with periods of *peak power demand*, proximity to power transmission lines, and access for heavy equipment and construction cranes.

State and local financial incentives, tax breaks, favorable approval procedures and zoning, or other incentives help project developers by shortening the approval process and allowing a rate of return sufficient to attract private capital investment. Wind power stations are usually located at uncomfortably windy sites that are normally away from residential areas, reducing concerns about turbine noise, electromagnetic interference, and visual impacts. Environmental issues involving bird and bat populations and migration can be a concern but must be considered along with environmental effects from other power generation sources.

The physical arrangement of the turbines within the station depends upon the terrain, wind directions and speeds, and turbine size. In general, if a site is flat terrain, with winds predominantly from one direction, the turbines are spaced from 1.5 to 3.0 rotor diameters in the cross-wind direction, in rows 8 to 10 rotor diameters apart, as shown in Figure 4-26. Spacing between turbines on ridgeline projects depends on terrain orientation with respect to the primary wind direction.

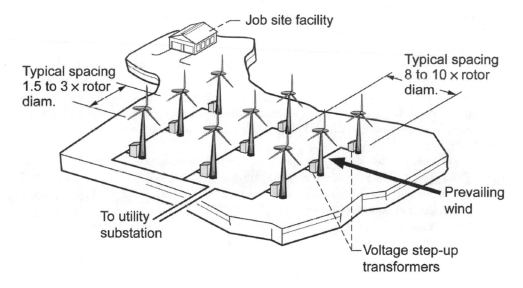

Figure 4-26. Typical arrangement of turbines in a wind power station on flat terrain.

Sizes of wind power stations vary widely. In the United States the average size has grown with the industry, from 10-MW plants in 1984, composed of 100-kW turbines, to most projects larger than 120 MW by 2007, with an average turbine size of 1.65 MW. Projects in the United States are continuing to grow in size and may exceed 1,000 MW. In Europe, land-based clusters of wind turbines are much smaller, typically from 10 to 20 units. However, much larger stations involving hundreds of turbines are being installed offshore in shallow water (less than 20 m depth), with some individual machine ratings as high as 5 MW (see Fig. 2-4).

Evolution of Wind Power Stations

Tax incentives and favorable power purchase rates in the late 1970s and early 1980s set the stage for an explosive growth in formation of *Independent Power Producers (IPP)* that sold wind turbines in a project to private owners. These individuals could use the state and federal *Investment Tax Credits (ITC)* available at the time. This incentive mechanism was subsequently replaced by a *Production Tax Credit* (PTC) that emphasized energy production rather than investment. Economic advantages of siting wind turbines near each other in wind power stations were found to be advantageous over installing many individual units, each with its own equipment for handling power and its own infrastructure for operations and maintenance. However, prior to the 1980s, most wind energy development focused on individual, often prototype, wind turbines. The fledgling wind industry was in the business of manufacturing turbines, not installing and operating large numbers of machines.

From 1982 through 1984, more than 40 firms developed wind power stations in California. In order to sell their machines, many of the turbine manufacturers also built power stations. Later as markets matured, turbine suppliers realized that they were competing with potential customers and shifted focus to supplying and maintaining turbine equipment. This business pattern was often repeated in new markets in other states or other countries, in which turbine manufacturers participated in the development of initial installations. By the end of 1986, the 10 largest wind turbine manufacturers installed over 80 percent of California's wind power capacity. The four largest of these (*United States Windpower, Micon, Fayette*, and *Vestas*) accounted for 55 percent of the projects. By 1987, however, only a half-dozen major companies remained.

Until the U.S. Department of Energy in the early 1980s sponsored the installation of a cluster of three 2.5-MW Mod-2 turbines (shown in Fig. 3-38), little attention was paid to the design of groups of machines and the potential impacts of upwind turbines on others operating nearby. Wake effects on downwind turbines were measured at this early cluster of turbines, including increases in turbulence-induced fatigue loads and decreases in energy production.

Energy generated by wind power stations in California quadrupled from 1983 to 1984, tripled from 1984 to 1985, and quadrupled again from 1985 through 1991. Early in 1991, installed capacity surpassed 1,500 MW, with more than 95 percent of the wind energy output coming from three areas in California: The Altamont Pass area (Fig. 4-21) accounted for 39 percent, the Tehachapi Pass area (Fig. 4-22) produced 38 percent, and the output from the San Gorgonio Pass area (Fig. 4-23) was 18 percent of the statewide total [CEC 1992]. Following expiration of the Federal tax credits in 1985 and the California tax credits in 1990, energy prices dropped and the wind power business virtually dried up in the United States.

Current Business Structure of Wind Power Stations

In October 1992 the *Federal Energy Policy Act* was made law and it included provisions that shaped the wind business for years to come. Title XIX, Subtitle A—Energy Conservation and Production Incentives, Section 45, established the renewable energy *Production Tax Credit (PTC)* that allowed a federal tax credit of $0.015/kWh (adjusted for inflation) for qualifying wind power plants that sell their energy. In addition, Title XII, Renewable Energy, Section 1212, established the *Renewable Energy Production Incentive* (REPI) that provided a similar inflation-adjusted $0.015/kWh payment to state-owned or municipal utilities that do not pay federal taxes. Factoring in inflation, the PTC and REPI were $0.02/kWh in 2008.

While new federal incentives were passed in 1992, it was not until 1999 that major project deployment began again. This rapid increase in the installation of wind turbines is shown graphically in Figure 4-27. The PTC proved to be effective at spurring growth in 2001, 2003, and 2005 to 2008, but the extensions of the tax provisions were for only a year or two and were passed too late to allow business continuity. Another 1- year extension of PTC was passed in October 2008 but some projects planned for 2009 had already been canceled. Despite these disruptive discontinuities in federal tax policies, by 2008 wind power installations had spread across the country with 16 states having more than 100 MW installed and six states with 1,000 to 4,500 MW. Total installed capacity in the United States in midyear passed 20,000 MW.

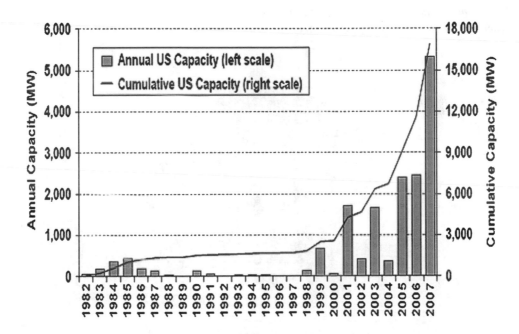

Figure 4-27. Wind energy capacity installed in the U.S. from 1982 to 2007. The bars indicate the capacity installed per year and the line shows the cumulative capacity.

The business structure of a wind power station project is a complex relationship among developers, investors, wind turbine manufacturers, utilities, insurers, and government entities which must approve many aspects of the project. By 2007, the structure of the wind business had changed substantially with increasing globalization. Companies like *General Electric (GE)* and the Spanish firm *Gamesa* were selling equipment but no longer developing power stations. GE had captured 45 percent of the U.S. market with their popular 1.5-MW turbine. Other leading turbine suppliers and their market shares in 2007 in the United States include *Vestas* at 18 percent; *Siemens*, 16 percent (following acquisition of *Bonus*); *Gamesa*, 11 percent; the remaining approximately 10 percent by *Mitsubishi*, *Suzlon*, *Clipper*, and *Nordex*.

Another new business trend began when large power companies started investing in the building of wind power stations. Examples of these companies are *Florida Power and Light* and *American Electric Power* in the United States, *Iberdrola* in Spain, *ENEL* in Italy, *Vattenfall* in Sweden, and *DONG* in Denmark. All these utility companies have projects in several countries.

The majority of the power sales from wind plants continue in 2007 to be through long-term *power purchase agreement (PPA)* sales contracts. Typically PPA contracts range from 15 to 25 years, with 20 years most common length. Most of the wind-generated power is sold through investor-owned utilities (IOUs) and publicly owned utilities (POUs), as illustrated in Figure 4-28. Some power companies and other investors are creating power marketing companies that buy and sell power and in addition can sell renewable energy credits and participate in the *green house gas cap-and-trade* business [Capoor and Ambrosi 2008]. *Merchant plants* are another category that sells power directly into the wholesale spot market when prices are high during peak demand periods.

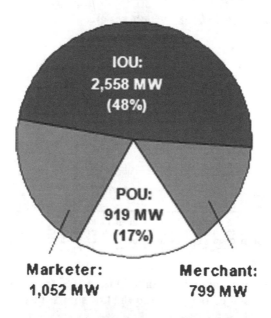

Figure 4-28. Relative amounts of wind-generated electricity sold by investor-owned utilities (IOUs), publicly owned utilities (POUs) and others in 2007. [DOE Berkeley Laboratory estimates based on the American Wind Energy Association database]

Distributed Wind Turbines on Utility Grids

Distributed wind turbines are dispersed across a utility's service area, as opposed to turbines in wind power stations that are bulk electricity producers concentrated in one location. Distributed turbines generally serve residential, farm, business, or light commercial loads with only excess power sold to a utility. Most of these turbines are small machines, with ratings ranging from a few hundred watts to 100 kW. They are usually connected to the distribution grid at low voltage (typically 220 to 4,160 VAC). At the beginning of 2008, there were about 25 companies manufacturing and selling small-scale wind turbines in the United States, and an estimated 85 similar companies in other countries, including American companies [Cohen and Wind 2001].

In Europe, most wind power installations are distributed applications. Denmark had 3,125 MW installed at the end of 2007, with nearly all in small clusters of 1 to 3 turbines. These units were supplying about 20 percent of that country's electricity.

U.S. Experience with Grid-Connected Distributed Turbines

The U.S. market for small-scale grid-connected wind turbines, designed to serve single-family homes and small businesses, has been hampered from the start by several factors:

- relatively high cost of small-scale turbines
- low power purchase prices and other financial incentives
- uncertainty about wind resources
- unproven turbine performance and reliability
- lack of distribution and service systems
- environmental concerns *vs.* potential benefits

In the early 1970s, many small-turbine manufacturers in the United States identified a market for machines in the 1 to 5 kW range. They believed this size was ample to meet most of the electrical needs of homes without electric heat, throughout most of the country. Prior to the passage of the *Public Utility Regulatory Policies Act (PURPA)*, and in areas where power purchase rates were low, it was desirable to use as much electricity on-site as possible and to minimize the excess that was sold to the utility. Moreover, the companies entering the wind turbine manufacturing business often did not have the capital or the engineering expertise to build machines much larger than 5 kW.

Largely as a result of commercial initiatives using venture capital, and with some support from the *U.S. Federal Wind Energy Program*, performance and reliability of small machines continued to improve and unit costs declined. As listed in Table 4-9, AWEA reported that sales of distributed wind turbines in 2007 in the United States totaled nearly 10 MW, with a slight majority used in grid-connected applications. Approximately 90 percent of this new

Table 4-9. Annual Sales of Small-Scale Wind Turbines in U.S. During 2007

Application	Number of Units	Capacity added (MW)	Sales (millions $)
Off Grid	7,800	4.0	14
On Grid	1,292	5.7	28
Total	9,092	9.7	42

capacity came from turbines manufactured by U.S. companies, including but not limited to *Southwest Windpower* (see Fig. 4-29), *Bergey Windpower, Wind Turbine Industries, Entegrity Wind Systems,* and *Distributed Energy Systems.* This was a 14 percent growth in annual sales—in capacity terms—relative to 2006, yielding a cumulative installed capacity of roughly 55 to 60 MW of distributed wind power in the United States in this small-scale size.

Figure 4-29. Small-scale wind power plant with four 400-W Southwest Windpower Air X turbines. (*Courtesy of National Renewable Energy Laboratory*)

Net metering, also called *net billing*, is a policy implemented by some states and electric utilities to ensure that any excess electricity produced by an on-site generator can be sent back into the utility system for fair credit. This excess generation can cause a homeowner's electric meter to spin backwards, indicating essentially negative electricity usage and effectively banking excess power production. Net metering allows the utility customer to be charged only for any net consumption of electricity. Since a single electric meter is used to measure in and out flows, the customer automatically receives compensation from the utility at the full retail electricity rate for any excess electricity produced.

Impediments to Commercial Small-Scale Wind Turbines

Initially, the manufacturers of small-scale wind turbines operated at a significant disadvantage in the United States because of competition with solar photovoltaic systems that had larger federal financial incentives for similar applications. That problem was resolved recently by the *Emergency Economic Stabilization Act* of 2008. This act contained provisions for an ITC of 30 percent of the cost of a small wind system (rated not more than 100 kW and up to $4,000) that is placed in service before December 31, 2016. Both wind and solar technologies now have comparable incentives.

Another impediment to small-scale wind turbine installations is the lack of reliable wind resource data. Furthermore, in many cases the timing of the available wind resource did not match that of the electrical demand. Interested buyers often found their sites had poor winds or were sheltered from high winds by buildings or trees. Even when the turbines operated properly, their energy production was too low to be cost effective. These productivity problems were in fact not the fault of the turbines but of incorrect wind estimates or siting.

Danish Experience with Grid-Connected Distributed Turbines

The Danish wind industry, like its counterpart in the United States, was first developed to serve a distributed market. Only years later did the industry begin to develop and sell large-scale turbines for wind power stations. For a variety of cultural and economic reasons, the Danes were very successful in serving this distributed market [Madsen 1987]. By 1993, about 3,500 wind turbines were operating in Denmark, mostly in distributed applications serving farms, homes, and small businesses. Although the total number of turbines in Denmark is only 22 percent, of those in California, the average number per unit of land area is twice as much (one per 12 square kilometers (km^2) in Denmark *vs.* one per 25 in California). While California wind turbines are generally concentrated in power stations, wind turbines in Denmark are generally dispersed, avoiding densely packed wind plants for aesthetic reasons.

The Danish Setting for Wind Energy

A nation of small cities, Denmark's population is distributed uniformly over the Jutland peninsula and the major islands of Zealand, Fyn, and Lolland. The uniform terrain is characterized by a flat, glaciated plain just above sea level with only a few hills and coastal dunes rising above the plain. Numerous inlets pierce the plain, creating a greatly indented coastline. There are 100 inhabited islands in Denmark. The indented coastline, the flat terrain, and the dominance of agriculture facilitate the installation of distributed wind turbines. Numerous bays and inlets provide an unobstructed fetch for winds sweeping across Denmark's open water. The flat terrain and the dominance of agriculture, with its dispersed housing and open fields, further contribute to a greater availability of good wind sites.

Like the Dutch to their southwest, the Danes have a long history of working with and using the wind, so the use of wind turbines seems more commonplace to them and less of a novelty. Another cultural characteristic in Denmark that is advantageous for the distributed-turbine market is the frequent use of *cooperative ventures*. Two or more homeowners or farmers will often buy a larger wind turbine together, taking advantage of the economy of scale and sharing the costs and benefits from production. Although there are no estimates of how many turbines have been installed in this way, cooperative ventures are significant not only as a means for obtaining a larger single turbine, but also because they are the model for acquiring multiple-unit installations.

There are many cases in Denmark where towns and villages have installed turbines to offset electricity consumption at municipal facilities (*e.g.*, sport stadiums, ferry terminals, and technical schools). About 50 percent of the projects using five or more turbines have been sponsored by municipal governments.

Development of the Danish Wind Turbine Industry

The early market for small wind turbines in Denmark developed like its counterpart in the United States, along two lines: Stand-alone systems to heat water, and grid-connected systems. But later, wind turbine development in Denmark took a different path that led to success in the distributed-system market. Influential factors were

- agricultural origin of turbine designs
- the well-established type of manufacturer
- the proximity of good sites to manufacturers
- the compact geographical size of the market served
- the development of a national certification program

Most Danish turbine designs grew out of the agricultural sector, and design development was both publically and privately supported. Because of their agricultural roots, early Danish turbine designs were simple, rugged, and used well-understood technology. The Danes paid a price for their use of heavier construction methods in terms of higher initial costs, higher transportation costs, and higher installation costs, but these penalties were offset by higher reliability. Most of the firms entering the wind market were well-established, midsized companies. Their ample size, coupled with their willingness to fund design development, enabled Danish firms to build first-generation turbines that were fairly reliable in the low-turbulence wind environment common in Denmark. Problems in the field and the cost of the solutions were consequently low. Consistent government policies supporting the domestic market and available financial resources for exports permitted leading Danish manufacturers of wind turbines to survive into the mid-1980s.

The distributed-turbine market in Denmark was aided by the close physical proximity of good sites to the machine manufacturers. For example, two of Denmark's largest wind power stations, the privately owned 2.3-MW *Tændpibe* and the utility-owned 10-MW *Velling Mærsk* wind plants, are only a few kilometers from the *Vestas* factory where their turbines were built. Close proximity of sites to manufacturing plants makes it much easier to move quickly through design iterations. Because of the small geographic area of Denmark, manufacturers were able to use in-house maintenance and repair teams to serve the entire country. Marketing was also beneficially affected. Because of the homogeneity of the market, the sales force of each manufacturer was able to call on customers directly, eliminating the need for an intermediate, cumbersome distribution system that relies on dealers who often have allegiance to more than one product.

Danish Approval and Certification System

Denmark was the first country to introduce design standards for wind turbines, and it did so through a government certification program. To qualify for federal tax credits and export loan guarantees, turbines have to be certified by the *Risø National Laboratory Station for Wind Turbines*. Risø personnel are aware that wind turbine design is complex and that manufacturers do not always fully understand the forces at work on the turbines. To address this problem, they originally set what they believed to be conservative design criteria for critical components. Later, Risø and the Danish industry learned that these criteria needed to be even more conservative. The result was Risø's imposition of design disciplines on the industry at an early stage. Although it may have delayed some creative new designs and eliminated variety in turbine configurations, government certification plays an important role in ensuring that Danish turbines performed reliably in the field.

In 1991 a new approval and certification system was introduced for the purpose of improving the overall quality of Danish wind turbines [Nielsen 1993]. It specifies very comprehensive requirements for documentation of all design criteria (*e.g.*, load cases and loads), fatigue evaluation, safety levels, power curves, and noise emissions, plus quality procedures for manufacturing, transporting, installing, and subsequently servicing the turbine. Many of these design standards are the basis for the *International Electrotechnical Standards* adopted in the late 1990s and now in use worldwide.

Status of Wind Energy Development in Denmark

By 2001, the total number of grid-connected wind turbines in Denmark peaked at about 8,000, with a total rated capacity of almost 3,000 MW. Energy production from wind turbines that year approached 5,000 GWh, which was 14 percent of the national electricity consumption. Private ownership accounted for 86 percent of the turbines, 76 percent of the installed power, and 79 percent of the energy production. The remainder came from utility-owned wind turbines. Some older machines have now been taken off-line and replaced, so the installed capacity at the end of 2007 was up slightly to 3,121 MW. That year, wind turbines produced 7,138 GWh, or 19.7 percent of electricity consumed in Denmark. Since 2001 most of the new plants, totaling 423 MW by 2007, have been offshore. More offshore wind power stations are planned, and the better winds offshore have helped increase station capacity factors.

Wind Turbines on Isolated or Utility Mini-Grids

The application of wind power to isolated- or mini-grids is a growing and important market for wind turbines. There are unique issues and constraints with isolated or small utility grids that depend on the conventional generators, the size of the network and the physical setting. There are a variety of grid applications that can be classified as isolated or small. For example,

-   islands that have no cable connection to mainland grid or where the island loads have exceeded cable capacity
-   isolated mainland communities
-   mainland communities that are not isolated, but choose to generate their own electricity

Island Applications

Some island communities use fossil-fuel thermal power plants to produce electricity that is distributed through an extensive grid network (*e.g.*, Tasmania in Australia and the Hawaiian Islands). Others employ *continuous-duty* diesel-electric generators, typically one to five units, to meet the electrical load. On some Caribbean islands, diesel generators are run intermittently to charge battery banks used for lighting and refrigeration. In all of these applications, wind power can supplement power production and save fuel.

There are many examples of island applications employing wind. In 1979, a 200-kW Mod-0A experimental HAWT was interconnected with the utility on Block Island, 10 miles off the coast of Rhode Island, as described in Chapter 3 [Hoff 2000]. The wind turbine was operated successfully with one to five of the diesel generators needed to meet the peak load in summer of 1,800 kW. During some low load periods in winter the turbine supplied more than 50 percent of the island's electricity, running with only a single diesel to control grid frequency. When these prototype tests were completed the turbine was disassembled, but in 2008 the Block Island power company is considering installing a combined wind, solar, and residential/commercial co-generation facility. The proposed wind turbines may be installed offshore on reefs near the island for aesthetics reasons.

Isolated Mainland Communities

Even in the developed world there are mainland communities isolated from regional electricity distribution systems. The first Mod-0A prototype HAWT was interconnected with the diesel-fired, municipal utility in Clayton, New Mexico. In the developing world, where regional networks are rudimentary, such applications are much more common. Islands linked to mainland networks by undersea cables are similar to remote mainland communities, since it is advantageous to use wind turbines for local generation whenever possible to offset power purchases from the mainland.

Kotzebue Electric Association in Alaska has led the way in a successful demonstration of using wind in an isolated community of about 3,000 people. Three 50-kW wind turbines were installed in 1994, and the wind plant now includes 17 machines supplying about 7 percent of the community's electricity. The latest addition was a 100-kW *Northern Power* turbine designed specifically to survive in the harsh arctic climate. More than a dozen remote villages along the Alaskan coast have begun using wind power.

Grid-Connected Communities

Some communities have chosen to generate their own electric power. Although they may be connected to a regional grid, their intent is to provide some or all of their electricity from local sources using the grid for backup. Such arrangements can present a difficulty for the regional power company, because their revenue base is reduced yet they must maintain sufficient power for backup and transmission lines to the community. On the other hand, in regions with load growth, the power companies can delay or cancel construction of new generating plants. Balancing these tradeoffs is done through the power purchase agreement.

Vashon Island located in Puget Sound near Seattle, Washington, is an example of a community planning an energy independence project. The local *Institute for Environmental Research and Education* conducted a study on sustainable energy for the island. All the renewable energy sources on Vashon-Maury Island were evaluated, from both engineering and an economic feasibility perspective. Possible energy sources include wind, solar, geo-thermal, tidal, biomass, and hydrogen. This study was the first step in a project whose goals are to move this community of 10,000 inhabitants, which currently imports essentially all its energy, to one that has no net energy importation and to minimize their *carbon footprint*. This would make Vashon-Maury Island fully energy sustainable and an example for other communities.

Stand-Alone Wind Turbine Applications

As the name implies, wind energy systems intended for stand-alone applications are designed as the sole or principal source of mechanical or electrical energy, and are not connected in any way to an electrical distribution grid. The major tasks of stand-alone turbines are those that have been performed throughout the history of wind power: pumping water and generating electricity for local consumption.

As shown in Figure 4-30, the electrical output from a stand-alone wind turbine can be DC for charging batteries to power low-voltage DC appliances, or DC for inversion to 110 VAC, or variable-frequency AC for *resistance heating*, or utility-grade AC to operate lights and motors directly. In practice, most of the commercial wind generators for remote applications are designed for charging batteries. No wind power system has been marketed successfully that is intended solely for producing heat.

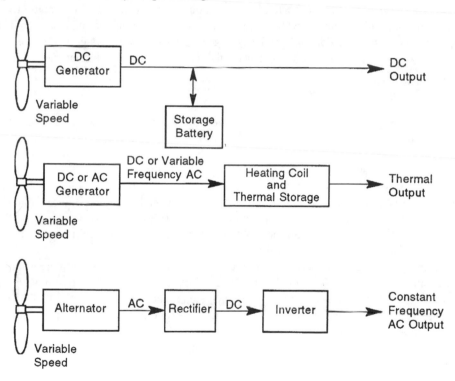

Figure 4-30. Output configurations for stand-alone wind turbine generators.

Residential and Agricultural Heating By Wind-Generated Electricity

Resistance heating, because of its insensitivity to the quality or frequency of electricity, allows the designer to tailor a relatively simple wind turbine for variable rotational speed, which results in more efficient operation. Controls necessary for the generations of utility-grade power are eliminated, along with most power-conditioning electronics. Researchers in the United States and Ireland have experimented with the variable-voltage, variable-frequency output from wind-driven AC alternators and DC generators for producing heat with resistance elements. Another concept for heating with wind power was the *wind furnace* [Cromack and Heronemus 1977], which used water heated by a wind driven mechanical churn. Several firms have tried to commercialize these concepts, but without success.

Stand-Alone Turbines for Remote Homesteads and Telecommunication Sites

The initial capital cost of a wind power system has been a major deterrent to homeowners but less of a concern to operators of remote telecommunications installations where reliable power is a necessity. Commercial wind turbines have been developed specifically for these remote applications since the late 1970s and have performed satisfactorily in harsh environments. Because of the need for storing excess energy to compensate for extended periods of low winds, nearly all stand-alone wind systems are designed to charge batteries. However, few turbines available today generate DC directly. Most drive alternators and rectify the AC output to DC.

In the developing world the market for small-scale turbines used in remote applications has been primarily for telecommunications, both commercial and military. Manufacturers who design products for the remote market have modified their turbines in response to customer needs, particularly their need for high reliability with minimal maintenance. In developing world communities, demand continues to grow for small wind systems of modest output that are used to power essential appliances, like refrigeration units in medical facilities.

Operational Issues on Small Grids and Isolated Applications

Because the generation of electricity from the wind varies with wind speed, other energy sources or storage are needed in applications requiring continuous electric power. Currently, wind turbines are designed to operate with some other generator setting the line voltage and frequency. In most small or isolated grids, wind generation is used as a fuel saver, reducing the need for diesel fuel, water impounded behind a dam, or some other controllable but intermittent source. Weather changes affect wind speed seasonally, hour-to-hour, and minute-to-minute, but these fluctuations are easily compensated by output from other sources. Wind power can be considered as backup for a hydroelectric plant during droughts and vice versa, as backup for diesels during fuel shortages.

Blending the outputs from wind, hydropower, and fuel cells will be an increasingly attractive and synergistic option in the future. In arctic regions the windiest season is typically in winter months when hydro resources are sequestered in ice and snow. Spring and summer melting and runoff are often the low wind periods. This inherent seasonal energy synergism could be enhanced with storage systems, either in diesel fuel today or in fuel cells in the future. The cost of fuel cells is declining and this quiet, clean energy source could be used to make hydrogen from excess wind or water power that could be stored for later use.

Penetration

There are a variety of operational issues involved in integrating wind turbines with other energy sources. One of these is *penetration*, a term which refers to the ratio of wind power to the utility's total power at any instant of time. If the amount of wind-generated power is small in relation to the total capacity of the utility, no *power curtailment* controls on the wind turbines (*i.e.*, methods for reducing maximum power) are needed. Recent wind power station experience indicates that wind turbines can contribute 50 percent or more of the grid power without upsetting power quality. Wind penetration in some areas of the Hawaiian Islands is already reaching this level during the night, when demand for electricity is low. When wind power generation is a large fraction of the load, the control system of the turbines must play an active part in regulating power characteristics. This control can be achieved by reducing power, dumping energy, or storing excess energy.

Wind system sizing and integration are more of a problem on small grids than on regional networks where power quality is negligibly affected by even large numbers of turbines. On a small grid, the degree of penetration often varies seasonally as a function of both the wind resource and the electrical demand. Wind turbines sized for modest penetration during peak summer loads will have a high penetration during the off season. This may require dumping of excess power that might otherwise disturb grid stability, or it may require the use of a fully controllable wind turbine, or the shutting down of other generators.

Limitations on Diesel Fuel Savings

One factor that affects maximum wind penetration on diesel-powered grids is how much the output of the diesel generators can be reduced safely and economically [Oei 1986, Qi 1993]. Diesel engines operate inefficiently at partial load, and their operating and maintenance costs may increase when they do not run at nearly full capacity. This inefficiency is caused by incomplete combustion, engine friction, generator excitation current losses, and losses in pumps, fans, and other auxiliary equipment. Partial load, suboptimal operation of diesels can cause excess smoke and result in carbon buildup in the exhaust system and stack fires. Therefore, fuel savings from reduced demand on a diesel generator are limited because of the machine's reduced performance under partial load.

Ideally, there should be a method to restrict partial-load operations, or to operate the grid with one or more diesels shut down but left in quick-start standby mode with oil heaters, to provide quick response to load changes while maximizing fuel savings and economic benefit to the system. Having different size diesels, an operator can dispatch the appropriate size engine to match the expected load and available wind energy. Operators should avoid frequent and damaging stop-start cycles of the engine as wind speed varies.

Grid Load Management

Load management is important in the successful integration of wind turbines into a small grid. Because power is expensive and often in short supply on small grids, users (unlike their counterparts on large grids) are accustomed to limiting their consumption to essential needs. Nonessential loads are either not met or delayed until the time of day when rates are lower. The effectiveness of wind power can be increased by regulating the grid load (a so-called *smart grid*) to reduce the degree of wind penetration. If nonessential loads are switched on as wind generation increases, the penetration will remain relatively constant, and it will be easier to maintain power quality. For example, the *dump circuit* illustrated in Figure 4-31 could direct excess wind power into a lower-priority heating load or dissipate it through a resistance load bank. With this load-management approach, fixed costs of operating the entire grid do not change, but fuel costs decrease as more wind energy becomes available.

Load management can be as simple as an on-off switch. On Caribbean islands it is common to connect a single diesel and wind turbine to charge batteries, run refrigeration, make ice, and provide light. During the night the entire system is shut down. These are *intermittent-duty* systems.

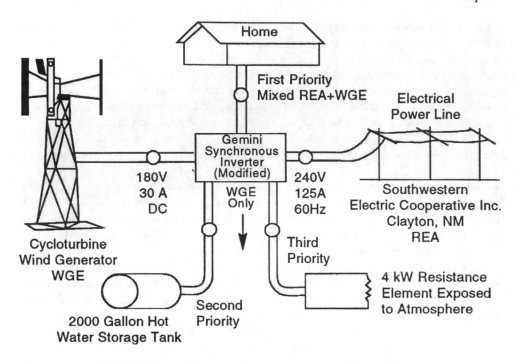

Figure 4-31. Schematic diagram of a dump circuit for switching excess wind power to nonessential loads on small grids.

Turbine Control

Variable-pitch turbines are more expensive but offer power and speed control advantages over fixed-pitch stall-regulated machines when it is necessary to match power from fluctuating winds to the fluctuating demand of the electrical grid. The potential for greater control with variable-pitch turbines allows higher penetration on small grids. Coupled with an asynchronous generator, blade pitch control can regulate frequency as well as power, acting more like a conventional power plant.

Recent advancements in variable-speed constant-frequency generating systems allow rotor speed to vary ± 50 percent as it tracks the wind speed while the generator produces a constant 60 Hz frequency. With variable speed asynchronous machines, the output voltage, frequency, and power factor to the grid can be managed to provide both real and reactive power as needed.

Factors Affecting the Market for Wind Turbines

Initial markets for wind power in the United States in the 1970s were driven by oil prices. At the time many thermal power plants used heavy oil for fuel. During the early 1980s the economic growth in California caused increasing demand for power that further fueled early wind markets. Following the oil embargos and to reduce atmospheric emissions, many plants shifted from oil to natural gas. This change and decreasing electricity consumption reduced the demand for wind power in the United States during the 1980s.

Incentive Factors

Since 1990, wind energy has again been recognized as a good source for meeting increasing demand for electric power. As the wind technology matured, turbine prices declined with increased production volume, and by the turn of the century wind power represented one of the largest new sources of electric capacity additions in the United States. It is clear that wind power plants can be installed quickly and in increments to match increasing demand for electricity. Following are several significant factors driving the current growth in wind energy.

Federal Renewable Production Tax Credit (PTC)

The PTC directly affects the economics of power projects by providing a tax credit to owners of *qualifying facilities*. The PTC provides a 2.0 cent per kWh tax credit (adjusted for inflation) for electricity generated in the first 10 years of operation to new projects brought on-line by the end of 2009. At that time the PTC is scheduled to expire again. A wind power plant that is producing electricity for sale to a third party is a *qualifying facility*.

Renewable Portfolio Standards (RPS) *and State Mandates*

Some states have established rules called RPS stating that a certain portion of the electricity that utilities sell must come from renewable sources. While the objectives and conditions of RPS and other state mandates vary widely among the some 26 states with such laws, these standards are providing an impetus to renewable energy development. This is particularly true for the stricter standards, such as those that make the RPS mandatory with specified renewable generation requirements and penalties for non-compliance. Differences in RPS provisions include variations such as

- what renewable energy sources will be counted
- whether power can come from existing renewable capacity or must be from new capacity
- what percentage of generation must be renewable and when
- how much of a challenge meeting that requirement will be from within the state
- generation base from which the state starts measuring compliance
- whether the provisions are mandatory or voluntary
- whether renewable energy credits, as established by many RPS programs, may be traded

The *North Carolina Solar Center* maintains the *Database of State Incentives for Renewables & Efficiency* (DSIRE), which contains summary information on portfolio standards for each state.

Rising Natural Gas Costs

Although the cost of natural gas may not be the single deciding factor in choosing to build a wind power station, the average cost of natural gas received by electric power plants has been in an upward trend since 1990. The average cost of natural gas reached a peak NYMEX futures cost of $15 per million Btu in 2005, and it has approached that level again in 2008. Higher electricity prices, driven by higher natural gas and other fossil fuel costs, improve wind's competitive position and make investment in wind power more profitable, particularly as developers speculate that future natural gas costs may rise further.

Water Savings

Water scarcity is a significant problem in many parts of the country. Few realize that fossil-fuel electricity generation now accounts for nearly half of all water withdrawals in the nation, with irrigation coming in second at 34 percent, according to the U.S. Geological Service in 2005. Water is used for cooling power plants fueled by natural gas, coal, and nuclear energy, and the availability of sufficient cooling water is an increasing challenge to further development of these energy sources.

Although a significant portion of the cooling water for electricity production is recycled back through the system, approximately 2 to 3 percent of this water is consumed through evaporative losses and must be replaced. Even this small fraction adds up to approximately 1.6 to 1.7 trillion gallons of water consumed for power generation each year. As additional wind generation displaces fossil-fuel generation, each megawatt-hour generated by wind could save as much as 600 gallons of water that would otherwise be lost in a fossil-fuel plant. Because wind energy generation uses a negligible amount of water, by 2030 annual water savings could be as much as 450 billion gallons. In addition, water pollution is reduced by decreasing coal washing effluent and chemical additives in feed water.

Emissions and Global Warming Concerns

Concerns over the potential impact of global warming have resulted in some states and regions establishing commitments to reduce greenhouse gas emissions. To illustrate, 10 northeastern states formed the *Northeastern States Regional Greenhouse Gas Initiative (RGGI)* with the nation's first multistate cap-and-trade system for carbon [Capoor and Ambrosi 2008]. RGGI raised $38.5 million in its first auction of carbon dioxide emissions credits held on September 25, 2008, with allowance being sold at a clearing price of $3.07 per ton on carbon monoxide. Also, California, Oregon and Washington have banded together to form the *West Coast Governors Global Warming Initiative* to reduce global warming. Development of wind power to meet electricity demand can help states and localities meet these commitments. The *European Union* emission trading scheme commenced operation in January 2005.

Wind power generation offsets or eliminates *air pollution* that otherwise would be caused by conventional power plants, a health and economic benefit often not accounted for quantitatively in the cost of energy. Most wind power stations offset generation from conventional plants fired with natural gas, reducing atmospheric emissions of nitrogen oxides, sulfur oxides, mercury, and carbon monoxide. A study performed at *Brookhaven National Laboratory* on the benefits of nuclear power in offsetting air pollution found that gas-fired plants caused 150 deaths per gigawatt-year of generation, coal plants 220 deaths, oil-fired plants 140 deaths, and wood-fired plants 57 deaths [Hamilton 1984]. Based on these findings, wind power stations are saving more than 600 lives each year of operation since 2006. This invaluable benefit is difficult to quantify in economic terms.

Financial Elements Needed for Development

From a financial perspective, the key elements that determine the viability of wind power development are *cost*, *revenue*, and the desired *rate of return* on investment. State and federal policies are often implemented through financial incentives such as tax credits or tax holidays. In some cases disincentives are reflected in the overall financial equation. The *cost element* includes the price of the turbine, the cost of installation (including permits, site plans and drawings, insurance, and financing charges), the infrastructure needed to operate the station (*e.g.*, roads, transmission lines, power handling devices, and service buildings), and operation and maintenance costs. The infrastructure costs are significantly influenced by the terrain, difficulty of access and construction (whether the land is flat or mountainous), and proximity to the existing utility grid. Operation and maintenance costs are determined by the station's size and the reliability of its equipment.

Revenue is determined by the wind resource, the turbine availability, turbine performance at individual sites throughout the station, and the rate paid by the off-taker for the energy generated. The desired *rate of return* is influenced by the investment community's perception of the *project risk*, the rate of return from competing investments of similar risk, the type of investment, investor needs, federal and state tax laws, and the political climate.

Wind Resource Assessment

The wind resource is the single most important factor in determining the economic viability of a wind power project. To determine the feasibility of wind energy development the Department of Energy (DOE) funded development of the *Wind Energy Resource Atlas for the United States* [Elliott 1987]. Wind sites were cataloged in seven wind power classes. To help power companies interpret data in the Atlas, DOE collaborated with the *Electric Power Research Institute* (EPRI) to produce a simplified version of the atlas. For example, EPRI assigned Class 5 to an excellent wind resource with an average annual speed greater than 6.0 m/s measured at a standard 10 m elevation above ground. With a typical Midwestern wind shear, this annual average wind speed becomes 7.5 m/s at the 50-m hub elevation of a typical medium-scale turbine and 8.3 m/s at the 100-m hub height of many large-scale turbines. There are vast areas of the country with this level of wind resource, especially in the Midwest, on ridgeline sites in the East, and offshore.

Measuring the wind at prospective sites is critically important. In the past, developers often underestimated the difficulty of characterizing the wind resource especially in hilly terrain. They also underestimated the decrease in wind speed within an array of turbines caused by the upwind rotors. In theory, the ideal resource assessment would include collecting wind speed, wind direction, turbulence intensity, and wind shear data for a 2-year period at representative locations within the proposed wind power station. Alternatively, at least 1 full year of wind measurements may be adequate with good correlation to a nearby long-term reference station with at least 10 years of data.

Regional wind maps are now publicly available for most parts of the country. These maps are based on atmospheric flow models that are tuned and adjusted using actual measurements where available. Despite improvements in *micro-siting technology* and procedures, energy estimates for a new wind power station in hilly terrain may still be inaccurate by as much as ±15 percent. Consequently, financial institutions will often apply the so-called *P-90* rule of thumb, which states that no matter how well the wind measurement campaign is run, the wind resource is probably 90 percent of the developers estimate.

Most small turbines are designed to start and operate at low wind speeds. Consequently, a Class 2 or 3 wind site may be adequate for an economically viable small-scale turbine application.

Land Requirements

Land area used for the turbines in a wind power station is small, with unoccupied land available for other compatible uses. Dedicated land actually occupied by the wind turbines, the connecting roads, and the plant infrastructure is only 2 to 5 percent of the total station area. The area surrounding the turbines is available for agriculture, forestry, fishing, or other commercial uses. Midwest farmers have found that revenues per acre from wind plants often exceed that for crops without effecting normal agricultural production.

In a DOE study to support a scenario in which 20 percent of the U.S. electricity demand would be supplied by wind power stations in 2030, a total wind power capacity of 305 GW was projected [DOE 2008]. Of this, 241 GW would be located on land-based stations and 54 GW in shallow water offshore. To meet that 20 percent goal, new land-based installations would require approximately 50,000 km^2 of land. However, the actual footprint of land-based turbines and related infrastructure would require only about 1,000 to 2,500 km^2 of dedicated land—slightly less than the area of Rhode Island. According to the conclusions of this study, available land area is clearly not a limiting factor in meeting and even doubling the 20 percent estimate.

Land costs are usually reflected in operating expenses. Land for wind power stations is generally leased to the developer or operator, although in some cases the developer may buy the site. Lease terms often include a one-time, advance payment of $200 to $2,000 per turbine (in 2008 dollars) plus a royalty or percentage of earnings from electricity sales. The royalty may be calculated on either gross or net earnings. Such fees have generally been 2 to 4 percent of net revenues.

Critics of wind power stations have raised concerns about visual pollution, noise, avian and bat impacts, and other land-use issues. While questions of aesthetic and noise impacts arise with the installation of all energy facilities, in other ways the land-use requirements for wind power stations are unique. Developers must place the turbines in geometric patterns to optimize the land use while minimizing the wake interference from upwind turbines on downwind machines. These patterns depend upon the terrain and the prevailing winds. In ridgeline installations turbines can be seen from greater distances, and often the land owners that have to look at the turbines are not those receiving revenues. Offshore installations can be seen for even greater distances and there is pressure to place installations out of sight of land, up to 20 km from shore. Experience has shown that aesthetic objections diminish as wind plants become more common and their environmental benefits are recognized.

Wind Turbine Size Factors

The trend in commercial sector has been to produce wind turbines with ratings of either less than 100 kW or more than 1 MW. For offshore applications, where transportation of very large and heavy components is not a limiting factor, turbines with rotors up to 120 m diameter and ratings of 7 MW are now being developed. Since 1998 average turbine size has increased 130 percent. In 2007, 40 percent of the turbines installed in wind power stations had rated capacities larger than 1.5 MW, compared to only 13 percent in 2002 and 2003. Machine sizes are expected to continue to increase, but at some point the transportation limitations on components for land-based systems are expected to become a firm barrier.

Variability of Electrical Output of Wind Power Stations

Grid integration studies have generally concluded that large-scale wind deployment does not have negative impacts at penetration levels less than 10 to 15 percent. Even at 30 percent or higher wind penetration power regulation, impacts are expected to be small. An example is a major study conducted by the *New York State Energy Research and Development Authority* [NYSERDA 2005, Piwko 2005]. Potential impacts on the existing grid systems were analyzed and this study concluded that the New York bulk power system can reliably accommodate at least 10 percent penetration, which would be 3,300 MW of wind generation, with only minor adjustments to its existing planning operational, and reliability practices.

Integrating the output from wind plants with other variable sources can even be beneficial. This is especially true in regions with hydro-based generation. Several project feasibility studies in the United States, Canada, and Russia have shown that windy periods in winter can provide power when most water resources are frozen and that the lower wind periods tend to occur during the spring and summer hydro runoff. A 10 year comparison of 800 MW of wind power from Vermont compared well with the *Hydro-Québec* load, as shown in Figure 4-32. This study also examined the economic impact on the value of wind by using hydro resources in Canada as storage. The value of wind could be up to 22 percent higher than selling it on the spot market [Ancona *et al.* 2003]. Similar results were found in Northwest Russia.

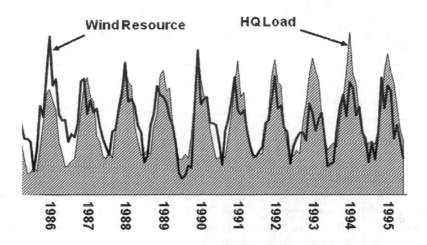

Figure 4-32. Beneficial match between seasonal variations in the Vermont wind resource and the Hydro-Quebec utility electric load over a 10 yr period.

Concerns about the variability of wind-generated power on grid control systems have largely been proven to be unfounded, but these concerns still persist. Several factors that tend to mollify impacts of variable output of a large wind power station include the following:

- slow variation of average wind speeds compared to normal grid power fluctuations
- spatial diversity of wind speeds over the station area
- geographic separation of stations within a utility's service area

Wind plant output does vary with time, but not as rapidly as might be expected from wind variations at a single location.

High-wind-penetration economic and technical issues have been studied in many cases in the United States and in Europe, and there are added costs when *spinning reserve* plant cycling is necessary to accommodate variations in wind power output. More than a dozen studies of different U.S. utility systems have shown that wind integration costs are well below $10/MWh and typically below $5/MWh, for wind capacity penetrations as high as 30 percent of peak load. The results of these studies are summarized in Table 4-10.

Table 4-10. Utility Cost Effects Resulting from Wind Power Penetration

Year	Utility study	Penetration of wind power	Costs ($/MWh)				
			Regulation	Load following	Unit commitment	Gas supply	Total
2006	MN-MISO [1]	32%	NA	NA	NA	NA	4.41
2007	Avista Utilities [2]	30%	1.43	4.40	3.00	NA	8.84
2003	We Energies	29%	1.02	0.15	1.75	NA	2.92
2007	Idaho Power	20%	NA	NA	NA	NA	7.92
2005	PacifiCorp	20%	0.00	1.60	3.00	NA	4.60
2006	Xcel-PSCo	15%	0.20	NA	3.32	1.45	4.97
2007	Arizona Public Service	15%	0.37	2.65	1.06	NA	4.08
2004	Xcel-MNDOC	15%	0.23	NA	4.37	NA	4.60
2007	Puget Sound Energy	10%	NA	NA	NA	NA	5.50
2006	CA RPS [3]	4%	0.45	Trace	Trace	NA	0.45
2003	Xcel-UWIG	3.5%	0.00	0.41	1.44	NA	1.85

[1] Highest costs over 3-yr evaluation period
[2] Unit commitment includes costs of wind forecast errors
[3] Regulation costs are 3-yr average

A study of the generation and transmission systems for Ireland examined energy portfolios that contained up to 59 percent renewable, mainly wind energy. This analysis showed that there was little difference in cost (less than 7 percent) among the various portfolios. For a portfolio with 36 percent renewable content, 70 percent to 80 percent of the investment cost could be recovered directly from the electricity market [Ireland 2008]. In addition, there could be revenue from carbon reduction credit sales.

Transmission System Expansion

The power grid in much of the United States today is characterized by mature, heavily loaded transmission systems. To move electricity to the major load centers from the projected new wind power plants will require new and expanded transmission lines. Both thermal- and voltage-related constraints affecting regional power deliveries have been well documented on systems operating at voltages up to and including 765 kV. In the long term, a mature system facing growing demands is most effectively strengthened by introducing a new, higher voltage class, both AC and DC. This approach can provide the transmission capacity and operating flexibility necessary to achieve the goal of a competitive electricity marketplace with wind power as a major contributor. A recent study by *American Electric Power* concluded that an additional 19,000 miles of high voltage lines will be required in order for wind power to generate 20 percent of the U.S. electrical load [AEP 2007].

Wind Power Station Costs

Wind power plant project costs include turbine hardware acquisition and installation, site preparation, balance of station equipment and facilities (often called hard costs), and costs to finance and legally structure the project (often called soft costs). The answer to the question "How much does a wind turbine cost?" is quite different from the question "How much does a wind power station cost?" Because the latter includes not only turbines but many other costs, only the wind plant cost is relevant to answering the question, "How much does wind-generated electricity cost?" In other words, "How much is the COE from a wind-powered plant?"

Breakdown of Wind Plant Costs

Turbines and other capital equipment are the largest element in wind plant costs. Turbine system costs are specified to include the following:

- rotor assembly
    - blades
    - aerodynamic control system
    - rotor hub
    - miscellaneous costs, including labor for assembly of rotor components
- nacelle assembly
    - low-speed shaft, bearings, and couplings
    - gearbox
    - generator
    - mechanical brake system
    - mainframe (chassis)
    - yaw system, including drives, dampers, brakes, and bearings
    - nacelle cover
    - work platform
    - miscellaneous costs, including labor for factory assembly
- tower (less on-site assembly costs included in "installation" below)
- control and electrical systems, including labor for factory assembly
- shipping costs, including permits and insurance
- warranty costs, including insurance
- markup, including royalties, profit, and overhead not included above

Balance-of-station cost is the other key element of project equipment costs that includes

- wind measurement anemometers and recorders
- site survey
- site preparation, including roads, grading and fences
- electrical collection system infrastructure
- substation
- foundations for the wind turbines
- operation and maintenance (O&M) facilities and equipment
- receiving, installation, checkout, and startup
- System Control and Data Acquisition System (SCADA)
- initial spare parts inventory
- permits and licenses
- legal expenses

- project management and engineering
- construction insurance
- construction contingency

In 1986, a study by the *Pacific Gas & Electric Company* concluded that a wind power station could be built for $1,050/kW, operated at a capacity factor of 27 percent, and maintained for $0.01/kWh [Smith 1986]. At a *fixed charge rate* of 10 percent, a 30-year life, and constant dollars (*i.e.*, no inflation), such a wind power plant could generate electricity for $0.054/kWh.

By 1987, wind turbine installed costs (power station turnkey costs less the *balance-of-station costs*) had declined to approximately $700 to $900 per kW. Further reductions were anticipated as wind turbine designs are refined and economies of scale took effect. Although costs appeared to be stabilizing from 1987 to 1990, this effect was actually caused by two opposing factors: (1) Wind turbine costs per installed kilowatt decreased by about 30 percent, and (2) the dollar weakened by approximately 35 percent during the same period.

Current Reference Costs in the United States

As a benchmark for estimating the cost of a wind power station, the DOE recently published *reference prices* for component parts of a typical 1.5-MW machine installed in a 100-MW wind plant [George and Schweitzer 2008]. These reference costs are listed in Table 4-11, in 2002 dollars. These cost estimates are being used for technology characterization and for tracking progress in cost-effective technology development. Reference component costs total to $614/kW for the turbine and $259 for the balance of station. With a market price adjustment at $108/kW to include the manufacturer's markup or profit margin, the reference wind power station has an initial cost of $981 per rated kilowatt.

Wind Plant Cost Trends

The costs of commercial wind turbines used in wind power stations have declined dramatically since 1980, bottoming at $1,400/kW and then increasing about $700/kW by 2007, as shown in Figure 4-33 by the trend in *turnkey system costs* (costs to investors). The installed cost of wind projects in the United States reached its lowest point in 2001. This was the result of a combination of larger projects, intense competition between developers, turbine manufacturers offering improved technology, rising fossil energy prices, state level financial incentives and the scheduled expiration of the PTC.

Many of these cost reduction drivers continued, but two changes caused prices to rise after 2001, for the first time in 20 years. First, the PTC was allowed to lapse or was renewed late in the year, disrupting project planning. Second, accelerating global demand for wind turbines and other products caused backlogs in orders for machines and shortages in raw materials, including steel for towers, copper for wire and generators, and cement for foundations, all of which drove up prices. It is important to note that material supply issues are not unique to wind projects and are increasing the cost of all new electricity generating systems.

In spite of these increases in installed project costs, the COE from wind projects was competitive with other wholesale power costs in all areas of the United States in the years 2000 to 2007. If the COE to the utility is in the range of $0.05/kWh to $0.06/kWh and a typical *capacity factor* is 0.31, turnkey station costs will have to be limited to approximately $900 to $1,200 per kW, to provide a fair rate of return for the station owners. Because the balance-of-station costs are approximately $400 per kW, turbines will have to be produced for less than $800/kW to support COEs in this range.

Table 4-11.
Reference Costs for a 1.5-MW Turbine in a 100-MW Wind Power Station
(in 2002 dollars) [George and Schweitzer 2008]

Component	Component Cost ($1,000)	Normalized component cost ($/kW)
Rotor		248	165
Blades	149		
Hub	64		
Pitch mechanism and bearings	36		
Drive train and nacelle		563	375
Gearbox	151		
Variable-speed electronics	101		
Generator	98		
Main frame	64		
Electrical connections	60		
Nacelle cover	36		
Low-speed shaft	20		
Bearings	12		
Yaw drive and bearing	12		
Hydraulic system	7		
Mechanical brake, high-speed coupling, *etc.*	3		
Tower		101	67
Control safety system		10	7
TURBINE CAPITAL COST		**921**	**614**
Electrical interconnection		127	
Roads and civil works		79	
Transportation		51	
Assembly and installation		51	
Foundations		49	
Permits and engineering		33	
BALANCE OF STATION COST		**388**	**259**
Market price adjuster		**162**	**108**
TOTAL INITIAL CAPITAL COST		**$1,471,000**	**981 $/kW**

A comprehensive study by the *World Bank* [2005] found that wind power is competitive with many other generating sources in a variety of applications in other countries, ranging from small isolated turbines and mini-grids to large grid-connected wind plants. Cost estimates are reported for 22 power generation technology configurations for global markets, and these relative costs are illustrated in Figure 4-34 [World Bank 2005].

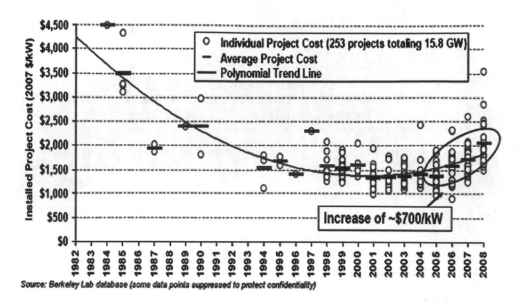

Figure 4-33. Trend of total normalized initial capital costs of wind power stations from 1982 to 2008. Costs are in $2007.

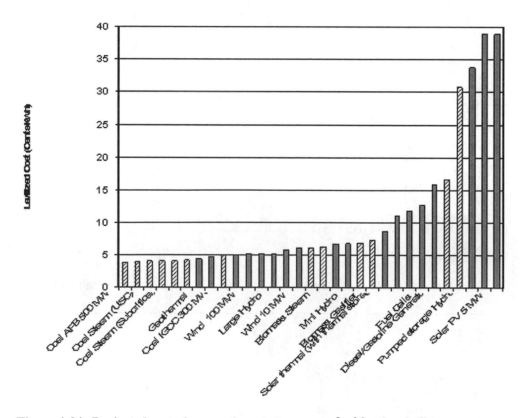

Figure 4-34. Projected cost of energy for wind compared with other bulk power generating technologies in 2010, in $2004. Cross-hatched bars denote technologies with atmospheric emissions. [World Bank 2005]

Wind Turbine Operation and Maintenance

Operation and maintenance (O&M) activities include all work necessary to *monitor and control* power output and keep the turbines *on-line* (available to generate electricity whenever the wind is within the operating range) through *scheduled* and *unscheduled maintenance*. These subjects will be discussed here primarily from the point of view of wind power stations composed of many turbines. Other O&M activities include contacts with government and utility personnel, and the public; general facility maintenance (*e.g.*, roads and buildings), and development and implementation of turbine improvements.

Performance Monitoring and Control Activities

Most modern wind power stations use a central control system from which the performance and operation of each wind turbine is continuously monitored.

Predicting and Assessing Energy Production

Figure 4-35 shows a flow chart for the most common method of predicting the *net annual energy production* from a wind power station. The level of uncertainty and the impact of errors on the final energy prediction are highest for the first step in the process (wind speed distribution) and decrease to the last step (miscellaneous losses).

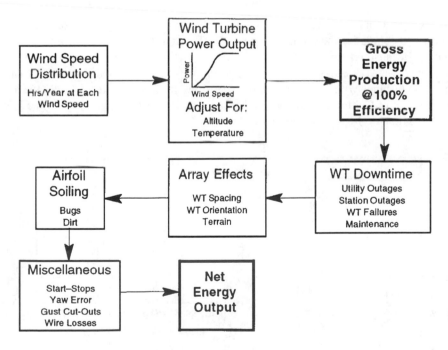

Figure 4-35. Flow chart for predicting net annual energy production from a wind power station.

Various *performance parameters* have been used to monitor and assess the net energy production from individual wind turbines and complete wind power stations. Two of these are *capacity factor* and *plant availability factor*, which are defined as follows:

Capacity Factor: The ratio of actual net energy production to the product of the power rating times the calendar time interval of interest expressed as a percentage.

Plant Availability Factor: The ratio of actual time that the plant or individual turbines are available for energy production regardless of the wind resource availability during the time interval of interest.

Turbines installed since the early 1990s have met or nearly met their performance goals, as a result of

- characterizing the wind resource with more accuracy
- more thorough accounting for losses within the wind power station
- significant upgrading of turbines by the power station operators
- better maintenance scheduling and procedures
- more accurate prototype performance testing by manufacturers of new turbines

Capacity Factor

Capacity factors for wind turbines vary widely as shown in Figure 4-36. Caution must be used when comparing the capacity factors of different turbines because wind regimes vary so much. Because energy capture is primarily a function of rotor diameter, turbines with higher *rated power densities* (rated power per unit of swept area) will have lower capacity factors than turbines with lower rated power densities at the same site. In 1992 the average capacity factor of California wind power stations was 18.5 percent, varying from a low of 2 percent to a high of 38 percent [Loyola 1993].

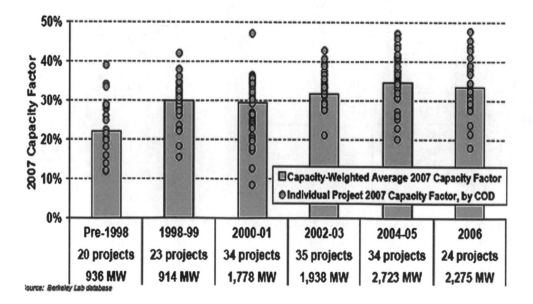

Figure 4-36. Trend in capacity factors of wind power stations from before 1998 through 2006.

As the result of many engineering improvements, the weighted-average 2007 capacity factor increased from 22 percent for projects installed prior to 1998, to 32 to 35 percent for machines installed from 2004 to 2006. Regional wind resources, siting, and turbine age are important considerations, but it is clear that machine performance is continuing to improve. Of the projects built before 2004 only 3.6 percent had capacity factors higher than 40 percent, compared to 25.9 percent for projects constructed in 2004 to 2006 timeframe.

Specific Energy Output

Another performance parameter that is directly related to the purpose of wind power stations is *specific energy output*, expressed in terms of kilowatt-hours per unit of rotor swept area during a specified time period. Table 4-12 is an example of the use of this parameter by the California Energy Commission to determine the effect of turbine size on wind energy production in 1991 [CEC 1992]. Capacity factors are also included for comparison.

Table 4-12.
Specific Energy Output of Early Wind Turbines in California during 1991,
According to Size [CEC 1992 & manufacturers literature]

Power rating (kW)	Specific energy output (kWh/m²/y)	Capacity factor
1 to 50	438	16%
51 to 100	670	19%
101 to 150	785	22%
151 to 200	833	22%
201 to 600 [1]	856	23%
1,500 [2]	947	34%

[1] Plus one 750-kW turbine
[2] For comparison; not in 1991 data set: 2007 commercial 1.5-MW turbine with 77 m rotor diameter, 80-m hub height, assuming 7 percent losses due to nonavailability, array wind losses, blade soiling, and electrical outages

Operation and Maintenance Costs

Operations and maintenance costs have decreased as operational environments are better understood and machine designs are improved. In 1987, the *Electric Power Research Institute* (EPRI) estimated O&M costs to be 0.8 to 1.2 cents per kWh (in 1987 dollars, not levelized). Assuming a capacity factor of 0.25, the annual O&M costs for a 50-kW system would range from $900 to $1,300, and from $3,200 to $4,800 for a 200-kW system. O&M costs for small-scale machines are higher per kilowatt-hour than for large-scale turbines, because many of the same activities have to be performed on each of many of the smaller units. Thus, there are economies of scale that can be realized in O&M costs.

The actual project O&M costs today are consistent with the early estimates. A survey done in 2007 showed that O&M costs averaged from $40/MWh for projects built in the early 1980s, to $9/MWh for projects installed in 2006. Data for projects built in the last 6 years are shown in Figure 4-37.

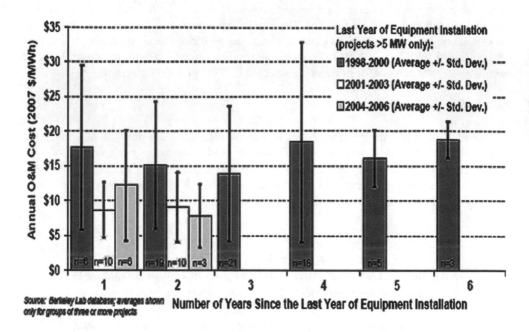

Figure 4-37. Trend of normalized operations and maintenance (O&M) costs for wind power stations.

According to EPRI estimates, O&M expenditures are distributed approximately as follows:

- labor   44 percent
- parts   35 percent
- lost operations   12 percent
- equipment   5 percent
- facilities   4 percent

Based on long-term experience with operating wind power station, annual O&M costs can be assumed to be 3 percent of the initial cost of the wind turbine or about $0.008/kWh to $0.015/kWh. It should be noted, however, that this percentage is for a mature turbine, whose design defects have already been corrected. The range is related to the severity of the environment where that turbine is installed. It does not provide for special inspections, product improvements, or other retrofit activities. Also, the costs of O&M can vary widely depending on equipment type and age. Plant size is also a key cost determinant for maintenance of wind plants as it is for other technologies.

Governmental and Business Environments Affecting Wind Power Development

The initial growth of the wind energy industry occurred in response to market and regulatory forces resulting from the 1973 Arab oil embargo that drastically increased the price of crude oil. Although future social and political considerations will have some effect, the worldwide prices of oil and natural gas will remain the dominant factors in the future of wind energy. Past experience has proven that future oil price fluctuations cannot be predicted reliably, and this situation is expected to continue indefinitely. Eventually oil and gas supplies will be depleted and the world most turn to other sources of energy.

The recent emergence of the global warming issue will change this picture, as nations move to decrease fossil-fuel burning. Therefore, the future rate of development of wind energy resources in the world remains uncertain, although the outlook is promising.

U.S. Wind Energy Industry

In the United States in 2007, 33 of the states had wind power installations (see Table 3-3). In 2008, six states have between 1,000 and 4,400 turbines, with the highest number in Texas. While some utilities are now getting more than 10 percent of their electricity from wind, one of the primary problems facing the U.S. wind industry is a constantly shifting policy toward renewable energy. Tax incentives for development have been complex and inconsistent. As an example, in 2008 there are 26 states with *Renewable Portfolio Standards (RPS)* that encourage the development of wind power plants and other renewable energy sources. However, half of these have been created since the beginning of 2004. A national RPS has been discussed, but its future is uncertain. Also, the goals, measurement criteria, qualification requirements, and financial incentives are different in each state.

Background on Federal and State Incentives in the United States

Nations responded to the sharp price increases and economic disruption of the early 1970s by instituting programs for reducing oil and natural gas consumption and for developing alternative sources of energy. U.S., federal tax credits of up to 15 percent spurred the domestic wind industry. Several states, including California, followed the federal example and created state tax incentives to increase the development of alternate energy sources of all types, including wind.

In 1978, the U.S. Congress passed the *Public Utility Regulatory Policies Act* (PURPA), which required every utility to buy electricity from independent producers at its *avoided cost*. Avoided cost is the cost per kilowatt-hour that the utility would have to pay for additional energy and capacity if it were to build new facilities. Although implementation of the new statute was delayed until the early 1980s, PURPA immediately established both a market and a pricing system for wind-generated power.

The California "Wind Rush"

Even with PURPA, many utilities objected to negotiating power sales contracts with independent energy producers. These objections dissipated just before the end of 1979 in California, when the *California Public Utilities Commission* fined the *Pacific Gas & Electric Company* $15 million for not considering conservation and alternative energy in its future generation plans. Major utilities in California began to actively support PURPA by giving more consideration to independent suppliers of renewable energy.

In the following year, 1980, California's governor organized a conference to attract financial interest in commercial wind power development, particularly in mountain passes that state-funded studies had identified as having excellent wind resources. In 1981 independent energy producers installed the first commercial wind power stations in the Altamont Pass east of San Francisco (Fig. 4-21) and the Tehachapi Mountains northeast of Los Angeles (Fig. 4-22). Several factors combined to attract initial development to these specific areas, factors that remain important today, as follows:

- excellent wind resources, with high average wind speeds that matched peak power demands
- abundant low-cost land with few land-use conflicts and proximity to load centers
- growing demand for electricity
- favorable power purchase rates (avoided costs), where a large percentage of the state's electricity was generated predominantly by gas- or oil-fired plants
- strong state regulatory support for developing alternative sources of energy
- lucrative financial incentives in the form of tax breaks
- liberal investment climate receptive to new ideas like commercialized wind power
- available funds for project financing

Between 1981 and 1985, the major source of financing for California's wind industry was private individuals who invested in *limited partnerships*. By the end of 1987, about $2.4 billion had been invested in California wind power stations. Since then, development has spread to the Midwest, the East Coast, and around the world.

Wind Energy Systems Act of 1980

The *U.S. Wind Energy Systems Act* (Public Law 96-345) was signed September 8, 1980. The goal of this Act was to establish an accelerated research and development program that would make wind competitive with other energy sources and, at the end of a 7-year program, have in place 100 MW of wind power capacity. At least 5 MW of the total capacity was to be in small-scale machines. The program projected increasing budgets for R&D and demonstration, reaching an annual level of $173 million in 1982. The Act also supported new technology, loans, grants, wind resource assessment, federal procurement of turbines, and other projects.

As a result of declining oil prices and plentiful energy supplies, federal funding was not available as envisioned, and consequently many provisions of this act could not be implemented. The Federal Wind Program budget declined from $63.4 million in 1980 to 34.4 million in 1982. Despite budget shortfalls, available state and Federal investment tax credits led to accelerated project deployments, and the 100 MW goal was reached in 1983, four years ahead of schedule.

Negative Impact of Lost Investment Tax Credits on the Wind Industry

The federal wind energy investment tax credits expired at the end of 1985 and were not renewed. The 10 percent federal investment tax credit, which could be applied to most equipment investments (whether wind turbines or industrial machinery) was repealed by the 1986 tax reform act. California's state tax credit decreased in 1986 from 25 to 15 percent and expired by July 1987. Following expiration of these tax credits, orders for new wind turbines fell in 1986 and 1987. At one time, as many as 200 to 300 firms professed interest in building wind turbines. By late 1986, however, less than a dozen major manufacturers, worldwide, were still building commercial wind turbines. This group shrank even further in

1987 as several major manufacturers faltered or went into bankruptcy. However, the existing California stock of more than 16,000 turbines assured a substantial business in wind turbine operations and maintenance.

Energy Policy Act of 1992

In 1992, comprehensive energy legislation called the *Energy Policy Act* (EPAct) contained a key provision for a PTC. This provision gave wind power producers a $0.015/kWh tax credit (adjusted for inflation) for energy produced during the first 10 years of operation. In 2008 the value of the PTC rose to $0.020/kWh. EPAct did little to foster new wind installations until just before its expiration in 1999. Nearly 700 MW of new wind generation were installed in that year, more than in any previous 12-month period since 1985. The PTC was later extended for two brief periods, ending in 2003, and then reinstated in late 2004 and extended again in 2008. This intermittent policy support led to only sporadic growth. In addition, business inefficiencies inherent in serving an intermittent market inhibited investment and increased the cost of project financing.

Current Incentives

In the United States a variety of additional incentives have been implemented. In many cases these incentives are additive, meaning that a project may benefit from a combination of federal, state, and local incentives. Table 4-13 shows a matrix of the various incentives and their beneficiaries. It should be noted that most energy technologies deployed in the United States now receive a plethora of incentives.

Table 4-13.
Beneficiaries of Government Financial Incentives for Developing Renewable Energy in the U.S.

Incentive	Beneficiary					
	Utilities	Independent Power Producers	Public Power	Rural Electric Coops	Manufacturers	Individuals and Businesses
Federal Incentives						
Production Tax Credit	X	X				X
Renewable Energy Production Incentive			X	X		
2002-8 Farm Bills						X
State and Local Incentives						
Green Power Purchasing Choices		X		X	X	
Green Tags		X		X	X	
Renewable Portfolio Standards	X	X	X	X	X	
Land Use and Tax Breaks [1]	X	X		X		X
Net Metering						X

[1] Property taxes, sale tax waivers, and tax holidays

Competitive Renewable Energy Zones

The cost of connection of wind plants and any renewable power source to the grid is an issue that can be solved in several ways. Some states have established *competitive renewable energy zones* (CREZ) connected to the existing grid by new high-voltage transmission lines. The cost of the new lines, at about $1.5 million per mile, will be paid by all the consumers across the state. This is similar to the German business model where the power companies were required to bring the power lines to new wind projects, again at no cost to the wind project developer.

Foreign Wind Energy Industries

On a global scale the wind power development picture is more complex, with different markets driven by various combinations of carbon reduction requirements, fuel supply shortages, and growing demands for electricity (especially in developing countries). Table 4-14 shows installed wind capacity in the top 10 countries at the end of 2007. Development of wind energy in the world did not take place in an economic or political vacuum. As in the United States, increases in the international price of oil spurred interest in wind technology. Future changes in the price of oil and other competing energy sources will determine the economic viability of wind power in the future.

Table 4-14.
Worldwide installation of wind energy generation capacity at the end of 2007.
[Global Wind Energy Council 2007]

Rank	Country	Wind power capacity (MW)	Relative capacity
1	Germany	22,247	23.7%
2	United States	16,818	17.9%
3	Spain	15,145	16.1%
4	India	7,845	8.4%
5	Peoples Republic of China	5,906	6.3%
6	Denmark	3,125	3.3%
7	Italy	2,726	2.9%
8	France	2,454	2.6%
9	United Kingdom	2,389	2.5%
10	Portugal	2,150	2.3%
	Total in top 10 countries	80,805	86.1%
	Other countries	13,060	13.9%
	Total worldwide	93,864	100%

A number of noneconomic factors will also influence the ability of wind power to compete with other energy sources. Foreign and domestic political considerations, international trade policies, and local and regional concerns are all expected to influence the future development of wind energy. European and some Middle Eastern countries also instituted financial incentives to develop alternate energy sources. Although most of these programs were associated with tax credits, other incentives were employed, such as partial government subsidies of early projects.

Denmark

Although Denmark did not have a comprehensive law equivalent to PURPA to encourage competitive sources of electric power, individuals are permitted to connect wind turbines to the utility grid and sell their excess electricity to the utility at 80 percent of retail prices. Alternatively, they could use this excess generation to offset their consumption, whether or not the wind turbines are installed at the owner's residence or business. Flexible tax incentives and utility power purchase provisions, among other factors, encouraged the growth of a powerful wind turbine manufacturing industry in Denmark. Total production reached over 200 MW per year, more than three-fourths of which went to California during 1984 and 1985. The Danish industry has the largest market share for dispersed wind turbine applications in the world.

In the early 1980s the Danish kroner was low in relation to the United States dollar, which provided a competitive advantage for sales in the United States. In addition the Danish government waived *Value Added Tax* (VAT) on products exported outside the European Union. But the Danes were not immune to the reduced world market for wind turbines. The expiration of the U.S. tax credits made wind power stations in this country more expensive to potential investors, and developers needed significant cost reductions to make new projects economically viable. Concurrently, the Danish kroner rose in value almost 40 percent against the dollar in the mid-1980s, making Danish exports more expensive. By 1987 only a few Danish manufacturers had escaped hardship or bankruptcy. Nevertheless, Denmark is still the world's leading producer of commercial wind turbines.

Germany

As listed in Table 4-10, Germany has more wind energy installed capacity than any other European Country or the United States, although U.S. wind power stations produced more energy by mid-2008. Usable wind resources in Germany are located primarily in the lowlands near the Baltic and North Sea, where more than 33 percent of electricity came from wind in 2006. Wind plants in Germany have the benefit of energy feed laws that provide lucrative, transparent, easily understood long-term tariffs for wind. As an example, Germany's *Renewable Energy Sources Law* required grid operators to pay 0.0819 Euros/kWh for 5 years beginning in 2007. This law requires the subsequent starting tariff be reduced by 2 percent yearly. However, offshore wind power installations are allowed higher tariffs and have the additional incentive requiring grid operators to pay the underwater line connection costs. The goal of these additional incentives is expansion in offshore wind power stations in German seas of 1,500 MW by 2011 and up to 25,000 MW by 2030.

Other European Countries

Businesses in European countries were initially slower to exploit the market for commercial wind turbines, but growing concerns about future energy supplies caused virtually every country in the European Union to begin investing in wind power. France was slow in joining this trend. Several initiatives were started there but stalled due to the dominant nuclear industry. However, by mid-2008 France had installed 2,454 MW of wind power, more than the United Kingdom that has vastly better wind resources.

Wind energy research activities in most European countries began in earnest by the early 1980s and accelerated after the Chernobyl nuclear accident in 1986. By then, 19 countries and territories had installed medium-scale wind turbines. The Netherlands, Greece, Spain, Germany, the United Kingdom, Italy, Norway, Sweden, Finland, and Portugal all have active research and development programs aimed at eventual commercial applications.

In 2007 Russia passed new legislation that will provide *Renewable Energy Certificates* that are expected to be sufficient incentive for wind power project financing. With the largest indigenous wind resources, the Russian wind power market is expected to be huge, freeing Russia's abundant oil and natural gas resources for export.

Asian Countries

Peoples Republic of China is now one of the fastest growing wind turbine markets, with 3,300 MW added in 2007. India has more installed capacity than China with 7,845 MW by the end of 2007.

Developing Countries

A critically important application for wind power is in developing countries. Deployment of wind and other renewable energy technologies in developing countries has been limited in spite of urgent need for clean energy, drawn from indigenous sources. Wind resources are widely available and wind technology is one of the lowest-cost options for generating electricity. Costs have declined to the point that, if reasonable resources are available, wind energy technologies can be the least-cost electric power option and should be part of the prefeasibility assessment process on all new power plant projects. In addition, wind plant construction can employ indigenous labor, and energy resources are critical to the balance-of-trade in developing countries.

Despite the importance of deploying wind power in developing countries, there are many obstacles. Technical, economic, and business issues are significant challenges in developing these applications. Primary difficulties include

- absence of wind resource data; wind speeds are normally recorded at weather stations and airports near urban areas and not at good wind sites
- short supply of wind turbine equipment that is consequently allocated to established and less-risky markets
- expensive transportation and logistics to develop remote sites
- lack of financial incentives
- costly financing because of higher risk
- absence of familiarity, knowledge, and experience with wind power

Distributed generation instead of large central power plants has advantages in developing countries and can be compared to proven markets in developed countries. Although many of the financial and economic assumptions would be similar, business structure and project costs can change significantly. The situation in each country will be different and many of the same assumptions and considerations apply equally to conventional fossil-fueled and wind energy plants. While the first installations of new technologies will be more expensive, costs will decline as the number of installations increases. In assessing the possible effect on both project and resulting energy cost, it is important to recognize that the individual effects are not necessarily additive and may be offsetting. In addition, government policies and commitment, business framework, joint venturing, project size and configuration can have a significant effect on project economics.

Offshore Installations

Offshore wind plants were first deployed in Europe and are increasingly attractive in the United States. Large urban areas are often located near bodies of water with good wind resources. In the United States, 28 percent of the coastal states use 78 percent of the electricity. Transmission lines run from offshore wind plants would be much shorter but are more costly than connections to most land-based sites. To reduce the higher cost of undersea cabling, consideration is being given to using DC links to larger projects. Wind resources are excellent off the East and West Coasts and on the Great Lakes. In order to achieve a 20 percent share of electricity generation, 54 GW of the planned 305 GW of wind power are estimated to be from offshore plants. Consideration of offshore resource potential is only beginning, and current estimates are probably low.

Offshore projects are being planned or are developing in eight states: Massachusetts, Rhode Island, New York, New Jersey, Texas, Ohio, Virginia, and Delaware. The first of these should be operating in 2010. These sites are in relatively shallow water (about 30 m or less) beyond the 3-nautical mile (nm) limit of state waters (Texas and the Gulf coast of Florida claim a 9-nm limit on state waters) and in areas regulated by the Minerals Management Service in the Department of the Interior. Far-offshore sites are attractive for aesthetic reasons and to avoid migratory bird routes. Protected waters, including the Chesapeake and Delaware Bays, may prove to have more cost-effective sites due to their shallower water and lower wave heights, provided that environmental concerns can be overcome. The configuration of a typical offshore wind plant is shown in Figure 4-38.

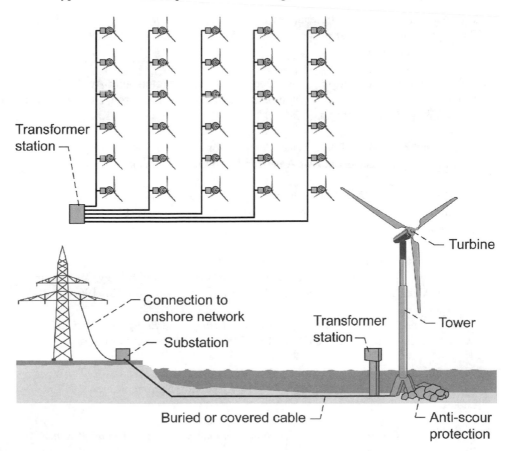

Figure 4-38. Schematic diagram of a typical offshore wind power station in relatively shallow water.

Offshore installations in Europe are the fastest growing market. The first offshore wind plant was commissioned in 1991 near Vindeby, Denmark. The plant consisted of eleven 450-kW turbines in two rows on concrete foundations in shallow water less than 10 m deep. By the close of 2006, more than 925 MW had been installed offshore in seven countries: Denmark, Germany, Ireland, Japan, The Netherlands, Sweden, and the United Kingdom.

References

AEP, 2007, *Interstate Transmission Vision for Wind Integration,* White Paper, American Electric Power, with American Wind Energy Association.

Ancona, D.F., G. Lafrance, S. Krau, P. Bezrukikh, 2003, "Operational Constraints and Economic Benefits of Wind-Hydro Hybrid Systems Analysis of Systems in the United States/Canada and Russia," *Proceedings, European Wind Energy Conference,* Madrid, Spain.

Capoor, Karan, P. Ambrosi, 2008, *State and Trends in the Carbon Market 2008,* Washington, DC: World Bank Institute.

CEC, 1992, *Results from the Wind Project Performance Reporting System:* 1991 *Annual Report,* Sacramento, California: California Energy Commission.

Clark, R. N., V. Nelson, R. E. Barieau, and E. Gilmore, 1981, "Wind Turbines for Irrigation Pumping," *Journal of Energy,* Vol. 5, No. 2: American Society of Mechanical Engineers, pp. 104-108.

Clark, R. N., and W. E. Pinkerton, 1988, "Operating Electric Motors with Autonomous Wind Power," *Proceedings, Windpower '88 Conference,* Washington, DC: American Wind Energy Association, pp. 313-323.

Clark, R. N., and K. E. Mulh, 1992, "Water Pumping for Livestock," *Proceedings, Windpower '92 Conference,* Washington, DC: American Wind Energy Association, pp. 284-290.

Cohen, J. M., and T. A. Wind, 2001, *Distributed Wind Power Assessment,* Washington, DC: National Wind Coordinating Committee, later renamed National Wind Coordinating Collaborative.

Cromack, D. E., and W. E. Heronemus, 1977, "Wind Power for Space Heating," *Proceedings, Third Biennial Conference and Workshop on Wind Energy Conversion Systems,* Vol. 1, Washington, DC: JBF Scientific Corporation, pp. 185-196.

DOE, 2008, *20 percent Wind Energy by 2030 – Increasing Wind Energy's Contribution to U.S. Electricity Supply,* DOE/GO-102008-2578: U.S. Department of Energy.

Cotrell, J., 2002, *The Mechanical Design, Analysis, and Testing of a Two-Bladed Wind Turbine Hub,* NREL TP-500-26645, Golden, Colorado: National Renewable Energy Laboratory.

Efiong, A., and A. Crispin, 2007, *Wind Turbine Manufacturers; Here Comes Pricing Power,* Merrill Lynch.

EIA, 2008, *Renewable Energy Consumption and Electricity Preliminary 2007 Statistics*, Washington, DC: Energy Information Administration.

Elliott, D.L., *et. al.*, 1987, *Wind Energy Resource Atlas of the United States*, DOE/CH10094-4, Richland, Washington: Pacific Northwest Laboratory.

Gaudiosi, G., and L. Pirazzi, 1987, "Wind Pumping Applications Meeting in Rome," *BWEA Windirections*, Vol. VI, No. 4, London: British Wind Energy Association, pp. 15-16.

George, K., and T. Schweizer, 2008, *Primer: The DOE Wind Energy Program's Approach to Calculating Cost of Energy*, NREL/SR-37653: Princeton Energy Resources International.

Griffin, D. A., 2001, *WindPACT Turbine Design Scaling Studies Technical Area 1 – Composite Blades for 80- to 120-Meter Rotor*, NREL/SR-500-29492, Golden, Colorado: National Renewable Energy Laboratory.

Griffin, D. A., 2002, *Blade System Design Studies Volume I: Composite Technologies for Large Wind Turbine Blades*, SAND2002-1879, Albuquerque, New Mexico: Sandia National Laboratory.

Hamilton, L. D., 1984, *Health and Environmental Risks of Energy Systems*, International Atomic Energy Agency Paper IAEA-SMM-273/51, Upton, New York: Brookhaven National Laboratory, p. 37.

Hansen, A. C. and C. P. Butterfield, 1993, "Aerodynamics of Horizontal-Axis Wind Turbines," *Annual Review of Fluid Mechanics*, Vol. 25, pp. 115-149.

Herrera, J. I., T. W. Reddoch, and J. S. Lawler, 1985, *Experimental Investigation of a Variable Speed Constant Frequency Electric Generating System from a Utility Perspective*, NASA CR-174950, DOE/NASA/4105-1, Cleveland, Ohio: NASA Lewis Research Center.

Hoff, T. E., 2000, *A Preliminary Analysis of Block Island Power Company's Use of Clean Distributed Resources to Provide Power to Its Customers*, NREL/SR520-27513, Golden, Colorado: National Renewable Energy Laboratory.

Ireland, 2008, *All Island Grid Study,* Report to the Department of Communications, Energy and Natural Resources and the Department of Enterprise, Trade and Investment, Report P4P601A-R003.

Jaras, T. F., 1987, *Wind Energy 1987: Wind Turbine Shipments and Applications*, Great Falls, Virginia: Stadia, Inc.

Johnson, W. R., and W. R. Young, 1985, "Status of the DOI SVU Wind Turbine Project," *Proceedings, Wind Power '85 Conference*, SERI/CP-217-2902, Washington, DC: Solar Energy Research Institute (now National Renewable Energy Laboratory), pp. 604-609.

Loyola, J., 1993, *Wind Project Performance: 1992 Summary*, Sacramento, California: California Energy Commission.

Lynette, R., 1986, *Wind Power Stations: 1985 Performance and Reliability*, Report for EPRI Project 1996-2, Palo Alto, California: Electric Power Research Institute.

Lynette, R., and P. Gipe, 1994, "Commercial Wind Turbine Systems and Applications," *Wind Energy Technology*, D.A Spera, ed., ASME Press, New York: American Society of Mechanical Engineers, pp. 139-214.

Madsen, B., 1987, "Analysis of a Success," *Windpower Monthly*, Vol. 3, No. 4, Knebel, Denmark: p. 4.

Moroz, E. M., 1993, "A Critical Review of "Water Pumping for Livestock" Presented at Windpower '92," *Proceedings, Windpower '93 Conference*, Washington, DC: American Wind Energy Association, pp. 551-558.

Nielsen, P., 1993, "Development of Wind Energy in Denmark," *Proceedings, Windpower '93 Conference*, Washington, DC: American Wind Energy Association, pp. 401-410.

NYSERDA, 2005, *The Effects of Integrating Wind Power on Transmission Planning, Reliability and Operations*, New York State Energy Research and Development Agency.

Oei, T. D., 1986, "Estimation of the Fuel Savings by Wind Energy Integration into Small Diesel Power Grids," *Proceedings, 1986 BWEA Wind Energy Conversion Conference*, London: British Wind Energy Association.

Piwko, R., 2005, *The Effects of Integrating Wind Power on Transmission Planning, Reliability and Operations*, for NYSERDA: GE Energy.

Qi, G., 1993, "Mathematic Programming Models for Economic Design and Assessment of Wind/Diesel Systems," *Proceedings, Windpower '93 Conference*, Washington, DC: American Wind Energy Association, pp. 501-507.

Smith, D., 1986, "Wind Energy at PG&E's Department of Engineering Research," *Proceedings, Fifth ASME Wind Energy Symposium*, New York: American Society of Mechanical Engineers.

Spera, D. A., and M. Miller, 1992, "Performance of the 3.2-MW Mod-5B Horizontal-Axis Wind Turbine During 55 Months of Commercial Service in Hawaii," *Proceedings, Windpower '92 Conference*, Washington, DC: American Wind Energy Association, pp. 231-238.

Williams, A, 2008, "A Strategic Review of Rotor Blades," *Proceedings, Windpower 2008 Conference*, Washington, DC: American Wind Energy Association.

Viterna, L. A., and R. D Corrigan., 1981, "Fixed Pitch Rotor Performance of Large Horizontal Axis Wind Turbines," *Proceedings, Workshop on Large Horizontal Axis Wind Turbines,* NASA CP-2230, DOE Publication CONF-810752, Cleveland, OH: NASA Lewis Research Center, pp. 69-85.

Wiser, R., and M. Bolinger, 2008, *Annual Report on United States Wind Power Installation, Cost, and Performance Trends: 2007*, NREL Report No. DOE/GO-102008-2590, Golden, Colorado: National Renewable Energy Laboratory.

World Bank, 2005, *Technical and Economic Assessment: Off Grid, Mini-Grid Grid Electrification Technologies*, Discussion Paper: Chubu Electric Power Co., Toyo Engineering and Princeton Energy Research International.

5

Wind Turbine Aerodynamics
Part A: Basic Principles

Robert E. Wilson, Ph.D.
Emeritus Professor of Mechanical Engineering
Oregon State University
Corvallis, Oregon

Introduction

Designing wind turbines starts with knowledge of the aerodynamic forces acting at the wind machine. This chapter concentrates on the basic principles of the aerodynamics of conventional horizontal- and vertical-axis wind turbines (HAWTs and VAWTs). Aerodynamic theories are discussed, starting from the *actuator disk model* of a HAWT, extending through the *Glauert optimum actuator disk model*, to the *strip theory*, which has been the mainstay of HAWT aerodynamic design and analysis. Various corrections, including *thrust coefficient modifications*, *tip-loss models*, and *gap corrections*, are developed. Comparisons are made between test data, strip theory, and *free vortex* calculations to evaluate the accuracy of strip theory.

HAWT operational and design features are presented, including the *teetered rotor, yawing* and *yaw stability, blade-* and *tip-pitch controls, ailerons, transient aerodynamics,* and *vortex generators. Power outputs* and *aerodynamic loads* of medium- and large-scale HAWTs are presented and compared with theory. The aerodynamic behavior of VAWTs is examined in a parallel fashion, starting with an analysis of limiting VAWT performance and then proceeding to a development of the *streamtube theory.* Comparisons are made between power output predictions and test results for a medium-scale research VAWT. The effects on VAWT performance of *rotor solidity, blade number, rotor shape,* and *Reynolds Number* are presented, along with a discussion of starting and stopping. Test data are used to demonstrate the shape of rotor power curves and the effects of *vortex generators.*

Translating Aerodynamic Devices

Perhaps the simplest type of wind power device is one that moves in a straight line under the action of the wind, like the ice-boat shown in Figure 5-1. Historically, these wind-driven *translating devices* have been used for propulsion rather than power extraction. Examination of translating *lift-* and *drag-driven* devices can be illuminating for the aerodynamic analysis of rotary machines, since a rotating blade element can be considered as instantaneously translating.

Drag Translator

First, consider a device driven only by drag forces. Figure 5-2 illustrates the action of an elementary drag device in which the power extracted is the product of the drag force and the translation velocity. Drag results from the *relative velocity* between the wind and the device, so that

Figure 5-1. An iceboat traveling at a speed of 27 m/s. *(Courtesy of Gougeon Brothers, Inc., Bay City, MI)*

$$P = D\,l\,v = C_D\,q\,A_p\,v = [0.5\,\rho\,(U - v)^2]C_D\,c\,l\,v \qquad (5\text{-}1)$$

where

$\qquad P$ = power extracted (W)
$\qquad D$ = drag force per unit length of device (N/m)
$\qquad l$ = length (spanwise) dimension of device (m)
$\qquad v$ = translation velocity (m/s)
$\qquad C_D$ = *drag coefficient*; function of device geometry
$\qquad q$ = *dynamic pressure* = $0.5\,\rho\,V_r^2$ (N/m^2)
$\qquad \rho$ = *air density* (kg/m^3)
$\qquad V_r$ = relative velocity (m/s)
$\qquad A_p$ = projected area of device (m^2)
$\qquad c$ = width (chordwise) dimension of device (m)
$\qquad U$ = steady free-stream wind velocity (m/s)

Thus, the velocity of the device must always be less than the wind velocity. The *power coefficient* of a wind power device is defined as the ratio of the power extracted to the wind power over an area equal to the projected area of the device, or

$$C_P = \frac{P}{0.5 \, \rho \, U^3 \, A_P} = \frac{v}{U} \left(1 - \frac{v}{U}\right)^2 C_D \tag{5-2a}$$

where C_p = power coefficient

The power coefficient for a drag-driven device is a maximum when $v/U = 1/3$. Thus,

$$C_{P,\,max} = (4/27) \, C_{D,\,max} \tag{5-2b}$$

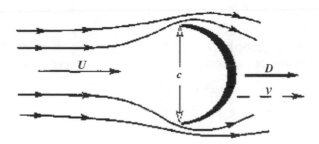

Figure 5-2. Translating drag device.

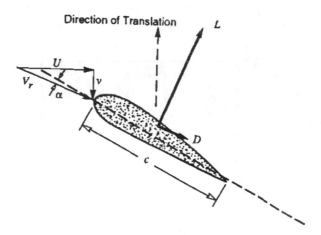

Figure 5-3. Translating airfoil with lift and drag forces acting.

Lifting Translator

By contrast, the lifting translator does much better. Figure 5-3 illustrates an airfoil that is translating at right angles to the wind direction and is subject to both lift and drag forces. This *lifting surface* sees a relative velocity that is the vector sum of the free-stream wind velocity and the wind speed induced by translation. The angle from the direction of the relative velocity to the *chord line* of the airfoil is termed the *angle of attack*. Lift and drag forces are perpendicular and parallel, respectively, to the relative wind and are given by

$$L = 0.5\,\rho\,V_r^2\,C_L\,c$$

$$D = 0.5\,\rho\,V_r^2\,C_D\,c \tag{5-3}$$

where

L = aerodynamic lift force per unit length of airfoil (N/m)
D = aerodynamic drag force per unit length of airfoil (N/m)
C_L, C_D = lift and drag coefficients, respectively; functions of airfoil shape and α
α = angle of attack

Analysis of the airfoil as a *free body* yields the power extracted as

$$P = 0.5\,\rho\,U^3\,A_P\,\frac{v}{U}\left(C_L - C_D\,\frac{v}{U}\right)\sqrt{1 + \left(\frac{v}{U}\right)^2} \tag{5-4}$$

At maximum power v/U approximately equals $(2/3)\,C_L/C_D$ (*i.e.* 2/3 of the *lift-to-drag ratio*). Therefore, the maximum power coefficient for an airfoil translating at right angles to the wind is given by

$$C_{P,\,\text{max}} \doteq (2/9)\,C_L\,(C_L/C_D)\sqrt{1 + (4/9)\,(C_L/C_D)^2} \tag{5-5}$$

Figure 5-4 is a comparison of lift and drag as mechanisms to extract power from the wind which readily shows the advantages or using lilting surfaces. Equations (5-2) and (5-4) are used to construct the curves in this figure. The aerodynamic properties are $C_L = 1.0$ and $C_D = 0.10$ for the airfoil, and $C_{D,\,\text{max}} = 2.0$ for the drag-driven device. The airfoil has a maximum

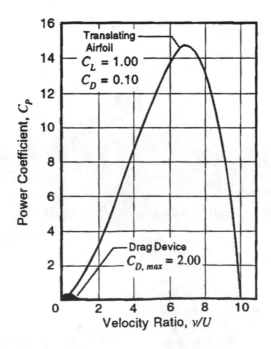

Figure 5-4. Comparison of typical power coefficients of a translating airfoil and a translating drag device. The airfoil is moving at right angles to the wind direction.

power coefficient of 15.0, compared with 0.3 for the drag device. Thus, lift devices can quite readily produce 50 times the power per unit of projected area than that produced by drag devices. Moreover, operating a lifting device at velocities well in excess of the wind velocity is easily achieved by rotating machines. It is further noted that the maximum power coefficient of any rotary machine using drag is also less than $(4/27)C_{D,max}$ based on the projected area of the drag elements.

With the superiority of the lifting translator established, the concept of placing lifting surfaces on a rotating machine to form a turbine is seen to be an obvious method of converting wind energy to useful work.

Performance Parameters

The *power performance* of a wind turbine can be expressed in dimensionless form in two ways. First, for a fixed wind speed, the *power coefficient*, C_p, and the *tip-speed ratio*, λ, are used. The power coefficient is defined in Equation (5-2), in which A_p is now the projected area of the *moving rotor* (called the *swept area*) and the tip-speed ratio is

$$\lambda = (v/U)_{max} = R\Omega/U \tag{5-6}$$

where λ = tip-speed ratio
R = maximum rotor radius (m)
Ω = rotor speed (rad/s)

The power in the Equation (5-2) can be either the rotor output, in which case we have the *rotor power coefficient*, $C_{P,r}$, or the system output power, in which case we have the *system power coefficient*, $C_{P,s}$. The difference between these two outputs is the power-train and electrical equipment losses.

The second dimensionless form for expressing performance is for a fixed rotor angular speed, in which the *advance ratio*, J, and a *rotor speed power coefficient*, K_p are used. These parameters are defined by

$$J = (U/v)_{min} = \frac{U}{R\Omega} = \frac{1}{\lambda} \tag{5-7a}$$

$$K_P = \frac{P}{0.5\,\rho\,R^3\Omega^3 A} = \frac{C_P}{\lambda^3} \tag{5-7b}$$

As before, K_p can be given for the rotor or for the entire system.

Figures 5-5 and 5-6 illustrate $C_{P,r}$ as a function of λ and $K_{P,r}$ as a function of J for a typical HAWT operating at fixed pitch. Operating points A, B, C, and D are shown in both figures. The left-hand side of Figure 5-5 (ABC) is controlled by *blade stall*. Local *angles of attack* (angles between the relative wind and the blade *chord line*) are relatively large as point A is approached. Changes in blade *pitch angle* have a great effect on power output along segment ABC. The right-hand side of Figure 5-5 (CD) is controlled by drag, particularly *skin friction*, because the angles of attack are small as point D is approached.

The plot of K_p versus J is a dimensionless plot of *power vs. wind speed*, or a *power curve*, at fixed rotor speed and blade pitch. Note that maximum power, B, does not occur at the maximum power coefficient point, C. At fixed pitch, maximum power occurs in the stall region when the lift coefficient is near its peak value over much of the blade.

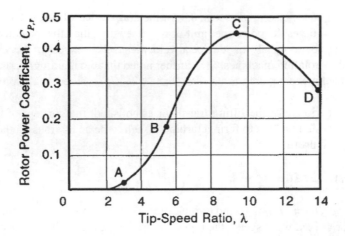

Figure 5-5. Typical plot of rotor power coefficient *vs.* tip-speed ratio for a HAWT with a fixed blade pitch angle.

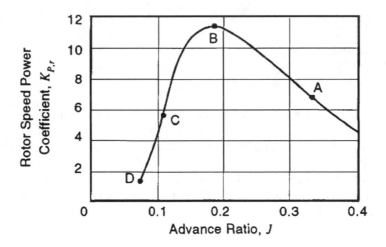

Figure 5-6. Rotor speed power coefficient *vs.* advance ratio for the same HAWT. Operating points A, B, C, and D are the same as in Figure 5-5.

Annual Energy Output

One of the most significant measures of the cost-effectiveness of a wind turbine is its annual production of energy. In the design and analysis of wind turbines, the *annual energy output* is calculated as illustrated by the flow chart in Figure 4-35. All of the steps in this process are normally the responsibility of the aerodynamic specialist. Calculation of annual energy output requires knowledge of the *wind speed frequency distribution* (probability that the wind speed will be within a given range) and the system power output of each turbine as a function of wind speed. Furthermore, every prediction of annual energy output is site-specific, depending on the local wind flow patterns and turbulence, the number and type of neighboring turbines, and the local air density.

The frequency distribution required is that of the wind speed at the elevation of the center of the rotor's swept area (*hub elevation* in a HAWT or *equatorial elevation* in a VAWT) at the site where the annual energy output is to be determined. When this frequency distribution is obtained from anemometer test data, it is frequently expressed in the form of a *histogram*, which gives the number of hours per year that the wind speed is within each given range or *bin*. Bin width is usually 0.5 to 1.0 m/s. System output power as a function of wind speed is also needed, calculated using the *mean air density* at the selected site. The general equation for calculating *gross annual energy output* (*i.e.* output exclusive of energy consumed for station-keeping, down-time losses, array effects, etc.) is

$$AEO_g = \sum_{k=1}^{K} \Delta E_{a,k} = \sum_{k=1}^{K} P_k \Delta t_k \tag{5-8}$$

where

AEO_g = gross annual energy output (kWh/y)
k = index of wind speed bin, from 1 to K
$\Delta E_{a,k}$ = gross annual energy output in the kth wind speed bin (kWh/y)
P_k = average power output over the wind speed range ΔU_k (kW)
Δt_k = cumulative time the wind speed at the elevation of the center of the swept area is in the range ΔU_k (h/y)

Table 5-1 illustrates the application of Equation (5-8) using design performance data for the Boeing Mod-2 2.5-MW HAWT [ASME 1988]. Analytical expressions for the system power and for the wind frequency distribution (*e.g.* the single-parameter *Rayleigh distribution* and the two-parameter *Weibull distribution*) can sometimes produce closed-form estimates for annual energy production, but the tabular approach shown in Table 5-1 recommended because of its clarity and flexibility. The accuracy obtained in representing actual power curves and wind distributions by analytical expressions is often poor.

Table 5-1. Typical Calculation of Gross Annual Energy Output [ASME 1988]

Bin k	Wind speed range ΔU_k (m/s)	Duration Δt_k (h/y)	Power output [1] P_k (kW)	Energy output $\Delta E_{a,k}$ (MWh/y)
1	0.0 - 6.25	2,147	0	0
2	6.25 - 6.75	416	175	73
3	6.75 - 7.25	440	318	140
4	7.25 - 7.75	458	460	211
5	7.75 - 8.25	468	603	282
6	8.25 - 8.75	470	745	350
7	8.75 - 9.25	466	949	442
8	9.25 - 9.75	453	1,153	522
9	9.75 - 10.25	435	1,316	572
10	10.25 - 10.75	410	1,479	606
11	10.75 - 11.25	381	1,642	626
12	11.25 - 11.75	349	1,805	630
13	11.75 - 12.25	314	1,968	618
14	12.25 - 12.75	278	2,120	589
15	12.75 - 13.25	242	2,263	548
16	13.25 - 13.75	208	2,385	496
17	13.75 - 14.25	175	2,500	438
18	14.25 - 22.40	648	2,500	1,620
19	>22.40	2	0	0
Totals:		8,760		8,763

AEO_g = 8,763 MWh/y

[1] at an air density of 1.15 kg/m³

Actuator Disk Theory for Horizontal-Axis Wind Turbines

The simplest aerodynamic model of a HAWT is the *actuator disk model*, in which the rotor becomes a homogeneous disk that removes energy from the wind. Originated by Rankine [1895], the actuator disk concept was motivated by the development of the marine propeller. The actuator disk provides a rational basis for illustrating that the flow velocity at the rotor is different from the free-stream velocity. However well the actuator disk theory provides an understanding of the flow field, this theory fails to provide a link between rotor geometry and rotor performance.

Rankine-Froude Theory

The *axial momentum theory* [Rankine 1865, W. Froude 1878, R. Froude 1889] idealizes flow through the rotor of a wind turbine as shown in Figure 5-7. The free-stream wind speed is U, which is slowed by the wind device. Applying requirements for *continuity*, momentum, and energy to the flow, we may determine the *thrust* and *power* if the flow is assumed to be entirely *axial*, with no rotational motion. Two expressions for thrust may be obtained. First, from the momentum theorem, the thrust is

$$T = \dot{M}_a (U - V_1) = \rho A V (U - V_1) \qquad (5-9)$$

where

$\quad T$ = thrust force on the disk (N)
$\quad \dot{M}_a$ = air mass flow rate through the disk (kg/s)
$\quad V$ = wind velocity at the disk (m/s)
$\quad V_1$ = wind velocity in the far wake (m/s)

Second, from consideration of the pressure drop caused by the wind machine,

$$T = A (p_u - p_d) \qquad (5-10)$$

where p_u, p_d = pressures upwind and downwind of the disk, respectively (N/m²)

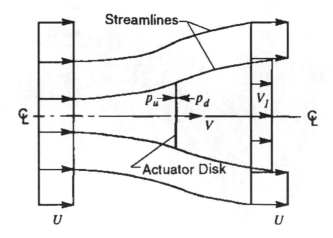

Figure 5-7. Idealized flow through a wind turbine represented by a non-rotating, actuator disk.

Now, the *Bernoulli equation* may be used between the free-stream and the upwind side of the turbine and again between the downwind side of the turbine and the far wake, so that Equation (5-10) becomes

$$T = 0.5\rho A \ (U^2 - V_1^2) \tag{5-11}$$

Combining Equations (5-9) and (5-11) we obtain

$$V = 0.5(U + V_1) \tag{5-12}$$

Thus, the wind velocity at the disk is the average of the free-stream and far-wake velocities, so the total velocity change from free-stream to far-wake is twice the change from free-stream to the disk. Let

$$U - V = aU \tag{5-13a}$$

Then
$$U - V_1 = 2aU \tag{5-13b}$$

The term a is known as the *axial induction factor* (or the *retardation factor*) and is a measure of the influence of the turbine on the wind. Because the minimum far-wake velocity is zero, the maximum value of the axial induction factor is 0.5. The thrust is not of immediate importance, but the power is. From the *first law of thermodynamics*, assuming isothermal flow and ambient pressure in the far wake, power is equal to

$$P = 0.5\rho A(U^2 - V_1^2)V = 0.5\rho AV(U + V_1)(U - V_1) \tag{5-14}$$

Combining Equations (5-13) and (5-14), the power coefficient for the actuator disk, according to the Rankine-Froude theory, is

$$C_P = \frac{P}{0.5\rho U^3 A} = 4a(1 - a)^2 \tag{5-15a}$$

which has a maximum when $a = 1/3$. Thus

$$C_{P,\,\text{max}} = (16/27) = 0.593 \tag{5-15b}$$

When examining Equation (5-15a), note that the denominator is the kinetic energy of the free-stream wind contained in a *streamtube* with an area equal to the disk area. However, Equation (5-15b) does not represent maximum *efficiency* (power output/power input), since the mass flow rate through the disk is not ρAU but ρAV. Instead,

$$\eta_d = \frac{P}{0.5\rho U^2 VA} = 4a(1 - a) \tag{5-16}$$

where η_d = actuator disk efficiency.

The maximum efficiency is 100 percent at $a = 0.5$, which yields a power coefficient of 0.5. The disk efficiency is 88.8 percent at the maximum power coefficient of 0.593.

General Momentum Theory with Wake Rotation

Further idealized-flow modeling can be accomplished with the additional consideration of wake rotation. While the initial free-stream wind is not rotational, interaction with a rotating wind machine will cause the wake to rotate. In the case of a propeller, the wake rotates in the same direction as the blades. In the case of an energy-extracting device (such as a HAWT), the wake rotates in the opposite direction. If there is rotational kinetic energy in the wake in addition to translational kinetic energy, we may expect (from thermodynamic considerations) less power extraction than if the wake has translational kinetic energy only.

Joukowski [1918] considered the effect of wake rotation in the analysis of propellers. Adopting his notation to the analysis of wind turbines, the effect of wake rotation on power removal may be estimated. Using a streamtube analysis, equations can be written that express the relation between the wake velocities (both axial and rotational) and the corresponding wind velocities at the rotor disk. In addition, for certain special cases, an expression for the power coefficient can be obtained. The main outcome of this approach is a measure of the effects of rotation on the relative values of the induced (retarded) velocities at the rotor and in the wake.

Figure 5-8 illustrates the geometry of the *streamtube model* of wind flow through a HAWT. Assuming fluid drag is zero, the resulting equations are

Continuity:
$$V r \, dr = V_1 \, r_1 \, dr_1 \tag{5-17a}$$

Moment of momentum:
$$r^2 \omega = r_1^2 \omega_1 \tag{5-17b}$$

Energy:
$$0.5(U - V_1)^2 = \left(\frac{\Omega + \omega_1/2}{V_1} - \frac{\Omega + \omega/2}{V} \right) V_1 \omega_1 r_1^2 \tag{5-17c}$$

where

r, r_1 = radial coordinates at the rotor and at the far wake, respectively (m)
dr, dr_1 = radial thickness of the streamtube at rotor and far wake, respectively (m)
ω, ω_1 = angular velocity in the fluid at rotor and far wake, respectively (rad/s)

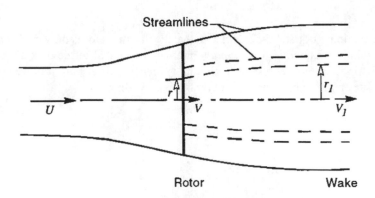

Figure 5-8. Geometry of the streamtube model of flow through a HAWT rotor.

Finally, an expression for the radial gradient in axial velocity may be obtained from Euler's equation:

$$\frac{d}{dr_1}\left(\frac{U^2 - V_1^2}{2}\right) = (\Omega + \omega_1)\frac{d}{dr_1}\left(\omega_1 r_1^2\right) \qquad (5\text{-}17\text{d})$$

These four equations may be used to obtain the relationships among thrust, torque, and flow in the wake. Closure cannot be obtained without specification of one of the variables, say ω. When this is done, the flow is found to have the following features:

-- The pressure varies across the wake due to the rotational velocity.
-- The rotor and wake axial velocities vary radially.
-- The angular velocity of the fluid, which is opposite in direction to the rotation of the rotor, changes discontinuously at the rotor.

In the expression for the wake radial velocity gradient, Equation (5-17d), let us assume that $r^2\omega$ is constant (*i.e.*, the rotor wake is an *irrotational vortex*). Then, the wake axial velocity is constant along a radius, because this equation's right-hand-side is zero. Defining

$$V = U(1 - a) \qquad (5\text{-}18\text{a})$$
$$V_1 = U(1 - b) \qquad (5\text{-}18\text{b})$$

where a = axial induction (retardation) factor at the disk
              b = axial induction (retardation) factor at the far-wake

We may obtain from Equation (5-17c)

$$a = \frac{b}{2}\left[1 - \frac{(1-a)\,b^2}{4\,\lambda^2\,(b-a)}\right] \qquad (5\text{-}19\text{a})$$

where, as before, λ is the tip-speed ratio. The power coefficient is given by

$$C_P = \frac{b^2\,(1-a)^2}{b-a}$$

Examination of Equation (5-19a) shows that the axial velocity reduction at the disk is always approximately one-half the reduction in the far-wake for tip-speed ratios above 2, which is the same result reached when wake rotation was neglected. The above equation for the power coefficient requires some modification, since the assumption that $r^2\omega$ is constant produces infinite velocities near the axis. In lieu of an irrotational vortex wake, we may substitute a *Rankine vortex wake*, which contains a rotational core with a constant angular velocity equal to the maximum specified for the rotor. This leads to

$$C_P = \frac{b\,(1-a)^2}{b-a}\left[b + (2a - b)\,\Omega/\omega_{\max}\right] \qquad (5\text{-}19\text{b})$$

where $\omega_{\max}$ = angular velocity of the wake core (rad/s)

The maximum power coefficient for a rotor with a Rankine vortex wake is shown in Figure 5-9. As would be expected, the power coefficient is insensitive to ω_{max} at high tip-speed ratios, where the torque and consequently the wake rotation are the least.

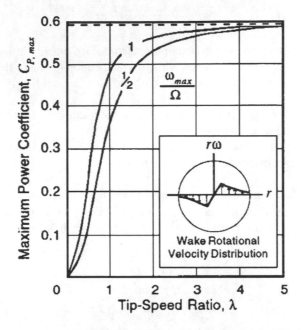

Figure 5-9. Maximum power coefficient for a HAWT rotor with a Rankine vortex wake, as a function of tip-speed ratio and the wake core rotational velocity.

Optimum Actuator Disk Theory

Glauert [1935] developed a simple model for the optimum HAWT rotor. The approach used is to treat the rotor as a rotating actuator disk (*i.e.*, a rotor with an infinite number of blades) and set up an integral for the power. The power integral is made stationary subject to an energy constraint and yields the maximum power output for a given tip-speed ratio. The power extracted from a streamtube (Fig. 5-8) is given by

$$dP = \Omega \, dQ = \Omega \, r \, \rho \, V \, (r \, \omega) \, (2 \, \pi \, r \, dr) \qquad (5\text{-}20)$$

where dP = increment of power extracted from the streamtube (W)
   Q = rotor torque (N-m)
   dQ = increment of rotor torque extracted from the streamtube (N-m)

The first term in parenthesis is the change in tangential velocity, and the second term in parenthesis is the streamtube annulus area. The power coefficient equations may be written

$$C_P = \frac{8}{\lambda^2} \int_0^\lambda (1 - a) \, a' \, x^3 \, dx$$

$$x = r \, \Omega / V \qquad (5\text{-}21)$$

$$a' = \frac{\omega}{2 \, \Omega}, \; rotational \; induction \; factor$$

Since this integral for the power involves two dependent variables, another relation is required. This is the *momentum equation*, as follows:

$$a' (1 + a') x^2 = a (1 - a) \tag{5-22}$$

A unique way of illustrating this relation is to consider the velocities at the rotor *plane of rotation* of a HAWT, as shown in Figure 5-10. The flow is assumed to be uniform in annular streamtubes with no circumferential variations. Under these conditions, two-dimensional flow may be assumed. In the absence of drag, the velocity *induced* at the rotor must be caused by lift and, hence, be perpendicular to the relative velocity. Two expressions for the tangent of the angle from the plane of rotation to the relative wind vector may be developed under the condition that the total induced velocity is perpendicular to the relative velocity. These are

$$\tan \phi = \frac{(1 - a) U}{(1 + a') r \Omega} = \frac{(1 - a)}{(1 + a') x} \tag{5-23a}$$

$$\tan \phi = \frac{a' r \Omega}{a U} = \frac{a'}{a} x \tag{5-23b}$$

Equating the right-hand-sides of Equations (5-23) also produces Equation (5-22). The *Calculus of Variations* can now be used to solve Equation (5-21) with the constraint of Equation (5-22), obtaining the following relationship between the rotational and axial induction factors:

$$a' = \frac{1 - 3a}{4a - 1} \tag{5-24}$$

so that

$$x = (4a - 1) \sqrt{\frac{1 - a}{1 - 3a}} \tag{5-25}$$

Hence, *1/3 > a > 1/4*. A tabulation of variations in the parameters x, a, a', ϕ, and C_p is given in Table 5-2. Since high-speed rotors easily reach tip-speed ratios of 7 or more, it can be seen that most of an ideal rotor will operate with $a = 1/3$ and a rotational velocity distribution in the form of an *irrotational vortex*. The power coefficient for various tip-speed ratios is also given in Table 5-2, by equating λ to x. At low tip-speed ratios, power coefficients are low because of the large rotational kinetic energy in the wake. At high tip-speed ratios, the power coefficient approaches 0.593 and the wake rotation approaches zero.

Further information may be obtained from the Glauert optimum disk model using the *blade-element theory*. Blade-element theory equates the thrust on a radial increment of blade of length dr to the momentum change in a flow annulus (streamtube) of area $2\pi r dr$. The blade torque is then equated to the moment of momentum in a similar fashion. As the quantities a and a' are known for each radial position, the relative velocity and the angle ϕ may be determined. Figure 5-10 illustrates the velocities and forces in relation to the blade configuration. Of course, since we have assumed that the drag is zero, the only force that acts on the blade is lift.

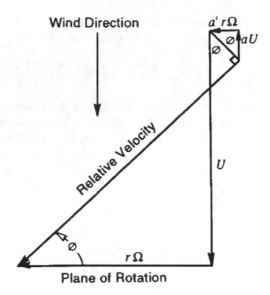

Figure 5-10. Flow velocity diagram at an annulus in a HAWT rotor disk.

Table 5-2.
Flow Conditions and Rotor Parameters for the Glauert Optimum HAWT Rotor

Velocity ratio x	Axial induction factor a	Rotational induction factor a'	Relative inflow angle ϕ (deg)	Rotor power coefficient $C_{p,\,r}$	Blade-element parameter $Bc\Omega C_L / 2\pi U$
0.25	0.280	1.364	50.6	0.176	0.3658
0.50	0.298	0.543	42.3	0.289	0.5205
0.75	0.310	0.294	35.4	0.364	0.5552
1.00	0.317	0.183	30.0	0.416	0.5359
1.25	0.322	0.124	25.8	0.451	0.4974
1.50	0.324	0.089	22.5	0.477	0.4551
1.75	0.326	0.067	19.8	0.496	0.4151
2.00	0.328	0.052	17.7	0.511	0.3791
2.50	0.330	0.034	14.5	0.532	0.3200
3.00	0.331	0.024	12.3	0.545	0.2750
3.50	0.331	0.018	10.6	0.555	0.2403
4.00	0.332	0.014	9.4	0.562	0.2129
4.50	0.332	0.011	8.4	0.566	0.1909
5.00	0.332	0.009	7.5	0.570	0.1729
5.50	0.332	0.007	6.9	0.573	0.1580
6.00	0.333	0.006	6.3	0.576	0.1453
7.00	0.333	0.004	5.4	0.580	0.1252
8.00	0.333	0.004	4.8	0.582	0.1099
9.00	0.333	0.003	4.2	0.584	0.0979
10.00	0.333	0.002	3.8	0.585	0.0883
11.00	0.333	0.002	3.5	0.586	0.0803
12.00	0.333	0.002	3.2	0.587	0.0737

The incremental thrust, dT, and torque, dQ, acting on an annulus are given by

$$dT = 0.5\,\rho\,V_r^2\,C_L c\,B\,\cos\phi\,dr \tag{5-26a}$$

$$dQ = 0.5\,r\,\rho\,V_r^2\,C_L c\,B\,\sin\phi\,dr \tag{5-26b}$$

where

V_r = relative wind velocity (m/s)
c = blade chord length at radius r (m)
B = number of rotor blades

Assuming that the far-wake axial induction is twice the axial induction at the rotor, momentum expressions yield

$$dT = 4\,\pi\,r\,\rho\,U^2\,(1 - a)\,a\,dr \tag{5-27a}$$

$$dQ = 4\,\pi\,r^3\,\rho\,U\,\Omega\,(1 - a)\,a'dr \tag{5-27b}$$

Combining Equations (5-26) and (5-27) then gives

$$\frac{a}{1 - a} = \frac{B\,c\,C_L\,\cos\phi}{8\,\pi\,r\,\sin^2\phi} \tag{5-28a}$$

$$\frac{a'}{1 + a'} = \frac{B\,c\,C_L}{8\,\pi\,r\,\cos\phi} \tag{5-28b}$$

Now, a, a', and ϕ are known as functions of x, so a dimensionless blade parameter equal to $(Bc\Omega C_L /2\pi U)$ may be determined. Results are listed in the right-hand column of Table 5-2. Note that an optimum blade for a given tip-speed ratio and constant C_L will have a chord that approaches a maximum at x approximately equal to 0.75.

To illustrate the use of Table 5-2 to determine an optimum blade shape, consider a two-bladed rotor designed to operate at $\lambda = 6$ with $C_L = 0.8$. The left-hand column of the table may be used to determine radial position along the blade, since

$$\frac{r}{R} = \frac{r\Omega /U}{R\Omega /U} = \frac{x}{\lambda} = \frac{x}{6} \tag{5-29a}$$

The required blade chord can be obtained from the right-hand column as follows:

$$\frac{c}{R} = \left(\frac{B\,c\,\Omega\,C_L}{2\,\pi\,U} \right) \frac{2\,\pi}{B\,C_L} \frac{U}{R\Omega} = 0.654 \left(\frac{B\,c\,\Omega\,C_L}{2\,\pi\,U} \right) \tag{5-29b}$$

The required twist angle, β, of the *zero-lift line* (Fig. 5-10) can be obtained from the relationship $\beta = \phi - \alpha$, where α is the angle of attack. Assuming *ideal lift*, for which $C_L = 2\pi \sin \alpha$, the angle of attack is 7.3 degrees for $C_L = 0.8$. Thus

$$\beta = \phi - 7.3° \tag{5-29c}$$

Thus, although the effects of finite blade number are not included, Table 5-3 gives an indication of the rotor geometry required to achieve maximum C_P at a tip-speed ratio of 6. The blade is noted to have *nonlinear taper and twist*. In order to determine the effects of finite blade number, drag, and operation at other tip-speed ratios, another approach is required.

Table 5-3.
Blade Chord and Twist Distribution for an Optimum Two-Bladed HAWT
(Lift coefficient = 0.8 and tip-speed ratio = 6)

Velocity ratio x	Radial position r/R	Chord ratio c/R	Twist angle β (deg)
0.75	0.125	0.363	28.1
1.50	0.250	0.298	15.2
3.00	0.500	0.180	5.0
4.50	0.750	0.125	1.0
6.00	1.000	0.095	-1.0

Flow States

In the previous analysis, it was tacitly assumed that the wind turbine is operating as a power-extraction device that produces a downwind force and retards the free-stream wind speed. This requires that the axial induction factor be between zero and unity. Figure 5-11(a) shows flow patterns or *flow states* and thrust force vectors T associated with a wide range of

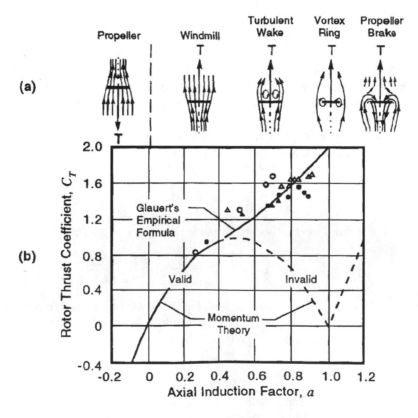

Figure 5-11. Relationship between the axial induction, flow state, and thrust of a rotor. [Lock *et al.* 1926, Wilson *et al.* 1976]

induction factors [Lock *et al.* 1926, Wilson *et al.* 1976]. A *thrust coefficient*, which can also be used to characterize the different flow states of a rotor, is shown in Figure 5-11(b) as a function of the axial induction factor, and is defined as

$$C_T = \frac{T}{0.5\,\rho\,A\,U^2} \tag{5-30a}$$

By Momentum Theory: $\qquad C_T = 4\,a\,|1-a| \tag{5-30b}$

where $\quad C_T$ = rotor thrust coefficient

Wind turbines will normally operate in the *windmill state*, with $0 \le a \le 0.5$. For negative inductions ($a < 0$) it is simple to continue the analysis to show that the device will act as a propulsor producing an upwind force (*i.e.* $C_T < 0$) that adds energy to the wake. This is typical of the *propeller state*.

When a wind turbine rotor operates at tip-speed ratios appreciably above its design value, blade tips may be driven into the *turbulent wake state*. As illustrated by the data in Figure 5-11(b) obtained on wind turbines, autogiros, and helicopters, rotor thrust increases with increasing induction in the turbulent wake state, instead of decreasing as predicted by Equation (5-30b). Thus, momentum theory is considered to be invalid for induction factors larger than about 0.4. Glauert's empirical formula [1926] and, more recently, Equation (2-16) [Spera 2008] are acceptable models of rotor thrust behavior for induction factors from 0.4 to 1.0, or, equivalently, $0.96 < C_T < 2.0$ [Dugundji *et al.* 1978].

When the induction factor is somewhat over unity, the flow regime is called the *vortex ring state*, a condition which is experienced by helicopters during powered (slow) descent. A particularly interesting case occurs for axial inductions greater than unity, where the rotor reverses the direction of flow. This may be physically modeled by considering a powered propeller with blades pitched so that they induce a forward flow. This is termed the *propeller brake state*, with power being added to the flow to create downwind thrust on the rotor. Further discussion of wind rotor states is given by Eggleston and Stoddard [1987].

Summary Comments on Actuator Disk Theory

The actuator disk models discussed in this section provide limiting values of rotor performance and a general understanding of the rotor configuration for a particular operating condition. For example, there is a maximum power coefficient of 16/27 for HAWTs, and this limit (called the *Betz* or *Lanchester-Betz* limit) can be approached when the wake rotation is low. Further, we noted that wind turbines operating at high values of induction develop forces that are considerably different from the axial forces predicted by the momentum relation, Equation (5-30). The assumptions of actuator disk theory, particularly the assumption of an infinite number of blades, restrict our understanding of the effect of rotor geometry (*i.e.* blade airfoil section, chord, and twist) on HAWT operation. Additionally, we find that the Betz limit is higher than the power coefficients achieved in practice, because actual rotors (1) have a finite number of blades and (2) are acted upon by drag forces.

Strip Theory for Horizontal-Axis Wind Turbines

In order to bridge the gap between actuator disk models of wind turbines and a rigorous *vortex theory*, an intermediate theory known as strip theory has been developed that relates rotor performance to rotor geometry. A particularly important prediction of strip theory is the effect of *finite blade number*. We find that performance-optimized HAWTs do have configurations similar to the Glauert-optimized rotor when finite blade number and drag are included. In the development of strip theory, the relation between the local induction and the local thrust coefficient in each streamtube is of paramount importance.

Background

A HAWT can be considered to be an *airscrew* that extracts kinetic energy from the driving air and converts it into mechanical energy. The similarity of a HAWT to a propeller (which puts energy into the air) enables the same theoretical development used for the propeller to be followed for the HAWT. Propeller theory was developed along two independent approaches: *Actuator disk theory* (which was discussed in the previous section) and *blade-element theory*. The strip theory presented here has been called *modified blade element theory*.

Blade-element theory was originated by Froude [1878] and later developed further by Drzewiecki [1892]. The approach of blade-element theory is opposite that of momentum theory since it is concerned with the forces produced by the blades as a result of the motion of the fluid. It was hampered in its original development by the lack of knowledge of sectional aerodynamics and the mutual interaction of blades. Modern rotor theory has developed from the concept of *free vortices* being shed from the rotating blades. These vortices define a *slipstream* and generate induced velocities. This rigorous theory can be attributed to the works of Lanchester [1907] and Flamm [1909], for the original concept; to Joukowski [1912], for induced velocity analysis; to Betz [1919], for optimization; to Prandtl [1919] and Goldstein [1929], for *circulation distribution* or *tip-loss* analysis; and to Glauert [1922a, 1922b, 1935], Pistolesi [1922], and Kawada [1926], for general improvements.

It has been found that strip-theory approaches are adequate for the analysis of wind machine performance. One reason is that a wind turbine wake expands rather than contracts. At high tip-speed ratios (*i.e.* low *advance ratios*), propellers and helicopter rotors have been observed to shed strong tip vortices. Since these wakes are contracting, the shed vortices are inboard of the tip and interact strongly with the flow through the rotor disk. The resulting radial distribution of aerodynamic forces is found to be appreciably different from that predicted by strip theory. Because most wind turbines operate at high tip-speed ratios, they might be expected to experience the same strong interaction. However, because of the expanding wake, the tip vortex moves outboard of the rotor, negating a strong interaction. From an outboard position, the tip vortex generates induced velocities that decrease local angles of attack and reduce aerodynamic loads.

Various forms of strip theory have been the standard methods of design and design analysis of HAWTs. Strip theories are easy to program, inexpensive to run, and readily adaptable to any size of computer. They are used with modest success to predict output power. However, it is important to note that the largest sources of error in power prediction have been in the airfoil lift and drag data, and these errors are frequently large enough to mask the inaccuracies of the theory.

Flow Model

The foremost assumption in strip theory is that individual *streamtubes* or *strips* (the intersection of a streamtube and the surface swept by the blades) can be analyzed independently of the rest of the flow. Such an assumption works well for cases where the circulation distribution over the blade is relatively uniform, so that most of the vorticity is shed at the blade root and the blade tip. The deployment of *control surfaces*, however, can violate this assumption. For example, the use of *partial-span pitch control* on a HAWT rotor introduces discontinuities in circulation, and appreciable vorticity can be shed near the junction between the inboard blade and the moveable tip. Fortunately, control surfaces are usually deployed to spill excess power at high wind speeds, when the induced velocity is relatively small. Thus, calculation errors that arise when strip theory methods are used to analyze rotors with deployed control surfaces are not usually significant.

A second assumption associated with the development of strip theory is that *spanwise flow* is negligible, and therefore airfoil data taken from two-dimensional section tests are acceptable. Strip theory does not predict any induced flows along the blades. However, when a blade is not perpendicular to the axis of rotation (*e.g.*, when the blade has a *coning angle*) the wind has a component that is directed along the span of the blade. This component is neglected and two-dimensional flow is assumed, adding some error to the airfoil data. A third assumption is that flow conditions do not vary in the circumferential direction. With this assumption, the "strip" to be analyzed is a uniform annular ring centered on the axis of revolution.

Figure 5-12 shows the strip geometry and the coordinate system used. Note that the coordinate s is measured along the blade and r is at right angles to the rotor axis. The wind velocity, U, is assumed to be constant in time and space and aligned with the axis of rotation, so that the *yaw error*, $\Delta\Psi$, is zero. When the rotor blades are coned, the velocity diagram in

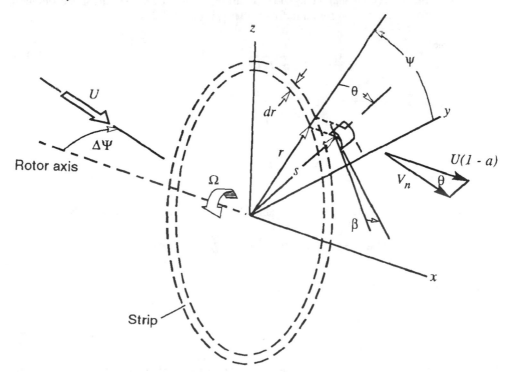

Figure 5-12. Strip geometry and coordinate system.

Figure 5-10 is altered somewhat, to that shown in Figure 5-13. The wind velocity normal to the blade reference plane, V_n, is equal to the free-stream wind velocity reduced by the axial induction factor, a, and the cosine of the *coning angle*, θ. The cosine of the coning angle also appears in the tangential velocity, v, because the local radial distance to the axis, r, is equal to $s \cos\theta$. In addition, a *drag force, D,* is now included.

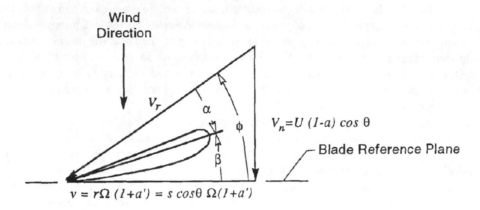

Figure 5-13. Flow velocity diagram at an annulus in a coned HAWT rotor disk.

Annulus Flow Equations

Momentum and moment of momentum relations are used to obtain equations with which to determine the induced axial and tangential (rotational) velocities. The thrust coefficient in Equation (5-30) can be written for an individual streamtube as follows:

$$C_t = \frac{dT}{0.5\,\rho\,U^2\,(2\,\pi\,r\,dr)} \tag{5-31a}$$

$$T = \int_0^R dT$$

where C_t = local value of the thrust coefficient, at radial coordinate r
        dT = increment of axial thrust on the blade area within the streamtube (N)

Referring to Figure 5-13, the thrust increment, dT, for a rotor with B blades of chord c is

$$dT = 0.5\,\rho\,V_r^2\,B\,c\,(C_L \cos\phi + C_D \sin\phi)\,dr \tag{5-31b}$$

Combining Equations (5-31a) and (5-31b),

$$C_t = \frac{B}{2\pi}\left(\frac{c}{r}\right)\left(\frac{V_r}{U}\right)^2 (C_L \cos\phi + C_D \sin\phi)$$

From Figure 5-13, we can express the velocity ratio in terms of the wind angle, ϕ, and the axial induction factor, a, and obtain

$$C_t = \frac{B}{2\pi} \left(\frac{c}{r}\right) (1-a)^2 \left(\frac{\cos^2\theta}{\sin^2\phi}\right) (C_L\cos\phi + C_D\sin\phi) \qquad (5\text{-}32a)$$

where θ = coning angle, measured from the plane of rotation (rad)

Turning our attention to the tangential direction, examination of the fluid torque about the axis of rotation for the streamtube and the increment of rotor torque from lift and drag forces on the blade segment within the streamtube, it can be shown that Equation (5-22) is applicable, with a minor change to account for coning, as follows:

$$a'\,(1+a')\,x^2 = a\,(1-a)\cos^2\theta \qquad (5\text{-}32b)$$

Thus, Equations (5-32) are the relations that determine the dimensionless induced velocities a and a'. Before these equations can be used, however, the local thrust coefficient, C_t, must be modified to account for two effects: the departure of the local thrust coefficient from the momentum relation (as discussed in the previous section), and the non-uniformity of the induced velocities in the flow, particularly near its outer edges. These so-called *tip losses* will be discussed first.

The Tip-Loss Factor

Strip theory, as previously developed, does not account for the interaction of shed vorticity with the blade's *bound vorticity*. This effect is usually greatest near the blade tip, although strong vortex interaction can occur at deployed control surfaces. Denoting the *bound circulation* of all the blades by Γ, a *tip-loss factor* F is defined as

$$F = \Gamma/\Gamma_\infty \qquad (5\text{-}33)$$

where Γ_∞ = bound circulation of a rotor with $B \rightarrow \infty$ and $c \rightarrow 0$ (m²/s)

Recalling that the induced rotational velocity is directly proportional to the local vorticity intensity, one can give a useful physical interpretation to F, which is that it is equal to the ratio of mean induced velocity in the flow annulus to the induced velocity at the blades.

The flow in the blade tip region significantly affects rotor torque and thrust. Focusing on these rather than the flow field, we can account for the diminished thrust and torque output of the tip regions by defining an *effective radius* of the rotor, which is about 3 percent smaller than the tip radius. In this empirical approach, we set

$$
\begin{aligned}
F = 1 \quad &if \quad 0 < r < r_e \\
F = 0 \quad &if \quad r_e \leq r \leq R \\
r_e \approx 0.97\,R &
\end{aligned}
\qquad (5\text{-}34)
$$

The interaction between shed and bound vorticity is like a light switch: either on or off.

Propeller development in the 1920s called for better flow models, and the behavior of the wake of the *optimum propeller* resulted in the development of two tip-loss models. The optimum propeller, as conceived by Betz [1919], has a *shed vortex sheet* that appears to move like a rigid body as it is convected away from the rotor. Prandtl [1919] noted that the flow at the edges of the Betz wake appeared to be like the flow over an infinite stack of round, parallel plates. He developed a simple and ingenious approximation for F expressed solely in terms of local flow parameters. While the Prandtl model was useful, questions as to its accuracy remained. Goldstein solved the problem of the flow about a lightly-loaded optimum rotor (with a constant wake diameter), and this "exact" solution was used as a test of the accuracy of the Prandtl model. Comparison showed that the Prandtl model gave results very close to the Goldstein solution at high tip-speed ratios, and qualitative agreement at all tip-speed ratios.

Comparisons between the Betz and Prandtl tip-loss models have been made for wind turbines [Wilson and Lissaman 1974], and agreement was also good. However, since an expanding wind turbine wake is different from the constant-diameter wake of an optimum propeller, it is not apparent which model is more accurate. In the absence of test data on local flow quantities near HAWT blade tips, the Prandtl tip-loss model is recommended. This recommendation is based on the following:

- The Prandtl model predicts a continuous change in circulation, in qualitative agreement with the behavior of wind turbine rotors.
- Calculations of rotor power and thrust made with the Prandtl model are in good agreement with test data.
- Strip theory calculations made with the Prandtl model show good agreement with calculations made with free wake vortex theory.
- The Prandtl model is easier to program and use.

It should be noted that, as shown in Figure 5-16, the agreement between test and calculations using the Prandtl tip-loss model is poorest at the blade tip.

In Prandtl's tip-loss model, the *vortex sheets* generated by the blades are replaced (for purposes of analysis) with a series of parallel planes at a uniform spacing equal to the normal distance between successive vortex sheets at the slipstream boundary. Thus

$$\Delta z = \frac{2\pi R}{B} \sin \phi_R$$

where Δz = axial spacing between planes representing vortex sheets (m)
         ϕ_R = angle between relative wind vector and plane of rotation at the tip (rad)

Using this spacing, the expression obtained by Prandtl for the tip-loss factor is

$$F = \frac{2}{\pi} \arccos \left(\exp \left[-\frac{B(R-r)}{2R\sin\phi_R} \right] \right)$$

In practice, Prandtl's expression is modified to be

$$F = \frac{2}{\pi} \arccos \left(\exp \left[-\frac{B(R-r)}{2r\sin\phi} \right] \right) \tag{5-35}$$

since, as the tip-loss factor F goes to zero at the tip, the tip flow angle ϕ_R goes to the blade pitch angle at the tip, which may be zero. Equation (5-35) avoids a potential tip singularity and, more significantly, produces good results when compared with theoretical approaches.

The tip-loss factor F enters the determination of the induced flow calculation through the local thrust coefficient, C_t. Recall from Figure 5-11 that the overall thrust coefficient departs from the momentum theory value of $4a(1-a)$ when a exceeds about 0.4. It has been shown [Wilson and Walker 1984] that the following formulation for the local thrust coefficient compares well with vortex theory calculations and produces good results in the correlation of calculated and measured performance and loads:

$$C_t = 4aF(1-a) \qquad\qquad if \ a \le a_c$$
$$C_t = 4F\left[a_c^2 + (1-2a_c)\,a\right] \qquad if \ a > a_c \qquad (5\text{-}36)$$
$$a_c \ge {}^1\!/_3$$

The second equation is a linear extrapolation of the first from the junction point $a = a_c$. The value of a_c should be equal or greater than 1/3 in order that the Betz Limit is not exceeded. In summary, the set of equations used to determine the induced velocities in accordance with strip theory is composed of Equations (5-23a), (5-32), (5-35), and (5-36).

Cascade Effects on Lift and Drag Coefficients

Lift and drag coefficients, C_L and C_D, are functions of the angle of attack, α, as follows:

$$C_L = C_L[\alpha]$$
$$C_D = C_D[\alpha] \qquad\qquad (5\text{-}37)$$
$$\alpha = \phi - \beta + \Delta\alpha_c$$

where $\Delta\alpha_c$ = cascade correction to angle of attack (rad)

At a given radial position on the blades, the circumference can be unrolled and represented as a sequence or *cascade* of airfoils, as shown in Figure 5-14. An expression for the so-called *cascade correction* will be derived using the flow geometry shown in this figure. In the strip theory developed to this point, we have used a *lifting line* approach, where the flow distortions caused by the rotor width and blade thickness are ignored. Near the root of a thick blade, however, flow distortion can be significant.

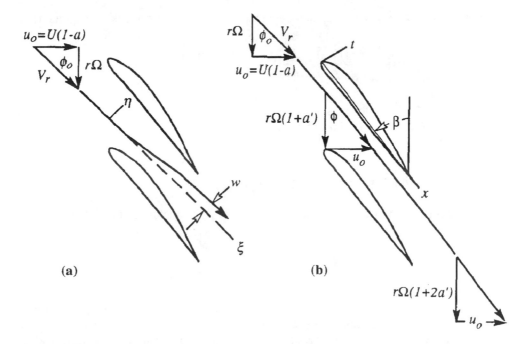

(a)   **(b)**

Figure 5-14. Cascade effects on angle of attack. (a) Streamline displacement caused by finite blade thickness (b) Streamline curvature caused by the finite blade width

The distance between the blades is the circumference divided by the number of blades. As the flow traverses the path shown, two distinct effects occur to change the flow pattern from that of the actuator disk which was used to develop strip theory: (1) An increase in tangential velocity occurs between the leading and trailing edges of the airfoil, so that the flow follows a curved path, and (2) the axial component of the flow increases because the *airfoil thickness* reduces the size of the flow path. We will treat these two effects separately, developing the equation

$$\Delta\alpha_c = \Delta\alpha_1 + \Delta\alpha_2 \tag{5-38a}$$

where $\Delta\alpha_1$ = effect of finite airfoil thickness (rad)
   $\Delta\alpha_2$ = effect of finite airfoil width (rad)

Because of *blockage* by the blade thickness, axial velocity between the blades is increased, which in turn increases the wind angle, ϕ. This causes flow streamlines to be displaced in a direction normal to the blade chord by an amount w, as shown in Figure 5-14 (a). The circumferential spacing between the blades in this figure is greatly reduced for purposes of illustration. The actual spacing-to-chord ratio is large, so we may employ a one-dimensional flow model. Using the requirement for flow continuity, an increment of the displacement, dw, is expressed in terms of the blade thickness as follows:

$$dw = \frac{B\cos\phi_o}{2\pi r}\frac{t}{\cos\alpha}d\xi = \frac{B\cos\phi_o}{2\pi r}t\,dx$$

where ϕ_o = flow angle prior to rotational induction (*i.e.*, $a' = 0$) (rad)
   t = local thickness of airfoil, normal to chord line (m)
   ξ = coordinate in the direction of the relative wind (m)
   x = airfoil chordwise coordinate, measured from leading edge (m)

The flow displacement w causes an increase in the apparent angle of attack of magnitude

$$\Delta\alpha_1 = \frac{w}{c} = \frac{1}{c}\int_0^c\left(\frac{dw}{dx}\right)dx = \frac{B\cos\phi_o}{2\pi rc}\int_0^c t\,dx$$

The latter integral is the airfoil cross-sectional area, A_a. Using the thickness distribution functions for the *NACA four-digit series* of airfoils,

$$A_a \approx 0.68\,c\,t_{max}$$

where t_{max} = maximum thickness of the airfoil (m)

Therefore, the increase in angle of attack caused by the finite thickness of the blades is

$$\Delta\alpha_1 = 0.11\,B\left(\frac{c}{R}\right)\left(\frac{t}{c}\right)_{max}\frac{\lambda}{\sqrt{(1-a)^2 + \left(\frac{r}{R}\right)^2\lambda^2}} \tag{5-38b}$$

The curvature of the flow caused by the finite blade width, illustrated in Figure 5-14(b), results in a change in the circulation developed by the blades. It is assumed that the induced tangential velocity is developed linearly along the blade, from a value of zero at the leading edge to its final value of $2\,a'r\Omega$ at the trailing edge. Note that blade coning is ignored in this development. The change in ϕ from a leading edge to a trailing edge is

$$\Delta\phi = \arctan\left[\frac{(1-a)\,x}{(1+2a')}\right] - \arctan\left[(1-a)\,x\right]$$

This change in flow angle occurs in a time period of $\Delta\tau = c/V_r$, so there is an *effective pitch rate* of the blades, $d\beta/dt$. Airfoil pitching is assumed to occur about an axis through the *1/4-chord point*, so the effective change in angle of attack from a finite blade width is

$$\Delta\alpha_2 = \arcsin\left(\frac{\Delta\tau}{4}\frac{d\beta}{dt}\right) \approx \frac{\Delta\phi}{4} \tag{5-38c}$$

Note that for the conditions illustrated in Figure 5-14, $\Delta\alpha_1 > 0$ and $\Delta\alpha_2 < 0$. The cascade corrections are most significant near the blade roots where they influence yaw loads.

Gap Correction

Another modification of the strip theory approach is the *gap correction* required to account for the effects of a space along the span of a HAWT blade. Such spaces occur, for example, on blades with *partial-span pitch control*, to accommodate the actuating mechanism. Gaps can affect both lift and drag forces, so corrections are expressed as

$$\Delta C_{P,G} = \Delta C_{P,GL} + \Delta C_{P,GD} \tag{5-39a}$$

where $\Delta C_{P,G}$ = total reduction in power coefficient caused by a gap in the airfoil
$\Delta C_{P,GL}$ = reduction in power coefficient caused by the loss of lift in the gap
$\Delta C_{P,GD}$ = reduction in power coefficient caused by the increased drag in the gap

Gap Lift Correction

A gap in the span of the blade causes the circulation to drop to zero locally. As a result, vorticity is shed in the wake, with vorticities of opposite sign being present. That is, the vorticity shed outboard of the gap is of the opposite sign of the vorticity shed inboard of the gap. The effect of this shed vorticity can be modeled as a *semi-infinite vortex doublet* extending downwind of the gap. A linear doublet is employed in lieu of a helical vortex because the extent of the velocity field of a doublet is smaller. The influence of the doublet falls off inversely as the square of the downwind distance, while the extent of the velocity field of a vortex diminishes inversely with distance.

The local circulation deficit from such a doublet can be calculated by considering its local *downwash* and employing linear aerodynamics. We find that the effect of the gap is approximated by deleting the lift contribution to the power coefficient for a distance along the span of the rotor of

$$s_{g,e} = 2c\sqrt{\frac{s_g}{c}}$$

where $s_{g,e}$ = effective spanwise width of the gap (m)
s_g = actual spanwise width of the gap (m)

We can approximate the effect of lost lift by assuming that no power is extracted from the streamtube enclosing the effective width of the gap, which gives

$$\Delta C_{P,GL} = 2B\cos^2\theta\left(\frac{r}{R}\right)\left(\frac{c}{R}\right)^2\sqrt{\frac{s_g}{c}}\frac{\lambda}{\pi}\frac{V_r}{U}C_L(1-a) \tag{5-39b}$$

where r and c are the radius and airfoil chord, respectively, at the gap.

Gap Drag Correction

Because no power is extracted from the streamtube containing the gap, the relative velocity causing drag is equal to the local tangential velocity, $r\Omega$. Neglecting the rotational induction factor, the gap drag effect on power coefficient is

$$\Delta C_{P,GD} = \frac{B\, C_{D,g}\, \lambda^3}{\pi}\left(\frac{r}{R}\right)^3 \left(\frac{s_g}{R}\right)\left(\frac{t_g}{R}\right)\cos^3\theta \qquad (5\text{-}39c)$$

where $C_{D,g}$ = drag coefficient of the structure in the gap
         t_g = average thickness of the structure in the gap (m)

An aerodynamic *fairing* usually covers some or all of the components in the gap, in order to reduce $C_{D,g}$ to a minimum. Otherwise, the lost power could be significant.

Comparison of Strip Theory with Free-Wake Analysis

Two methods of predicting HAWT performance will now be compared. The *strip theory* (modified blade-element theory) approach developed previously is compared with *lifting line, free-wake vortex* calculations performed with the *VORTEX* code, a steady-state analysis developed at the Massachusetts Institute of Technology under sponsorship of the U.S. Department of Energy [Gohard 1978]. The sample cases also offer an opportunity for a controlled comparison of the flow fields predicted by both methods. The rotor configuration for this comparative analysis has two, untwisted, constant-chord blades with a 20percent *root cutout* (*i.e.*, no airfoil inboard of $r/R = 0.2$) and no hub interference. The aerodynamic lift and drag properties of the blade airfoil are simple linear and quadratic functions of the angle of attack, respectively, with approximations for the effects of stall.

Table 5-4 gives rotor thrust and power coefficients calculated using the two analysis methods for four cases that include two tip-speed ratios and three blade pitch angles. Only eight *nodes* (*i.e.*, blade stations) were used in the free-wake analysis, and this does not really warrant reporting the results with three significant figures. Nevertheless, the free-wake calculations do represent the best theoretical predictions currently available of a HAWT flow field. Examination of these results shows that the strip theory predictions of thrust are about one percent lower and predictions of power coefficients are about four percent lower than coefficients calculated using the VORTEX free-wake analysis code.

Table 5-4.
Comparison of HAWT Rotor Coefficients Calculated by Two Analytical Methods

Case	Tip-Speed Ratio, λ	Pitch Angle β (deg)	Thrust Coefficient, C_T		Power Coefficient, C_P	
			Strip Theory [1]	Free-Wake Analysis [2]	Strip Theory	Free-Wake Analysis
1	6.49	0	1.051	1.077	0.421	0.443
2	"	2	0.910	0.920	0.430	0.443
3	"	4	0.758	0.769	0.406	0.422
4	9.52	2	1.036	1.042	0.322	0.338

[1] [Wilson and Walker 1984]   [2] [Gohard 1978]

The circulation developed by the blade is a key item associated with the flow field. Figure 5-15 illustrates the circulation distributions along the span of the blade as calculated according to the strip theory and the free-wake methods for Cases 1 and 4. These two cases had similar thrust coefficients but different power coefficients. The graphs show good agreement, with both methods predicting a fairly uniform circulation distribution in both cases. The two methods agree on the circulation in the stalled inboard region in Case 1.

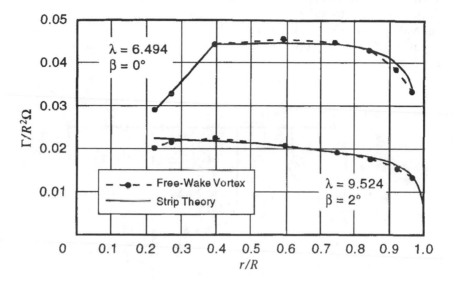

Figure 5-15. Circulation distributions calculated by two different analytical methods. [Wilson and Walker 1984, and Gohard 1978]

Comparison of Strip Theory with Test Results

Anderson *et al.* [1982] conducted wind tunnel tests of a 3-m diameter HAWT at a variety of wind speeds over a wide range of rotor speeds, with tip-speed ratios from 8 to 14. The rotor had two tapered and twisted blades with an NACA 4412 airfoil section and a solidity of 4.58percent. Test data included local axial and tangential velocities near the rotor, measured using *hot-film anemometry*, and rotor power and thrust. Angles of attack experienced by the blades where below those for stall, so *linear aerodynamics (i.e.,* lift coefficient directly proportional to angle of attack) could be employed. *Reynolds numbers*, N_R for the tests of this turbine at a free-stream wind speed of 10 m/s ranged from 240,000 at a tip-speed ratio of 8 to 420,000 at a tip-speed ratio of 14. (Because of the taper in the blade planform, Reynolds number is almost constant over the outer 70percent of these blades.) Lift and drag test data for the NACA 4412 airfoil below stall over this Reynolds-number range have been reported by Jacobs and Sherman [1936].

The turbine tests were performed in a wind tunnel with a blockage ratio of 0.12. Corrections were made to the test data to account for blockage due to wake expansion. At higher tip speeds, blockage corrections were very sensitive to the induction ratio used. This was not the case, however, at lower tip speeds. Since the blockage correction has a very large effect at the higher tip-speed ratios, comparison of experimental and theoretical results will be limited to those obtained at a tip-speed ratio of 8. The slope of the lift curve ($dC_L/d\alpha$) was modified by the *Prandtl-Glauert rule* to account for local *Mach number*, which reached 0.26 at this tip-speed ratio.

Figure 5-16 shows the local axial induction factors measured behind the rotor at various radial stations at a tip-speed ratio of 8, compared with factors predicted using modified strip theory. Predicted induction factors show good agreement with test data for the inboard stations out to about 80 percent of the blade's span, but they are higher near the tip. However, numerical integration of the local axial induction factors yields an overall rotor thrust coefficient that is consistent with the measured thrust coefficient, as shown in Table 5-5. Differences between measured and predicted power coefficients listed in this table fall within the reported range of experimental error ($\pm$ 2.5percent). On the basis of wind tunnel tests such as these on small-scale HAWTs and field tests of larger turbines, we conclude that performance calculations based on strip theory adequately model test results. The tip loss model requires further refinement.

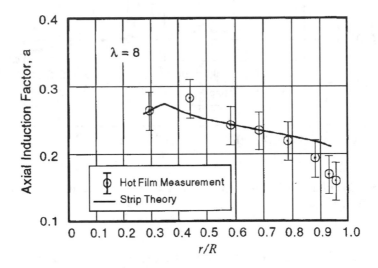

Figure 5-16. Comparison of experimental and theoretical axial induction factors for a 3.5-m, two-bladed HAWT tested in a wind tunnel. [data from Anderson *et al.* 1982]

Table 5-5.
Comparison of Predicted and Measured Rotor Parameters for a 3-m HAWT[1]

Tip-speed ratio λ	Axial Induction Factor, a [2]		Thrust Coefficient, C_T		Power Coefficient, C_P	
	Strip theory	Wind tunnel test	Strip theory	Wind tunnel test	Strip theory	Wind tunnel test
8.0	0.24	0.23	0.69	0.70	0.44	0.42

[1] Test data from [Anderson *et al.* 1982]   [2] Average for 0.25 < r/R < 0.95

Other Aerodynamic Analyses

While the Blade Element Momentum (BEM) approach has been a useful tool, the inability of BEM to accurately treat design conditions such as operation at large yaw angles and wind turbine response during short-term transients creates the need for a better wind turbine analysis technique. Along with the search for a better analysis technique it should be noted that all of potential improved techniques under consideration use a lifting line model in which the flow conditions are evaluated along the axis of the blade (usually at 1/4 chord) and these conditions are used to determine the angle of attack which in turn is used as an entry to a table of two-dimensional airfoil characteristics. Neither the span-wise nor the chord-wise variations in airfoil performance are modeled although Schreck *et al.* [2007] have examined the influence of rotation effects on wind turbine airfoils.

The need for a more accurate wind turbine analysis technique has produced several candidates that incorporate analysis of the wind turbine wake. There are two principle approaches, use of a *vortex wake* (either prescribed or free) and the *generalized dynamic wake* (GDW) method. Both of these methods claim to meet the challenge of modeling transient HAWT behavior in a turbulent wind.

GDW Method

Currently, GDW analysis is being used to determine loads for wind turbine certification. Adopted from a technique developed for helicopter analysis [Pitt and Peters 1981 and Peters and He 1989], the GDW method is based on a potential flow solution of Kinner [1937]. Using the acceleration potential, Kinner obtained a general solution for flow about a circular disk by considering the nonlinear convective flow acceleration to be based on the uniform free-stream wind speed, thus linearizing the Euler Equations.

While Kinner used boundary conditions on the disk for an airfoil, the Pitt and Peters solution used boundary conditions that allow flow through the disk such as occurs for a rotor. When using the acceleration potential, the linearized Euler equations are transformed into a LaPlace equation with the pressure as the dependent variable. This formulation of the flow about a rotor has a cylindrical, skewed wake that accounts for the trailing wake and does not require a tip-loss model. The GDW approach has thus far failed to give accurate results near the tips of helicopter blades. However, because the wake from a wind turbine expands rather than contracts (as is the case with a helicopter), there is no reversal of the induced velocity near the tips of wind turbine blades. Thus, the use of the GDW approach could be more accurate for wind turbines than for helicopters.

Advantages claimed for the GDW method include the ability to model transient aerodynamics, tip losses and operation at large yaw angles. However, there are also disadvantages associated with the GDW method when compared to strip theory. These are that the GDW method

- is poorest at wind speeds where induced velocities are large compared to the free stream wind velocity
- does not determine the induced rotational wind speed
- does not allow for blade coning or the deflection of the blades
- has so far failed to determine conditions near the blade tips

Vortex Wake Models

A variety of vortex wake models have been employed for wind turbines in axial flow. Coton and Wang [1999] extended the earlier work of Robinson *et al.* [1994, 1995] on the use of a *prescribed wake* to treat yawed flow. As the name implies, prescribed wake methods simplify the location of the wake and use the wake locations and the Biot-Savart relation to determine the influence of each trailing vortex on each blade station. Currin *et al.* [2007a, 2007b] have extended the prescribed wake approach to create a *dynamic wake model* which has been incorporated into a structural code. In this approach, the wake is divided into three near-wake regions where wake expansion is occurring and a far-wake region where radial expansion no longer occurs. Comparisons with test data have shown good agreement.

Rotor Configuration Effects

A large number of variables exist in the design of a HAWT rotor, including *taper*, *twist*, *solidity*, and *number of blades*. As taper and twist are intimately associated with and often controlled by manufacturing techniques, we will consider the aerodynamic effects on performance of only solidity and blade number.

Rotor Solidity and Number of Blades

Solidity, defined as the *total blade planform area* divided by the *swept area* of the rotor is typically less than 0.10 for modern wind turbines. The rotor solidities of the NASA/DOE Mod-0, Mod-1, and Mod-2 two-bladed experimental HAWTs were 0.029, 0.042, and 0.036, respectively, while typical three-bladed HAWTs have solidities between 0.070 and 0.080. (See Chapters 3 and 4 for descriptions of these wind turbines.)

As the solidity of a wind turbine rotor is increased, the tip-speed ratio for maximum power coefficient is reduced. This effect can be explained by considering a rotor of a given solidity at its peak power coefficient and associated tip-speed ratio, producing its optimum axial induction. If the solidity of this rotor is increased and all other parameters are held constant, both lift and drag forces will increase and, as a result, the axial induction factor will increase beyond its optimum value. To restore optimum axial induction, aerodynamic forces on the blade must be decreased, which is accomplished by reducing the tip-speed ratio. Additionally, the *peak power* of a fixed-pitch rotor increases as solidity is increased. These effects of solidity on power and power coefficient occur regardless of whether the solidity is changed by increasing the number of blades or by increasing the chord dimensions of each blade or both.

A different situation occurs when solidity is held constant and the number of blades is changed. In this case, performance changes are caused by two opposing effects, which are the effect of blade *Reynolds number* and the effect of *tip losses*. As the number of blades is increased, chord dimensions decrease (to keep solidity constant), which reduces the blade's Reynolds number. Rotor power decreases, as a result, since modern wind turbines usually have *turbulent flow* over their blades, which is a regime in which a lower Reynolds number produces lower lift and higher drag. The effect of increasing the number of blades on tip losses is the opposite: Reducing the tip chord dimensions decreases the total tip loss, and rotor power is increased.

Empirical Equation for Maximum Power Coefficient

Because of the many configuration variables that must be considered in the design of a wind turbine rotor, it is often useful to be able to quickly estimate the *maximum* or *peak power coefficient* that can be achieved with a potential rotor configuration. This was done previously for a drag-driven translator (Eq. (5-3)), a lifting translator (Eq. (5-5)), the Rankine-Froude actuator disk (Eq. (5-15b)), and the Glauert optimum HAWT rotor (Table 5-2). With the added effects of finite blade number, B, and realistic airfoil lift and drag properties, C_L and C_D, estimates of maximum power coefficient are best done with empirical equations such as the following [Wilson *et al.* 1976]:

$$C_{P,\max} = 0.593 \left[\frac{\lambda\, B^{0.67}}{1.48 + (B^{0.67} - 0.04)\,\lambda + 0.0025\,\lambda^2} - \frac{1.92\,\lambda^2\, B}{1 + 2\,\lambda B}\, D/L \right] \quad (5\text{-}40)$$

where D/L = ratio of C_D to C_L at the design angle of attack; drag-to-lift ratio

Figure 5-17 illustrates the application of Equation 5-40. The Glauert ideal HAWT performance ($B \rightarrow \infty$ and $D/L = 0$; data from Table 5-2) forms an upper bound.

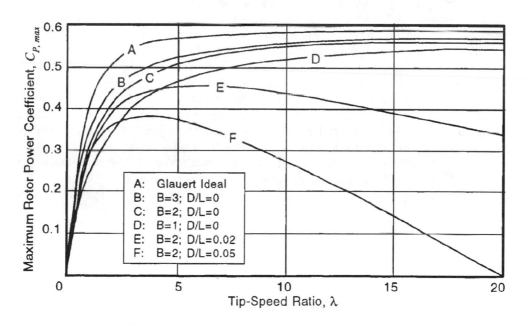

Figure 5-17. Typical effects of the number of blades and the design drag-to-lift ratio on the maximum power coefficient of a HAWT. Equation (5-40); [Wilson *et al.* 1976]

Aerodynamic Behavior of HAWTs in Operation

The operation of modern HAWTs includes a number of important features, such as the use of a *teetered rotor*, *yaw response*, *aerodynamic controls*, *transient aerodynamics*, and *vortex generators*. The aerodynamic qualities of these features are discussed in this section, along with examples of HAWT performance and aerodynamic loads.

Teetered Rotors

The rotor of an operating HAWT is continually subjected to wind conditions that cause significant variations in the local air loads along its blades. These conditions include *wind shear* (vertical and/or horizontal gradients in wind speed), *yaw error* (rotor disk not perpendicular to the wind), *tower shadow* (reduction in wind speed caused by tower blockage), and *small-scale turbulence* (perturbations above and below the mean wind speed over areas smaller than the swept area).

Wind shear, for example, can cause the blade at the topmost position to experience a higher wind speed than the blade closest to the ground. If the blade is held rigidly at the hub, this difference in wind speed will induce a cyclic variation in local angles of attack. As a result, each blade will experience *cyclic flatwise bending* (also called *flapwise* or *out-of-plane* bending), with a dominant frequency of once per rotor revolution or *1P*. Similar cyclic loading at a 1P frequency occurs with yaw error. The effects of tower shadow and turbulence are more complex, causing significant fluctuations in the aerodynamic loading on the blades not only at the *1P* frequency but also at higher harmonic frequencies (*i.e. 2P, 3P*, etc.).

Cyclic flatwise bending moments on the blades of a two-bladed HAWT are greatly reduced by allowing the rotor to oscillate like a see-saw about a *teeter axis* or *teeter hinge* through the outboard end of the turbine shaft, as illustrated in Figure 5-18. The teeter hinge prevents unbalanced out-of-plane aerodynamic loads from being transmitted to the turbine shaft, while still allowing in-plane torque loads to be carried. The effectiveness of teetering in reducing blade bending loads can be shown by the following analysis.

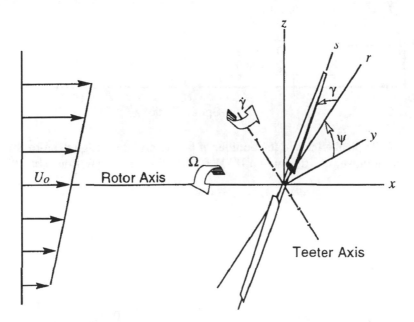

Figure 5-18. Schematic diagram of a teetered, two-bladed HAWT rotor.

Equilibrium of out-of-plane moments about the teeter axis requires that

$$0 = \int_{-R}^{R} p_{a,x}\, r\, dr - \gamma'' \int_{-R}^{R} m_b\, r^2\, dr - \gamma \Omega^2 \int_{-R}^{R} m_b\, r^2\, dr \qquad (5\text{-}41a)$$

where $p_{a,x}$ = axial air pressure loading per unit of blade span (N/m)
   m_b = blade mass per unit of span (kg/m)
   γ, γ'' = teeter deflection (rad) and acceleration (rad/s²), respectively

The limits -R to R indicate that the integration is over both blades. The first term is the net (unbalanced) aerodynamic moment on the rotor, the second is the inertial moment, and the third is the moment of centrifugal forces which oppose the teetering motion. We can analyze the *rigid-body motion* of the rotor by rewriting Equation (5-41a) as an *equation of motion* and setting the aerodynamic forces to zero (*i.e.* teetering in a vacuum), which gives

$$I_{tt}\, \gamma'' + I_{tt}\, \Omega^2\, \gamma = M_{a,net} = 0 \qquad (5\text{-}41b)$$

where I_{tt} = mass moment of inertia of the rotor about the teeter axis (kg-m²)
   $M_{a,net}$ = unbalanced aerodynamic out-of-plane moment on the rotor (N-m)

Equation (5-41b) describes simple harmonic motion, with a frequency equal to Ω (*1P*), so

$$\gamma = \gamma_1 \sin \Omega t + \gamma_2 \cos \Omega t = \gamma_1 \sin \psi + \gamma_2 \cos \psi \qquad (5\text{-}41c)$$

where γ_1, γ_2 = teeter amplitudes determined by the wind speed spatial distribution (rad)
   ψ = blade azimuth (rad)

Returning to a consideration of air loads, the teeter motion will cause a wind velocity component *opposing the motion*, which can reduce or eliminate variations in wind speed across the rotor. To illustrate this, assume a vertical, linear wind shear gradient. The axial wind velocity, V_x, seen by the teetering blade is

$$V_x = \left(U_0 + \frac{\Delta U}{2R}\, r \sin \psi \right)(1 - a) - r\Omega\,(\gamma_1 \cos \psi - \gamma_2 \sin \psi) \qquad (5\text{-}42a)$$

where U_0 = free-stream wind velocity at hub elevation (m/s)
   ΔU = vertical change in wind velocity across the rotor diameter (m/s)

Variations in wind speed from U_0 are eliminated at the rotor by a small teetering motion that is 90 degrees out-of-phase with the wind shear gradient, with the component amplitudes

$$\gamma_1 = 0 \qquad \gamma_2 = -\frac{(1-a)}{2\lambda}\, \frac{\Delta U}{U} \qquad (5\text{-}42b)$$

In practice, teeter deflections are restrained by *teeter stops* that prevent the blade from hitting the tower during extreme gusts or during starting and stopping. Teetered rotors may also be *coned*, and sometimes the teeter axis is not perpendicular to the blade axis (the angle of inclination is called the δ_3 angle). While these modifications complicate the aerodynamic analysis, the benefits of teetering remain the same: reductions in cyclic loads on the blades and, most importantly, on the nacelle and tower. Analysis of rotor loads must include teeter motion, but mean power output can be determined without including teetering.

Yaw of HAWT Rotors

All HAWTs operate for some fraction of the time with their rotor axis misaligned with the wind azimuth. It is common for a rotor to encounter yaw angles of 30 deg for periods as long as several minutes. Active-yaw-controlled rotors typically respond to a time-averaged wind direction with averaging times of several minutes. Some free-yaw rotors which have been designed to operate with the rotor downwind of the tower have been observed with the rotor operating upwind of the tower for extended periods. Many HAWTs operate continuously with a small yaw error. Others have experienced high rates of yaw rotation which have caused fatigue failures induced by gyroscopic loadings.

Historically, yaw-related problems have been one of the leading causes of turbine downtime in California wind power stations. These problems are of two types: loss of *yaw stability*, and excessive *yaw-induced loads*. Yaw stability is particularly important for free-yaw HAWTs which rely on aerodynamic forces to achieve proper orientation with respect to the wind. The wind turbines of the 1930s used a tail vane in order to obtain yaw stability. However, a modern free-yaw HAWT, with its rotor downwind of the tower, obtains its yaw stability from the aerodynamic forces on the rotor.

The presence of a yaw error causes changes in the aerodynamic forces normal to and tangential to the airfoils as the blades rotate. In a rigid-hub wind turbine, both the normal and tangential forces can contribute to a moment about the yaw axis. In a HAWT with a teetered rotor, the tangential forces are dominant in producing this yaw moment. A yaw moment arises from blade-to-blade differences in aerodynamic loading. Thus, when making analytical estimates of the yaw moment, greater accuracy is required than in the determination of rotor thrust or rotor torque, because the aerodynamicist must now evaluate small differences between large numbers.

Aerodynamic Analysis of a Yawed Rotor

The aerodynamic behavior of a yawed wind turbine has been examined by Glauert [1935a], Miller [1979], de Vries [1985], Hansen [1992], and Croton *et al.* [1999]. Additionally, a review article by Hansen and Butterfield [1993] contains a discussion of yaw aerodynamics. Aerodynamic analysis of the power extracted from the wind stream by a yawed rotor is complicated by the fact that the wake is skewed, so that wake cross-sections are elliptical rather than round. This modifies the induced velocity at the rotor. Large cyclic changes also occur in blade angles-of-attack because of yaw error, and this has been shown to cause dynamic stall [Hansen 1992]. Yaw moments are particularly sensitive to dynamic stall because of the asymmetry produced by stall hysteresis. As noted by Ribner [1948], most of the yaw force is generated at the inboard stations of the rotor, near the hub. Additionally, tower shadow, and wind shear (both vertical and horizontal) are known to contribute to the differential blade loading that causes yaw misalignment.

Glauert's contribution to the aerodynamic analysis of a yawed rotor is expressed as a modification of the momentum equation, Equation (5-9), as follows:

$$dT = 2\,a\,\rho\,U\,U'\,dA$$

$$U' = \sqrt{(U\cos\Delta\Psi - a\,U)^2 + U^2\,\sin^2\Delta\Psi}$$

(5-43)

where U' = component of the free-stream wind speed shown in Figure (5-19) (m/s)

Yaw Error and Power Output

It has been observed that even relatively small yaw errors ($\pm$ 10 deg or less) can reduce the power output of a HAWT significantly. Anderson *et al.* [1982] measured the power

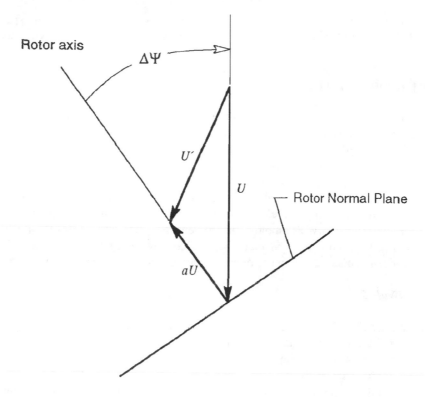

Figure 5-19. **The wind velocity component U' in the Glauert modification of the momentum equation accounting for yaw heading error.** [Glauert 1935a]

output of a 3-m, two-bladed HAWT in a wind tunnel at various angles of yaw. The results are shown in Figure 5-20, in the form of ratios of power with yaw error to power without yaw error. Tests were run at three different tip-speed ratios. For comparison, power reductions predicted in accordance with strip theory are also shown in the figure. Generally, strip theory predicts too large a loss in power from yaw error. Performance was found to vary as

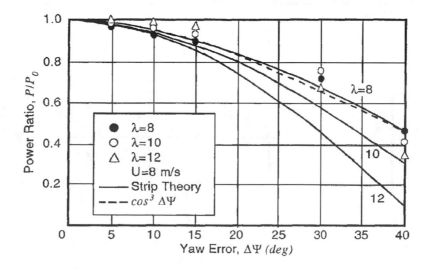

Figure 5-20. **Measured and calculated effects of yaw error on the power output of a 3-m HAWT.** Test data from [Anderson *et al.* 1982]

the cube of the cosine of the yaw error, which means the effective normal wind speed is U $cos\ \Psi$. It should be noted that turbulence can also result in yaw error and a corresponding power loss.

Aerodynamic Controls

The operation of a HAWT involves starting the wind turbine from rest, stopping the wind turbine under a wide range of normal and abnormal conditions, and modulating system power and loads while the turbine is running. The starting of many current wind turbines is accomplished by running the generator as a motor. For large-scale HAWTs aerodynamic control surfaces are employed to assist in delivering the large *starting torques* needed. *Loss-of-load emergencies* and wind speeds above the operating range require a reliable means of stopping the turbine. Aerodynamic controls are particularly attractive for stopping the rotor, and almost all current HAWTs employ some type of aerodynamic control to prevent *rotor overspeed*. A third control function, that of regulating power output and/or system loads, has been accomplished historically by use of aerodynamic control surfaces.

Full-Span Blade Pitch Control

First on the list of aerodynamic controls is full-span blade pitch control. Figure 5-21 shows a typical blade pitch angle schedule required to produce a *ramp-plateau power curve*. The power plateau, at the turbine's *rated power*, has the advantage of restricting the power and loads at high wind speeds to cost-effective levels. The ramp portion of this power curve represents operations at below-rated wind speeds, during which blade pitch is usually held constant. The control systems in some HAWTs change pitch angle slightly along the ramp (either continuously or by a mid-ramp change) in order to optimize energy capture. In practice, the blade pitch changes required at the "knee" of the power curve represent a problem area, particularly under gusty wind conditions that can cause power "spikes," since the sharp change in the power curve is accomplished by corresponding sharp changes in the blade pitch rate. Control strategies are often used that "round off" these sharp knees.

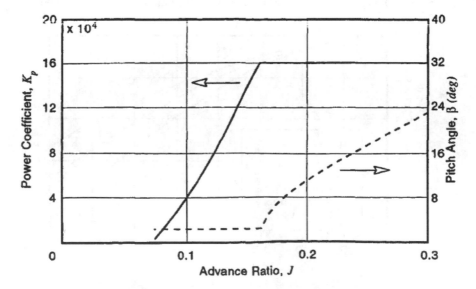

Figure 5-21. Dimensionless power curve and associated full-blade pitch angles for a typical HAWT. Blade pitch is held constant until rated power is produced.

Partial-Span Pitch Control

Pitching the entire blade requires large bearings and substantial actuators, both of which tend to increase rotor weight and cost. One approach that retains the advantages of full-span blade-pitch control and yet decreases the complexity of the hub is partial-span pitch control. This is particularly advantageous for a teetered rotor, permitting designers to separate the teeter-axis bearings and supporting structure from the pitch-axis bearings and pitch actuators. The multimegawatt Boeing Mod-2 and Mod-5B HAWTs (Figs. 3-39 and 4-5) were examples of turbines with partial-span blade pitch control for starting, power regulation, and stopping. Typically, only the outer third of the blade span is pitched on this type of rotor. The gap between the movable tip and the fixed inboard portion of the blade can cause power losses, as discussed earlier. Fairings are used when possible to minimize these losses.

Passively-deployed pitchable tips are used for overspeed control in a number of normally fixed-pitch HAWTs (see Chapter 4). Centrifugal forces are used to deploy short tips at angles up to 90 deg when the rotor speed exceeds its operating limit. Data on the effectiveness of these so-called *air brakes* have been presented by Jensen *et al.* [1986]. *Tip plates*, also deployed by centrifugal force, have been employed for overspeed control on several medium-scale HAWTs, including the *ESI* and *Enertech* machines.

Aileron Control

Aileron control surfaces (sometimes referred to as *flaps*) are a potentially cost-effective means of rotor control. These devices have been investigated both theoretically and experimentally at the NASA Lewis Research Center for application to medium- and large-scale HAWTs [Miller and Sirocky 1985]. This work began in 1979, when aileron control was first evaluated as an alternative to blade-pitch control on DOE/NASA experimental HAWTs [Wentz *et al.* 1980]. The concept selected for study consisted of incorporating a movable control surface into the trailing edge of the rotor blade, in a manner analogous to the placement of ailerons and flaps in the trailing edge of an airplane wing. Control surfaces on an aircraft wing usually deflect toward the high-pressure (lower) surface to increase lift during takeoff and landing. On a wind turbine blade, ailerons deflect toward the low-pressure (downwind) surface of the airfoil to reduce lift and produce a braking effect.

Figure 5-22 shows cross-sectional diagrams of two typical aileron configurations evaluated for HAWT rotor control. The hinge line is on the downwind side of the blade, so that the

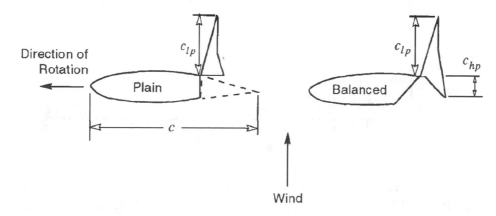

Figure 5-22. Schematic diagrams of typical aileron configurations for HAWT aerodynamic control. Ailerons are shown in the fully-deflected position for braking the rotor. [Miller and Sirocky 1985]

aileron can be deflected toward the low-pressure surface. The primary difference between the two configurations illustrated is the chord length of the high-pressure control surface, c_{hp}. The configuration without a high-pressure control surface, referred to as a *plain aileron*, was found to have the best aerodynamic braking characteristics of all of the configurations tested. If the aileron has both a low-pressure and a high-pressure control surface it has been referred to as a *balanced aileron*.

Ailerons change the lift and drag characteristics of the basic blade airfoil as a function of the *aileron deflection angle*, producing corresponding changes in rotor torque and thrust loads. The changes in torque can be controlled by regulating the deflection angle, which in turn enables the regulation of rotor speed and/or rotor power output. A considerable amount of information on the aerodynamic behavior of HAWT rotors with ailerons has been obtained from full-scale turbine tests and wind tunnel tests of airfoils with ailerons [*e.g.* Miller and Corrigan 1984, Savino *et al.* 1985, Ensworth 1985].

One question of considerable importance is the potentially detrimental effect of *aileron gaps* on the aerodynamic performance of a rotor blade when the ailerons are not deflected. Since ailerons are located in the most productive spanwise portion on the blade, the question arises as to performance penalties caused by leakage from high-pressure to low-pressure surfaces through gaps between the aileron and the and the main structure of the blade. Figure 5-23 [Savino *et al.* 1985] shows the measured effect of aileron gap on the lift coefficient of a 7.3-m segment of a medium-scale HAWT blade with an undeflected aileron 6.1 m long. Normal gaps were found to reduce the chordwise (in-plane) force coefficient by about 0.01 to 0.02 at angles of attack below stall and to have no measurable post-stall effect. Drag coefficients below stall were not affected by normal aileron gaps. Some power loss may result when an aileron is deployed and a rapid negative gust occurs.

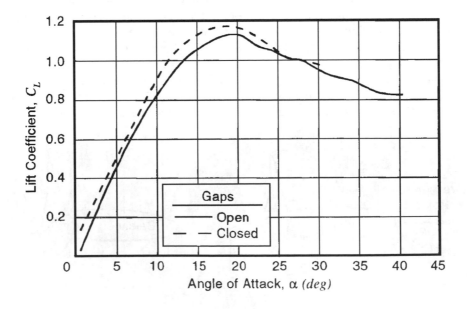

Figure 5-23. Effect of normal aileron gaps on the lift coefficient of a medium-scale HAWT blade when the aileron is not deflected. Wind tunnel test data. [Savino *et al.* 1985]

Transient Aerodynamics

Current state-of-the-art approaches use *steady-state aerodynamics* to determine aero-dynamic forces, although the transient processes are well known. When *aerodynamic transients* occur during blade pitch changes, large-scale gusts or coherent turbulence (enveloping the whole rotor), or small-scale turbulence (smaller than the swept area) the question arises as to the time history of aerodynamic forces experienced by the rotor. If the initial force is F_1 and the final force is F_2, do transient forces ever fall outside the range between F_1 and F_2? The answer this question if yes, because several processes are at work. First, pitch motion and/or gusts will induce a circulation (and, therefore, a force) which varies with the *pitch rate* and/or the *gust strength*. Second, during a transient, vorticity is shed parallel to the blade that interacts with the blade to dampen the forced response. Finally, there is vorticity trailing the blade, perpendicular to the span, which induces a flow that alters the angle of attack experienced by the blade.

Time Scales

Some insight into the relative importance of these force-producing processes can be inferred from their time scales. Let τ_g, τ_β, τ_{sv}, and τ_{tv} be the time scales associated with *gusts*, *pitch changes*, *shed vorticity*, and *trailing vorticity*, respectively. The gust time scale, τ_g, is a function of the site (through its *turbulence scale*) and the size of the rotor. The pitch-change time scale, τ_β, is specific to a particular control system. The change in lift coefficient induced by a pitch change at the rate $d\beta/dt$ is as follows:

$$\Delta C_{L,B} \approx \left(\frac{\pi}{2} \right) \left(\frac{c}{r} \right) \left(\frac{d\beta/dt}{\Omega} \right)$$

While the time scale τ_β is a significant parameter, the induced change in lift coefficient appears to be small, since $c \ll r$ and $d\beta/dt \ll \Omega$.

The third time scale, τ_{sv}, is associated with the shed vorticity that is convected downstream. This vorticity is initially parallel to the blade-pitch axis and yields a force response of the type $[1 - exp(- t/\tau_{sv})]$, as exhibited by the classical *Wagner gust function*. In broad terms, $\tau_{sv} = 5c/V_r$, so it can be seen that it varies with rotor size, tip-speed ratio, and position on the blade, since blades are usually tapered and

$$V_r \approx R\Omega$$

The fourth time scale, τ_{tv}, is associated with vorticity trailing downstream from the rotor. A simple actuator disk model of this process yields $\tau_{tv} = R/V$, where V is the wake velocity near the rotor.

Sample calculations of τ_{sv} and τ_{tv}, will illustrate their relative durations. Assume a HAWT of radius R operating in a free-stream wind speed of 20 m/s, with a tip-speed ratio of 5.7, an axial induction factor of 0.2, and a chord-to-radius ratio of 0.045. Evaluating the shed vorticity time scale at 3/4 span, the results are as follows:

Shed vorticity: $\quad \tau_{sv} = 0.0026\,R$
Trailing vorticity: $\quad \tau_{tv} = 0.0625\,R$

where τ is in seconds and R is in meters. It can be seen that the time scale associated with shed vorticity is much shorter than that associated with trailing vorticity, and that both scales increase linearly with rotor size. From this example, it is evident that the first focus on aerodynamic transients should be on the trailing vorticity process. Further, the second focus should be to determine the nature of τ_g, and τ_β.

The way in which an aerodynamic force approaches its steady-state value during a transient can be discussed qualitatively using an example in which a blade initially at pitch angle β_1 is suddenly brought to pitch angle $\beta_2 < \beta_1$. Since decreasing pitch increases angles of attack, the aerodynamic force then goes from one positive value to a higher positive value. The shed vorticity effect dies out in a time period of $2\tau_{sv}$ to $3\tau_{sv}$. This leaves the blade with the final pitch angle but the *initial* induced velocity, since τ_{tv} (which is also the time scale of induction changes) is much longer than τ_{sv}. Hence, the aerodynamic forces will first exceed the final forces and then decrease to them asymptotically as the added induced velocity develops from the wake vorticity. In every case of sudden pitch change, transient forces will go outside the range of the initial and final forces. Similar arguments can be used to examine gust responses.

Dynamic Stall

Transient aerodynamics has another facet called dynamic stall, which is related to changes in the lift curve near its peak and in the first stages of stall, resulting from oscillations in the angle of attack. These changes often produce a *hysteresis loop* in the lift coefficient which can, in turn, lead to cyclic pressure loadings that are not predictable from conventional lift and drag data obtained at steady angles of attack. Increased excitation of blade structural-dynamic modes becomes a possibility during dynamic stall. Analytical investigations of the potential scope of these cyclic loads have been conducted using the *Gormont* model [1973] and more recently the Leishman-Beddoes [1986, 1989] method. Dynamic stall effects have been measured and analyzed for HAWT rotors by Hansen [1992] and Hansen and Butterfield [1993], and these have also been analyzed by Berg [1983] for VAWT rotors. *The AeroDyn Theory Manual* [Moriarty *et al.* 2005] discusses the treatment of dynamic stall used in current *aero-structural dynamic codes*.

Vortex Generators

Faced with a tradeoff between thin airfoils with superior performance but inferior strength and moderately thick airfoils with superior strength but inferior performance, wind turbine designers have recognized the benefit from some sort of *boundary-layer control* to increase the resistance of the boundary layer on a thicker airfoil to *adverse pressure gradients*. This delays separation and stall, and allows a moderately-thick airfoil to reach a higher lift coefficient without a large drag penalty.

Re-energizing the boundary layer on a wind turbine blade can be achieved by mixing the boundary layer air with faster-moving air from the free stream. This mixing occurs naturally in a *turbulent boundary layer*, and one simple "fix" used to extend the performance of a number of airfoils is to add a *trip strip* (a small step or band of roughness perpendicular to the flow direction) to the low-pressure surface, to force the transition of a previously laminar boundary layer to a turbulent one. However, the rate of momentum transfer in a turbulent boundary layer is limited by the magnitude of the turbulence. If the turbulence could be increased, the mixing rate would also increase, and there would be the potential for performance improvements greater than those possible with trip strips.

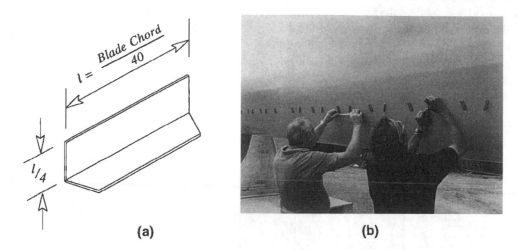

Figure 5-24. Vortex generators on the 3.2-MW Mod-5B HAWT. (a) Typical vane dimensions (b) Installing counter-rotating pairs of VGs on the low pressure side of the blade, with ± 20 deg angles of incidence at 10 percent of chord.

Vortex generators provide a simple method of increasing this mixing rate. These devices were first developed at the *United Aircraft Corporation* and are credited largely to Taylor [1947] and Bruynes [1951]. In their simplest form, they consist of small vanes projecting normal to the low-pressure surface of the airfoil at an angle of attack to the incoming flow (Fig. 5-24). In this configuration they behave like half-wings and generate trailing vortices from their tips. These vanes can be parallel to one another or arranged with alternate positive and negative angles of incidence, producing *corotating* or *counterrotating* vortices, respectively. They have been used to suppress or delay separation in diffusers, pipe bends, and wings. It should be noted that by creating an energetic vortex, vortex generators necessarily incur a drag penalty when separation is not imminent.

The first application of vortex generators to wind turbine blades was performed by the Boeing Engineering and Construction Company on the 2.5-MW Mod-2 HAWT [Boeing 1982]. This installation and that on the later 3.2-MW Mod-5B HAWT were of the counter-rotating type. Wind tunnel tests verified that vortex generators would delay stall and increase the maximum lift coefficient of the moderately-thick airfoils used in these rotors. In addition, vortex generators helped solve a control instability problem in the Mod-2 caused by intermittent stalling of the airfoil sections near the gap between the fixed and moveable sections of the rotor.

Shown in Figure 5-25 are power curves for the Mod-2 HAWT with and without vortex generators on the blades [Sullivan 1984]. Without vortex generators, control instability forced a non-optimum pitch setting during operations in below-rated winds, particularly at the "knee" of the power curve. This resulted in a substantial loss in power when wind speeds were between 11 and 15 m/s. Vortex generators over the fixed sections of the rotor (the inboard 70 percent of span) eliminated the control instability and permitted a more-optimum pitch setting, reducing the rated wind speed by almost 2.5 m/s, compared with previous operations. When vortex generators were also added to the tip sections, improvements in the lift coefficient further increased power in below-rated winds.

Similar tests were also made on the 25-kW Carter Model 25 HAWT [Gyatt 1986]. In this rotor the blades are highly twisted, so at low wind speeds the angles of attack along the inboard segments of the blade are well below stall. Thus, while vortex generators increased the maximum power of the Carter Model 25, power output at low wind speeds was reduced by the additional drag of the vortex generators.

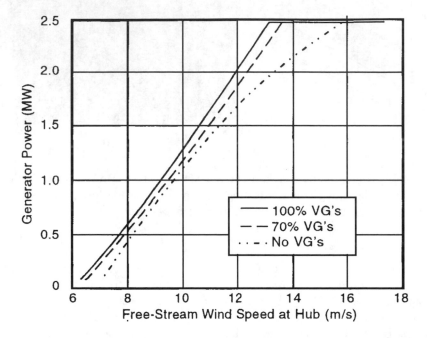

Figure 5-25. Test power curves for the 2.5-MW Mod-2 HAWT with and without vortex generators on the rotor airfoils. [Sullivan 1984]

HAWT Aerodynamic Performance Tests

Testing Methods

One of the aerodynamicist's few windows for observing and understanding the physical phenomena of wind turbine aerodynamic behavior comes from performance data obtained during field testing of HAWTs. There are three basic methods of testing wind turbine rotors, each method having its own advantages and disadvantages.

Wind Tunnel Testing

Wind tunnel testing, the mainstay of the aircraft industry, has been of limited value in wind turbine rotor development, although it is useful for obtaining basic two-dimensional airfoil lift and drag data. While tests in a facility such as the NASA Ames Research Center's *80-ft by 120-ft Wind Tunnel* have provided useful data [Simms *et al.* 2001 and Schreck 2002], financial limits have restricted the extent of wind tunnel tests. *Scaling* and *blockage* problems occur in most wind turbine tests in wind tunnels. For the test and full-scale *Reynolds Numbers* to be approximately equal, a large test model is wanted. However, because the rotor wake expands, the *blockage ratio* (swept area/cross-sectional area of the wind tunnel) should be small, preferably less than 1/10.

Tow Testing

Tow testing, whereby the wind turbine is pulled or pushed through static air, can relieve the scaling and blockage problems of wind tunnels, but the rotor size is limited. The tow testing of a 15-m diameter wind turbine, for example, would be a formidable task. Tow testing also shares with wind tunnel testing the problem of failing to subject the wind turbine to the unsteady nature of the wind.

Field Testing

Field testing presents the proper wind environment, but it brings new challenges in measuring and recording test data. Field test data acquisition and recording vary with the type of test to be made. For *load tests*, it is desirable to sample the wind at several circumferential stations upwind of the rotor (*e.g.* with an array of anemometers as shown in Fig. 3-25), with time resolution sufficient to distinguish *small-scale turbulence*. Load measurement is made with *strain gages*, using a sampling rate sufficiently fast to determine the *frequency response* of the rotor blades. *Performance tests*, in which the power output of the wind turbine is determined, use time-averaged measurements. Discussion of testing in this chapter focuses on performance data, since loads data are covered in later chapters.

Test Data Acquisition

Performance data from wind turbines are stored and plotted using the well-proven *method of bins* [Akins 1978]. Output power and free-stream wind speed are sampled over periods of time, on the order of 5 minutes, and average values for each period are stored in wind-speed "bins." The bins are on the order of 1 m/s in range. The power and wind speed samples in each bin are averaged for plotting. This method greatly smooths the resulting graph of power *vs.* wind speed or the *power curve*.

As a result of using time-averaged data, performance measured by field testing lacks the potential accuracy of wind tunnel or tow testing. At the same time, it contains the response of the turbine to turbulence, wind shear, and terrain features. The *reference wind speed* for performance testing is the wind speed measured at the elevation of the center of the swept area of the rotor. Because a rotor retards the wind flowing through it, the reference wind speed must be measured at a distance upwind, usually a minimum of about 1.5 to 2.0 rotor diameters. Performance test data based on a single reference wind speed measurement is the current industry standard, but two anemometers are usually used for redundancy [ASME 1988].

Examples of Field Tests of HAWT Performance

Numerous field tests have been conducted on three generations of medium- and large-scale prototype HAWTs, from the 100-kW 38-m diameter Mod-0 to the 3.2-MW 97.5-m diameter Mod-5B (see Chapter 3). Performance test data from three of these evolutionary turbines will be presented here and compared with performance predictions made in accordance with modified strip theory. The specific examples are these:

-- The Mod-0 HAWT (38-m diameter; 100-kW rating) operating at *fixed pitch*. Under high wind conditions the flow over the rotor blades was well into the stall region.
-- The MOD-1 HAWT (61-m diameter; 2-MW rating), which operated in the *vortex ring/windmill brake* state under low wind conditions.
-- The MOD-2 HAWT (91.4-m diameter; 2.5-MW rating) using *partial-span pitch* control as well as *teetering*.

As discussed in the introduction to this chapter, performance data taken during a field test are stored and plotted using the *method of bins* [Akins 1978], producing a graph of output power versus wind speed or a *power curve*. Comparisons with theory will be made on the basis of the mean value of the test data in each bin. The *standard deviation* will also be shown when available.

Mod-0 HAWT Performance

This rotor had tapered and twisted blades with airfoil sections from the NACA 23000 series. Airfoil thickness-to-chord ratios varied ranging from 0.12 at the tip to more than 0.40 at the blade root, with a rotor solidity of 2.9percent. During this series of tests, the MOD-0 wind turbine was operated at fixed pitch, with a tip speed of 55 m/s [Viterna and Corrigan 1981]. Results from these tests are shown in Figure 5-26 and are typical of the performance of fixed-pitch HAWTs. Power increases with increasing wind speed, holds constant, and then decreases somewhat. This process is referred to as *stall regulation*. Strip theory predictions of power agree with the test results over the entire range of wind speeds, including post-stall operations for which an empirical model was developed by the investigators.

Mod-1 HAWT Performance

The Mod-1 rotor had tapered and twisted blades with a solidity of 4.2 percent, half again as much as the Mod-0 rotor. The higher solidity of the Mod-1 turbine resulted in the development of large axial inductions at low wind speeds, so wake expansion effects were considerable. The NACA 4400 series of airfoils was employed for the two rotor blades, with thickness-to-chord ratios varying from 0.09 at the tip to 0.33 at the root. Performance test data [Spera and Janetzke 1981] were measured at a rotor speed of 3.63 rad/s, which produced

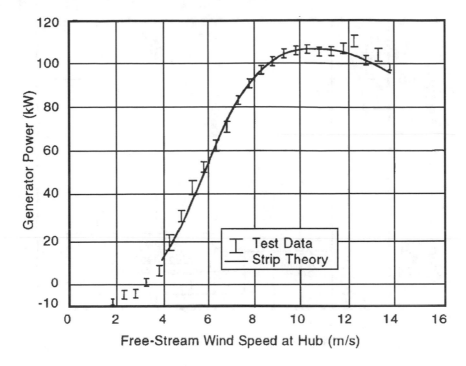

Figure 5-26. Power curve of the 38-m diameter Mod-0 HAWT with fixed blade pitch.

a relatively high tip speed of 111 m/s. Full-blade pitch was used to hold power constant at 2 MW in high winds, and pitch was held constant in below-rated winds.

Figure 5-27 shows measured and predicted power output for the Mod-1 HAWT as a function of free-stream wind speed. Strip theory power predictions are shown based on both *smooth* and *half-rough airfoil* lift and drag data. Half-rough data are obtained by averaging reference smooth and *rough* airfoil data [Abbot *et al.* 1945]. (Rough airfoil performance is unrealistically low for wind turbine blades.) Smooth airfoil data leads to predictions which closely match measured power at levels greater than one-half of rated power, where wind turbines produce the majority of their energy output (see Table 5-1).

Mod-2 HAWT Performance

The 2.5-MW Mod-2 turbine design had a two-bladed teetering rotor that used airfoils in the NACA 23000 series. It had a rotor solidity of 0.036 and tip speed of 84 m/s, which places these parameters midway between those of the Mod-0 and the Mod-1 rotors. The outboard 30percent of each blade had variable pitch for starting, stopping, and controlling power output. In contrast to the wind turbines examined previously, the Mod-2 control system employed pitch change during operation in below-rated winds. The test data presented here are for the rotor without vortex generators, and in this configuration the pitch schedule below rated power was as follows (tips are fully-feathered at a pitch angle of -90 deg): For power outputs between 0 and 1.0 MW, the tip pitch varied linearly with power, from -5 deg to -2 deg; from 1.0- to 2.5-MW the turbine was operated at a fixed blade pitch of -2 deg.

Figure 5-28 is a plot of test data reported by Boeing [1982] compared with predicted power output obtained using modified strip theory [Wilson and Walker 1984]. A correction for the 0.3-m gaps between the fixed- and variable-pitch portions of the rotor (developed earlier in this chapter) is included in the predicted power. At the "knee" of the power curve,

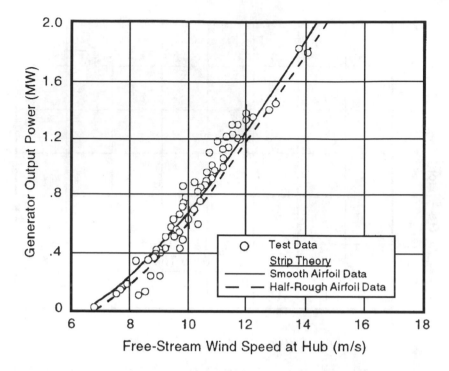

Figure 5-27. Power curve of the 61-m diameter Mod-1 HAWT.

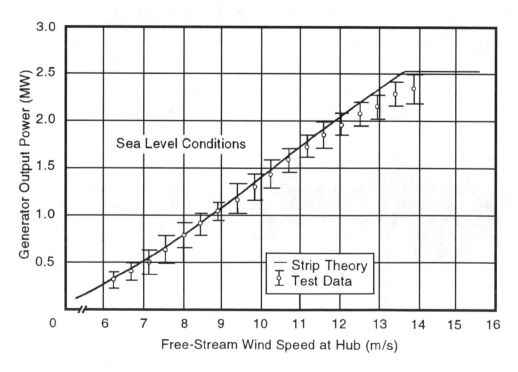

Figure 5-28. Power curve of the 91.4-m diameter Mod-2 HAWT. Test data from [Boeing 1982]; theory from [Wilson and Walker 1984]

wind turbulence and control system actions generally cause the mean test power of HAWTs to fall below theoretical predictions, unless the effects of these two factors are specifically added to the performance model. Below the knee of the power curve, strip theory predicts the system power with acceptable accuracy.

Dimensionless Performance Parameters

Shown in Figure 5-29 is a dimensionless plot of power coefficient K_P versus advance ratio J, displaying performance test data for all three of the wind turbines in these examples. Although the three turbines vary in rated capacity from 100 kW to 2.5 MW, their dimensionless performance characteristics are seen to be remarkably similar. The Mod-1 HAWT, with its highly-twisted blades, high tip speed, low solidity, and NACA 44XX-series airfoils, is observed to have superior performance in below-rated winds.

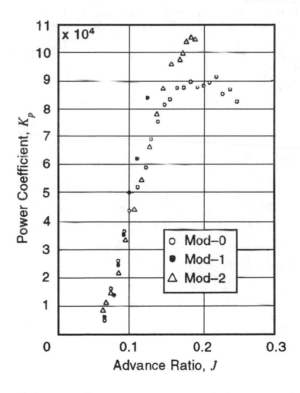

Figure 5-29. Dimensionless performance curves, permitting comparison of test data for the Mod-0, Mod-1, and Mod-2 HAWTs. [Wilson and Walker 1984]

HAWT Aerodynamic Loads

In general, the determination of the loads on a HAWT rotor involves the complex interaction of aerodynamic forces, structural deformations, and wind variability in both time and space. However, the aerodynamic moment obtained from a rigid-body analysis (the first integral in Equation (5-41a)) is a fair estimate of the *steady blade bending load*. The peak (maximum) value of this steady load occurs at rated power for a pitch-controlled rotor. Using the *Glauert optimum rotor* model, upper bounds can be established for steady bending loads, but these will greatly overestimate the actual aerodynamic loadings when the blade design departs appreciably from the optimum configuration. For the Glauert optimum rotor, the circulation is constant along the blade, approaching the following value at higher tip-speed ratios:

$$\Gamma \rightarrow \frac{8\pi}{9} \frac{U^2}{B\Omega} \tag{5-44}$$

The following axial and tangential air loadings are easily obtained from this circulation:

$$p_{a,x} = 0.5 \, \rho \, U^2 \, \frac{16\pi}{9\,B} \, r \tag{5-45a}$$

$$p_{a,y} = 0.5 \, \rho \, U^3 \, \frac{32\pi}{27\,B\,\Omega} \left(1 - \frac{3\,r\,\Omega}{2\,U} \frac{C_D}{C_L} \right) \tag{5-45b}$$

where x, y = subscripts designating axial and tangential air loadings, respectively

Drag has less effect on the axial loading, and Equation (5-45a) contains only the lift contribution. The aerodynamic flatwise bending moment on a Glauert optimum blade is

$$M_y = \int_s^R p_{a,x} \, r \, dr = 0.5 \, \rho \, U^2 \, \frac{8\,\pi\,R^3}{27\,B} \left[2 - 3\frac{s}{R} + \left(\frac{s}{R}\right)^3 \right] \tag{5-46}$$

where s = blade station, measured from the rotor axis (m)

Again, Equations (5-45) are applicable only at higher tip-speed ratios and can be expected to overestimate air loadings on blades significantly different in shape from the Glauert optimum configuration. Figure 5-30 illustrates the relative sizes of steady flatwise bending moments measured on a Mod-2 HAWT blade [Boeing 1982], an upper bound calculated in accordance with Equation (5-46) for tip-speed ratios greater than 10, and steady bending loads predicted using strip theory. This blade design is far from the Glauert performance-optimized configuration, and Equation (5-46) significantly overestimates the bending load.

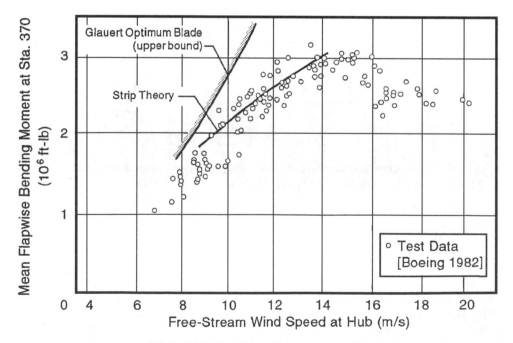

Figure 5-30. Measured and calculated steady flatwise bending moments *vs.* free-stream wind speed at 20 percent span in a Mod-2 HAWT blade.

Aerodynamic Analysis of Vertical-Axis Wind Turbines

Modern VAWTs are almost exclusively of the curved-blade design patented by G. Darrieus [1931]. However, this type of wind turbine did not see extensive development until the 1970s when it was re-invented at the National Research Council of Canada [South and Rangi 1972]. Engineers there and at the NASA Langley Research Center [Muraca *et al.* 1975] and the Sandia National Laboratories [Blackwell and Reis 1974] undertook the analysis, design, construction, and testing of Darrieus VAWTs. In the course of subsequent development projects, many aerodynamic models for the Darrieus rotor have been proposed, and these may be classified into the following three groups:

-- *Streamtube models* [Templin 1974, Wilson and Lissaman 1974, Muraca *et al.* 1975, Shankar 1975, Strickland 1975, Wilson and Walker 1981, Parashiviou 1981].
-- *Fixed-wake vortex models* [Holme 1976, Wilson 1978, McKie *et al.* 1978]
-- *Free-vortex models* [Fannuci and Walters 1976, Strickland *et al.* 1980, Wilson *et al.* 1983]

These models vary considerably in their treatment of the flow, and no single approach is available that covers all significant effects over the entire operating range of a VAWT, such as *dynamic stall, variable induced flow,* and *wake crossing* by the downwind blade. The history and a description of the many analysis techniques used for the Darrieus rotor is covered in a book about Darrieus Rotor design [Paraschiviou 2002].

In streamtube models the induced axial velocity is calculated at the rotor by equating the *time-averaged force* on the blades to the *mean momentum flux* through a streamtube of fixed location and dimensions. Single, tandem, and multiple streamtubes have been used, as illustrated in Figure 5-31 [Touryan *et al.* 1987]. Streamtube models, with some exceptions [Wilson and Walker 1981, Paraschiviou 1981], predict fore-and-aft symmetry of flow quantities. Aerodynamic forces are calculated from local angles of attack and local relative velocities, using static airfoil lift and drag coefficients which may include post-stall behav-

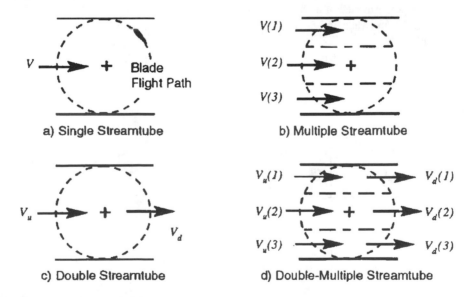

Figure 5-31. Plan view sketches of various streamtube models used for aerodynamic analysis of VAWT rotors. (a) Single streamtube (b) Multiple streamtube (c) Double streamtube (d) Double-multiple streamtube [Touryan *et al.* 1987]

ior. Streamtube approaches have met with modest success in predicting the overall torque and thrust loads on Darrieus rotors. Additionally, the computer time needed for streamtube computations is much less than that for any other approach.

Fixed-wake vortex models use a vortex sheet wake that is locally independent of time. Forces are determined from circulation using the Kutta-Joukowski law. The fixed-wake vortex analysis has a distinct advantage over streamtube methods below stall, in that it has the ability to identify the *crosswind-induced flow*. Its computational time requirements exceed those of streamtube models, but are at least of the same order.

The most complete and accurate method of analysis that has been developed for predicting the aerodynamic behavior of a Darrieus rotor is the *free-vortex* approach. The rotor wake is modeled by discrete, force-free vortices, each of which is convected downstream at a velocity determined by induced velocities from the rest of the system. Unfortunately, the free-vortex approach is also the most complex and has several other disadvantages:

-- Power and load values are approached asymptotically from above as the wake length is increased towards infinity. As a result, the practical requirement of a finite-length wake results in predictions that are too high.
-- Since the wake moves very slowly at high tip-speed ratios, it is always easier to analyze rotors operating at low tip-speed ratios.
-- Overall, free-vortex analysis is quite expensive to perform.

Limiting VAWT Performance

We will first consider the performance limit of a *giromill rotor*, a design also proposed by Darrieus. A giromill consists of straight, vertical, rectangular blades rotating about a vertical axis. The blades are *articulated* in pitch to maximize the extraction of energy from the wind. A necessary condition for maximum energy extraction is that the far-wake velocity be uniform. By virtue of its articulated blades, a giromill will have higher performance than a fixed-pitch Darrieus rotor. Therefore, a performance limit for a giromill is also an upper bound for the performance of a Darrieus VAWT.

Figure 5-32 illustrates the development of an actuator disk model of a giromill. In this schematic plan view, B blades ($B \gg 1$) of chord c are situated around the circumference of a rotor of radius R. As a blade moves around the circle, its pitch angle is modulated to keep the

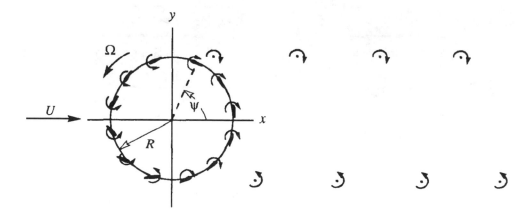

Figure 5-32. Schematic plan view of a vortex-sheet model of a giromill flow field. [Wilson 1978]

circulation constant. At the azimuthal positions of $\psi = 90$ deg and $\psi = 270$ deg, each blade is "flipped" (*i.e.*, its pitch angle changes sign) to produce a circulation of equal magnitude but opposite sign. This change of sign in circulation is necessary to maintain a positive force in the tangential direction. The wake consists of concentrated vortices moving downwind in vortex-street fashion. As $B \to \infty$ and $c \to 0$ while the product Bc is held constant, the discontinuous vortex streets become continuous *vortex sheets*, and the entire flow field becomes steady.

The velocity field of the rotor consists of three parts that can be superimposed to obtain the flow field: the free-stream velocity, U; the wake-vortex sheet velocity, v_w; and the bound-vortex sheet velocity, v_b. Analysis of this vortex sheet model of the flow field about a giromill shows that the power and thrust coefficients of a VAWT have the same limits as those of a HAWT, given by Equations (5-15a) and (5-30) [Wilson 1978]. In addition, this analysis also yields a *shear (crosswind) force coefficient*, C_S, as follows:

$$C_S = \frac{S}{0.5 \, \rho \, U^2 \, R \, H} = -\frac{\pi}{\lambda} \, a^2 \, (1 - a) \tag{5-47}$$

where H = height of the giromill rotor (m)

Fixed-Wake Streamtube Analysis

In order to analyze the aerodynamic behavior of a VAWT in accordance with the streamtube approach, first consider an airfoil traversing the idealized path shown in Figure 5-33(a). This airfoil is assumed to be symmetrical in cross-section and have negligible drag. Between points A and B the airfoil moves parallel to the free stream and generates no force.

At point B, the airfoil changes direction and sheds vorticity. It then moves across the wind for a distance w, generating both lift and circulation. Again at point C, the direction of motion is changed and vorticity is shed, opposite in sign to that shed at B. From point C to point D no force is generated. Finally, on the path from D to A, the airfoil once again generates lift and circulation, the latter opposite in direction to the circulation along path BC. The wake system that is generated is illustrated in Figure 5-33(b), in which the local circulation direction is indicated by the arrows and where

Γ_f, Γ_r = magnitudes of front and rear circulation, respectively (m²/s)

Axial Induction of a VAWT Rotor

We may take certain liberties with the idealized path illustrated in Figure 5-32(a) without changing the wake pattern. In Figure 5-33(c), the width w has been reduced relative to the distance AB, and the paths BC and DA have been made to conform to the path of a Darrieus rotor blade. Thus, we have a streamtube of a Darrieus rotor. Considering the flow in other streamtubes to be similar, we arrive at several significant observations:

--   Since flow along BC is influenced only by flow inside the vortex street shed by BC, forces on an airfoil traversing BC are not influenced by adjacent streamtubes.
--   As the streamtube gets smaller in width, the wake from BC appears as a semi-infinite vortex street. The resulting induced velocity at the front is due only to the semi-infinite wake of the front caused by Γ_f.
--   An airfoil on the rear path DA "sees" an infinite wake caused by the front circulation, Γ_f, and a semi-infinite wake caused by the rear circulation, Γ_r. Thus, the rear induced velocity depends on both the front and rear circulations.

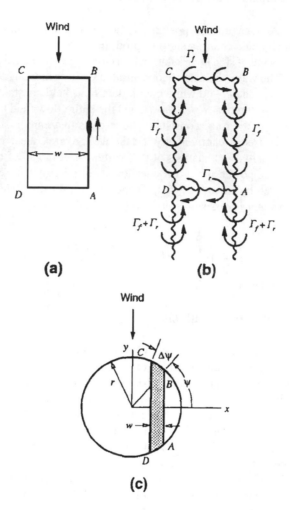

Figure 5-33. Development of a VAWT streamtube model. (a) Plan view of an idealized blade path (b) Resulting circulation in the wake (c) VAWT streamtube

These observations lead to the following equations for the induced velocities:

$$a = a_r - a_f \tag{5-48a}$$

$$\frac{a_r}{a_f} = 2 - \frac{\Gamma_r}{\Gamma_f} \tag{5-48b}$$

where a_f, a_r = front and rear axial induction factors, respectively

In order to determine the three local induced velocities (a, a_f and a_r), a third equation is required. Let us use the momentum theorem and equate the windwise momentum change to the mean downwind blade force, which results in

$$a(1-a) = \frac{B\Omega}{4\pi U^2}[\Gamma_f - \Gamma_r] \tag{5-48c}$$

The circulations in Equation (5-48b) do not include bound circulation. For a symmetrical airfoil with its chord tangent to the path of rotation, the only bound circulation is that from pitching. Using linear aerodynamics with a lift curve slope of C_L' and noting that $(sin\psi)_{rear}$ equals $- (sin\psi)_{front}$, Equations (5-48) may be reduced to the following:

$$a^3 - 2a^2 (1 + G) + a \left(1 + 4G + G^2\right) - 2G = 0 \qquad (5\text{-}49a)$$

$$G = \frac{\lambda B c C_L'}{8 \pi R} \cos\theta \sin\psi \qquad (5\text{-}49b)$$

$$a_f = \frac{1 - \sqrt{1 - 2a}}{2} \qquad a_r = a_f + a \qquad (5\text{-}49c)$$

where θ = inclination of blade norm1 from the horizontal (rad)

Explicit solutions of the cubic equation may be obtained. For nonlinear aerodynamics, the local circulation (exclusive of bound circulation) is $0.5C_LcV_r$, and iteration is required.

Torque of a VA WT Rotor

Consider an element of a Darrieus rotor blade, as shown in Figure 5-34, acted upon by the local effective wind. The airfoil is assumed to be symmetrical, with its chord line tangent to the circumferential path. As was true for the coned HAWT rotor, the spanwise component of the relative wind does not contribute to lift or drag, so

$$V_e = \sqrt{V_\psi^2 + V_n^2} \qquad (5\text{-}51a)$$

$$\alpha = \text{arc sin} \left(V_n / V_e\right) \qquad (5\text{-}51b)$$

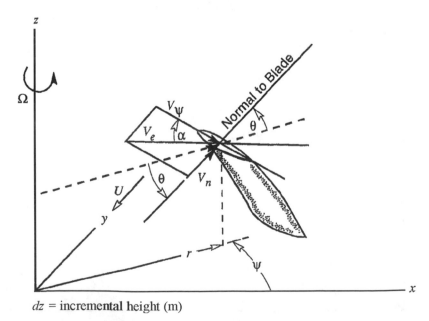

dz = incremental height (m)

Figure 5-34. Geometry of a Darrieus blade element and the effective wind velocity.

where V_e = effective wind speed (m/s)
V_ψ = circumferential (chordwise) wind speed (m/s)
V_n = wind speed normal to the element (m/s)

Referring to Figure 5-34, the circumferential and normal wind speeds can be calculated from the free-stream wind speed, the rotor speed, and the axial induction factors as follows:

$$For\ 0 \leq \psi \leq \pi : \quad V_\psi = r\Omega + U\left(1 - a_f\right)\cos\psi$$
$$V_n = -U\left(1 - a_f\right)\sin\psi\cos\theta \tag{5-51c}$$

$$For\ -\pi \leq \psi \leq 0 : \quad V_\psi = r\Omega + U\left(1 - a_r\right)\cos\psi$$
$$V_n = -U\left(1 - a_r\right)\sin\psi\cos\theta \tag{5-51d}$$

As shown in Figure 5-35, the lift and drag forces on an element of blade of length ds produce a differential torque dQ about the axis of rotation that is given by

$$dQ = dQ_{KJ} + dQ_D \tag{5-52a}$$

$$dQ_{KJ} = 0.5\,\rho\,V_e^2\,c\,C_L\,\frac{\sin\alpha}{\cos\theta}\,r\,dz \tag{5-52b}$$

$$dQ_D = -0.5\,\rho\,V_e^2\,c\,C_D\,\frac{\cos\alpha}{\cos\theta}\,r\,dz \tag{5-52c}$$

where dQ_{KJ} = *Kutta-Joukowski* (lift) torque contribution (N-m)
dQ_D = drag torque contribution (N-m)
dz = incremental height (m)

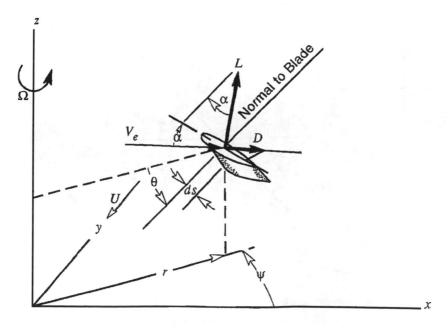

Figure 5-35. Wind velocity and force diagram for a Darrieus airfoil segment.

The torque of a VAWT rotor varies with blade azimuthal position (the so-called *torque ripple*), so rotor power is defined as the average power over one revolution, as follows:

$$P = \frac{B\Omega}{2\pi} \int\limits_{0}^{2\pi} \int\limits_{-H/2}^{H/2} (dQ_{KJ} + dQ_D)\, d\psi \qquad (5\text{-}53)$$

where H = height of the area swept by the airfoils (m)

Airfoil Data for VAWT Blades

When using streamtube analysis, the determination of aerodynamic forces is accomplished using available static lift and drag data. Despite the amount of airfoil section properties compiled by the NACA and more recent work by NASA, extensive gaps remain in our knowledge of wind turbine airfoils. Aircraft airfoils are designed to operate below stall, and so the data base is concentrated in the below-stall region. A VAWT blade, however, operates at angles of attack considerably beyond stall, particularly when it is producing peak power. Another variance between VAWT and aircraft airfoils is that a VAWT blade follows a circular path with significant cyclic variations in relative velocity and angle of attack.

The applicability of steady-state airfoil data to the unsteady operating conditions of a VAWT may be questioned. Below stall, however, the use of static lift data appears to be valid. Shown in Figure 5-36 is the variation in lift coefficient during a typical rotor revolution, calculated for a section on a VAWT blade. The VAWT lift curve remains approximately parallel to the theoretical static lift curve for chord-to-radius ratios of 0.10. Typical ratios for current VAWTs are less than 0.08.

A third factor affecting airfoil data is the pitch rate of a VAWT blade, which is equal to the rotor speed Ω. In Figure 5-36, the offset angle between the static and dynamic lift curves, α_p, is caused by pitching circulation. A first-order analysis of the effect of pitching on drag coefficient of a pitching *NACA 0015* airfoil below stall indicates that the offset angle for drag is 40percent to 50percent less than that for lift [Wilson and Neff 1985].

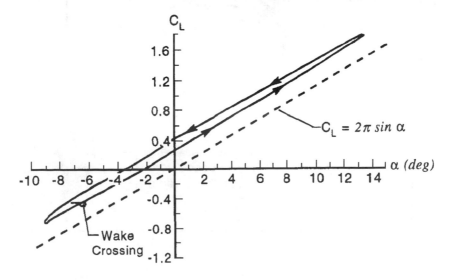

Figure 5-36. Calculated lift coefficient *vs*. angle of attack at a VAWT blade section during one rotor revolution. Tip-speed ratio is 3.5 and the chord-to-radius ratio is 0.2. [Wilson *et al.* 1983]

Comparison of Theoretical VAWT Performance with Test Results

Theoretical predictions of the performance of the *Sandia/DOE 17-m VAWT* (similar to that in Fig. 3-24) will now be compared with measured performance data [Worstell 1980]. The principal parameters of the Darrieus rotor on this machine are as follows:

Airfoil = NACA 0015; lift and drag data from [Sheldal and Klimas 1981]

     $R = 8.28$ m (at $z = 0$) $c = 0.61$ m

     $H = 17.0$ m $A = 183$ m^2

     $B = 2$ $\Omega = 4.0$ to 5.3 rad/s

     $N_R = 1.5$ to 2.0×10^6 at the rotor equator

$$r = 2.790 + \sqrt{30.100 - z^2} \quad 0 \le |z| \le 4.548 \; m$$

$$r = 12.60 - 1.4826 \, |z| \quad 4.548 \le |z| \le 8.090 \; m$$

Test power coefficients are plotted versus tip-speed ratio in Figure 5-37, together with vortex-model predictions [Strickland *et al.* 1979]. It can be seen that this type of aerodynamic model is quite accurate. However, since an aerodynamic analysis based on vortex theory is the most complex and time-consuming of the methods for predicting VAWT performance, it is generally reserved for validating simpler models.

The test data and vortex theory predictions in Figure 5-37 are plotted in a dimensionless power-curve form in Figure 5-38, as rotor-speed power coefficient, K_P, *vs.* advance ratio, J. We note that the region of maximum power coefficient ($J \approx 0.18$ or $\lambda \approx 5.5$) occurs at low power levels, indicating that $C_{P, \, max}$ is not a critical parameter in determining the annual

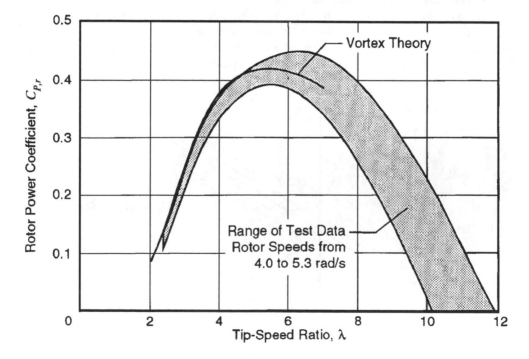

Figure 5-37. Comparison of experimental and theoretical power coefficients for the Sandia/DOE 17-m VAWT. Test data from [Worstell 1980]; theory from [Strickland *et al.* 1979]

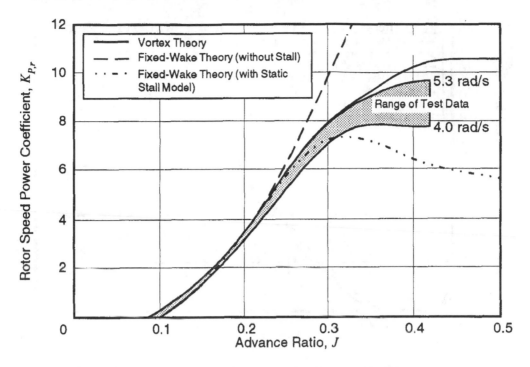

Figure 5-38. Experimental and theoretical dimensionless power curves for the Sandia/ DOE 17-m VAWT. Test data and vortex theory are the same as in Figure 5-37. Fixed-wake theory from [Wilson and Walker 1981]

energy output of this VAWT or in evaluating various aerodynamic theories. Of more importance is performance at higher advance ratios, or (equivalently) tip-speed ratios less than that for $C_{P,\text{max}}$. It can be seen in Figure 5-38 that the power curves flatten at higher advance ratios, an effect attributed to increasing amounts of stall. Power levels trend upward with increasing rotor speed, reflecting the benefits of higher Reynolds Number and lower angles of attack.

Two additional theoretical predictions are shown in Figure 3-38. These are based on *fixed-wake theory, with and without static stall* in the airfoil properties [Wilson and Walker 1981]. At high advance ratios (*i.e.* high wind speeds), fixed-wake theory without stall over-predicts the measured power by large amounts. Note, however, that measured peak power levels trend toward the theoretical unstalled performance with increasing rotor speed. When static stall properties are added to the fixed-wake theory, the predicted power curve flattens at higher advance ratios and then drops below the test data. This latter behavior emphasizes the role of stall in reducing VAWT power at high wind speeds, and at the same time indicates that the transient stall experienced by the rotating blade is not as fully-developed as static stall.

A final comparison between experimental and theoretical VAWT performance is shown in Figure 5-39. Here, the 17-m test data are compared with predictions made with three streamtube theories: *multiple streamtube with static stall* [Strickland 1975], *double-multiple streamtube with static stall* [Parashiviou 1981], and *double-multiple streamtube with dynamic stall* [Berg 1983]. When static stall data are used, the streamtube and fixed-wake theories give much the same predictions of power, and these are in good agreement with test data at the most energy-productive advance ratios of a VAWT. With the addition of a dynamic stall model, correlation between test data and streamtube theory at high advance ratios becomes equivalent to that of the more-accurate vortex theory.

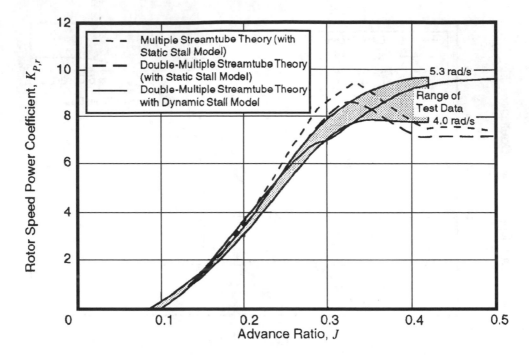

Figure 5-39. Experimental dimensionless power curves for the Sandia/DOE 17-m VAWT compared with streamtube theories. Test data are the same as in Figure 5-37. Theoretical predictions from [Strickland 1975, Parashiviou 1981, Berg 1983].

Dynamic-Stall Factors

Relatively early in the development of VAWTs, differences between performance test behavior and theory were attributed to dynamic stall [Strickland *et al.* 1980], which has a measurable effect on HAWTs but appears to be even more important to VAWT aerodynamics. Simply stated, dynamic stall on a blade can produce lift and pitching-moment values that are much larger than static values. These additional forces are brought on by a strong vortex that forms at the leading edge and is quickly convected over the blade and into the wake. Its formation and movement strongly depend on the following factors:

-- type of airfoil, leading edge radius, and thickness distribution;
-- initial angle of attack;
-- rate of increase of angle of attack;
-- excursion of angle of attack past the static stall angle;

The effect of this vortex on airfoil properties begins with a rapid increase in forces and ends with full flow separation and a large drop in lift. It is clear that the dynamic stall process depends both on the amplitude and the history of the angles of attack on the airfoil. There is also a *hysteresis effect*, involving a delay in re-attachment of the flow and recovery of blade forces to their static levels after the event. Stall time delays related to rapid increases in angle of attack have been used to reproduce analytically the flattening of VAWT power curves at high advance ratios [Massé 1984].

Future Developments of VAWT Aerodynamic Theory

Continued comparison of theoretical and experimental VAWT performance has focused current research on the following:

-- improving the accuracy of predicting peak power output;
-- understanding dynamic stall at high wind conditions;
-- including stochastic wind models in order to predict fatigue loads on blades;
-- correcting the tendency to over-predict power output for operations between maximum rotor power coefficient and the onset of dynamic stall;
-- introducing interactions between adjacent flow regions;
-- eliminating assumptions that are known to be mathematically incorrect.

Local circulation methods [Azuma *et al.* 1983, Massé 1984] are candidates for improved models of the performance of VAWTs.

Aerodynamic Behavior of VAWTs in Operation

More than 600 VAWTs were placed in commercial service, mostly in California, with generator capacities that range from 120 to 300 kW. All were two-bladed machines. Despite their lack of wind direction sensitivity, yaw control equipment, aerodynamic control surfaces, and teetered hubs, VAWTs still have several operational characteristics that are unique, such as rotor blade shape and torque ripple. Other characteristics, such as starting and stopping, the role of transient aerodynamics, the use of vortex generators, and the effects of rotor solidity are subjects that VAWTs have in common with HAWTs.

Rotor Blade Shape

The blade shape of a Darrieus blade is patterned after the *troposkien* ("spinning rope"), a configuration in which every section of an ideal blade is locally in tension under action of the centrifugal acceleration (including a small correction for gravity loads), without bending stresses. Practical Darrieus blade shapes are composed of mixtures of circular arcs and straight sections that approximate the ideal shape, in order to reduce bending stresses and yet have a shape that can be easily manufactured. The *Sandia/DOE 17-m* research VAWT (as well as several commercial VAWTs that are derivatives of it) uses the blade configuration shown in Figure 5-40. It consists of a circular arc, *AB*, and a straight section, *BC*, instead of the continuously changing curvature of a true troposkien. The ratio of height to diameter, *H/2R*, is very close to unity in this rotor. U.S. commercial VAWTs have employed *aspect ratios* between 1.2 and 1.3, and the Canadian *Etolé* VAWT (Fig. 3-36) had a ratio of 1.5.

On the circular arc or *equatorial section* of the blade, local angles of attack are found to decrease as the radius *r* decreases. By contrast, angles of attack on the straight blade section increase from the junction with the arc toward the axis of rotation. Some of the operational characteristics of a VAWT rotor may be inferred by an examination of these variations in the local angle of attack.

First, consider the performance of a two-bladed Darrieus rotor in high winds (lower tip-speed ratios), during which power output is produced almost exclusively by the circular arc sections of the rotor. Since angle of attack decreases along the circular arc, and since a decrease in angle of attack reduces the amount of stall caused by high winds, the power output of the rotor will increase as the central angle Θ increases (holding blade chord constant).

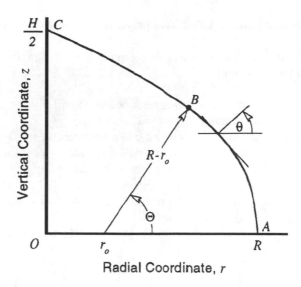

Figure 5-40. General configuration of a Darrieus VAWT rotor blade.

Increasing Θ in turn decreases the rotor aspect ratio. Thus, in broad terms one can say that decreasing the height-to-diameter ratio of a VAWT rotor will increase $K_{P,\,max}$. Carried to the limit, a VAWT with a circular rotor will have a higher $K_{P,\,max}$ than one with straight blades of the same length. However, a high value of $K_{P,\,max}$ is not particularly desirable, since it may force the power train to be too large to be cost-effective.

Next, consider the effect of rotor shape on VAWT performance in low winds (higher tip-speed ratios), when power is obtained from the straight sections of the blades. In low winds equatorial sections have very low angles of attack, to the point where power produced by lift is insufficient to overcome power lost by drag. As the angle Θ is increased, the straight segments of the rotor become longer (with a corresponding increase in aspect ratio) and the low-wind power output of the rotor increases. When the chord varies over different sections of the blade, the previous observations will hold as long as the blade chord remains fixed on each section.

Torque Ripple

Each section of a VAWT blade sees a wide variation in angle of attack during each rotor revolution. As a result, the torque produced by a blade, even in steady winds, is characterized by a cyclic time history termed *torque ripple*. Figure 5-41 shows typical variations in *torque coefficient* with blade azimuthal position, for a 17-m Darrieus operating at several tip-speed ratios. The torque coefficient is defined by

$$C_{Q,b} = \frac{Q_b}{0.5\,\rho\,U^2 A R} \tag{5.53}$$

where $C_{Q,b}$ = blade torque coefficient
    Q_b = time-varying torque produced by one blade (N-m)

Over the front portion of the path, for about 40 percent of the revolution (ψ from about 30 to 160 deg), a blade produces positive torque. However, torque production on the remainder of the circuit is low or negative, mostly because of the lower angles of attack experienced on the rear path. At a tip-speed ratio of 11, the average torque is negative in this example. The

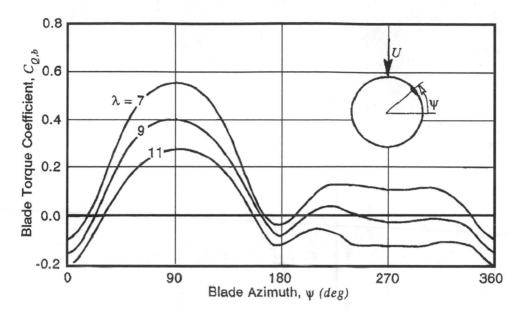

Figure 5-41. Typical variations in torque coefficient with blade azimuth or "torque ripple" produced by a single VAWT blade.

torque ripple is smoothed significantly when the individual torques of two or more blades are added together, because of their phase differences.

Rotor Solidity Effects

The solidity of a VAWT rotor can be changed by changing the number of blades or the blade chord dimensions. Current VAWTs are two-bladed machines, the two-bladed configuration being lighter and having fewer joints than a three-bladed rotor. Little has been done to determine the effect of blade number while holding other parameters constant. The most significant difference between two- and three-bladed rotors is that the rotor torque ripple, which has a distinctive twice-per-revolution oscillation for a two-bladed VAWT, is almost eliminated in a three-bladed rotor. With little else examined for the aerodynamics of three-bladed rotors, the subject of the effects of solidity on rotor aerodynamics is focused here on changes in blade chord.

For a VAWT, solidity is expressed in two different ways. The first is the traditional way, as the ratio of the blade planform area to the rotor swept area. Such a determination requires detailed knowledge of blade chord *vs.* blade length and the swept area of the rotor, quantities which are tiresome to determine. A second expression for solidity, not equal to the first, is the term Bc_{max}/R, where c_{max} is the blade equatorial chord. Numerically, the second parameter is usually within 20percent of the conventional solidity.

Figure 5-42 shows the effect of changes in Bc_{max}/R on the dimensionless power of a VAWT for various advance ratios. The geometry of the VAWT is that of the Sandia 17-m rotor described earlier, except for the chord dimension which is varied. The changes in rotor solidity (as measured by Bc_{max}/R) affect rotor performance in two ways. Largest of the effects is the fact that the increased chord results in larger aerodynamic forces and therefore in higher power. Although the axial induction increases and reduced local wind speeds offset some of the force increase caused by larger chords, the net effect of increasing the chord is an increase in power coefficient, except at very low advance ratios.

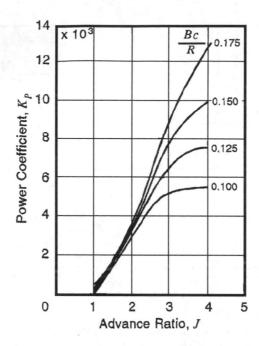

Figure 5-42. Typical effect of rotor solidity on the power coefficient of a VAWT. Rotor shape is that of the Sandia 17-m rotor, but blade chord is a variable.

A second (and smaller) effect is caused by changes in *Reynolds number*. At a constant rotor speed, increasing the blade chord directly increases local Reynolds numbers. When the Reynolds number is less than one million, increases in performance with increases in Reynolds number are most noticeable.

Vortex Generators

Vortex generators have been employed on VAWTs as well as HAWTs to improve performance. The test power curves shown in Figure 5-43 are from three tests run on the *DAF Indal* 50-kW VAWT [Quinlan 1986]. The blades on this rotor were approximately 9 m long. The first test, without vortex generators, was made to establish the basic power curve of the turbine. A second test was made with vortex generators installed over 3 m of the equatorial section of each blade, on both surfaces. Finally, a third test was run with vortex generators placed over the entire blade length, again on both surfaces.

As shown in Figure 5-43, the basic power curve (*i.e.*, without vortex generators) exhibits the same flattening at high wind speeds as was observed during the slowest test on the Sandia 17-m VAWT discussed earlier (Fig. 5-38). As a result of placing vortex generators only on the equatorial sections of the blades, there was a reduction in power output over the entire range of wind speeds. At the highest wind speed in the test (22 m/s) the reduction reached over 20 percent. Utilization of vortex generators over the entire blade resulted in a small decrease in power output under low wind conditions, but a substantial increase in power output at high winds. The tip-speed ratio at the transition from power loss to power gain was about 5.

The conclusion drawn from these tests is that the vortex generators on the equator sections of a VAWT rotor will reduce power output, while vortex generators on the upper and

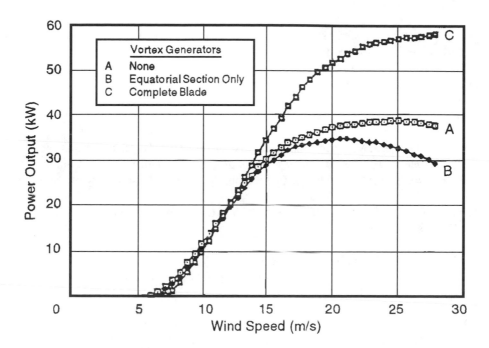

Figure 5-43. Effect of vortex generators on the power output of the DAF Indal 50-kW VAWT. [Quinlan 1986]

lower portions of the blades (which are the regions of highest angles of attack) can increase power output by counteracting the "flattening" of the power curve at high wind speeds.

Starting and Stopping

Darrieus rotors require mechanical power input for starting, although they have been known to start by themselves (several VAWTs have been lost this way). The usual method of starting a VAWT is to run the generator as a motor. Motoring the rotor is required only to bring the VAWT up to a fraction of its rated speed, after which aerodynamic forces are sufficient to bring it up to operating speed. Auxiliary rotors, such as Savonius turbines, were tried on early VAWTs for starting, but this practice was discarded in favor of motor starting.

The process of stopping a VAWT in the absence of aerodynamic control surfaces is presently accomplished almost entirely by means of mechanical brakes. Aerodynamic brakes were attempted on the Canadian Eolé VAWT, but these were not successful, primarily because of their complexity and the fact that they had to operate reliably in the high acceleration environment of the rotor equator. Aerodynamic control surfaces such as flaps and air brakes have not been employed on U.S. VAWTs to date.

References

Abbot, I. H., A. E. von Doenhoff, and L. S. Stivers, Jr., 1945, *Summary of Airfoil Data*, NACA Report 824.

ASME, 1988, *Performance Test Code for Wind Turbines*, ASMEIANSI PTC 42-1988, New York: American Society of Mechanical Engineers.

Akins, R. E., March 1978, *Performance Evaluation of Wind Energy Conversion Systems Using the Method of Bins--Current Status*, SAND 77-1375, Albuquerque, New Mexico: Sandia National Laboratories.

Anderson, M. B., D. J. Milborrow, and N. J. Ross, April 1982, *Performance and Wake Measurements on a 3-m Diameter Horizontal Axis Wind Turbine*, DOE Contract Report No. E/5A/CON/1090/177/020, Cambridge, England: University of Cambridge.

Azuma, A., K. Nasu, and T. Kayashi, 1983, "An Extension of the Local Momentum Theory to Rotors Operating in a Twisted Flow Field," *Vertica*, Vol. 7, No. 1, pp. 45-49.

Berg, D. E., 1983, "Recent Improvements to the VDARTS VAWT Code," *Proceedings, 1983 Wind and Solar Energy Technology Conference*, Albuquerque, New Mexico: Sandia National Laboratories, pp. 31-41.

Betz, A., 1919, "Schraubenpropeller mit geringstem Energieverlust," *Gottinger Nachrichten, mathematisch-physikalische Klasse*: pp. 193-213.

Blackwell, B. F., and G. E. Reis, 1974, *Blade Shape for a Troposkien Type of Vertical-Axis Wind Turbine*, SLA74-0154, Albuquerque, New Mexico: Sandia National Laboratories.

Boeing, 1982, *Mod-2 Wind Turbine System Development Final Report, Vol. II Detailed Report*, NASACR-168007, DOE/NASA/0002-8212, Cleveland, Ohio: NASA Lewis Research Center.

Bruynes, H., 1951, *Fluid Mixing Device*, U.S. Patent No. 2,558,816, Washington, DC: U.S. Patent Office.

Corbus, D., A. Hansen, and J. Minnema, 2006, "Effect of Blade Torsion Effects on Modeling Results for the Small Wind Research Turbine (SWRT)," *44th AIAA Aerospace Sciences Meeting*, American Institute of Aeronautics and Astronautics.

Coton, F. and T. Wang, 1999, "The Prediction of Horizontal Axis Wind Turbine Performance in Yawed Flow Using an Unsteady Prescribed Wake Model," *Proceedings of the Institution of Mechanical Engineers*, Vol. 213, Part A: pp. 33-43.

Currin, H. D., F. N. Coton, 2007a, "Validation of a Dynamic Prescribed Vortex Wake Model for Horizontal Axis Wind Turbines," *Proceedings of FEDSM07 5th Joint ASME/JSME Fluid Engineering Summer Conference*, July 30-August 2, 2007, San Diego, CA.

Currin, H. D., F. N. Coton, B. Wood, 2007b, "Dynamic Prescribed Vortex Wake Model for AeroDyn," AIAA 2007-421, *45th AIAA Aerospace Sciences Meeting and Exhibit*.

de Vries, O., 1985, "Comment on the Yaw Stability of a Horizontal-Axis Wind Turbine at Small Angles of Yaw," *Wind Energy*, Vol. 9, No. 1: pp. 42-49.

Darrieus, G. J. M., 1931, *U.S. Patent No. 1,835,018*, Washington, DC: U.S. Patent Office.

Drzewiecki, S., 1892, "Sur une méthode pour la détermination des élements mécaniques des propulseurs hélicoidaux," *Comptes rendus de l'Académie des Sciences*, Paris, 114: pp. 820-822.

Dugundji, J., E. E. Larrabee, and P. H. Bauer, 1978, "Experimental Investigation of a Horizontal Axis Wind Turbine," *Wind Energy Conversion*, Vol. V, ASRL TR-184-11, Cambridge, Massachusetts: Massachusetts Institute of Technology.

Eggleston, D. M., and F. S. Stoddard, 1987, *Wind Turbine Engineering Design*, New York: Van Nostrand Reinhold. pp. 30-35.

Ensworth, C. B., 1985, "Comparison of Blade Loads for Aileron and Tip Controls," *Proceedings, Fourth ASME Wind Energy Symposium*, A. H. P. Swift, ed., New York: American Society of Mechanical Engineers, pp. 115-123.

Fanucci, J. B., and R. E. Walters, 1976, "Innovative Wind Machines: The Theoretical Performance of a Vertical Axis Wind Turbine," *Proceedings, Vertical-Axis Wind Turbine Technology Workshop*, SAND76-5586, Albuquerque, New Mexico: Sandia National Laboratories, pp. **III** 61-95.

Flamm, 1909, *Die Schiffschraube*, Berlin.

Froude, R. E., 1889. *Transactions, Institute of Naval Architects*, Vol. 30: p. 390.

Froude, W., 1878, "On the Elementary Relation between Pitch, Slip, and Propulsive Efficiency," *Transactions, Institute of Naval Architects*, Vol. 19: pp. 47-57.

Glauert, H., 1922a, *An Aerodynamic Theory of the Airscrew*, Reports and Memoranda, AE. 43, No. 786, London: Aeronautical Research Committee.

Glauert, H., 1922b, *Notes on the Vortex Theory of Airscrews*, Reports and Memoranda, No. 869. London: Aeronautical Research Committee.

Glauert, H., 1926, *The Analysis of Experimental Results in the Windmill Brake and Vortex Ring States of an Airscrew*, Reports and Memoranda, No. 1026, London: Aeronautical Research Committee.

Glauert, H., 1935a, *Aerodynamic Theory*, Vol. 6, Div. L, W. F. Durand, ed., Berlin: Julius Springer, p. 324.

Glauert, H., 1935b, "Airplane Propellers," *Aerodynamic Theory*, Vol. 4, Div. I, W. F. Durand, ed., Berlin: Julius Springer, pp. 169-360.

Gohard, J. D., Sept. 1978, *Free Wake Analysis of Wind Turbine Aerodynamics*, TR 184-14, Cambridge, Massachusetts: Massachusetts Institute of Technology.

Goldstein, S., 1929, "On the Vortex Theory of Screw Propellers," *Proceedings, Royal Society*, A 123: pp. 440-465.

Gormont, R. E., 1973, *A Mathematical Model of Unsteady Aerodynamics and Radial Flow for Application to Helicopter Rotors*, USAAMRDL TR 72-67.

Gyatt, G. W., 1986, *Development and Testing of Vortex Generators for Small Horizontal Axis Wind Turbines*, NASA CR- 17951 4, DOE/NASA/0367- 1, AV-FR-861822, Cleveland, Ohio: NASA Lewis Research Center.

Hansen, A. C., 1992, *Yaw Dynamics of Horizontal-Axis Wind Turbines: Final Report*, NREL Technical Report 442-4822, Golden, Colorado: National Renewable Energy Laboratory.

Hansen, A. C., and C. P. Butterfield, 1993, "Aerodynamics of Horizontal-Axis Wind Turbines," *Annual Reviews of Fluid Mechanics*, Vol. 25, Palo Alto, California: Annual Reviews Inc., pp. 115-149.

Holme, O., 1976, "A Contribution to the Aerodynamic Theory of the Vertical-Axis Wind Turbine," *Proceedings, International Symposium on Wind Energy Systems*, Cambridge, England: St. John's College, pp. c4-55 - c4-72.

Jacobs, E. N., and A. Sherman, 1936, *Airfoil Section Characteristics as Affected by Variations of the Reynolds Number*, NACA Report 586, Hampton, Virginia: NASA Langley Research Center.

Jensen, S. A., P. Ingham, and P. A. Moller, 1986, "Aerodynamic Characteristics for Wind Turbine Blades and Air Brakes," *Proceedings, European Wind Energy Association Conference*, Rome, Italy.

Joukowski, N. E., 1912, *Soc. Math. Moscow*, reprinted in Theorie Tourbillonnaire de l'Helice Propulsive, Paris.

Joukowski, N. E., 1918, *Travanx du Bureau des Calculs et Essais Aeronautiques de l'Ecole Superieure Technique de Moscou*.

Kawada, S., 1926, Aeronautical Research Institute Report No. 14, Tokyo: Tokyo Imperial University.

Kinner, W., 1937, "Die kreisförmige Tragfläche auf potentialtheoretischer Grundlage," *Ing. Arch*, Vol. 8, No. 1: pp. 47-80.

Lanchester, F. W., 1907, *Aerodynamics*, London; see Bergey, K. H., 1980, "The Lanchester-Betz Limit," *Journal of Energy*, Vol. 3, No. 6: pp. 382-384.

Leishman, J. G. and T. S. Beddoes, 1986, "A Generalized Model for Airfoil Unsteady Behavior and Dynamic Stall Using the Indicial Method," *Proceedings from the 42nd Annual Forum of the American Helicopter Society*, Washington, DC: pp. 243-266.

Leishman, J. G. and T. S. Beddoes, 1989, "A Semi-Empirical Model for Dynamic Stall," *Journal of the American Helicopter Society*, Vol. 34, No. 3: pp. 3-17.

Lock, C. N. H., H. Batemen, and H. C. H. Townsend, 1926, *An Extension of the Vortex Theory of Airscrews with Applications to Airscrews of Small Pitch, Including Experimental Results*, Aeronautical Research Committee Reports and Memoranda, No. 1014, London: Her Majesty's Stationery Office.

Massé, B., 1986, "A Local Circulation Model for Darrieus Vertical-Axis Wind Turbine," *Journal of Propulsion and Power*, Vol. 2, New York: American Society of Mechanical Engineers, pp. 135-141.

McKie, W. R., R. E. Wilson, and P. B. S. Lissaman, 1978, *Analytical Investigation of Darrieus Rotor Aerodynamics*, Chapter 2, Corvallis, Oregon: Oregon State University.

Miller, D. R., and P. J. Sirocky, 1985, "Summary of NASAIDOE Aileron-Control Development Program for Wind Turbines," *Proceedings, Windpower '85 Conference*, SERYCP-217-2902, Washington, DC: American Wind Energy Association, pp. 537-545.

Miller, D. R., and R. D. Corrigan, 1984, *Shutdown Characteristics of the Mod-0 Wind Turbine with Aileron Controls*, NASA TM-869 18, DOE/NASA/20320-61, Cleveland, Ohio: NASA Lewis Research Center.

Miller, R. H., 1979, "On the Weathervaning of Wind Turbines," *AIAA Journal of Energy*, Vol. 3, No. 5.

Moriarty, P. J. and A. C. Hansen, 2005, *AeroDyn Theory Manual*, NREL/TP-500-36881, Golden, CO: National Renewable Energy Laboratory, January.

Muraca, R. J, S Stephen, V. Maria, and R. J. Dagenhart, 1975, *Theoretical Performance of Vertical Axis Windmills*, NASA TM TMX-72662, Hampton, Virginia: NASA Langley Research Center.

Parashiviou, I., 1981, "Double-Multiple Streamtube Model for Darrieus Wind Turbines," *Proceedings, Wind Turbine Dynamics Workshop*, NASA CP-2185, DOE Publication CONF-810226, SERIICP-635-1238, R.W. Thresher, ed., Cleveland, Ohio: NASA Lewis Research Center, pp. 19-25.

Parashiviou, I., 2002, *Wind Turbine Design with Emphasis on Darrieus Concept*, Polytechnic International Press, Montreal.

Peters, D. A., and C. J. He, 1991, "Correlation of Measured Induced Velocities with a Finite-State Wake Model," *Journal of American Helicopter Society*, July.

Pistolesti, E., 1922, *Vortrage aus dem Gebiete der Hydro- und Aerodynamik*, Innsbruck, Austria.

Pitt, D. M. and D. A. Peters, 1981, "Theoretical Prediction of Dynamic-Inflow Derivatives," *Vertica*, Vol. 5, No. 1: March.

Prandtl, L., 1919, Appendix to [Betz 1919]: pp. 213-217.

Quinlan, P. J., 1986, Private Communication.

Rankine, W. J. M., 1865, "On the Mechanical Principles of the Action of Propellers," *Transactions, Institute of Naval Architects*, Vol. 6: pp. 13-30.

Reuss, R. L., M. J. Hoffman, and G. M. Gregorek, 1995, *Effects of Surface Roughness and Vortex Generators on the NACA 4415 Airfoil*, NREL/TP-442-6472, Golden, CO: National Renewable Energy Laboratory, December.

Ribner, N. S., 1948, *Propellers in Yaw*, NACA Report 820, Washington, DC.

Robinson, D., F. Coton, R. Galbraith, and M. Vezza, 1994, *The Development of a Prescribed Wake Model for the Prediction of the Aerodynamic Performance of Horizontal Axis Wind Turbines in Steady Axial Flow*, Tech. Rep. No. 9403, Department of Aerospace Engineering, University of Glasgow.

Robinson, D., F. Coton, R. Galbraith, and M. Vezza, 1995, "Application of a Prescribed Wake Aerodynamic Prediction Scheme to Horizontal Axis Wind Turbines in Axial Flow," *Wind Engineering*, Vol. 19: pp. 41-51.

Savino, J. M., T. W. Nyland, and A. G. Birchenough, 1985, *Reflection Plane Tests of a Wind Turbine Blade Tip Section with Ailerons*, NASA TM-87018, DOE/NASA 20320-65, Cleveland, Ohio: NASA Lewis Research Center.

Shankar, P. N., 1975, *On the Aerodynamic Performance of a Class of Vertical Axis Windmills*, AE TM-13-15, Bangalore, India: National Aeronautical Laboratory.

Sheldal, R. E., and P. C. Klimas, 1981, *Aerodynamic Characteristics of Seven Symmetrical Airfoil Sections Through 180-Degree Angle of Attack for Use in Aerodynamic Analysis of Vertical Axis Wind Turbines*, SAND80-2114, Albuquerque, New Mexico: Sandia National Laboratories.

Schreck, S., 2002, "The NREL Full-Scale Wind Tunnel Experiment," *Wind Energy*, Vol. 5: pp. 77-84.

Schreck, S. J., N. N. Sorensen, M. C. Robinson, 2007, "Aerodynamic Structures and Processes in Rotationally Augmented Flow Fields," *Wind Energy*, Vol. 10, No. 2: pp. 159-178.

Simms, D., S. Schreck, M. Hand, and L. J. Fingersh, 2001, *NREL Unsteady Aerodynamics Experiment in the NASA-Ames Wind Tunnel: A Comparison of Predictions to Measurements*, NREL/TP-500-29494, Golden, CO: National Renewable Energy Laboratory.

South, P., and R. S. Rangi, 1972, *A Wind Tunnel Investigation of a 14-ft Diameter Vertical Axis Windmill*, NRC Laboratory Technical Report LTR-LA-105, Ottawa: National Research Council of Canada.

Strickland, J. H., 1975, *The Darrieus Turbine: Performance Prediction Model Using Multiple Streamtubes*, SAND75-0431, Albuquerque, New Mexico: Sandia National Laboratories.

Strickland, J. H., B. T. Webster, and T. Nguyen, 1980, *A Vortex Model of the Darrieus Turbine: An Analytical and Experimental Study*, SAND79-7058, Albuquerque, New Mexico: Sandia National Laboratories.

Spera, D. A., and D. C. Janetzke, 1981, "Performance and Load Data from Mod-OA and Mod-1 Wind Turbine Generators," *Proceedings, Workshop on Large Horizontal-Axis Wind Turbines*, NASA CP-2230, R.W. Thresher, ed., Cleveland, Ohio: NASA Lewis Research Center, pp. 447-467.

Spera, D. A., 2009, "Introduction to Modern Wind Turbines," *Wind Turbine Technology, Second Edition*, Chapter 2, ASME Press, D. A. Spera, ed., New York: American Society of Mechanical Engineers.

Sullivan, T. L., 1984, *Effect of Vortex Generators on the Power Conversion Performance and Structural Dynamic Loads of the Mod-2 Wind Turbine*, NASA TM-83680, DOE/NASA/20320-59, Cleveland, Ohio: NASA Lewis Research Center.

Taylor, H. D., 1947, *The Elimination of Diffuser Separation by Vortex Generators*, Report R-4012 3, East Hartford, Connecticut: United Aircraft Corporation.

Templin, R. S., 1974, *Aerodynamic Performance Theory for the NRC Vertical-Axis Wind Turbine*, LTR-160, Ottawa: National Research Council of Canada.

Touryan, K. J., J. H. Strickland, and D. E. Berg, 1987, "Electric Power from Vertical-Axis Wind Turbines," *Journal of Propulsion*, Vol. 3, No. 6: pp. 481-493.

Viterna, L. A., and R. D. Corrigan, 1981, "Fixed Pitch Rotor Performance of Large Horizontal Axis Wind Turbines," *Proceedings, Workshop on Large Horizontal-Axis Wind Turbines*, NASA CP-2230, R. W. Thresher, ed., Cleveland, Ohio: NASA Lewis Research Center, pp. 69-85.

Wentz, W. H. Jr., M. H. Snyder, and J. T. Calhoun, 1980, *Feasibility Study of Aileron and Spoiler Control Systems for Large Horizontal Axis Wind Turbines*, NASA CR-159856, DOElNASN3277- 1, WER-10, Cleveland, Ohio: NASA Lewis Research Center.

Wilson, R. E., 1978, "Vortex Sheet Analysis of the Giromill," *Transactions, ASME Journal of Fluids Engineering*, Vol. 100.

Wilson, R. E., and P. B. S. Lissaman, 1974, *Applied Aerodynamics of Wind Power Machines*, Corvallis, Oregon: Oregon State University.

Wilson, R. E., P. B. S. Lissaman, M. James, and W. R. McKie, 1983, "Aerodynamic Loads on a Darrieus Rotor Blade," *ASME Journal of Fluid Engineering*, Vol. 105: pp. 53-58.

Wilson, R. E., P. B. S. Lissaman, and S. N. Walker, 1976, *Aerodynamic Performance of Wind Turbines*, ERDA/NSF/04014-7611, Washington, DC: U.S. Department of Energy.

Wilson, R. E., and S. N. Walker, 1981, *Fixed Wake Analysis of the Darrieus Rotor*, SAND81-7026, Albuquerque, New Mexico: Sandia National Laboratories.

Wilson, R. E., and S. N. Walker, 1984, *Performance Analysis of Horizontal Axis Wind Turbines*, Corvallis, Oregon: Oregon State University.

Wilson, R. E., and J. A. Neff, 1985, *A First Order Analysis of the Effect of Pitching on the Drag Coefficient*, SAND85-7003, Albuquerque, New Mexico: Sandia National Laboratories.

Wilson, R. E., L. N. Freeman, and S. N. Walker, 1994, "Parametric Analysis of a Teetered Rotor Model," *Proceedings, Thirteenth ASME Wind Energy Symposium*, W. Musial, ed., New York: American Society of Mechanical Engineers.

Worstell, M. H., 1979, *Aerodynamic Performance of the 17-Meter-Diameter Darrieus Wind Turbine*, SAND78-1737, Albuquerque, New Mexico: Sandia National Laboratories.

Worstell, M. H., 1980, "Measured Aerodynamic and System Performance of the 17-Meter Research Machine," *Proceedings, Vertical Axis Wind Turbine Design Technology Seminar for Industry*, S. Johnston, ed., SAND80-0984, Albuquerque, New Mexico: Sandia National Laboratories, pp. 233-258.

5

Wind Turbine Aerodynamics
Part B: Turbine Blade Flow Fields

Scott Schreck, Ph.D.
National Renewable Energy Laboratory
Golden, Colorado

Introduction

As summarized previously in Part A of this chapter, field testing of the Mod-0, Mod-1, and Mod-2 horizontal-axis wind turbines (HAWTs) during the 1980's produced unique measurements of structural and aerodynamic loads not previously available. Analyses of Mod-series data in connection with contemporary aerodynamic models suggested that anomalous wind turbine aerodynamics events were prominent. However, the aerodynamics models then available were not capable of resolving turbine flow field details to the extent required to dissect these phenomena. Sufficiently detailed aerodynamics measurements were impractical because of the large scale of the Mod-series turbines, which encompassed rotor diameters from 38.1 m (Mod-0) to 97.5 m (Mod-5B).

To address these limitations, efforts began in the early 1990s to acquire highly resolved structural and aerodynamic load measurements in field tests of turbines having rotor diameters in the 10 m to 20 m range. Concurrent testing at the Netherlands Energy Research Foundation and the Delft University of Technology (The Netherlands), Mie University (Japan), National Renewable Energy Laboratory (United States), Risø Wind Turbine Test Station (Denmark), Imperial College (United Kingdom), and the Center for Renewable Energy

Sources (Greece) greatly advanced the state of the art for performing research grade measurements in the field environment. Data acquisition and documentation was carried out under the auspices of International Energy Agency Wind R&D Annexes XIV [Schepers *et al.* 1997] and XVIII [Schepers *et al.* 2002].

Need for Inflow Control and Wind Tunnel Testing

However, improved field measurement systems and enhanced data quality accentuated a long standing dilemma. Large scale turbines could be thoroughly instrumented, but size constraints required testing in the atmosphere. There, wind inflow fluctuations introduced uncertainties that precluded unambiguous characterization of key aerodynamic phenomena. Alternatively, wind tunnels offered controlled inflows, but severely limited test turbine size. Disparities between wind tunnel and field turbine sizes led to uncertainties regarding the validity of extrapolating wind tunnel measurements to larger-scale turbines.

In the U.S., resolving this dilemma led to the design and construction of an NREL test turbine designated as the *Unsteady Aerodynamics Experiment Phase VI HAWT* (UAE Phase VI), Fig. 5-44(a), which was tested in the NASA Ames 24.4 m x 36.6 m wind tunnel [Hand *et al.* 2001]. The rotor configuration of the UAE Phase VI turbine was two-bladed, constant speed (72 rpm), mounted upwind of its support pylon, and stall regulated. Its 5.03-m (16.5-ft) rotor radius included a tapered, twisted blade and hub. A 21percent thick, NREL S809 airfoil was used along the blade except for the cuff region, which transitioned from the circular root-end fitting to the S809 airfoil at 25 percent span [Tangler and Kocurek 2004].

(a)   (b)

Figure 5-44. Wind turbines specifically designed for detailed aerodynamic testing in wind tunnels. (a) UAE Phase VI turbine mounted in NASA Ames 24.4 m x 36.6 m wind tunnel. (b) MEXICO turbine mounted in DNW LLF 9.5 m x 9.5 m wind tunnel.

In Europe, the three-bladed 4.5 m diameter *MEXICO* (Model Rotor Experiments In Controlled Conditions) HAWT shown in Figure 5-44(b), sponsored by the European Union (EU), was tested in the DNW an open section of the *Large Low Speed* (LLF) 9.5 m x 9.5 m wind tunnel to address the same set of issues [Snel *et al.* 2007]. The MEXICO blades included 3 different airfoil sections, with a *Delft* section near the root, an *NACA* section near the tip, and a *Riso* section in between.

The U.S. and EU tests had in common several high level test goals and approaches, which were balanced by subtle distinctions that enabled each experiment to compensate for limitations in the other.

Evaluation of Aerodynamic Computer Codes

Data from the NREL/NASA Ames test of the UAE Phase VI wind turbine were employed as a reference standard in a blind code comparison exercise designed to evaluate the accuracy and robustness of various wind turbine aerodynamics codes [Simms *et al.* 2001]. Participating in this exercise were over 30 aerodynamic specialists from 18 organizations, including commercial companies, universities, and government laboratories in both Europe and the U.S. The range of computer models encompassed *blade element momentum* (BEM) models, *prescribed wake* models, *free wake* models, and *Navier-Stokes* codes.

Figure 5-45 shows the range of low-speed shaft (rotor) torques predicted by the participants compared to the actual measured torques, which is typical of results of this code comparison exercise. In general, unexpectedly large margins of disagreement were found between code predictions and measured data, particularly in the post-stall operating regime of the blades, at wind speeds above 10 m/s. Moreover, no consistent trends were apparent regarding the magnitudes or the directions of these deviations. These discrepancies highlighted a pressing need for better understanding of the details of turbine blade aerodynamics, in order to provide a solid physical basis for improving the accuracy of current rotor aerodynamic codes in both the pre-stall and post-stall regimes.

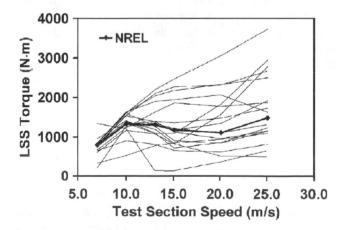

Figure 5-45. Typical comparison of predicted and measured rotor low-speed shaft (LSS) torques for the NREL UAE Phase VI wind turbine *vs.* airspeed in the test section. Over 30 aerodynamic specialists participated in this computer code evaluation exercise [Simms *et al.* 2001].

Data Analyses for Physical Comprehension

In-depth analysis of the UAE Phase VI test data produced new findings regarding HAWT aerodynamic phenomena, such as the following:

–   Blade rotation can amplify maximum normal forces at inboard span stations, to nearly three times the levels observed under nonrotating conditions.
–   These amplified normal forces are associated with unconventional blade surface pressure distributions that have been observed infrequently in previous experiments.
–   Dynamic stall vortex structure and dynamics varied significantly along the blade span, being highly three-dimensional under yawed operating conditions.

It is important to note that the following discussion concentrates solely on wind turbine blade aerodynamics, and does not consider the wind turbine wake. Admittedly, the wake plays a strong role in overall turbine aerodynamics, and wake modeling has advanced significantly in recent years. Unfortunately, measured data containing detail comparable to that available for the turbine blade flow field do not exist at present.

Measurement of Local Inflow Angles and Normal Force Coefficients

Determination of angles of attack and lift coefficients for rotating blades using measurements on or near the blade itself remains a challenging and essential activity [Brand 1994; Schepers 1995; Shipley *et al.* 1995b]. For the UAE Phase VI turbine, these difficulties were deferred in order to simplify physical relationships and concentrate on rotational modifications to the flow field. This was accomplished by analyzing the *local inflow angle*, *LFA*, as a surrogate for the angle of attack, and the normal force coefficient, C_n, in lieu of the lift coefficient. *LFA* is defined as the angle between the local inflow vector and the local blade sectional chord. *LFA* was measured by five-hole probes located 0.80 chord width ahead of the blade leading edge, at normalized radial positions of $r/R = 0.34, 0.51, 0.67, 0.84$, and 0.91.

The normal force coefficient, C_n, is equal to the sectional aerodynamic force *intensity* (force per unit of span length) at right angles to the blade chord divided by the product of the local dynamic pressure and the local chord width. Normal force intensities were determined by integrating surface pressures measured along the blade chordlines at $r/R = 0.30, 0.47, 0.63, 0.80$, and 0.95. Detailed information regarding instrumentation and procedures can be found in Hand *et al.* [2001].

Rotational Augmentation of Aerodynamic Properties

Background

Prior research concerning rotational augmentation of airplane propeller and helicopter rotor aerodynamics helped guide early work specifically directed at wind turbines. Augmentation of rotating blade aerodynamic properties, including stall delay and lift enhancement, was first observed for airplane propellers and qualitatively explained in terms of centrifugal and Coriolis accelerations [Himmelskamp 1950]. Later, analytical modeling quantitatively accounted for key elements of the rotating blade flow field [Banks and Gadd 1963]. Analytical modeling of helicopter rotors determined that rotationally induced cross flows played an important role in blade lift production [McCroskey and Yaggy 1968]. Experimental research [McCroskey 1971] suggested that centrifugal forces are important in the presence of flow separation, but of limited influence otherwise.

Early wind turbine field testing aimed at understanding rotational augmentation affirmed the importance of blade geometry with respect to rotational influences [Madsen and Christensen 1990]. A wind tunnel experiment showed that blade geometry coupled with blade rotation maintained blade lift under conditions in which lift otherwise would significantly decline [Barnsley and Wellicome 1992]. Subsequent wind tunnel research determined that rotational augmentation was most active at the inboard portion of the turbine blade [Ronsten 1992]. Analytical modeling of rotational augmentation furnished better comprehension of the aerodynamics underlying this phenomenon, and provided foundational predictive capabilities for design and analysis [Eggers and Digumarthi 1992, Snel *et al.* 1994, Du and Selig 2001; Corten 2001, Tangler and Selig 1997, Corrigan and Schillings 1994].

Inboard Locations on Turbine Blades

As shown in Figures 5-46 and 5-47, inboard radial locations exhibit rotational augmentation more strongly than do radial locations farther outboard on the blade. In Figure 5-46, C_n at 0.30R is plotted as a function of *LFA*, for both the stationary (parked) and rotating UAE Phase VI blade with its NREL S809 airfoil. All data correspond to an axisymmetric operating state, with the rotor plane normal to the wind vector.

With the blade held stationary and a free-stream wind speed, U_∞, equal to 20 m/s, the Reynolds number, *Re*, at station 0.30R equals 0.98 x 10^6. At a U_∞ of 30 m/s Re equals 1.46 x 10^6. These Reynolds numbers encompass the range for the 11 fastest rotating blade data points, thus establishing a consistent baseline against which the rotating blade data can be compared. For the stationary blade, C_n increases progressively with *LFA*, reaches a relative maximum of 0.94 at *LFA* = 32.2°, and thereafter remains approximately level.

To disclose hysteresis effects, if any, the stationary blade normal force curve for each wind speed consists of two data sets: One corresponding to increasing *LFA*, and the other for decreasing *LFA*. These two curves exhibited close agreement, testifying to the absence of hysteresis effects for stationary operation, and disqualifying hysteresis as an explanation for rotational augmentation.

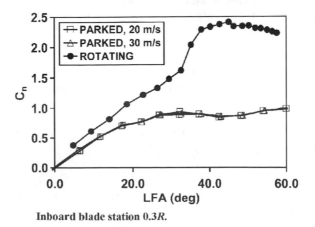

Inboard blade station 0.3R.

Figure 5-46. Comparison of sectional normal force coefficients, C_n, at an inboard station for stationary (parked) and rotating operations of the UAE Phase VI wind turbine, as functions of the local inflow angle, *LFA*. Differences between rotating and stationary C_n values were all consistently larger for inboard locations

Plotted in Figure 5-46 is C_n *vs. LFA* for the inboard portion of the rotating blade at wind speeds from 5 m/s to 25 m/s, with a fixed blade pitch angle and constant rotor speed. The C_n curve rises monotonically until *LFA* = 45 deg, where it attains a stalled C_n of 2.41. Thereafter, the normal coefficient undergoes a gentle decline with increasing inflow angle. Note that this rotating blade C_n represents an amplification of nearly threefold relative to the baseline stationary blade C_n for the inboard station.

Outboard Locations on Turbine Blades

Rotational augmentation undergoes progressive attenuation at radial locations farther outboard on the blade. In Figure 5-47, C_n at station 0.80R is plotted versus *LFA*, for both the stationary and rotating blades. Again, all data correspond to an axisymmetric operating state, with the rotor plane orthogonal to the wind vector. At 0.80R, 20 m/s yields a Reynolds number, *Re*, of 0.63 x 10[6], while at 30 m/s the *Re* is 0.94 x 10[6]. Normal force hysteresis was found to play a role in shaping the post stall C_n curve for the stationary blade.

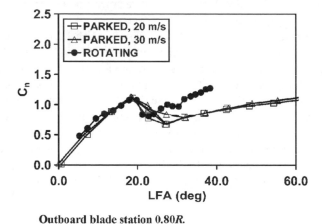

Outboard blade station 0.80R.

Figure 5-47. Comparison of sectional normal force coefficients, C_n, at an outboard station for stationary (parked) and rotating operations of the UAE Phase VI wind turbine.

Up to a local inflow angle of approximately 25 deg, the normal force coefficients of the rotating blade closely track those of the stationary blade. However, at higher *LFA* in the post-stall regime, the rotating blade coefficients at this outboard blade station exceed the stationary blade coefficients by as much as 45 percent.

Significant stall *LFA* delay and stall C_n amplification were observed for the rotating blade relative to the stationary blade at all blade span locations. Stall *LFA*, stall C_n, and differences between rotating and stationary C_n values were all consistently larger for inboard span locations.

Surface Pressure Distributions

To understand the physical mechanisms responsible for rotational stall delay and C_n amplification relative to stationary conditions, surface pressure distributions were compared for the stationary and rotating operating conditions. Figure 5-48 shows the results of this comparison of surface *pressure coefficients, c_p,* which are equal to the measured local pressure divided by the section dynamic pressure. These c_p distributions correspond to C_n at stall on

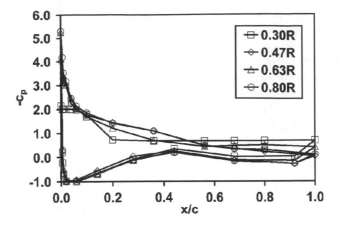

Figure 5-48. Surface pressure coefficient distributions measured on a stationary blade at stall. The horizontal axis is the normalized distance aft of the leading edge. Note that negative pressure is plotted above the horizontal axis.

the stationary blade. Note that negative or suction pressures are plotted above the horizontal axis. All four of these distributions contain suction peaks on the *suction surface* (or *upper* or *low-pressure surface*) of the blade, near the leading edge.

Extending aft of the leading edge, also on the blade's suction surface, are *adverse pressure gradients* (*i.e.* pressures that retard the blade's forward motion) of progressively decreasing magnitude. On the blade *pressure surface* (*lower* or *high-pressure surface*) the c_p distributions are remarkably similar. Overall, these surface pressure distributions strongly resemble those commonly observed on two-dimensional airfoils in the neighborhood of stall.

The rotating blade stall c_p distributions shown in Figure 5-49 were characterized by mild pressure gradients extending over significant portions of the suction surface chord. The rotating blade distributions at $0.30R$, $0.47R$, and $0.63R$ all differ considerably from those on the stationary blade at the same spanwise locations. However, the rotating blade c_p distribution at $0.80R$ closely resembles that shown in Figure 5-48 at the same spanwise location on the stationary blade. This is consistent with the strong similarity between the stationary and rotating blade C_n vs. *LFA* curves at $0.80R$ near stall. The magnitude of the rotating blade suction c_p levels, coupled with the mild gradients, is not consistent with conventional two-dimensional lifting surface flows, thus pointing to three-dimensional mechanisms.

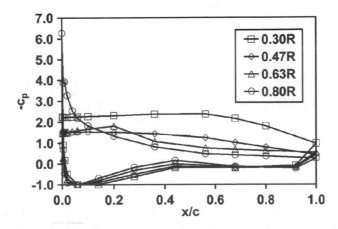

Figure 5-49. Stall surface pressure coefficient distributions measured on a rotating blade.

Previous wind tunnel experiments employing subscale HAWT models have revealed similarly augmented surface pressure distributions on rotating blades. These surface pressure profiles took on approximately trapezoidal [Barnsley and Wellicome 1992] or triangular [Ronsten 1992] profiles. In addition, large scale HAWT field measurements [Robinson, *et al.* 1995] have detected trapezoidal surface pressure distributions.

Clearly, stall c_p distributions on inboard locations of the rotating blade differ dramatically from those observed farther outboard under rotating conditions. Inboard rotating blade stall c_p distributions differ in equally dramatic fashion from the stall c_p distributions observed anywhere on the stationary blade. This enhanced suction at elevated *LFA* has been attributed to various mechanisms mediated by centrifugal or Coriolis forces [Himmelskamp 1950, Banks and Gadd 1963, McCroskey and Yaggy 1968, McCroskey 1977, Madsen and Christensen 1990, Barnsley and Wellicome 1992, Eggers and Digumarthi 1992, Snel *et al.* 1992] and to the presence of a stationary energetic vortex structure [Eggleston 1990].

Flow Field Structure

The pressure instrumentation on the blades of the UAE Phase VI wind turbine provided accurate portrayals of flow field states at the blade surface, but was incapable of discerning off-surface flow field structure. To overcome this limitation, the EllipSys3D computational fluid dynamics code was used [Michelsen 1992, Michelsen 1994, Sørensen 1995]. Code results were first validated against UAE Phase VI surface pressure measurements. The flow field above the blade surface was then extracted from these computations, revealing flow structures and processes responsible for rotational augmentation [Schreck *et al.* 2007].

Figure 5-50 contains a two-dimensional section of the computed streamlines in the stalled flow field at a blade radius of 0.30R, for U_∞ = 15 m/s. This computation corresponds to the measured data in Figure 5-46, for *LFA* = 40.2° and C_n = 2.33. Under these conditions, separation occurs at the leading edge. The resulting shear layer arches over the blade suction surface and finally impinges on the aft portion of the blade chord. Between separation and

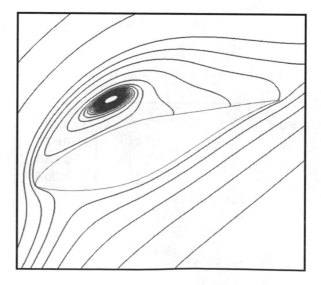

Figure 5-50. Two-dimensional section of computed UAE Phase VI blade flow field showing a region of intense recirculation above the blade's suction surface (U_∞ = 15 m/s, r/R = 0.30). [Computation courtesy of N. Sørensen, Risø DTU and Aalborg University]

impingement, the shear layer surrounds a region of intense recirculation strongly reminiscent of a vortical structure.

Other analyses of time variations in measured surface pressure data have enabled detection and tracking of separation and impingement locations, enabling more rigorous validation of computational results and direct inferences regarding flow field topology [Schreck and Robinson 2003, Schreck *et al.* 2007].   Together, these data and analyses make it clear that rotational augmentation is mediated by prominent concentrations of vorticity (if not vortices) present on the blade suction surface.

In addition to the predominantly two-dimensional components shown in Figure 5-50, strongly three-dimensional influences are also present, as illustrated in Figure 5-51.   The computed *separation line* now extends along the blade's leading edge from $0.20R$ to $0.85R$. From $0.85R$, the separation line travels outboard and aft, demarcating a small triangular region containing chordwise streamlines and two-dimensional flow.   Impingement is now prominent on the blade surface, signified by an aggregation of limiting streamlines that cross the blade from midchord at $0.30R$ to the leading edge at $0.65R$.   Within the triangular region between separation and impingement, the streamlines proceed outboard and forward toward the leading edge, suggesting the presence of a recirculating vortical region over this blade surface area.

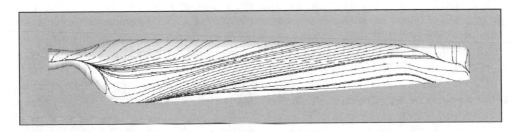

Figure 5-51. Planform view of the computed UAE Phase VI blade flow field, showing suction surface limiting streamlines in the stall regime. $U_\infty = 15$ m/s. [Computation courtesy of N. Sørensen, Risø DTU and Aalborg University]

Rotational Augmentation Modeling

Rotational augmentation models must account for both centrifugal and Coriolis influences on blade aerodynamics.   The combination of these effects frequently produces regions of chordwise and radial flow that coexist on the turbine blade.   Different rotational augmentation models can be classified as predominantly empirical, theoretical, or computational.

One rotational augmentation model typical of some empirical approaches is that of Corrigan and Schillings [1994].   Originally intended for use with rotorcraft, this model exploited theoretical research by Banks and Gadd [1963] as an initial guide.   This previous work suggested that the principal effects of rotational augmentation could be accounted for using a single parameter relationship.   Ultimately, Corrigan and Schillings arrived at a simple functional relationship, basing their model on correlations with extensive rotorcraft test data. These data inherently included centrifugal and Coriolis effects.   This model was later adapted for wind turbine power predictions and validated by Tangler and Selig [1997].

A theoretical approach was pursued by Snel [1991], one in which model development began by applying an order of magnitude analysis to the Navier-Stokes equations for the blade flow field.   This procedure yielded a reduced equation set that retained centrifugal and Coriolis effects. This equation was then solved using a finite difference algorithm [Snel *et al.* 1994]. Later, Du and Selig [2001] solved a similar equation set in alternative fashion, and captured key parameter variations in phenomenological relationships. In similar fashion, Eggers and Digumarthi [1992] began with the Navier-Stokes equations, but applied different criteria to arrive at a reduced equation set, which was then solved analytically. The approach adopted by Corten [2001] also began with the Navier-Stokes equations, but diverged from prior approaches by excluding the effects of viscosity. Separation over the aft blade surface was introduced via prescription, permitting emergence and characterization of pronounced radial flows.

Shen and Sørensen [1999] approximated the governing equations for flow over a rotating blade using an order of magnitude analysis, to efficiently account for rotational and spanwise flow effects. Resulting quasi-three-dimensional laminar and turbulent computations showed that rotational influences became more pronounced at locations nearer the rotation axis, reducing the chordwise extent of separation and amplifying aerodynamic forces. Chaviaropoulos and Hansen [2000] constructed a quasi-three-dimensional computational model, and then used it in a parametric study that showed orderly trends in three-dimensionality and rotational effects in response to blade taper ratio and twist distribution. Sorensen *et al.* [2002] solved the full Navier-Stokes equations for the Phase VI flow field, without reduction or empiricism.

Dynamic Stall

Dynamic stall initiates when the aerodynamic surface incidence angle (local inflow angle or angle of attack) dynamically exceeds the static stall angle. Soon thereafter, unsteady boundary layer separation gives rise to a small but energetic dynamic stall vortex, which appears near the leading edge. This vortex quickly grows, convects rapidly downstream toward the trailing edge, and soon sheds from the lifting surface.

During this process, the dynamic stall vortex generates a region of low pressure on the lifting surface, causing dramatic lift amplification beyond static levels, followed by abrupt deep stall at vortex shedding [McCroskey 1977, Carr 1988]. Surface pressure signatures confirm that dynamic stall occurs on horizontal-axis wind turbine blades [Shipley *et al.* 1995, Robinson *et al.* 1995, Huyer *et al.* 1996] and contributes significantly to rotor loads and yaw dynamics [Hansen *et al.* 1990].

The complexity of dynamic stall is compounded by three-dimensionality, or nonuniformity of the vortex structure along the length of the vortex. The three-dimensionality of a dynamic stall vortex has been visualized for a rectangular wing pitching in a wind tunnel [Freymuth 1988]. Visualizations have been corroborated with time-varying surface pressure measurements, which also were acquired using rectangular wings pitching in wind tunnels [Robinson and Wissler 1988, Schreck *et al.* 1991, Lorber *et al.* 1992, Schreck and Helin 1994, Piziali 1994]. More recently, dynamic stall vortex three-dimensionality was observed during yawed turbine operation via analyses of time varying surface pressure data [Schreck *et al.* 2000 and 2001].

Inflow Angle Oscillation

With the turbine rotor disc yawed relative to the wind inflow vector, blade incidence angle (local inflow angle or angle of attack) oscillates in pseudo-sinusoidal fashion as the blade rotates in azimuth about the rotor axis. Figure 5-52 shows *LFA* (local inflow angle) as a function of Ψ, the blade azimuth angle, at four five-hole probe locations, for $U_\infty = 13$ m/s and a rotor yaw angle of 40 deg. At $\Psi = 0$ deg, the instrumented blade is situated at the 12 o'clock position, and the period of one rotor revolution is 0.838 s.

These pseudo-sinusoidal oscillation amplitudes with a dominant frequency of one-per-rotor revolution are largest at $0.34R$ and decrease progressively at radial stations farther out-

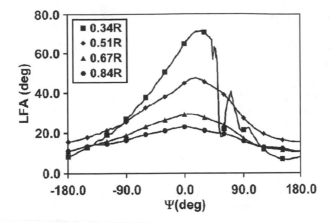

Figure 5-52. Cyclic variation of local inflow angle, *LFA*, with blade azimuth angle, Ψ, at four radial locations, for a rotor yaw angle of 40 deg. U_∞ = 13 m/s. Azimuth angle is zero when the blade is at the 12 o'clock position.

board on the blade. However, even at 0.84*R*, the *LFA* oscillation amplitude remains substantial. Notably, all four *LFA* cyclic amplitudes exceed the S809 stall *LFA* of approximately 21 deg, thus initiating dynamic stall.

Aerodynamic Force Amplification

An example of the aerodynamic response to the cyclic flow field is shown in Figure 5-53, in which mean normal force coefficients along the blade, C_n, are plotted as a function of blade azimuth angle, Ψ. Again, the rotor yaw angle is 40 deg and the test airspeed is 13 m/s. Prior research has shown that dynamic stall forces similar to those shown in this figure are common for a broad range of airspeeds and yaw angles [Schreck *et al.* 2005].

At 0.30*R*, C_n reached a maximum of 3.18 at Ψ approximately equal to zero. C_n maxima at the remaining three spanwise locations decreased progressively with distance outboard. Nevertheless, at 0.80*R* the C_n maximum of 1.31 still significantly exceeded the stall C_n of 1.00 for two-dimensional flow over a static NREL S809 airfoil, testifying to the presence and influence of dynamic stall.

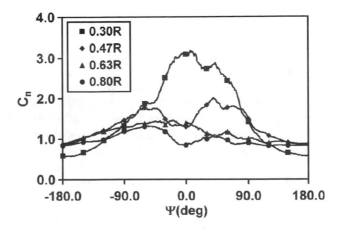

Figure 5-53. Variation of mean normal force coefficient, C_n, with blade azimuth angle, Ψ, at four radial locations, for a rotor yaw angle of 40 deg. U_∞ = 13 m/s.

Aerodynamic Force Unsteadiness and Repeatability

Mean aerodynamic force records exhibit prominent time variations, which prompt questions regarding repeatability and statistical significance. Figure 5-54 shows a superposition of UAE Phase VI C_n records for 36 blade rotation cycles, at 0.30R and 0.80R, for $U_\infty = 13$ m/s and a rotor yaw angle of 40 deg. The mean C_n data for these two radial locations also are shown, as thin white traces in the approximate center of the multiple overlapping black traces.

These plots reveal the presence of C_n oscillations having frequencies of approximately 30 to 50 Hz. Peak-to-peak amplitudes of these higher frequency components correspond to approximately 50 to 70 percent of mean levels. Both the once-per-revolution and the higher-frequency time variations of C_n associated with dynamic stall constitute significant components of the aerodynamic load spectrum.

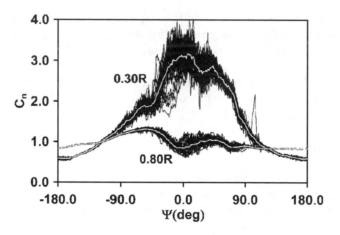

Figure 5-54. Instantaneous C_n records for 36 blade rotation cycles at two radial locations. $U_\infty = 13$ m/s and yaw angle = 40 deg.

Flow Field Structure

Visualization of dynamically stalled flow fields reveals flow field structures responsible for the dramatic C_n amplification. Unfortunately, dynamic stall visualization is prohibitively difficult on large-scale rotating turbine blades, and successful execution continues to elude turbine aerodynamicists. However, analogous dynamically stalled flow fields occur on non-rotating aerodynamic surfaces subjected to dynamic pitching, and are readily visualized.

Figure 5-55 shows smoke flow visualization of a two-dimensional airfoil with a 0.15 m chord, pitching from 0 to 60 deg at a constant rate of 115 deg/s. The airfoil pitch axis is located at the quarter-chord point. Air flows from left to right at 6.1 m/s. The three panels show instantaneous angles of attack of 20°, 25°, and 30°, from top down. The upper surface of the airfoil is visible in each panel as a brightly lit arc inclined with respect to the panel borders.

In the upper panel of Figure 5-55, the disordered smoke filaments immediately above the airfoil indicate that separation has enveloped the entire upper surface. However, the smoke line emanating from the leading edge is well defined, curving back toward the airfoil, and impinging on the surface at approximately 20 percent of the chord width from the leading edge. Subsequent panels show this structure to be an energetic *dynamic stall vortex*. The dynamic stall vortex produces a surface pressure minimum of significant magnitude on the

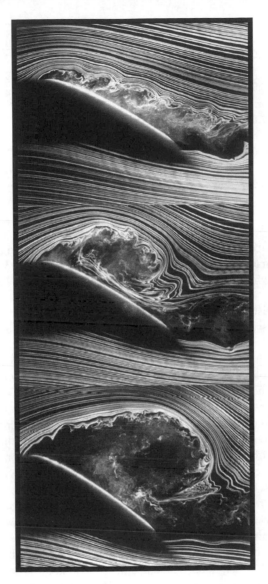

Figure 5-55. Smoke flow visualization of dynamic stall on a two-dimensional airfoil pitching in a wind tunnel.

airfoil surface beneath the vortex. This surface pressure minimum is responsible for the strongly amplified C_n levels observed during dynamic stall.

The middle panel, for a 25 deg angle of attack, shows that the dynamic stall vortex has grown substantially and has convected aft on the airfoil chord, residing approximately over midchord. The smoke line defining the outer perimeter of the vortex almost completely encircles the vortex, indicating the strongly rotational and energetic nature of the vortex. Consistent with vortex size, the surface pressure minimum below it has grown and C_n has increased.

In the lower panel, the vortex has grown still larger, and continues to exhibit a strongly rotational appearance, indicative of the energy contained in the vortex. However, the dynamic stall vortex has now reached the airfoil trailing edge, and has begun to lift away from the airfoil surface. As the vortex sheds from the airfoil surface, the surface pressure minimum weakens and C_n undergoes a catastrophic decrease, or stall.

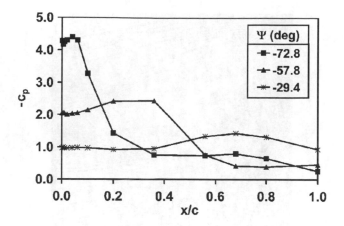

Figure 5-56. **Signatures of** c_p **across a blade chord at** *0.47R* **tracking the moving location of a dynamic stall vortex above the suction surface of a UAE Phase VI wind turbine blade at three successive azimuthal positions.** U_∞ = 13 m/s and rotor yaw angle = 40 deg.

Again, no flow visualization has been accomplished as yet for the UAE Phase VI turbine blade. However, pressure signatures elicited by a dynamic stall vortex have been identified and tracked on a UAE Phase VI blade surface. These c_p data are shown in Figure 5-56 for a station at 0.47R, with U_∞ = 13 m/s and a 40 deg yaw angle. The dynamic stall vortex was first detected just aft of the leading edge (*x/c* = 0.04), as indicated by a high, narrow surface pressure minimum (suction peak) centered at, when the blade was at an azimuth of -72.8 deg. Then, 35 msec later, when Ψ = -57.8 deg, the suction peak had moved aft, broadened, and decreased in magnitude to -2.42. Finally, after an additional 66 msec had passed and Ψ = -29.4°, the suction peak had continued aft, broadened further, and decreased in magnitude to -1.42. Movement of the peak aft corresponds to vortex convection and peak broadening is associated with vortex growth. Dynamic stall vortex convection was rapid, traversing 75 percent of the chord width in approximately 0.1 sec.

Using this suction peak tracking methodology, a detailed pressure history can be assembled at each spanwise location of pressure taps, documenting vortex chordwise position as a function of time. Then, for each combination of airspeed and yaw angle, *vortex convection topologies* can be mapped, producing diagrams such as that illustrated in Figure 5-57 [Schreck *et al.* 2001 and 2005]. Detailed interpretation of these topologies shows that the vortex structure evolves rapidly and dramatically, undergoing complex structural deformations.

The dynamic stall vortex exhibited restrained convection at 0.30R, implying the existence of pinning influences at the inboard vicinity of the vortex [Freymuth 1988; Robinson and Wissler 1988; Lorber *et al.* 1992; Schreck *et al.* 1991; Schreck and Helin 1994; Schreck *et al.* 2001; Coton and Galbraith 1999]. Similarly suppressed vortex convection was observed at the outboard extreme of the vortex, again consistent with pinning interactions. Vortex convection was most active near the central region of the blade.

Dynamic Stall Modeling

The fluid dynamic structures and processes described in connection with Figure 5-55 are common to dynamically-stalled flow fields in the qualitative sense. However, quantitative aerodynamic force production can vary radically in response to subtle input parameter variations. Thus, several dynamic stall models have been developed, with most intended originally for rotorcraft applications. Some of these dynamic stall models have been adapted

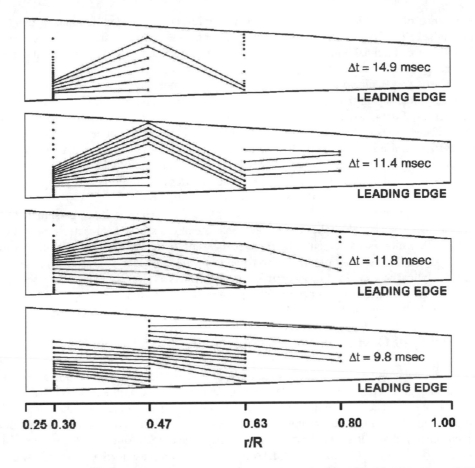

Figure 5-57. Examples of dynamic stall vortex topologies, illustrating the type of diagram developed specifically for tracking the evolution of vortices with time. Yaw angles are 20, 30, 40, and 50 deg (top planform to bottom) for $U_\infty = 13$ m/s. [Schreck *et al.* 2001 and 2005]

for wind turbine applications, and are summarized briefly below. Because all such models are semi-empirical in nature, these should be considered to be reconstructions rather than predictions of turbine aerodynamic response.

An early dynamic stall model was developed by Johnson [1980] based upon experimental aerodynamic data. This model assumed that a dynamic stall vortex generates a large lift increase, having a brief rise time to maximum lift and an equally brief decay time back to static lift levels. Required inputs are few and easily acquired, consisting of static stall angle of attack and lift for the blade airfoil section, blade chord length, rate of angle of attack change, and local inflow speed. However, depending on the application, limitations in specifying inputs may impose corresponding limitations in model generality.

The *Boeing-Vertol model* traces its origins to work by Harris *et al.* [1970]. In this method, unsteady inviscid stall angle is incremented by a factor that includes the a parameter termed the *reduced frequency* and an empirical constant that is intended to approximate the time lag in flow field development physically imposed by viscosity. Explicit inputs consist of static lift curve characteristics, and the empirical constant accounts for effects of airfoil thickness, Mach number, and Reynolds number.

The *ONERA dynamic stall model* [Tran and Petot 1981] superimposes two distinct components to obtain dynamic stall aerodynamic loading. The first component corresponds to the absence of dynamic stall, while the second represents the vortex-dominated, dynamically-stalled flow field. Both of these loading components are quantified by means of a set of nonlinear differential equations containing a significant number of coefficients that need to be extracted from unsteady experimental measurements.

Clearly, the Boeing-Vertol model formulation is mathematically simpler and requires fewer empirical inputs than the ONERA model. However, comparisons of model outputs with measured data have yielded apparently inconclusive results regarding relative model fidelity. Bierbooms [1992] reported that ONERA model fidelity was superior to that of other models that required fewer inputs. In contrast, Yeznasni *et al.* [1992] found that the ONERA model generally deviated more from measured results than the Boeing-Vertol model. It should be noted that both researchers were careful to stress the importance of experimental data quality, as well as correct analysis and application of these data.

The *Leishman-Beddoes dynamic stall model* [Pierce and Hansen 1995, Leishman 2006, Gupta and Leishman 2006] takes advantage of a more physics-based characterization of dynamic stall, which is implemented in three parts. The first part uses inviscid thin airfoil theory to quantify low angle of attack aerodynamics, while the second part superimposes different analytical relationships to account for viscous effects as angle of attack increases. The third part of the Leishman-Bedoes model replicates the formation, convection and growth, and ultimate shedding of the dynamic stall vortex, as illustrated in Figure 5-55.

Frequency of Occurrence of Dynamic Stall

Rotational augmentation and dynamic stall, as described above, are not anomalous events. Though operational turbines lack instrumentation that would enable specific detection, both events take place with some regularity during wind turbine operation, across the operating envelope. Figure 5-58 illustrates the operating conditions during which these phenomena occurred for the UAE Phase VI turbine during the NASA Ames wind tunnel test. The square marked "Nom" (nominal) near the origin of the plot represents the range of yaw error and test airspeed, U_∞, that was not influenced by rotational augmentation or dynamic stall. Rotational augmentation predominates for yaw error and U_∞ corresponding to the rectangular region labeled "Rot Aug", which lies immediately above the "Nom" square. Dynamic stall is dominant for conditions demarcated by the large rectangular region marked "Dynamic Stall" that fills the upper right portion of Figure 5-58.

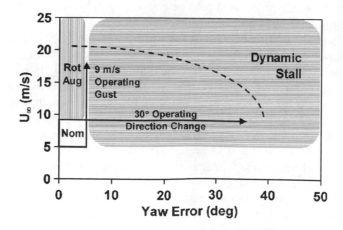

Figure 5-58. UAE Phase VI wind turbine operating envelope, showing the observed dynamic stall regime and typical gust and direction change equivalents.

Wind gusts and direction changes encountered during routine operations are sufficient to drive turbine operating state out of the nominal region and into rotational augmentation or dynamic stall. These departures are shown by the two vectors originating at the upper right corner of the nominal region. One of these vectors points to the right and is labeled "30 deg Operating Direction Change", while the other points upward and is labeled "9 m/s Operating Gust". Wind turbine certification standards were used to establish these vector magnitudes for operating direction change and normal operating gust [Germanischer-Lloyd 1998].

It may be noted that the UAE Phase VI machine was a stall controlled turbine and was intentionally operated at large yaw errors. Admittedly, these two characteristics differ from state-of-the-art utility class machines. However, wind gusts and direction changes that out-pace turbine pitch- and yaw-actuation rates are likely lead to similar circumstances. It also is important to remember that the current work relies on wind tunnel measurements and computational predictions that purposely exclude turbulent inflow. Thus, current work on modeling dynamic stall cannot quantify the extent to which turbulent inflow may compound adverse loading effects.

Conclusions

Wind turbine aerodynamics remains a challenging and crucial research area for wind energy, especially as turbines grow larger and aerodynamics and structural design become more closely coupled. Clear understanding of turbine blade flow physics and model numerics reveals analogies between the two and points toward attributes that yield accurate, reliable models. Both turbine blade aerodynamics and model construction can be subdivided consistent with turbine operating state and routinely occurring flow phenomena.

Under axisymmetric operating conditions and at low yaw angles, rotational augmentation delays turbine blade stall and significantly amplifies blade aerodynamic loads. This phenomenon depends on viscous flow effects in the presence of centrifugal and Coriolis influences, and occurs routinely during turbine operation. Current rotational augmentation models blend theory with empiricism to achieve acceptable model accuracy and efficiency.

During yawed operation, dynamic stall takes place when blade angle of attack dynamically exceeds the static stall threshold and a dynamic stall vortex initiates. As this vortex grows in size and convects aft over the blade, it significantly amplifies blade aerodynamic loads. Load amplification is fleeting, and ceases abruptly as the vortex departs the blade surface. The physical complexity of dynamic stall requires a semi-empirical approach that balances model fidelity with requirements for empirical input information.

Refinement of blade aerodynamics comprehension and modeling methodologies offers potential benefits for future wind energy machines, by facilitating the creation and validation of more accurate turbine aerodynamics design models. Enhanced aerodynamic design capabilities enable the development of larger, more efficient, and more cost effective turbines, thereby ensuring the continued growth and success of wind energy generation.

References

Banks, W., and G. Gadd, 1963, "Delaying Effect of Rotation on Laminar Separation," *AIAA Journ.*, v. 1, n. 4, pp. 941-942.

Barnsley, M., and J. Wellicome, 1992, "Wind Tunnel Investigation of Stall Aerodynamics for a 1.0 m Horizontal Axis Rotor," *Journ. Wind Engineering and Industrial Aerodynamics*, v. 39, pp. 11-21.

Bierbooms, W., 1992, "A Comparison Between Unsteady Aerodynamic Models," *Journ. Wind Engineering and Industrial Aerodynamics*, v. 39, n. 1-3, pp. 23-33.

Brand, A., 1994, *To Estimate the Angle of Attack of an Airfoil from the Pressure Distribution*, ECN Technical Report, ECN-R-94-002.

Carr, L. W., 1988, "Progress in Analysis and Prediction of Dynamic Stall", *Journ. Aircraft*, v. 25, n. 1, pp. 6-17.

Chaviaropoulos, P., and M. O. L. Hansen, 2000, "Investigating Three-Dimensional and Rotational Effects of Wind Turbine Blades by Means of a Quasi-3D Navier-Stokes Solver," *Journ. of Fluids Engineering*, v. 122, pp. 330-336.

Corrigan, J., and J. Schillings, 1994, *Empirical Model for Stall Delay Due to Rotation*, American Helicopter Society Aeromechanics Specialist Conf.

Corten, G., 2001, "Inviscid Stall Model," *Proceedings, European Wind Energy Conference, July 2001*, pp. 466-469.

Coton, F., and R. Galbraith, 1999, "An Experimental Study of Dynamic Stall on a Finite Wing," *The Aeronautical Journal*, v. 103, no. 1023, pp. 229-236.

Du, Z., and M. Selig, 2001, *A 3-D Stall-Delay Model for Horizontal Axis Wind Turbine Performance Prediction*, AIAA-98-0021: American Institute of Aeronautics and Astronautics.

Eggers, A., and R. Digumarthi, 1992, "Approximate Scaling of Rotational Effects on Mean Aerodynamic Moments and Power Generated by CER Blades Operating in Deep-Stalled Flow," *Proceedings, 11th ASME Wind Energy Symposium*: American Society of Mechanical Engineers.

Eggleston, D., 1990, "New Results in Flow Visualization for Wind Turbines," *Proceedings, AWEA Windpower Conference, September, 1990*, Washington, DC: American Wind Energy Association.

Freymuth, P., 1988, "Three-Dimensional Vortex Systems of Finite Wings," *Journ. Aircraft*, v. 25, n. 10, pp. 971-972.

Germanischer-Lloyd, 1998, *Rules and Regulations, IV-Non-Marine Technology, Part 1-Wind Energy Converters*, Chapter 4, p. 2-3.

Gupta, S., and J. G. Leishman, 2006, "Dynamic Stall Modeling of the S809 Aerofoil and Comparison with Experiments," *Wind Energy*, v. 9, n. 6, pp. 521-547.

Hand, M., D. Simms, L. Fingersh, D. Jager, J. Cotrell, S. Schreck, and S. Larwood, 2001, *Unsteady Aerodynamics Experiment Phase VI: Wind Tunnel Test Configurations and Available Data Campaigns*, NREL/TP-500-29955, Golden, Colorado: National Renewable Energy Laboratory.
Available at http://www.nrel.gov/docs/fy02osti/29955.pdf

Hansen, A., C. Butterfield, and X. Cui, 1990, "Yaw Loads and Motions of a Horizontal Axis Wind Turbine," *Journ. of Solar Energy Engineering*, v. 112, pp. 310-314.

Harris, F., F. Tarzanin, and R. Fisher, 1970, "Rotor High Speed Performance, Theory vs. Test," *Journ. American Helicopter Society*, v. 15, n. 3, pp. 35-44.

Himmelskamp, H, 1945, *Profiluntersuchungen an einem umlaufenden Propeller"* (*Profile Investigations on a Rotating Airscrew*), Dissertation, Gottingen, Mitt. Max-Planck-Institut fur Stromungsforschung Gottingen Nr. 2, 1950.

Huyer, S., D. Simms, and M. Robinson, 1996, "Unsteady Aerodynamics Associated with a Horizontal Axis Wind Turbine," *AIAA Journ.*, v. 34, n. 7, pp. 1410-1419.

Johnson, W., 1980, *Helicopter Theory*, Princeton University Press, Princeton, New Jersey, pp. 893-894.

Leishman, J. G., 2006, *Principles of Helicopter Aerodynamics*, Cambridge University Press, New York, New York.

Lorber, P., F. Carta, and A. Covino, 1992, *An Oscillating Three-Dimensional Wing Experiment: Compressibility, Sweep, Rate, Waveform, and Geometry Effects on Unsteady Separation and Dynamic Stall*, UTRC Report R92-958325-6, East Hartford, Connecticut: United Technology Research Center.

Madsen, H., and H. Christensen, 1990, "On the Relative Importance of Rotational, Unsteady and Three-Dimensional Effects on the HAWT Rotor Aerodynamics," *Wind Engineering*, v. 14, n. 6, pp. 405-415.

McCroskey, W. J., 1971, *Measurements of Boundary Layer Transition, Separation and Streamline Direction on Rotating Blades*, NASA TN D-6321: National Aeronautics and Space Administration.

McCroskey, W. J., 1977, "Some Current Research in Unsteady Fluid Dynamics – The 1976 Freeman Scholar Lecture," *Transactions of the ASME, Journ. of Fluids Engineering*, Mar.: American Society of Mechanical Engineers, pp. 8-39.

McCroskey, W., and P. Yaggy, 1968, "Laminar Boundary Layers on Helicopter Rotors in Forward Flight," *AIAA Journ.*, v. 6, n. 10: American Institute of Aeronautics and Astronautics, pp. 1919-1926.

Michelsen J.A., 1992, *Basis3D - A Platform for Development of Multiblock PDE Solvers*, Technical Report AFM 92-05, Technical University of Denmark.

Michelsen J.A., 1994, *Block Structured Multigrid solution of 2-D and 3-D Elliptic PDE's*, Technical Report AFM 94-06, Technical University of Denmark.

Pierce, K., and A. C. Hansen, 1995, "Prediction of Wind Turbine Rotor Loads Using the Beddoes-Leishman Model for Dynamic Stall," *Journ. of Solar Energy Engineering*, v. 117, n. 3, pp. 200-204.

Piziali, R., 1994, *2-D and 3-D Oscillating Wing Aerodynamics for a Range of Angles of Attack Including Stall*, NASA TM 4632: National Aeronautics and Space Administration.

Robinson, M., R. D. Galbraith, D. Shipley, and M. Miller, 1995, "Unsteady Aerodynamics of Wind Turbines," *Proceedings, AIAA 33rd Aerospace Sciences Meeting and Exhibit*, AIAA 95-0526: American Institute of Aeronautics and Astronautics.

Robinson, M., and J. Wissler, 1988, "Unsteady Surface Pressure Measurements on a Pitching Wing," *Proceedings, AIAA 26th Aerospace Sciences Meeting*, AIAA 88-0328: American Institute of Aeronautics and Astronautics.

Ronsten, G., 1992, "Static Pressure Measurements on a Rotating and a Non-Rotating 2.375 m Wind Turbine Blade. Comparison with 2D Calculations," *Journ. of Wind Engineering and Industrial Aerodynamics*, v. 39, pp. 105-118.

Schepers, J., 1995, *Angle of Attack in Aerodynamic Field Measurements on Wind Turbines*, ECN Technical Report.

Schepers, J., A. Brand, A. Bruining, J. Graham, M. Hand, D. Infield, H. Madsen, and D. Simms, 1997, *Final Report of IEA Annex XIV: Field Rotor Aerodynamics*, ECN-C--97-027, Petten, The Netherlands: Netherlands Energy Research Foundation.

Schepers, J., A. Brand, A. Bruining, J. Graham, M. Hand, D. Infield, H. Madsen, T. Maeda, J. Paynter, R. van Rooij, Y. Shimizu, D. Simms, and N. Stefanatos, 2002, *Final Report of IEA Annex XVIII: Enhanced Field Rotor Aerodynamics Database*, ECN-C--02-016, Petten, The Netherlands: Netherlands Energy Research Foundation.

Schreck, S., G. Addington, and M. Luttges, 1991, "Flow Field Structure and Development Near the Root of a Straight Wing Pitching at Constant Rate," *Proceedings, AIAA 22nd Fluid Dynamics, Plasma Dynamics and Lasers Conference*, AIAA 91-1793: American Institute of Aeronautics and Astronautics.

Schreck, S., and H. Helin, 1994, "Unsteady Vortex Dynamics and Surface Pressure Topologies on a Finite Pitching Wing," *Journ. of Aircraft*, v. 31, n. 4, pp. 899-907.

Schreck, S. and M. Robinson, 2003, "Boundary Layer State and Flow Field Structure Underlying Rotational Augmentation of Blade Aerodynamic Response," *Journ. of Solar Energy Engineering*, v. 125, pp. 448-456.

Schreck, S., M. Robinson, M., Hand, M., and Simms, D., "Blade Dynamic Stall Vortex Kinematics for a Horizontal Axis Wind Turbine in Yawed Conditions," *Journ. of Solar Energy Eng.*, v. 123, Nov. 2001, pp. 272-281.

Schreck, S., N. Sørensen, and M. Robinson, "Aerodynamic Structures and Processes in Rotationally Augmented Flow Fields," *Wind Energy*, v. 10, n. 2, pp. 159-178.

Shen, W., and J. Sørensen, 1999, "Quasi-3D Navier-Stokes Model for a Rotating Airfoil," *Journ. of Computational Physics*, v. 150, pp. 518-548.

Sheng, W., R. Galbraith, and F. Coton, 2007, "A Modified Dynamic Stall Model for Low Mach Numbers," *Proceedings, 45th AIAA Aerospace Sciences Meeting and Exhibit*, AIAA-2007-0626: American Institute of Aeronautics and Astronautics.

Shipley, D., M. Miller, and M. Robinson, 1995, *Dynamic Stall Occurrence on a Horizontal Axis Wind Turbine Blade*, NREL/TP-442-6912, Golden, Colorado: National Renewable Energy Laboratory.

Shipley, D., M. Miller, M. Robinson, and M. Luttges, 1995, *Techniques for the Determination of Local Dynamic Pressure and Angle of Attack on a Horizontal Axis Wind Turbine*, NREL TP-442-7393: National Renewable Energy Laboratory.

Simms, D., S. Schreck, M. Hand, and L. J. Fingersh, 2001, *NREL Unsteady Aerodynamics Experiment in the NASA-Ames Wind Tunnel: A Comparison of Predictions to Measurements*, NREL/TP-500-29494, Golden, Colorado: National Renewable Energy Laboratory. Available at http://www.nrel.gov/docs/fy01osti/29494.pdf

Snel, H., 1991, "Scaling Laws for the Boundary Layer Flow on Rotating Wind Turbine Blades," *Proceedings, 4th IEA Symposium on the Aerodynamics of Wind Turbines*, ETSU-N-118, United Kingdom: ETSU Dept. of Energy.

Snel, H., R. Houwink, and J. Bosschers, 1994, *Sectional Prediction of Lift Coefficients on Rotating Wind Turbine Blades in Stall*, ECN-C-93-052, The Netherlands: Netherlands Energy Research Foundation.

Snel, H., R. Houwink, and W. Piers, 1992, "Sectional Prediction of 3D Effects for Separated Flow on Rotating Blades," *18th European Rotorcraft Forum*.

Snel, H., J. Schepers, and B. Montgomerie, 2007, "The MEXICO Project (Model Experiments in Controlled Conditions): The Database and First Results of Data Processing and Interpretation," The Science of Making Torque from Wind Conference, *Journal of Physics: Conference Series 75*. 012014 doi:10.1088/1742-6596/75/1/012014

Sørensen, N. N., 1995, *General Purpose Flow Solver Applied to Flow over Hills*, Risø-R-827 EN, Roskilde, Denmark: Risø National Laboratory.

Sørensen, N., J. Michelsen, and S. Schreck, 2002, "Navier-Strokes Predictions to the NREL Phase VI Rotor in the NASA Ames 80 ft × 120 ft Wind Tunnel," *Wind Energy*, v. 5, pp. 151-169.

Tangler, J., and M. Selig, 1997, *An Evaluation of an Empirical Model for Stall Delay due to Rotation for HAWTS*, NREL/CP 440-23258, Golden, Colorado: National Renewable Energy Laboratory.

Tran, C., and D. Petot, 1981, "Semi-Empirical Model for the Dynamic Stall of Airfoils in View of the Application to the Calculation of Responses of a Helicopter Blade in Forward Flight," *Vertica*, v. 5, n. 1, pp. 35-53.

Yeznasni, A., R. Derdelinckx, and C. Hirsch, 1992, "Influence of dynamic stall in the aerodynamic study of HAWTs," *Journ. of Wind Engineering and Industrial Aerodynamics*, v. 39, n. 1-3, pp. 187-198.

6

Wind Turbine Airfoils and Rotor Wakes

Peter B. S. Lissaman, Ph.D.
Adjunct Professor of Aerospace Engineering
University of Southern California
Los Angeles, California

Introduction

In this chapter we will discuss the detailed characteristics of the *lifting surfaces* which we know as *airfoils*, in order to provide guidance in the selection of airfoil shapes that will perform satisfactorily over the broad operating environment of a wind turbine. In addition, the fundamental fluid-dynamic principles involved in modeling the structure of rotor wakes and integrating wake-induced effects over an array of turbines are presented in this chapter.

Wind turbines operate by the action of the *relative wind* (the natural wind plus wind caused by rotor motion and rotor-induced flow), which creates aerodynamic forces on the rotating blades. These can normally be grouped into *lift-like forces* and *drag-like forces*. Lift forces operate through the generation of *circulation* and do not involve large viscous losses in the flow and the associated loss of *total head*, while drag forces function through *flow separation* on the blade and the loss of total head. In the previous chapter, the relative merits of lift and drag power devices were discussed, with lift being the clear preference for wind turbine rotors. When considering the potential impact of wakes on the total output of a wind power station, a lift-type device is again preferred, since a drag-type unit develops a greater wake and less energy is available to downwind units. Thus, the array efficiency of a cluster of drag-type units will be lower than that of lift-type turbines.

Designing lift-type wind energy conversion systems depends on a knowledge of the properties of airfoils. There is extensive experience in airfoil technology, derived primarily from airplane wing design. It is normally assumed that the properties of airfoils that are desirable for wings will also be desirable for wind turbines. This assumption is only superficially valid, for reasons described later. It must be noted that fixed-pitch wind turbine airfoils generally operate over angles of attack ranging from 0 deg to 90 deg, and thus their *stall* and *post-stall behavior* are important. This situation does not apply for aircraft wings, which seldom require operation at angles in excess of 30 deg.

An excellent review of wind turbine airfoils and their properties is given by Miley [1980]. Included here are many of the airfoils that wind turbine designers have borrowed from the aircraft industry. More recently, Tangler and Somers [1995] have described the development of several families of airfoils designed specifically for HAWTs.

Airfoils are designed to generate lift; that is, to create a force normal to the incident flow when immersed in this flow at a small angle to the airfoil's *chord line*. The geometry of this flow and the forces created are shown in Figure 6-1. The angle from the *relative velocity* vector, V_r, to the chord line is the *angle of attack*, α. The lift force is normal to V_r. Generally this lift is produced simultaneously with a streamwise force at right angles to itself, called the drag. These forces are normalized by dividing the force per unit span by the *dynamic pressure* and the chord length to obtain *lift* and *drag coefficients*, C_L and C_D, in accordance with Equations (5-3).

Characteristic Behavior of Airfoils

Reynolds Number

The most significant flow factor influencing the behavior of low-speed airfoils is that of *viscosity*, which indirectly causes lift and directly causes drag and *flow separation*. This influence is characterized by the *Reynolds number* of the airfoil/fluid combination. For the airfoil in Figure 6-1, Reynolds number can be calculated as follows:

$$N_R = \frac{V_r c}{\nu} \approx \left(\frac{r\Omega}{30\ m/s}\right)\left(\frac{c}{0.5\ m}\right) \times 10^6 \approx \left(\frac{r\Omega}{110\ mph}\right)\left(\frac{c}{1.0\ ft}\right) \times 10^6 \qquad (6\text{-}1)$$

where

$$
\begin{aligned}
\nu &= \text{kinematic viscosity of air (m}^2\text{/s)} \\
V_r &= \text{relative velocity (m/s)} \\
r\Omega &= \text{local tangential velocity (m/s or mph)} \\
c &= \text{chord length (m or ft)}
\end{aligned}
$$

Airfoils in use on modern turbines range in representative chord size (typically at 3/4 span) from about 0.3 m (1 ft) on a small-scale turbine to over 2 m (7 ft) on a megawatt-scale rotor. Tip speeds typically range from approximately 45 to 90 m/s (100 to 200 mph), so tangential velocities at the 3/4-span of a HAWT blade can be estimated to range from about 34 to 68 m/s (75 to 150 mph). For turbine airfoils, then, Reynolds numbers range from about 10 million down to 0.7 million. This implies that turbine airfoils generally operate beyond the sensitive, low Reynolds number range (often taken to be below 0.5 million) in which extreme and unusual behavior is caused by anomalous *transition, separation*, and *bubble formation* phenomena. In this sensitive range, very large changes in airfoil behavior

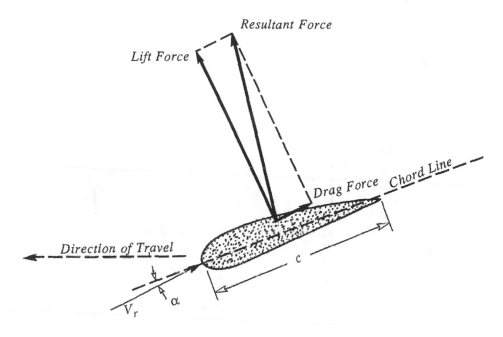

Figure 6-1. Typical airfoil geometry and aerodynamic forces.

can be induced by minor events, like changes in incoming flow turbulence, vibrating the airfoil itself, or roughness on the surface [Lissaman 1983]. Fortunately, this type of irregular airfoil behavior is not usually encountered with wind turbines, so it will not be necessary to discuss it further.

Typical Airfoil Lift and Drag Behavior

Generally, airfoil behavior is characterized by three flow regimes, as illustrated by the wind tunnel test data in Figure 6-2 for a representative quadrant of angle of attack, α. The first of these is the *attached regime*, for angles of attack from about -15 to +15 deg; the second is the *high-lift/stall-development* regime, for angles of attack between about 15 and 30 deg; and, thirdly, the *flat-plate, fully-stalled*, or *deep-stall* regime with attack angles between 30 and 90 deg. These regimes are repeated in the other three quadrants, with approximate symmetry for drag and anti-symmetry for lift.

Attached Flow Regime

In the attached flow regime, the general airfoil behavior is well understood. Although it can be complicated and significantly affected by geometrical and viscous parameters like *thickness, camber, nose radius, trailing edge angle, surface roughness*, and *Reynolds number*, airfoil behavior in the attached flow regime can be very accurately estimated by the wealth of theory and data accumulated in a half-century of refining two-dimensional airfoil theory for application to wings. Methods exist for accurate analytical estimation of the lift and drag forces in this regime, and there is also an extensive literature presenting experimental data. The upper end of the attached flow regime, around an angle of attack

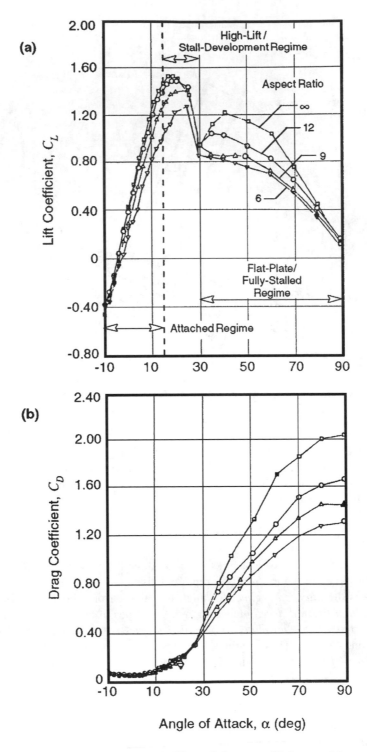

Figure 6-2. Typical variation of airfoil lift and drag coefficients with angle of attack and aspect ratio, illustrating the flow regimes used to characterize airfoil behavior. (a) Lift coefficients for an *NACA 4415* airfoil tested at a Reynolds number of 0.5 million. (b) Drag coefficients for the same airfoil [Ostowari and Naik 1984]

of 15 deg, is the region where separation and incipient stall generally commence. While the initial stages of stall are quite sensitively connected to details of the airfoil shape, the features of incipient stall are well-documented in the standard literature.

A basic insight into the effect of drag in the attached-flow regime on the power of a HAWT can be obtained by taking an approach similar to that used in the fundamental analysis of the performance of monoplane wings, that is, by calculating the ideal power without viscous or *profile drag* losses and then estimating the power losses caused by drag alone. A simple analysis of a constant chord HAWT rotor with a constant drag coefficient at zero lift coefficient indicates that

$$\delta P_V \approx \frac{\rho B c \Omega^3 R^4}{8} C_{D,0} \tag{6-2}$$

where

$$
\begin{aligned}
\delta P_V &= \text{viscous power loss (W)} \\
\rho &= \text{air density (kg/m}^3) \\
B &= \text{number of blades in the rotor} \\
\Omega &= \text{rotor speed (rad/s)} \\
R &= \text{rotor radius (m)} \\
C_{D,0} &= \text{drag coefficient at zero lift coefficient; profile drag coefficient}
\end{aligned}
$$

For tapered blades, the effective chord length is approximately that at about 80 percent of the rotor radius. As an example, consider a representative small-scale turbine, the *Carter 25* HAWT shown in Figure C-2, for which $B = 2$, $\Omega = 12.6$ rad/s, $R - 4.9$ m, $c - 0.31$ m, and $C_{D,0} = 0.009$. This provides a representative viscous power loss of 1.0 kW at all wind speeds. If we compare this with the rated power of 25 kW, we note that it is about 4 percent of that value. This estimate can be refined by using a state-of-the-art rotor performance code, taking into account details of chord and twist distributions and the air loading. When the actual drag coefficient is reduced by $C_{D,0}$ at all angles of attack, the improvement in power output is predicted to be roughly constant at 1.0 kW over the operating range, which is equal to the crude estimate above.

At a site with a typical *Weibull wind speed histogram* (Table 2-1), this represents an 8 percent energy loss from viscous effects. It should be noted that most well-sited turbines produce the majority of their energy when operating near their maximum power coefficient, so this energy loss estimate is accurate to a first order. For turbines generating power in a low wind regime, energy production will be more heavily impacted by airfoil drag.

High-Lift/Stall-Development Regime

This next regime involves airfoil behavior at angles of attack from approximately 15 to 30 deg. Designers of fixed-pitch turbines pay special attention to lift and drag coefficients in this regime, since these have a dominant effect on the peak power produced by the rotor. Here the flow state ranges from initial incipient separation near the trailing edge of the airfoil to massive separation over its entire low-pressure surface. The phenomenon of *dynamic stall* [Hibbs 1986] also takes place in this regime.

Flat-Plate/Fully-Stalled Regime

Airfoil behavior at angles of attack from approximately 30 to 90 deg is similar to that of a simple flat plate. At 45 deg, lift and drag coefficients are approximately equal, and lift approaches zero near 90 deg. Like a flat plate, the drag coefficients in Figure 6-2(b) show a significant effect of *aspect ratio* (span length divided by chord width), with the low aspect ratio case significantly lower in drag than the two-dimensional (*i.e.*, infinite length) situation.

It is further noted that the post-stall flow over the airfoils of an actual rotor experiences two additional fluid-dynamic effects that are not represented in typical wind-tunnel tests. One is the effect of *spanwise flow* caused by rotational effects. Generally, this will produce a flow towards the blade tip, and the normal expectation is that this postpones stall nearer the axis of rotation. Swept aircraft wings show a similar effect, with the spanwise component of the flow causing the wing tips to stall prematurely. No rational quantitative analysis is available to account for this. The second difference relates to the generally *non-uniform spanwise loading* on an actual rotor blade. A spanwise-constant lift coefficient is seldom achieved, so stall will develop differently at different radial stations along the blade. Again, no acceptable procedure is available to handle the influence of spanwise loading variations on stall. The usual procedure is to assume each section behaves independently, which effectively ignores spanwise interaction of sections.

Aspect Ratio Effects

The data in Figure 6-2 clearly indicate that lift and drag characteristics show a significant aspect-ratio dependence at angles of attack larger than 30 deg. In the fully-attached regime, airfoil section characteristics are not greatly affected by aspect ratio, so that two-dimensional (*i.e.* infinite aspect ratio) data can be used in predicting performance at low angles of attack. However, when two-dimensional data are used, a *tip-loss factor* must be added, as described in Equations (5-35) and (5-36).

The size of the aspect-ratio effects on airfoil coefficients in the attached regime can be estimated using the classical equations for correcting wind tunnel test data measured on a finite-span airfoil, from the work of Munk, Glauert, and Prandtl [Jacobs and Abbot 1932]. In this case we are using these equations in reverse, starting from infinite-span data and obtaining lift and drag curves for a finite aspect ratio. These formulas are as follows:

$$C_L = C_L' \tag{6-3a}$$

$$C_D = C_D' + \frac{C_L^2}{\pi\mu} \tag{6-3b}$$

$$\alpha = \alpha' + \frac{57.3\,C_L}{\pi\mu} \tag{6-3c}$$

where
C_L', C_D' = lift and drag coefficients for an infinite aspect ratio
C_L, C_D = lift and drag coefficients for a finite aspect ratio
μ = aspect ratio

In Equations (6-3b) and (6-3c), minor corrections for the shape of the pressure distribution on the airfoil (rectangular *vs.* elliptical) have been eliminated for convenience. Inspection of these equations shows that finite length increases the angle of attack and the drag coefficient for a given lift coefficient. Conversely, the lift coefficient is reduced for the same angle of attack. From Figure 6-2, it is found that stall occurs at a lower lift coefficient for low aspect ratios (which is not predicted by Equations (6-3)), and that the same situation (*i.e.*, a lower separated-lift coefficient) persists as the angle of attack increases beyond the stall point.

Post-Stall Modeling of Lift and Drag Coefficients

As previously noted, the airfoils in *fixed-pitch rotors* (used in *stall-controlled* HAWTs and most VAWTs) will operate over all three of the flow regimes in Figure 6-2. In order to accurately predict the peak power of a fixed-pitch rotor, it is particularly important to know the details of lift and drag behavior in the high lift/stall development regime. An empirical model for modifying two-dimensional airfoil data in all three regimes to more accurately represent wind turbine rotor behavior has been developed by Viterna and Corrigan [1981]. This model is based on the following three assumptions:

-- In the attached flow regime, Equations (6-3) adequately model end effects in terms of the blade aspect ratio, and no tip- or hub-loss models are needed.
-- In the high-lift/stall-development regime, the *torque force* (sometimes called the *suction force*) on the element, acting in the plane of rotation, does not decrease with increasing angle of attack; rather, it is independent of angle of attack.
-- In the flat-plate/fully-stalled regime, the dominant parameter is the maximum value of the drag coefficient, and this is determined by the blade aspect ratio.

The equations which implement these assumptions in the *Viterna-Corrigan post-stall model* are as follows:

$$\alpha \geq \alpha_s:$$

$$C_L = \frac{C_{D,\max}}{2} \sin 2\alpha + K_L \frac{\cos^2 \alpha}{\sin \alpha} \tag{6-4a}$$

$$C_D = C_{D,\max} \sin^2 \alpha + K_D \cos \alpha \tag{6-4b}$$

$$K_L = (C_{L,S} - C_{D,\max} \sin \alpha_S \cos \alpha_S) \frac{\sin \alpha_S}{\cos^2 \alpha_S} \tag{6-4c}$$

$$K_D = \frac{C_{D,S} - C_{D,\max} \sin^2 \alpha_S}{\cos \alpha_S} \tag{6-4d}$$

$$\mu \leq 50: \quad C_{D,\max} = 1.11 + 0.018 \, \mu$$
$$\mu > 50: \quad C_{D,\max} = 2.01 \tag{6-4e}$$

where

$C_{D,max}$ = maximum drag coefficient in the fully-stalled regime

To illustrate the application of Equations (6-3) and (6-4), Viterna and Corrigan [1981] analyzed the power output of the historic *Gedser* wind turbine, shown in Figure 3-1. The blades of this rotor are *Clark Y* airfoils with an aspect ratio of six. Figure 6-3 shows the modification of the unstalled two-dimensional lift and drag curves for this airfoil [Silverstein 1934] to represent an aspect ratio of six, in accordance with Equations (6-3). At the stall angle of 15.9 deg, these curves are coupled with the Viterna-Corrigan model of post-stall behavior. The power curves in Figure 6-4 show the effect of these changes on calculated power, compared to test data from [Lundsager *et al.* 1980].

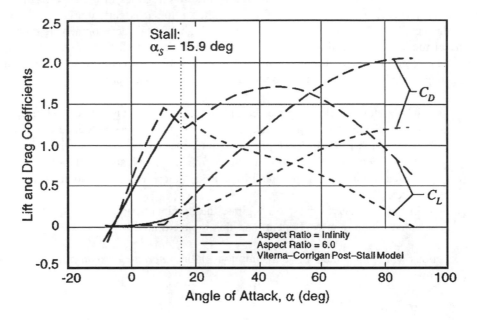

Figure 6-3. Two-dimensional and modified lift and drag curves for Clark Y airfoils.

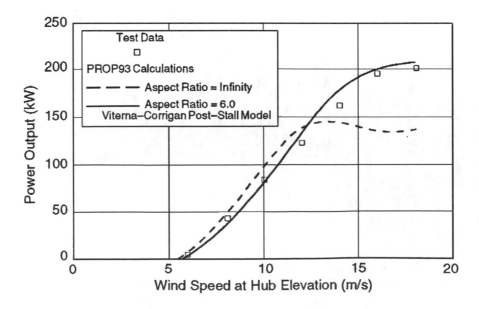

Figure 6-4. Measured output power of the Gedser 200-kW wind turbine compared to power curves calculated by two methods. [Viterna and Corrigan 1981]

Rotor power was calculated using a computer model derived from the original *PROP* code for predicting wind turbine aerodynamic performance [Wilson *et al.* 1976]. A later version of this is the *PROP93* computer code [McCarty 1993]. Power train losses were estimated using the following general loss model [Spera and Janetzke 1981] and deducted from the rotor power:

$$P_{PT} = -aP_{G,R} - (b + s)P_R \qquad\qquad \text{(6-5a)}$$

For the Gedser Wind Turbine:

$$a \approx 0.055$$
$$P_{G,R} = 200 \ kW \qquad\qquad \text{(6-5b)}$$
$$b \approx 0.040$$
$$s \approx 0.050$$

where

P_{PT} = power-train loss (kW)
a, b = empirical constants from tests on the 200-kW Mod-OA wind turbine
$P_{G,R}$ = rated power of the generator (kW)
s = slip in the generator
P_R = rotor power (kW)

The dashed power curve in Figure 6-4 is the result of calculations based on two-dimensional airfoil data and *Prandtl* tip- and hub-loss models. The solid power curve illustrates how the correlation between the calculated and measured peak powers of a fixed-pitch rotor can often be significantly improved by modifying lift and drag curves in accordance with Equations (6-3) and (6-4). This is an important achievement, since the accurate determination of this peak power controls the design of the gearbox and generating equipment and is thus a primary driver of the cost of a fixed-pitch turbine.

An alternative set of empirical equations for modeling lift and drag coefficients in the pre- and post-stall regimes has recently been developed, extending the Viterna-Corrigan model [Spera 2008, Fig. 2-16].

Airfoil Aerodynamic Requirements

There are evidently many engineering requirements entering into the selection of a wind turbine airfoil. These include primary requirements related to *aerodynamic performance, structural strength and stiffness, manufacturability*, and *maintainability*. Requirements related to other rotor characteristics like electromagnetic interference, acoustic noise generation, and aesthetic appearance are generally assumed to be of secondary importance. Here we refer only to aerodynamic aspects, although we note that the critical wind turbine performance and reliability features associated with *aeroelastic behavior* (changes in angle of attack caused by blade deflections) introduce a strong coupling between aerodynamic and structural requirements.

Lift and Drag Requirements

The usual assumption, historically established in airplane lifting surface theory, is that high lift and low drag are desirable for an airfoil, and that the *lift-to-drag ratio* (often abbreviated as *L/D*) is a critical consideration. For wind turbine rotors this point of view is not of the same importance as it is for aircraft wing design. General analysis of rotor performance shows that the primary factor is the product of the chord and the lift coefficient, or cC_L. Thus, when other characteristics like tip-speed ratio and diameter are held constant, operating at a higher lift coefficient will permit the use of narrower blades. Generally, this will not result in lower viscous power losses, since viscous torque is controlled more by the L/D ratio of the airfoil than the actual value of lift.

The principal factor controlling the lift-to-drag ratio of a given airfoil section is the Reynolds number. Since a smaller chord reduces Reynolds number in accordance with Equation 6-1, this is an aerodynamic reason to avoid narrow-chord blades. Another factor relating to the use of narrower blades is the negative effect on structural stiffness, which reduces rapidly as thickness decreases. Stiffness is approximately proportional to the square or the cube of the thickness, and thickness is proportional to the chord.

Drag coefficient is of limited importance in determining turbine performance in the usual operating range of wind speeds and at normal tip-speed ratios. Generally, its effect on the power coefficient varies directly with the cube of the tip-speed ratio. When this ratio is high (in excess of 10, typically) the drag coefficient does become somewhat more significant. This situation occurs in lower wind ranges when it may be more important to maximize energy extraction.

Surface Roughness Effects

An important operating requirement that relates to a wind turbine airfoil is its ability to perform when the smoothness of its surfaces has been degraded by dust, dirt, rain, or insect debris. Field experience with small- and medium-scale units has indicated that very severe performance degradation can occur in these circumstances. Figure 6-5 shows field test power curves for a small-scale (10-m dia.) HAWT with both clean and dirty blades. It appears that this deleterious effect is most pronounced in fixed-pitch stall-controlled

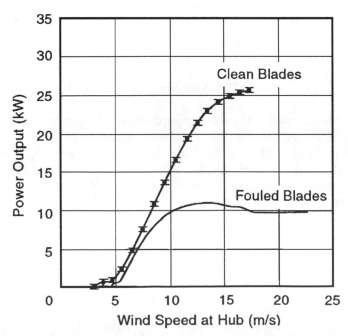

Figure 6-5. Effect of blade fouling by dust, dirt, and insects on the performance of a small-scale HAWT 10-m in diameter.

rotors, and that it is caused by the surface roughness precipitating blade stall at a relatively low lift coefficient. Consequently, peak power occurs at a lower tip-speed ratio than that for which the stall control was designed. Large-scale rotors are less affected because of their higher elevation (above most insects and dust particles) and because debris thicknesses are smaller fractions of the leading-edge radius and the airfoil thickness.

Airfoil Selection

In the present state of the art there is no unambiguous, rational procedure to determine the ideal airfoil for a given wind turbine rotor, or even for a given radial station on a blade. This is the result of both the very wide range of angles of attack over which a blade operates and the widely different geometrical combinations of airfoil section, chord, and twist possible in the blade design. A further factor relates to the differing spanwise aerodynamic requirements of the general rotor. For these reasons the current approach in airfoil selection is to use a rotor performance computer code [*e.g.*, Wilson *et al.* 1976, Tangler 1987, McCarty 1993, Buhl *et al.* 1998] that takes into account (as adequately as possible) all the forces on the rotor that vary with tip-speed ratio and radial position. The aerodynamicist can then vary the rotor geometry until an acceptable performance results. Inherent in this process is the assumption that the analytical models in the code properly account for the effects of these geometric changes on performance.

Standard Aircraft Airfoils

Many different standard airfoils developed for aircraft have been used on wind turbines with no special advantages from any particular choice when only the impact of the basic airfoil properties on power output is considered. Five of these are shown in Figure 6-6. The *NACA 230XX series* and the *NACA 44XX series* airfoils (where the XX stands for the thickness-to-chord ratio, in percent) have been used on many modern HAWT units, with thickness-to-chord ratios varying from about 28 percent at the root to about 12 percent at the tip. In some respects, these standard airfoils have unsatisfactory characteristics. For example, airfoils in the NACA 230XX series have maximum lift coefficients that are very sensitive to surface fouling, and their performance deteriorates with increased thickness more rapidly than that of other airfoils.

NACA 63-2XX series airfoils have demonstrated the best overall performance characteristics of the NACA families, and they provide reasonable resistance to roughness losses. Airfoils in the *LS(1)-04XX series* were designed to tolerate surface fouling, but HAWTs with these airfoils (*e.g.*, the *ESI 80* and *Carter 300*) have experienced large power losses induced by roughness. This airfoil series also has a very high nose-down pitching moment which can result in excessive elastic blade twist in thin blades and undesirable performance changes.

For most VAWTs, a symmetrical airfoil such as the four-digit series *NACA 00XX* is normally used, with thickness ratios varying from 12 percent to 15 percent.

Current Designs of Airfoils for Wind Turbines

Special airfoils designed for wind turbine applications have been developed since the mid-1980s. Classes of these for HAWTs are described by Tangler and Somers [1985, 1995] and Tangler [1987], and are shown in Appendix D. Some airfoils specially tailored for VAWTs are described by Klimas [1984]. Typical design goals for new HAWT airfoils are

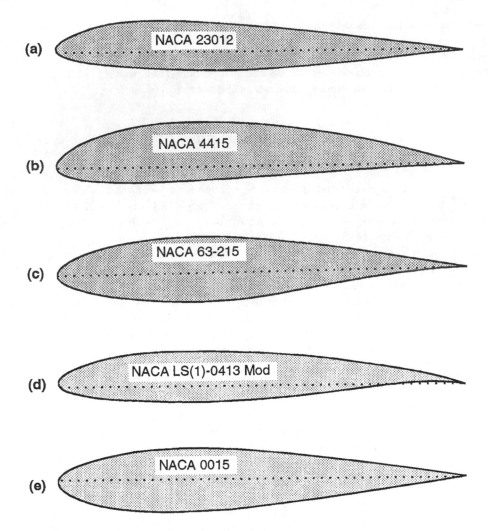

Figure 6-6. Common wind turbine airfoils, designed originally for aircraft.
(a) NACA 23012 (b) NACA 4415 (c) NACA 63-215 (d) NASA LS(1)-0413 Modified (e) NACA 0015

-- a stable $C_{L,\,max}$ at stall, despite blade fouling
-- larger lift-to-drag ratios
-- limited $C_{L,\,max}$ at outboard sections, for control of peak power in high winds

As we noted previously, the numerous different demands made of airfoil sections along the span of a wind turbine blade preclude the possibility of the design of a single airfoil which might be "ideal" in all senses. However, if a limited class of wind turbines is defined, it then becomes possible to specify airfoil requirements for the rotors of this class. The spanwise variation in requirements may be accounted for by specifying different lift and drag properties (and consequently different airfoil sections) at different radial locations. The definition of a rotor class provides us with input for which rotor performance codes may be run, so that the relative performance with different airfoil sections may be calculated and evaluated. Airfoil contours are shaped to produce the desired flow characteristics at selected radial stations using analysis tools such as the *Eppler code* [Eppler and Somers 1980], an early *computational fluid dynamics* (CFD) code.

HAWT Rotors 10 m to 30 m in Diameter

Tangler and Somers [1995] have applied this concept to the development of airfoils for a large class of modern wind turbines with the following characteristics:

--   system designs optimized for a site with an annual average wind speed between 4.5 and 6.2 m/s at an elevation of 10 m;
--   rotor diameters between 10 and 30 m;
--   fixed-pitch stall-controlled blades;
--   peak power coefficients at tip-speed ratios of approximately 8.

The contours of three airfoils designed for specific locations in the blades of this class of HAWTs are shown in Figure 6-7.

Assuming that the main power generating region of the blade is centered on a station at about 75 percent of the radius (*i.e.*, $r/R \approx 0.75$) we desire the airfoil at this primary outboard station to have a relatively high lift-to-drag ratio (to maximize the power coefficient), a limited maximum lift coefficient (to assure reliable stall control, particularly at a low-wind site), a low sensitivity of stall to surface roughness (to ensure that stall control behavior remains constant), and an appreciable thickness-to-chord ratio (to retain desirable structural stiffness and weight). An airfoil designated the *SERI S805A* has been designed to meet these requirements, with the contour shown in Figure 6-7(b) [Tangler and Somers 1986, Tangler 1987].

In Table 6-1 the basic properties of this new airfoil are compared with those of three conventional wing airfoils of comparable thickness, for a representative Reynolds number of one million. It is noted from these data that the lift-to-drag ratio (and hence the performance) of the SERI S805A is comparable to that of the high-performance standard

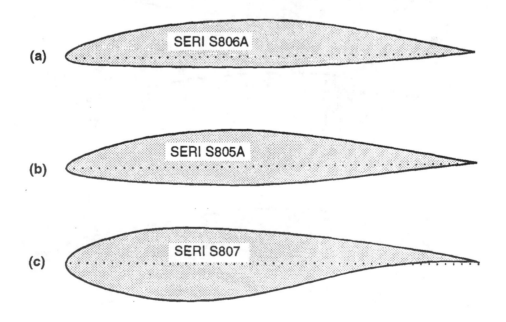

Figure 6-7. Contours of the SERI S805A/S806A/S807 series of airfoils designed specifically for fixed-pitch HAWTs with diameters from 10 m to 20 m. (a) Tip region ($r/R \approx 0.95$) (b) Primary outboard airfoil ($r/R \approx 0.75$) (c) Root region ($r/R \approx 0.40$) [Tangler and Somers 1995]

airfoils such as the NACA 4415 and the LS(1)-0413 Mod. The SERI S805A also has the lowest $C_{L, max}$, to prevent overpowering the generator of a fixed pitch turbine in high winds. Unlike the NACA 4415, the $C_{L,max}$ of the new airfoil has minimal sensitivity to surface roughness. This can be a pronounced performance advantage, as illustrated in Figure 6-5.

Table 6-1.
Properties of the *SERI S805A* Airfoil Compared to Standard Wing Airfoils
[Tangler and Somers 1986]

Airfoil $N_R = 1 \times 10^6$	Thickness ratio	Max. lift coefficient	Min. drag coefficient	Lift-to-drag ratio at $C_L = 0.7$	Moment coef. at $\alpha = 0$
SERI S805A	0.135	1.2	0.007	90	-0.05
NACA 4415	0.150	1.4	0.009	85	-0.08
NACA 23012	0.120	1.3	0.009	70	-0.01
LS(1)-0413 Mod	0.130	1.5	0.008	86	-0.10

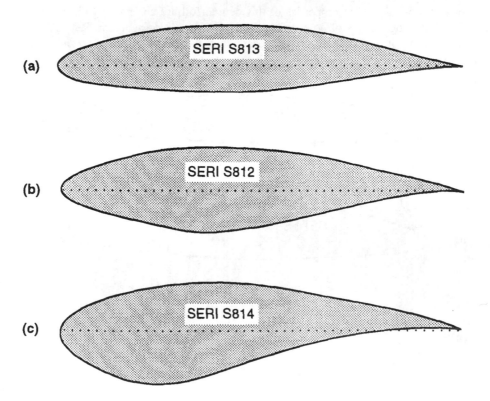

Figure 6-8.   Contours of the SERI S812/S813/S814 series of airfoils designed specifically for fixed-pitch HAWTs with diameters from 21 m to 35 m. (a) Tip region ($r/R \approx 0.95$)   (b) Primary outboard airfoil ($r/R \approx 0.75$)   (c) Root region ($r/R \approx 0.40$) [Tangler 1987, Tangler and Somers 1995]

Airfoils designed for the root and tip sections must not only address local aerodynamic requirements, but it is also desirable that they produce a monotonic change in aerodynamic properties from root to tip and a "faired" blade surface. The root section should be thicker for structural reasons, and it can accept a higher $C_{L,max}$. The tip section needs to be thinner, with lower values of both minimum drag and $C_{L,max}$. Two airfoil shapes shown in Figure 6-7, the *SERI S807* and *S806A*, have been designed for the root ($r/R \approx 0.40$) and tip ($r/R \approx 0.95$) regions, respectively, to meet these requirements.

HAWT Rotors 21 m to 35 m in Diameter

It should be noted that designing airfoils specifically for wind turbines may obtain benefits not only by improved aerodynamic performance but also by improved structural performance and lower blade costs. The thickness-to-chord ratio of an airfoil plays a very significant role in its strength and stiffness, and this is particularly important for longer HAWT blades. For this reason, a family of thicker airfoils has been designed for rotors from 21 to 35 m in diameter, on lines similar to those of the SERI S805A/806A/807 family, with the designations *S812, S813,* and *S814* (Fig. 6-8). The thickness-to-chord ratio of the S812 is 0.210, compared to 0.135 for the S805A. Wind tunnel tests confirm that these thicker airfoils also have a limited $C_{L,max}$ (in the range of 1.00) which is insensitive to surface roughness, and low drag. The latter is accomplished by maintaining large areas of *laminar flow.*

Figure 6-9 shows a third set of airfoils designed for HAWTs with diameters of 36 m and larger, tailored for an optimum combination of aerodynamic performance and strength. These are designated as *S816, S817,* and *S818* airfoils.

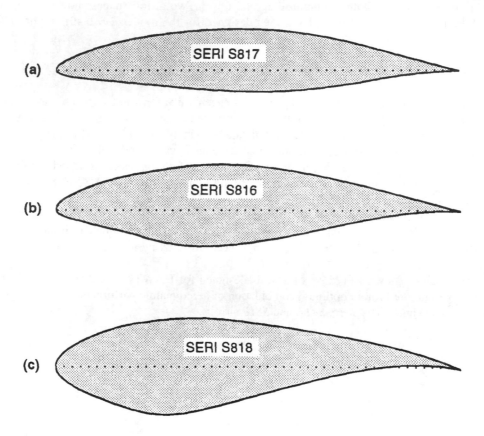

Figure 6-9. Contours of the SERI series of airfoils designed specifically for fixed-pitch HAWTs 36 m in diameter and larger. (a) Tip region ($r/R \approx 0.95$) (b) Primary outboard airfoil ($r/R \approx 0.75$) (c) Root region ($r/R \approx 0.40$) [Tangler and Somers 1995]

Field Performance Testing of Wind Turbine Airfoils

While idealized, two-dimensional wind tunnel tests are necessary for developing airfoil contours and obtaining lift and drag data with which to predict rotor power output, final evaluation of a wind turbine airfoil must be made on the basis of full-scale field performance tests. Only in the field can the combined effects of spatial and temporal turbulence, variability in steady wind speed, airfoil rotation, hub/tip losses, surface roughness, and manufacturing tolerances be measured. Of course, the variability of the wind complicates the testing, and airfoil-to-airfoil comparisons are more difficult to make. Standardized procedures have therefore been developed for the purpose of increasing the reliability and reducing the uncertainty in field tests of wind turbine performance, emphasizing *power curves* [AWEA 1988] and *energy output* [ASME 1988].

Comparative Power Performance

As an example of a field test conducted to obtain comparative power curves, let us examine the side-by-side performance tests conducted on two stall-controlled, three-bladed rotors installed on identical medium-scale turbines [Tangler *et al.* 1990]. The wind turbines in these tests were *Micon 65/13* HAWTs, rated at 65 kW. One turbine had its original-equipment *Aerostar* blades, with NACA 4415-24 airfoils (Fig. 6-6) and the planform shown in Figure 6-10(a). Candidate replacement blades on the second turbine [Jackson and Migliore 1987] utilize the SERI airfoils in Figure 6-7, and their planform is shown in Figure 6-10(b). Table 6-2 lists comparative data for the two sets of blades. The restrained $C_{L,max}$ in the tip region of the SERI blade permits the length extension and allows the use of 14 percent more swept area for the same generator rating.

With respect to the prevailing wind at the site, the test turbines had a crosswind spacing of 2.2 diameters, and a meteorological tower with an anemometer at hub height was located 2.0 diameters upwind. Based on the relative positions of the anemometer and the test rotors, data were collected at a sampling rate of 1 Hz, and then averaged for 30 sec, in compliance with standard procedures [AWEA 1988]. Both clean and "dirty" blades were tested. Dirty blade roughness was simulated by strips of tape on upper and lower surfaces near the leading edge. Over 100 hours of operational data were collected for each cleanliness condition, which is a typical duration for a reliable performance test.

The results of these comparative power performance tests are shown graphically by the power curves in Figures 6-11 and 6-12. The significant performance improvements exhibited by the SERI blades compared to the Aerostar blades are attributed to the following three factors, in the order of their relative contributions [Tangler *et al.* 1990]:

-- larger swept area of the SERI blades allowed by their restrained $C_{L,max}$;
-- less sensitivity of $C_{L,max}$ to leading-edge roughness in the outboard SERI airfoils;
-- improved aerodynamic performance at low and moderate wind speeds of the inboard SERI airfoils.

Figure 6-10. Planforms of two blade designs used in comparative field performance tests on *Micon 65/13* HAWTs. (a) Original-equipment *Aerostar* blade with *NACA 4415-24* airfoils (b) Candidate replacement blade with *SERI* airfoils [Tangler *et al.* 1990]

Table 6-2. Blade Data for Side-by-Side Performance Tests on a Micon 65/13 HAWT [Tangler *et al.* 1990]

Parameter	Original Blade	SERI Blade
Rotor diameter	16.0 m	17.0 m
Blade length	7.41 m	7.96 m
Planform area	5.97 m²	6.17 m²
Airfoil sections:		
Tip	NACA 4415	S806A
3/4 span		S805A
Root	NACA 4424	S807
Pitch angle	4.2 deg	0.5 deg
Tilt angle	4 deg	4 deg
Cone angle	4 deg	4 deg
Blade material	GFRP	GFRP
Blade weight	3.6 to 3.8 kN	2.8 kN
Natural frequencies:		
1st flapwise	4.05 hz	3.16 hz
1st edgewise	5.80 hz	7.20 hz
Rotor speed	48 rpm	48 rpm

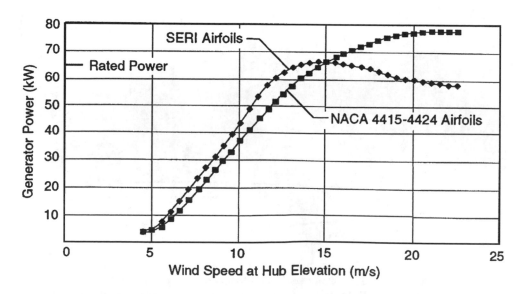

Figure 6-11. Power curves from field tests of clean blades. [Tangler *et al.* 1990]

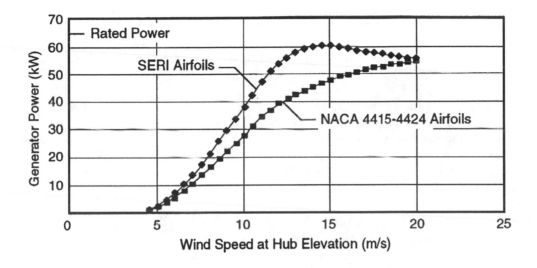

Figure 6-12. Power curves from field tests of dirty blades. [Tangler *et al.* 1990]

Comparative Energy Production

Comparative *annual energy outputs* are calculated by combining power curves, such as those in Figures 6-11 and 6-12, with a specified *wind speed histogram*, as illustrated by the example in Table 5-1. Tests on wind turbines conducted in accordance with standardized ASME procedures [ASME 1989] seek to define relative energy production in terms of an *annual energy ratio*, which is the ratio of the annual energy output of the test turbine to that a specified reference energy production. The latter could be a calculated value (*e.g.* during the design phase) or the result of previous tests.

To illustrate performance comparisons on the basis of annual energy ratios, we will assume (1) the clean original rotor is our reference, and (2) the annual wind speed histogram has a *Weibull distribution* with parameters $C = 8.51$ m/s and $k = 2.48$. The latter are taken from Table 2-1 for an elevation of 25 m, the entry nearest the 23-m hub elevation of the Micon 65/13 HAWTs. Table 6-3 lists the results of annual energy calculations obtained using this wind histogram and the power curves in Figures 6-11 and 6-12.

Table 6-3. Sample Comparative Annual Energy Ratios

Test Rotor Configuration	Gross Annual Energy Output[1] (MWh/y)	Annual Energy Ratio
Original: Clean	194	1.00
Original: Dirty	139	0.72
SERI: Clean	225	1.16
SERI: Dirty	184	0.95

[1] 7.54 m/s mean wind speed at hub elevation, with a Weibull wind speed distribution

The relative energy productions of the two rotors will depend on site conditions, such as the rate of blade soiling by dirt and insects and the frequency of washing. For example, is we assume that the clean condition is present 40 percent of the time and the dirty condition 60 percent, a rotor with SERI blades can be expected to produce about 24 percent more energy under the specified wind regime. This is calculated using the annual energy ratios in Table 6-3, as follows:

$$\frac{SERI - Rotor\ Energy}{Original - Rotor\ Energy} = \frac{0.40 \times 1.16 + 0.60 \times 0.95}{0.40 \times 1.00 + 0.60 \times 0.72} = 1.24 \qquad (6\text{-}6)$$

Actual field test comparisons in Tehachapi and San Gorgonio wind power stations in California have demonstrated improvements in annual energy output from 25 percent to 30 percent [Tangler 1993]. In addition to the improvements calculated in accordance with the annual energy ratios in Table 6-3, the SERI airfoils produce more energy because of easier startup of the turbine and longer run times in low winds. These are additional benefits of the lower sensitivity of the SERI airfoils to roughness. At a site with a low annual average wind speed, improvements in annual energy output as high as 40 percent have been projected.

Tip Shapes of HAWT Blades

One of the many distinctions between HAWT and VAWT blades is that the former have a tip whose effects on lift and drag forces must be taken into account. Two fundamental fluid-dynamic effects occur here. One is the purely *inviscid* effect of the termination of the lifting *vorticity*. This results in a shedding of vorticity near the tip which causes a streamwise vorticity sheet of high intensity in the vicinity of the tip. This inviscid effect modifies the induced *downwash field* in this region. The second effect is a viscous one associated with the highly three-dimensional flow around the tip. Normally, the inviscid and viscous effects interact, with the trailing vortex system inducing flow around the tip while flow separation near the tip (resulting from blade separation) may change the tip air loading and hence affect the trailing vorticity.

These effects in planar wings have been studied extensively. Many wing-tip shapes have been tested in attempts to improve aerodynamic performance, and advantages have been claimed for a number of these. Studies of tip shapes for HAWT rotors have also been made, dating back at least to those that resulted in the semi-circular tapered tips on the

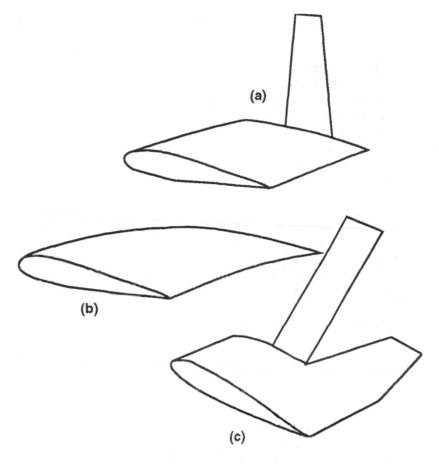

Figure 6-13. Experimental tip devices for improving HAWT rotor performance. (a) Single winglet (b) Shark's fin (c) Double winglet   [Gyatt and Lissaman 1985]

pioneering *Gedser HAWT* (Fig. 3-1). An early theoretical study [VanKuik 1986] indicates that tip effects are much more pronounced on wind turbine rotors than on airplane wings and suggests that modifications to HAWT performance models are required.

Gyatt and Lissaman [1985] report field tests on a number of tip shapes intended to improve performance by controlling the shedding of the tip vortex. The test configurations are shown in Figure 6-13 and include one planar swept-back tip and two non-planar tip geometries. These shapes were selected based their promising performance on planar wings, as determined by wind tunnel tests, flight tests, and analytical predictions.

The leading edge of the single winglet, Figure 6-13(a), was positioned normal to the chord plane of the blade on the low-pressure (downwind) side. The "shark's fin," Figure 6-13(b), is a planar tip with a strongly curved leading edge and a swept-back trailing edge. The double winglet, Figure 6-13(c), has two inclined surfaces, with the forward element mounted approximately in a plane at 45 deg to the plane of the blade and the aft element in the plane of the blade. As in the case of the single winglet, the double winglets extended above the low-pressure surface of the main blade.

Comparative performance tests, with and without these tip devices, were conducted on a two-bladed commercial HAWT, 10 m in diameter. Each of the tip devices replaced a length of blade equal to approximately 5 percent of the rotor radius, so the swept area remained constant during these tests. Modest performance improvements of the order of 3 percent were predicted for these tips, using the best available non-planar aerodynamic model. However, a small but significant reduction in power of about 3 percent to 6 percent was measured. While the theoretical improvement may still be achievable by proper adjustment of the non-planar elements, these tip devices do not appear promising for significantly increasing the power of a HAWT rotor.

Figure 6-14 illustrates a tip shape that has been found to be more desirable for HAWT blades [Tangler 1993]. Its characteristics are

- -- *rounded leading-edge corner*, to eliminate the start of separation caused by large, three-dimensional adverse pressure gradients at a sharp corner;
- -- *finite tip chord* (approx. $c/2$), to maintain a good Reynolds number and lift-to-drag ratio at the tip;
- -- *straight trailing edge*, to provide a pitching moment that is dynamically stable (*i.e.* nose-down).

This tip shape has been used successfully on SERI airfoil blades (Fig. 6-10(b)), both thin and thick.

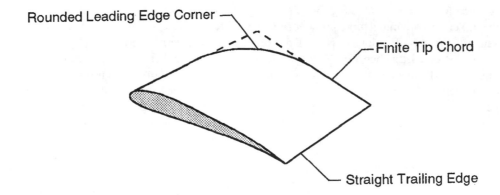

Figure 6-14. Aerodynamically desirable tip shape for a HAWT blade. [Tangler 1993]

Wind Turbine Wake Effects

Modern *wind power stations* consist of numbers of individual turbines arranged on a given site so as to best utilize the local wind energy. This appears to call for the placement of most of the units at the locations of strongest flow. However, such concentrations of turbines will cause shielding of neighboring units, so that downwind turbines are exposed to a lower wind speed. Clearly, the station designer is seeking an optimal situation, in which the most energetic regions of the site are exploited without crowding these desirable areas with so many units that *array interference* (or *wake interference*) prevents them from achieving the best energy capture. It has also been suggested that, in complex terrain, turbines on upwind slopes or near crests of hills and ridges may actually precipitate separation in the lee flow, significantly lowering the surface wind energy on the downwind slope and causing even larger wake deficits.

The extent of the loss due to array interference can be a significant factor in the economics of a wind power station. A representative picture of the loss situation for a 6x6 array of wind turbines is shown in Figure 6-15 [Lissaman and Zalay 1982]. In this hypothetical example, 36 turbines are arranged in a rectangular grid pattern, spaced 10 diameters apart in the direction of the prevailing wind and at various crosswind distances. An important parameter in determining the severity of wake effects on downwind machines is the *ambient turbulence intensity* of the wind, σ_0/U, which will be discussed later. Figure 6-15 gives the total array energy loss as a function of turbulence and crosswind spacing,

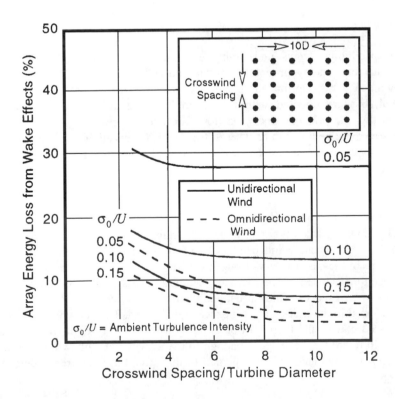

Figure 6-15. General effect of spacing and ambient turbulence intensity on the wake energy losses in an array of wind turbines on flat ground. [Lissaman and Zalay 1982]

as calculated for two wind regimes: all winds from the prevailing direction, and wind flow equally distributed in all directions.

A typical case, $\sigma_0/U = 0.15$ and crosswind spacing of 3 diameters, gives a wake loss estimate of 10 percent to 12 percent for this idealized array. As shown in Figure 4-26, extensive field experience in California wind power stations indicates that array losses are typically about 10 percent, but can vary from 2 percent to almost 30 percent depending on the terrain, the concentration of turbines, and the wind turbulence.

Another factor is introduced by array interference which can be as significant as the direct energy loss, although it is less quantifiable. This is the extra turbulence generated by upwind rotors, which will increase the turbulence to which downwind units are exposed [*e.g.*, Kelley 1989]. This so-called *generated turbulence* can affect the operation of a downwind turbine by reducing its fatigue life, increasing the probability of catastrophic failure caused by strong gusts, and by exciting blade vibrations and unfavorable control system responses. This is likely to increase the operating and maintenance costs of these machines and possibly cause premature wear-out and/or failure.

Physical Factors Controlling Wake Interference

It is clear that the dominant parameters are the *downwind distance* between units (usually defined in terms of turbine diameters and normally between *6D* and *12D*), the amount of *power extracted* from the wind stream by the turbine units (defined in terms of the power coefficient and normally about 0.40, maximum) and the *turbulence* in the wind stream (both ambient and generated). The qualitative effects of these parameters are described below.

Downwind Spacing

Close spacing is always undesirable, but this must be viewed in the context of increasing the total energy production of the site and utilizing the directional nature of seasonal wind patterns. Optimized spacing may call for arrays that are not orthogonally symmetrical, but are oriented with respect to both prevailing and local wind conditions. Typically, the principal energetic flows are from a prevailing direction, so that an average crosswind direction can be defined from which energetic flows are quite rare. This permits closer spacing crosswind than downwind, as pictured in Figure 4-21.

Power Extracted

The higher the output of a turbine, expressed in terms of its power coefficient, the greater is the downstream wake interference. Interference, as a percent of potential energy production, is approximately independent of wind speed, provided the wind is not very strong or light. In the case of very strong winds (above the turbine's rated speed), upwind turbines operating at rated power will be shedding excess wind power by stall or pitch control. This in turn reduces rotor thrust and the *axial induction factor* (see Chapter 5), so the retarded speed in the wake may still exceed rated speed. Thus downwind units will also operate at rated power, and there will be no array loss. For very light winds, modest axial inductions will retard the wake speed enough so that downwind units will experience winds below the *cut-in speed* and will not operate. In this case, the interference percentage will be very large.

Turbulence Intensity

Turbulence intensity is a measure of the relative unsteadiness of the wind, defined as follows:

$$\sigma/U = \frac{\sigma_0 + \sigma_G}{U} \tag{6-7a}$$

$$\sigma_0 = RMS(u - U) \tag{6-7b}$$

where

σ/U = turbulence intensity
σ_0 = ambient turbulence (m/s)
σ_G = turbulence generated by the rotor (m/s)
$RMS(\ )$ = root-mean-square of ( ) during a given time interval (typically 6 to 12 min)
u = instantaneous free-stream wind speed (m/s)
U = steady (mean) free-stream wind speed during the time interval (m/s)

Ambient turbulence is normally about 0.12 U. Generated turbulence within the first few diameters of distance downstream of a turbine is normally about 0.08 U.

Turbulence causes two opposing effects on the amount of wake interference. It tends to increase *entrainment* of air from the free stream surrounding the wake, thus re-energizing the wake and reducing the *velocity deficit*. At the same time turbulence increases the wake diameter, which causes a given turbine to affect more downwind units. Thus, turbulence spreads the energy loss over a wider area. As shown in Figure 6-15, array energy losses generally decrease with increasing turbulence intensity, indicating that the positive re-energizing effect is dominant. At the same time, however, fatigue loads increase with increasing turbulence.

Atmospheric Stability

Recent research has shown that *atmospheric stability* (see Chapter 8) may be a major parameter in the determination of wake size and structure [Kelley 1994]. Atmospheric stability controls the size of *eddies* within the general wind flow, and therefore the rate of entrainment of air from the free stream and diffusion of turbulence in the wake.

Effect of Turbine Configuration

The description presented so far of wake effects indicates that the interference of a given wind turbine is essentially a consequence of its global or external fluid mechanics and is thus not strongly dependent on the details of the turbine design. This can also be inferred by noting that the principal wake effects continue to be experienced 10 or more diameters downwind of a rotor. At this distance the rotor wake has started to develop the general structure common to all deficit wakes. Although there do appear to be some small distinctions (*e.g.*, in the amount of *generated turbulence*), we need not distinguish between HAWTs and VAWTs. Instead, we can discuss wake interference as a general phenomenon.

Development of Wind Turbine Wake Models

The earliest work on modeling wake interference appears to be by Templin [1974]. This was followed by a paper by Craaford [1975], who handled the subject by considering the effect of the turbines as equivalent to a *ground surface roughness* and estimating the flow retardation in the boundary layer caused by this effect. This approach gives a powerful overview with interesting global results, but it does not provide enough detail of the wake structure for use in the practical design of wind power stations.

There are at least three general types of wake models, which are described in detail by Luken and Vermeulen [1986]. The first is the *semi-empirical type* [Lissaman and Bate 1976]. Here a simple analytical relation is used to establish an invariant for the fundamental wake scales of speed and radius. Experimental data are used for modeling the flow profiles in various regions of the wake, and an empirical wake growth rate is used that depends on turbulence. The second type of model [Ainslie 1985] utilizes an axisymmetric boundary-layer *approximation of the Navier-Stokes equations*, with an eddy viscosity model to determine shear stress. Finally, a *full mixing length/eddy viscosity* (K,E) model [Crespo *et al.* 1985] is an advanced type that takes into account surface roughness as well as atmospheric stability. While this last procedure is valuable for analytical research, it is too complicated for design purposes. Various approximations are employed in all three types of models to reduce computer demands.

At the other end of the spectrum from the K,E model is an extremely simple analysis developed by Katic *et al.* [1986]. In spite of its simplicity, wind tunnel tests of a cluster of 37 model turbines indicate that its predictions are as accurate as those made with more-complex models. As noted by these authors, good correlation between model predictions and array test data is a consequence of satisfying the *conservation of wake momentum*, which is the dominant invariant. Overall, a semi-empirical model contains practical manifestations of all the significant fluid mechanic phenomena. Therefore, the basic principles of wake modeling will be discussed here using the semi-empirical approach. Wake models developed after 1994 are not covered.

Fluid-Dynamic Principles

Considering the turbine to be an *actuator disk*, we note that it essentially removes *total head* from the wind stream in the form of a pressure drop across the rotor disk. Because there is no means of establishing the strong flow curvature required to maintain a pressure difference between the wake and the surrounding free stream, the flow returns very rapidly to the ambient static pressure. As a consequence, the loss of head is manifested in a reduction in speed a short distance downwind of the disk. This is the so-called *velocity deficit*. Standard turbine theory operates under the implied assumption that this deficit extends unchanged from the disk to infinitely far downwind, which is known as the *Trefftz plane concept* (Fig. 5-7). This concept is not adequate for determining flow details in the wake itself, so we will construct a scheme for modeling the wake development downwind.

First, we define the wake as that portion of the flow that has a total head different from the free stream. Because the wake velocity deficit exists within the free stream, there is the characteristic axisymmetric *shear layer* between them, with turbulent transfer tending to reduce velocity differences across the layer. As a consequence, the more-energetic free stream is entrained into the wake. This increases both the mass flow and the total energy in the wake. The net result is an asymptotic return of the wake flow to the undisturbed conditions. However, since the flow in the wake just downwind of the rotor was of lower energy, this return to the free stream state can occur only when the mass of air that passed through the disk has been infinitely diluted, or when the wake has grown to an infinite diameter with a negligible velocity deficit. It is important to establish an *invariant* for this process, noting that neither mass nor energy flux is conserved in the wake.

If one considers the situation of a single turbine in a uniform flow on flat terrain, it can be seen that the thrust of the rotor must be manifest in the wake, assuming that the turbulent skin friction on the ground plane is not greatly affected by the wake flow. The thrust is equal to the rate of change of the momentum deficit in the wake, or

$$ T = \frac{d}{dt} \left[2\pi\rho \int_0^\infty (U - V) dx \, r \, dr \right] = 2\pi\rho \int_0^{R_W} (U - V) V \, r \, dr \tag{6-8} $$

where

T = rotor thrust force (N)
r, x = radial and downwind coordinates measured from the rotor center (m)
U = free stream wind speed at the elevation of the rotor center (m/s)
V = local velocity in the wake; function of the radial coordinate r (m/s)
R_W = effective radius of the wake (m)

Equation (6-8) can be recognized as the general form of Equation (5-9). Any effect of the ground in creating a non-axisymmetric wake has been neglected here and will be accounted for later. Equation (6-8) provides the basic conservation law for the wake development. We now require a "shape" law to define the velocity profile $V = V(r)$ and a "scale" law that will define the effective wake radius, R_W.

Wake Geometry Models

It has been observed that wind turbine wakes develop according to several fairly-well defined regimes at different downwind distances, and these can be idealized as shown in Figure 6-16. First, there is the flow emerging from the rotor disk itself at section *A*. For well-designed blades, air loading produces an almost uniform velocity deficit, removing the same energy at all radii. The wake velocity is assumed to be constant across this section and equal to $(1 - a)U$, where, as before, a is the axial induction factor. The initial wake is considered to behave like an inviscid slug of uniformly-reduced velocity submerged in an outer flow, forming a so-called *co-flowing jet* or *potential core*.

Proceeding downwind from the rotor to section *B*, the velocity deficit profile is "top hat" shaped, and an intense shear layer is developed, attenuating the velocity discontinuity. It is assumed that this shear layer extends itself outward and inward until the inviscid core region is eliminated at section *C*. The wake from *A* to *C* is defined as being in the *initial potential core regime*.

Farther downwind at section *D*, it is assumed that the wake has adopted its asymptotic "bell-shaped" profile, like the radial velocity distribution far downstream of a body of revolution. This is the beginning of the *fully-developed wake regime*. The profile of the wake at *C*, where the shear layer has penetrated to the flow center line, is not the same as the far-wake profile at *D*, so it is necessary to assume that there is a *transitional regime* of a certain streamwise extent that joins these two sections. It now remains to establish simple analytical models for intermediate wake profiles, between the rotor and far downwind, and the transitions from one to another [Lissaman 1976, 1979].

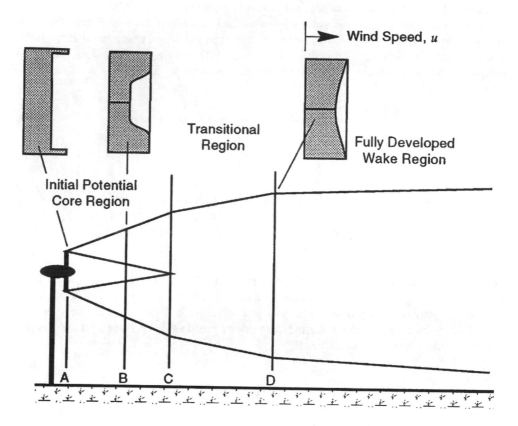

Figure 6-16. Regimes of wake development and associated radial profiles of the velocity deficit.

Initial Potential Core Regime

In this regime, it is assumed that the shear layer, having a profile given by a standard radially-symmetric shear layer model, develops according to the normal shear law where the growth rate of a layer thickness is proportional to the speed difference across it. This will define a profile shape for any downwind station in the regime. The basic conservation of drag is then maintained by selecting the effective radius in Equation (6-8), R_w, to conserve the momentum deficit in the wake. Axisymmetric shear layers like this are shown in Abramovitch [1963] and referenced in Lissaman [1976].

Fully-Developed Wake Regime

A suitable model for the flow profile in the far wake, well-supported by experimental results, is also provided by Abramovitch [1963]. This profile is expressed as a function of radial distance, r, normalized by a radial scale factor. To establish a growth law for the width of the wake in this regime, we assume that the effective wake radius develops temporally and spatially according to the following laws:

$$\frac{dR_W}{dt} = k_W(\sigma_0 + \sigma_G) \tag{6-9a}$$

$$\frac{dR_W}{dx} = k_W \frac{\sigma_0 + \sigma_G}{U} \tag{6-9b}$$

where k_w = empirical constant

This model produces a turbulent energy level which increases from free stream to a higher level upon passage through the rotor disk, is reinforced by the shear layers immediately downwind of the disk, and then decays to return to the ambient level. Details are given in [Lissaman 1976].

Transitional Regime

The transition law describing flow from the inviscid core to the far wake is based simply on establishing a geometric relationship that will provide a smooth transition from the profile at section C to that at section D in Figure 6-16.

Ground Effects

A refinement of Equations (6-9) takes into account differences between vertical and lateral ambient turbulence near the ground and accounts for differences in vertical and lateral growth rates. This causes the wake to adopt an elliptical cross-section downwind of its original circular shape. Another ground effect is taken into account by the standard method of *reflection*, so that the actual wake perturbation at a given downwind station now becomes the sum of that in the direct flow and that in the image flow. If the perturbations are linearized, then the thrust invariant is now associated with the thrust of the rotor plus its image. Thus, conservation of the integral for a single wake automatically takes into account the ground effect by the presence of the image perturbation. In other words, the momentum defect "removed" by the ground is "restored" by the addition to the actual wake of the ground perturbation flow of the image.

Integration of Wakes for Array Effect

In most models it is assumed that the wakes of an array of turbines may be directly superimposed. This is a linearizing assumption, based on the physical fact that the wake perturbations caused by a single turbine are relatively small. Normally, velocity deficits are less than 5 percent of the free-stream velocity by a distance of five diameters downwind of a rotor. Thus it is a good approximation to disregard any interaction between wakes.

The normal procedure for calculating the wake interference for a given array of turbines is therefore straightforward and as follows: For the given wind azimuth the most-upwind unit is selected, and its wake geometry and velocity deficits are calculated for a specified wind speed and turbulence intensity, progressing downwind through the array. The deficit at each turbine is tabulated. Turbine control parameters (such as cut-in and rated wind speeds) may be introduced into the model, as well as different rotor areas and elevations. Then the most-upwind of the remaining turbines is selected and its inflow velocity determined. In general, this will be the vector sum of the free-stream flow and the wake velocity deficit of the leading upwind turbine. The development of the wake of the second unit is then calculated and its velocity deficits at the locations of all other units determined. These are tabulated with the wake deficits from the most-upwind turbine.

This procedure is repeated until the most-downwind turbine has been reached, and results in the power output of the array for a given combination of wind azimuth, speed, and turbulence. The calculation must be repeated for differing azimuths (to account for the annual distribution of wind directions), wind speeds (to account for the annual wind speed histogram and turbine control characteristics), and ambient turbulence levels (to account for the varying wake expansion). Conceptually, this procedure provides the performance of all the turbines in the entire array viewed as a single wind power system. Output power of the array will be defined as a function of wind speed and turbulence, as it is for a single turbine. Unlike a single turbine, array output power is also a function of wind direction and speed variations across the site, taking into account the array geometry and terrain features.

Repeating this process for each wind azimuth, speed, and turbulence level is a formidable (but not complicated) computational task. The problem of completing the computation for an array within a reasonable time indicates the merits of a simple, linearized physical wake model like the one described. It is believed that models with more fundamental fluid-dynamic features, using more complex rational turbulence models, and employing finite difference techniques would be prohibitively complex for analyzing an array of practical size. Such models can, however, be used to validate the simple wake model for the case of a single turbine and assist in determining any empirical constants.

Analytical Results Obtained with Wake Models

Flat Terrain

From the numerous wake-interference calculations which have been made for the case of a uniform flow over flat terrain, it has been found that the level of ambient turbulence has a very large effect on the array energy loss. Typical of the analytical results for simple arrays on flat terrain are those shown in Figure 6-15. In the course of numerical investigations of this type, it has been determined that the wake structure within its first four of five diameters of downwind length does not have a strong effect on its velocity deficit in regions further downwind, where turbines are likely to be located. Thus, the modelling of the initial potential core regime (Fig. 6-16) does not need to be very refined, provided it gives the proper initial state for the fully-developed wake regime.

In the design of a wind power station the normal method is to start with a reasonable layout for the array, maximizing turbine spacing in the principal flow direction at the expense of closer spacing in the crosswind direction. This assumes that the case of wind at right angles to the prevailing wind direction, although involving large losses, does not occur frequently. Next, the initial layout is examined using the wake interference model to identify turbines that have particularly poor production because they are in sheltered positions and to find more favorable locations for these units. Spacing distances are then perturbed until the maximum (or near-maximum) annual energy production is determined. Thus the array design is approximately optimized by an iterative process.

Some effort has been devoted to defining a theoretically-exact, optimal arrangement for simple arrays on flat terrain. Such arrays could not normally be used directly in design, since each site has its own specific distributions of wind speed and direction. However, the value of optimal-spacing studies lies in determining the sensitivity of array energy output to changes in spacing from the theoretical optimum.

Two examples will illustrate the investigation of this sensitivity to spacing. In the first, turbines are placed in a single windwise row, with the constraint that a given number of units has to be placed on a strip of a given length. This is sufficiently simple that a *variational process* can be used to determine an ideal spacing. The optimal distribution is symmetrical about the center of the strip, with the spacing somewhat larger for the units near the center [Lissaman *et al.* 1982]. This implies that increasing the spacing in the center serves to provide a region of re-energization for the flow, a conclusion supported by Kaminsky *et al.* [1987]. The second example is a square array with a constant wind occurring with equal frequency from all directions. An optimal-spacing study of this case also indicates that it is slightly advantageous to have larger spacing in the interior region, with a heavier concentration of the turbines near the periphery of the square.

The most important conclusion from these optimal spacing studies of simple arrays is that uniform spacing gives an array energy output that is only insignificantly less than that for an optimum spacing, for normal spacings in the range of 6 to 10 diameters. Moreover, this energy difference is smaller than the implied accuracy associated with the approximations in the mathematical models. This provides the valuable practical result that a uniform spacing, at least in the prevailing wind direction, is an excellent initial arrangement which can then be optimized to account for the actual site features and wind characteristics.

Complex Terrain

In the case of a non-uniform wind field caused by complex terrain it is not possible to use the basic invariant of the momentum deficit, which is connected to the thrust of the turbine. In complex terrain the thrust in a plane downwind of the turbine is manifested in pressure perturbations on the sloping ground as well as in the velocity profile in the wake. Thus, the momentum deficit in the wake is not conserved for non-uniform flows. To handle this case we will use the concept of *invariance of the total energy* in the flow after dissipative losses due to entrainment have been taken into account [Lissaman *et al.* 1986].

If there were no losses, as the flow moves into areas of different pressure the wake speed would change in accordance with the *Bernoulli equation* connecting fluid speed and pressure. However, entrainment causes dissipation losses in total head, so that speed cannot be connected to pressure alone. As a first approximation, it can be assumed that dissipation (which is a function of shear gradients and flow distortion) is the same for non-uniform and uniform flows. Thus, our modeling approach for varying ambient pressure is to take the wake field which would exist over flat terrain, calculate the dissipation for this constant-pressure case, and then apply this dissipation to the wake flow speed at each turbine location on complex terrain, the varying pressure case.

For example, consider identical turbines at the two sites shown in Figure 6-17, where site (a) is flat and site (b) has complex terrain. The wake velocities can be expressed as

$$V_f = U_f(1 - d_f) \qquad\qquad\qquad (6\text{-}10a)$$

$$V_c = U_c(1 - d_c) \qquad\qquad\qquad (6\text{-}10b)$$

where

$\qquad\qquad V_f$ = wake speed at a given location over flat terrain (m/s)
$\qquad\qquad V_c$ = wake speed at the same location over complex terrain (m/s)
$\qquad U_f, U_c$ = ambient wind speeds for flat and complex terrain, respectively (m/s)
$\qquad\qquad d_f$ = known wake deficit factor, calculated using flat terrain model
$\qquad\qquad d_c$ = unknown wake deficit factor for complex terrain

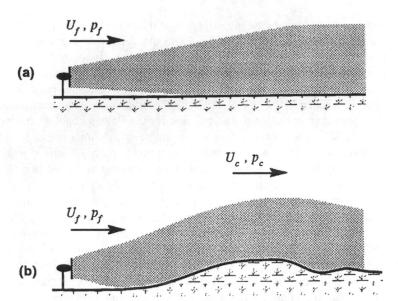

Figure 6-17. Comparison of wake flow conditions over flat and complex terrain.
(a) Flat terrain (b) Complex terrain, where the wind speed and static pressure outside the wake are changed by a hill

The total head at the same downwind distance in each site can be written as

$$h_f = p_f + 0.5\rho U_f^2 (1 - d_f)^2 \qquad\qquad (6\text{-}11a)$$

$$h_c = p_c + 0.5\rho U_c^2 (1 - d_c)^2 \qquad\qquad (6\text{-}11b)$$

where

h_f, h_c = total head in the wake, for flat and complex terrain, respectively (N/m^2)
p_f, p_c = local static pressure, for flat and complex terrain, respectively (N/m^2)

Now, outside the wake, the wind speeds and static pressures are coupled by the nondissipative Bernoulli equation to give

$$p_f + 0.5\rho U_f^2 = p_c + 0.5\rho U_c^2 \qquad\qquad (6\text{-}12)$$

If we assume that the dissipation at a given downwind distance is the same for both the flat and complex terrains, then the total head in each wake is also the same, and $h_c = h_f$. This provides a simple result:

$$U_c^2(-2d_c + d_c^2) = U_f^2 (-2d_f + d_f^2)$$

which, for small values of d_c and d_f, linearizes to

$$d_c = d_f \left(\frac{U_f}{U_c} \right)^2 \qquad\qquad (6\text{-}13)$$

This is an approximation that converges properly for uniform flow with dissipation as well as for non-uniform flow without dissipation, and thus it may be expected to be a reasonable model for the case of non-uniform flow with dissipation.

Measurement of Wake Effects

Wind Tunnel Wake Tests

Wind tunnel tests have been conducted to measure wake deficits behind single turbines. These have established the strong dependence of the wake deficit on the tunnel turbulence, as predicted by the theory expressed in Equations (6-9). They have also shown reasonable correlation with the semi-empirical model in decay law (the functional dependence of deficit with downwind distance) and the actual rate of the decay (the magnitude of the decay constant). A typical example of measured deficits compared with model predictions is shown in Figure 6-18. Both the data and the model indicate that the centerline velocity deficit $(U - V_{min})$ decays at a rate inversely proportional to the downwind distance x. Turbulence intensities in wind tunnels are generally low, even when they are artificially increased by the use of wire grids and boundary layer devices, as was done during these tests. However, it can be seen that even for these low turbulences an increase in the turbulence intensity of only 0.01 significantly increases the wake decay rate.

Because of the difficulty of reproducing the actual nature and scales of the ambient turbulence and other full-scale factors relating to the turbine rotor, the prospects of validating an analytical wake model in a wind tunnel to a higher degree than that shown in Figure 6-18 are quite low.

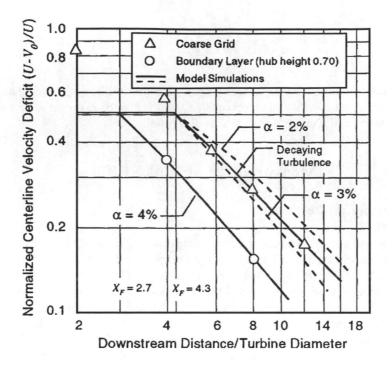

Figure 6-18. Comparison of wind tunnel test data and model predictions of the decay in wake velocity deficit.

Field Tests on a Single Turbine

Comprehensive field tests of the individual factors that cause array interference are extremely difficult to conduct. It is necessary, in principle, to check the different effects separately. These include at least the wake development (velocity deficit, wake radius, and turbulence) downwind of a single turbine as well as the performance of a downwind unit while experiencing wake flow. Further refinements in field testing would include measurements on the wake of one turbine operating within the wake of another. Certain fundamental features of the nature of wind turbine wakes have been established by full-scale field tests. The results of measurements by Faxen [1978] of the velocity profile behind an operating HAWT are shown in Figure 6-19 and confirm the expected, approximately *Gaussian* profile of the deficit in the far wake.

We can extract some additional information about HAWT rotor wakes from the data in Figure 6-19 if we integrate the deficit volumes under the two profiles and normalize the results. Table 6-4 lists various parameters obtained by this procedure. We note that the ratios of deficit-to-wind speed are not constant for the two test cases, but increase with increasing power. The normalized parameters that are found to be approximately equal for the two cases are the average-to-centerline deficit ratios (0.28, compared to 0.22 for an ideal Gaussian distribution), and the deficit-to-power density ratios. The latter are approximately 0.042 (m/s)/(W/m^2) for the average deficit and 0.0142 (m/s)/(W/m^2) for the maximum.

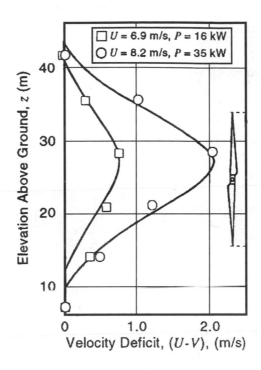

Figure 6-19.   Field measurements of the velocity profile downwind of an operating medium-scale HAWT. [Faxen 1978]

Field Tests on Multiple Turbines

Wind speed measurements have been made in the wake of a Mod-2 HAWT (Fig. 3-38) using instruments attached to tethered balloons [Zambrano *et al.* 1982]. Unfortunately, the results of these tests are too scattered to provide definitive corroboration of wake models. To overcome problems with data scatter, the absence of steady-state conditions, and the expected power differences between widely-spaced turbines, Neustadter and Spera [1984] developed an energy method for measuring wake interference losses. In this procedure, the energy output of a test turbine, with and without wake effect, is normalized by the energy output of a nearby control turbine during the same test periods. This corrects for changes in the free-stream wind speed prior to calculating interference losses.

The required arrangement of wind turbines is illustrated in Figure 6-20, using the 3-unit Mod-2 HAWT cluster as an example. The test and control turbines (Units 1 and 3, respectively) run continuously while the wake-producing turbine (Unit 2) is operated intermittently. Test data are the energy production in 10-min segments for the three turbines over a cumulative operating time of 27 hours. Only the time segments during which the wake envelops the test turbine (wind azimuths from 250 to 280 deg in this case) are used. For the 7-diameter downwind spacing of these tests, the energy deficit was found to be 10.4 percent, which is compatible with model predictions. The equivalent average wind velocity deficit was only 0.3 m/s, which emphasizes the difficulties of measuring instantaneous power and wind speed and the advantages of measuring energy output.

Table 6-4. Wake Deficit Parameters Derived from HAWT Field Test Profiles
[data from Faxen 1978]

Parameter	Test No. 1	Test No. 2
Turbine:		
Free-stream wind speed (m/s)	6.9	8.2
Power output (kW)	16	35
Power density [1] (W/m^2)	60	132
Wake:		
Wake-to-rotor area ratio	≈ 4.0	≈ 4.0
Centerline velocity deficit (m/s)	0.84	2.09
Average velocity deficit (m/s)	0.23	0.58
Normalized Wake:		
Deficit-to-free stream wind speed ratios		
Centerline	0.122	0.255
Average	0.034	0.071
Average-to-centerline deficit ratio	0.28	0.28
Deficit-to-power density ratios		
Centerline (m/s)/(W/m^2)	0.0139	0.0155
Average (m/s)/(W/m^2)	0.0039	0.0044

[1]Rotor swept area = 270 m^3

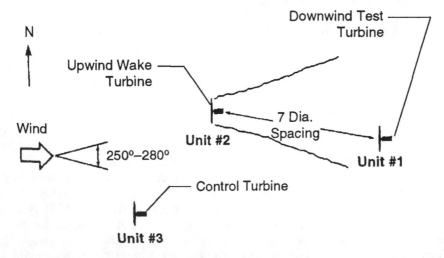

Figure 6-20. Schematic plan view of the Mod-2 HAWT test site near Goldendale, Washington, during wake effect tests using an energy method. Unit 2 is operated intermittently to change the energy output of Unit 1. Energy output of Unit 3 is used as a control, to eliminate wind variability effects. [Neustadter and Spera 1984]

Summary Comments on Wind Turbine Wake Effects

As the demand for wind turbine installations increases and the availability of large sites with high energy flow decreases, it will be necessary to employ arrays of higher density in the good wind areas as well as to install arrays on sites with lower wind speeds. In both cases, an understanding of array effects is critical to maximizing energy production. Since 1976 there has been an extensive research effort devoted to predicting wake interference effects. For uniform flows corresponding to flat terrain, the numerical modeling has reached an adequate state, particularly with semi-empirical models in which constants may be adjusted to conform with the accumulating data base.

For non-uniform flows, however, the situation is still relatively unquantified. Reliable wake-effect data are still needed from commercial wind power stations located in complex terrain. It is noted that modeling the array interference in this case must be based on an adequate modeling of the wind flow itself over complex terrain, and this topic is generally considered to be incompletely understood. One must have an acceptable wind flow model without wind turbines before one can reliably superimpose the turbine array effects. Simple approximate models for wake interference are probably as accurate as the complex-terrain flow models now available. Refinements are required for both, in order to produce an acceptable operational model for the micrositing of wind turbines.

References

Ainslie, J. F., 1985, "Development of an Eddy-Viscosity Model for Wind Turbine Wakes," *Proceedings, 7th BWEA Conference*, British Wind Energy Association, London: Multi-Science Publishing Co.

Buhl, M. L., Jr., A. D. Wright, and J. L. Tangler, 1998, *Wind Turbine Design Codes: A Preliminary Comparison of the Aerodynamics*, NREL/CP 23975, Golden, Colorado: National Renewable Energy Laboratory.

Craaford, J., 1975, (personal communication).

Crespo A., F. Manuel, D. Moreno, E. Fraga, and J. Hernandez, 1985, "Numerical Analysis of Wind Turbine Wakes," *Proceedings, Workshop on Wind Energy Application*, Delphi, Greece.

Eppler, R., and K. M. Somers, 1980, *A Computer Program for the Design and Analysis of Low-Speed Airfoils*, NASA TM-80210, Hampton, Virginia: NASA Langley Research Center.

Faxen, T., 1978, "Wake Interaction in an Array of Windmills," *Proceedings, 2nd International Symposium on Wind Energy Systems*, Amsterdam.

Gyatt, G. W., and P. B. S. Lissaman, 1985, *Development and Testing of Tip Devices for Horizontal Axis Wind Turbines*, NASA CR 174991, Cleveland, Ohio: NASA Lewis Research Center.

Hibbs, B. D., 1986, *HAWT Performance with Dynamic Stall*, SERI/STR-217-2732, Golden, Colorado: National Renewable Energy Laboratory.

Jackson, K. L., and P. G. Migliore, 1987, "Design of Wind Turbine Blades Employing Advanced Airfoils," *Proceedings, Wind Power '87 Conference*, SERI/CP-217-3315, Washington, DC: American Wind Energy Association, pp. 106-111.

Jacobs, E. M., and I. H. Abbot, 1932, *The NACA Variable-Density Wind Tunnel*, NACA TR 416, Hampton, Virginia: NASA Langley Research Center.

Katic, I., J. Hojstrup, and N. O. Jensen, 1986, "A Simple Model for Cluster Efficiency," *Proceedings, EWEC '86 Conference*, Rome: European Wind Energy Committee.

Klimas, P. C, 1985, *Tailored Airfoils for Vertical Axis Wind Turbines*, SAND 84-0941, Albuquerque, New Mexico: Sandia National Laboratories.

Lissaman, P. B. S., 1976, *Energy Effectiveness of Arrays of Wind Energy Collection Systems*, Report No. AV-R-6110, Monrovia, CA: AeroVironment, Inc.

Lissaman, P. B. S., A. Zalay, G.W. Gyatt, 1982, "Critical Issues in the Design and Assessment of Wind Turbine Arrays," *Proceedings, 4th International Symposium on Wind Energy Systems*, ISBN 0 906085 772, Stockholm.

Lissaman, P. B. S., D. R. Foster, and B. D. Hibbs, 1986, *Operational Model for the Design of Optimum Wind Farm Arrays*, Report No. AV-FR-86/837, Monrovia, CA: AeroVironment, Inc.

Luken, E., and P. E. J. Vermuelen, 1986, "Development of Advanced Mathematical Models for the Calculation of Wind Turbine Wake-Interaction Effects," *Proceedings, European Wind Energy Conference '86*, Rome.

Lundsager, P., S. Frandsen, and C. J. Christensen, 1980, *Analysis of Data from the Gedser Wind Turbine, 1977-1979*, Risø-M-2242, Roskilde, Denmark: Risø National Laboratory Station for Wind Turbines.

Kelley, N. D., 1989, *An Initial Look at the Dynamics of the Microscale Flow Field within a Large Wind Farm in Response to Variations in the Natural Inflow*, SERI/TP-257-3591, Golden, Colorado: National Renewable Energy Laboratory.

Kelley, N. D., 1994, *The Identification of Inflow Fluid Dynamics Parameters That Can Be Used To Scale Fatigue Loading Spectra of Wind Turbine Structural Components*, NREL/TP-442-6008, Golden, Colorado: National Renewable Energy Laboratory.

Miley, J., 1980, *A Catalog of Low Reynolds Number Airfoil Data for Wind Turbine Applications*.

Neustadter, H. E., and D. A. Spera, August 1985, "Method for Evaluating Wind Turbine Wake Effects on Wind Farm Performance," *Journal of Solar Energy Engineering*, Vol. 107: pp. 240-243.

Ostowari, C., and D. Naik, 1984, *Post-Stall Wind Turbine Studies of Varying Aspect Ratio Wind Tunnel Blades with NACA 44XX Series Airfoil Sections*, Golden, Colorado: National Renewable Energy Laboratory.

Silverstein, A., 1934, *Scale Effect on Clark Y Airfoil Characteristics from N.A.C.A. Full-Scale Wind Tunnel Tests*, NACA Technical Report No. 502, Hampton, Virginia: NASA Langley Research Center.

Spera, D. A., 2008, *Models of Lift and Drag Coefficients of Stalled and Unstalled Airfoils in Wind Turbines and Wind Tunnels*, NASA-CR-215434, Cleveland, Ohio: NASA Glenn Research Center.

Spera, D. A., and D. C. Janetzke, 1981, "Performance and Load Data from Mod-0A and Mod-1 Wind Turbine Generators," *Proceedings, Workshop on Large Horizontal-Axis Wind Turbines*, NASA CP-2230, DOE Publication CONF-810752, Cleveland, OH: NASA Lewis Research Center, pp. 447-467.

Tangler, J. L., and D. M. Somers, 1985, "Advanced Airfoils for HAWTS," *Proceedings, Wind Power '85 Conference*, SERI/CP-217-2902, Washington, DC: American Wind Energy Association, pp. 45-51.

Tangler, J. L., and D. M. Somers, 1986, "A Low Reynolds Number Airfoil Family for Horizontal Axis Wind Turbines," *Proceedings, International Conference on Aerodynamics at Low Reynolds Numbers*, London.

Tangler, J. L., 1987, *Status of Special-Purpose Airfoil Families*, SERI/TP-217-3264, Golden, Colorado: National Renewable Energy Laboratory.

Tangler, J. L., 1993, (personal communication), Golden, Colorado: National Renewable Energy Laboratory.

VanKuik, G. A. M., 1986, "The Physics and Mathematical Description of the Achilles Heel of Stationary Wind Turbine Aerodynamics: The Tip Flow Process," *Proceedings, European Wind Energy Conference '86*, Rome.

Viterna, L. A., and R. D. Corrigan, 1981, "Fixed Pitch Rotor Performance of Large Horizontal Axis Wind Turbines," *Proceedings, Workshop on Large Horizontal Axis Wind Turbines*, NASA CP-2230, DOE Publication CONF-810752, Cleveland, OH: NASA Lewis Research Center, pp. 69-85.

Zambrano, T. G., and G. W. Gyatt, 1982, *Wake Structure Measurements at the Mod-2 Cluster Test Facility at Goodnoe Hills*, Report No. AV-QS-A-82/608, Monrovia, CA: AeroVironment, Inc.

7

Wind Turbine Acoustics

Harvey H. Hubbard
Distinguished Research Associate
NASA Langley Research Center

and

Kevin P. Shepherd, Ph.D.
NASA Langley Research Center
Hampton, Virginia

Introduction

Wind turbine generators, ranging in size from a few kilowatts to several megawatts, are producing electricity both singly and in wind power stations that encompass hundreds of machines. Many installations are in uninhabited areas far from established residences, and therefore there are no apparent environmental impacts in terms of noise. There is, however, the potential for situations in which the radiated noise can be heard by residents of adjacent neighborhoods, particularly those neighborhoods with low ambient noise levels. A widely publicized incident of this nature occurred with the operation of the experimental Mod-1 2-MW wind turbine (Fig. 3-31), which is described in detail in [Kelley *et al.* 1985]. Pioneering studies which were conducted at the Mod-1 site on the causes and remedies of noise from wind turbines form the foundation of much of the technology described in this chapter.

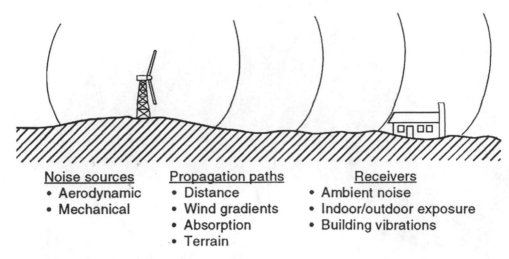

Noise sources
- Aerodynamic
- Mechanical

Propagation paths
- Distance
- Wind gradients
- Absorption
- Terrain

Receivers
- Ambient noise
- Indoor/outdoor exposure
- Building vibrations

Figure 7-1.   Factors contributing to wind turbine noise.

Significant factors relevant to the potential environmental impact of wind turbine noise are illustrated in Figure 7-1. All acoustic technology is built on an understanding of three primary elements: *Noise sources, propagation paths*, and *receivers*. The purpose, therefore, of this chapter is to describe in quantitative terms the specific wind turbine factors that characterize each of these elements. The most important of these are listed in Figure 7-1.

The noise produced by wind turbines ranges in frequency from low values that are sometimes inaudible to higher values in the normal audible range [Kelley *et al.* 1985]. Although increased distance is beneficial in reducing noise levels, the wind can enhance noise propagation in certain directions and impede it in others. A unique feature of wind turbine noise is that it can result from essentially continuous periods of daytime and nighttime operation. This is in contrast to the more common aircraft and road traffic noises that vary markedly as a function of time of day.

This chapter summarizes available information on the physical characteristics of the noise generated by wind turbines and includes example *sound pressure time histories, narrow-band* and *broadband frequency spectra*, and *noise radiation patterns*. Also reviewed are noise measurement standards, analysis technology, and methods for characterizing and predicting the intensity of noise from wind turbines, both singly and in clusters. Atmospheric propagation data are included that illustrate the effects of distance and the effects of refraction caused by a vertical gradient in the mean wind speed. Perception thresholds for humans are defined for both narrow-band and broadband spectra from systematic tests in the laboratory and from observations in the field. Also summarized are structural vibrations and interior sound pressure levels, which could result from the low-frequency noise excitation of buildings.

For more detailed information, a bibliography is available that lists technical papers on all aspects of wind turbine acoustics [Hubbard and Shepherd 1988].

Characteristics of Wind Turbine Noise

Noise from wind turbines may be characterized as aerodynamic or mechanical in origin. Aerodynamic noise components are either narrow-band (containing discrete harmonics) or broadband (random) and are related closely to the geometry of the rotor, its blades, and their aerodynamic flow environments. The low-frequency, narrow-band rotational components typically occur at the blade passage frequency (the rotational frequency times the number of blades) and integer multiples of this frequency. Of lesser importance for most configurations are mechanical noise components from operating bearings, gears, and accessories.

An example of a spectrum of wind turbine noise is shown in Figure 7-2. These data, which were measured 36 m downwind of a vertical-axis wind turbine (VAWT), show the decrease of sound pressure levels with increasing frequency (a general characteristic of wind turbines). All sound pressure levels presented in this chapter are based on root-mean-square (RMS) values of pressure; they are referenced to 2 x 10^{-5} Pa and are averaged over 30 to 180 seconds, depending on the frequency bandwidth. The spectrum generally contains broadband random noise of aerodynamic origin, although discrete components identified as mechanical noise from the gearbox are also evident. The blade passage frequency is readily apparent in the time history illustrated in Figure 7-2, as is the random nature of the emitted sounds.

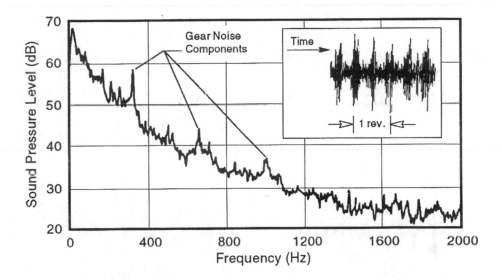

Figure 7-2. Typical narrow-band noise spectrum of a wind turbine, measured 36 m from a VAWT generating 185 kW at a wind speed of 16.5 m/s (bandwidth = 2.5 Hz)

The many analytical and experimental acoustical studies conducted on horizontal-axis wind turbines (HAWTs) indicate that for given geometrical and operational characteristics (such as power output, rotor area, and tip speed) HAWTs with downwind rotors (downwind of the tower) will generate more noise than will those with upwind rotors. This is because an additional noise source in downwind rotors is introduced when the rotating blades interact with the aerodynamic wake of the supporting tower.

Because little information on the acoustics of VAWTs is currently available, it is difficult to directly compare the noise-generation characteristics of HAWTs and VAWTs. Example VAWT spectra, levels, and directivity data are contained in Kelley, Hemphill, and Sengupta [1981] and Wehrey *et al.* [1987]. The blades of a VAWT interact with the aerodynamic wake of the rotor's central column in a manner similar to the way that a downwind HAWT rotor interacts with its tower wake, but at a greater distance relative to the column diameter. Thus, the magnitude of the noise from a VAWT caused by this interaction is expected to be less than that of an equivalent downwind HAWT rotor and greater then that of an upwind HAWT rotor. There is currently no detailed information available describing other aerodynamic noise sources associated with VAWTs. Thus, to gain an understanding of the acoustics of this type of turbine, additional studies are needed.

Blade Impulsive Noise

Impulsive noise is often associated with downwind rotors on HAWTs; in many cases, it is the dominant noise component for that configuration. Figures 7-3(a) and (b) show example sound pressure time histories for two different HAWTs with downwind rotors [Shepherd, Willshire, and Hubbard 1988; Hubbard and Shepherd 1982]. Figure 7-3(a) relates to the 4-MW WTS-4 HAWT (Fig. 4-24), with its 78.2-m-diameter rotor supported downwind of a twelve-sided shell tower. Strong impulses are superposed on less-intense broadband components. The impulse noise arises from the blade's interaction with the aerodynamic wake of the tower. As each blade traverses the tower wake, it experiences short-duration load fluctuations caused by the velocity deficiency in the wake. These load fluctuations lead directly to the radiated acoustic pulses. The acoustic pulses are all of short duration and vary in amplitude as a function of time. This variation in amplitude is believed to result from variations in the blade loadings caused by detailed differences in the time-varying structure of the aerodynamic wake [Kelley *et al.* 1985].

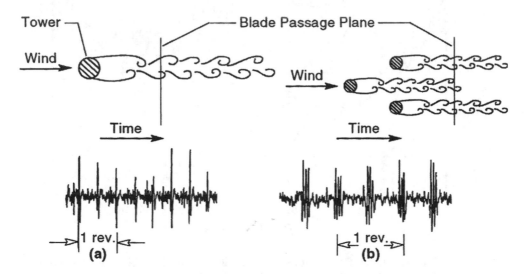

Figure 7-3.   Sound pressure time histories from two downwind-rotor HAWTS. [Shepherd, Willshire, and Hubbard 1988; Hubbard and Shepherd 1982]. (a) 78.2-m-diameter rotor, 2 blades, 2,050-kW output, 30-rpm rotor speed, 200-m distance (b) 17.6-m-diameter rotor, 3 blades, 50-kW output, 72-rpm rotor speed, 30.5-m distance

The same phenomena, differing only in detail, are illustrated in Figure 7-3(b). These data relate to a small-scale turbine with a 17.6-m-diameter rotor supported downwind of a three-legged open truss tower [Hubbard and Shepherd 1982]. Each blade passage produces a three-peaked pulse as the blade interacts with the wakes of the three tower legs. Experimental studies by Hubbard and Shepherd [1982] and Greene [1981] showed that the character of the wake of a tower element can be altered to various degrees by adding such modifications as strakes, screens, and vanes. Because some velocity deficiency remains in the lee of the tower, it is inevitable that such modifications can ameliorate but not eliminate the impulsive noise components.

Figure 7-4 compares narrow-band spectra for upwind- and downwind-rotor HAWTs, along with their typical sound pressure time histories. The upwind HAWT is the NASA/DOE Mod-2 HAWT (Fig. 3-37), 91 m in diameter and operating at a speed of 17.5 rpm. The downwind HAWT is the WTS-4 HAWT. Note that the upwind-rotor spectrum shows an

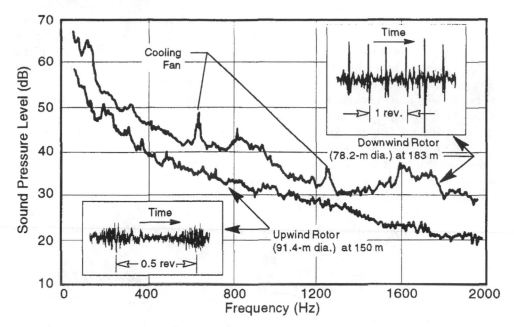

Figure 7-4. Narrow-band noise spectra from large-scale HAWTs with upwind and downwind rotors. (bandwidth = 2.5 Hz)

amplitude-modulated time history, but without the sharp pressure peaks that are evident for the downwind rotor. The two spectra have essentially the same shape, but the downwind-rotor spectrum shows generally higher noise levels because of its higher blade tip speed.

An expanded frequency scale is shown in Figure 7-5, in which the lower-frequency portions of the spectra in Figure 7-4 were analyzed with a narrower effective bandwidth resolution. Impulsive noises such as those illustrated in Figures 7-3(a) and 7-4 can be

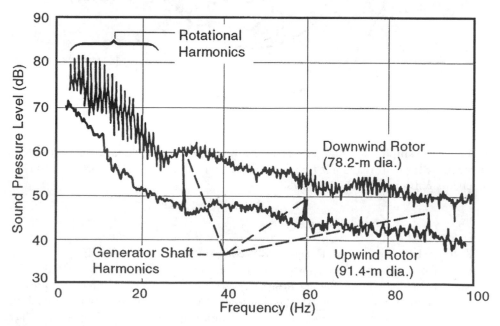

Figure 7-5. Low-frequency, narrow-band noise spectra from large-scale HAWTs with upwind and downwind rotors. (bandwidth = 0.25 Hz, distance = 150 m)

resolved into their *Fourier components*, which are pure tones at the blade passage frequency and integer harmonics of this frequency. These components are evident in the lower-frequency portion of the downwind-rotor spectrum of Figure 7-5, which shows identifiable rotational components out to about 30 Hz. The spectrum indicates a peak near 5 Hz and then a general decrease as the frequency increases [Shepherd and Hubbard 1983].

Figure 7-6 illustrates the nature of the noise radiation patterns for low-frequency rotational noise components. Shown are the results of simultaneous measurements of sound pressure levels at a frequency of 8 Hz; the measurements were taken at a distance of 200 m around the turbine. Acoustic radiations upwind and downwind are about equal and are greater than that in the crosswind direction. The two patterns in Figure 7-6 provide a direct comparison of measurements made at the same nominal wind conditions for daytime and nighttime operation. The nighttime levels are generally lower than the daytime levels, and the resulting radiation pattern generally appears as an acoustic dipole. The lower levels are believed to result from a different atmospheric turbulence structure during the night.

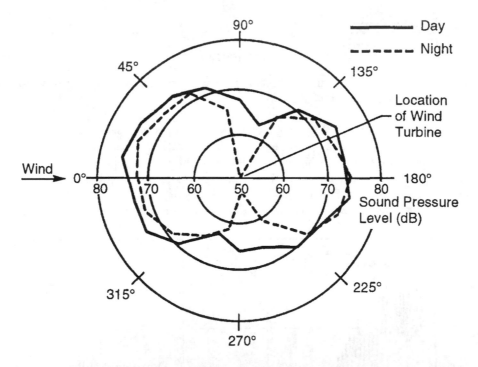

Figure 7-6. Example radiation patterns for low-frequency rotational noise 200 m from a large-scale HAWT. (harmonic frequency = 8 Hz, wind speed = 7.2 m/s, power = 100 kW) [Shepherd, Willshire, and Hubbard 1988]

Kelley, Hemphiil, and McKenna [1982] compare characteristic low-frequency noise emissions from upwind-rotor HAWTs, downwind-rotor HAWTs, and a VAWT. These comparisons are based on joint probability distributions of octave-band sound pressure levels. The authors conclude that a downwind-rotor HAWT presents the highest probability of emitting coherent low-frequency noise, while an upwind-rotor HAWT appears to have the lowest probability of emitting such noise. The probability associated with a VAWT providing coherent noise was found to be between the two HAWT probabilities.

Blade Broadband Noise

Broadband noise arises as the rotating blades interact with the wind inflow to the rotor. It is a significant component for all configurations of rotors, regardless of whether the low-frequency impulsive components are present. Broadband noise components are characterized by a continuous distribution of sound pressure with frequency, and they dominate a typical wind turbine acoustic spectrum at frequencies above about 100 Hz.

Example broadband-noise radiation patterns for a large-scale HAWT are shown in Figure 7-7. Data are included for one-third-octave bands with center frequencies of 100, 200, and 400 Hz. The band levels in the upwind and downwind directions are comparable but generally higher than those in the crosswind direction. The general shapes of these patterns are similar to those in Figure 7-6 for the low-frequency, rotational noise components during the daytime.

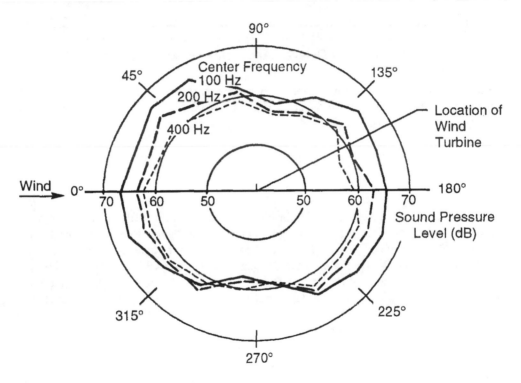

Figure 7-7. Example radiation patterns for broadband noise 200 m from a large-scale HAWT. (one-third-octave bands, wind speed = 12.1 m/s, power = 2050 kW) [Shepherd, Willshire, and Hubbard 1988]

The one-third-octave band spectra of Figure 7-8 were obtained for wind speeds varying by a factor of two. At lower frequencies, dominated by the rotational harmonics, the highest levels are shown to be associated with the highest wind speeds and the highest power outputs. At higher frequencies, dominated by broadband components, there is no clear trend in relation to wind speed. This result is in contrast to a scaling law given in Sutherland, Mantey, and Brown [1987], in which A-weighted sound pressure levels increase in proportion to the logarithm of the wind speed, and this contrast is verified by data from a group of several small wind turbines.

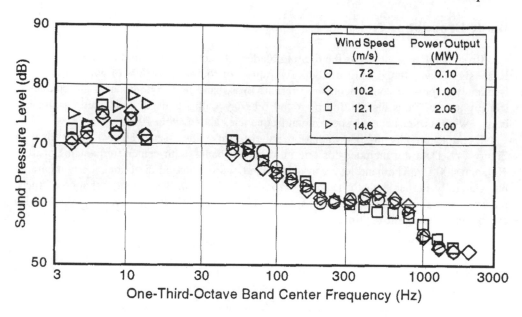

Figure 7-8. Typical variation in noise spectra with power output and wind speed, measured 200 m from a large-scale HAWT. (78.2-m diameter, downwind rotor) [Shepherd, Willshire, and Hubbard 1988]

Figure 7-8 and the upper spectrum in Figure 7-4 both represent the acoustic output of the same wind turbine. The higher sound pressure levels in Figure 7-8 are the typical result of increasing the frequency bandwidth selected for analyzing the acoustic output.

Figures 7-9 and 7-10 contain measured data for several HAWTs of various sizes and configurations [Shepherd, Willshire, and Hubbard 1988]. In Figure 7-9 [Hubbard and Shepherd 1984], measured far-field data for several upwind-rotor turbines are adjusted to a distance 2.5 rotor diameters from the base of the tower and are plotted as one-third-octave band spectra. The disk power densities (in W/m^2) and tip speeds for all of these machines are comparable, and the spectra (adjusted for distance) are in general agreement except at the lower frequencies. Comparable data are presented in Figure 7-10 for several downwind rotors [Shepherd *et al.* 1988; Hubbard and Shepherd 1982; Shepherd and Hubbard 1981; Lunggren 1984]; the results are similar. The variations in noise levels in Figure 7-10 can be related to the variations in rotor tip speed noted in the legend. A reference gradient of -10 dB per decade is included to indicate roughly the rate at which the broadband noise levels decrease as frequencies increase.

Effects of HAWT Yaw Error

Horizontal-axis turbines sometimes operate such that the wind direction is not aligned with the rotor axis. The effects of nonalignment, or *yaw error*, on the generated noise have been evaluated for a large-scale HAWT with a downwind rotor. In Figure 7-11, data are shown for yaw errors of 0, 20, and 31 deg. The band levels plotted are arithmetic averages of measured values in the upwind and downwind quadrants. The obvious result is that sound pressure levels at low frequencies are reduced as yaw error increases. This would

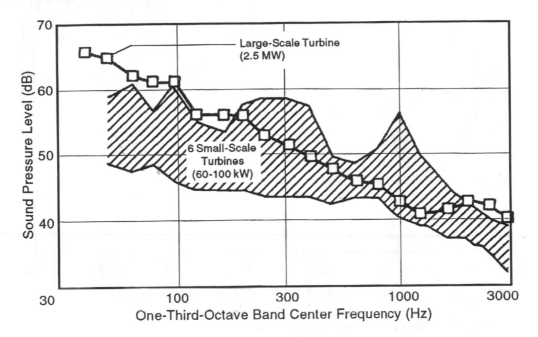

Figure 7-9. Noise spectra from small- and large-scale HAWTs with upwind rotors. (downwind distance = 2.5 rotor diameters)

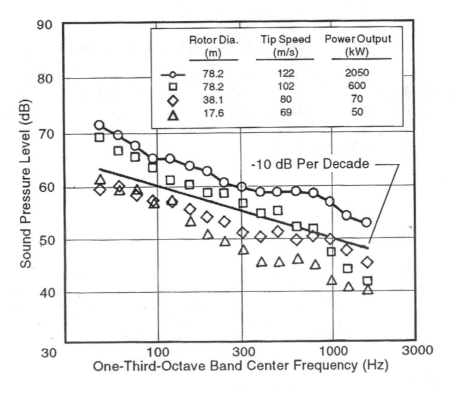

Figure 7-10. Noise spectra from small-, intermediate-, and large-scale HAWTs with downwind rotors. (downwind distance = 2.5 rotor diameters) [Shepherd *et al.* 1988; Lunggren 1984; Shepherd and Hubbard 1981; Hubbard and Shepherd 1982]

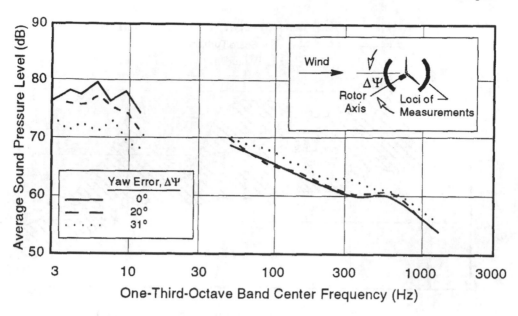

Figure 7-11. Effect of yaw error on the noise spectra of a large-scale HAWT with a downwind rotor. [Shepherd, Willshire, and Hubbard 1988]

be expected because of the reduced aerodynamic loading associated with a yaw error. At higher frequencies yaw errors cause some small increases in the sound pressure levels.

Noise from Aileron Control Surfaces and Vortex Generators

Some experimental HAWT blades have contained *ailerons* for speed and power control. Data for two different aileron configurations are given in Shepherd and Hubbard [1984]. The unusually high noise levels observed in these tests are believed to result from the excitation of internal cavity resonances in the blades by the external flow. Well-designed aileron systems with sealed bulkheads would not have this problem.

In some situations, small tabs or *vortex generators* are installed on the low-pressure surfaces of both HAWT and VAWT blades to delay local stall and generally improve aerodynamic performance. Studies to evaluate the effects of vortex generators on noise radiations show these effects to be insignificant [Hubbard and Shepherd 1984].

Machinery Noises

Most of the acoustic noise associated with the wind turbines studied to date has been aerodynamic in origin. Potential sources of mechanical noise, such as gears, bearings, and accessories, have not been important for large-scale HAWTs. Narrow-band analyses of noise from two large-scale HAWTs (Figures 7-4 and -5) show some cooling-fan noise and some identifiable components at the shaft speed of the generator (30 Hz) and harmonics of this speed. Because these mechanical components generally radiate in the crosswind direction and are not normally heard, they are of only secondary importance.

For some of the smaller HAWTs and some VAWTs, however, gear noise can be an important component of the total acoustic radiation. Some straightforward approaches to controlling gear noise are to include noise and vibration limits in the design specifications and to apply noise insulation around the gearbox.

Predicting Noise from a Single Wind Turbine

Extensive research studies have been conducted to predict noise from isolated airfoils, propellers, helicopter rotors, and compressors. Many of those findings have helped identify the significant noise sources of wind turbines and have helped develop methods for noise prediction. This section summarizes the technology available for predicting the sound pressure levels radiating from known sources of wind turbine noise, particularly from the aerodynamic sources which are believed to be the most important.

Rotational Harmonics

Impulse noises like those shown in Figures 7-3 and 7-4 can be resolved into their Fourier components (Figure 7-5), which are at the blade passage frequency and its integer multiples. Acoustic pulses arise from rapidly-changing aerodynamic loads on the blades as they routinely encounter localized flow deficiencies which result in momentary fluctuations in lift and drag. Airfoil lift and drag coefficients can be transformed into thrust and torque coefficients, and these can be used to determine the unsteady blade loads associated with periodic variations in the wind velocity. These variations may occur within the tower wake, as indicated schematically in Figure 7-3, or within the swept area of the rotor, through wind shear and small-scale turbulence.

Variations in blade loading can be represented by *complex Fourier coefficients* modified by the *Sears function* to determine the effects of unsteady aerodynamics on the airfoil. The Sears function represents aerodynamic loading on a rigid airfoil passing through a sinusoidal gust [Sears 1941]. Following the model presented in Viterna [1981], a general expression for the RMS sound pressure level of the nth harmonic can be derived in the following form:

$$P_n = \frac{\sqrt{2}\sin\gamma}{4\pi R_e d} M_n^2 \sum_m e^{im(\phi - \pi/2)} J_x \left(a_m^T \cos\gamma - \frac{nB - m}{M_n} a_m^Q \right) \tag{7-1a}$$

$$M_n = nB \frac{R_e \Omega}{a_0} \tag{7-1b}$$

where

P_n = RMS sound pressure for the *n*th harmonic (N/m²)
n = sound pressure harmonic number (n = 1, 2,...)
γ, ϕ = azimuth and altitude angles to the listener, respectively, referred to the rotor thrust vector (rad)
R_e = effective blade radius ≈ 0.7 x tip radius, R (m)
d = distance from the rotor to the listener (m)
M_n = Mach Number factor for the *n*th harmonic
m = blade loading harmonic index (m = ..., -2, -1, 0, 1, 2,...)
J_x = Bessel function of the first kind and of order $x = nB - m$
a_m^T, a_m^Q = complex Fourier coefficients for the thrust and torque forces acting at R_e, respectively (N)
B = number of blades
Ω = rotor speed (rad/s)
a_0 = speed of sound (m/s)

Note that each blade loading harmonic *m*, caused by fluctuating air loads, gives rise to more than one sound harmonic *n* in the radiation field.

For the special case in which the inflow wind to the rotor disk is uniform and the listener is located in the plane of the axis (*i.e.*, $\phi = 0$), Equation (7-1a) reduces to

$$P_n = \frac{\sqrt{2}\sin\gamma}{4\pi R_e d}\left(T\cos\gamma - \frac{Qa_0}{R_e^2\Omega}\right)M_n^2 J_{nB} \tag{7-2}$$

where

T = rotor shaft thrust (N)
Q = rotor shaft torque (N-m)

Example Rotational Noise Calculations

Examples of sound pressure levels calculated by means of Equation (7-1) for the Mod-1 HAWT are presented in Viterna [1981] and are included in Figures 7-12 and 7-13. The calculations relate to the following geometric and operating conditions:

R = 30.5 m   hub height, H = 46 m
B = 2   power output, P = 1.500 kW
Ω = 34.6 rpm   wind speed, U = 13.4 m/s
d = 79 m and 945 m downwind   deficit behind tower = 20 percent over a 20 deg sector

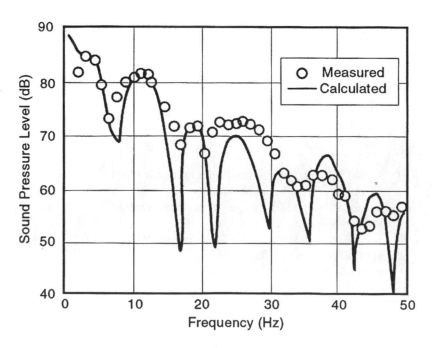

Figure 7-12. Measured and calculated rotational noise spectra 79 m downwind of the Mod-1 HAWT. (rotor diameter = 61 m, wind speed = 13.4 m/s, power output = 1,500 kW) [Viterna 1981]

Figure 7-12 compares calculated and measured sound pressure levels of the first 50 rotational harmonics for the Mod-1 downwind rotor. The calculations predict the maximum levels quite well, as well as the general shape of the spectrum. Other calculations [Viterna 1981] suggest that the maximum levels of the rotational harmonics occur in the upwind and downwind directions, while the minimum levels occur in the crosswind directions. Note that the calculation procedure presented in Equation (7-1) has been validated for the Mod-1 and WTS-4 machines. Alternative methods for predicting the magnitude of rotational harmonics are discussed, and pertinent results are presented in Kelley *et al.* [1985]; Meijer and Lindblad [1983]; Green and Hubbard [1980]; Martinez, Widnall, and Harris [1982]; George [1978]; and Lowson [1970].

Calculations made with Equation (7-2) were compared with those for a nonuniform wind inflow, and the results are shown in Figure 7-13. For a uniform flow field, the fundamental rotational harmonic is relatively strong, but all higher harmonics are weak. A similar result is obtained when the rotor operates in a shear flow that produces a once-per-revolution variation of inflow velocity at each blade. When a tower wake deficit is added, however, the levels of the higher frequencies are greatly enhanced.

These results suggest that both configuration and siting effects are significant in the rotational noise generation of wind turbines. For example, the tower wake of both VAWTs and downwind-rotor HAWTs can greatly enhance the strength of the rotational noise harmonics. Other deviations in wind inflow from a uniform velocity over the disk may also enhance the strength of the rotational harmonics for all rotor configurations. Flow deviations may be caused by the vertical wind velocity gradient in the earth's boundary layer and may be exaggerated by atmospheric turbulence or terrain features that can impose additional velocity gradients on the inflow.

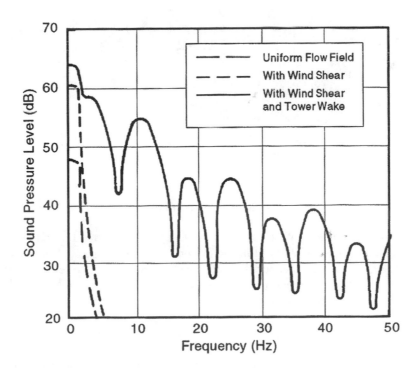

Figure 7-13. Calculated envelopes of rotational noise spectra for various wind inflow conditions 945 m downwind of the Mod-1 HAWT. (rotor diameter = 61 m, wind speed = 13.4 m/s, power output = 1,500 kW) [Viterna 1981]

Broadband Noise Components

Extensive research on propellers, helicopter rotors, compressors, and isolated airfoils has provided a wealth of background information and experience for predicting broadband noise for wind turbine rotors. The main noise sources have been identified, prediction techniques have been described, and comparisons have been made with available experimental data [George and Chou 1984; Glegg *et al.* 1987; Grosveld 1985]. Measurements to date indicate three main sources of broadband noise are as follows:

-- *aerodynamic loading fluctuations* caused by inflow turbulence interacting with the rotating blades

-- *turbulent boundary-layer flow* over the airfoil surface interacting with the blade trailing edge

-- *vortex shedding* caused by the bluntness of the trailing edge

These sources of broadband noise are illustrated in Figure 7-14, along with their sound power dependencies, definitions of critical dimensions, and flow velocities [Grosveld 1985].

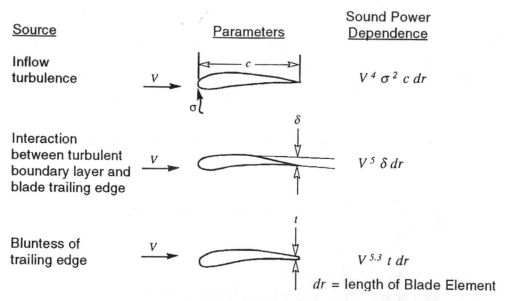

Figure 7-14. Sources of wind turbine broadband noise. [Grosveld 1985]

Another possible source of broadband noise is that of *tip vortex* formation. Based on the experimental data of isolated airfoils and rotors [George and Chou 1984; Brooks and Marcolini 1986], this source is expected to be of secondary importance relative to the three listed. However, unusual geometries, such as those associated with some tip brakes and deflected tip control surfaces, could result in significantly more radiated noise.

Inflow Turbulence Noise

As the wind turbine blades move through the air, they encounter atmospheric turbulence that causes variations in the local angle of attack, which in turn causes fluctuations in the lift and drag forces. The length scales and intensities are a function of local atmospheric and site conditions and are different at different heights above the ground [Kelley *et al.*

1987]. The following expression for HAWT rotor noise induced by inflow turbulence is based on the work presented in Grosveld [1985]:

$$SPL_{1/3}(f) = 10 \log_{10}[B \sin^2 \theta \, \rho^2 \, c_{0.7} R \sigma^2 V_{0.7}^4 / (d^2 a_0^2)] + K_a \qquad (7\text{-}3a)$$

$$V_{0.7} = 0.7 \, R \, \Omega \qquad (7\text{-}3b)$$

$$f_{peak} = S_0 \, V_{0.7} / (H - 0.7 \, R) \qquad (7\text{-}4)$$

where

$SPL_{1/3}$ = one-third octave band sound pressure level (dB)
f = band center frequency (Hz)
θ = angle between the hub-to-receiver line and its vertical projection in the rotor plane (rad)
ρ = air density (kg/m³)
$c_{0.7}$ = blade chord at a radius = 0.7 R (m)
σ^2 = mean square of turbulence (m²/s²)
$V_{0.7}$ = blade forward speed at 0.7 radius (m/s)
K_a = frequency-dependent scaling factor (dB; Figure 7-15)
f_{peak} = frequency at which K_a is maximum (Hz; Figure 7-15)
S_0 = constant Strouhal number = 16.6
H = hub elevation above ground (m)

A peak in the frequency domain is obtained when $f = f_{peak}$, which corresponds to the maximum value of K_a in Figure 7-15. Inherent in the derivation of Equation (7-3) are the assumptions that the *turbulence is isotropic* and the *atmosphere is neutrally stable* within the vertical layer occupied by the rotor. In addition, the noise source is considered to be a *point dipole* at hub height, and the wavelength of the radiated sound is much shorter

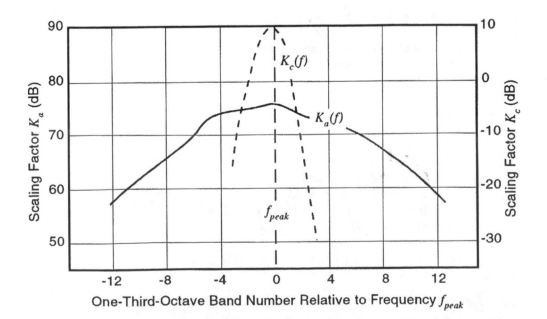

Figure 7-15. Predicted frequency-dependent scaling factors for broadband noise.

than d, the distance to the receiver. The frequency-dependent scaling factor K_a, in Figure 7-15, has been determined empirically from measured frequency spectra of rotor noise caused largely by inflow turbulence.

Noise from the Interaction of the Turbulent Boundary Layer and the Blade Trailing Edge

Noise is generated by the convection of the blade's attached turbulent boundary layer into the wake of the airfoil. This is a major noise source for helicopter rotors, and the studies on this subject by Schlinker and Amiet [1981] have been adapted to wind turbine rotors. The resulting expression [Grosveld 1985] for one blade airfoil is as follows:

$$SPL_{1/3}(f) = 10\log_{10} \int_0^R \Phi_b \, dr + K_b \tag{7-5a}$$

$$\Phi_b = V_r^5 \, B \, D \, \frac{\delta}{d^2} \left(\frac{S}{S_{max}} \right)^4 \left[\left(\frac{S}{S_{max}} \right)^{1.5} + 0.5 \right]^{-4} \tag{7-5b}$$

$$D = \frac{\sin^2(\theta/2)}{(1 + M \cos \theta)[1 + (M - M_c) \cos \theta]} \tag{7-5c}$$

$$M = V_r/a_0 \tag{7-5d}$$

$$\delta = 0.37c/N_R^{0.2} \tag{7-5e}$$

$$N_R = V_r c / \nu \tag{7-5f}$$

$$S = f\delta/V_r \tag{7-5g}$$

where

V_r = resultant velocity at a blade segment (m/s)
D = directivity factor
θ = angle between the segment-to-receiver line and its vertical projection in the rotor plane (rad)
M = blade segment Mach number
M_c = convection Mach number = 0.8 M
δ = boundary layer thickness (m)
c = segment chord (m)
N_R = segment Reynolds number
ν = kinematic viscosity (m²/s)
d = distance from the segment to the receiver (m)
S = segment Strouhal number
S_{max} = 0.1
dr = spanwise length of the blad segment (m)
K_b = constant scaling factor = 5.5 dB

Sound pressure levels for the rotor are obtained by integrating contributions of all acoustic sources over the length of each blade and adding the results.

Noise from Vortex Shedding at the Trailing Edge

Another broadband noise source is associated with vortex shedding caused by the bluntness of the trailing edge. This phenomenon is analogous to the shedding noise from wings with blunt trailing edges, as well as from flat plates, and struts [Schlinker and Amiet 1981; Brooks and Hodgson 1980]. The expression derived in Grosveld [1985] for the noise from the blunt trailing edge of one blade is

$$SPL_{1/3}(f) = 10 \log_{10} \int_0^R \Phi_c \, dr + K_c \tag{7-6a}$$

$$\Phi_c = \frac{B \, V_r^{5.3} \, t \, \sin^2(\theta/2) \, \sin^2\psi}{(1 + M \cos \theta)^3 [1 + (M - M_c) \cos \theta]^2 \, d^2} \tag{7-6b}$$

$$f_{peak} = \frac{0.1 \, V_r}{t} \tag{7-7}$$

where

  t = trailing edge thickness (m)
  ψ = angle between the segment-to-receiver line and its horizontal projection in the rotor plane (rad)
  K_c = frequency-dependent scaling factor (dB; Figure 7-15)

The corresponding K_c has its maximum value when f reaches f_{peak} (Figure 7-15). Once again, sound pressure levels for the rotor are obtained by integrating the contributions of all acoustic sources over the length of each blade and adding the results.

Example Calculations and Measurements of HAWT Broadband Noise

Figure 7-16 illustrates the relative contributions of the broadband noise components calculated by using Equations (7-3) to (7-7) for a large-scale HAWT with an upwind rotor. The calculations are in the form of one-third-octave band spectra for each of the broadband components identified. Also included is the summation of the components. As shown in Figure 7-16, inflow turbulence contributes noise over the whole frequency range and dominates the spectrum at frequencies below about 500 Hz. Effects of boundary-layer interaction also contribute noise over a wide frequency range but are most significant at higher frequencies. On the other hand, the noise spectrum of the trailing edge wake is sharply peaked; the maximum for the example turbine is near 1,250 Hz.

Figure 7-17 presents sound pressure levels calculated by using the methods of Grosveld and compares them with acoustic far-field measurements for a large, upwind-rotor HAWT and two different downwind-rotor HAWTs. Good agreement is shown in all cases. Note that the validation of Equations (7-3) to (7-7) has been limited to acoustic radiation in the upwind and downwind directions only.

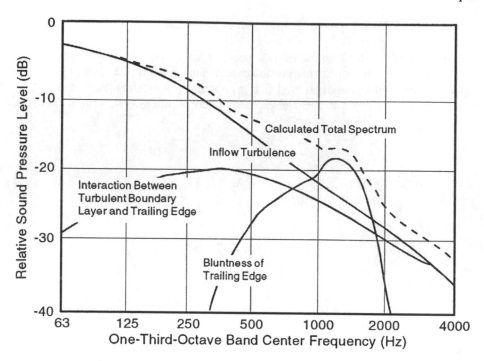

Figure 7-16.   Relative contributions of broadband noise sources to the total noise spectrum calculated for a large-scale HAWT. [Grosveld 1985]

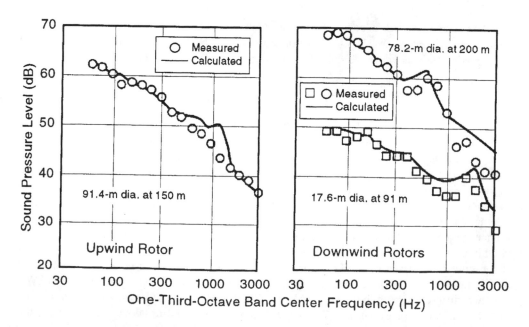

Figure 7-17. Measured and calculated broadband noise spectra downwind of various HAWTs. [Grosveld 1985]

An alternative broadband-noise prediction scheme is proposed in Glegg *et al.* [1987] and includes noise from *unsteady lift, unsteady thickness, trailing edges,* and *separated flows.* Inflow turbulence at the rotor must be specified to predict unsteady lift and thickness noises. Using the turbulence data associated with the atmospheric boundary layer as input yielded poor agreement between calculated and measured noise levels. Thus, the authors hypothesized that there was an additional source of turbulence: that each blade ran into the tip vortex shed by the preceding blade. Note that Grosveld [1985] also used atmospheric boundary layer turbulence but found that better agreement with acoustic measurements required an empirical turbulence model. The two approaches share the same theoretical background and therefore should give the same results.

Noise Propagation

A knowledge of the manner in which sound propagates through the atmosphere is basic to the process of predicting the noise fields of single and multiple machines. Although much is known about sound propagation in the atmosphere, one of the least understood factors is the effect of the wind. Included here are brief discussions of the effects of distance from various types of sources, the effects of such atmospheric factors as absorption in air and refraction caused by sound speed gradients, and terrain effects.

Distance Effects

Point Sources

When there is a nondirectional point source as well as closely-grouped, multiple point sources, *spherical spreading* may be assumed in the far radiation field. Circular wave fronts propagate in all directions from a point source, and the sound pressure levels decay at the rate of -6 dB per doubling of distance, in the absence of atmospheric effects. The latter decay rate is illustrated by the straight line in Figure 7-18. The dashed curves in the figure represent increased decay rates associated with *atmospheric absorption* at frequencies significant for wind turbine noise.

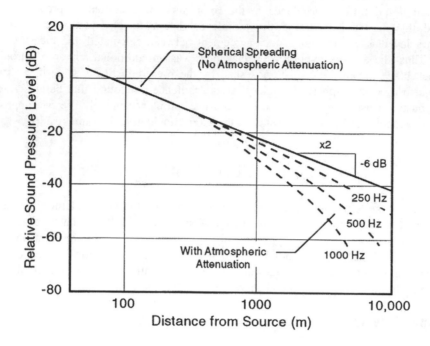

Figure 7-18. Decrease in the sound pressure levels of pure tones as a function of distance from a point source. [ANSI 1978]

Line Sources

For an infinitely long line source, the decay rate is only -3 dB per doubling of distance, compared with the -6 dB per doubling of distance illustrated in Figure 7-18. Such a reduced decay rate is sometimes observed for sources such as trains and lines of vehicles on a busy road. Some arrays of multiple wind turbines in wind power stations may also behave acoustically like line sources.

Atmospheric Factors

Absorption in Air

As sound propagates through the atmosphere, its energy is gradually converted to heat by a number of molecular processes such as shear viscosity, thermal conductivity, and molecular relaxation, and thus atmospheric absorption occurs. The curves in Figure 7-19 were plotted from ANSI values [1978] and show changes in atmospheric absorption as a function of frequency, at typical ambient temperatures and relative humidity levels. Atmospheric absorption is relatively low at low frequencies and increases rapidly as a function of frequency. Atmospheric absorption values for other temperature/humidity conditions can be obtained from the ANSI tables.

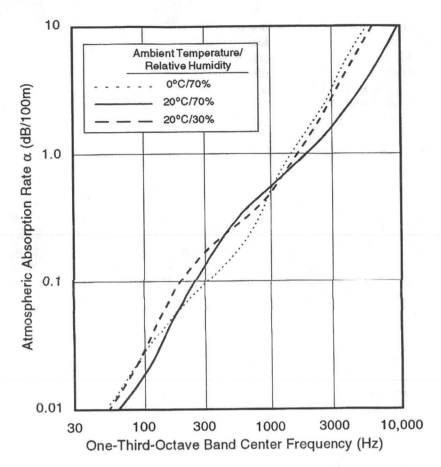

Figure 7-19. Standard rates of atmospheric absorption. [ANSI 1978]

Refraction Caused by Wind and Temperature Gradients

Refraction effects arising from the sound speed gradients caused by wind temperature can cause nonuniform propagation as a function of azimuth angle around a source. A simple illustration is shown in Figure 7-20 of *atmospheric refraction* (*i.e.*, bending) of sound rays, caused by a vertical wind-shear gradient over flat, homogeneous terrain around an elevated point source. Note that in the downwind direction the wind gradient causes the

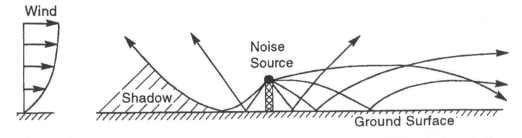

Figure 7-20. Effects of wind-induced refraction on acoustic rays radiating from an elevated point source. [Shepherd and Hubbard 1985]

sound rays to bend toward the ground, whereas in the upwind direction the rays curve upward away from the ground. For high-frequency acoustic emissions, this causes greatly increased attenuation in a *shadow zone* upwind of the source, but little effect downwind. The attenuation of a low-frequency noise, on the other hand, is reduced by refraction in the downwind direction, with little effect upwind.

The distance from the noise source to the edge of the shadow zone is related to the wind speed gradient and the elevation of the source. In a 10- to 15-m/s wind, for a source height from 40 to 120 m above flat, homogeneous terrain, the horizontal distance from the source to the shadow zone was calculated to be approximately five times the height of the source [Shepherd and Hubbard 1985].

Attenuation exceeding that predicted by spherical spreading and atmospheric absorption can be found in the shadow zone. This attenuation is frequency-dependent, and the lowest frequencies are the least attenuated. Figure 7-21 presents an empirical scheme for estimating attenuation in the shadow zone, based on information in Piercy *et al.* [1977]; SAE [1966]; and Daigle *et al.* [1986]. The estimated extra attenuation is assumed to develop from zero to a maximum of A_E over a distance equal to twice that from the source to the edge of the shadow zone. The predicted decay in the sound pressure level from the source to the edge of the shadow zone is caused by atmospheric absorption [ANSI 1978] and spherical spreading. Within the shadow zone, extra attenuation should be added to these two effects, estimated according to Figure 7-21.

Note that vertical temperature gradients, which are also effective sound speed gradients, will normally be present. These will add to or subtract from the effects of wind that are illustrated in Figure 7-20. Effects of the wind gradient will generally dominate over those of temperature gradients in the propagation of noise from wind power stations.

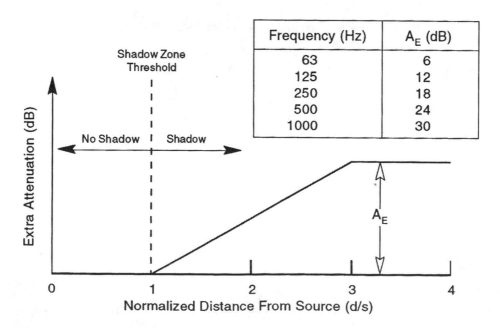

Figure 7-21. Empirical model for estimating the extra attenuation of noise in the shadow zone upwind of an elevated point source. (s = 5h, 40 ≤ h ≤ 120 m, where h = source elevation) [Shepherd and Hubbard 1985]

Distributed Source Effects

Because of their large rotor diameters, some wind turbines exhibit distributed source effects relatively close to the machines. Only when listeners are at distances from a turbine that are large in relation to its rotor diameter does the rotor behave acoustically as a point source. As indicated in Figure 7-22, distributed source effects are particularly important in the upwind direction. In this figure, sound pressure levels in the 630-Hz, one-third-octave band are presented as a function of distance in the downwind, upwind, and crosswind directions. In the downwind and crosswind directions, the measured data agree well with the solid curves representing spherical spreading and atmospheric absorption. In the upwind direction, however, the measured data fall below the solid curve, which indicates the presence of a shadow zone.

An improvement in predicting upwind sound pressure levels is obtained when the noise is modeled as being distributed over the entire rotor disk. Each part of the disk is then considered to be a point source, and attenuation is estimated by means of the empirical model shown in Figure 7-21. The resulting predictions are shown as the dashed curve of Figure 7-22 and are in good agreement with the sound measurements upwind of the turbine. In the downwind and crosswind directions, point-source and distributed-source models result in identical calculations of sound pressure levels.

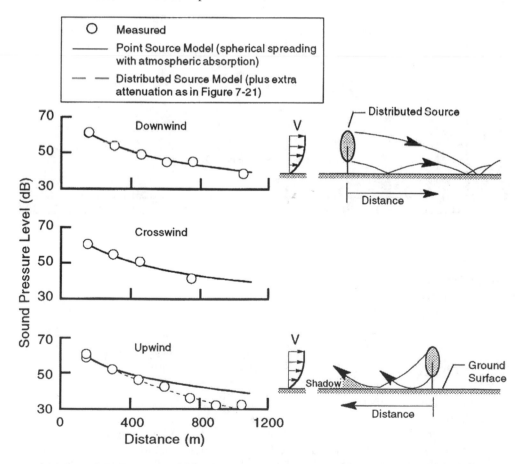

Figure 7-22. Measured and calculated sound pressure levels in three directions from a large-scale HAWT. (one-third-octave band = 630 Hz, rotor diameter = 78.2 m) [Shepherd and Hubbard 1985]

Figure 7-23 illustrates the special case of propagation of low-frequency rotational-harmonics when the atmospheric absorption and extra attenuation in the shadow zone are very small. Measured sound pressure levels are shown as a function of distance for both the upwind and downwind directions. For comparisons, the curves representing decay rates of -6 dB and -3 dB per doubling of distances are also included. Note that in the upwind case the sound pressure levels tend to follow a decay rate of -6 dB per doubling of distance, which is equal to the rate of spherical spreading. No extra attenuation from a shadow zone was measured.

In the downwind direction, the sound pressure levels tend to follow a decay rate of -3 dB per doubling of distance, similar to that for cylindrical spreading. This reduced decay rate in the downwind direction at very low frequencies is believed to result from atmospheric refraction which introduces a channeling sound path in the lower portions of the earth's boundary layer [Willshire and Zorumski 1987, Thomson 1982, Hawkins 1987].

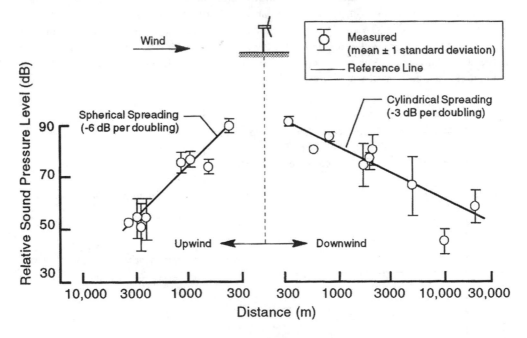

Figure 7-23. Measured effect of wind on the propagation of low-frequency rotational harmonic noise from a large-scale HAWT. (Harmonics with frequencies from 8 to 16 Hz, rotor diameter = 78.2 m) [Willshire and Zorumski 1987]

Terrain Effects

Terrain effects include ground absorption, reflection, and diffraction. Furthermore, terrain features may cause complex wind gradients, which can dominate noise propagation to large distances [Kelley *et al.* 1985, Thompson 1982]. Wind turbines are generally located in areas devoid of trees and other large vegetation. Instead, ground cover usually consists of grass, sagebrush, plants, and low shrubs, which are minor impediments to noise propagation except at very high frequencies. At frequencies below about 1,000 Hz, the ground attenuation is essentially zero.

Methods are available for calculating the attenuations provided by natural barriers such as rolling terrain, which may interrupt the line of sight between the source and the receiver [Piercy and Embleton 1979]. However, very little definitive information is available regarding the effectiveness of natural barriers in the presence of strong, vertical wind gradients. Piercy and Embleton [1970] postulate that the effectiveness of natural barriers in attenuating noise is not reduced under conditions of upward-curving ray paths (as would apply in the upwind direction) or under normal temperature-lapse conditions. However, under conditions of downward-curving ray paths, as in downwind propagation or during temperature inversions (which are common at night), the barrier attenuations may be reduced significantly, particularly at large distances.

Predicting Noise from Multiple Wind Turbines

Methods are needed to predict noise from wind power stations made up of large numbers of machines, as well as for a variety of configurations and operating conditions. This section reviews the physical factors involved in making such predictions and presents the results of calculations that illustrate the sensitivity of radiated noise to various geometric and propagation parameters. A number of valid, pertinent, simplifying assumptions are presented. A logarithmic wind gradient is assumed, with a wind speed of 9 m/s at hub height. Flat, homogeneous terrain, devoid of large vegetation, is also assumed. Noises from multiple wind turbines are assumed to add together incoherently, that is, in random phase.

Noise Sources and Propagation

Reference Spectrum for a Single Wind Turbine

The most basic information needed to predict noise from a wind power station is the noise output of a single turbine. Its noise spectrum can be predicted from knowledge of the geometry and operating conditions of the machine [Viterna 1981; Glegg *et al.* 1987; Grosveld 1985], or its spectrum can be measured at a reference distance. Figures 7-9 and 7-10 are examples of spectral data for HAWTs. The solid line shown in Figure 7-10 is a hypothetical spectrum used in subsequent example calculations to represent a HAWT with a 15-m rotor diameter and a rated power of approximately 100 kW. The example spectrum has a decrease of 10 dB per decade in sound pressure level with increasing frequency. This spectral shape is generally representative of the aerodynamic noise radiated by wind turbines. However, predictions for a specific wind power station should be based, if possible, on data for the particular types of turbines in the station.

Directivity of the Source

Measurement of aerodynamic noise for a number of large HAWTs [*e.g.*, Kelley *et al.* 1985, Hubbard and Shepherd 1982] indicate that the source directivity depends on specific noise-generating mechanisms. For broadband noise sources, such as inflow turbulence and interactions between the blade boundary layer and the blade trailing edge, sound pressure level contours are approximately circular. Lower-frequency, impulsive noise, which results when the blades interact with the tower or central column wake, radiates most strongly in the upwind and downwind directions. Furthermore, while there is one prevailing wind direction at most wind turbine sites, it is not uncommon for the prevailing wind vector to vary ± 45 deg in azimuth angle during normal operations. Therefore, one of the simplifying assumptions made in the calculations that follow is that each individual machine behaves like an *omnidirectional source.*

Considerations for Frequency Weighting

A-*weighted* sound pressure levels, expressed in dB(A), are in widespread use for evaluating the effects of noise on communities [Pearsons and Bennett 1974]. This particular weighting emphasizes the higher frequencies and de-emphasizes the lower frequencies, according to the sensitivity of the human ear. Figure 7-24 shows the results of applying this descriptor. The solid line is the assumed single-turbine reference spectrum, at a distance of 30 m from the machine, and the upper dashed curve is the equivalent A-weighted spectrum at the same distance. As distances increase, atmospheric absorption causes the levels of the higher-frequency components to decay faster than those of the lower frequencies (see Figure 7-19).

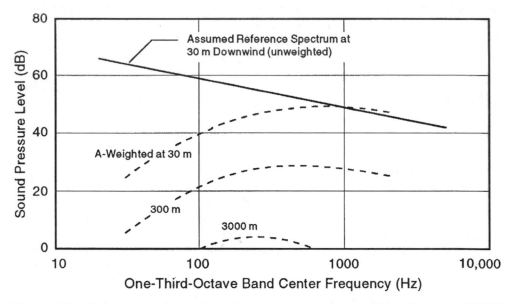

Figure 7-24. Reference and A-weighted noise spectra from a 15-m-diameter HAWT with a rated power of 100 kW. Assumed for example calculations of noise from a wind power station

The result of the combination of A-weighting and atmospheric absorption is that the midrange frequencies (100 to 1000 Hz) tend to dominate the audible spectrum at long distances. Frequencies higher than 1000 Hz will generally not be important considerations at long distances because of the effects of atmospheric absorption. Frequency components below about 100 Hz may not be significant in terms of audible noise, but they can be significant in terms of such indirect effects as noise-induced building vibrations.

Representative Wind Power Station

The basic geometric arrangement of wind turbines shown in Figure 7-25 is assumed to represent an example wind power station. The station consists of four rows, with 31 100-kW, 15 m-diameter turbines per row. The spacing between turbines is 30 m, the row length is 900 m, and the spacing between rows is 200 m. This basic four-row configura-

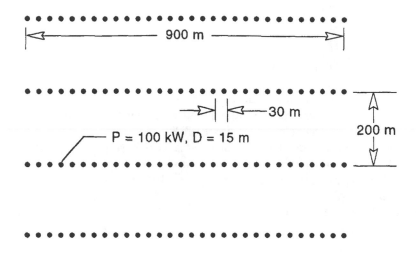

Figure 7-25. Layout of 124 wind turbines in an representative wind power station. [Shepherd and Hubbard 1986]

tion can be perturbed to investigate the effects of such variables as the number of rows, row spacing, turbine spacing, row length, and turbine power rating.

Absorption and Refraction

The example calculations that follow assume an ambient temperature of 20° C and a relative humidity of 70 percent. From the data in Figure 7-19, assumed values of atmospheric absorption of 0, 0.10, 0.27, and 0.54 dB per 100 m then correspond roughly to one-third-octave band center frequencies of 50, 250, 500 and 1,000 Hz, respectively. These frequencies were chosen because they encompass the range of frequencies considered important in evaluating the perception of wind turbine noise in adjacent communities [Shepherd and Hubbard 1986].

Calculation Methods

The method presented here for calculating the sound pressure level from incoherent addition is a sum of the random-phase multiple noise sources at any arbitrary receiver distance. This method assumes that each source radiates equally in all directions, and attenuation caused by atmospheric absorption is included. Propagation is over flat, homogeneous terrain, and there is a logarithmic wind-speed gradient. The method has no limitations on the number of wind turbines or their geometric arrangement.

The required input is a reference sound-pressure-level spectrum, $L_0(f)$, either narrow-band or one-third-octave band, for a single wind turbine. This spectrum can be either measured or predicted, and should represent the radiated noise at a reference turbine-to-receiver distance of approximately 2.5 times the rotor diameter. The sound pressure level received from any individual wind turbine in the array in a given frequency band can then be calculated with the following equation:

$$L_n(f_i) = L_0(f_i) - 20 \log_{10}(d_n/d_0) - \alpha\,(d_n - d_0)/100 \tag{7-8}$$

where

$L_n(f_i)$ = sound pressure level in the *i*th frequency band from the *n*th turbine (dB)
n = wind turbine index = 1,2...,*N*
N = number of wind turbines in the power station
f_i = center frequency of the *i*th band (Hz)
$L_0(f_i)$ = sound pressure level from the reference wind turbine in the *i*th frequency band at the reference distance (dB)
d_n = distance from the nth turbine to the receiver (m)
d_0 = reference turbine-to-receiver distance (m)
α = atmospheric absorption rate (dB per 100 m)

The total sound pressure level in the *i*th frequency band, from all wind turbines in the array, is then calculated as follows:

$$SPL_{total}\,(f_i) = 10 \log_{10} \sum_n 10^{L_n(f_i)/10} \tag{7-9}$$

This procedure is repeated for all frequency bands to provide a predicted spectrum of sound pressure levels at the receiver location. Noise measures such as the A-weighted sound pressure level may also be calculated by adding the A-weighting corrections at each frequency to the values of $L_n(f_i)$ or $SPL_{total}(f_i)$ in Equations 7-8 and 7-9. If the sources are arranged in rows, the required computations can be reduced by using the simplified procedures of Shepherd and Hubbard [1986]

Examples of Calculated Noise from Wind Power Stations

A series of parametric calculations of unweighted sound pressure levels has been performed based on the array of Figure 7-25 and systematic variations of that array [Shepherd and Hubbard 1986]. The receiver is assumed to be on a line of symmetry either in the downwind, upwind, or crosswind direction.

Effect of Distance from a Single Row

Figure 7-26 shows calculated sound pressure levels from one row of the example wind power station, as a function of downwind distance for various rates of atmospheric absorption. Also shown are reference decay rates of -3 dB and -6 dB per doubling of distance. For an atmospheric absorption rate of zero, the decay rate is always less than that for a single point source (Figure 7-18). At intermediate distances, the row of turbines acts as a line source, for which the theoretical decay rate is -3 dB per doubling of distance (or -10 dB per decade of distance). Only at distances greater than one row length (900 m in this case) does the decay rate approach the single-point-source value of -6 dB per doubling of distance (-20 dB per decade). Decay rates increase as the absorption coefficient increases.

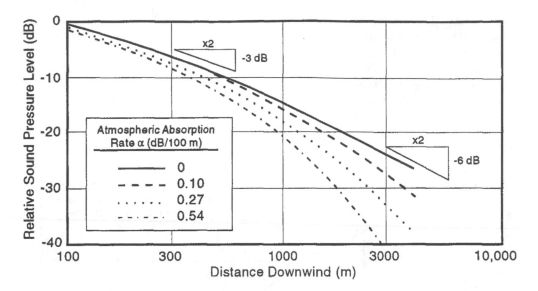

Figure 7-26. Calculated noise propagation downwind of a single row of wind turbines in the example array for four atmospheric absorption rates. [Shepherd and Hubbard 1986]

Effect of Multiple Rows

Figure 7-27 presents the results of sound-pressure-level calculations that were made for one, two, four, and eight rows of wind turbines; this illustrates the effects of progressively doubling the number of machines for a constant turbine spacing. At zero atmospheric absorption, and at receiver distances that are large relative to the array dimensions, a doubling of the number of rows results in an increase of 3 dB in the sound pressure level. This simply reflects a doubling of acoustic power. At shorter distances, the closest machines dominate and the additional rows result in only small increments in the sound

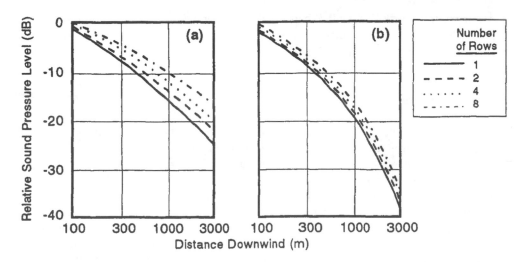

Figure 7-27. Calculated noise propagation downwind of various numbers of rows of wind turbines in the example array. [Shepherd and Hubbard 1986] (a) Without atmospheric absorption. (b) Absorption coefficient $\alpha = 0.54$ dB/100 m

pressure level. For nonzero atmospheric absorption, the effect of additional rows is less significant at all receiver distances. Doubling the number of rows results in an increase in the sound pressure level of less than 3 dB.

Figure 7-28 shows similar data for two different row lengths. For these comparisons, the turbine spacing is constant and the row lengths are doubled by doubling the number of machines per row. When the receiver is at shorter distances, the predicted sound pressure levels are equal because of the equal turbine spacing. At longer distances, the levels for the double-length row are higher by 3 dB because the acoustic power per row is doubled.

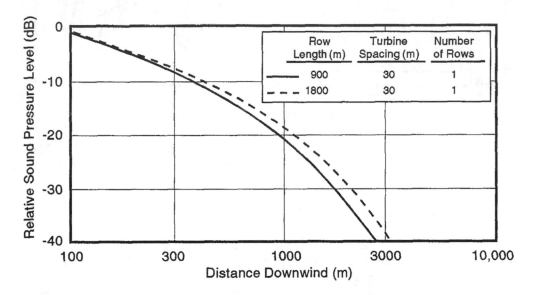

Figure 7-28. Calculated noise propagation downwind of wind turbines in rows of two different lengths. (α = 0.54 dB/100 m) [Shepherd and Hubbard 1986]

Computations were also made [Shepherd and Hubbard 1986] for a configuration similar to that of Figure 7-25, except that the row spacing was reduced from 200 m to 100 m. At all distances to the receiver, the computed sound pressure levels were higher for this more compact array.

Effect of Turbine Rated Power

Shepherd and Hubbard [1986] calculated the effect of the turbine's rated power on noise emissions by increasing the power of each turbine and the total station power. The turbine and row spacings were adjusted from those of Figure 7-25 to more appropriate values for larger machines (four times the rated power). Sound pressure levels from rows of 16 400-kW wind turbines were compared with levels from the same number of rows of 31 100-kW machines. This approximately doubled the rated power of the station. The reference spectrum for the larger turbines was assumed to have the same shape as that of the smaller turbines (Figure 7-10), although the levels were all 6 dB higher. This implies four times the acoustic power for four times the rated power. The computed sound pressure levels are 3 dB higher for the array of larger turbines because the acoustic power is doubled for each row of the array. Different results would be obtained if the reference spectra of the two sizes of turbines had different shapes.

Directivity Considerations for a Wind Power Station

Although individual turbines have been treated as if they radiate sound equally in all directions, an array of such sources may not have uniform directivity characteristics. Figure 7-29 compares the predicted sound pressure levels for two array configurations as received from two different directions. Calculations are presented for a receiver located downwind on the line of symmetry perpendicular to the rows and for a receiver located crosswind on the line of symmetry parallel to the rows. For the case of one row of turbines, the cross-wind sound pressure level is predicted to be about 5 dB lower than the downwind level near the turbines, and only about 2 dB lower in the far field. For an array with eight rows, the crosswind sound pressure level is only 3 dB lower near the turbines, and there is little directivity once the receiver distance exceeds 300 m. Downwind levels are higher close to the eight-row array, because the turbine spacing in the row is less than the row spacing.

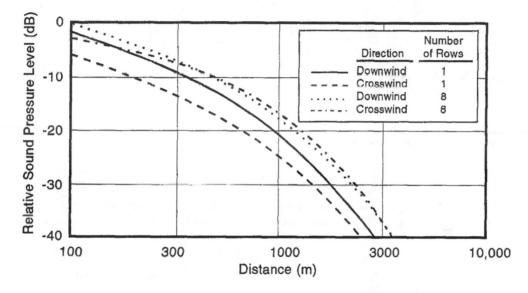

Figure 7-29. Calculated noise propagation downwind and crosswind of single and multiple rows of wind turbines in the representative wind power station. ($\alpha = 0.54$ dB/100 m)[Shepherd and Hubbard 1986]

Estimates of complete contours of sound pressure level around a wind power station are shown in Figure 7-30. The array geometry in this case consists of five rows of 31 machines each, spaced as shown in Figure 7-25. This gives an approximately square array. Figure 7-30 shows predicted contours for sound pressure levels of 40, 50, and 60 dB for an atmospheric absorption rate of 0.54 dB/100 m (which corresponds to a frequency of 1000 Hz, at 20°C and 70 percent relative humidity). Assuming a hub-height wind speed of 9 m/s, the distances to contours in the upwind direction are greatly reduced. These upwind contours are derived from computed distances to the acoustic shadow zone and the extra attenuation that occurs within this zone (see Figure 7-21).

An acoustic shadow zone forming upwind of the array results in greatly reduced distances to particular noise level contours (*i.e.*, greatly reduced noise propagation) for all frequencies above about 60 Hz. The dashed curve in Figure 7-30 shows the location of the 40-dB contour in the absence of a shadow zone.

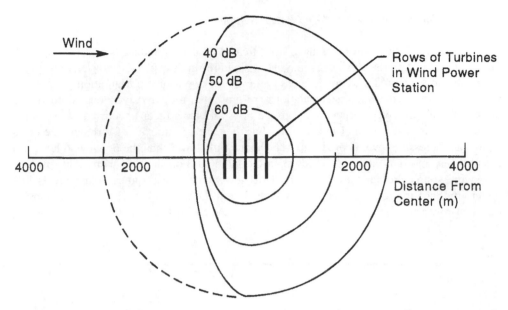

Figure 7-30. Calculated contours of sound pressure level around a five-row example array for the one-third-octave band at 1000 Hz. (α = 0.54 dB/100 m) [Shepherd and Hubbard 1986]

A-Weighted Composite Spectra

Figure 7-31 illustrates the effects of A-weighting the composite sound spectrum from the representative wind power station. Predicted sound spectra for the array are compared with equivalent spectra for a single machine (Figure 7-24). At large distances, the midrange frequencies dominate the A-weighted spectrum for both the single turbine and the array.

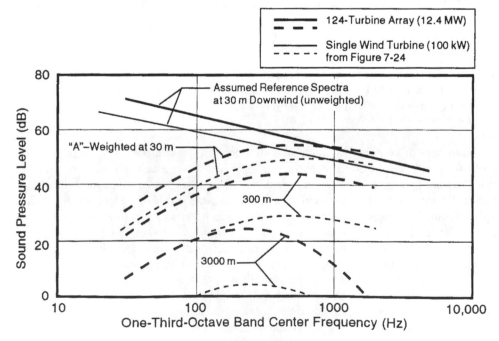

Figure 7-31. **Reference and A-weighted noise spectra for the representative 124-turbine wind power station and for a single wind turbine** [Shepherd and Hubbard 1986]

Receiver Response

Evaluating the effect of receivers' exposure to noise at various locations involves determining people's responses to direct acoustic radiation as well as the acoustic and vibrational environments inside buildings. The factors involved in such an evaluation are diagrammed in Figure 7-32 and are explained in detail in Stephens *et al.* [1982]. Noise radiated by the wind turbine is propagated through the atmosphere to a receiver (a person or a building), and the characteristics of that receiver then determine the acoustic and vibration effects of the noise. The broadband and impulsive components of the acoustic response are treated separately, and either may be significant. Background noise and building vibrations must also be considered in evaluating people's responses to wind turbine noise.

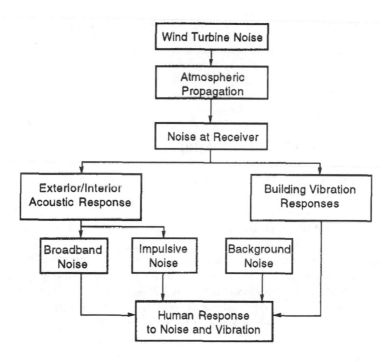

Figure 7-32. Factors and interactions to be considered in evaluating human response to wind turbine noise [Stephens *et al.* 1982]

If wind turbine noise levels are below the corresponding background noise levels, they will generally not be perceived; therefore, no adverse human response is expected. When any noise level exceeds the *threshold of perception*, however, there is the potential for community response, as indicated in Table 7-1 [ISO 1971]. The data in this table were derived from responses to noise sources other than wind turbines. Because there has been little experience to date with community responses to wind turbines, the applicability of Table 7-1 is tentative. The substantial variations in background noise in terms of time-of-day and location are complicating factors.

Perception thresholds for acoustic noise and structural vibrations have been derived separately. There are no known threshold criteria for combined effects, except in terms of the quality of the ride in transportation vehicles [Stephens 1979].

Table 7-1. Estimated Community Response to Noise (ISO 1971)

Amount by which received noise exceeds threshold level (dB)	Estimated community response	
	Category	Description
0	None	No observed reaction
5	Little	Sporadic complaints
10	Medium	Widespread complaints
15	Strong	Threats of community action
20	Very Strong	Vigorous community action

Perception of Noise Outside Buildings

Evaluating people's responses to wind turbine noise outside buildings involves the physical characteristics of the noise of the machines, the pertinent atmospheric phenomena, and the ambient or outdoor background noise at the receiver's location. Both broadband and narrow-band noise components must be considered if they are present in the noise spectrum.

In Figure 7-33, a one-third-octave band spectrum of broadband wind turbine noise is compared with a one-third-octave band spectrum of the typical background noise in a residential neighborhood. In this case, the background noise is a combination of noises from numerous distant sources, with no dominant specific source. Wind effects are also absent. Note that the turbine noise levels are generally lower than the background noise levels, except at 1000 Hz, where they are about equal. In the laboratory, human subjects exposed to the spectra of Figure 7-33 can just perceive the wind turbine noise. High-frequency wind turbine noise is generally not perceived in laboratory tests when the turbine's one-third-octave band levels are below the corresponding levels of background noise (which, in this case, had small temporal fluctuations).

The same general findings apply to the perception of low-frequency impulsive noise. A series of laboratory tests [Stephens *et al.* 1982; Shepherd 1985] were conducted to determine the detection thresholds of impulsive wind turbine noises in the presence of ambient noise with a spectral shape similar to that in Figure 7-33. In contrast to the relatively simple detection model for higher-frequency noises, understanding the perception of low-frequency impulsive noise requires that a full account be taken of the blade passage frequency of the wind turbine, the ambient noise spectrum, and the absolute hearing threshold. The latter is important because the human ear is relatively insensitive to the low frequencies that characterize impulsive wind turbine noise.

In addition to laboratory tests with sample spectra, field tests can be used to determine thresholds of perception around wind turbines, including directivity effects. For example, aural (hearing) detectability contours were determined for two large-scale HAWTs surrounded by flat terrain. The results are shown in Figure 7-34.

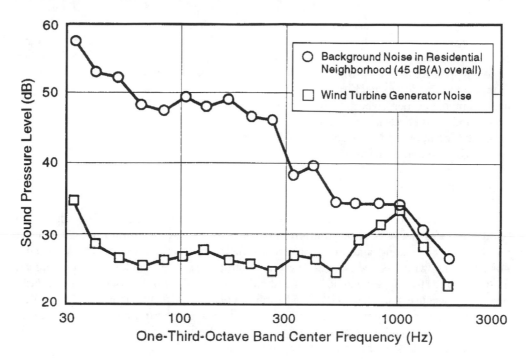

Figure 7-33. Example broadband noise spectrum at the perception threshold in the presence of the given background noise spectrum. [Stephens *et al.* 1982]

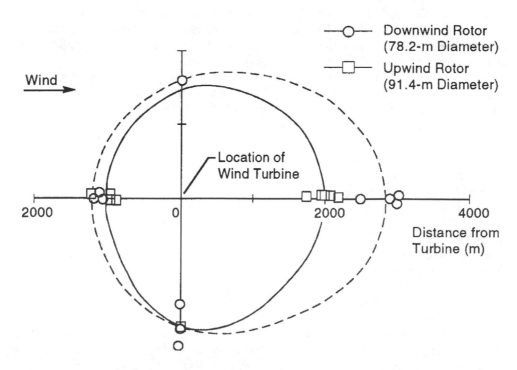

Figure 7-34. Perception thresholds for large-scale HAWTs with downwind and upwind rotors.

In Figure 7-34, each data point represents observations by one or two people and defines the distance at which the wind turbine noise is heard intermittently. The two aural curves in the figure are then estimated from these observations and from a limited number of sound pressure measurements. Both curves are foreshortened in the upwind direction and elongated in the downwind direction. With one exception, broadband noise was the dominant component perceived for both HAWTs. The exceptional case is that of noise downwind of the downwind-rotor machine, for which low-frequency impulses are the dominant component. This accounts for its longer downwind detection distance as compared with that of the upwind-rotor turbine.

Background Noise

Because background noise is an important factor in determining people's responses to wind turbine noise, it must be carefully accounted for by site measurements without the wind turbines operating, and preferably prior to their construction. Sources of background noise are the wind itself; its interaction with structures, trees, and vegetation; human activities; and, to a lesser extent, birds and animals. Natural wind noises are particularly important because they can mask wind turbine noise, as a result of the fact that their broadband spectra are similar to those of wind turbines. Measuring background noise, at the same locations and with the same techniques used for measuring wind turbine noise, is an integral part of assessing receiver response.

Noise Exposure Inside Buildings

People who are exposed to wind turbine noise inside a building experience a much different acoustic environment than do those outside. The noise transmitted into the building is affected by the mass and stiffness characteristics of the structure, the dynamic response of structural elements, and the dimensions and layouts of rooms. People may actually be more disturbed by the noise inside their homes than they would be outside [Kelley *et al.* 1985]. Indoor background noise is also a significant factor.

Data showing the reductions in outdoor noise provided by typical houses are given in Figure 7-35 as a function of frequency. The hatched area shows experimental results obtained from a number of sources [Stephens *et al.* 1983]. The noise reduction values of the ordinate are the differences between indoor and outdoor levels. The most obvious conclusion here is that noise reductions are larger at higher frequencies. This implies that a spectrum measured inside a house will have relatively less high-frequency content than that measured outside. These data are derived from octave-band measurements but are generally not sensitive to frequency bandwidth.

Very few data are available on outdoor-to-indoor noise reduction at the lowest frequencies (*i.e.*, below 50 Hz). In this range, wavelengths are comparable to the dimensions of rooms, and there is no longer a diffuse sound field on the inside of the building. Other complicating factors are low-frequency building resonances and air leaks. The inside distribution of sound pressure can be nonuniform because of structure-borne sound, standing wave patterns, and cavity resonances in rooms, closets, and hallways.

Data relating to the noise-induced vibration responses of houses are summarized in Figure 7-36, in which RMS acceleration levels are plotted as a function of external sound pressure level. The trend lines for windows, walls, and floors are averaged from a large number of test measurements of aircraft and helicopter noises, sonic booms, and wind turbine noise.

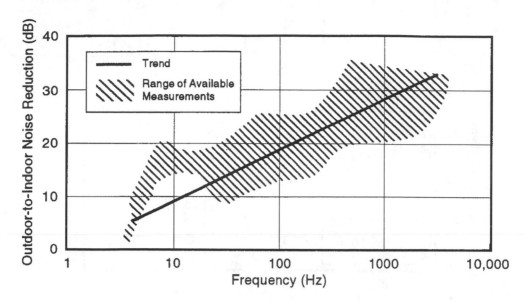

Figure 7-35. **Outdoor-to-indoor noise reduction in a typical house with closed windows.** [Stephens *et al.* 1982]

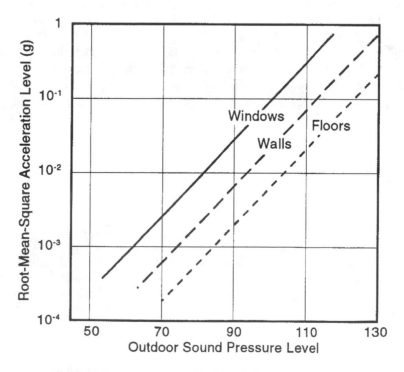

Figure 7-36. **The noise-induced acceleration of the typical structural elements of a house, as a function of outdoor sound pressure level.** [Stephens *et al.* 1982]

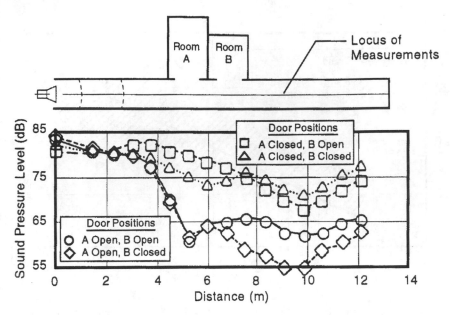

Figure 7-37. Sound-pressure-level gradients in a hallway excited by a pure-tone (21 Hz), constant-power loudspeaker. [Hubbard and Shepherd 1986]

Gradients and Resonances for Indoor Sound Pressure Levels

Large spatial variations in sound pressure level may occur within a house from a uniform external noise excitation. People moving within the house could be sensitive to these variations. Figure 7-37 illustrates the sound-pressure level gradient in a hallway with various combinations of open and closed doors. Noise was produced by a loudspeaker at a discrete frequency of 21 Hz. This frequency represents the low-frequency noise components from wind turbines that would propagate efficiently through buildings. When doors to Rooms A and B are both closed, there is a general decrease in the sound pressure level with distance up to the end of the hallway. When doors are opened in various combinations, the hallway pressure levels can be raised substantially. The changes in level that occur when room doors are open are similar to what might occur for side-branch resonators in a duct.

Because of the way rooms are arranged in houses, it is possible that *Helmholtz resonances* (cavity resonances) may be excited at certain frequencies, depending on the volumes of the rooms and whether doors are open or closed [Davis 1957, Ingard 1953]. Hubbard and Shepherd [1986] present results of sound-pressure-level surveys conducted inside a room during resonance. For this condition, the inside pressures were everywhere in phase and tended to maintain a uniform level. This is in contrast to the large gradients observed in the excitation of normal acoustic modes in a room [Knudsen and Harris 1978]. The latter modes are excited at frequencies for which the acoustic half-wavelengths are comparable to or less than the room dimensions, whereas Helmholtz resonance wavelengths are characteristically large compared with the room dimensions. Rooms A and B in Figure 7-37 both exhibit Helmholtz resonance behavior at an excitation frequency of 21 Hz.

Coupling Noise Fields in Adjacent Rooms

As sound-pressure-level gradients change in a hallway outside rooms according to whether doors are open or closed (Figure 7-37), so also do the levels inside the rooms.

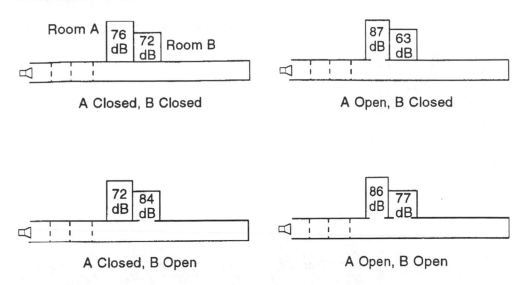

Figure 7-38. Effect of door positions on the maximum sound pressure levels in rooms adjacent to a hallway excited by a loudspeaker at 21 Hz. [Hubbard and Shepherd 1986]

Figure 7-38 illustrates the manner in which these changes can occur for various door positions. Variations in sound pressure level are as high as 20 dB for a steady noise input, both inside the rooms and in the hallway, as shown previously in Figure 7-37. This implies that a person might experience a change in levels of this order of magnitude at a particular location, depending on the doors, or as a function of location for a particular door arrangement. During tests, the highest sound pressure levels of Figure 7-38 could be readily heard, but the lowest levels were not audible.

Mechanical coupling between adjacent rooms can also excite acoustic resonances, as indicated by data in Hubbard and Shepherd [1986]. One wall of a test room was mechanically excited, and two response peaks were noted. One peak corresponded to the Helmholtz mode of the room, and the other was the first structural mode of the wall. Measured sound-pressure-level gradients were small in both cases.

Perception of Building Vibrations

One of the common ways that a person senses noise-induced excitation of a house is through structural vibrations. This mode of observation is particularly significant at low frequencies, below the threshold of normal hearing.

No standards are available for the threshold of perception of vibration by occupants of buildings. Guidelines are available, however, that cover the frequency range from 0.063 to 80 Hz [Hubbard 1982, ISO 1987]. The appropriate perception data are reproduced in Figure 7-39. The hatched region in this figure shows the perception threshold data obtained in a number of independent studies. Different investigators, using different measurement techniques, subjects, and subject orientations, have obtained perception levels extending over a range of about a factor of 10 in vibration amplitude. The composite guidelines of Figure 7-39 are judged to be the best representation of the most sensitive cases from the available data on *whole-body vibration perception*.

The two small shaded regions in Figure 7-39 are from the data of Kelly *et al.* [1985]. These are estimates of levels of vibration perceived in two different houses excited by noise from the Mod-1 HAWT. The latter data fit within the body of test data on which the *International Standards Organization* composite guideline is based. Therefore, they generally confirm the applicability of this guideline for structural vibrations induced by wind turbine noise.

Note that if measured vibration levels are not available they can be estimated for typical house building elements from Figure 7-36, provided the external noise excitation levels are known.

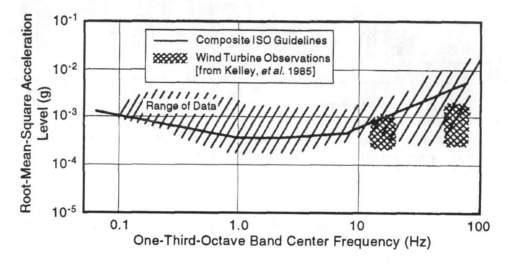

Figure 7-39. Most sensitive threshold of perception of vibratory motion. [Stephens *et al.* 1982]

Measuring Wind Turbine Noise

Wind turbine noise is measured to define source characteristics, to provide acoustic information for environmental planning, and to validate compliance with existing ordinances. It is important to use the appropriate equipment and measurement procedures and to acquire data under appropriate test conditions. Measuring noise from wind turbine generators is particularly difficult because of the adverse effects of the wind [Andersen and Jakobsen 1983, Jakobsen and Andersen 1983]. As a result, a number of special considerations are involved in selecting measurement locations and equipment and in recording and analyzing data. This section presents some guidelines on each of these subjects.

To make meaningful comparisons of the noise outputs of different wind turbines and evaluations of environmental noise control, it is necessary to have generally accepted standards of measurement. AWEA [1988], IEA-WECS [1988], and IEC [1998] contain the results of work in the wind energy community to develop such standards. These documents address significant issues in the measurement of wind turbine noise.

To interpret acoustic measurements, it is usually necessary to simultaneously record various non-acoustic quantities. Among these are wind speed and direction, ambient temperature and relative humidity, rotor speed, power output, time-of-day and date, type of terrain and vegetation, and amount of cloud cover. Atmospheric turbulence (which is often difficult to measure directly) may be inferred from this information.

Measuring Points

Most noise measurements (other than those for research purposes) are made to characterize the radiated noise of a particular machine. This infers that all data should be obtained far enough from the machine to be in the acoustic far-field. For practical applications, the reference distance d_0 illustrated in Figure 7-40 should be approximately equal to the total height of a HAWT or, in the case of a VAWT, the total height plus the rotor equatorial radius [IEA-WECS 1988]. The choice of a much greater distance may not be acceptable because of the reduced signal-to-noise ratio and because atmospheric attenuation and refraction effects can complicate interpretation of the data. To ensure the best possible signal-to-noise ratio, measurement points should be as close to the source as possible without being in the acoustic near-field.

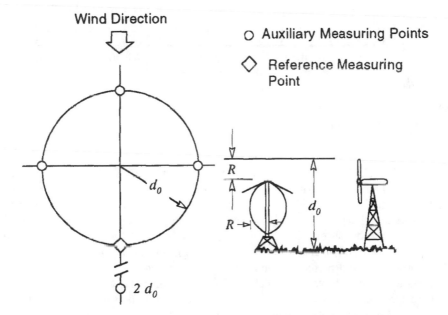

Figure 7-40. Recommended patterns of measuring points for acoustic surveys of wind turbines. [IEA-WECS 1988]

The number of measurement points needed can be determined by inspecting the polar diagrams in Figures 7-6 and 7-7. The aerodynamic noise sources in wind turbines are not highly directional, but the highest levels are usually in the upwind and downwind quadrants. A rather coarse azimuthal spacing seems adequate to define these aerodynamic radiation patterns around a HAWT, because they are generally symmetrical about its axis of rotation. If a particular turbine produces significant mechanical noise, however, that radiation pattern may be asymmetrical and highly directional.

Microphone Positions

An important consideration in laying out a measurement program is defining microphone height above the ground. Placing the microphone at ear level is conceptually attractive because it should record what people hear. The disadvantage of this height is that the data are more difficult to interpret. Figure 7-41 illustrates how sound pressure level data can be affected by microphone height, and compares sound pressure levels near the ground

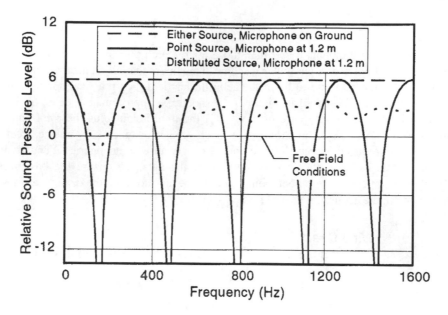

Figure 7-41. Calculated effect of microphone height on the measured noise spectra for point and distributed sources.

to results in free-field conditions (*i.e.* away from all reflecting surfaces). The solid curve represents a calculated spectrum from a point source (such as a gearbox) located 20 m above hard ground, as received at a microphone position 1.2 m above ground and 40 m from the source. The peaks and valleys represent interference patterns caused by differences between the distances traveled by sounds coming directly from the source to the receiver and those reflected from the ground surface. Under ideal conditions (with no mean wind or turbulence and perfect ground reflection), the levels vary alternately from 6 dB above free-field values to very low values. For an assumed incoherent ring source with a 20-m diameter (which represents the broadband aerodynamic noises) the distributed-source curve applies. For a microphone height of 1.2 m, the variation of sound pressure level with frequency from a distributed source is less than that for a point source, but it is still significant.

A measurement at the ground surface, however, gives a constant enhancement above free-field values that is 6 dB over the entire frequency range, as indicated by the horizontal line in Figure 7-41. Thus, it is common practice to place microphones at ground level on a hard, reflecting surface (such as plywood) and then to deduct 6 dB from all measured sound pressure levels. When there are very low-frequency components to be measured, calculations suggest that microphone placement is not critical. The first dip in the spectrum occurs at a frequency well above that associated with low-frequency rotational harmonics, as shown in Figure 7-5.

Acoustic Instrumentation

The requirements for acoustic instrumentation are derived from the type of measurements to be performed and most directly from the frequency range of concern. For the frequency range of 20 to 10,000 Hz, standardized equipment is available for detecting, recording, and analyzing acoustic signals. A number of different microphones with flat frequency responses are available. Likewise, sound-level meters that meet existing acoustic standards are available for direct readout or for recording them. Either frequency-modulated or direct-record tape systems are acceptable.

For cases where the frequency range of measurements must extend below 20 Hz, some special items of equipment may be required. Although standard microphone systems can be used, their frequency response is poor at the lowest frequencies. Special microphone systems will increase signal fidelity, along with special procedures to minimize wind noise problems.

Windscreen Applications

Noise measured in the presence of wind is contaminated by various types of wind-related noises. These include *natural wind noise*, from atmospheric turbulence; *microphone noise*, caused by the aerodynamic wake of the microphone or its windscreen; *vegetation noise*, caused by nearby trees, bushes, and ground cover; and *miscellaneous wake noise*, from the aerodynamic wakes of accessories (such as a tripod) or a nearby structure.

Because of the deleterious effects of the wind, windscreens are recommended to reduce microphone noise for all measurements of wind turbine and site background noise. Commercial windscreens of open-cell polyurethane foam are usually adequate for routine measurements at or near ground level, where wind speeds are relatively low. For measurements above ground, larger windscreens of custom design may be necessary [Sutherland, Mantley, and Brown 1987]. It is essential that the acoustic insertion loss of any windscreen be either zero or known as a function of frequency, so that appropriate corrections can be made to the data.

Wind noise is a particularly severe problem at the lowest frequencies. The ambient (*i.e.*, wind-only) noise spectrum increases as frequency decreases and may submerge some low-frequency wind turbine noise components in the ambient noise at the microphone location. In such situations, customized windscreens may help reduce the low-frequency wind noise. Some special cross-correlation analysis techniques have also been applied that use measurements from pairs of microphones [Bendat and Piersol 1980].

Little can be done to reduce noise from vegetation, other than to locate microphones away from significant sources. Noise generated by the aerodynamic wakes of accessories such as tripods can often be reduced by streamlining.

Data Analysis

The data analysis required depends on the types of acoustic information desired. If A-weighted data are needed, they can be obtained directly from a sound-level meter or from tape recordings and A-weighting filters. Statistical data can also be obtained directly by means of a *community noise analyzer*. Broadband data are routinely produced from one-third-octave band analysis such as that illustrated in Figure 7-8. Narrow-band analysis can be obtained with the aid of a wide range of filter bandwidths, the main requirement being that the bandwidth is small compared to frequency intervals between discrete frequency components.

Measured Sound Power Levels

There are two important measures of the magnitude or intensity of sound that can easily be confused. These are the sound *pressure* level and the sound *power* level.

Sound Pressure Level

Sound pressure level is a measure of the intensity of sound at a listener's location, and as such it is a combination of the radiated acoustic power of the noise source and the propagation of that power from the source to the listener. Thus the sound pressure level at an observer location will depend on factors such as the distance from the source, the source directivity, and the propagation path. The latter includes effects due to the atmosphere (absorption, refraction) and terrain (absorption, shielding and diffraction). As explained previously, all sound pressure levels presented in this chapter are based on root-mean-square (RMS) values of pressure and are averaged over 30 to 180 seconds, depending on the frequency bandwidth. Pressure levels in decibels are referenced to the threshold of hearing, typically defined as 20 micro-Pascals, as follows:

$$SPL = 20 \log_{10} (p / p_0) \qquad (7\text{-}10)$$

where
$\quad SPL$ = sound pressure level (dB or dBA)
$\quad\quad p$ = RMS acoustic pressure (Pa)
$\quad\quad p_0$ = reference acoustic pressure, 2 x 10^{-5} Pa

A sound pressure level can be measured with a microphone or microphone/meter at the listener's location. Sound pressure levels can vary considerably with location and time, because they depend not only on the generation of sound but also on its propagation. Therefore, sound pressure level alone is not a convenient parameter for assessing and comparing the acoustic outputs of wind turbines. Another measure is needed, and that is sound power level.

Sound Power Level

Sound power level is a characteristic of the noise source, and as such it is independent of the environment around the source and the location of the listener. It is expressed in decibels and referenced to a power of 1 picowatt, as follows:

$$PWL = 10 \log_{10} (P_A / P_0) \qquad (7\text{-}11)$$

where
$\quad PWL$ = sound power level (dB)
$\quad\quad P_A$ = acoustic power of the sound source (W)
$\quad\quad P_0$ = reference acoustic power, 1×10^{-12} W

The sound power level of a wind turbine is a noise characteristic of that turbine that does not depend on either the environment in which it is located or the relative location of a listener. For example, Søndergaard and Plovsing [2005] measured the sound power level of a 2 MW wind turbine situated in interior Danish waters, and found that the offshore levels were 1 dB to 3 dB above the onshore levels measured for the same model of wind turbine located on land. This difference is consistent with expected machine-to-machine variation when both are located on land.

While a sound power level of a wind turbine cannot be measured directly with a meter, it can be determined from sound pressure levels measured nearby, by assuming (a) the wind turbine is a point noise source and (b) that its sound power spreads outward over a spherical surface. Sound power level is related to sound pressure level as follows:

$$PWL = SPL + 10 \log_{10}(a\pi R^2/A_0) + \alpha(R/100) + b \qquad (7\text{-}12)$$

where

a = propagation shape factor
     = 4 for spherical spreading *free-field* or above an ideally absorbing or anechoic surface
     = 2 for hemispherical spreading above an ideally reflecting (hard) surface
R = slant distance from rotor center to microphone or meter (m)
A_0 = reference area of sound source, 1 m²
α = atmospheric absorption rate (dB per 100m)
b = empirical microphone correction constant (dB or dBA)
     = - 6 dB recommended for a microphone located on a ground board [IEC 1998]

Figure 7-26 illustrates the relative decrease in sound pressure levels assuming four different atmospheric absorption rates, α, varying from 0 to 0.54 dB per 100m.

Sound Power of Utility-Scale Wind Turbines

According to Howe *et al.* [2007], modern wind turbines are considerably quieter than earlier versions, with some investigators finding a reduction in recent years of about 10 dB. While different models and different manufacturer's systems have their own acoustic characteristics, various investigators indicate that sound power levels of 105 dBA are typical for modern wind turbines in the 1 to 2 MW range at moderate wind speeds (Fig. 7-42).

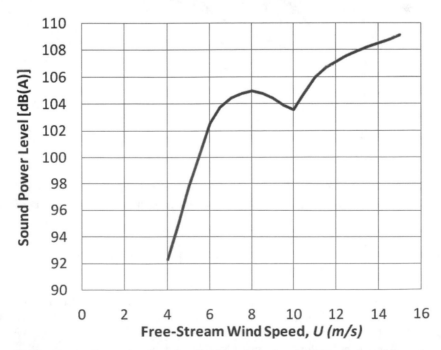

Figure 7-42. Typical sound power levels of modern 1 to 2 MW wind turbines as a function of wind speed. [Howe *et al.* 2007]

Sound Power Levels of Small-Scale Wind Turbines

Sound power levels of a variety of 3-bladed small-scale wind turbines were measured by Migliore *et al.* [2003]. Rated powers of these turbines ranged from 1.0 to 100 kW, with rotor diameters of 2.1 to 19 m. A graphical summary of the results of this series of acoustic tests is shown in Figure 7-43. The authors refer to these power levels as "apparent", because they are not measured directly but inferred through the use of Equation (7-12). The constant a in this equation was set equal to 4, and the constant b was taken equal to -6 dB, because sound pressure measurements were made with microphones mounted on a ground board. The atmospheric absorption rate α was assumed to be zero.

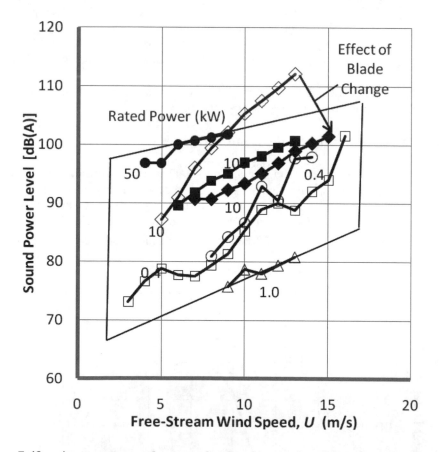

Figure 7-43. Apparent sound power levels of several small-scale, 3-bladed wind turbines with at least 6 dB(A) separation from background noise. [Migliore *et al.* 2003]

It can be seen from Figure 7-43 that sound power levels from these small-scale wind turbines were found to increase with increasing wind speed at an average rate of about 1 dB per meter per second. Sound power also increased significantly with the rated power of the turbines. All else being equal, one would expect a ten-fold increase in rated power to result in an approximately 10 dB increase in sound power. The range of the rated powers in the figure (1-100 kW) indicates an expected 20dB range in sound power, which is approximately consistent with the measured levels.

The scale of the reduction in sound power that can be achieved by altering the shape of a rotor blade is indicated by the diamond symbols in Figure 7-43. The open diamonds show the sound power of the rotor with blades of the original design that was found to be quite noisy in higher winds. The closed diamonds indicate the much lower sound power coming from the redesigned blades on the same wind turbine. At wind speeds from 10 to 15 m/s, sound power was reduced 12 to 15 dB by the blade design change.

Sound Sources on Wind Turbines

Acoustic tests have been conducted on an operating intermediate-scale wind turbine to determine the specific locations from which sound power is emitted. Oerlemans *et al.* [2005] conducted a detailed series of noise measurements on a pitch-controlled, three-bladed *GAMESA G58* wind turbine, which has a rotor diameter of 58 m, a tower height of 53.5 m, and a rated power of 850 kW. Using an elliptical array of 152 microphones mounted on a wooden platform on the ground, these researchers were able to obtain a profile of the noise sources on the operating turbine.

Figure 7-44. Noise sources measured on a Gamesa G58 850 kW wind turbine by means of a microphone array on the platform in the foreground. The downward-moving blades emit the majority of the noise, as indicated by the sound power contours whose range is 12 dB. [Oerlemnes 2005]

Figure 7-44 shows a downwind view of the test turbine onto which contours of down-ward-radiated noise levels have been projected, both from the blades and from the nacelle. The range of power levels on these contours is 12dB. The most striking phenomenon is that practically all downward radiated blade noise (as measured by the microphone array on the platform in the foreground) is produced during the downward movement of the blades. Since the range of the contours shown is 12 dB, this means that the downward-radiated noise produced during the upward movement of a blade (on the left of the tower) is at least 12 dB less than during its downward movement on the right of the tower.

Referring to Figure 7-14, the noise emission pattern illustrated in Figure 7-44 indicates that trailing edge noise is the leading source of broadband noise from this rotor. Furthermore, it can be seen that the majority of the blade noise is produced by the outboard section of the blades, but not by the very tip in this blade design. The authors note that for a different observer location the pattern of sound emission will be different. For example, an observer located further away would hear noise being generated over the full-blade rotation, not just over one side as indicated in this figure.

A second important observation is that the noise from the blades clearly dominates the noise from the nacelle. The difference between the overall sound pressure levels from the nacelle and those from the blades was found to increase with increasing wind speed, from about 8 dB(A) at 6 m/s to about 11 dB(A) at 10 m/s.

Survey of Community Response to Wind Turbine Noise

Studies of the reactions of nearby residents to the sight and sound of wind power stations have been conducted as part of the planning and zoning approval processes. One of the most detailed is a recent study in the Netherlands sponsored by the *European Union* [van den Berg *et al.* 2008]. The study population was selected from all residents in the Netherlands living within 2.5 km from a wind turbine. As the study emphasized modern wind power stations, wind turbines were selected with a rated power of 500 kW or more and one or more turbines within 500 m from the first. Excluded were wind turbines that were erected or replaced in the year preceding the survey. Residents lived in the countryside with or without a busy road close to the turbine(s), or in built-up areas (villages, towns). Excluded were residents in mixed and industrial areas. The survey was conducted by mail.

The sound level at the residents' dwellings was calculated according to the international ISO standard for sound propagation, the almost identical Dutch legal model, and a simple (non-spectral) calculation model. The indicative sound level used was the sound level when the wind turbines operated at a wind speed of 8 m/s in the daytime, at high but not maximum power. Respondents were exposed to levels of wind turbine sound between 24 and 54 dB(A) and wind turbines at distances from 17 m to 2.1 km. The angular elevations of the largest wind turbines ranged from 2 degrees to 79 degrees, with an average value of 10 degrees. Wind turbines occupied about 2 percent of the space above the horizon, on average.

Summary of the Main Conclusions of the Survey

With Respect to Hearing Wind Turbines

- Not having wind turbines visible from the dwelling and high levels of background (road traffic) sound decreased the probability of hearing wind turbine sound, though the influence of background sound was found to be small.

- Wind turbines were perceived as louder when the wind was blowing from the wind-turbine towards the dwelling and less loud vice versa.
- Wind turbines were perceived as louder when the wind was strong and less loud with a weak or no wind. However, more respondents thought it was louder than less loud at night, even though wind speeds were lower at night, on average.

With Respect to Annoyance from Wind Turbine Sound

- Of the potential annoyances from wind turbines, noise was the most annoying.
- The most common description of annoying wind turbine sound was swishing; annoyance was more probable for respondents who gave this description than for those who did not.
- Benefiting economically from wind turbines decreased the probability of being annoyed by wind turbine sound.
- Location near a main road (in comparison with a built-up area) decreased the probability of being annoyed by wind turbine sound.
- Although the presence of background sound from road traffic made wind turbine sound less noticcable, higher levels of background sound did not reduce the probability of being annoyed.
- Annoyance with wind turbine noise was associated with a negative attitude towards wind turbines in general and the perceived negative visual impact of wind turbines on the landscape.

With Respect to Other Health Effects Associated with Wind Turbines

- The risk for sleep interruption by noise was higher at levels of wind turbine sound pressure levels above 45 dB(A) and lower at levels below 30 dB(A).
- Annoyance with wind turbine noise was associated with complaints of psychological distress, stress, difficulties in falling asleep, and sleep interruption.

References

American National Standards Institute (ANSI), 1978, *Method for the Calculation of the Absorption of Sound in the Atmosphere*, ANSI SI.26, New York: Acoustical Society of America.

American Wind Energy Association (AWEA), 1988, *Standard Performance Testing of Wind Energy Conversion Systems*, Standard AWEA 1.2-1988, Washington, DC: American Wind Energy Association.

Andersen, B., and J. Jakobsen, 1983, *Noise Emissions from Wind Turbine Generators: A Measurement Method*, Danish Acoustical Institute Technical Report No. 109, Lyngby, Denmark: Danish Academy of Technical Sciences.

Bendat, J. S., and A. G. Piersol, 1980, *Engineering Applications of Correlation and Spectral Analysis*, New York: John Wiley and Sons.

Brooks, T. F., and T. H. Hodgson, 1980, *Prediction and Comparison of Trailing Edge Noise Using Measured Surface Pressures*, Preprint 80-0977, American Institute of Aeronautics and Astronautics.

Brooks, T. F., and M. A. Marcolini, 1986, "Airfoil Tip Vortex Formation Noise," *AIAA Journal*, Vol. 24, No. 2.

Daigle, G. A., T. F. W. Embleton, and J. E. Piercy, 1986, "Propagation of Sound in the Presence of Gradients and Turbulence Near the Ground," *Journal of the Acoustical Society of America*, Vol. 79, No.3.

Davis, D. D. Jr., 1957, "Acoustical Filters and Mufflers," *Handbook of Noise Control*, C. M. Harris, ed., New York: McGraw Hill.

George, A. R., 1978, "Helicopter Noise State-of-the-Art," *Journal of Aircraft*, Vol. 15, No. 11: pp. 707-711.

George, A. R., and S. T. Chou, 1984, "Comparison of Broadband Noise Mechanisms, Analyses, and Experiments on Rotors," *Journal of Aircraft*, Vol. 21.

Glegg, S. A. L., S. M. Baxter, and A. G. Glendinning, 1987, "The Prediction of Broadband Noise for Wind Turbines," *Journal of Sound and Vibration*, Vol. 118, No. 2: pp. 217-239.

Greene, G. C, and H. H. Hubbard, 1980, *Some Effects of Non-Uniform Inflow on the Radiated Noise of a Large Wind Turbine*, NASA TM-8183, Hampton, Virginia: NASA Langley Research Center.

Greene, G. C, 1981, "Measured and Calculated Characteristics of Wind Turbine Noise," *Proceedings*, *Wind Turbine Dynamics Conference*, R. W. Thresher, ed., NASA CP-2185, DOE Publication CONF-810226, Cleveland, Ohio: NASA Lewis Research Center, pp. 355-362.

Grosveld, F. W., 1985. "Prediction of Broadband Noise from Horizontal Axis Wind Turbines," *AIAA Journal of Propulsion and Power*, Vol. 1, No. 4.

Hawkins, J. A., 1987, *Application of Ray Theory to Propagation of Low Frequency Noise from Wind Turbines*, NASA CR-178367, Hampton, Virginia: NASA Langley Research Center.

Howe, B, W. Gastmeier, and N. McCabe, 2007, "Wind Turbines and Sound: Review and Best Practice Guidelines," report submitted by HGC Engineering, Mississauga, Ontario, to the Canadian Wind Energy Association, Ottawa, Ontario.

Hubbard, H. H., 1982, "Noise Induced House Vibrations and Human Perceptions," *Noise Control Engineering Journal*, Vol. 19, No. 2.

Hubbard, H. H., and K. P. Shepherd, 1982, *Noise Measurements for Single and Multiple Operation of 50-kW Wind Turbine Generators*, NASA CR-166052, Hampton, Virginia: NASA Langley Research Center.

Hubbard, H. H., and K. P. Shepherd, 1984, *The Effects of Blade Mounted Vortex Generators on the Noise from a Mod-2 Wind Turbine Generator*, NASA CR-172292, Hampton, Virginia: NASA Langley Research Center.

Hubbard, H. H., and K. P. Shepherd, 1986, *The Helmholtz Resonance Behavior of Single and Multiple Rooms*, NASA CR-178173, Hampton, Virginia: NASA Langley Research Center.

Hubbard, H. H., and K. P. Shepherd, 1988, *Wind Turbine Acoustics Research Bibliography with Selected Annotation*, NASA TM 100528, Hampton, Virginia: NASA Langley Research Center.

IEA-WECS, 1988, *Recommended Practices for Wind Turbine Testing 4. Acoustics Measurement of Noise Emission from Wind Turbines*, International Energy Agency Programme for Research and Development on Wind Energy Conversion Systems, Second Edition, Bromma, Sweden: Aeroacoustical Research Institute of Sweden.

Ingard, U., 1953, "On the Theory and Design of Acoustic Resonators," *Journal of the Acoustical Society of America*, Vol. 25, No. 6.

IEC, 1998, "Wind Turbine Generator Systems - Part 11: Acoustic Noise Measurement Techniques," *International Standard* IEC 61400-11 (First ed.): International Electrotechnical Commission.

International Standards Organization (ISO), 1971, *Community Response to Noise*, ISO Standard R 1996-1971 (E), New York: Acoustical Society of America.

International Standards Organization (ISO), 1987, *Evaluation of Human Exposure to Whole Body Vibration - Part 2: Human Exposure to Continuous and Shock-Induced Vibrations in Buildings (1 Hz to 80 Hz)*, Draft ISO Standard 2631-2.2, New York: Acoustical Society of America.

Jakobsen, J., and B. Andersen, 1983, *Wind Noise Measurements of Wind-Generated Noise from Vegetation and Microphone System*, Danish Acoustical Institute Technical Report No. 108, Lyngby, Denmark: Danish Academy of Technical Sciences.

Kelley, N. D., R. R. Hemphill, and D. L. Sengupta, 1981, "Television Interference and Acoustic Emissions Associated with the Operation of the Darrieus VAWT," *Proceedings, 5th Biennial Wind Energy Conference and Workshop, Vol. I,* I. E. Vas, ed., SERI/CP-635-1340, Golden, Colorado: National Renewable Energy Laboratory, pp. 397-413.

Kelley, N. D., R. R. Hemphill, and H. E. McKenna, 1982, "A Comparison of Acoustic Emission Characteristics of Three Large Wind Turbine Designs," *Proceedings, Inter-Noise '82*, New York: Noise Control Foundation.

Kelley, N.D., H.E. McKenna, R.R. Hemphill, C.L. Etter, R.C. Garrelts, and N.C. Linn, 1985, *Acoustic Noise Associated with the MOD-1 Wind Turbine: Its Source, Impact and Control*, SERITR-635-1166, Golden Colorado: National Renewable Energy Laboratory.

Kelley, N.D., H.E. McKenna, E. W. Jacobs, R.R. Hemphill, and N.J. Birkenheuer, 1987, *The MOD-2 Wind Turbine: Aeroacoustical Noise Sources, Emissions, and Potential Impact*, SERI/TR-217-3036, Golden, Colorado: National Renewable Energy Laboratory.

Knudsen, V.O., and CM. Harris, 1978, *Acoustical Designing in Architecture*, New York: American Institute of Physics.

Lowson, M.V., Nov. 1970, "Theoretical Analysis of Compressor Noise," *Journal of the Acoustical Society of America*, Vol. 47, No. 1 (Part 2).

Lunggren, S., 1984, *A Preliminary Assessment of Environmental Noise from Large WECS, Based on Experience from Swedish Prototypes*, FFA TN 1984-48, Stockholm, Sweden: Aeronautical Research Institute of Sweden.

Martinez, R., S.E. Widnall, and W.L. Harris, 1982, *Prediction of Low Frequency Sound from the MOD-1 Wind Turbine*, SERI TR-635-1247, Golden, Colorado: National Renewable Energy Laboratory.

Meijer, S. and I. Lindblad, 1983, *A Description of Two Methods for Calculation of Low Frequency Wind Turbine Noise, Including Applications for the Swedish Prototype WECS Maglarp*, FFA TN 1983-31, Stockholm, Sweden: Aeronautical Research Institute of Sweden.

Migliore, P., J. van Dam, and A. Huskey, 2003, "Acoustic Tests of Small Wind Turbines," NREL/CP-500-34662, National Renewable Energy Laboratory, Golden, Colorado.

Oerlemans, S., and B. Méndez López, 2005, "Acoustic Array Measurements on a Full-Scale Wind Turbine," NLR-TP-2005-336, National Aerospace Laboratory NLR.

Pearsons, K.S., and R.L. Bennett, 1974, *Handbook of Noise Ratings*, NASA CR-2376, Hampton, Virginia: NASA Langley Research Center.

Piercy, J.E., T.F.W. Embleton, and L.C. Sutherland, 1977, "Review of Noise Propagation in the Atmosphere," *Journal of the Acoustical Society of America*, Vol. 61, No. 6.

Piercy, J.E., and T.F.W. Embleton, 1979, "Sound Propagation in the Open Air," *Handbook of Noise Control 2nd Edition*, edited by CM. Harris, New York: McGraw Hill.

SAE, 1966, *Method for Calculating the Attenuation of Aircraft Ground to Ground Noise Propagation During Takeoff and Landing*, SAE Aerospace Information Report AIR 923, Warrendale, Pennsylvania: Society of Automotive Engineers.

Schlinker, R.H, and R.K. Amiet, 1981, *Helicopter Rotor Trailing Edge Noise*, NASA CR-3470, Hampton, Virginia: NASA Langley Research Center.

Sears, W.R., 1941, "Some Aspects of Non-Stationary Airfoil Theory and Its Practical Application," *Journal of Aerospace Sciences*, Vol. 8, No. 3.

Shepherd, K. P., and H. H. Hubbard, 1981, *Sound Measurements and Observations of the Mod-OA Wind Turbine Generator*, NASA-CR-165752, Hampton, Virginia; NASA Langley Research Center.

Shepherd, K. P., and H. H. Hubbard, 1983, *Measurements and Observations of Noise from a 4.2 Megawatt (WTS-4) Wind Turbine Generator*, NASA CR-166124, Hampton, Virginia: NASA Langley Research Center.

Shepherd, K. P., and H. H. Hubbard, 1984, *Acoustics of the Mod-0/5A Wind Turbine Rotor with Two Different Ailerons*, NASA CR-172427, Hampton, Virginia: NASA Langley Research Center.

Shepherd, K. P., 1985, *Detection of Low Frequency Impulsive Noise from Large Wind Turbine Generators*, NASA CR-172511, Hampton, Virginia: NASA Langley Research Center.

Shepherd, K. P., and H. H. Hubbard, 1985, *Sound Propagation Studies for a Large Horizontal Axis Wind Turbine*, NASA CR-172564, Hampton, Virginia: NASA Langley Research Center.

Shepherd, K. P., and H. H. Hubbard, 1986, *Prediction of Far Field Noise from Wind Energy Farms*, NASA CR-177956, Hampton, Virginia: NASA Langley Research Center.

Shepherd, K. P., W. L. Willshire, Jr., and H. H. Hubbard, 1988, *Comparisons of Measured and Calculated Sound Pressure Levels around a Large Horizontal Axis Wind Turbine Generator*, NASA TM 100654, Hampton, Virginia: NASA Langley Research Center.

Søndergaard, B. and B. Plovsing, 2005, "Noise from Offshore Wind Turbines," Danish Ministry of the Environment Report, Environmental Project No. 1016 2005.

Stephens, D. G., 1979, "Developments in Ride Quality Criteria," *Noise Control Engineering Journal*, Vol. 12, No. 1.

Stephens, D. G., K. P. Shepherd, H. H. Hubbard, and F. W. Grosveld, 1982, *Guide to the Evaluation of Human Exposure to Noise from Large Wind Turbines*, NASA M-83288, Hampton, Virginia: NASA Langley Research Center.

Sutherland, L. C, R. Mantey, and R. Brown, 1987, *Environmental and Cumulative Impact of Noise from Major Wind Turbine Generator Developments in Alameda and Riverside Counties — Literature Review and Measurement Procedure Development*, WR87-8, El Segundo, California: Wylie Laboratories.

Thomson, D. W., 1982, *Analytical Studies and Field Measurements of Infrasound Propagation at Howard's Knob, North Carolina*, SERI/TR-635-1292, Golden, Colorado: National Renewable Energy Laboratory.

Thomson, D. W., 1982, "Noise Propagation in the Earth's Surface and Planetary Boundary Layers," *Proceedings, Internoise-82 Conference*, New York: Noise Control Foundation.

van den Berg, F., E. Pedersen, J. Bouma, R. Bakker, 2008, *WINDFARMperception: Visual and Acoustic Impact of Wind Turbine Farms on Residents*, FP6-2005-Science-and-Society-20 Specific Support Action Project No. 044628 Final Report.

Viterna, L. A., 1981, *The NASA-LeRC Wind Turbine Sound Prediction Code*, NASA CP-2185, Cleveland, Ohio: NASA Lewis Research Center.

Wehrey, M. C, T. H. Heath, R. J. Yinger, and S. F. Handschin, 1987, *Testing and Evaluation of a 500-kW Vertical Axis Wind Turbine*, AP-5044, Palo Alto, California: Electric Power Research Institute.

Willshire, W. L. Jr., and W. E. Zorumski, 1987, "Low Frequency Acoustic Propagation in High Winds," *Proceedings, Noise-Con 87 Conference*, New York: Noise Control Foundation.

8

Characteristics of the Wind

Walter Frost, Ph.D.
*The University of Tennessee Space Institute
and FWG Associates, Inc.
Tullahoma, Tennessee*

and

Carl Aspliden, Ph.D.
*Battelle Pacific Northwest Laboratory
Richland, Washington*

Introduction

Wind is air in motion relative to the surface of the earth. For purposes of wind turbine design, the wind vector is considered to be composed of a *steady wind* plus *fluctuations about the steady wind.* This chapter deals with the characteristics of both the steady and fluctuating components of the wind, as an energy source and as aerodynamic forcing functions on wind turbine rotors. U.S. and world-wide wind resources are reviewed, including major methods of assessing these resources. Standard methods for measuring and describing wind turbulence are discussed, supplemented by the more recent work in measuring turbulence from a rotational frame of reference. Finally, siting techniques are described for individual wind turbines and clusters of turbines.

Causes of Wind

The primary cause of air motion is uneven heating of the earth by solar radiation. By and large, the air is not heated directly, but solar radiation is first absorbed by the earth's surface and is then transferred in various forms back into the overlying atmosphere. Since the surface of the earth is not homogeneous (land, water, desert, forest, *etc.*), the amount of energy that is absorbed varies both in space and time. This creates differences in atmospheric temperature, density, and pressure, which in turn create forces that move air from one place to another. For example, land and water along a coastline absorb radiation differently, as do valleys and mountains, and this creates breezes.

During the year, tropical regions receive more solar energy than they radiate to space, and polar regions receive less. Since the tropics do not get hotter continuously from year to year nor do the poles get colder, there is an exchange of thermal energy across latitudes with wind as the medium of convection. In addition to uneven heating, a second important factor in large-scale air movement is the *earth's rotation*, which gives rise to two effects. First, the rotation results in *Coriolis forces* which accelerate each moving particle of air. This acceleration moves an air particle to the right of its direction of motion in the Northern hemisphere, and to the left in the Southern hemisphere [Plate 1971], When air movements reach a steady state, Coriolis forces balance pressure gradient forces, leaving a resultant motion approximately along *isobars* (lines of equal pressure) which is called the *geostrophic wind.* Actual air movement is close to this idealized motion at altitudes of 600 m or more.

The second effect of the earth's rotation on wind flow becomes apparent in the mid-latitudes. By virtue of the rotation, each air particle in the atmosphere has an angular momentum directed from west to east. As a particle moves towards the poles (remaining at approximately the same altitude) it draws closer to the axis of rotation. Conservation of angular momentum requires an increase in the component of its velocity in the west-to-east direction. This effect is small near the equator, but in the temperate zones it accounts for the *Westerlies* which are in a direction opposite to the general flow, in both hemispheres.

Solar heating and rotation of the earth establish certain semi-permanent patterns of circulation in the atmosphere, as illustrated in Figure 8-1. From the above discussion, it is evident that regions of the earth in the Westerlies are preferable to others for extracting power from the wind in the boundary layer of the atmosphere. In addition to these major wind-creating factors, local topographical features may alter the energy distribution considerably, so that many exceptions from this general picture exist.

Power in the Wind

How much power is contained in the wind? The basic unit of measurement for this resource is *wind power density*, or power per unit of area normal to the wind *azimuth* (the direction *from* which the wind is blowing), calculated as follows:

$$p_W = \frac{d}{dt}(q_d x) = 0.5\rho U^2 \frac{dx}{dt} = 0.5\rho U^3 \tag{8-1}$$

where

p_W = wind power density (W/m²)
q_d = dynamic pressure of the wind (N/m²)
x = run of wind past a given point (m)
ρ = air density (kg/m³)
U = horizontal component of the mean free-stream wind velocity (m/s)

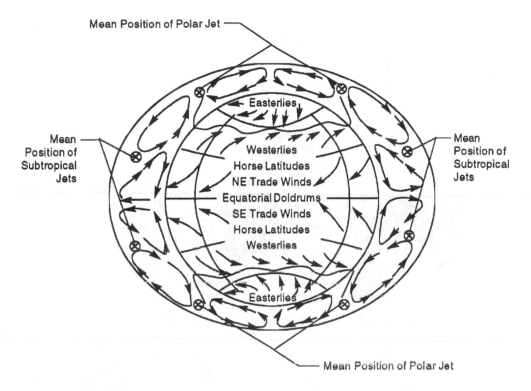

Figure 8-1. Semi-permanent global wind patterns. Trade winds and Westerlies are particularly well-suited for wind energy conversion.

Since we are interested in capturing energy from the wind and not just producing power, our best indication of the size of the local wind energy resource is its *annual average wind power density*, or

$$p_{W,a} = \frac{0.5\,\rho}{8760} \int\limits_{year} U^3\, dt \tag{8-2}$$

where

$\quad p_{W,a}$ = annual average wind power density (W/m²)
$\quad\quad t$ = time (h)

Wind resource maps generally estimate the potential for wind energy conversion in terms of *wind power classes*, as illustrated in Figure 8-2. Each class represents a range of annual average wind power densities and equivalent mean wind speeds (Table 8-1). Areas designated as Class 3 or higher are suitable for most wind turbine applications, and it is the challenge of the wind power engineer to capture this energy in a safe, cost-effective manner.

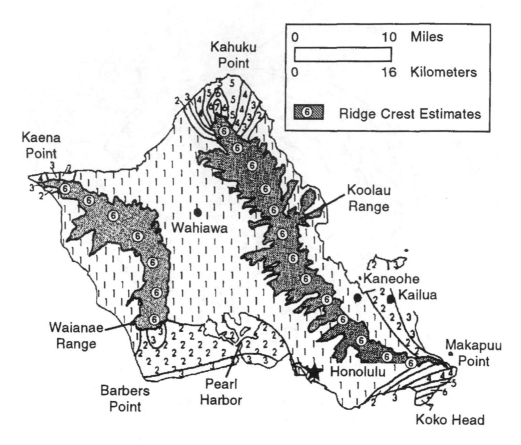

Figure 8-2. Typical wind energy resource map using the wind power classes defined in Table 1 to show estimated annual average wind power densities. [Elliott *et al.* 1986]

Table 8-1.

Wind Classification System based on Annual Average Wind Power Density[1]

Wind power class	Annual average wind power density (W/m²)		Equivalent mean wind speed (m/s)	
	10-m elevation	**50-m elevation**	**10-m elevation**	**50-m elevation**
1	0 - 100	0 - 200	0.0 - 4.4	0.0 - 5.6
2	100 - 150	200 - 300	4.4 - 5.1	5.6 - 6.4
3	150 - 200	300 - 400	5.1 - 5.6	6.4 - 7.0
4	200 - 250	400 - 500	5.6 - 6.0	7.0 - 7.5
5	250 - 300	500 - 600	6.0 - 6.4	7.5 - 8.0
6	300 - 400	600 - 800	6.4 - 7.0	8.0 - 8.8
7	400 - 1000	800 - 2000	7.0 - 9.4	8.8 - 11.9

[1]A Rayleigh frequency distribution is assumed

Scales of Motion in the Atmosphere

In order to understand the wind energy resource and its distribution in time and space, it is helpful to look at the different scales of air flow that occur on our rotating globe. While the ultimate energy source for atmospheric flow is the sun, wind is not driven directly by solar radiation but by energies derived through different kinds of conversion. Thus it is beneficial for the wind turbine engineer to be acquainted with the physical processes (*i.e.*, *dynamic, thermodynamic*, and *radiative*) that take place simultaneously in the atmosphere and cause the wind vector.

The atmosphere operates on many time and space scales, ranging from seconds and fractions of a meter to years and thousands of kilometers. Time and space scales of the atmospheric motions and their importance to wind energy utilization are summarized in Figure 8-3. The very large or *climatic scale* includes seasonal and annual fluctuations in the wind, which are useful for assessing regional wind resources. On a scale comparable with the weather maps seen in the press or used by airlines there are *large-scale synoptic fluctuations* identified by the patterns of *isobars* moving across a country. These large-scale fluctuations influence the output of wind power stations, so effective site selections depend on a knowledge of regional atmospheric motions. *Small-scale fluctuations*, which are even more local in size and higher in frequency, are best observed on local anemometer records, providing data for wind turbine design and *micrositing* (*i.e.*, siting of individual turbines).

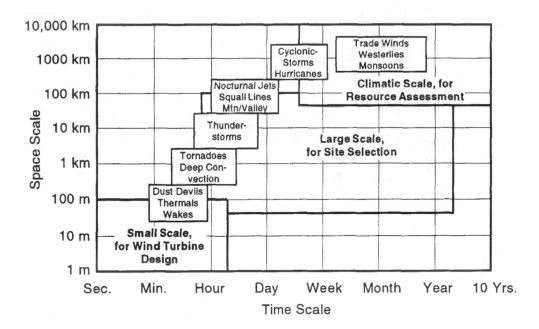

Figure 8-3. Temporal and spatial scales of atmospheric motion.

Given a record of a meteorological variable, such as the horizontal wind speed, we can always decompose it into components representing a time or spatial average and fluctuations superimposed on the average. After doing this, we view the average value of the variable as a *deterministic quantity* and assign *statistical characteristics* to only the fluctuations. Wind loadings on the turbine can also be classified into two parallel groupings: those associated with the mean wind speed, which are conveniently described as *quasi-steady* or *time-averaged* loadings; and those associated with the gustiness or turbulence of the wind, which are predominantly *dynamic* in character. Thus, for the horizontal component of the free-stream wind velocity we obtain

$$u(t) = U + g(t) \tag{8-3a}$$

$$\int_0^{\Delta t} g(t) \, dt = 0 \tag{8-3b}$$

$$RMS[u(t) - U] = \sqrt{\frac{1}{\Delta t} \int_0^{\Delta t} g^2(t) \, dt} = \sigma_0 \tag{8-3c}$$

where

$u(t)$ = instantaneous horizontal free-stream wind velocity (m/s)
U = steady horizontal free-stream wind velocity (m/s)
$g(t)$ = fluctuating wind velocity; instantaneous deviation from U (m/s)
Δt = averaging time interval (h)
$RMS[\ ]$ = root-mean-square average of [ ]
σ_0 = *ambient* or *natural* turbulence (m/s)

The characteristics of both the steady wind and the fluctuating wind depend strongly on the principal scale of interest. On the climatic or *macro-meteorological* scale, changes in wind speed will be referred to as fluctuations in the steady wind, and g is neglected. In the large-scale regime, U represents the field that appears when weather observations are plotted and smoothed and is approximately equivalent to a one-hour average. In this regime g includes all microscale motions. Small-scale or *micro-meteorological* fluctuations, such as those appearing in the anemometer record, will be referred to as *gusts* or *turbulence*.

Figure 8-4 is a *wind energy spectrum* developed by Van Der Hoven [1957] which shows that the majority of the fluctuating energy is contained at the macro- and micro-meteorological scales, and that a region of low energy exists between them. The spectral gap for periods between 0.1 and 5 hours defines a convenient range of averaging periods, Δt, to which a steady wind speed can be referenced [Davenport 1967]. The period chosen should be long enough to minimize *non-stationarities*, and short enough to reflect short-term storm activity. Periods from 10 minutes to one hour have been found to be suitable for defining the steady wind speed, giving it only a weak dependence on the averaging period. Panofsky and Dutton [1984] provide a clear discussion of different averaging techniques.

Steady wind speeds required for wind resource assessments may vary throughout each day and from day to day, but they are still essentially constant relative to the dynamic response frequencies of typical wind turbines. Much of the wind data used in assessment methodologies, however, is derived from one-minute averages centered on the hour and sampled once an hour or once every three hours (*one-hour* and *three-hour readings*, respectively). Figure 8-4 clearly shows that a one-minute average wind (*i.e.*, $\Delta t \approx 0.02$ h) contains a high level of fluctuation energy to which a wind turbine might be sensitive. However, since the most extensive sources of wind data provide information in this for-mat, one-minute averages are commonly used to formulate statistical data on the *frequency distribution* (cumulative duration within a given wind speed range per year) and the *persistence* (duration within a wind speed range per occurrence) of steady winds.

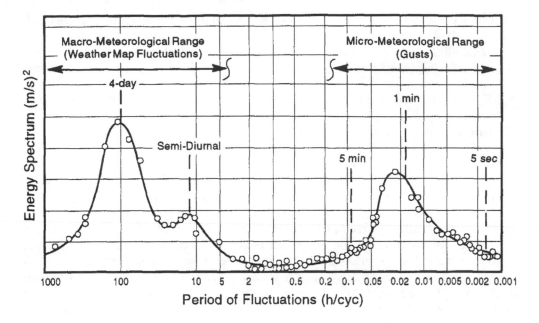

Figure 8-4. Energy spectrum of wind speed fluctuations in the atmosphere. [Van Der Hoven 1957]

Spatial Variations

Disregarding climatic scales, the planetary scale illustrated in Figure 8-1 is the largest in size and longest in period, while the turbulence scale is the smallest and shortest. On a planetary scale, semi-permanent features are part of the general circulation of the earth's atmosphere and are found at every latitude. Some of these features are associated with useful wind energy potential and lend themselves to energy extraction, while others do not. Turbulence, for example, cannot be considered as a potential wind energy resource that can be converted into useful energy. However, turbulence does play a significant role in wind turbine structural design and will be discussed in detail later in this chapter.

The *trade winds* that emerge from subtropical, anticyclonic cells in both hemispheres are known for being the steadiest wind systems in the lower atmosphere. They blow throughout the year and, on the average, are best developed and strongest in the winter hemisphere. *Traveling-wave* perturbations are superimposed on the trades, causing them to vary in time and space. Most parts of the trade wind regions, however, have good wind energy potential throughout the year.

Monsoons are seasonal winds that last for a number of months. Monsoons are caused by the larger annual range of air temperatures over large land masses compared with those over neighboring ocean and sea surfaces. These temperature differentials create pressure gradients which move the monsoon winds. Many monsoons are strong enough for wind energy use, especially where topography and other features (such as upwelling of cool water in coastal areas) further increase temperature differentials between land and water areas.

Broad belts of the *Westerlies* extend across the mid-latitudes in both hemispheres. They blow throughout the year, but are quite variable in space and time, depending on many factors such as season, topography, and configuration of land masses. The Westerlies and *sub-polar flows* are regions of very disturbed weather. Mid-latitude cyclonic storms frequently come in series and have a periodicity of 4 to 6 days. When a wave passes over a location, large variations are experienced in both wind speed and wind direction. Despite this variability, however, most regions of the middle-latitude belts have Westerlies that can be used for energy conversion, with peaks occurring most commonly in spring and winter and lulls in summer.

Synoptic-scale motions (*i.e.* correlated over a wide geographical region) are associated with periodic systems, such as traveling waves in the tropical *Easterlies* or the temperate Westerlies, or at temperate latitudes. Some parts of these waves have very good wind potential (Class 5 or better). Typically the area of influence of a travelling wave is of the order of 1,000 to 1,500 km, with a time scale of about 2 to 4 days.

Mesoscale wind systems can be associated either with traveling disturbances (such as squall lines) or with topographical features (such as valleys and coastal areas). Squall lines are generally convective systems that consist of several convective cells of the cumulo-nimbus type. Squall-line winds can be very violent and destructive, and may not always be of value for wind energy conversion. Mesoscale winds caused by differential heating of topographical features are generally referred to as *breezes*. A breeze is similar to a monsoon, but it operates on much smaller scales, typically a few hundred kilometers and a few hours. In many areas breezes are a regular daily occurrence and, therefore, are of great value as a wind energy resource, especially when they enhance the existing basic wind.

Convective-scale motion is associated with vertical activity in the lower atmosphere, especially in connection with cumulus clouds. Since the scales of convective flow are a few kilometers and minutes to a couple of hours, this motion of the air does not contribute significantly to the wind energy resource. An exception to this is the condition where topographic lifting occurs on the windward side of a mountain. The convective activity that may result could keep local circulations going for several hours. In regions where the winds are prevailing from one direction (*e.g.*, the trades) this phenomenon may repeat itself from day to day, and the enhanced low-level winds may contain sufficient energy to be extracted and used.

Time Variations

It has been pointed out that the motions of the atmosphere vary over a wide range of time scales (seconds to months) and space scales (meters to thousands of kilometers), and that these time and space scales are related. In this section, time variation of the wind is discussed in general terms. Statistical methods and data are described later.

Long-Term Variability

The first concern about a site that is under consideration for a wind power station is with the long-term mean wind speed. Can the winds be counted on to "fuel" cost-effective power production over many years? What is the year-to-year variability at this site? What period of wind speed measurement at the site is adequate to establish a reliable estimate of the long-term mean wind speed? Wind power station operators have indicated that the ability to estimate the interannual variability at a site is almost as important as estimating its annual mean wind speed. The complexity of the interactions of the meteorological and topographical factors that cause the mean wind to vary from one year to the next have hampered the development of a reliable prediction method.

One statistically-developed rule of thumb is that one year of record is generally sufficient to predict long-term seasonal mean wind speeds to within an accuracy of 10 percent with a confidence level of 90 percent [Corotis 1977]. A study that compared seven different methods of estimating long-term average wind speeds from short-term data samples showed that the accuracy of more-sophisticated methods (including *principal component analysis* and *weather pattern classification*) was not significantly higher than that of simpler, linear statistical methods [Barchet and Davis 1983]. Accuracy was measured by the degree of correlation between each of the estimates and actual long-term wind speed data.

Seasonal and Monthly Variability

Significant variations in wind speeds from season to season are common over most of the world. The degree of seasonal variation in the wind at a given site depends on latitude and position with respect to specific topographic features such as land masses and water. In general, mid-latitude continental locations that are well-exposed will experience higher winds in winter and spring, primarily because of large-scale storm activity. However, mountain passes in coastal areas may experience strong winds in the summer when cool maritime air moves into a hot interior valley [Elliott 1979].

Within a given season, time variations in the wind over periods of one to several days can be caused by disturbances in the overall flow pattern such as *cyclonic storms* (in temperate latitudes) and *traveling wave systems* (in the tropics). These disturbances are quite capable of causing the output of a wind power station to cycle between zero and rated power several times a month. This type of wind variability is illustrated in Figure 8-5 for a mid-latitude site. Here the fluctuations in wind speed have several different time periods, but three storms approximately 10 days apart are dominant.

In tropical latitudes, pronounced wind changes from season to season are well recognized. A belt of feeble winds named the *Intertropical Convergence Zone* or ITCZ

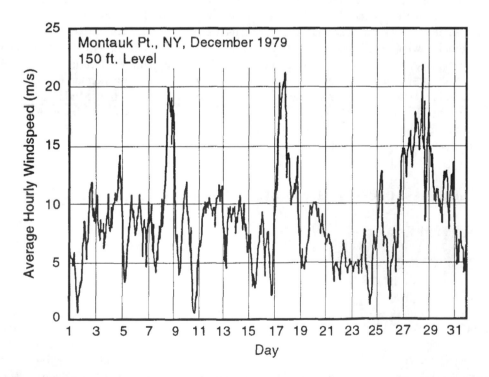

Figure 8-5.   Example of wind speed variations caused by a series of cyclonic storm disturbances at a mid-latitude site. (*Courtesy of the U.S. Department of Energy*)

(commonly referred to as the *doldrums*) moves north and south within the tropics, following the annual march of the sun. The trades or monsoons on either side of the ITCZ are also affected. Therefore, large seasonal time variations of the wind regime occur at most latitudes in the tropics, and they are most pronounced over continents. During a year in the tropics, it is reasonable to expect three or four months of low-to-mediocre winds and eight or nine months of good-to-excellent wind energy conditions.

Skill in reliable forecasting of wind variability on the time scale of a day or two has been shown to be valuable in the operation of a wind power station [Goldenblatt *et al.* 1982]. However, there is little evidence that the appropriate level of reliability in wind forecasting is available through public or private organizations.

Diurnal Variations

In both tropical and temperate latitudes, large wind variations can occur on the *diurnal* (daily) time scale. In the tropics, these variations are most pronounced over land areas and during dry seasons, when the humidity content of the air is very low and the skies are often cloudless or almost so. Variations in radiation flux during the day enhance momentum transfer in the vertical direction during daytime and inhibit such transfer during night. As a result, wind speeds are a maximum during the afternoon and a minimum during the early morning. In extreme cases this diurnal range of wind speed may approach 10 m/s. Such wind variations could not be detected from monthly and seasonal averages, but require sampling with time resolutions of one hour or less.

Daily variations in solar radiation are also responsible for diurnal wind variations in temperate latitudes over relatively flat land areas. The largest diurnal changes generally occur in spring and summer, and the smallest occur in winter. *Low-level jets* may be induced at night in some parts of the Great Plains of the United States [Mahrt and Heald 1979]. In some cases they extend downward to low enough elevations (30 m) to benefit large-scale HAWTs, but they disappear during the day.

Finally, wind speed and direction variations with periods on the order of a few minutes are important to turbines with active pitch and/or yaw controls. Wind variability on this time scale is evidence of turbulence in the flow that may have been generated by upstream topographic features, surface roughness elements, or thermal stratification. Modeling studies have shown that operational strategies involving wind measurements and the logic of turbine start-up, shut-down, and yaw control actions can play an important role in optimizing performance and reducing control-generated fatigue loads.

Wind Resources in the United States and Around the World

U.S. Wind Resources

The *National Renewable Energy Laboratory* (NREL) maintains a current wind energy resource atlas of the U.S. This atlas can be accessed on-line at http://www.nrel.gov/gis/wind_maps.html.

A preliminary wind resource assessment of the U.S. and its territories was carried out under the Federal Wind Energy Program described in Chapter 3, producing twelve regional wind energy *atlases*, which are listed in Table 8-2. These atlases map annual and seasonal average wind resources on regional and state levels, based on data collected before 1979. They include *certainty ratings* of the estimated resource and *areal distributions* (*i.e.* percentages of land area suitable for wind energy development). The complexity of the

Table 8-2. Early Regional Wind Energy Resource Atlases

	Volume No. and Title	Authors	Date	Report No.
1.	The Northwest Region	Elliott, D. L., *et al.*	1980	PNL-3195 WERA-1
2.	The North Central Region	Freeman, D. L., *et al.*	1981	PNL-3195 WERA-2
3.	The Great Lakes Region	Paton, D. L., *et al.*	1981	PNL-3195 WERA-3
4.	The Northeast Region	Pickering, K. E., *et al.*	1980	PNL-3195 WERA-4
5.	The East Central Region	Brode, R., *et al.*	1980	PNL-3195 WERA-5
6.	The Southeast Region	Zabransky, J., *et al.*	1981	PNL-3195 WERA-6
7.	The South Central Region	Edwards, R. L., *et al.*	1981	PNL-3195 WERA-7
8.	The Southern Rocky Mountain Region	Anderson, S. P., *et al.*	1981	PNL-3195 WERA-8
9.	The Southwest Region	Simon, R. L., *et al.*	1980	PNL-3195 WERA-9
10.	Alaska	Wise, J. L., *et al.*	1980	PNL-3195 WERA-10
11.	Hawaii and Pacific Trust Territories	Schroeder, T. A., *et al.*	1981	PNL-3195 WERA-11
12.	Puerto Rico and U.S. Virgin Islands	Wegley, H. L., *et al.*	1981	PNL-3195 WERA-12

topography and the availability of reliable measurements in the vicinity determine the *certainty rating* assigned to the wind resource estimates for exposed locations. Ratings range from 1 (lowest degree of certainty) to 4 (highest degree of certainty). The percentage of land area represented by a specified wind power class is also of interest in interpreting wind power resource estimates. As the ruggedness of the terrain increases, the percentage of land area well-exposed to the wind decreases dramatically. Much of the data used in these early atlases was not collected for wind energy assessment purposes. As a result, the certainty ratings are low of many areas estimated to have excellent wind resources, because few (if any) actual measurements were available for these locations. Since the late 1970's, however, many new sites have been instrumented specifically for wind energy assessment.

Barchet [1981] describes a wind resource data base produced specifically for this application, containing detailed wind statistics from 975 meteorology stations in the U.S. This expanded data base has been employed by Elliott *et al.* [1986] to produce a revised *Wind Energy Resource Atlas of the United States.* New site data were identified and obtained for approximately 270 new sites across the U.S., the majority of which were identified as promising in the earlier regional atlases. Approximately 200 of these new sites were instrumented specifically for wind energy assessment purposes. As before, the revised atlas maps the wind resource in terms of *wind power classes* (Table 8-1 and Figure 8-2). Areas designated Class 3 or higher are considered suitable for wind turbines, Class 2 areas are marginal, and Class 1 areas are generally unsuitable. Isolated sites (*e.g.* exposed hilltops not shown on the maps) with adequate winds may exist within some Class 1 areas.

Wind power estimates in regional and state atlases always apply to well-exposed areas free of local obstructions to the wind, such as open plains, tablelands, and hilltops. In mountainous areas, wind resource estimates apply to exposed ridge crests and summits. Although wind maps identify many areas estimated to have a high wind resource, these maps do not depict variability caused by local terrain features that can cause the available wind energy to vary considerably over short distances, especially in areas of coastal, hilly, and mountainous terrain.

World-Wide Wind Resources

NREL has also produced international wind resource maps available on-line at http://www.nrel.gov/wind/international_wind_resources.html. Wind energy resource atlases for many parts of the world are either not available or not nearly as comprehensive as those for the U.S. A preliminary world-wide wind resource assessment based on statistical and subjective analyses is given by Elliott *et al.* [1981]. Global pressure and wind patterns, upper-air wind data, boundary layer meteorology, and assumed annual frequency distributions were used to obtain consistent estimates of the wind energy resource on a planetary scale. By necessity, such assessments are limited in detail. However, this survey does give a rough indication of the world's mean annual wind energy resource. The reference elevation used for this assessment was 50 m, which is high enough above the ground for the wind to be relatively independent of minor surface features and directly useable for medium- and large-scale turbines. At the same time, it is close enough to the ground to allow meaningful interpolations for small-scale machines.

The results of this preliminary world-wide wind energy resource assessment have been summarized on a *Molleweide map projection* that shows the estimated distribution of the seven wind energy classes defined in Table 8-1. As expected, the map shows greater complexity over land than over sea and generally reflects only large-scale patterns. Much greater spatial variability exists than is depicted, and the same is true for variations in time, since the map only gives annual mean estimates. Because of these limitations, this map should be used only for its intended purpose: to provide wind power project planners with rough estimates of the mean annual wind energy resource in countries other than the U.S.

Analysis and Assessment Methodologies

Three basic methods have been used in wind energy resource assessments: (1) statistical and subjective analysis of existing wind measurements, other meteorological data, and topographical information [Elliott *et al.* 1986]; (2) qualitative indicators of long-term wind speed levels [*e.g.* Putnam 1948]; and (3) application of boundary layer similarity theory and the use of surface pressure observations [Petersen *et al.* 1981].

Statistical and Subjective Analysis Methodology

This method depends on the availability of wind measurements representing a wide variety of geographical, topographical, and climatological conditions. As an example, the surface wind data which are the basis of the Wind Energy Resource Atlas of the United States [Elliott *et al.* 1986] were obtained from a wide variety of sources including the National Climatic Data Center (NCDC), the U.S. Forest Service, university research projects, and power plant sites. Data sets from almost 7,000 recording stations were screened, with almost half retained for further analysis. In unpopulated areas, adequate summarized data are often not available, and an effort must be made to identify other sources, such as private organizations and other government agencies. These data frequently exist as unreduced strip-chart records or as partial compilations of hourly data records collected for very specialized purposes. Formats are not always suitable for a wind energy assessment, and the adequacy of the data set should be reflected in its certainty rating.

In general, wind data in summarized or digitized formats are preferred. For stations where both are available, the digitized data can be used to improve the existing summaries. Analyses of wind speed records for periods with constant anemometer elevation, locations, and observation frequency are more useful than routine summaries. For stations having several different types of summarized wind data covering various time periods, one or two of the better summaries for those stations should be selected considering

-- the most suitable format for wind power assessment;
-- the longest record;
-- the least change in anemometer elevation and exposure;
-- the most frequent daily observations.

The same screening criteria should be applied when assessing areas with a high density of meteorological stations.

Two other types of wind data that may be of value are *coastal marine data* (*i.e.* ship observations and offshore "fixed" stations) and *upper-air data.* In coastal regions where very few land stations with good exposure are available, the marine data can be a very useful supplement. Upper-air wind data are useful in estimating the wind resource on mountain summits and ridge crests, where existing surface station data are sparse. While a strong correlation exists between mountain-top and free-air speeds [Wahl 1966], there is currently no universal procedure for reliably estimating the wind energy potential over mountainous areas. In particular, a procedure which applies to one season may not apply to the others.

Time scales involved in wind resource analysis include annual, seasonal, monthly, and (to a lesser extent) diurnal. Annual mean values are generally based on an average of the one- or three-hour observations of wind speed, and a complete calendar year of data is needed. Data from stations with less than 24 one-hour observations (or 8 three-hour observations) per day should be used only as a last resort when calculating annual mean wind speeds. For purposes of calculating seasonal mean wind speeds, the months in each of the four seasons in the northern hemisphere are generally divided as follows:

-- Winter: December, January, and February
-- Spring: March, April, and May
-- Summer: June, July, and August
-- Autumn: September, October, and November

Procedures for calculating wind power density from various types of wind data records are described by Elliott [1979] and Wegley *et al.* [1980], including adjusting the data for differences in elevation and accounting for differences in air density. Quite often in assessing a stations's seasonal and annual mean wind speeds, a visual examination of the data provides a rough but fast and inexpensive means of making a preliminary estimate of its wind power class. A subjective estimate of this type depends on the skill and experience of the observer, but it often provides the only timely information on the wind resource in many areas.

Qualitative Indicators of the Wind Resource

In many remote areas wind data may be sparse or non-existent, and evaluation of the wind resource may have to rely on qualitative rather than quantitative methods. For example, there are *topographic/meteorologic indicators* of both high and low wind power classes. The following are some indicators of a potentially high wind power class:

-- gaps, passes, and gorges in areas of frequent strong pressure gradients;
-- long valleys extending down from mountain ranges;
-- plains and plateaus at high elevations;
-- plains and valleys with persistent downslope winds associated with strong pressure gradients;

--   exposed ridges and mountain summits in areas of strong upper-air winds;
--   exposed coastal sites in areas of strong upper-air winds or strong thermal pressure gradients.

Features generally indicative of low mean wind speeds are as follows:

--   valleys perpendicular to the prevailing winds aloft;
--   sheltered basins;
--   short and/or narrow valleys and canyons;
--   areas of high surface roughness (*e.g.*, forested, hilly terrain).

Evidence of strong, persistent winds can be found from *wind-deformed vegetation* [Putnam 1948]. Hewson *et al.* [1978, 1979] suggest methods by which mean wind speeds can be deduced from the extent of wind deformation on trees and shrubs. It should be noted that although *wind-flagged trees* (*i.e.*, trees with stunted branches on the side toward the prevailing wind) may indicate that the annual average wind speed is stronger than 4 m/s, unflagged trees do not necessarily indicate that the winds are light. These trees may be exposed to strong winds from all directions, with insufficient persistence in any one direction to cause flagging. Methods of identifying areas of wind-deformed vegetation may include aerial surveys, ground investigations, and public questionnaires.

The removal and deposition of surface materials by the wind to form playas, sand dunes, and other types of *eolian landforms* indicate strong winds from a nearly constant direction. However, correlation of the characteristics of eolian terrain features with annual average wind speeds has proven difficult [Marrs and Gaylord 1979].

Boundary Layer Similarity and Surface Pressure Methodology

A team of Danish scientists devoted much time and effort over several years to develop a wind atlas for Denmark based on a wind resource assessment method that uses boundary layer similarity theory in combination with surface pressure measurements [Petersen *et al.* 1981]. The basic data used to map wind speeds at a given elevation over terrain of a specified roughness are ordinary surface ambient pressure readings from synoptic observations. Existing direct wind measurements were used to validate this procedure, and the best agreement between measurements and predictions was found where the terrain is the least complicated. It is reasonable to conclude that this methodology has promise for estimating wind power classes with sufficient accuracy for identifying areas worthy of more detailed assessment.

Characteristics of the Steady Wind

Throughout this book, time-varying wind speed is considered to be made up of a *steady* value plus *a fluctuation* about this steady value, as expressed in Equations 8-3. Moreover, the steady value is assumed to be *quasi-static*, so that its time variation is negligible for the purpose at hand. It is generally the steady wind which is referred to when discussing wind energy resources and wind turbine siting. The fluctuating component of the wind is referred to when discussing turbulence effects on the turbine structure and controls. Obviously, the characteristics of both the steady and fluctuating components of the wind will depend on the time and space scales selected for averaging.

Two parameters commonly used to characterize the steady wind at a given elevation are *frequency distribution of wind speed* on an annual basis and *persistence*. "Frequency" indicates the *cumulative time* the wind blows at a prescribed value as distinct from "persistence" which provides statistics on the *continuous time* the wind maintains that speed. For example, a frequency distribution might indicate that the summer wind is below the turbine cut-in speed 25 percent of the time, while persistence analysis will indicate how this "downtime" is distributed in terms of periods of different lengths.

Frequency distribution and persistence are important factors in both the design and siting of a wind turbine generator. The energy input to a wind turbine can be calculated from the frequency distribution of the wind. Turbulence usually increases directly with steady wind speed, so frequency distribution is also a significant factor in the structural fatigue life of turbine components. The persistence of wind is important in the assessment of wind energy potential, since *dependability* of generated power, required *storage levels, capacity credits* from the user, and the design of *hybrid systems* depend on this information. The higher the persistence, the more uniform and dependable is the wind energy production.

Wind Speed Frequency Distribution

Equation (8-2) for the annual average wind power density can be re-written in terms of a wind speed *frequency distribution function* as follows:

$$w_a = \frac{0.5\,\rho}{8,760} \int_{year} U^3\,dt = \frac{0.5\,\rho}{8,760} \int_0^\infty U^3 f_U\,dU \tag{8-4a}$$

$$f_U = \left(\frac{dt}{dU}\right)_{year} = \frac{d}{dU}[F(U_1 \geq U)] \tag{8-4b}$$

where

f_U = frequency distribution function of the steady wind speed [(h/y)/(m/s)]
$F(\ )$ = annual time that ( ); cumulative frequency distribution function (h/y)
U_1 = arbitrary value of U (m/s)

The frequency distribution function, f_U, is expressed as a function of the steady wind speed. It must start at zero for a wind speed of zero, rise to at least one maximum value, and then decrease to zero as the wind speed becomes large. Several *non-Gaussian* distributions have been suggested as appropriate models for f_U. These distributions include the *gamma distribution* [Putnam 1948 and Sherlock 1951], the *lognormal distribution* [Luna and Church 1974], the *inverse Gaussian distribution* [Bardsley 1980], the *squared normal distribution* [Carlin and Haslett 1982], and the *Weibull distribution* [e.g., Davenport 1965, Justus *et al.* 1976a and 1978]. Of these distributions the Weibull has received the most use in wind energy assessment analyses and wind load studies. Therefore, the following analysis will use the Weibull distribution, although the same general procedures would apply to other mathematical models.

Weibull Wind Speed Distribution Functions

In Equation (8-4b) *F* is the *cumulative distribution function* which defines the so-called *wind duration curve.* The Weibull equation for the duration curve is

$$F(U_1 \geq U) = 8{,}760 \exp\left[-(U/C)^k\right] \tag{8-5}$$

where

$$\exp[\ ] = \text{exponential function of } [\ ]$$
$$C = \text{empirical Weibull scale factor (m/s)}$$
$$k = \text{empirical Weibull shape factor}$$

Duration curves for several values of the shape factor *k* are shown in Figure 8-6. The range from 1.5 to 3.0 for *k* includes most site wind conditions. The *Rayleigh distribution* is a special case of the Weibull distribution in which the shape factor is 2.0. Cliff [1977] suggests that sites with annual average wind speeds greater than 4.5 m/s tend to have a near-Rayleigh cumulative wind distribution.

Substituting Equation (8-5) into (8-4b), the Weibull frequency distribution function is

$$f_U = (8{,}760/C)\, k\, (U/C)^{k-1} \exp\left[-(U/C)^k\right] \tag{8-6}$$

from which we can calculate the annual average wind speed to be

$$U_a = C\, \Gamma(1 + 1/k) \approx (0.90 \pm 0.01)C \tag{8-7}$$

where

$$U_a = \text{annual average wind speed (m/s)}$$
$$\Gamma(\ ) = \text{gamma function of } (\ )$$

Figure 8-7 shows the Weibull frequency distribution curves associated with the duration curves in Figure 8-6 and the annual average wind speed for each value of *k*. Note that the annual average wind speed is higher than the most-frequent wind speed, but that these two parameters approach equality at higher values of *k*.

Reference Wind Speed Distribution

Early in the U.S. Federal Wind Energy Program a reference annual wind speed distribution was defined to serve as a uniform basis for research and development projects. To be representative of a large portion of the worldwide wind resource, the annual average wind speed was selected to be 6.26 m/s (14 mph) at an elevation of 9.1 m (30 ft) above level terrain. Weibull factors for this reference distribution are $C = 7.17$ m/s and $k = 2.29$. The resulting duration and frequency curves are shown in Figures 8-6 and 8-7. Additional details on the reference wind speed distribution are given in Chapter 2.

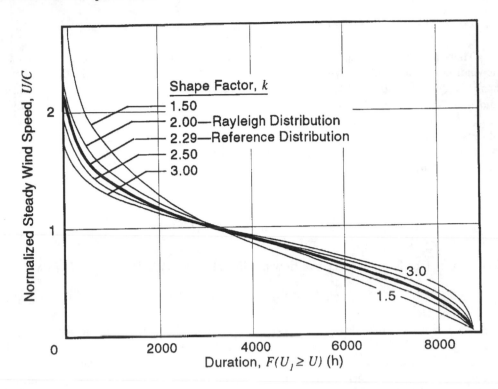

Figure 8-6. Wind speed duration curves according to the Weibull distribution model.

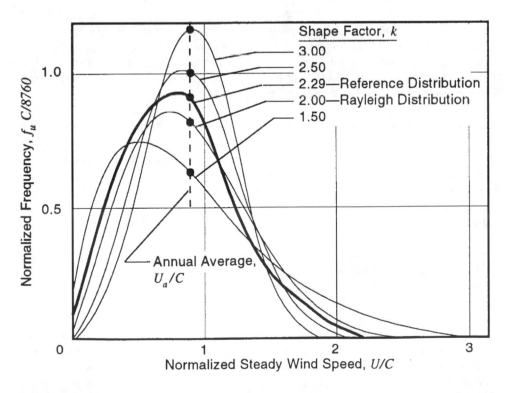

Figure 8-7. Wind speed frequency distributions in dimensionless form, based on the Weibull duration curves. Annual average wind speeds are indicated.

Methods for Estimating Weibull Distribution Factors

There are several methods which can be used to estimate the Weibull factors C and k, depending on the available wind statistics and the desired level of sophistication in data analysis [Justus *et al.* 1978]. These methods are (1) the *least-squares curve-fit*, (2) *median and quartile*, (3) *annual average and standard deviation*, (4) *annual average and fastest mile*, and (5) *variance vs. annual average trend*. The least-squares method requires an observed *wind speed histogram*, which is frequency data for n speed intervals or *bins*. A duration curve is then constructed from the histogram and the results plotted in the linearized form

$$y = y_0 + m\,x$$

$$x = \ln\,(U) \quad y = \ln\,[-\ln\,(F/8{,}760)]$$

(8-8a)

The data are then least-squares curve-fit with a line of slope m and intercept y_0, from which

$$k = m \quad C = \exp\,(-y_0/m)$$

(8-8b)

The median and quartile method is useful if a complete histogram is not available, since it requires only the wind speeds at $F = 2{,}190$, $4{,}380$, and $6{,}579$ hours. The third method requires the annual average wind speed, U_a, and the annual standard deviation from this average, σ_a. The fourth method uses the publication *Local Climatological Data* (from the National Climatic Data Center in Ashville, North Carolina) which lists monthly mean wind speeds and the monthly fastest mile, U_{max} (average speed associated with the most rapid one mile run of wind), for many locations in the U.S. The parameter k is determined from U_{max} and then C can be calculated using Equation (8-7). Finally, Justus *et al.* [1976a] identified a general trend between the annual *variance* of the wind distribution (equal to the square of the standard deviation) and the annual average wind speed, so that a qualitative estimate of the average wind speed is sufficient to make a rough estimate of the Weibull factors. The references cited should be consulted for more details.

Wind Speed Distribution Data

The results of work by Justus *et al.* [1976b] on wind speed distributions have been incorporated into a graphical and analytical format to provide for the rapid construction of duration curves for geographic locations throughout the U.S. [Frost *et al.* 1978]. Values of C and k, adjusted to a 10-m reference elevation, are given for 138 geographical locations in the U.S., based on data with averaging times of one minute [Justus *et al.* 1976b] and two minutes [Doran *et al.* 1977]. It must be borne in mind that wind speed duration curves are quite sensitive to location because of *surface roughness* conditions, and are valid only for relatively flat terrain. Steady wind speed data include the influence of *atmospheric stability* and *terrain features* peculiar to the site at which they were measured, but the influence of these factors is normally small enough under higher wind conditions that the available data are a good representation of the wind for design purposes.

The factors in the Weibull distribution are elevation-specific and must be adjusted to account for *wind shear*. One method for adjusting the C and k values for changes in elevation uses power law equations [Justus *et al.* 1976a, Spera and Richards 1979]. These equations are valid for elevation corrections over a fairly wide range of surface roughnesses, provided the terrain is fairly level. Wade and Walker [1988] confirmed the adjustment of C for elevation, but found it unnecessary to adjust k.

Persistence of Wind Speeds

Persistence is the duration of the wind speed within a given range. Sigl *et al.* [1979] used hourly wind speed records to develop a model for the probability distribution of wind speed persistence above and below fixed reference speeds, which represented the operating range of a wind turbine. Examination of duration histograms from 19 sites for records varying from 5 to 24 years in length led to the development of a simple *composite distribution* model. As shown in Figure 8-8, the probability that a run duration will exceed a given time period is modeled by a power function for the shorter runs and an exponential function for longer runs. The transition between the two functions occurs at a run duration of t_p, referred to as the *partition parameter*. Equations for these probability functions are

$$F(t_r \leq t) = 1 - (1 - F_p)(t/t_p)^{-\mu}, \quad t_o \leq t \leq t_p \tag{8-9a}$$

$$F(t_r \leq t) = 1 - (1 - F_p) \exp[-\mu(t/t_p - 1)], \quad t > t_p \tag{8-9b}$$

$$t_p = t_o(1 - F_p)^{-1/\mu} \tag{8-9c}$$

where

$F(\ )$ = cumulative probability of ( )
t_r = run duration (h)
t = specified time (h)
t_p = partition parameter (h)
F_p = empirical probability parameter; cumulative probability at $t_r = t_p$
μ = empirical shape parameter, greater than zero
t_o = shortest observable run duration (h)

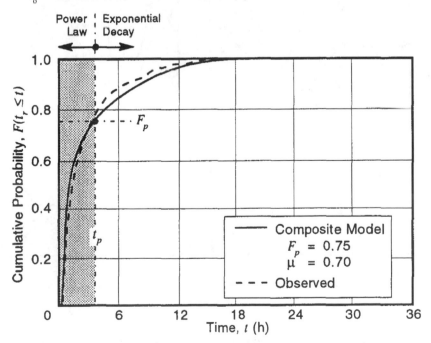

Figure 8-8. Composite distribution model for estimating probabilities of wind speed persistence, with comparison to sample test data. Data are for runs at wind speeds below 4 m/s during five winters at Cheyenne, Wyoming. [Sigl *et al.* 1979]

The size of the partition probability, F_p, varies somewhat with the level of wind speed during the run, but for practical speed levels it can be taken equal to 0.75. Thus, t_p is the duration for which 75 percent of the runs are shorter, for a particular site and wind speed level. Comparison of observed and calculated mean run lengths has shown differences of less than 10 percent when $F_p = 0.75$, for wind speeds up to roughly 2.0 to 2.5 times the seasonal mean speed at the site.

To determine recommended values for the shape parameter μ, hourly wind speed records from 15 sites in the U.S. were analyzed, with a cumulative record length of over 122 years. The shortest observable run duration, t_0, was taken as half the sampling period, or 0.5 h. With the adoption of 0.75 for the partition probability, the fitting of the composite distribution model to test data reduces to calculating μ from the observed mean logarithm of the run durations above and below selected wind speeds or *run levels*. A high degree of correlation between the shape parameter and the run level was found when the run level was normalized by the seasonal average wind speed. Figure 8-9 shows the results of this analysis in the form of curves of μ *vs.* wind speed ratios, with scatter bands of ± one standard deviation indicated.

The parameter values given in Figure 8-9 are considered to be adequate for wind turbine design purposes when only the seasonal average wind speeds are known. If a complete persistence analysis is available, best-fit values for the parameters F_p and μ can be selected.

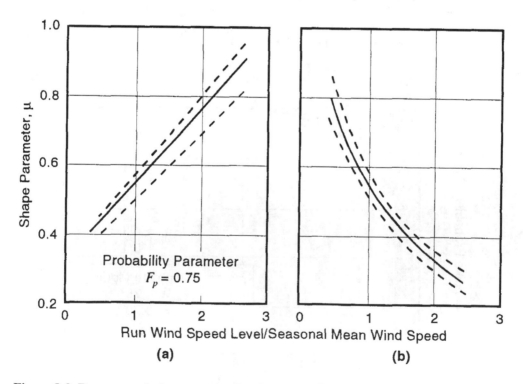

Figure 8-9. Recommended parameters for the composite persistence probability model, derived from wind speed records at 15 U.S. sites. Bands indicate ± one standard deviation of scatter, (a) Runs with wind speeds above a selected run level (b) Runs with wind speeds below a selected run level [Sigl *et al.* 1979]

Vertical Profiles of the Steady Wind

The mean horizontal wind speed is zero at the earth's surface and increases with altitude in the atmospheric boundary layer. Instantaneous measurements of the horizontal wind speed at various altitudes, $u(t,z)$, would typically appear as illustrated in Figure 8-10. The variation of wind speed with elevation is referred to as the *vertical profile of the wind speed* or the *wind shear*. This instantaneous profile shows a number of peaks associated with gusts or turbulent "eddies" of different sizes, whose locations and strengths are time-dependent. Averaging is generally necessary to obtain the steady wind speed profile, $U(z)$.

Variations in the steady wind speed also occur in a horizontal direction, but these are generally not significant over length scales important to wind turbines, except in complex terrain. Turbulent fluctuations, on the other hand, can create essentially instantaneous, short-lived variations in wind speed in both the horizontal and vertical directions, and these will be presented later. Detailed discussions of spatial variations in wind speed within the atmospheric boundary layer are given by Haugen [1973], Lumley and Panofsky [1964], and Panofsky and Dutton [1984].

The variation of wind speed with elevation above ground has important influences on both the assessment of wind energy resources and the design of wind turbines. Assessment of wind energy resources over a wide geographical area normally requires that anemometer data from a variety of sources be corrected to a common elevation. To do this, a model of the variation of wind speed with altitude must be used. Rotor blade fatigue life will be influenced by the cyclic loads resulting from rotation through a wind field that varies in the vertical direction, especially when the rotor is not teetered. Power output is affected by the height of the rotor above ground, so turbine designers must be able to trade off potential gains in energy output against the costs of taller towers, for example.

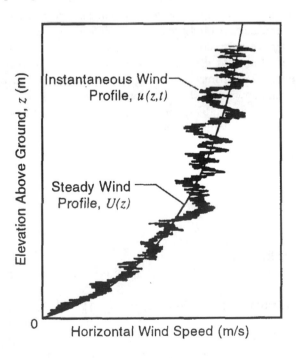

Figure 8-10. Typical vertical profiles of the wind speed, both instantaneous and steady, in the atmospheric layer where wind turbines operate. The steady profile shown represents "positive" wind shear, in which wind speed increases monotonically with altitude.

Two mathematical models or "laws" are commonly used to quantify the vertical profile of wind speed over regions of homogeneous, flat terrain (*e.g.*, fields, deserts, and prairies). These are the *logarithmic/linear law* and the *power law*. The former can be derived theoretically from basic principles of fluid mechanics. It is valid over large ranges of altitude and incorporates the phenomenon of *atmospheric stability*. By contrast, the power law is empirical, and its validity is generally limited to the lower elevations of the atmosphere. Because of its simplicity, however, the power law is the engineering model most commonly used to describe wind speed variations with elevation above ground.

Logarithmic/Linear Law for Vertical Profiles of Wind Speed

There are many detailed derivations and discussions of this model in the literature [*e.g.* Plate 1971, Haugen 1973, Panofsky and Dutton 1984]. Therefore, only a brief description of the principle features of the logarithmic/linear law is given here. The basic equation is

$$U = \frac{U_*}{\kappa}[\ln(z/z_0) + \psi_s(z/L_s)], \quad z \gg z_0 \tag{8-10a}$$

where

U_* = friction velocity (m/s)
κ = von Karman constant, approximately equal to 0.4
z = elevation above ground level (m)
z_0 = empirical surface roughness length (m)
$\psi_s(\ )$ = atmospheric stability function dependent on z/L_s (m/s)
L_s = Monin-Obukhov stability length [see, *e.g.*, Mikhail 1984] (m)

The surface roughness length, z_0, is an empirical parameter which characterizes the influence of surface irregularities on the vertical wind speed profile. The rougher the terrain (*i.e.*, the larger the surface obstructions that oppose the flow of the wind) the thicker will be the affected layer of air and the more gradual will be the increase of velocity with height. In the absence of experimental data, z_0 must be selected on the basis of visual inspection of the terrain upwind of the turbine and reference to tables such as Table 8-3 [*e.g.*, Counihan 1975, Kaufman 1977, Frost *et al.* 1978]. The parameter U_*, the friction velocity at the ground, is a function of the surface shear stress and the air density. Since U_* is not easily evaluated, the ratio U_*/κ is generally calculated using a reference wind speed at a specified reference elevation. Thus

$$\frac{U_*}{\kappa} = \frac{U_r}{\ln(z_r/z_0) + \psi_s(z_r/L_s)}, \quad z \gg z_0 \tag{8-10b}$$

where

z_r = reference elevation = 10 m
U_r = reference steady wind speed at the reference elevation (m/s)

Atmospheric Stability and Wind Shear

The stability of the atmosphere is governed by the vertical temperature distribution resulting from radiative heating or cooling of the earth's surface and the subsequent convective mixing of the air adjacent to the surface. Atmospheric stability states are classified as *stable, neutrally stable,* or *unstable.* These states are important to our modeling of the vertical profile of the steady wind speed because of the different amounts

Table 8-3.
Surface Roughness Parameters for Various Types of Terrain

Type of terrain	Typical surface roughness length, z_o (m)		
Urban and suburban areas	3.0	to	0.4
— cities with very tall buildings		3.0	
— cities and large towns		1.2	
— small towns		0.55	
— outskirts of towns		0.40	
Woodlands and forest	1.2	to	0.4
Farmland and grassy plains	0.30	to	0.002
— many trees and hedges, and a few buildings		0.30	
— scattered trees and hedges		0.15	
— many hedges		0.085	
— few trees; summer		0.055	
— crops; tall grass		0.050	
— isolated trees		0.025	
— uncut short grass		0.020	
— few trees; winter		0.010	
— cut grass		0.007	
— snow covered cultivated farmland	0.002		
Large expanses of water	0.001	to	0.0001
— coastal areas with off-sea winds		0.001	
— calm open sea		0.0001	
Flat desert	.0001	to	0.0001
Snow-covered flat ground	0.0001		
Mud flats; ice	0.00003	to	0.00001

of *atmospheric mixing* which are characteristic of each. A review of the literature through 1980 on the relation of atmospheric stability to the siting of wind turbines is given in [Frost and Shieh 1981].

The concept of atmospheric stability is best illustrated by considering the upward displacement of a small element of air to an altitude with a lower ambient pressure. Assuming a rapid displacement, there will be no time to lose or gain heat and the element will expand *adiabatically*. If the expanded element is less dense than the surrounding air, it will continue to rise due to *buoyancy* and will not return to its original location. This *unstable* atmospheric state is characterized by significant mixing that tends to decrease vertical gradients of wind speed. If our expanded element of air has the same density as the air at its new elevation it will not move further, and the atmosphere is termed *neutrally stable*. There is little mixing in a neutrally-stable atmosphere, and vertical gradients in wind speed tend to remain constant. Finally, if the element is more dense than its new surroundings it will sink back to its original location. This *stable* atmospheric state is also characterized by very little mixing between layers at different altitudes, so vertical gradients tend to be larger.

Typical Diurnal Cycle of Wind Shear

Because atmospheric stability is governed by solar radiation with its diurnal (daily) cycle, we can expect that wind shear will also exhibit a diurnal cycle. This is typically the case, as illustrated in Figure 8-11 by wind shear variations with time-of-day, measured for a year at Clayton, New Mexico [Spera 1991]. Here wind shear is given by the difference between hourly-average wind speeds at elevations of 9.1 and 45.7 m, normalized by the wind speed at 30.7 m, and then averaged over 12 months. The highest wind shears occur during stable conditions at night, and the lowest during unstable and neutral atmospheric conditions near mid-day.

The atmospheric behavior that leads to these changes in wind shear can be explained in qualitative terms by following a typical diurnal cycle of heating and cooling. Before sunrise, heat flux is negative and the air temperature increases with elevation. An element of air forced upward will sink back to its original elevation, so the atmosphere is stable. Without mixing, wind shears can be high. A positive heat flux occurs at sunrise from solar thermal radiation, and the temperature gradient near the ground reverses, so that air temperature decreases with increasing elevation up to a level referred to as the *inversion height*. Between sunrise and noon the inversion height grows rapidly in response to the steadily increasing surface heat flux. In our typical diurnal cycle, we will classify this period as unstable, because of the large amount of mixing that often occurs accompanied by decreasing wind shear.

Growth of the so-called *convective boundary layer* below the inversion height slows down between 1300 and 1600 hours (when the solar heat flux reaches it maximum value) and levels off to a value on the order of 1 to 2 km, which is maintained even after the earth's surface begins to cool by radiation. From 1200 to about 1800 hours in a typical day, this layer of the atmosphere can be classified as neutral. Approximately one hour prior

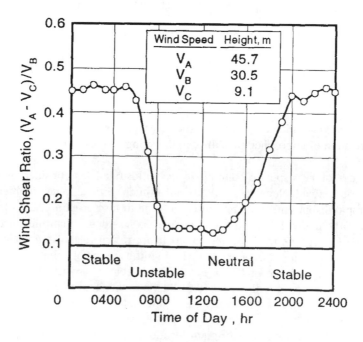

Figure 8-11. Diurnal cycle of wind shear at Clayton, New Mexico, during 1979, showing the effect of atmospheric stability conditions. [Spera 1991]

to sunset, dissolution of the convective boundary layer begins abruptly, and the heat flux over the entire layer turns negative within minutes. With the surface temperatures dropping rapidly after sunset, an inversion layer often develops near the ground. This *nocturnal boundary layer* continues to deepen throughout the evening, reaching a few hundred meters in thickness about six hours after sunset. The atmosphere becomes stable and remains so until sunrise.

It is important to note that on cloudy or windy days, neutral or near-neutral conditions may prevail throughout the entire day. Neutral conditions are also observed to prevail at high wind speeds and during heavy cloud cover.

Empirical Equations for the Atmospheric Stability Function

The influence of atmospheric stability on the wind speed profile is given in the logarithmic/linear law by the term ψ_s, which is a function of the ratio of the elevation to the *Monin-Obukhov stability length*, or z/L_s. The stability length L_s is a measure of the ratio of mechanical shear forces to the thermal buoyant forces causing atmospheric motion [Mikhail 1984]. It is difficult to predict L_s quantitatively, so we will treat it as an empirical constant in the same way as z_0. The different functions reported in the literature for ψ_s can be simplified for engineering purposes to the following:

$$\textit{Neutral atmosphere:} \quad \psi_s = 0 \qquad\qquad\qquad\qquad\qquad (8\text{-}10\text{c})$$

$$\textit{Stable atmosphere:} \quad \psi_s = +4.5\, z/L_s \qquad\quad , z \le L_s \qquad (8\text{-}10\text{d})$$

$$\psi_s = +4.5[1 + \ln(z/L)] \quad , z > L_s \qquad (8\text{-}10\text{e})$$

$$\textit{Unstable atmosphere:} \quad \psi_s = -0.5\, z/L_s \qquad\quad , z \le L_s \qquad (8\text{-}10\text{f})$$

$$\psi_s = -0.5[1 + \ln(z/L_s)] \quad , z > L_s \qquad (8\text{-}10\text{g})$$

Equations (8-10e and -10g) effectively uncouple the wind shear at elevations above L_s from z_0 and extend the logarithmic/linear gradient at L_s to higher elevations. In the terminology of atmospheric science, L_s is negative for an unstable atmosphere, but this is not significant here since we are using it as an empirical constant.

Figure 8-12 illustrates the application of Equations (8-10). The test data in this figure are the average and range of steady wind speeds measured at 15 elevations above flat, homogeneous terrain at 0500 and 1200 hours on 16 consecutive days [Sisterson and Frenzen 1978]. At noon (Fig. 8-12(a)) the atmosphere is well-mixed and neutral or near-neutral, so ψ_s is zero. Fitting a straight line to the average wind speed data produces a value of 0.025 for z_0. At the reference elevation of 10 m, the reference wind speed is 3.5 m/s (by interpolation), and the coefficient U_*/κ is calculated to be equal to 0.58 m/s.

Figure 8-12(b) shows the increased wind shear and an uncoupled high-level jet that can occur under stable atmospheric conditions during the early morning hours. At an elevation of 10 m the reference wind speed is 2.1 m/s. The curve-fit parameters for this condition are 15 m for L_s and 0.25 m/s for U_*/κ. In accordance with Equation 8-10e, the extended logarithmic/linear line above L_s is tangent to the curve of the data at $z = L_s$, and its slope is 5.5 times the slope at $z = z_0$. This linear extension also indicates the uncoupled nature of the jet centered at an elevation of approximately 150 m.

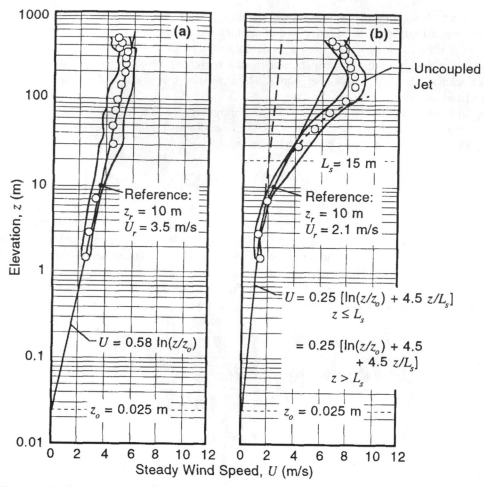

Figure 8-12. Examples of vertical profiles of the steady wind speed modeled by the logarithmic/linear law. Data are averages and ranges of measurements on 16 consecutive days, (a) Well-mixed neutral or near-neutral atmospheric conditions at noon (b) Uncoupled stable conditions at 0500 hours   [Data from Sisterson and Frenzen 1978]

Power Law for Vertical Profiles of Steady Wind

A *power law* is commonly used in wind engineering for defining vertical wind profiles because it is simple and direct.   The basic equation of the wind shear power law is

$$U = U_r(z/z_r)^{\alpha} \tag{8-11}$$

where α = empirical wind shear exponent

In general, the exponent α is a highly variable quantity, often changing from less than 1/7 during the day to more than 1/2 at night over the same terrain, as is the case for the profiles in Figure 8-12. In early wind energy work, it was recognized that α varies with elevation, time of day, season of the year, nature of the terrain, wind speed, temperature, and various thermal and mechanical mixing parameters [Golding 1955, 1977]. Implicit in α must be a dependence on atmospheric stability [DeMarrais 1959, Cramer *et al.* 1972, Touma 1977, Panofsky and Dutton 1984, Wade and Walker 1988]. Relationships have been suggested for calculating α from the parameters in the logarithmic/linear law, which is physically more correct [*e.g.* Plate 1971, Counihan 1975, Justus 1978, Simiu and Scanlan 1978]. These usually complicated approximations reduce the simplicity and directness of the power law. Generally, it is more useful if the wind turbine engineer accepts the empirical nature of the power law and chooses values of α that best fit available wind shear data.

Spera and Richards [1979] proposed equations for α based on both the surface roughness length, z_0, and the wind speed at the reference elevation, U_r. These are shown graphically in Figure 8-13. Following work by Justus *et al.* [1976a], this model decreases the wind shear exponent with increasing wind speed until the wind profile becomes vertical or "homogeneous" at a very high speed (67 m/s). The Spera-Richards model was used extensively by researchers at the NASA Lewis Research Center and its contractors to design large-scale HAWTs. For high wind speeds of short averaging times, Kaufman [1977] gives

$$\alpha = 0.55\, U_r^{-0.77} \tag{8-12}$$

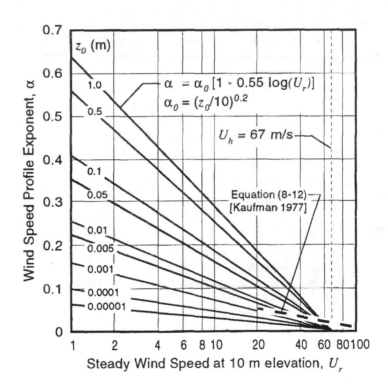

Figure 8-13. Graph of equations for calculating wind shear power-law exponents from surface roughness lengths and steady wind speeds. [Spera and Richards 1979]

Heister and Pennell [1981], however, argue that any apparent dependence of α on wind speed is really a reflection of the variation of atmospheric stability with wind speed. In a very comprehensive study of wind profiles over a wind power station near Whiskey Run, Oregon, Wade *et al.* [1986] identified the power output from the station as a factor which influences the value of α. Wind shear increased after the wind turbines were installed. Additional information useful in adjusting α to account for significant atmospheric and topographic factors can be found in Frenkiel [1962], Fichtl and Smith [1977], Peterson and Hennessy [1978], Elliott [1979], Heald and Mahrt [1981], and Simon [1982].

Extreme Winds

Design of a wind turbine for structural integrity requires that the support structures, (both above and below ground) and the rotor, pitch, and yaw assemblies be of sufficient strength to withstand the *extreme wind* loading that is likely to be encountered during its design life. Generally, the turbine is shut down under extreme wind conditions, so dynamic loading of the structure caused by rotor rotation is not usually an additive factor. Thus, the discussion in this section pertains to *maximum static wind loading* of the same type as that experienced by buildings, towers, water tanks, and signs.

Measuring Extreme Winds

Two measures of extreme wind speed are currently used in wind engineering, namely the *fastest-mile wind speed* and the *peak gust*. The first of these is recorded by a special anemometer called a *1-mile contact anemometer*, in which an electrical contact causes a transverse mark to be made on a rotating drum chart at the completion of each whole mile of wind movement past the instrument. The shortest time interval between two consecutive marks in a wind record gives the fastest-mile wind speed. This measure of extreme wind is therefore the maximum one-hour steady wind speed.

Peak wind statistics (like the peak gust) have an advantage over mean wind statistics (like the fastest-mile wind speed) in that they do not depend upon an averaging operation that can vary from day to day and from observer to observer. Hourly peak wind speed readings avoid this sometimes subjective averaging process. The time duration of a peak wind speed, however, is an important factor in wind loading, and this duration is a function of the anemometer sensitivity. Standard weather service anemometers measure gusts with approximately 3-second durations, whereas speeds recorded by research anemometers can normally be resolved to 0.1 second.

Recurrence Intervals

The general concept of determining an extreme design wind speed is to select the most extreme value of the wind that the structure will probably experience in a given number of years of exposure. Extreme wind values are therefore tabulated according to percent probability of occurring at least once in a given *recurrence interval*. Thorn [1961, 1968] has published *isotach maps* (lines of constant wind speed) of the fastest-mile wind speed, based on U.S. Weather Bureau measurements over a 21-year period. Typical mean recurrence intervals graphed on these maps are 2, 10, 25, 50, and 100 years. The accuracy of the speeds on these maps is given as approximately ± 15 percent. Extreme wind speed values are normally reported at a given reference elevation (usually 9.1 m) over smooth, uniform terrain that is typical of airports. These speeds must be corrected for elevation, including the effects of surface roughness (see Figure 8-13), and to account for the dynamic response time of the structure.

Risk of Exceedance

Since there is a finite probability that the actual extreme wind speed will be higher than the selected design value during the turbine's design life, the design engineer must select an acceptable degree of *risk of exceedance*. The relationship between risk of exceedance, mean recurrence interval, and design life is relatively straightforward, and is shown graphically in Figure 8-14. Two different *design criteria* for selecting an acceptable risk of exceedance have been used: An *aerospace vehicle criterion*, which is a 10 percent risk of exceedance [Kaufman 1977]; and a *building code criterion*, which is a 63 percent risk for a design life of 50 years [British Standard Code of Practice CP3 1972]. The latter criterion is equivalent to specifying a mean recurrence interval of 50 years.

The aerospace criterion has been used for experimental large-scale HAWTs (*e.g.*, the 2.5-MW Mod-2 turbines), but it is considered to be too conservative for commercial wind power stations. Risk of exceedance for these is usually determined by building design practices and state and local building codes. Lower (more conservative) risks of exceedance may be selected for commercial turbines on the basis of economic trade-off studies.

The following example will illustrate the selection of a design extreme wind speed:

-- design life . . . . 30 yr
-- acceptable risk of exceedance . . . 10 percent (aerospace criterion)
-- site location . . . . 150 km east of Portland, Oregon
-- surface roughness length . . . . 0.05 m
-- reference elevation . . . . 10.0 m
-- elevation of center of swept area . 61.0 m

The first two specifications locate point **A** in Figure 8-14, from which the minimum mean recurrence interval is determined to be 280 years. Using isotach map data for Portland

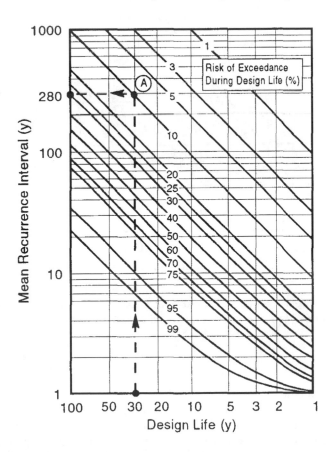

Figure 8-14. Relationship between risk of exceedance, mean recurrence interval, and design life used to determine the design extreme wind speed. [Frost *et al.* 1978]

[Thorn 1968], a graph of mean recurrence interval *vs.* fastest-mile wind speed can be plotted for the reference elevation of 10 m (ignoring the slight difference between 10 m and the 9.1 m elevation of the reference maps), as shown in Figure 8-15. The scales in this figure follow a *Fisher-Tippet Type II distribution*, which permits straight-line extrapolation of available data. For a 280-year interval, the extreme wind speed at 10 m is 53.5 m/s (point B). The wind shear exponent α is calculated to be 0.015 from the equations in Figure 8-13 and 0.026 from Equation 8-12. Using the higher value for conservatism, the extreme wind speed at 61.0 m that meets the design requirements is determined to be 56.1 m/s.

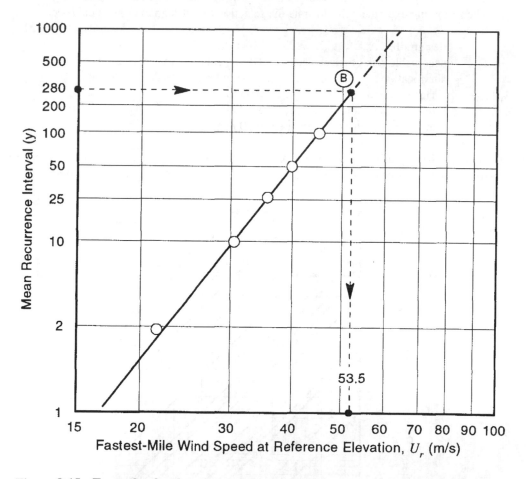

Figure 8-15.   Example of a plot of mean recurrence interval *vs.* fastest-mile wind speed, using Fisher-Tippet Type II distribution coordinates. [data from Thom 1968]

Detailed computational procedures and data from which basic values of extreme wind speeds for wind turbine design can be estimated are given in Frost *et al.* [1978], including mathematical expressions for adjusting these basic speeds for elevation, terrain roughness, and structural-dynamic response times of different-sized turbine components. Additional references on extreme wind loads are Hollister [1970], Sachs [1972], Canadian Structural Design Manual [1975], Simiu and Scanlan [1978], and Mayne [1979].

Turbulence

Referring once again to the energy spectrum of the wind shown in Figure 8-4, all of the time variations in wind speed and direction with periods less than about 1/10 hour are generally considered to be turbulence. These turbulent fluctuations in the *micro-meteorological range* create unsteady, non-uniform aerodynamic forces on the wind turbine that must be taken into account in the design of its structure and controls. By way of introduction to the subject of turbulence, let us review the concept introduced earlier, that wind loads on the turbine can be classified into two groups: those associated with the steady wind speed, which are described as *quasi-static* or *time-averaged*; and those associated with the gustiness or turbulence of the wind, which are predominantly *dynamic*.

Expanding Equations (8-3) to represent a wind field instead of a streamtube, we have

$$u(y, z, t) = U(y, z) + g(y, z, t) \tag{8-13a}$$

$$\int_{A_r} \int_0^{\Delta t} g(y, z, t)\, dt\, dA = 0 \tag{8-13b}$$

$$RMS[u(y, z, t) - U(y, z)] = \sqrt{\frac{1}{A_r \Delta t} \int_{A_r} \int_0^{\Delta t} g^2(y, z, t)\, dt\, dA} = \sigma_0 \tag{8-13c}$$

where

$$y, z = \text{lateral and vertical coordinates, respectively (m)}$$
$$u(y,z,t) = \text{instantaneous horizontal free-stream wind velocity field (m/s)}$$
$$U(y,z) = \text{steady horizontal free-stream wind velocity field (m/s)}$$
$$g(y,z,t) = \text{fluctuating wind velocity field; instantaneous deviation from } U(y,z) \text{ (m/s)}$$
$$A_r = \text{swept area of the rotor (m}^2\text{)}$$
$$\Delta t = \text{averaging time interval (h)}$$
$$RMS[\ ] = \text{root-mean-square average of [ ]}$$
$$\sigma_0 = \text{ambient or natural turbulence (m/s)}$$

Equation (8-13c) states the common assumption that the turbulence is *homogeneous* (*i.e.*, has the same structure over the swept area of the rotor). Wind models which describe various fluctuating velocity field functions $g(y,z,t)$ that create dynamic wind forces are discussed in this section. The structural response to these dynamic forces is treated in Chapters 10, 11, and 12.

Types of Turbulence Models

Mathematical models of the fluctuating or turbulent wind field at the rotor of a wind turbine can be grouped into four categories of increasing complexity and realism [Stoddard and Potter 1986], as follows:

1. Non-uniform in space, steady in time . . . . . . $g = g(y,z,0)$
2. Uniform in space, unsteady in time . . . . $g = g(0,0,t)$
3. Non-uniform in space, unsteady in time . . . . $g = g(y,z,t)$
4. Stochastic . . . . $g = g(power spectrum vs.$
   discrete gust parameters)

Turbulence Models Non-Uniform in Space and Steady in Time

The earliest models of wind inflow to a rotor were of this type and were developed primarily to represent *tower shadow* (flow distortion to a downwind rotor caused by the tower wake), not turbulence [Spera 1978]. However, the free-stream wind field itself can be highly non-uniform, as illustrated in Figure 8-16. These variations have been measured in the field by *rotational sampling* methods [*e.g.*, Verholek 1978, Connell 1981], which will be discussed later. Rotational sampling is a procedure that uses an array of anemometers distributed along the hypothetical path of a section of a turbine blade. These anemometers record the time history of wind speed that would be experienced by that blade section.

In order to use a non-uniform, steady wind model to represent rotationally-sampled turbulence, the assumption is made that spatial variations in wind speed are *quasi-static*, changing slowly in comparison with the rotational period of the rotor. Any mathematical function with periodic frequency might be used to convert the time series of wind speed at each rotor station to (*y,z*) coordinates. The swept area of the rotor is then simply divided into sectors, and a wind speed is specified for each sector. Presently, many turbulence inflow models uses this procedure.

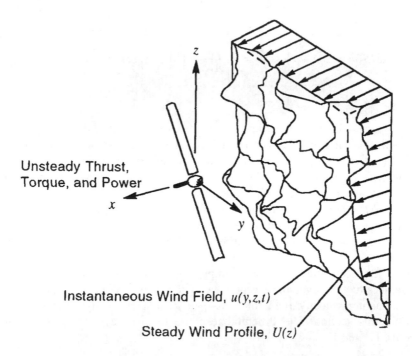

Figure 8-16. Schematic diagram of the non-uniform wind field that typically flows into the swept area of a wind turbine rotor. [Sundar and Sullivan 1981]

Models Uniform in Space and Unsteady in Time

This type of model may also be called a *planar gust front* or a *discrete gust model*. The gust profile (in the coordinates *y* and *z*) is constant and envelopes the rotor uniformly, but its level changes with time. However, a given turbulence eddy may appear uniform to a small-scale rotor, but non-uniform to a larger one. This illustrates the need to model gust *amplitude*, duration, number of occurrences, and dimensions together. Generally, the larger the scale of the rotor, the larger the amplitude, the longer the duration, and the lower the frequency of occurrence of the planar gust fronts that envelop it.

The *ultimate strength* design of a wind turbine and the design of shutdown and safety controls are typically driven by extreme, somewhat isolated, gusts which are embedded in the turbulent winds. For analysis of extreme values, one generally resorts to a *probability of maximum value* or *external gust* model [Coleman and Meyers 1982]. Discrete gust models are used routinely in aircraft design [*e.g.*, Bisplinghoff and Ashley 1957, Babister 1980].

Models Non-Uniform in Space and Unsteady in Time

The actual inflow to a turbine is both unsteady and nonuniform, so this type of model is the most realistic and, of course, the most complex. It considers a nonuniform flow front, that is perhaps two-dimensional, which varies with time. The need for this level of complexity will differ for different rotor sizes. For example, both spatially and temporally varying inflow models may be required for large-scale rotors, whereas temporal variations alone may suffice for small rotors.

Stochastic Models

The final turbulence model type is stochastic (sometimes called *probabilistic*). The three types of inflow models described previously contain *deterministic* descriptions of the wind fluctuations. Stochastic models, on the other hand, are based on the concept that turbulence is made up of sinusoidal waves or eddies with many periods and random amplitudes. The *spectrum* of the turbulence, like that shown in Figure 8-4, is typically used to describe the frequency of occurrence of fluctuations with different periods, and is given as a graph of the average kinetic energy associated with eddies or disturbances which have a common period but time-random amplitudes. Stochastic models may also use probability distribution or other statistical parameters.

Stoddard and Potter [1986] point out that stochastic analysis can be invaluable in developing models of the wind inflow from measurements of the response of an operating wind turbine, because it facilitates the following critical tasks:

-- representation of many data points taken in field tests;
-- rapid evaluation of fatigue loads;
-- comparison of model predictions with historical field data.

The *method of bins* [Akins 1978] is a simple application of stochastic methods to power performance testing of wind turbines. Another application of a stochastic approach is to represent the timewise variation of a wind "front" with probabilistic formulations. This might be the probability of experiencing a given shape and magnitude of non-uniformity in the inflow. In this case the spatial variation is still deterministic. The number of times the wind speed exceeds a prescribed value (*i.e., exceedance statistics*) is another example of a stochastic model. Such a model is applicable to fatigue analysis. There are many other applications of stochastic modeling in wind engineering.

Dimensions of Turbulence Models

Turbulence models of all types can be classified as one-, two-, or three-dimensional, depending on the number of spatial coordinates (*i.e., x, y,* and *z*) needed to describe the wind velocity field. One-dimensional or *large-scale turbulence* models assume that the rotor is entirely engulfed by each gust, so only the longitudinal coordinate (*x*) is required. Note that the dimensional classification does not reflect the number of components of the wind velocity vector that are included. Two- and three-dimensional (*i.e.*, *small-scale turbulence*) models define fluctuations in wind speed with length scales smaller than the diameter of the rotor, and the lateral and/or vertical coordinates (*y*, *z*) must be used.

Spectral Models of Continuous Turbulence

Structural fatigue life and the control of power output and HAWT rotor heading are typically sensitive to *continuous turbulence* (*i.e.*, wind fluctuations during routine operations) which causes dynamic forces to act continuously on the rotor and its supporting structure. Methods of analyzing continuous turbulence generally rely on *spectral models* or *deterministic gust models*, combined with statistical exceedance values [Frandsen and Christensen 1980, Raab 1980, Thresher *et al.* 1981a, Sundar and Sullivan 1981, Thresher *et al.* 1981b, Waldon and Hansen 1983].

Basic Equations

The basic relationships between the spectra of atmospheric turbulence and the spectra of structural dynamic responses (such as motions, deformations, and loads) are contained in the following general equation, which is written here for one component of the wind velocity:

$$S_k(n) = \phi_{11}(n)H_1(n)H_1^*(n) + \phi_{22}(n) H_2(n) H_2^*(n) + \cdots$$
$$+ Re[\phi_{12}(n) H_1^*(n) H_2(n) + \phi_{13}(n) H_1^*(n) H_3(n) + \cdots] \qquad (8\text{-}14)$$

where

n = circular frequency (rad/s)
S_k = power spectrum of the dynamic response of the turbine parameter k which has units of K (K^2/rad/s)
ϕ_{ij} = two-point power spectrum of the wind velocity component acting at points i and j (m^2/s)
H_i^* = complex conjugate of H_i (K/m/s)
H_i = turbine parameter response transfer function (K/m/s)
$Re[\ ]$ = real part of [ ]

This expression can be extended directly to two or three components of velocity and is general within linear theory. It can be reduced to a simpler form depending on whether the assumed turbulence is classified as one- or two-dimensional. If the turbulence can be modeled as one-dimensional (*i.e.*, large-scale turbulence relative to the rotor diameter) Equation 8-14 can be simplified to the following form:

$$S_k = \phi_x(n) H_x(n) H_x^*(n) + \phi_y(n) H_y(n) H_y^*(n) + \phi_z(n) H_z(n) H_z^*(n) \qquad (8\text{-}15)$$

The wind characteristics required to solve Equations (8-14) and (8-15) are contained in the turbulence power spectra $\phi_{ij}(n)$, and these will now be discussed. The primary focus of this discussion is on spectra for wind over flat, homogeneous terrain, although a few general comments will be made about spectra over complex terrain.

Wind Turbulence Spectra in a Neutral Atmosphere

Analytical models for spectra of the longitudinal, lateral, and vertical components of atmospheric turbulence are discussed in a number of references. Haugen [1973], and Panofsky and Dutton [1984] give detailed physical and mathematical descriptions, whereas Counihan [1975] and Frost *et al.* [1978] use an engineering handbook format. A modified version of a spectrum model developed by Kaimal [1973] is as follows:

$$\frac{n\phi_\alpha}{\sigma_{0,\alpha}^2} = \frac{0.164\,\eta/\eta_\alpha}{1 + 0.164(\eta/\eta_\alpha)^{1.667}} \tag{8-16a}$$

$$\eta = \frac{nz}{U} \tag{8-16b}$$

where

ϕ_α = power spectrum of wind fluctuations in the α direction at elevation z (m^2/s)

α = x (longitudinal), y (lateral), or z (vertical) direction

$\sigma_{0,\alpha}$ = ambient turbulence in the α direction (m/s)

η = normalized frequency

η_α = reference normalized frequency, dependent on elevation and stability

Figure 8-17 is a graph of Equation (8-16a) for conditions of neutral stability, with the following recommended values of the elevation-dependent reference normalized frequencies:

$$\eta_x = 0.0144(z/30)^{0.78} \quad \eta_y = 0.0265(z/30)^{0.78} \quad \eta_z = 0.0962 \tag{8-16c}$$

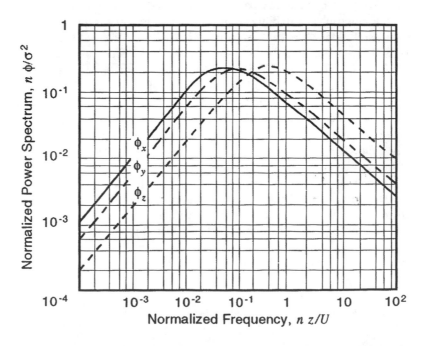

Figure 8-17. Modified Kaimal model of turbulence spectra at an elevation of 30 m.

Other models of turbulence spectra have been used for wind turbine design and analysis, such as the *Davenport spectra* [1961; see Sundar and Sullivan 1981], the *Dryden spectra* [see Holley *et al.* 1981], and the *von Karman spectra* [see Connell 1981]. The von Karman spectra, which depend upon integral turbulence length scales, agree well with atmospheric data in the high frequency range. They are, however, based on the assumption that the turbulence is *isotropic* (equal in all directions), which is not realistic near the ground. Moreover, turbulence length scales are difficult to estimate for general design purposes, and the influence of atmospheric stability cannot be readily incorporated. Therefore, Equation 8-16 and others which include effects of boundary layer stability are recommended for representing atmospheric turbulence spectra.

Turbulence and Turbulence Intensity

Turbulence is not only a qualitative term for wind fluctuations, but it is also a fundamental, quantitative measure of the unsteadiness of the wind. Adapting Equation (8-3c) so that it represents the components of turbulence in three directions, we have

$$\sigma_{0,\alpha} = RMS[u_\alpha(t) - U_\alpha] \tag{8-17}$$

where, as before, the subscript α refers to direction x, y, or z, with x parallel to the steady wind. Hence, $U_x = U$ and $U_y = U_z = 0$. The value of the turbulence is a function of the time period selected for averaging the wind speed to determine U.

Turbulence intensity (also known as *relative turbulence intensity*) is defined as the ratio of the turbulence to the steady wind speed. Turbulence intensity is typically measured when characterizing the wind regime at a turbine site, using the same wind speed data recorded for the purpose of calculating seasonal and annual average wind speeds. Normalizing the turbulence by the steady wind speed tends to produce a characteristic range of intensities for a given site, but these intensities are by no means constant. Turbulence intensity has been found to vary with the same parameters as wind shear: surface roughness, wind speed, elevation, atmospheric stability, and topographic features. Equations for predicting turbulence intensity in the absence of measurements will be presented here, but only direct measurements are recommended for the final stages of site selection and turbine design.

Predicting turbulence intensity generally begins with estimating the ratio of the vertical component of turbulence to the *friction velocity*, $\sigma_{0,z}/U_*$ [Barr *et al.* 1974 and Panofsky and Dutton 1984]. For neutral atmospheric stability this ratio is approximately 1.3 [Frost 1980]. It has been observed that under neutral conditions $\sigma_{0,z}$ is dependent upon the steady wind speed and the surface roughness length, z_o. Experimental results indicate that vertical gusts are primarily a function of small-scale roughness features, whereas lateral and longitudinal gusts are influenced by large-scale surface features.

Because of the lack of mathematical models for the influence of terrain features and atmospheric stability on $\sigma_{0,x}$ and $\sigma_{0,y}$, it has been proposed that the ratios of these components to the vertical turbulence be treated as functions of altitude only, according to the following relationships for elevations less than 600 m [Frost *et al.* 1978]:

$$\sigma_{0,x}/\sigma_{0,z} = (0.177 + 0.00139\,z)^{-0.4} \tag{8-18a}$$

$$\sigma_{0,y}/\sigma_{0,z} = (0.583 + 0.00070\,z)^{-0.8}, \quad z < 600\ m \tag{8-18b}$$

Assuming $\sigma_{0,z}/U_* = 1.3$, $U_*/U = \kappa\,/\ln(z/z_0)$, and $\kappa = 0.4$, the turbulence intensities for neutral atmospheric conditions at elevations less than 600 m can be expressed as

$$\frac{\sigma_{0,x}}{U} = \frac{0.52}{\ln(z/z_0)}(0.177 + 0.00139\,z)^{-0.4} \tag{8-19a}$$

$$\frac{\sigma_{0,y}}{U} = \frac{0.52}{\ln(z/z_0)}(0.583 + 0.00070\,z)^{-0.8} \tag{8-19b}$$

$$\frac{\sigma_{0,z}}{U} = \frac{0.52}{\ln(z/z_0)}, \quad z_0 << z < 600\ m \tag{8-19c}$$

The effect of surface roughness on longitudinal turbulence intensity is shown in Figure 8-18 where the elevation is held constant at 30 m. Both the test data (compiled by Counihan [1975]) and Equation (8-19a) indicate a strong effect of surface roughness on turbulence intensity. Test data showing the dependence of turbulence intensity on elevation have been correlated with Equations (8-19), and the general trend and magnitude are in agreement, although the data scatter is relatively large.

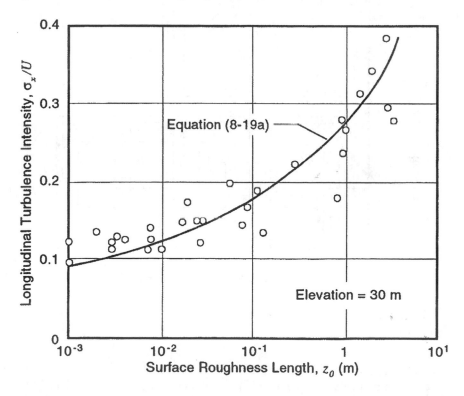

Figure 8-18. Effect of surface roughness on turbulence intensity. [data: Counihan 1975]

Influence of Atmospheric Stability on Turbulence Spectra

In general, the critical wind forces on a wind turbine are those that occur at high wind speeds for which neutral atmospheric conditions can normally be assumed. However, winds containing significant turbulent energy may also persist for long hours under both stable and unstable atmospheric conditions at many turbine sites. Good design practice would therefore call for structural response studies with turbulence representative of a range of stability conditions. Panofsky and Dutton [1984] provide details of the physical principles and/or experimental results on turbulence spectra for stable and unstable atmospheric conditions. Frost *et al.* [1978] recommend that Equation 8-16 be used to mathematically describe the spectra for a stable atmosphere with η_α expressed as a function of *Richardson's number* as well as elevation.

Mathematical models of the influence of atmospheric stability on turbulence are still being developed. It has been observed that spectra of horizontal turbulence in unstable air consist of a lower-frequency portion that depends on the stability length, L_s, and a higher-frequency portion that depends of the elevation [Hojstrup 1982]. Figure 8-19 illustrates the lower-frequency effect on the longitudinal spectrum for unstable conditions ($L_s < 0$), compared with the spectrum for neutral stability (Eq. (8-16a)). The peaks observed at $\eta \approx 0.004$ are associated with coherent, large-scale eddy structure in unstable stratified atmospheric boundary layers [Wilczak 1984, Antonia *et al.* 1982 and 1983, Khalsa 1980, Van Atta 1977]. This topic continues to be a major area of meteorological research.

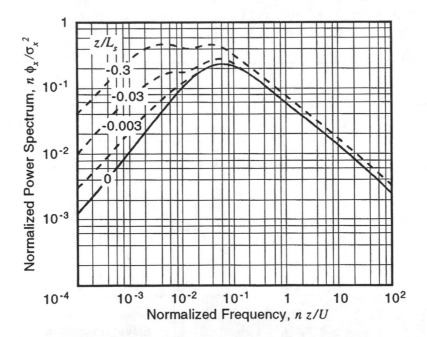

Figure 8-19. Effect of unstable atmospheric conditions on the spectrum of longitudinal turbulence. [based on Hojstrup 1982].

Large-scale eddies are observed as distinctive ramp-like signals in temperature records. As an example, a particular ramp with a horizontal length slightly greater than 500 m was observed at the 50 m level of the *Boulder Atmospheric Observatory* tower in Colorado, at a wind speed of 11 m/s and a stability length of -70 m [Wilczak 1984]. Characteristic velocity fields are associated with these temperature ramps, and these may be of interest to the wind turbine engineer. The interior ramp region generally has an upward vertical velocity, while the surrounding quiescent region has downward vertical velocity. The longitudinal velocity component attains its minimum value in the ramp interior and its maximum value immediately behind the ramp's trailing edge. These velocity fluctuations are generally embedded in the along-wind speed/time records.

Turbulence Spectra Over Complex Terrain

No well-formed theory is currently available with which to predict quantitatively the effects of complex terrain on turbulence spectra. Panofsky *et al.* [1982] provide the following general observations:

-- Lower frequencies of the spectrum of horizontal turbulence (roughly $\eta < 0.1$) are affected by the upwind terrain. In this respect, there are important differences among spectra associated with various types of terrain. For example, smaller "eddies" from upwind terrain features appear to persist longer than larger eddies.

-- In the surface boundary layer, higher-frequency spectra (wavelengths much less than the fetch length over the changed terrain) are in local equilibrium, so methods for estimating spectral characteristics over uniform terrain apply.

-- Since vertical velocity fluctuations (in contrast to horizontal fluctuations) have most of their energy at relatively high frequencies, their spectra over complex terrain closely resemble those over uniform terrain.

Discrete Gust Models

A discrete gust is typically defined in terms of a *magnitude* (maximum change from the steady wind speed), a *duration* (period of time during which the wind speed differs from the steady wind speed), and a *shape function* which gives the variation of wind speed with time during the gust. The shape function defines the rate of build-up and decay of the gust. A classic example is the *one-minus-cosine* shape function shown in Figure 8-20, in which

$$u(t) = U + 0.5\Delta u[1 - \cos(2\pi t/\tau)] \qquad (8\text{-}20)$$

where

Δu = magnitude of the gust; positive *(up gust)* or negative *(down gust)* (m/s)
t = elapsed time from start of gust (s)
τ = gust duration or period (s)

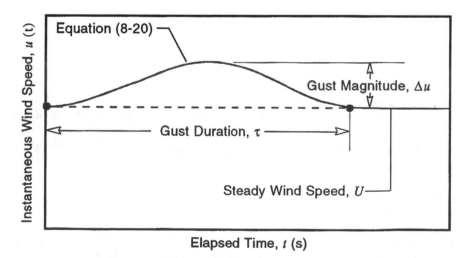

Figure 8-20. The idealized "one-minus-cosine" model of a discrete gust.

One of the most detailed sets of experimental data on gust shapes is that given by Camp [1968]. These data suggest that a typical gust has an essentially exponential rise with a long dwell and an exponential decay. This is a more severe gust than that given by Equation (8-20) and may be preferred for predicting extreme wind loads. On the other hand. Equation (8-20) may be preferable for fatigue load analysis. Other research indicates that *gust energy* is the important parameter to preserve in any model used for dynamic analysis.

Coherence Modeling of the Gust Environment

A model of the total gust environment experienced by a wind turbine includes *the joint probability distribution* of magnitude and duration and *frequency of occurrence*. The discrete gust environment of most interest here is that actually experienced by a wind turbine. This environment is not identical to the typical time history of wind speed measured by an anemometer, since the turbine responds to some spatially-averaged wind field whose extent is considerably larger than that to which an anemometer responds, and the turbine may be capable of changing pitch angle, teeter angle, and/or rotor speed to "absorb" some parts of the gust environment. Therefore, selection of meaningful values of gust magnitude and duration depends not only on statistical wind data from isolated anemometers, but also on turbine size and dynamic response.

While discrete gust models are highly idealized representations of the actual wind, they are quite useful for wind turbine analysis when the size of the gust is large enough to engulf the entire rotor. For such cases, the wind speed may be assumed to be changing uniformly across the rotor. Gust size is generally considered to be related to the gust duration, τ. The longer the duration, the larger the spatial dimensions of the gust. To estimate the duration of a gust which will engulf a turbine rotor, it is convenient to use the so-called *coherence function* [Frost *et al.* 1978].

Coherence is a dimensionless quantity between zero and unity that represents the degree to which two events, separated in space, are alike in their time histories. If the two time histories are identical their coherence is unity, and if they are completely unrelated their coherence is zero. The coherence of two wind speeds measured at points separated in space during a gust may be expressed empirically by the following equation:

$$coh_\alpha = \exp\left(-d_\alpha \Delta l_\alpha / U_r \tau\right) \tag{8-21}$$

where

$$
\begin{aligned}
coh_\alpha &= \text{coherence between stations separated in the } \alpha \text{ direction} \\
\alpha &= x \text{ (longitudinal), } y \text{ (lateral), or } z \text{ (vertical) direction} \\
d_\alpha &= \text{decay coefficient in the } \alpha \text{ direction} \\
\Delta l_\alpha &= \text{distance between measurement stations separated in the } \alpha \text{ direction (m)}
\end{aligned}
$$

A physical explanation of Equation 8-21 is that for long durations or small separations the coherence approaches unity, while for short durations or large separations there is no time correlation between fluctuations "felt" at the two measurement stations.

Tentative values of the decay coefficients are $d_x = 4.5$, and $d_y = d_z = 7.5$, based on averages of coefficients reported in the literature [*e.g.*, Davenport 1961, Ropelewski *et al.* 1973]. It is known that d is dependent on terrain roughness, atmospheric stability, and spatial separation. Panofsky and Dutton [1984] discuss research which provides additional insight on decay coefficients. Figure 8-21 shows the coherence model for lateral separation compared with test data exhibiting typical scatter [Frost and Lin 1981].

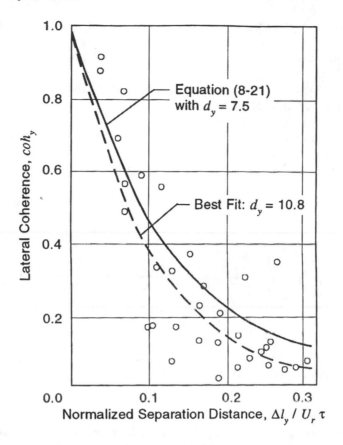

Figure 8-21. Typical lateral coherence data modeled by exponential decay functions. [data from Frost and Lin 1981].

Because the coherence function is useful for estimating the sizes of gusts of different durations, Equation (8-21) provides a method for calculating the minimum duration of gusts that are large enough to engulf a turbine rotor. Solving for the gust duration gives

$$\tau = -\frac{d_\alpha \Delta l_\alpha}{U_r \ln(coh_\alpha)} \tag{8-22}$$

As an example, consider a HAWT with a rotor diameter of 30 m, operating in a wind that is 15 m/s at the reference elevation of 10 m. Assuming a coherence of 0.5 (a high coherence for wind-related events) and a decay coefficient of 7.5, the minimum duration of a gust that can be assumed to be constant over the entire area swept by this rotor is 22 seconds, according to Equation (8-22). Additional information on spatial correlation of gusts, derived from the *von Karman correlation model*, is given in Frost *et al.* [1987].

Gust Factor Model for Peak Horizontal Gust Magnitudes and Durations

As discussed previously, a gust model useful to wind turbine engineers correlates magnitude, duration, and frequency of occurrence. Several such models have been derived from published gust statistics, such as those given in a number of reports reviewed by Powell and Connell [1980]. Doran and Powell [1980], Kaimal *et al.* [1981], and Akins [1981] have also compiled statistics on gust characteristics. Both the magnitude and duration of gusts vary randomly, but magnitude generally increases with decreasing duration. The total gust environment can be defined in terms of the joint probability distribution of magnitude and duration for gust events, together with some basic frequency-of-occurrence data.

On the basis of 17 years of measurements of peak horizontal wind speeds at Cape Kennedy, Florida, reported by Kaufman [1977], a gust model has been developed that consists of two empirical factors: a *mean gust factor*, which is defined as the ratio of the average wind speed during the gust to the steady wind speed averaged over a 10-minute period or longer; and a *statistical factor*, to account for deviations from the mean gust. Steady wind speeds for averaging periods longer than 10 min showed little variation from the 10-min average. These factors are expected to be representative of winds over reasonably flat, uniform terrain. In this model, the gust magnitude in Equation (8-20) is given by

$$\Delta u = (F_g F_s - 1)\, U \qquad\qquad (8\text{-}23)$$

where

F_g = empirical mean gust factor from Table 8-4
F_s = empirical statistical factor from Table 8-5

Table 8-4.
Mean Gust Factors, F_g, for Calculating Extreme Gust Magnitudes
[Frost *et al.* 1978; data from Kaufmann 1977]

Ref. wind speed, U_r (m/s)	Duration of gust, τ (s)	Elevation, z (m)					
		10	20	40	80	160	300
5	1	1.450	1.398	1.350	1.307	1.270	1.239
	5	1.380	1.336	1.297	1.263	1.232	1.207
	10	1.335	1.300	1.266	1.235	1.208	1.186
	50	1.219	1.197	1.176	1.158	1.141	1.127
	100	1.160	1.144	1.129	1.116	1.105	1.094
	300	1.059	1.054	1.049	1.044	1.040	1.036
10	1	1.390	1.331	1.282	1.239	1.203	1.175
	5	1.322	1.276	1.236	1.201	1.172	1.149
	10	1.288	1.248	1.211	1.179	1.153	1.132
	50	1.186	1.163	1.141	1.121	1.104	1.086
	100	1.135	1.119	1.105	1.090	1.077	1.066
	300	1.049	1.043	1.038	1.047	1.038	1.025
≥ 20	1	1.382	1.322	1.271	1.227	1.192	1.162
	5	1.315	1.269	1.228	1.192	1.162	1.139
	10	1.279	1.238	1.201	1.171	1.145	1.125
	50	1.178	1.153	1.132	1.112	1.096	1.082
	100	1.128	1.111	1.096	1.082	1.070	1.060
	300	1.047	1.041	1.036	1.031	1.027	1.023

Table 8-5. Statistical Factors, F_s, for Calculating Extreme Rust Magnitudes
[Frost *et al.* 1978; data from Kaufmann 1977]

Ref. wind speed, U_r (m/s)	Standard deviations above mean	Elevation, z (m)					
		10	20	40	80	160	300
5	1	1.19	1.25	1.31	1.37	1.43	1.49
	2	1.37	1.52	1.68	1.85	2.04	2.24
	3	1.64	1.88	2.16	2.52	2.91	3.32
10	1	1.37	1.40	1.45	1.48	1.53	1.57
	2	1.48	1.56	1.64	1.73	1.83	1.93
	3	1.60	1.73	1.87	2.01	2.17	2.34
15	1	1.07	1.09	1.12	1.14	1.16	1.19
	2	1.15	1.20	1.25	1.30	1.36	1.41
	3	1.23	1.31	1.39	1.48	1.57	1.67
25	1	1.05	1.06	1.07	1.09	1.11	1.12
	2	1.11	1.15	1.20	1.24	1.29	1.33
	3	1.17	1.23	1.29	1.36	1.43	1.49

Discrete Gusts for Fatigue Analysis

The gust magnitudes described in Equation (8-23) are based on statistics of hourly peak wind speeds. In this sense they are expected to be extreme values and most useful for the analysis of ultimate loads. For fatigue analysis, however, one is interested in smaller but more numerous gusts which occur routinely throughout the life of the structure. Analysis of gust fatigue loadings on the NASA/Boeing 2.5-MW Mod-2 HAWT was performed with what has been termed the *NASA Lewis gust model* [Spera and Richards 1977, Powell and Connell 1980]. In this model, fatigue gust amplitudes are determined by first computing the wind turbulence intensity at a given elevation and steady wind speed for a given wind turbine by means of the following equation:

$$\sigma_{0,x}(z, U) = \sqrt{\int_{n_{min}}^{n_{max}} \phi_x(z, U, n)\, dn} \qquad (8\text{-}24)$$

where the frequency limits, n_{min} and n_{max}, are obtained from known or assumed dynamic response characteristics of the turbine being analyzed.

The turbulence of the wind is then assumed to consist of a set of discrete gusts with *Gaussian random amplitudes* but with the deterministic shape given by Equation 8-20. Thus, the population of gust magnitudes, Δu, has a *normal distribution* based on $\sigma_{0,x}$ in this model. To complete this definition of the set of discrete gusts, a duration must be specified, and this is selected as follows: Let $\tau_0(\Delta u)$ be the most probable period of a gust with magnitude Δu, determined from the data in Table 8-4. Then,

$$\tau = 0.5/n_{max} \quad , \tau_0 \leq 0.5/n_{max} \tag{8-25a}$$

$$\tau = 0.5/\tau_0 \quad , 0.5/n_{max} < \tau_0 < 0.5/n_{min} \tag{8-25b}$$

$$\tau = 0.5/n_{min} \quad , \tau_0 \geq 0.5/n_{min} \tag{8-25c}$$

If it is desired to utilize a discrete gust approach for fatigue load calculations, it is necessary to estimate the number of times that the turbine will be exposed to gusts exceeding a given magnitude. A model for estimating the number of times per hour, averaged over a year, that the gust speed exceeds a prescribed value is given in Frost *et al.* [1978] and Frost and Turner [1982]. This estimate is based on the well-known "number of crossings" model of Rice [1944, 1945] and the assumption of a Weibull distribution of steady wind speeds during a year. Validation of this model with test data is limited.

A number of studies on gust statistics and other gust characteristics have been carried out in an effort to shed more light on gust phenomena and to validate discrete gust models References in addition to the sources mentioned earlier are Ramsdell [1978], Cliff and Fichtl [1978], and Huang and Fichtl [1979]. The general approach used in these studies is to spatially-average gusts by *band-pass filtering* of the data. Various high- and low-pass filter combinations are reported throughout these works. Doran and Powell [1980] provide some of the more complete statistics, including the following:

--   likelihood of occurrence of gusts of a given magnitude regardless of duration;
--   likelihood of occurrence of gusts of a given duration regardless of magnitude;
--   statistical relationships between gust magnitude and gust duration (Fig. 8-22).

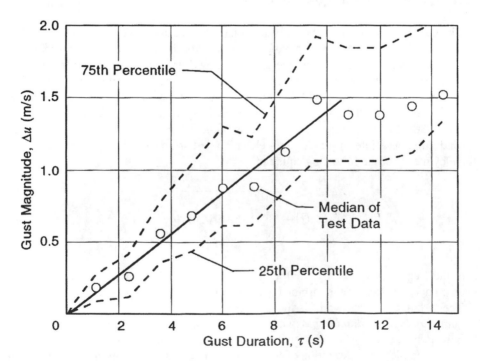

Figure 8-22.   **Typical statistical relationships between gust magnitude and duration.**
[Doran and Powell 1980]

Spatial Turbulence Models

Spatial turbulence models are mathematical descriptions of the fluctuating wind field in two- and three-dimensions. These models are important to the design of wind turbine control systems, to the experimental verification of power output, to the prediction of asymmetric forces on rotors resulting from nonuniform gusts over the swept area, and many other responses of the turbine to unsteady, non-uniform wind conditions.

Information about spatial wind fields is normally based on a time-history of wind speed measured by an isolated anemometer measurement and the application of *Taylor's hypothesis* [Hinze 1975]. Taylor's hypothesis assumes the fluctuating velocity of the wind at a fixed point, $g(x_0, t)$, can be converted using the relationship $x = x_0 + Ut$, producing a spatial relationship at a fixed time, $g(x, t_0)$. This is also referred to as the *frozen turbulence* concept, since the spatial variation is assumed to remain unchanged during the averaging time of U. To illustrate the relation between time and distance inherent in Taylor's hypothesis, consider two points in space, A and B, in which B is a distance Δx downwind of A. According to the frozen turbulence concept

$$g_\alpha(B, t) = g_\alpha(A, t - \Delta x/U)$$

where, as before, the subscript α represents each of the three components of the wind.

Spatial variation at two or more points in space can be modeled without recourse to Taylor's hypothesis using wind speed data from an array of anemometers. Spatial variations in the vertical direction can be measured with a single tower instrumented at different levels, and in both the vertical and lateral directions with multiple towers supporting anemometers in a two-dimensional pattern called a *vertical plane array* (Figure 2-24).

Statistical Parameters

Quantitative estimates of spatial variations in wind fields are provided by the following three statistical parameters:

-- correlation coefficient
-- two-point spectrum (or two-point correlation)
-- coherence function

Full mathematical descriptions of these parameters are given in a number of references [*e.g.*, Papoulis 1965, Bendat and Piersol 1971, Panofsky and Dutton 1984]. Brief physical descriptions of some models will be given here.

Correlation Coefficient

A correlation coefficient is a measure of how well fluctuations in a wind speed component measured at one position in space correspond or correlate with fluctuations in a wind speed component at a second point. The full three-dimensional correlation coefficient, correlating each of three components at the first point with each of the three at the second, is a *nine-component tensor*. However, for the conditions of isotropic, homogeneous turbulence normally assumed in wind turbine analysis, the correlation coefficient reduces to two components: a *longitudinal correlation coefficient* for wind fluctuations parallel to a line joining the two points in space, and a *transverse correlation coefficient* for wind fluctuation components perpendicular to this line. Assuming a separation distance of ζ, some examples of these two types of correlation are as follows:

Longitudinal Correlation:

   $g_x(x,y,z)$ with $g_x(x + \zeta, y, z)$ $g_y(x,y,z)$ with $g_y(x, y + \zeta, z)$

Transverse Correlation:

   $g_x(x,y,z)$ with $g_x(x, y + \zeta, z)$ $g_y(x,y,z)$ with $g_y(x + \zeta, y, z)$

A correlation coefficient of unity indicates that wind fluctuations are identical at the two points in question, while a negative correlation coefficient suggests structured reverse flow.

The calculation of correlation coefficients can be illustrated by considering a HAWT with a rotor diameter D, hub elevation h, and time-histories of the longitudinal wind speed at three locations in space: the center of the rotor, ($x = y = 0$, $z = h$); upwind from the rotor center a distance $2D$; and laterally outward from the rotor center a distance $0.5D$. Assuming isotropic turbulence (*i.e.*, equal in all directions), the correlation coefficients are

$$\text{Longitudinal:} \quad \kappa_L(-2D) = \overline{g_x(0,0,h)\, g_x(-2D,0,h)} / \sigma_{0,x}^2 \qquad (8\text{-}26a)$$

$$\text{Transverse:} \quad \kappa_T(0.5D) = \overline{g_x(0,0,h)\, g_x(0,0.5D,h)} / \sigma_{0,x}^2 \qquad (8\text{-}26b)$$

where

$$\kappa_L = \text{longitudinal correlation coefficient}$$
$$\kappa_T = \text{transverse correlation coefficient}$$
$$\text{overbars} = \text{time averages}$$

Three mathematical forms of the correlation coefficients are the *von Karman* [Hinze 1975, p. 247], the *Dryden* [*loc. cit.*, p. 58], and the *Batchelor and Townsend* [*loc. cit.*, p. 202]. All can be written in terms of a parameter called the *integral length scale, Z,* as follows:

$$\text{von Karman:} \quad \kappa_L = 0.593 \left(\frac{\zeta}{Z}\right)^{0.333} K_{1/3}(\zeta/Z) \qquad (8\text{-}28a)$$

$$\kappa_T = 0.593 \left(\frac{\zeta}{Z}\right)^{0.333} \left[K_{1/3}(\zeta/Z) - \frac{\zeta}{Z} K_{-2/3}(\zeta/Z) \right] \qquad (8\text{-}28b)$$

$$\text{Dryden:} \quad \kappa_L = \exp(-1.50\,\zeta/Z) \qquad (8\text{-}29a)$$

$$\kappa_T = (1 - 1.50\,\zeta/2Z)\,\exp(-1.50\,\zeta/Z) \qquad (8\text{-}29b)$$

$$\text{Batchelor–Townsend:} \quad \kappa_L = \exp[-(\zeta/Z)^2] \qquad (8\text{-}30a)$$

$$\kappa_T = [1 - (\zeta/Z)^2]\exp[-(\zeta/Z)^2] \qquad (8\text{-}30b)$$

where

$$\zeta = \text{separation distance between two points in space (m)}$$
$$Z = \text{integral length scale of isotropic turbulence (m)}$$
$$K_\nu(\,) = \text{modified Bessel function of the second kind of ( ), of fractional order } \nu$$

The coefficient of 1.50 in Equations (8-29) has been selected in order to fit the simpler Dryden functions to the von Karman functions, eliminating the need for the more-complex Equations (8-28). The three pairs of correlation functions are plotted in Figure 8-23. In Figure 8-23(a), the data points represent experimental longitudinal correlation coefficients computed from wind speeds measured simultaneously along a line of anemometer towers [Steely and Frost 1981]. Additional experimental data and discussion of spatial correlations and integral length scales are given in Frost and Lin [1983] and Frost *et al.* [1985]. The integral length scale, Z, is defined as the integral of the correlation coefficient over all

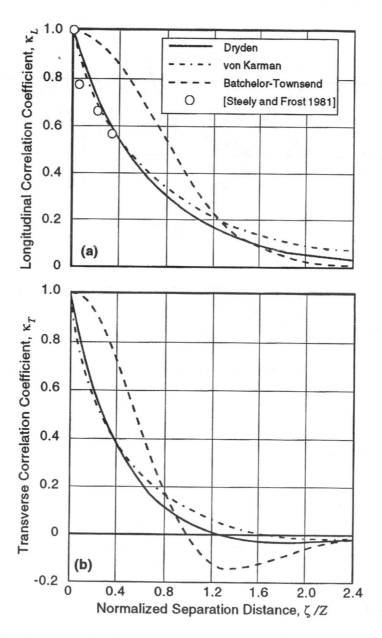

Figure 8-23. Three models of turbulence correlation coefficients. The Dryden model has been fitted to the von Karman. (a) Longitudinal correlation coefficients compared to test data (b) Transverse correlation coefficients

separation distances. Empirical relationships for non-isotropic integral length scales are plotted in Figure 8-24, as functions of surface roughness length, elevation, and direction [Counihan 1975].

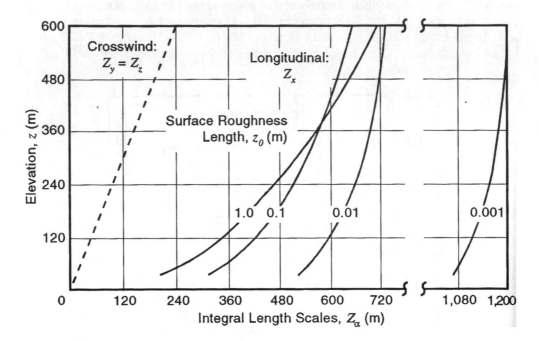

Figure 8-24. Empirical models for calculating non-isotropic integral length scales. [Counihan 1975]

Two-Point Spectrum

The correlation coefficient provides a relationship between the fluctuations in the wind averaged over all values of gust sizes. In many cases, however, the correlation between fluctuations of a prescribed frequency is needed. In these cases, the statistical parameter known as the *two-point* or *cross spectrum* is useful. The two-point spectrum is computed by *Fourier transforms* of the two-point correlation discussed previously. A convenient theoretical model of the two-point spatial correlation can be derived from the von Karman model [Houbolt and Sen 1972, Frost *et al.* 1987]. The Fourier transforms of these intermediate two-point correlation equations then gives the two-point spectra.

Coherence

The coherence parameter is defined as the absolute value of the normalized two-point spectrum, and it is often more useable than the spectrum itself. Coherence is expressed empirically by Equation (8-21).

Small-Scale Wind Shear Fluctuations

Wind shear fluctuations across the swept area of a wind turbine rotor are another form of turbulence that can have a pronounced effect on rotor blade loads, fatigue life, and the output power quality. Ramsdell [1978] established standard deviations and other statistics of vertical and lateral gradients in wind speed, between two points separated by distances the size of a rotor, treating these gradients as random variables. Correlation of test data on the basis of the steady wind speed at mid-elevation on the rotor, U, surface roughness length, z_0, and rotor diameter, D, yields the empirical models shown in Figure 8-25 for the standard deviations of the average vertical and lateral gradients.

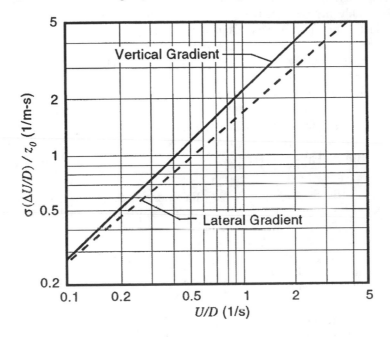

Figure 8-25. Empirical models for estimating standard deviations of the vertical and lateral wind gradients across the swept area of a rotor of diameter D. [Ramsdell 1978]

Elliott [1984] analyzed wind shear data from three sites in the U.S. with tall towers supporting anemometers at various elevations, under day, night, summer, and winter conditions. A two-slope system was developed for categorizing the observed wind shear profiles (Fig. 8-26), and 30 of the most common were analyzed statistically for *frequency of occurrence, mean duration* (*i.e.* persistence), and magnitude of *fluctuations in steady wind speed* associated with each profile. Of particular interest here is the relatively short persistence of many of the wind shear profiles observed. For example, during a winter test at one site, there was a 71 percent probability that the wind shear profile would change to a different pattern in less than one minute. In the summer this probability rose to 96 percent. Much additional information on wind shear as a dynamic phenomenon can be obtained from this study.

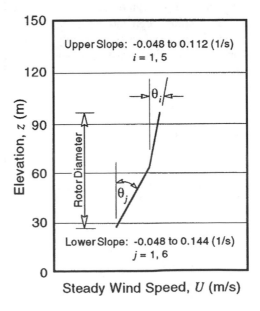

Figure 8-26. Two-slope system for categorizing vertical wind shear profiles across a HAWT rotor. [Elliott 1984]

Turbulence Modeled From a Rotational Frame of Reference

Previous discussions have dealt with turbulence measured at a single point or a multiple of fixed points along a line in space. This type of turbulence modeling, in which the reference coordinate axes *x-y-z* are fixed in space, is also called *Eulerian*. Since the late 1970s, considerable study has been carried out by wind energy researchers on measuring and modeling fluctuations in the wind from a rotating or *Lagrangian* frame of reference. Starting in 1977, personnel at the Battelle Pacific Northwest Laboratories in Richland, Washington, began studies of the wind as seen by a rotating turbine blade passing repeatedly through a wind field with spatial and temporal fluctuations. Measurements have been made using *fixed arrays* of towers and anemometers [Verholek 1978, Connell 1981, George 1984, George and Connell 1984], a *rotating-boom hot-wire anemometry* system [Sanborn and Connell 1984], and a *lidar scanning* system [Hardesty *et al.* 1984].

Vertical Plane Arrays

The test installation shown schematically in Figure 8-27 [Verholek 1978] is typical of the so-called *vertical plane arrays*. Instrument towers along a line perpendicular to the prevailing wind support a circular pattern of anemometers at equal intervals. This particular anemometer circle represents the path followed by a blade section 12.2 m from the rotor axis. The wind velocity sensed by each anemometer is continuously recorded. By joining segments of wind speed records taken sequentially from consecutive anemometers around the circle, the time-history of wind velocity experienced by a section on a rotating blade can be *synthesized*. The length of the segment from each anemometer depends on the tangential speed of the hypothetical blade section.

Figure 8-28 illustrates typical rotationally-sampled wind speed data. In Figure 8-28(a), the dominant fluctuations with a period equal to the period of rotation (0.8 s) are caused by the mean wind shear profile. Also evident are higher-frequency fluctuations which indicate the non-uniform structure of the small-scale turbulence measured at the various anemometers along the circle of rotation. Such

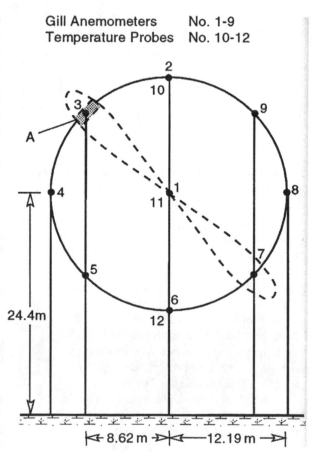

Figure 8-27. Schematic diagram of a vertical-plane array of anemometers for measuring turbulence experienced by a rotating HAWT blade. Sampling of signals from anemometers 2 to 9 is synchronized with the passage of blade section **A**. [Verholek 1978]

fluctuations are indicative of cyclic wind forces imposed on a turbine blade as it rotates through a quasi-static but non-uniform wind field.

In Figure 8-28(b) the time-series of wind speed measured by the central or "hub" anemometer is compared to the array-average time series. The figure shows that the array-average is much smoother than that of the single wind speed measurement at the center of the array, because array-averaging acts as a *low-pass spatial filter*. The fluctuations in the array-averaged wind are the average of the instantaneous fluctuations at each anemometer. An examination of 10-min averages demonstrates that the steady wind speed at the hub is representative of the array-average steady wind speed.

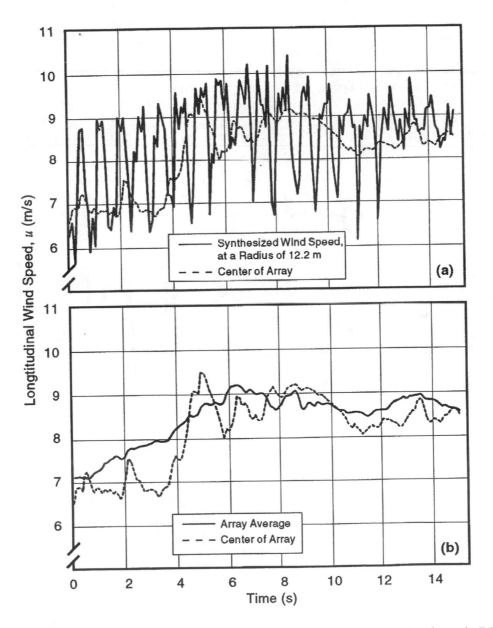

Figure 8-28. Examples of rotationally-sampled wind speed data. Rotational rate is 7.8 rad/s. (a) Synthesized wind speed compared to speed at the center of the array (b) Array-average wind speed compared to speed at the center of the array [Verholek 1978]

Spectral Analysis of Rotationally-Sampled Wind

Spectral analysis of the three time-series illustrated in Figure 8-28 gives us further information on the distinctions between the *Eulerian* (fixed coordinate) and *Lagrangian* (rotating coordinate)[1] representations of the same wind field, and these spectra are shown in Figure 8-29. Here the power amplitude, $\phi(n)$, has been multiplied by the circular frequency, n, to accentuate the higher-frequency portion of the spectrum. The spectrum of the center anemometer provides a reference in the Eulerian system. The rotationally-sampled spectrum contains several large "spikes", the largest of which occurs at the simulated rotor speed of 7.8 rad/s. The average magnitude of the Lagrangian spectrum in the frequency range of the spikes is much greater than that in the Eulerian spectrum.

Conversely, there is a considerable loss of energy in the decade of frequency below the simulated rotor speed, compared to the Eulerian spectrum. Connell [1981] points out that rotational sampling transfers energy from intermediate frequencies to energy at higher frequencies and collects a portion of this energy into narrow frequency bands at harmonics of the rotational speed. The total turbulent energy, of course, cannot be changed by a change of coordinate system. As expected, the spectrum of the array-average wind speed contains less energy than that of the center anemometer, because of the low-pass filtering mentioned earlier.

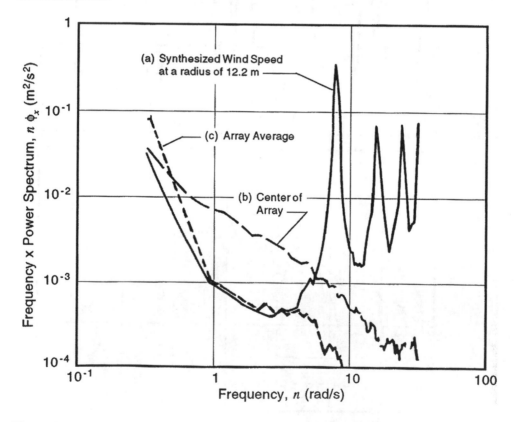

Figure 8-29. Power spectra of the time-series wind speeds illustrated in Figure 8-28. (a) Synthesized wind speed in the Lagrangian (rotating) coordinate system (b) Center anemometer in the Eulerian (fixed) coordinate system (c) Array-averaged wind speed [Verholek 1978]

---

[1] Note that these are not the same definitions as used in classical fluid mechanics.

Lagrangian Mathematical Models of Turbulence

Several mathematical models of turbulence spectra seen by a section of a rotating turbine blade have been developed [Rosenbrock 1955, Holley *et al.* 1981, Connell 1981, Kristensen and Frandsen 1982, Powell and Connell 1986a]. These models are similar in approach to that of Houbolt and Sen [1972] for describing two-point spatial spectra. The basic spectral theory consists of the Fourier transform of a two-point spatial correlation function for turbulence as experienced at a rotating point. The point rotates either in a vertical plane, as on a HAWT blade, or in a horizontal plane, as on a VAWT blade.

Empirical equations developed by Spera [1995] model rotationally-sampled turbulence for input into structural-dynamic computer codes for calculating fatigue loads.

HAWT Rotationally-Sampled Turbulence Model

Consider the plane of revolution for a HAWT blade as shown in Figure 8-30. The distance ζ represents the separation distance between the two fluctuating velocities in the turbulent flow field which will correlate at the plane of revolution on the circular path of a section on the moving blade. That is, at time zero the blade section is at point A and t seconds later it is at B. However, since the turbulence is assumed to drift without change at the steady wind speed, U, the turbulent fluctuation seen by the blade section at B was upwind a distance $x = Ut$ at time zero, at point B'. The two-point correlation we seek, between the longitudinal wind speeds at points A and B' that are separated by a distance ζ, is given by Equations (8-31), which are derived in detail in [Frost *et al.* 1987]. These follow the approach in Rosenbrock [1955], Connell [1981], and Kristensen and Frandsen [1982] and assume the turbine shaft is aligned with the wind and turbulence is isotropic.

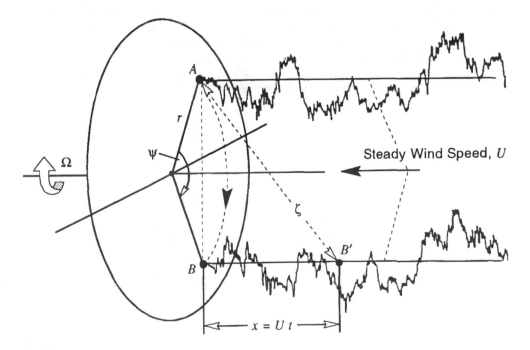

Figure 8-30. Rotating (Lagrangian) coordinate system for calculating two-point correlation coefficients for a HAWT.

$$R_{xx}(\zeta) = [\kappa_L(\zeta/Z)(x^2/\zeta^2) + \kappa_T(\zeta/Z)(1 - x^2/\zeta^2)]\sigma_{0,x}^2 \qquad (8\text{-}31a)$$

$$x = \psi U/\Omega \qquad (8\text{-}31b)$$

$$\zeta = \sqrt{4 r^2 \sin^2\psi/2 + x^2} \qquad (8\text{-}31c)$$

where

$R_{xx} =$ *two-point autocorrelation function* for longitudinal winds at A and B (m^2/s^2)

$\kappa_L(\ ), \kappa_T(\ ) =$ longitudinal and transverse correlation coefficients evaluated at ( ), referenced to A-B' (Eqs. (8-28) to (8-30))

$\psi =$ azimuthal coordinate from A to B (rad)

$\zeta =$ distance from A to B' (m)

Inspection of Equations (8-32) reveals that the Eulerian autocorrelation function is obtained by setting r equal to zero, and the *frozen turbulence* case is obtained with U equal to zero. For non-isotropic turbulence, we can define an equivalent isotropic integral scale length as

$$Z = Z_x(x/\zeta) + Z_y(1 - x/\zeta) \qquad (8\text{-}31d)$$

where

Z_x, Z_y = longitudinal and lateral integral scale lengths, respectively (m)

Powell *et al.* [1985] found that the best correlation between calculated and measured Lagrangian spectra from the Clayton, New Mexico, vertical-plane array was obtained when the integral length scales Z_x and Z_y were both set equal to about 1.7 times the mid-elevation of the array. Therefore, a reasonable approach would be to assume *isotropic turbulence* and treat Z as an empirical parameter, to be evaluated on the basis of available test data. Powell and Connell [1986b] provide a computer model for simulating rotational data for both HAWTs and VAWTs and compare calculated spectra to experimental data.

To illustrate the use of Equations 8-31, we will calculate horizontal autocorrelation functions for a Mod-OA HAWT, using the Dryden correlation coefficients for simplicity. The configuration parameters for this sample case are as follows:

R = rotor radius = 19.05 m		Ω = rotor speed = 4.19 rad/s	
h = hub elevation = 30.5 m		Z = 51 m	
U = steady wind speed = 8.21 m/s		σ = 0.9 m/s	

Figure 8-31 shows typical behavior of the autocorrelation function (normalized by the square of the turbulence or the *variance*) with increasing azimuthal angle, at four radial locations. The valleys are regions of lower correlation which produce the characteristic peaks in the Lagrangian power spectrum shown in Figure 8-32. This spectrum is obtained from the autocorrelation function by a standard *fast Fourier transform (FFT)* analysis.

VAWT Rotational Sampling

The two-point spatial correlation for a VAWT is developed in the same manner as that for the HAWT [Powell and Connell 1986a]. Comparisons of simulated Lagrangian spectra for VAWTs with equivalent spectra for a HAWT [George 1984] indicate much lower turbulent energy for VAWTs in flow fields with strong vertical gradients, as would be expected.

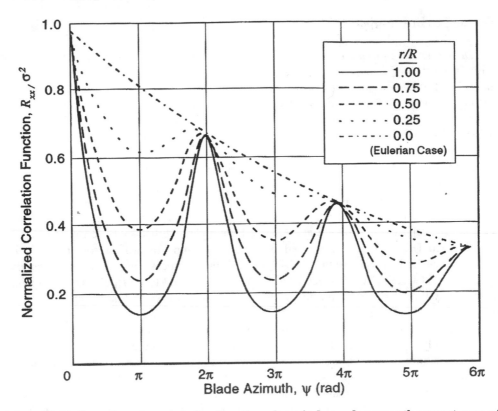

Figure 8-31. Sample autocorrelation functions for wind speeds across the swept area of the Mod-OA HAWT, calculated using Equations 8-31 and the Dryden correlation coefficients. Rotationally-sampled turbulence spectra are obtained from these functions.

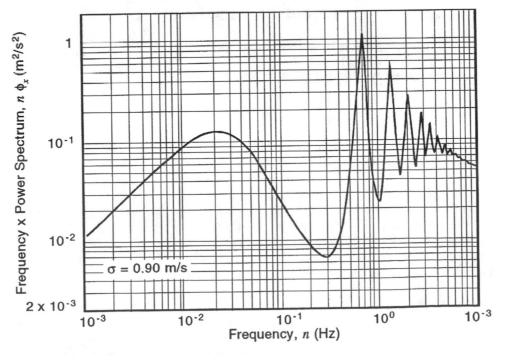

Figure 8-32. Sample Lagrangian turbulence spectra calculated from the autocorrelation function in Figure 8-31, at a radius of 19.05 m. [Powell *et al.* 1985]

Wake Turbulence

Mathematical models of wind turbine wake geometries are discussed in Chapter 6. In addition to predicting reductions in steady wind speed, a model of the turbulence within the wake of a wind turbine is important in the design of wind power stations. As noted earlier in Equations (6-7), wake turbulence is a combination of *ambient* and *rotor-generated* components. The total turbulence in the wake has a fairly significant effect on its crosswind size and downwind persistence. The turbulence intensity in the wake of a single turbine has also been simulated in wind tunnels [*e.g.* Blackwell and Sheldahl 1977, Boschloo 1977, Vermeulen 1978, Builtjes 1979]. However, there are various limitations in all of the existing models in such areas as scaling, modeling of the interaction between the wind and the turbine, and modeling of the boundary layer of the atmosphere.

One of the more complete field studies of the turbulent structure of a wind turbine wake is that reported by Connell and George [1982]. The vertical-plane array at Clayton, New Mexico, (Fig. 8-33) was employed to rotationally-sample the wake turbulence two diameters downwind of the Mod-OA HAWT. The multiple-peak shape of the Lagrangian spectra in the wake at this distance from the rotor was found to be similar to that upwind of the turbine, for the same atmospheric stability. Turbulence was measured at 17 locations in the wake and one location at the wake's edge. The *ambient turbulence intensity* (ratio of turbulence to steady wind speed upwind of the rotor and outside the rotor wake) was approximately 0.15. By deducting the ambient turbulence from the measured wake turbulence, the pattern of *generated turbulence* shown in Figure 8-33 was obtained. The change in turbulence is noted at each anemometer location in the array. Turbulence increased within the main power-extraction region, while behind the tower and near the ground turbulence stayed the same or decreased. Overall, the turbulence two-diameters downwind of the turbine was increased about 0.24 m/s or 15 percent by the action of the rotor.

Turbulence Simulation

Simulated turbulence is a computer-generated analog or digital signal which has statistical properties equivalent to that of the turbulent wind. The resulting signal thus resembles a time history of wind speed fluctuations. Turbulence simulation, when carried out completely, can provide all three gust components as functions of time and position in space for input into equations of motion (linear or non-linear) for structural dynamic analysis. The output of these equations gives the dynamic response of the system in analog or digital form, from which parameters such as the RMS response of the system, extremes or peaks in response, and correlations between the response output and the turbulence input can be calculated. Moreover, 3-dimensional turbulence simulation automatically includes the effects of wind direction fluctuations on the power output of a HAWT.

Turbulence simulation procedures generally require considerable computer memory and lengthy computational time. On the other hand, turbulence simulation provides the most complete analysis of the effects of an unsteady, non-uniform wind field on the performance, loads, and control of a wind turbine. In principle, one could simply use measured time histories of wind speed and direction rather than simulated ones. However, this generally results in lengthy and costly field projects with multi-anemometer arrays in the uncontrolled environment provided by nature. Moreover, simulations can contain random or chance events that do not necessarily occur during a specific measurement program, and statistical and atmospheric properties (such as power spectrum, turbulence intensity, stability, and surface roughness) can be systematically varied to study their effects on the turbine.

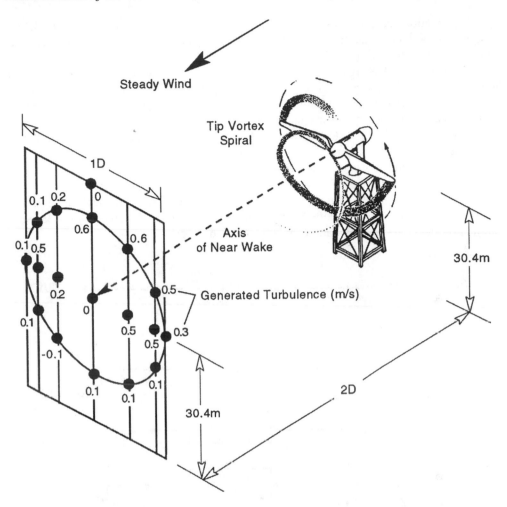

Figure 8-33. Pattern of generated turbulence in the wake of an operating 200-kW Mod-OA HAWT, two rotor diameters downwind. Changes from the ambient turbulence are noted at each anemometer location. [Connell and George 1982]

Methods of simulating turbulence can be grouped into three generic techniques: *transformation, correlation*, and *harmonic series*. Simulated turbulence signals are frequently superimposed on a signal representing the steady, non-uniform wind field to give the instantaneous wind.

Transformation Techniques

In the transformation technique a *Gaussian white noise signal* is passed through a filter or series of filters, as illustrated in Figure 8-34. The filters tailor the input signal so that the output signal has the statistical properties of the wind field being simulated. The number of statistical properties which are matched depends on the sophistication of the filtering system. In the simple system shown in Figure 8-34(a), the one-dimensional wind speed fluctuations have a Gaussian probability distribution, and the filter function h(t) is designed to provide a time history with a prescribed turbulence intensity and power spectrum. The procedure for determining filter characteristics is given in Frost and Moulden [1977] and Frost and Wang [1980]. Many of the early simulations used a *Dryden spectrum* for simplicity, but recent models use spectra more representative of atmospheric turbulence.

The complex simulation system shown in Figure 8-34(b) generates time histories for all three wind velocity components and is capable of matching considerably more statistical properties than the simple system. Filter functions are selected so that the spectrum of each component simulates that of a wind which contains large eddy fluctuations. *Vertical coherence* requirements can also be satisfied by a simulation of this complexity.

It is apparent that the completeness of the turbulence simulation produced by the transformation technique is basically dependent on the degree of mathematical sophistication represented by the filters. Table 8-6 lists a number of reported turbulence transformation simulations and the statistical parameters incorporated in each.

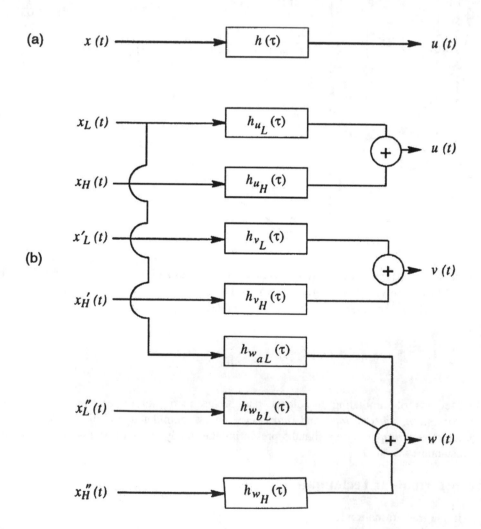

Figure 8-34. Schematic diagrams of turbulence simulations using the transformation technique. (a) Simple one-dimensional simulation (b) Three-dimensional complex system

Table 8-6.

Turbulence Simulations Using the Transformation Technique

Reference	No. of components	Spectrum model	Probability distribution	Coherence
Neuman and Foster [1970]	1	Dryden	Gaussian	NA
Reeves *et al.* [1976]	1	Dryden	Non-Gaussian	NA
Perlmutter *et al.* [1977]	3	Dryden	Gaussian	Vertical
Campbell [1984]	3	von Karman	Gaussian	NA
Frost and Wang [1980]	1	Dryden, von Karman, and Kaimal	Gaussian	Vertical
Arman and Frost [1986a]	3	Kaimal	Gaussian	Vertical
Arman and Frost [1986b]	3	Hojstrup	Gaussian	Vertical, plus cross-correlation

Correlation Techniques

Most applications of the *correlation technique* to turbulence simulation are based on the *Markov hypothesis* [e.g. Cliff and Hall 1980, Lee and Stone 1983, Legg and Raupach 1982]. The velocity of a small volume of air at a future time is represented by the sum of a velocity correlated with the velocity at the present time and a purely random, uncorrelated increment of velocity. For a stationary, homogeneous turbulent flow the correlated part is given mathematically by the *Lagrangian velocity auto-correlation function.* An application of this technique to wind turbine design is that by Cliff and Hall [1980].

Harmonic Series Techniques

Two turbulence simulations which use harmonic series have been reported in the wind energy literature. That of Sundar and Sullivan [1981] simulates a random process by a series of cosine waves at almost evenly spaced frequencies and with amplitudes weighted according to the spectral energy at the wave frequency, and the authors report that with as few as 50 frequencies one can obtain a fairly good representation of the spectrum. Their procedure has been expanded to three-dimensions using a composite of the *Davenport spectrum* and the *Fourier transform of the coherence function.* Because direct evaluation of the three-dimensional spectrum is time consuming, a fast Fourier transform technique was applied to decrease the computational time. A second harmonic series technique is that developed by Powell and Connell [1986a].

Turbulence Simulation in Lagrangian (Rotating) Coordinates

Powell and Connell [1986b] provide a comparison of the following three turbulence simulation models based on rotating coordinates: Connell [1982], Sundar and Sullivan [1983], and Veers [1984]. All three models are based on the harmonic series technique, which the authors recommend over the transformation (white noise) technique. However, comparisons of the time series produced by the different models showed that significant differences can occur, and recommendations were made for improving simulation methods. More recent models proposed by Veers [1988] and expanded by Kelley [1993] are now widely used and reported.

Wind Turbine Siting

The major goal in siting wind turbines is to maximize energy capture and thereby reduce the unit cost of generating electricity. Maximum energy capture can be difficult to achieve because of the wide variability of the wind over both space and time. The first step in site analysis is to examine all available information, such as the Wind Energy Resource Atlas of the United States [Elliott *et al.* 1986]. Although wind resource maps identify many promising areas, they do not depict variability caused by local terrain features. Siting of a wind turbine relative to local factors, which is termed *micrositing*, is described in this section. The scale of the turbines to be sited can determine the specific procedures used.

Siting Small-Scale Wind Turbines

Small-scale wind turbines (rotor diameters up to about 12 m) are typically used to generate power for consumption near the turbine by a private household or a small business. This restricts the location of the machine to a relatively small area. The siting approach under these circumstances must address two basic problems: finding the best (or at least the most acceptable) site for the turbine in a given area, and accurately estimating the wind characteristics at the selected site. Picking the initial location can be done using empirical guidelines formulated for evaluating the effects of the local topography and roughness elements on the wind, including the effects of trees and buildings [*e.g.* Frost and Nowak 1979, Wegley *et al.* 1980, Frost and Shieh 1981].

Depending on the time and funds available, there are three possible approaches for estimating the wind characteristics at the site: first, using wind data from a nearby site, which is the quickest and least costly approach; second, correlating limited on-site wind measurements with data from a nearby site; and third, collecting a representative data sample at the proposed site, which is the most accurate approach, but also the most time consuming and most expensive. The success of the first and second approaches depends very heavily on the complexity of the terrain between the proposed site and site where data were collected and on whether the two locations are similarly exposed to the prevailing wind in the area.

Siting Medium- and Large-Scale Wind Turbines for Utility Applications

The meteorological aspects of siting medium- and large-scale wind turbines have been documented by Hiester and Pennell [1981]. These methods also apply to wind power stations composed of large numbers of medium-scale machines. Siting activities for utility applications include more than just the identification of a large, windy site. Economic viability, compatibility with wind turbine design capabilities, and accept-ability to the public are also important. Formal stepwise procedures have been established to help wind power developers through the decision process involved in locating large, viable wind turbine sites [Pennell 1982]. This process combines the meteorological aspects of siting (*e.g.*, annual average wind speed, frequency distribution, persistence, turbulence, and wind shear) with the non-meteorological aspects (economic potential, safety, and environmental considerations) that are equally essential to a successful wind power project. Only the meteorological aspects of siting will be discussed here.

The spacing between turbines in a wind power station is an important factor in extracting maximum energy from the wind without significantly affecting turbine life. Optimal spacing requires knowledge of the size and intensity of wakes downwind of each type of turbine. The wake analysis problem is complicated by the terrain-induced variability of the ambient wind over the station. The nature of this variability and preliminary attempts to simulate it with a *mass-consistent numerical flow model* have been reported by Phillips [1979], and Wendell [1984], The subject of wake modeling is treated in more detail in Chapter 6.

Veenhuizen and Lin [1987a] have proposed a set of computer models that estimate wind flow over complex terrain and account for wake interference on the productivity of individual turbines. For verification, measured energy output from 50 turbines in a 275-turbine wind power station was compared with predicted output based on an earlier micro-siting study in which *kite anemometers* were used to survey the wind field over the station. Monthly energy production was used to calculate an effective monthly mean wind speed for each turbine site, and these were found to be about 10 percent less than the wind speeds determined by the micrositing study. Errors in power output were found to be about twice the size of errors in estimating wind speed, on a percentage basis.

Characterizing Wind Flow Over a Large Area

The so-called *field measurement method* of determining the spatial variation of the wind resource across a wind power station is currently the procedure generally followed by private wind power developers. This method consists of subjectively interpolating between on site measurements of the wind speed at several locations to estimate wind speeds at potential turbine sites. Two alternative approaches are *wind tunnel physical models* and *computerized numerical models*, but these have not yet been used extensively by private industry.

The Field Measurement Method

A very complete description of the field measurement method for micrositing of wind turbines is given by Wade and Walker [1988]. A basic assumption for determining spatial variations in the wind resource is that power-producing winds are associated with a persistent set of characteristic meteorological conditions such that when these conditions occur, the wind field is spatially well-correlated. It is operationally assumed that the wind flow pattern is fixed and that the temporal variation in speed can be scaled locally with speed, once the wind direction at a given reference point is known from the prevailing wind direction. Ratios of wind speed measured simultaneously at separate locations on the terrain are constants under this assumption.

The general procedure of the field measurement method is to establish the *wind speed ratios* between the short-term (4 to 12 months) data from on-site anemometers and an existing long-term data base (several years) from a nearby reference station. Typically, two to three short-term anemometer stations are located per square kilometer, although the trend recently has been toward twice that density. Correlation of wind speeds between short- and long-term data records is obtained by *linear regression analysis.* For large-scale turbines, it is economically justified to utilize one anemometer per turbine site for at least one year.

Kite anemometer measurements can be made with commercially available equipment to provide additional definition of the spatial variation of the wind resource to supplement the fixed anemometer data. Kite anemometers typically add approximately 60 to 80 estimates of wind speed and direction to the data already available from fixed anemometers. About one kite measurement per 20,000 to 60,000 m^2 is recommended, depending on the steepness of the terrain. Several kite anemometers are normally flown simultaneously in relatively close proximity, and one is always located near a fixed anemometer which acts as a reference for calibration. Data records are about 20 min long, and three or more flights are generally made per kite location on different days. Kite elevation is usually the same as that of the center of the swept area of the prospective turbines.

Physical Modeling

Physical modeling consists of obtaining velocity, direction, and turbulence measurements over a scale model of the selected terrain in a wind or water tunnel. Data from the tunnel model may be acquired in sufficient detail to determine the spatial variation in the wind flow field. One wind tunnel study of a wind power station site [Davis 1984] was able to accurately predict the sign of the change in local wind speed from the average (*i.e.* whether a specific turbine site would receive more or less wind than a reference site) and it also did well in defining areas with low speeds. However, the magnitude of wind speed variations caused by complex terrain was underestimated. Testing of a contoured model of the Kahuku Point area on the north coast of Oahu (Fig. 8-2) produced wind speeds about 5 percent higher than those measured at 18 field anemometer data stations [Chien *et al.* 1980]. The opposite result was obtained with a physical model of the region of the Rakaia Gorge in New Zealand [Meroney *et al.* 1978].

Physical modeling of the flow field over a planned wind power station can be very useful in the final stages of wind turbine siting. It can be used to locate individual machines and wind monitoring equipment to within a small area. It does, however, require large specialized facilities [Hunt and Fernholz 1975, Riley and Delisi 1977], and only neutral stability conditions are easily simulated in a wind tunnel. References for more detailed descriptions of the theoretical principles of physical modeling of the atmospheric boundary layer are Snyder [1972], and Hiester and Pennell [1981].

Numerical Modeling

Numerical flow modeling uses digitized and gridded terrain elevations as the lower boundary of the model and produces *vector* and *contour plots* of predicted local wind speeds and directions for specified boundary conditions. Complex terrain in a 2-km square section of a wind power station was numerically modeled using nodal points on a 40-by-40 grid [Wegley and Barnard 1986]. For 8 different flow conditions, values of RMS error between calculated and observed data at 28 short-term anemometer stations ranged from 5 percent to 18 percent, and errors were less than 8 percent for six of these flow conditions. A numerical model of a 1.6-by-3.7 km wind power station with a 30.5-m grid spacing [Veenhuizen and Lin 1987b] accurately simulated the combined energy production of 93 turbines and gave the undisturbed wind flow at 18 measuring stations within an RMS error of 7 percent.

The accuracy of numerical flow models is significantly improved when measurements from a few locations are used to adjust the direction of the initial flow and the model parameters that simulate the enhancement or suppression of vertical motion due to thermal stratification. Measurements for this purpose can be made over a very short time period when atmospheric conditions are climatologically representative of the power-producing winds. The few measurements recommended to help determine the spatial variability of the wind over an area should not be confused with measurements made to estimate the long-term average of the wind speed. Under most conditions, a minimum of one year of data is required to estimate a long-term mean wind speed to an accuracy of 10 percent with a confidence level of 90 percent [Corotis 1977].

Anemometry

Anemometry for wind turbines is discussed in Wegley *et al.* [1980], Ramsdell and Wetzel [1981], AWEA [1986], and ASME [1989]. Anemometer systems consist of three major components: *sensors, signal conditioners*, and *recorders*. According to ASME [1989], sensors can be classified into the following categories:

-- *momentum transfer*, cups, propellers, and pressure plates
-- *pressure* on stationary sensors: pitot tubes and drag spheres
-- *heat transfer*, hot wires and hot films
-- *Doppler effects:* acoustic and laser
-- *special methods:* ion displacement, vortex shedding, etc.

Cup and propeller anemometers are currently used in conjunction with wind turbines more often than the other types. Two surveys of wind-measuring instruments that provide information on principles of operation, specifications, and expected performance are those by Stone and Bradley [1977] and Moroz and Brousaides [1980].

For accurate results, sensors should be mounted on free-standing, low-blockage masts or towers at appropriate distances from interfering objects, including the mast itself. It is very important that sequential wind data be retained. Most *data loggers* record the peak gust during a certain interval and store information about turbulence over the same period.

Anemometers are subject to a variety of errors in the determination of true wind speed, and equations which may be used to estimate the size of these errors are given in Justus [1978]. When anemometers are calibrated in steady air flows in a wind tunnel, they may measure the true wind within ± 1 percent. In gusty winds, however, anemometers as a rule speed up faster than they slow down, so that accuracies of ± 5 percent may be more realistic. Wind speed measurements are the principal contributors to uncertainty in performance tests [Vachon 1985], because of factors such as

-- anemometer instrument errors
-- measuring wind speed at a point rather than over an area
-- separation between the anemometer and the turbine
-- topography and turbulence effects

The number of anemometers required for measuring wind speed during a performance test of a wind turbine is addressed in ASME [1989]. Duplicate anemometers at a single elevation are recommended, with guidelines given for determining acceptable anemometer elevations with respect to the swept area of the turbine and the profile of the terrain between the anemometer and the turbine.

A *wind rose* is a convenient tool for displaying anemometer data for siting analysis. A wind rose is defined as any one of a class of diagrams showing the temporal distribution of wind direction and azimuthal distribution of wind speed at a given location. Figure 8-35 illustrates the most common form, which consists of several equally-spaced concentric circles with 16 equally-spaced radial lines, each representing a *compass point*. Line length is proportional to the frequency of the wind *from* the compass point, with the circles forming a scale. The frequency of calm conditions is entered in the center. The longest lines identify the prevailing wind directions. Average wind speeds are entered at the ends of the lines. Table 8-10 lists the data contained in the sample wind rose. Wind roses may represent annual, seasonal, monthly, or time-of-day data.

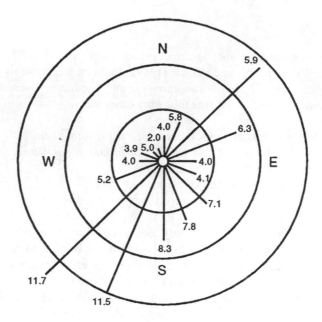

Figure 8-35. Sample wind rose diagram. Line lengths show time distribution, and average wind speeds are noted for each direction.

Table 8-10. Sample Wind Rose Data in Tabular Format

Direction	Percent frequency per wind speed interval (mph)						Totals (%)	Average speed
	0-3	4-7	8-12	13-18	19-24	25-31		
N	1	1					2	4.0
NNE	1	2	1				4	5.8
NE	3	8	3				14	5.9
ENE	1	5	2				8	6.3
E	1	2					3	4.0
ESE	1	2					3	4.1
SE	1	3	2				6	7.1
SSE		3	2	1			6	7.8
S	1	3	3	1			8	8.3
SSW	1	3	5	5	1		15	11.5
SW	1	4	5	5	2		17	11.7
WSW	2	2	1				5	5.2
W	1	1					2	4.0
WNW	1	1					2	3.9
NW		1					1	5.0
NNW	1						1	2.0
Calm	3						3	0.0
Totals (%)	20	41	24	12	3	0	100	7.6 mph

References

Akins, R. E., 1978, *Performance Evaluation of Wind Energy Conversion Systems Using the Method of Bins--Current Status*, SAND 77-1375, Albuquerque, New Mexico: Sandia National Laboratories.

Akins, R. E., 1981, *Gust Structure Analysis for WECS Design and Performance Analysis*, DOE/ET/23007-80-2, Richland, Washington: Battelle Pacific Northwest Laboratory.

ASME, 1989, *Performance Test Code for Wind Turbines*, ASME/ANSI PTC 42-1988, New York: American Society of Mechanical Engineers.

AWEA, 1986, *Standard Procedures for Meteorological Measurements at a Potential Wind Turbine Site*, D-FC02-86CH10302, Washington, DC: American Wind Energy Association.

Arman, E. F., and W. Frost, 1986a, *Documentation of the Monte Carlo Particle Dispersion (MoCaPD) Model*, Final Report of Contract DAAG29-81-D-0100, DO 1885, Tullahoma, Tennessee: FWG Associates, Inc.

Arman, E. F., and W. Frost, 1986b, *Large Eddy Turbulence Simulation Model*, Final Report of Contract DAAG29-81-D-0100, Tullahoma, Tennessee: FWG Associates, Inc.

Barchet, W. R., 1981, "Wind Energy Data Base," *Proceedings, Fifth Wind Energy Conference*, I.E. Vas, ed., SERI/CP-635-1340, Vol. II, Golden, Colorado: National Renewable Energy Laboratory.

Barchet, W. R., and W. E. Davis, 1983, *Estimating Long-Term Mean Winds from Short-Term Data*, PNL-4875, Richland. Washington: Battelle Pacific Northwest Laboratory.

Bardsley, W. E., 1980, "Note on the Use of the Inverse Gaussian Distribution for Wind Energy Applications," *Journal of Applied Meteorology*, Vol. 19: pp. 1126-1130.

Barr, N. M., G. Dagfinn, and D. R. Schaeffer, 1974, *Wind Models for Flight Simulator Certification of Landing and Approach Guidance and Control Systems*, Report No. FAA-RD-74-206, Washington, DC: Federal Aviation Administration.

Bendat, J. S., and A. G. Piersol, 1971, *Random Data: Analysis and Measurement Procedures*, New York: John Wiley & Sons, Inc.

Bisplinghoff, R. L., and H. Ashley, 1957, *Aeroelasticity*, Menlo Park, CA: Addison-Wesley Publishing Co.

Blackwell, B. F, and R. E. Sheldahl, 1977, "Selected Wind Tunnel Test Results for the Darrieus Wind Turbine," *Journal of Energy*, Vol. 1, No. 6, New York: American Society of Mechanical Engineers, pp. 382.

Boschloo, G., 1977, *Wake Structure of a Darrieus Rotor*, TNO Report 77-07244, Netherlands: Netherlands Organization for Applied Scientific Research.

British Standard Code of Practice CP3, 1972, *Code of Basic Data for the Design of Buildings*, Chapter V, Part 2: Wind Loads, London: British Standards Institution.

Builtjes, P., 1979, *Wind Turbine Wake Effects*. TNO Report 79-08375, Netherlands: Netherlands Organization for Applied Scientific Research.

Camp, D. W., 1968, *Low Level Wind Gust Amplitude and Duration Study*, NASA TMX-53771, Huntsville, Alabama: NASA Marshall Space Flight Center.

Campbell, C. W., 1984, *A Spatial Model of Wind Shear and Turbulence for Flight Simulation*, NASA TP 2313, Hampton, Virginia: NASA Langley Research Center.

Canadian Structural Design Manual, 1975, *National Building Code of Canada*, Supplement No. 1, Ottawa: Assoc. Committee on National Building Code and National Research Council of Canada.

Carlin, J., and J. Haslett, 1982, "The Probability Distribution of Wind Power from a Dispersed Array of Wind Turbine Generators," *Journal of Applied Meteorology*, Vol. 21: pp. 303-313.

Chien, H. C, R. N. Meroney, and V. A. Sandborn, 1980, "Sites for Wind Power Installations: Physical Modeling of the Wind Field of Kahuku Point, Oahu, Hawaii," *Proceedings, Third International Symposium on Wind Energy Systems*, Lyngby, Denmark: pp. 75-90.

Cliff, W. C, 1977, *The Effect of Generalized Wind Characteristics on Annual Power Estimates from Wind Turbine Generators*, PNL-2436, Richland, Washington: Battelle Pacific Northwest Laboratory.

Cliff, W. C, and G. H. Fichtl, 1978, *Wind Velocity-Change (Gust Rise) Criteria for Wind Turbine Design*, PNL-2526, Richland, Washington: Battelle Pacific Northwest Laboratory.

Cliff, W. C, and D. L. Hall, 1980, *Simulation of Turbulence Using Conditional Sampling Techniques*, PNL-SA 8338, Richland, Washington: Battelle Pacific Northwest Laboratory.

Connell, J. R., 1981, *The Spectrum of Wind Speed Fluctuations Encountered by a Rotating Blade of a Wind Energy Conversion System: Observations and Theory*, PNL-4083, Richland, Washington: Battelle Pacific Northwest Laboratory.

Connell, J. R., 1982, "The Spectrum of Wind Speed Fluctuations Encountered by a Rotating Blade of a Wind Energy Conversion," *Solar Energy*, Vol. 29: pp. 363-375.

Connell, J. R., and R. L. George, 1982, *The Wake of the Mod-OAl Wind Turbine at Two Rotor Diameters Downwind on December 3, 1981*, PNL-4210, Richland, Washington: Battelle Pacific Northwest Laboratory.

Corotis, R. B., 1977, *Stochastic Modeling of Site Wind Characteristics*, RLO/2342-77/2, Evanston, Illinois: Northwestern University.

Counihan, J., 1975, "Adiabatic Atmospheric Boundary Layers: A Review and Analysis of Data from the Period 1880-1972," *Atmospheric Environment*, Vol. 9: Pergamon Press, pp. 871-905.

Davenport, A. G., 1961, "The Spectrum of Horizontal Gustiness Near the Ground in High Winds," *Quarterly Journal of the Royal Meteorological Society*, Vol. 87: pp. 194-211.

Davenport, A. G., 1965, "Relationship of Wind Structure to Wind Loading," *Proceedings, Symposium on Wind Effects on Structures*, Paper No. 2, London: HMSO.

Davenport, A. G., 1967, "The Dependence of Wind Loads on Meteorological Parameters," *Proceedings, International Conference on Wind Loads on Buildings and Structures*, Toronto, Canada: University of Toronto Press, pp. 19-81.

Davis, E. L., 1984, "Wind Energy Meteorology for a 190-MW Windpower Facility," *Proceedings, European Wind Energy Conference.*, Hamburg, Germany.

Doran, J. C, J. A. Bates, P. J. Liddell, and T. D. Fox, 1977, *Accuracy of Wind Power Estimates*, PNL-2442, Richland, Washington: Battelle Pacific Northwest Laboratory.

Doran, J. C., and D. C. Powell, 1980, *Gust Characteristics for WECS Design and Performance Analysis*, PNL-3421, Richland, Washington: Battelle Pacific Northwest Laboratory.

Elliott, D. L., 1979, "Meteorological and Topographical Indicators of Wind Energy for Regional Assessments," *Proceedings, Conference on Wind Characteristics and Wind Energy Siting*, Boston: American Meteorological Society, pp. 273-283.

Elliott, D. L., N. J. Cherry, and C. I. Aspliden, 1981, "World-Wide Wind Energy Resource Assessment," *Energy and Special Applications Programme Report No. 2 (1981)*, Geneva: Secretariat of World Meteorological Organization.

Elliott, D. L., 1984, *Wind Shear for Large Wind Turbine Generators at Selected Tall Tower Sites*, PNL-4895, Richland, Washington: Battelle Pacific Northwest Laboratory.

Elliott, D. L., C. G. Holladay, W. R. Barchet, H. P. Foote, and W. F. Sandusky, 1986, *Wind Energy Resource Atlas of the United States*, DOE/CH 10093-4, Richland, Washington: Battelle Pacific Northwest Laboratory.

Fichtl, G. H., and O. E. Smith, 1977, "Wind," *Terrestrial Environment (Climatic) Criteria Guidelines for Use in Aerospace Vehicle Development*, 1977 Revision, J.W. Kaufman, ed., 1977 Revision, NASA TM-78118, Huntsville, Alabama: NASA Marshall Space Flight Center, pp. 8.15-8.17.

Frandsen, S., and C. J. Christensen, 1980, "On Wind Turbine Power Measurements", *Proceedings, Third International Symposium on Wind Energy Systems,* Lyngby, Denmark.

Frenkiel, J., 1967, "Wind Profiles Over Hills in Relation to Wind-Power Utilization," *Quarterly Journal of the Royal Meteorological Society*, Vol. 88: pp. 281-283.

Frost, W., and T. H. Moulden, 1977, *Handbook on Turbulence*, New York: Plenum Press.

Frost, W., B. H. Long, and R. E. Turner, 1978, "*Engineering Handbook on the Atmospheric Environmental Guidelines for Use in Wind Turbine Generator Development*," NASA TP-1359, Cleveland, Ohio: NASA Lewis Research Center.

Frost, W., and D. K. Nowak, 1979, *Guidelines for Siting Wind Turbine Generators Relative to Small-Scale, Two-Dimensional Terrain Features*, Report RLO/2443-77/1, Richland, Washington: Battelle Pacific Northwest Laboratory.

Frost, W., 1980, *Standard Deviations and Confidence Intervals for Atmospheric Criteria Used in WECS Development*, Final Report of DOE Contract B-23461-A-P, Tullahoma, Tennessee: FWG Associates, Inc.

Frost, W. and C. F. Shieh, 1981, *Wind Turbine Generator Siting Guidelines Relative to Terrain Features*, Final Report, DOE Contract AC06076ET20242, Tullahoma, Tennessee: FWG Associates, Inc.

Frost, W., and M. C. Lin, 1981, "Two-Dimensional Turbulence Models," *Proceedings, Wind Turbine Dynamics Conference*, R. W. Thresher, ed., NASA CP 2185, DOE Publication CONF-810226, Cleveland, Ohio: NASA Lewis Research Center, pp. 155-162.

Frost, W., and R. E. Turner, 1982, "A Discrete Gust Model for Use in the Design of Wind Energy Conversion Systems", *Journal of Applied Meteorology*, Vol. 21: p. 770.

Frost, W., and M. C. Lin, 1983, *Statistical Analysis of Turbulence Data from the NASA Marshall Space Flight Center Atmospheric Boundary Layer Tower Array Facility*, NASA CR-3737, Huntsville, Alabama: NASA Marshall Space Flight Center.

Frost, W., M. C. Lin, H. P. Chang, and E. A. Ringnes, 1985, *Analysis of Data from NASA B-57B Gust Gradient Program*, Interim Report, Contracts NAS8-35347/NAS8-36177 Tullahoma, Tennessee: University of Tennessee Space Institute.

Frost, W., M. C. Lin, H. P. Chang, and E. A. Ringnes, 1987, *Analyses and Assessments of Data from NASA B-57B Aircraft*, Final report, Contracts NAS8-35347/NAS8-36177 Tullahoma, Tennessee: FWG Associates, Inc.

George, R. L., 1984, *Simulation of Winds as Seen by a Rotating Vertical Axis Wind Turbine Blade*, PNL-4914, Richland, Washington: Battelle Pacific Northwest Laboratory.

Goldenblatt, M. K., H. L. Wegley, and A. H. Miller, 1982, *Analysis of the Effects of Integrating Wind Turbines Into a Conventional Utility*, PNL-3962, Richland, Washington: Battelle Pacific Northwest Laboratory.

Golding, E. W., 1955, *The Generation of Electricity by Wind Power*, New York: Philosophical Library, Inc.

Golding, E. W., 1977, *The Generation of Electricity by Wind Power*, London: E. & F.N. Spon Ltd.

Hardesty, R. M., J. A. Korrell, and F. F. Hall, 1981, *Lidar Measurement of Wind Velocity Turbulence Spectra Encountered by a Rotating Turbine Blade*, Final Report, U.S. DOE Contract AC06-80RL10236, Boulder, Colorado: NOAA/ERLAVave Propagation Laboratory.

Haugen, D. A., ed., 1973, *Proceedings, Workshop on Micrometeorology*, Boston, MA: American Meteorological Society.

Heald, R. C. and L. Mahrt, 1981, "The Dependence of Boundary-Layer Shear on Diurnal Variation of Stability," *Journal of Applied Meteorology:* pp. 859-867.

Heister, T. R. and W. T. Pennell, 1981, *The Meteorological Aspects of Siting Large Wind Turbines*, PNL 2522, Richland, Washington: Battelle Pacific Northwest Laboratory.

Hewson, E. W., J. E. Wade, and R. W. Baker, 1978, *Vegetation as an Indicator of High Wind Velocity*, RLO/2227-T24-78-2, Corvallis, Oregon: Oregon State University.

Hewson, E. W., J. E. Wade, and R. W. Baker, 1979, *A Handbook on the Use of Trees as an Indicator of Wind Power Potential*, RLO/2227-79/3, Corvallis, Oregon: Oregon State University.

Hiester, T. R. and W. T. Pennell, 1981, *Meteorological Aspects of Siting Large Wind Turbines*, PNL-2522, Richland, Washington: Battelle Pacific Northwest Laboratory.

Hinze, J. O., 1975, *Turbulence*, New York: McGraw-Hill.

Hojstrup, J., 1982, "Velocity Spectra in the Unstable Boundary Layer," *Journal of Atmospheric Sciences.*, Vol. 39: pp. 2239-2248.

Holley, W. E., R. W. Thresher, and S. R. Lin, 1981, "Wind Turbulence Inputs for Horizontal-Axis Wind Turbines", *Proceedings, Wind Turbine Dynamics Conference*, R. W. Thresher, ed., NASA CP-2185, DOE Publ. CONF-810226, Cleveland, Ohio: NASA Lewis Research Center, pp. 101-112.

Hollister, S. C, 1970, "The Engineering Interpretation of Weather Bureau Records for Wind Loading on Structures," *Building Science Series 30*, Washington, DC: U.S. Department of Commerce.

Houbolt, J. C, and A. Sen, 1972, *Cross-Spectral Functions Based on von Karman's Spectral Equation*, NASA CR-2011, Hampton, Virginia: NASA Langley Research Center.

Huang, C. H., and G. H. Fichtl, 1979, *Gust-Rise Exceedance Statistics for Wind Turbine Design*, PNL-2530, Richland, Washington: Battelle Pacific Northwest Laboratory.

Hunt, J. C. R., and H. Fernholz, 1975, "Wind-Tunnel Simulation of the Atmospheric Boundary Layer: A Report on Euromech 50," *Journal of Fluid Mechanics*, Vol.70: pp. 543-559.

Justus, C. G., W. R. Hargraves, and A. Mikhail, 1976a, *Reference Wind Speed Distributions and Height Profiles for Wind Turbine Design and Performance Evaluation Applications*, ORO/5108-76/4, Atlanta, Georgia: Georgia Institute of Technology.

Justus, C. G., W. R. Hargraves, and A. Yacin, 1976b, "Nationwide Assessment of Potential Output from Wind-Powered Generators," *Journal of Applied Meteorology*, Vol. 15: pp. 673-678.

Justus, C. G., 1978, *Winds and Wind System Performance*, Philadelphia: Franklm Institute Press.

Justus, C. G., W. R. Hargraves, A. Mikhail, and D. Graber, 1978, "Methods for Estimating Wind Speed Frequency Distributions," *Journal of Applied Meteorology*, Vol.17: pp. 350-353.

Kaimal, J. C, 1973, "Turbulence Spectra, Length Scales and Structure Parameters in the Stable Surface Layer", *Boundary Layer Meteorology*, Vol. 4: pp. 289-309.

Kaimal, J. C, J. E. Gaynor, and D. E. Wolfe, 1981, *Gust and Gust-Rise Statistics of Wind Speed and Direction for Two Strong Mountain Downslope Wind Cases for Design of Wind Turbines*, DOE/ET/23115-80-1, Boulder, Colorado: Environmental Research Laboratories.

Kaufman, J. W., ed., 1977, *Terrestrial Environment (Climatic) Criteria Guidelines for Use in Aerospace Vehicle Development*, 1977 Revision, NASA TM X-78118, Huntsville, Alabama: NASA Marshall Space Flight Center.

Kelley, N. D., 1993, *Full Vector (3-D) Inflow Simulation in Natural and Wind Farm Environments Using an Expanded Version of the SNLWIND (Veers) Turbulence Code*, NREL/TP-442-5225, Golden, Colorado: National Renewable Energy Laboratory.

Kristensen, L., and S. Frandsen, 1982, "Model for Power Spectra Measured From the Moving Frame of Reference of the Blade of a Wind Turbine," *Journal of Wind Engineering and Industrial Aerodynamics*, Vol. 10: pp. 249-262.

Lee, J. T., and G. L. Stone, 1983, "The Use of Eulerian Initial Conditions in a Lagrangian Model of Turbulence Diffusion," LA-UR-82-3034, *Proceedings, Sixth Symposium on Turbulence and Diffusion*, Boston: American Meteorological Society.

Legg, B. J., and M. R. Raupach, 1982, "Markov-Chain Simulation of Particle Dispersion in Inhomogeneous Flows: The Mean Drift Velocity Induced by a Gradient in Eulerian Velocity Variance," *Boundary-Layer Meteorology*, Vol. 24: pp. 3-13.

Lumley, J. L., and H. A. Panofsky, 1964, *The Structure of Atmospheric Turbulence*, New York: Interscience-Wiley Publications.

Luna, R. E., and H. W. Church, 1974, "Estimation of Long-Term Concentrations Using a "Universal" Wind Speed Distribution," *Journal of Applied Meteorology*, Vol. 13: pp. 910-916.

Mahrt, L., and R. C. Heald, 1979, *Analysis of Strong Nocturnal Wind Shears for Wind Machine Design*, DOE/ET23116-79/1, Corvallis, Oregon: Oregon State University.

Marrs, R. W., and D. R. Gaylord, 1979, *A Guide to Interpretation of Windflow Characteristics From Eolian Landforms*, RLO/2343-79/2, Corvallis, Oregon: Oregon State University.

Mayne, J. R., 1979, "The Estimation of Extreme Winds," *Journal of Industrial Aerodynamics*, Vol. 5: pp. 109-137.

Meroney, R. J., A. J. Bowen, D. Lindley, and J. Pearce, 1978, *Wind Characteristics Over Complex Terrain: Laboratory Simulation and Field Measurements at Rakaia Gorge, New Zealand*, RLO/2438-77/2, Corvallis, Oregon: Oregon State University.

Mikhail, A., 1984, "Height Extrapolation of Wind Data," PNL 4367, Richland, Washington: Battelle Pacific Northwest Laboratory.

Moroz, E. Y., and F. J. Brousaides, 1980, *Survey of Sensors for Automated Tactical Weather Observations*, AFGL-TR-80-0195, Washington, DC: U.S. Air Force.

Neuman, F., and J. Foster, 1970, *Investigation of a Digital Automatic Aircraft Landing System in Turbulence*, NASA TN-D-6066, Moffett Field, California: NASA Ames Research Center.

Panofsky, H. A., J. A. Dutton, D. Larko, R. Lipschutz, and G. Stone, 1982, *Spectra Over Complex Terrain in the Surface Layer*, PNL-3745, Richland, Washington: Battelle Pacific Northwest Laboratory.

Panofsky, H. A., and J. A. Dutton, 1984, *Atmospheric Turbulence*, New York: John Wiley & Sons.

Papoulis, A., 1977, *Signal Analysis*, New York: McGraw-Hill Book Co.

Pennell, W. R., 1982, *Siting Guidelines for Utility Application of Wind Turbines*, AP-2795, Palo Alto, California: Electric Power Research Institute.

Perlmutter, M., W. Frost, and G. H. Fichtl, 1977, "Three Velocity Component, Nonhomogeneous Atmospheric Boundary-Layer Turbulence Modeling", *AIAA Journal*, Vol. 15, No. 10: pp. 1444-1454.

Petersen, E. L., I. Troen, S. Frandsen, and K. Hedegaard, 1981, *Danish Wind Atlas: A Rational Method of Wind Energy Siting*, Roskilde, Denmark: Risø National Laboratory.

Peterson, E. W. and J. P. Hennessey, Jr., 1978, "On the Use of Power Laws for Estimates of Wind Power Potential," *Journal of Applied Meteorology*, Vol. 17: pp. 390-394.

Phillips, G. T., 1979, *A Preliminary User's Guide for the NOABL Objective Analysis Code*, DOE/ET/20280-T1, LaJolla, California: Science Applications, Inc.

Plate, E. J., 1971, *Aerodynamic Characteristics of Atmospheric Boundary Layers*, Critical Review Series, Washington, DC: U.S. Atomic Energy Commission.

Powell, D. C, and J. R. Connell, 1980, *Definition of Gust Model Concepts and Review of Gust Models*, PNL-3138, Richland, Washington: Battelle Pacific Northwest Laboratory.

Powell, D. C, J. R. Connell, and R. L. George, 1985, *Verification of Theoretically Computed Spectra for a Point Rotating in a Vertical Plane*, PNL-5540, Richland, Washington: Battelle Pacific Northwest Laboratory.

Powell, D. C, and J. R. Connell, 1986a, *A Model for Simulating Rotational Data for Wind Turbine Applications*, PNL-5857, Richland, Washington: Battelle Pacific Northwest Laboratory.

Powell, D. C, and J. R. Connell, 1986b, *Review of Wind Simulation Methods for Horizontal-Axis Wind Turbine Analysis*, PNL-5903, Richland, Washington: Battelle Pacific Northwest Laboratory.

Putnam, P. C, 1948, *Power from the Wind*, New York: Van Nostrand-Reinhold Publishing Co.

Raab, A., 1980, "Combined Effects of Deterministic and Random Loads in Wind Turbine Design," *Proceedings*, *Third International Symposium on Wind Energy Systems*, Lyngby, Denmark: pp. 169-182.

Ramsdell, J. V., 1978, *Estimates of the Number of Large Amplitude Gusts*, PNL-2508, Richland, Washington: Battelle Pacific Northwest Laboratory.

Ramsdell, J. V., and J. S. Wetzel, 1981, *Wind Measurement Systems and Wind Tunnel Evaluation of Selected Instruments*, PNL-3435, Richland, Washington: Battelle Pacific Northwest Laboratory.

Reeves, P. M., R. G. Joppa, and V. M. Ganger, 1976, *A Non-Gaussian Model of Continuous Atmospheric Turbulence for Use in Aircraft Design*, NASA CR-2639, Moffett Field, California: NASA Ames Research Center.

Rice, S. O., 1944 and 1945, "Mathematical Analysis of Random Noise", *Bell System Tech. Journal:* Vol. 23, pp. 282-332; Vol. 24, pp. 46-156.

Riley, J. J., and D. P. Delisi, 1977, *Survey of Laboratory Modeling of Plume Dynamics*, EPRI EA-323, Palo Alto, California: Electric Power Research Institute.

Ropelewski, C. F., H. Tennekes, and H. A. Panofsky, 1973, "Horizontal Coherence of Wind Fluctuations," *Boundary Layer Meteorology*, Vol. 5.

Rosenbrock, H. H., 1955, *Vibration and Stability Problems in Large Turbines Having Hinged Blades*, C/T 113, Surrey, England: E.R.A.

Sachs, P., 1972, *Wind Forces in Engineering*, Oxford, England: Pergamon Press.

Sandborn, V. A., and J. R. Connell, 1984, *Measurement of Turbulent Wind Velocities Using a Rotating Boom Apparatus;* PNL-4888, Richland, Washington: Battelle Pacific Northwest Laboratory.

Sherlock, R. H., 1951, "Analyzing Wind for Frequency and Duration," *On Atmospheric Pollution*, Meteorology Monogram, No. 4: American Meteorological Society, pp. 42-49.

Sigl, A. B., R. B. Corotis, and D. J. Won, 1979, "Run Duration Analysis of Surface Wind Speeds for Wind Energy Application," *Journal of Applied Meteorology*, Vol. 18: pp. 156-166.

Simiu, E., and R. H. Scanlan, 1978, *Wind Effects on Structures: An Introduction to Wind Engineering*, New York: John Wiley & Sons.

Simon, R., 1982, "Potential Errors in Using One Anemometer to Characterize the Wind Power Over an Entire Rotor Disk," *Proceedings, Workshop on Large Horizontal-Axis Wind Turbines*, R. W. Thresher, ed., NASA CP 2230, DOE Publ. CONF-810752, Cleveland, Ohio: NASA Lewis Research Center, pp. 427-444.

Sisterson, D. L. and P. Frenzen, 1978, "Nocturnal Boundary-Layer Wind Maxima and Problem of Wind Power Assessment," *Environmental Science and Technology*, Vol. 12: pp. 218.

Snyder, W. H., 1972, "Similarity Criteria for the Application of Fluid Models to the Study of Air Pollution Meteorology," *Boundary-Layer Meteorology*, Vol.3: pp. 113-134.

Spera, D. A., 1995, "A Model of Rotationally-Sampled Wind Turbulence for Predicting Fatigue Loads in Wind Turbines," *Collected Papers on Wind Turbine Technology*, D. A. Spera, ed., NASA CR-195432, Cleveland, Ohio: NASA Lewis Research Center, pp. 17-26.

Spera, D. A., and T. R. Richards, 1977, "NASA Lewis Gust Model", Appendix to *Definition of Gust Model Concepts and Review of Gust Models*, by D. C. Powell and J. R. Connell, PNL-3138, 1980, Richland, Washington: Battelle Pacific Northwest Laboratory.

Spera, D. A., 1978, *Comparison of Computer Codes for Calculating Dynamic Loads in Wind Turbines*, NASA TM-73773, DOE/NASA/1028-78/16, Cleveland, Ohio: NASA Lewis Research Center.

Spera, D. A. and T. R. Richards, 1979, *Modified Power Law Equations for Vertical Wind Profiles*, NASA TM-79275, DOE/NASA/1059-79/4, Cleveland, Ohio: NASA Lewis Research Center.

Spera, D.A., 1991, *Analysis of the Diurnal Cycle of Wind Shear at Clayton, New Mexico*, (personal correspondence), Cleveland, Ohio: DASCON Engineering.

Steely, S., and W. Frost, 1981, *Statistical Analysis of Atmospheric Turbulence About a Simulated Block Building*, NASA CR-3366, Huntsville, Alabama: NASA Marshall Space Flight Center.

Stoddard, F. S. and B. K. Porter, 1986, *Wind Turbine Aerodynamics Research Needs Assessment*, DOE/ER/30075-H1, Washington, DC: U. S. Department of Energy.

Stone, R. J., and J. T. Bradley, 1977, *Survey of Anemometers*, FAA-RD-77-49, Washington, DC: U. S. Federal Aviation Administration.

Sundar, R. M., and J. P. Sullivan, 1981, "Performance of Wind Turbines in a Turbulent Atmosphere," *Proceedings, Wind Turbine Dynamics Conference*, R. W. Thresher, ed., NASA CP 2185, DOE Publ. CONF-810226, Cleveland, Ohio: NASA Lewis Research Center, pp. 79-86.

Sundar, R. M., and J. P. Sullivan, 1983, "Performance of Wind Turbines in a Turbulent Atmosphere," *Solar Energy*, Vol. 31: pp. 567-575.

Thresher, R. W., W. E. Holley, and N. Jafarey, 1981, "Wind Response Characteristics of Horizontal Axis Wind Turbines," *Proceedings, Wind Turbine Dynamics Conference*, R. W. Thresher, ed., NASA CP 2185, DOE Publ. CONF-810226, Cleveland, Ohio: NASA Lewis Research Center, pp. 87-99.

Thom, H. C. S., 1961, "Distributions of Extreme Winds in the United States," *Trans. ASCE*, Vol. 126.

Thom, H. C. S., 1968, "New Distributions of Extreme Winds in the United States," *Jour. ASCE Structural Division:* Paper No. 6038.

Touma, J. W., 1977, "Dependence of the Wind Profile on Stability for Various Locations," *Journal of Air Pollution Control Association*, Vol. 27: pp. 863-866.

Vachon, W., 1985, "Critical Issues Involved in Making Proper Wind Measurements for Wind Energy Production Estimates," *Proceedings, Windpower '85 Conference*, SERI/CP-217-2902, Washington, DC: American Wind Energy Association, pp. 1-7.

Van Der Hoven, I., 1957, "Power Spectrum of Horizontal Wind Speed in the Frequency Range From 0.0007 to 900 Cycles Per Hour," *International Meteorology*, Vol. 4: pp. 160-164.

Veenhuizen, S. D., and J. T. Lin, 1987a, "Wind Turbine Micrositing Status," *Proceedings, Windpower '87 Conference*, SERI/CP-217-3315, Washington, DC: American Wind Energy Association, pp. 168-173.

Veenhuizen, S. D., and J. T. Lin, 1987b, "Numerical Model for Predicting Wind Turbine Array Performance in Complex Terrain," *Proceedings, Windpower '87 Conference,* SERI/CP-217-3315, Washington, DC: American Wind Energy Association, pp. 393-397.

Veers, P. S., 1984, *Modeling Stochastic Wind Loads on Vertical Axis Wind Turbines*, SAND 83-1909, Albuquerque, New Mexico: Sandia National Laboratories.

Veers, P. S., 1988, *Three-Dimensional Wind Simulation*, SAND88-0132, Albuquerque, New Mexico: Sandia National Laboratories.

Verholek, M. G., 1978, *Preliminary Results of a Field Experiment to Characterize Wind Flow Through a Vertical Plane*, PNL-2518, Richland, Washington: Battelle Pacific Northwest Laboratory.

Vermeulen, P., 1978, *A Wind Tunnel Study of the Wake of a Horizontal-Axis Wind Turbine*, Netherlands Organization for Applied Scientific Research, TNO Report 78-09674.

Wade, J. E, S. N. Walker, and R. W. Baker, 1986. "*Wind Resource Characterization and Performance Modeling at the Whisky Run Windfarm*," PWE, 86-2, Corvallis, Oregon: Oregon State University.

Wade, J. E., and S. N. Walker, 1988, *Local Flow Measurements for Micrositing*, Draft Report, U.S. Department of Energy Contract DE-FC02-86CH10251, Golden, Colorado: National Renewable Energy Laboratory.

Wahl, E. W., 1966, *Windspeed on Mountains*, No. AF19(628)-3873, Bedford, Massachusetts: USAF-CRL.

Waldon, C. A., and A. C. Hansen, 1983, *Fatigue Modeling for Small Wind Systems - Basic Theory*, Contract No. DE-AC04-76DP03533, RFP-3564, Golden, Colorado: National Renewable Energy Laboratory.

Wang, S. T., and W. Frost, 1980, *Atmospheric Turbulence Simulation Techniques with Application to Flight Analysis*, NASA CR 3309, Huntsville, Alabama: NASA Marshall Space Flight Center.

Wegley, H. L., J. V. Ramsdell, M. M. Orgill, and R. L. Drake, 1980, *A Siting Handbook for Small Wind Energy Conversion Systems*, PNL-2521, Rev. 1, Richland, Washington: Battelle Pacific Northwest Laboratory.

Wegley, H. L., and J. C. Barnard, 1986, *Using the NOABL Flow Model and Mathematical Optimization as a Micrositing Tool*, PNL-6070, Richland, Washington: Battelle Pacific Northwest Laboratory.

Wendell, L. L., 1984, "An Overview of Wind Turbine Siting Considerations," *Proceedings, Energy Technology XL Applications and Economics*, Rockville, Maryland: Government Institutes, Inc., pp. 485-496.

Wilczak, J. M., 1984, "Large-Scale Eddies in the Unstably-Stratified Atmospheric Surface Layer - Part I: Velocity and Temperature Structure", *Journal of the American Meteorological Society*, Vol. 24: pp. 35-37.

9

Electromagnetic Interference
from Wind Turbines

Dipak L. Sengupta, Ph.D.
Professor of Electrical Engineering
University of Detroit Mercy
Detroit, Michigan

and

Thomas B. A. Senior, Ph.D.
Professor of Electrical Engineering
The University of Michigan
Ann Arbor, Michigan

Introduction

It is well known that any large structure, whether stationary or moving, in the vicinity of a receiver or transmitter of electromagnetic signals may interfere with those signals and degrade the performance of the transmitter/receiver system. Under certain conditions, the rotor blades of an operating wind turbine may passively reflect a transmitted signal, so that both the transmitted signal and a delayed *interference signal* (varying periodically at the blade passage frequency) may exist simultaneously in a zone near the turbine. The nature and amount of electromagnetic interference (EMI) in this zone depend on a number of parameters, including location of the wind turbine relative to the transmitter and receiver, type of wind turbine, physical and electrical characteristics of the rotor blades, signal frequency and modulation scheme, receiver antenna characteristics, and the radio wave propagation characteristics in the local atmosphere. When the influence of these parameters on EMI is understood, wind turbines can usually be designed and sited so that any interference with communication signals may not exceed allowable levels.

Figure 9-1 illustrates the field conditions under which a wind turbine may cause EMI. A transmitter, *T*, sends a *direct signal* to two receivers, *R*, and a wind turbine, *WT*, that may be of either the horizontal- or vertical-axis configuration (HAWT or VAWT). The rotating blades of the turbine produce and transmit a *scattered signal.* Thus, the receivers may acquire two signals simultaneously, with the scattered signal causing EMI because it is delayed in time or distorted. Signals reflected in a manner analogous to mirror reflection are termed *back-scattered.* As shown in the figure, about 80 percent of the region around the turbine is the *backward-scatter zone.* On the other hand, signal scattering that is analogous to shadowing is termed *forward scattering*, and about 20 percent of the region around a turbine is the *forward-scatter or front-scatter zone.*

Other wind turbine components which have been considered to be potential causes of EMI are towers and electrical systems. However, neither of these has been found to be a significant source of interference. Modern wind turbine towers, unlike buildings and water towers that sometimes cause stationary "ghost" images on television screens, are usually slender cylinders or trusses which are not efficient reflectors of electromagnetic radiation. *Radio frequency interference* (RFI) caused by signals emitted by generators and motors has not been observed near wind turbines. Thus, moving rotor blades are the components of most importance in determining EMI levels, and these are the only components which will be discussed here.

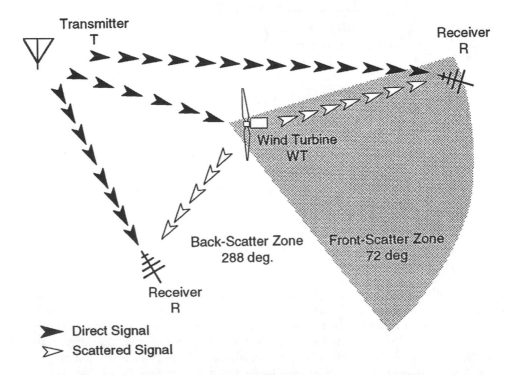

Figure 9-1. Schematic plan view of the relative positions of a transmitter, receivers, and a wind turbine that may produce EMI. Interference is caused by the simultaneous reception of both the direct and the scattered signals, the latter being delayed or distorted.

The purpose of this chapter is to summarize the current state of understanding of the potential EMI effects of operating wind turbines on various communication systems and to discuss the analysis and measurement of these effects in specific cases. Topics which are discussed include EMI mechanisms, mathematical models for predicting EMI effects, and the results of laboratory and full-scale field measurements.

Types of Observed and Potential EMI Effects

Television Interference (TVI)

TVI effects from wind turbines were first measured in 1976 in the vicinity of the Mod-0 100-kW experimental HAWT near Sandusky, Ohio, [Sengupta and Senior 1977 and 1979a]. Later, such effects were also observed on Block Island, Rhode Island, near the Mod-0A 200-kW HAWT [Sengupta *et al.* 1980], at Boone, North Carolina, around the Mod-1 2-MW HAWT [Sengupta *et al.* 1981a], at the Goldendale, Washington, cluster of three Mod-2 2.5-MW HAWTs [Sengupta *et al.* 1983a], and at Albuquerque, New Mexico, near an experimental Darrieus VAWT [Sengupta *et al.* 1981b]. Laboratory studies [Sengupta *et al.* 1983b] indicate that small-scale turbines can also cause TVI in their immediate vicinity. Methods for assessing TVI potential in the field are given in [Senior and Sengupta 1983].

TVI from wind turbines is characterized by *video distortion* that generally occurs in the form of a jittering of the picture that is synchronized with the blade passage frequency (*i.e.* the rotor speed times the number of blades). In the worst case, a strong interference signal may cause complete break-up of the picture. No audio distortion has been observed.

FM Radio Interference

Effects on FM broadcast reception have been observed only in laboratory simulations [Sengupta and Senior 1978]. These appear as a background hissing noise pulsating in synchronism with the rotor blades.

Interference to Navigation Systems

The effects on VOR (DVOR) and LORAN-C systems have only been studied theoretically. A VOR study has indicated that a stopped wind turbine may produce *scalloping errors* in the azimuthal bearings provided by the VOR. However, when the wind turbine is operating, potential interference effects are much less than those under stopped conditions [Sengupta and Senior 1978]. A LORAN-C study indicates that under normal siting conditions (*i.e.*, the wind turbine is not in close proximity to the transmitter or receiver) no degradation in communication performance is likely to occur [Sengupta and Senior 1979b].

Interference to Microwave Links

Theoretical studies indicate that the effects tend to smear out the modulation used in typical microwave transmission systems [Sengupta and Senior 1978]. In Denmark, severe problems with *frequency-shift-keying modulation* have been reported on microwave links operating at a frequency of 7.4 GHz with an 8-Mbit/sec data rate [IEA 1986].

Analysis of Signals with Electromagnetic Interference

Interference in the Ambient Signal Field

A general mathematical model has been developed of the essential mechanism by which a wind turbine can produce electromagnetic interference [*e.g.*, Senior *et al.* 1977]. Referring to Figure 9-1, the total *electric field strength* of the ambient signal, *E*, surrounding a receiving point, *R*, can be written as

$$E_R(\Pi, t) = E_{R,D}(\Pi, t) + E_{R,S}(\Pi, \omega t, B\Omega t) \tag{9-1}$$

where

$$
\begin{aligned}
E_R &= \text{field strength at the receiver of the total signal (mV/m)} \\
E_{R,D} &= \text{field strength at the receiver of the direct (transmitter) signal (mV/m)} \\
E_{R,S} &= \text{field strength at the receiver of the scattered (wind turbine) signal (mV/m)} \\
\Pi(r,\theta,\phi) &= \text{relative positions of the transmitter, wind turbine, and receiver (m, rad, rad)} \\
\omega &= \text{frequency of the transmitted signal (rad/s)} \\
t &= \text{time (s)} \\
B &= \text{number of blades} \\
\Omega &= \text{rotor speed of the turbine (rad/s)}
\end{aligned}
$$

Field strengths will be treated as scalar quantities with assumed time dependence *exp(jωt)*, which has been omitted from Equation (9-1) for simplicity. All of the quantities in this equation are complex. The time dependence of the scattered signal (and, hence, of the total signal) is caused by the blade rotation. Further insight into the interference mechanisms can be obtained by considering the general nature of the total signal given by Equation (9-1). We can separate out the time dependence in the *scattered field* by writing

$$E_{R,S}(\Pi, \omega t, B\Omega t) = |E_{R,S}(\Pi)| f_m(t) \exp[j\delta_s(t)] \tag{9-2}$$

where

$$
\begin{aligned}
|E_{R,S}(\Pi)| &= \text{maximum amplitude of the scattered field during a rotor revolution (mV/m)} \\
f_m &= \text{time-varying modulation shape function; } -1 \le f_m \le 1 \\
\exp[\ ] &= \text{exponential of [ ]} \\
j &= \text{square root of -1} \\
\delta_s &= \text{time-varying radio-frequency (RF) phase of the scattered signal (rad)}
\end{aligned}
$$

If we also express the *direct field* as

$$E_{R,D}(\Pi, t) = |E_{R,D}(\Pi)| \exp(j\delta_D) \tag{9-3}$$

where

$$
\begin{aligned}
|E_{R,D}(\Pi)| &= \text{amplitude of the direct field at the receiver (mV/m)} \\
\delta_D &= \text{RF phase of the direct signal (rad)}
\end{aligned}
$$

the *total field* at a receiver then becomes

$$E_R(\Pi, t) = |E_{R,D}(\Pi)|\{1 + m_E f_m(t) \exp[j\delta(t)]\} \tag{9-4a}$$

$$m_E = |E_{R,S}(\Pi)| / |E_{R,D}(\Pi)| \tag{9-4b}$$

where

$$\delta = \delta_S - \delta_D$$
$$m_E = \text{ambient field modulation index}$$

Assuming $m_E \ll 1$, Equation (9-4a) can be approximated by

$$E_R = |E_{R,D}|[1 + m_E f_m \cos(\delta)] \exp\{j[\delta_D + m_E f_m \sin(\delta_S)]\} \tag{9-5}$$

eliminating the functional notation (Π,t) and (t) for simplification. Equation (9-5) shows that, in general, the total signal is both amplitude and frequency modulated, the former being dominant. For later use, we write the amplitude of the total field at a receiver R as

$$|E_R| = |E_{R,D}|[1 + m_E f_m \cos(\delta)] \tag{9-6}$$

displaying the *amplitude modulation* (AM) resulting from rotation of the blades. In the non-rotating case, $f_m = 1$. Equation (9-7) indicates that any electromagnetic communication system using AM for transmission could be vulnerable to interference produced by rotating wind turbine blades. The severity of interference is measured by the modulation index, m_E, and the nature of the interference effects is described by the modulation shape function, f_m.

To interpret Equation (9-6) in terms of observable quantities, we note that $\cos(\delta)$ is a rapidly-varying function of time compared to f_m. It attains its extreme values (± 1) many times during a single cycle of f_m which occurs in the time period $2\pi/B\Omega$. The *envelope* of Equation (9-6), therefore, represents the field of the total signal that is actually observed and is given by

$$|E_R|_{envelope} = |E_{R,D}|(1 + m_E f_m) \tag{9-7}$$

Thus, the modulation shape function represents the time dependence of the envelope of the scattered signal introduced by the blade rotation.

Signal Power Interference

Up to this point, the discussion of signal interference has centered on the ambient signal field around the receiver. However, EMI *perception* depends not only on the modulation of the signal field but also on the degree of modulation of the signal power at the input terminals of the receiver. This involves the *receiving antenna orientation and response*. Using a TV signal as an example, signal power and signal field are related as follows:

$$P_R = G_O F_A (\lambda/693\pi)^2 E_R^2 \qquad (9\text{-}8)$$

where

P_R = signal power *input* at the receiver location (mW)
G_o = effective gain of the receiving antenna pointed at the transmitter mW/mV²)
F_A = azimuthal response factor of the receiving antenna, dependent on ϕ_A;
   $F_A \le 1$, with $F_A(0) = 1$ (mW/mW)
ϕ_A = azimuthal angle from the receiving antenna beam to the signal source (rad)
λ = signal wave length = 299.8/f (m)
f = carrier signal frequency (MHz)

Signal power is usually expressed in *dBm* or *dB above 1 mW*, for which the definition is

$$P_R(dBm) = 10\log_{10}[P_R(mW)] \qquad (9.9)$$

Effective gains for TV antennas range from about 4 to 16 mW/mV² (*i.e.* 6 to 12 dB). Figures 9-2 and 9-3 show typical azimuthal response functions for a directional antenna used extensively for field evaluation of TV interference. Figure 9-2 presents the results of a laboratory test in the form of a polar diagram for one signal frequency. The azimuthal response function F_A (in dB) is equal to the difference between the reading at a given antenna direction and that at zero degrees. In Figure 9-3, the results of a field test of the same antenna at the same frequency are compared to the laboratory data, showing the effects of local terrain and atmospheric conditions.

Figure 9-4 shows typical modulations of an audio power signal that can be caused by the addition of a secondary signal scattered by a HAWT rotor. In this case the turbine is the 2.0-MW Mod-1 machine with two steel blades rotating at 10 rpm, which produces a modulation wave form with a period of 3.0 s [Sengupta *et al.* 1981a]. In Figure 9-4(a), the antenna beam is pointed at the wind turbine, at an azimuth of 288 deg. The direct (desired) signal on Channel 3 is received from the transmitter at an azimuth of 154 deg. Thus,

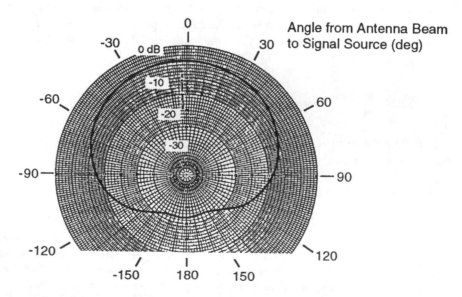

Figure 9-2. Typical azimuthal response factors for a directional TV antenna, as measured in the laboratory. Frequency = 63 MHz (Ch. 3) [Sengupta *et al.* 1981a]

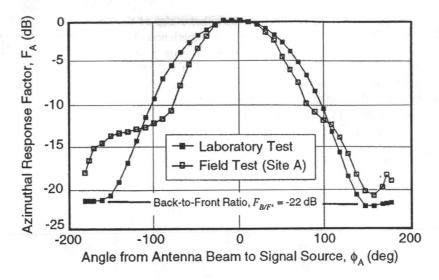

Figure 9-3. Azimuthal response factors measured in the field compared to those measured in the laboratory. Frequency = 63 MHz (Ch. 3). [data: Sengupta *et al.* 1981a]

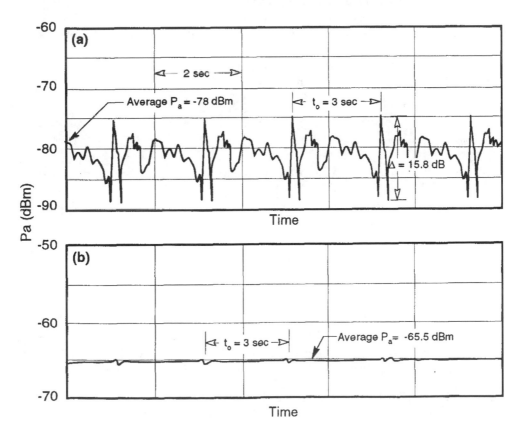

Figure 9-4. Typical modulation of a TV power signal by a secondary signal scattered by an operating wind turbine. The turbine is the two-bladed 2.0-MW Mod-1 HAWT. Modulation frequency is twice the rotor speed of 10 rpm. (a) Antenna pointed at the wind turbine (b) Antenna pointed at the transmitter [Sengupta *et al.* 1981a].

the antenna angle for the direct signal is -134 deg, for which the site response factor is -13 Db (Fig. 9-3). This greatly increases the *relative size* of the modulation compared to the direct signal and, therefore, the interference experienced by a viewer.

Figure 9-4(b) shows that the opposite is true when the antenna is aimed at the transmitter. Here the direct signal is received at full strength and the scattered signal is reduced by -12.5 dB. Thus, the modulation-to-direct signal ratio is about 25 dB lower when the antenna is aimed at the transmitter than when it is aimed at the turbine. These figures illustrate how potential electromagnetic interference can be avoided by the use of a properly-oriented directional antenna.

Combining Equations (9-7) and (9-8) gives

$$|P_R|_{envelope} = |P_{R,D}|(1 + m_R f_m)^2 \qquad (9\text{-}10a)$$

$$m_R = \sqrt{F_{A,W}/F_{A,T}} \; m_E \qquad (9\text{-}10b)$$

where

$|P_{R,D}|$ = amplitude of direct signal power input at the receiver location (mW)
m_R = receiver input modulation index
$F_{A,W}$ = antenna response factor for a signal from the wind turbine (mW/mW)
$F_{A,T}$ = antenna response factor for a signal from the transmitter (mW/mW)

Because the maximum magnitude of f_m is always unity, the maximum and minimum departures (in dBm) from the level of the direct signal are

$$\Delta_1 = 20\log_{10}(1 + m_R) \qquad (9\text{-}11a)$$

$$\Delta_2 = 20\log_{10}(1 - m_R) \qquad (9\text{-}11b)$$

giving rise to

$$\Delta = \Delta_1 - \Delta_2 = 20\log_{10}\left(\frac{1 + m_R}{1 - m_R}\right) \qquad (9\text{-}11c)$$

where

$\Delta = P_{R,\,max} - P_{R,\,min}$ = signal power modulation range (dBm)
$\Delta_1 = P_{R,\,max} - P_{R,\,mean}$ (dBm)
$\Delta_2 = P_{R,\,min} - P_{R,\,mean}$ (dBm)
$P_{R,\,mean} = |P_{R,\,D}|$ (dBm)

Figure 9-5 is a graphical solution of this equation which shows the receiver input modulation index as a function of the signal power modulation range. A convenient empirical equation for this relationship is

$$m_R = 0.0620\,\Delta(1 - 0.0169\,\Delta) \qquad (9\text{-}12)$$

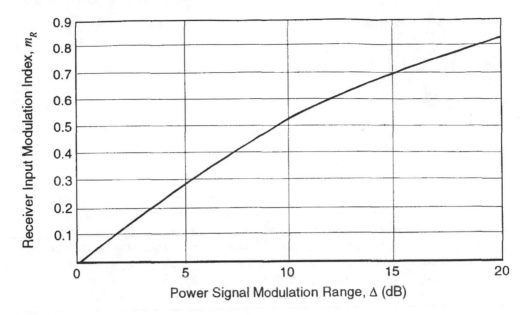

Figure 9-5. Receiver modulation index as a function of the power signal modulation range.

It is clear now that knowledge of both the direct and scattered signals at the receiver's input terminals is necessary for the determination of m_R, the receiver modulation index, and f_m, the modulation shape function, associated with the total signal. The maximum expected value of m_R is the basic parameter that is used to judge the acceptability of the potential EMI with a particular receiver. Thus

$$m_{R,\max} < m_{P,T} : Imperceptible\ EMI \qquad (9\text{-}13)$$
$$m_{R,\max} \geq m_{P,T} : Perceptible\ EMI$$

where

$m_{R,max}$ = maximum expected receiver modulation index, for all possible wind turbine rotor and blade orientations

$m_{P,T}$ = modulation threshold of perception, from studies of test subjects

Thresholds of perception will differ for different electromagnetic systems and periods of exposure. Some guideline values for modulation perception indices will be discussed later.

Signal Scatter Ratio

The strength of the scattered signal field at a receiver location caused by a wind turbine can be conveniently expressed as a *signal scatter ratio*, which is the ratio of the amplitude of the scattered signal field *at the receiver* to the direct signal field *at the turbine*, or

$$Z = \frac{|E_{R,S}|}{|E_{WT,D}|} \tag{9-14}$$

where

Z = signal scatter ratio

$|E_{WT,D}|$ = amplitude of the direct field at the wind turbine (mV/m)

The ratio Z is a characteristic of the turbine and its location relative to the transmitter and receiver, and it is independent of the receiver antenna response or the ambient signal fields at either the receiver or turbine locations. We can use the signal scatter ratio to predict the receiver input modulation index (the measure of EMI potential) using Equations (9-4b) and (9-10b), as follows:

$$m_R = Z \sqrt{\frac{F_{A,W}}{F_{A,T}}} \frac{|E_{WT,D}|}{|E_{R,D}|} \tag{9-15}$$

Equation (9-15) shows clearly that EMI potential depends on the combination of turbine and site characteristics (Z), relative antenna characteristics ($F_{A,W}/F_{A,T}$), and relative direct field strengths ($|E_{WT,D}|/|E_{R,D}|$).

Signal scatter ratios actually observed during EMI tests can be determined from records of signal power versus time, like those shown in Figure 9-4. Data reduction equations, derived from Equations (9-8), (9-11), (9-12), and (9-14), are as follows:

$$Z_O = \frac{|E_{R,S}|_O}{|E_{WT,D}|} = \frac{m_R(\Delta)}{\sqrt{F_{A,W}}} \frac{\sqrt{P_{R,mean}}}{|P_{WT,D}|} \tag{9-16a}$$

$$P_{R,mean} = \frac{P_{R,\max}}{(1 + m_R)^2} \tag{9-16b}$$

$$F_{A,W} = \frac{[P_{R,mean}]_W}{P_{R,mean}} \tag{9-16c}$$

where

Z_O = observed signal scatter ratio

$|E_{R,S}|_O$ = observed amplitude of the scattered signal at the receiver (mV/m)

$|P_{WT,D}|$ = amplitude of the direct signal power at the wind turbine rotor (mW)

$[P_{R,mean}]_W$ = average signal power with the antenna beamed at the wind turbine (mW)

The following calculations of the observed signal scatter ratios for the signal records in Figures (9-4a) and (9-4b) will illustrate the use of Equations (16) to analyze test data:

From preliminary measurements at the wind turbine site, $P_{WT,D}$ = -21.0 dB.

(a)   For the antenna aimed at the wind turbine, Figure (9-4a), let $Z_o = Z_{O,W}$ and

$$\Delta = P_{R,max} - P_{R,min} = -74.5 - (-89.5) = 15.0 \text{ dB}$$

$$m_R = 0.0620\,\Delta(1 - 0.0169\,\Delta) = 0.695$$

$$P_{R,mean} = P_{R,max}/(1 + m_R)^2 = 10\log_{10}[10^{-7.45}]/(1.695)^2] = -79.1 \text{ dB}$$

$$F_{A,W} = [P_{R,mean}]_W/P_{R,mean} = -79.1 - (-79.1) = 0\,\text{dB}$$

from which

$$Z_{O,W} = 10\log_{10}(m_R) + 0.5(P_{R,mean} - F_{A,W} - P_{WT,D} = -27.6\,\text{dB} = 0.0017)$$

(b)   For the antenna aimed at the transmitter, Figure (9-4b), let $Z_o = Z_{O,T}$ and

$$\Delta = -64.7 - (-65.2) = 0.5\,\text{dB}$$

$$m_R = 0.0620 \times 0.5(1 - 0.0169 \times 0.5) = 0.031$$

$$P_{R,mean} = 10\log_{10}[10^{-6.47}/(1.031)^2] = -65.0\,\text{dB}$$

$$F_{A,W} = -79.1 - (-65.0) = -14.1\,\text{dB}$$

from which

$$Z_{O,T} = 10\log_{10}(0.031) + 0.5[-65.0 - (-14.1) - (-27.0)] = -27.1\,\text{dB} = 0.0020$$

Thus, as expected, $Z_{O,W} \approx Z_{O,T}$ and Z_o is independent of the aiming point of the antenna. In this case, the turbine scatters about 0.2 percent of its incident field onto the receiver.

Idealized Scattering Models of Wind Turbine Rotors

The potential scattering of an electromagnetic signal by wind turbine blades necessarily depends on many factors, such as the three-dimensional geometry and the material of the blades, the blade rotation frequency, polarization of the direct signal, the directions and elevations of the transmitter and receiver relative to the blades, the terrain, and the medium of transmission. An exact mathematical description of the scattered signal would therefore require the solution of a rather complicated boundary-value problem. In principle, it would be feasible to obtain a numerical solution for a sequence of incremental positions of the blade, but it is doubtful whether this type of solution would bring out the features that are common to all wind turbines. Moreover, the scattering of electromagnetic waves by even a stationary blade is a complicated problem by itself.

Instead, we will develop simplified, idealized models of HAWT and VAWT rotors and compare model predictions of signal scattering with measured scattering. This approach will illuminate the basic principles of EMI by wind turbines and provide us with empirical equations useful for estimating the magnitude of potential interference in practical situations.

Flat Plate Model of a Single HAWT Blade

It has been found that the main scattering characteristics of a rotating HAWT blade can be adequately analyzed with the help of an idealized model consisting of a rotating flat plate [*e.g.*, Sengupta *et al.* 1981a]. Consider the rectangular flat plate of length l and width w shown in Figure 9-6, lying in the *y-z* (vertical) plane of a Cartesian coordinate system and rotating about the *x* axis. The origin of the x-y-z coordinates is also the origin of a spherical coordinate system (r, θ, ϕ). The transmission medium is assumed to be free space

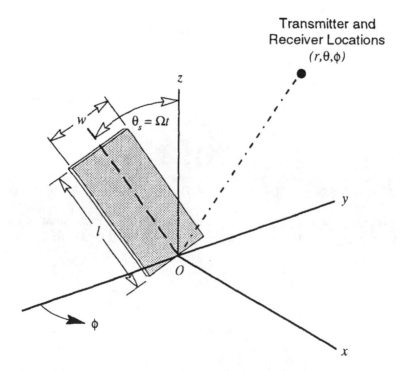

Figure 9-6. Flat plate model of a rotating HAWT blade and its position relative to the transmitter and the receiver.

without any *ground reflection*, and the plate is assumed to be of constant thickness, planar (*i.e.* untwisted), and metallic in order to produce maximum scattering.

When the direct signal is a *plane electromagnetic wave* with its electric field vector in the x-y plane, the horizontal component of the idealized scattered signal at the receiver is

$$E_{R,P} = j|E_{WT,D}| \frac{lw}{\lambda\zeta} \exp[-j2\pi\zeta/\lambda]|\sin(\phi_R)| \cdot$$

$$\cdot sinc \left\{ \frac{\pi l}{\lambda}[p\sin(\Omega t) + q\cos(\Omega t)] \right\} \cdot$$

$$\cdot sinc \left\{ \frac{\pi w}{\lambda}[p\cos(\Omega t) + q\sin(\Omega t)] \right\} \tag{9-17a}$$

$$p = \sin(\theta_T)\cos(\phi_T) + \sin(\theta_R)\cos(\phi_R) \tag{9-17b}$$

$$q = \cos(\theta_T) + \cos(\theta_R) \tag{9-17c}$$

where

$E_{R,P}$ = scattered signal field at the receiver, from a rotating plate at the wind turbine representing one rotor blade (mV/m)

$|E_{WT,D}|$ = amplitude of the direct signal field at the wind turbine (mV/m)

l, w = length and width of the plate, respectively (m)

λ = wave length of the direct signal (m)

ζ = distance from the receiver to the wind turbine (m)

$sinc\{\ \}$ = $|\sin\{\ \}/\{\ \}|$ (1/rad)

r_T, θ_T, ϕ_T = spherical coordinates of the distant transmitter; $r_T \rightarrow \infty$ (m, rad, rad)

ζ, θ_R, ϕ_R = spherical coordinates of the receiver (m, rad, rad)

As a result of the assumption of a plane wave, the direct (or *incident*) signal at the receiver has the same amplitude as that of the signal incident at the plate. Actually, these two quantities may differ, as indicated in Figure 9-6. To maximize the signal scattered by the plate, we first assume that $\Omega t = \theta_R = \theta_T = \pi/2$. This gives an idealized amplitude of

$$|E_{R,P}| = |E_{WT,D}| \frac{lw}{\lambda\zeta} \sin(\phi_R) \, sinc \left\{ \frac{\pi lp}{\lambda} \right\} \tag{9-18a}$$

$$p = \cos(\phi_T) + \cos(\phi_R) \tag{9-18b}$$

In modelling a wind turbine rotor, l is much larger than w, and the time dependence of the modulation shape function is primarily determined by the first *sinc* factor in Equation (9-17a). This sinc function is a maximum (unity) when $p = 0$, which is the case for *specular* or *backward scattering* ($\phi_R = \pi - \phi_T$) and *forward scattering* ($\phi_R = \pi + \phi_T$). For these two directions the idealized scattered signal is independent of time. In directions near these, however, the modulation function is a time-varying sinusoid with frequency twice the rotation frequency. In receiving directions that are well away from it $\pi \pm \phi_T$, measurements show that the waveform consists of sine-like pulses with a width proportional to l repeating at twice the rotation frequency [LaHaie and Sengupta 1970]. The fact that there is no time-varying modulation in the directions of maximum scattering is a consequence of modeling the turbine blade as a flat plate rotating in its own plane, which is not quite true for an actual blade that has an airfoil contour and may be twisted and *coned* (*i.e.*, inclined downwind) out of its plane of rotation.

Back-Scatter and Front-Scatter Zone Shapes

Referring to Equation (9-18a), the azimuthal shape of contours of equal scattering intensity around a wind turbine are defined by the product of the sinc and sine terms. The *back-scatter zone* is the region in which scattering is specular, with $\phi_R = \pi - \phi_T$ and $p = 0$. Thus

$$|\sin(\phi_R)| \, sinc \left\{ \frac{\pi l p}{\lambda} \right\} = \cos(\phi_{RT}/2) \; Back-scatter \tag{9-19a}$$

where $\phi_{RT} = \phi_R - \phi_T$ (rad)

In the *front-scatter zone* $\phi_R = \pi + \phi_T - \delta\phi$, $\phi_T = \pi/2$ for maximum effect, and $\delta\phi$ is a small angle measured from the direction of the receiver. The zone shape function becomes

$$|\sin(\phi_R)| \, sinc \left\{ \frac{\pi l p}{\lambda} \right\} = \cos(\delta\phi) \, sinc \left\{ \frac{\pi l \sin(\delta\phi)}{\lambda} \right\}$$

which drops rapidly to zero with increasing $\delta\phi$, at a rate dependent on the ratio l/λ. As a result, the front-scatter zone is a narrow spike behind a HAWT aligned with the transmitter. For our purposes we can approximate the shape of this spike by

$$|\sin(\phi_R)| \, sinc \left\{ \frac{\pi l p}{\lambda} \right\} \approx \cos(2\phi_{RT}), \; 0.8\,\pi < \phi_{R\,T} < 1.2\,\pi \; Front-scatter \tag{9-19b}$$

Figure 9-7 shows the azimuthal shapes of the two scatter zones, plotted in accordance with Equations (9-19). The back-scatter zone, in the shape of a *cardioid*, is seen to dominate.

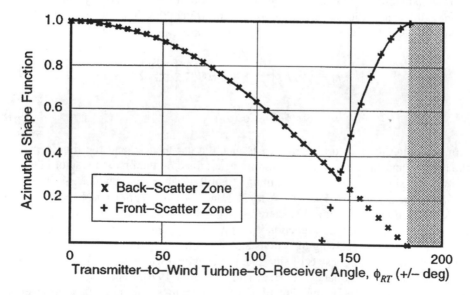

Figure 9-7. Idealized interference zone shapes around a HAWT oriented for maximum signal scatter. Elevations of the transmitter, turbine rotor, and receiver are the same.

Multiple Blades and Coning

For an idealized rotor with B blades, we assume again that $\theta_R = \theta_T = \pi/2$ and modify Equation (9-18a) as follows:

$$|E_{R,S}|_I = |E_{WT,D}| \frac{B_E lw}{\lambda \zeta} \sin(\phi_R) \tag{9-20a}$$

$$B_E = \sum_{i=1}^{B} sinc \left\{ \frac{\pi l}{\lambda} p_i \sin(\Omega t)_i \right\} \tag{9-20b}$$

where

$|E_{R,S}|_I$ = idealized amplitude of the scattered signal field at the receiver (mV/m)
B_E = effective number of blades, $\leq B$
i = blade index number, from 1 to B

For a two-bladed rotor, blade 1 is placed in the horizontal, specular position as before, for maximum scattering. The azimuthal angles ϕ_R and ϕ_T of blade 2 will then differ from those of blade 1 by twice the coning angle, so that

$$B_E = 1 + sinc \left\{ \frac{2\pi l}{\lambda} \sin(2\theta) \cos(\phi_{RT}/2) \right\} \tag{9-20c}$$

where θ = coning angle between the blade axis and the plane of revolution (rad)

In an idealized three-bladed rotor, blades 1 and 2 are placed at $\Omega_t = \pi/6$ and $\Omega t = 5\pi/6$ and are assumed to be in a specular position with negligible coning effect. Blade 3 is then in the same position as blade 2 in the idealized two-bladed rotor. Substitution of these angles into Equation (9-20b) shows that Equation (9-20c) also applies to a three-bladed rotor.

Signal Scattering Efficiency of a HAWT Blade

Because the dimensions l and w in Equation (9-18) are those of an ideally-flat metal plate, we can expect that their product will be less than the *planform area* of a general blade airfoil (*i.e.*, its maximum projected area) with the same signal-scattering capability. To account for this, a *scatter efficiency factor* is included in our idealized HAWT rotor scattering model. Laboratory tests [*e.g.*, Sengupta and Senior 1978, 1980] have shown that l is very nearly equal to the physical length of an airfoil, so the following equations can be used to convert plate dimensions to blade dimensions:

$$lw = \eta_s A_P \tag{9-21a}$$

$$l \approx L \quad w \approx \eta_S(A_P/L) \tag{9-21b}$$

where

η_S = signal scattering efficiency of a blade compared to a flat metallic plate
A_P = planform area of one blade (m²)
L = length of one blade (m)

Scale-model blade scattering tests made with microwave signals in an *anechoic chamber* indicate the relative effects of *airfoil contour, material*, and *total twist* on back-scatter efficiency. The results of these laboratory experiments are shown in Figure 9-8 and lead to the following empirical equations:

$$\eta_{S,H} = \eta_A \eta_M \exp(-2.30 \Delta\beta) \tag{9-22a}$$

$$\eta_A = 0.80 \tag{9-22b}$$

$$\eta_M = \begin{cases} 1.00 \; for \; metallic \; blades \\ 0.41 \; for \; non-metallic \; blades \end{cases} \tag{9-22c}$$

where

$\eta_{S,H}$ = signal scattering efficiency of a HAWT blade
η_A = airfoil contour factor
η_M = material factor
$\Delta\beta$ = total blade twist from root to tip (rad)

If it is necessary to provide lightning protection in the form of metallic strips along a non-metallic blade, this may increase the scattering efficiency almost to the value of the corresponding metallic configuration [Sengupta and Senior 1978]. It may be possible to reduce the scattering efficiency of a metallic blade by coating its surfaces with some radar-absorbing material or by judicious shaping of the blade itself.

Signal Scattering Efficiency of a Darrieus VAWT Blade

A general scattering model for VAWT rotors has yet to be developed, although some preliminary results have been obtained for a *Darrieus* (curve-bladed) turbine based on signal scattering tests around the DOE/Sandia 17-m Darrieus VAWT (Fig. 3-20) and a very simple hoop model [Sengupta *et al.* 1981b]. During these tests, it was observed that the strength of the scattered field was relatively independent of the carrier wave length λ. Referring to Equation (9-18a), this suggests that the scatter area lw and, therefore, η_S, must be proportional to λ for a Darrieus VAWT. Signal scattering tests in which the receiver was placed directly between the transmitter and the VAWT at a distance of about 2 rotor diameters confirmed this suggestion. For these tests, $\phi_{RT} = 0$ and $B_E = 1$ (because the second blade is shadowed by the first as well as the central column of the rotor). Combining Equations (9-16a), (9-18a), and (9-21a), the observed signal scattering efficiency, $\eta_{S,O}$, is

$$\eta_{S,O} = \frac{Z_O \lambda \zeta}{A_P} \tag{9-23}$$

Figure 9-9 shows observed scattering efficiencies of Darrieus blades as a function of the wave length normalized by the blade length, which is 24.1 m for the 17-m VAWT. A linear relationship is found which leads to the following empirical equation:

$$\eta_{S,V} = \eta_A \eta_M \lambda / L \tag{9-24}$$

where $\eta_{S,V}$ = signal scattering efficiency of a Darrieus VAWT blade

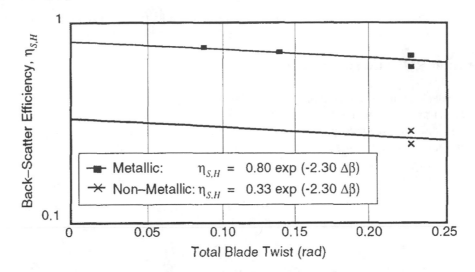

Figure 9-8. Signal scattering efficiency of scale-model HAWT blades compared to flat metal plates. [data: Sengupta and Senior 1978, 1980]

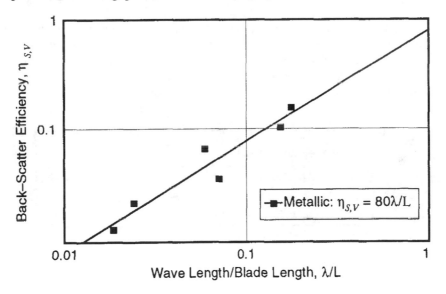

Figure 9-9. Signal scattering efficiency of a Darrieus VAWT blade compared to a flat metal plate of the same planform area. [data from Sengupta *et al.* 1981b]

The airfoil and material factors, η_A and η_M, are the same as those for a HAWT blade and are given in Equations (9-22b) and (9-22c). Equation (9-24) should be considered to be preliminary until verified by scattering tests on Darrieus blades of other lengths.

Effective Number of VAWT Blades

Equation (9-20c) can be applied to VAWT rotors by assuming that coning angles are determined by the angles between blade chords at the rotor equator. Thus, the coning angle is zero for a two-bladed VAWT and 60 deg for a VAWT with three blades. In the infrequent case of a receiver directly between the transmitter and a two-bladed VAWT, only one blade is effective because the second is shadowed by the first blade and the central column.

Ground Reflection of the Scattered Signal

Under certain terrain conditions, signals scattered by a wind turbine rotor could reach a receiver after reflection from the ground and interact with the scattered signal transmitted directly through the atmosphere. Assuming a homogeneous flat earth between the wind turbine and the receiver, a *ground effect factor* can be calculated as follows [Kerr 1964]:

$$F_G = \sqrt{1 - \Gamma^2 + 4\Gamma^2 \sin^2\left(\frac{2\pi}{\lambda\zeta} z_W z_R\right)} \qquad (9\text{-}25)$$

where

$\quad F_G$ = ground reflection amplification factor
$\quad \Gamma$ = amplitude of the reflection coefficient for grazing incidence on the ground
$\quad z_R, z_W$ = elevations of the receiver and the wind turbine rotor, respectively (m)

The coefficient Γ can range from 0 (no reflection) to 1.00 (ideal reflection). Ground effects can alternate between attenuation and amplification when Γ is not zero, as illustrated in Figure 9-10. For our idealized plate model, we will assume there is no ground reflection and F_G is always unity. However, actual ground reflection coefficients greater than zero can be a significant cause of deviation between observed and idealized signal scattering.

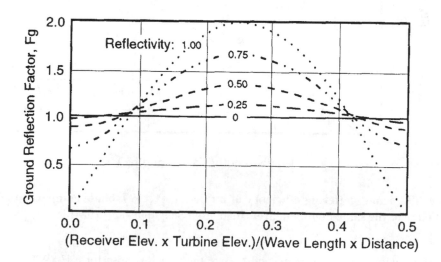

Figure 9-10. Theoretical ground reflection amplification factors for reflectivity coefficients from 0 (no reflection) to 1.00 (ideal reflection). [based on Kerr 1964]

Idealized Signal Scatter Ratio

Using the concepts developed with idealized plate models, we can now define and calculate *idealized signal scatter ratios* for HAWT and VAWT rotors as follows:

$$Z_I = \frac{|E_{R,S}|_I}{|E_{WT,D}|} \tag{9-26a}$$

where

Z_I = idealized signal scatter ratio

$|E_{R,S}|_I$ = idealized amplitude of the scattered signal field at the receiver (mV/m)

Receiver Located in the Backward Scattering Zone

$$Z_I = \eta_S \frac{B_E A_P}{\lambda \zeta} \cos(\phi_{RT}/2), \quad -08\pi \le \phi_{RT} \le 0.8\pi \tag{9-26b}$$

$$\eta_S = \begin{cases} 0.80\exp(-2.30\Delta\beta) & for\ metallic\ HAWT\ blades \\ 0.33\exp(-2.30\Delta\beta) & for\ non-metallic\ HAWT\ blades \\ \\ 0.80\,\lambda/L & for\ metallic\ Darrieus\ blades \\ 0.33\,\lambda/L & for\ non-metallic\ Darrieus\ blades \end{cases} \tag{9-26c}$$

$$B_E = 1 + sinc\left\{\frac{2\pi R}{\lambda}\sin(2\theta)\cos(\phi_{RT}/2)\right\} \le B_{E,\max} \tag{9-26d}$$

$$B_{E,\max} = \begin{cases} \lambda R/A_P & for\ HAWT\ rotors \\ 2 & for\ Darrieus\ rotors \end{cases} \tag{9-26c}$$

where the condition $B_E \le B_{E,max}$ ensures that the amplitude of the scattered signal is less than that of the direct signal at distances greater than the rotor radius R.

Receiver Located in the Forward Scattering Zone

$$Z_I = \eta_S \frac{B_E A_P}{\lambda \zeta} \cos(2\phi_{RT}), \quad 0.8\pi < \phi_{RT} < 1.2\pi \tag{9-26f}$$

$$B_E = 1 + sinc\left\{\frac{2\pi R}{\lambda}\sin(2\theta)\cos(2\phi_{RT})\right\} \le B_{E,\max} \tag{9-26g}$$

Multiple Wind Turbines

Measurements in the vicinity of three Mod-2 wind turbines indicated that TVI effects are enhanced when several turbines operate in *synchronism* (with blades always parallel). The amplitude of the interference pulses produced by two synchronized turbines is about twice that for a single wind turbine. Without synchronism, modulation pulses caused by each machine can be identified individually and produce interference singly [Sengupta *et al.* 1983a]. For a large number of units, interferences may add randomly, but this is yet to be verified. Assuming synchronism for our idealized cluster scatter model we obtain

$$Z_{I,C} = \frac{|E_{R,S}|}{|E_{C,D}|} = \sum_{j=1}^{N} Z_{I,j} \qquad\qquad (9\text{-}6h)$$

$$|E_{C,D}| = \frac{1}{N} \sum_{j-1}^{N} |E_{WT,D}|_j \qquad\qquad (9\text{-}26i)$$

where

$Z_{I,C}$ = idealized cluster scatter ratio

$|E_{C,D}|$ = average amplitude of the direct signal incident on the cluster (mV/m)

N = number of wind turbines in the cluster

Field Measurements of Television Interference

Measurements of television signals in the regions around full-scale operating wind turbines are used to determine the extent to which the idealized signal scattering just discussed is modified by local conditions of topography, direct signal strength, transmitter and wind turbine directions and elevations, and by wind turbine rotor orientation, size, speed, material, and blade pitch. Background information on field measurement equipment and observed TVI effects is presented in this section, followed by a comparison of measured signal scattering data with calculated idealized scatter ratios.

Field Test Equipment

Systematic procedures and equipment for conducting TVI tests in the field are discussed in several references [*e.g.*, Sengupta *et al.* 1981a and 1983a]. Two types of measurement systems are described here, for attended and unattended testing.

Attended Testing

A schematic block diagram of a typical experimental arrangement for attended TVI testing is shown in Figure 9-11, where only those components pertinent to data collection have been included. With any given TV transmission a portion of the signal is scattered by the wind turbine. This portion, together with the desired signal from the transmitter, is picked up by a *rotatable antenna*, fed to a *spectrum analyzer*, and recorded on a *paper chart recorder* for later evaluation. The combination of the spectrum analyzer and the chart recorder is used to (a) measure the ambient levels of the video and audio carrier signals (*i.e.* without the wind turbine operating), and (b) to record the total signal received as a function of time, including any modulation produced by moving wind turbine blades. Two-way radios provide for communication between the TV test site and the turbine control room.

The general quality of the ambient TV reception and the existence of any wind turbine-produced video distortion are observed on the screen of the *TV receiver*. The *video recorder* is employed for documentation of interference phenomenon for correlation with quantitative measurements. It is often convenient to have two TV receivers, with one displaying the received program while the other feeds the video recorder. The equipment is usually powered from the 60 Hz utility system. In the absence of such power at the test site, a portable AC generator is necessary. All of these test instruments can be fitted

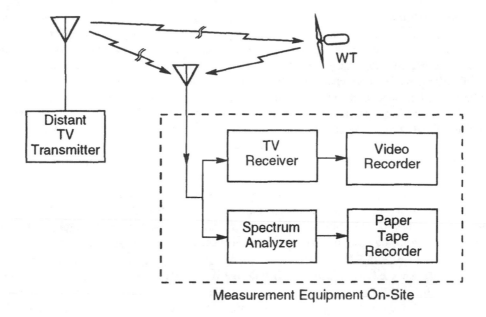

Figure 9-11. Schematic block diagram of a measurement system used for attended field observation of TV signals in the vicinity of wind turbines. [Sengupta *et al.* 1981a]

comfortably inside a station wagon or van, with the receiving antenna mounted on a rack on the roof as shown in Figure 9-12. A typical antenna height is 6 m, and the tower is usually hinged or telescoped to facilitate moving from one site to another.

Figure 9-12. TVI test equipment vehicle with a roof-mounted rotatable antenna. The antenna tower is hinged for transportation from site to site. [Sengupta *et al.* 1981a]

Unattended Testing

A TV signal measuring system that will operate unattended is coupled to the same antenna as before and has a *programmable computer* controlling a spectrum analyzer and a *plotter/printer*. For a given orientation of the antenna, the total signal can be sampled at incremental frequencies spanning the band of frequencies within the TV channel of interest, and the results can be plotted as a function of frequency. Peak signals exceeding a predetermined strength level can be printed out, along with the frequencies at which they occurred. Video and audio carrier signal strengths can be found from these data. The process is repeated for different antenna azimuths, to determine maximum signal strengths for each TV channel.

An unattended system of this type typically updates the printed information about every 15 minutes. It has been observed that peak audio frequencies are often missed on some channels because of the incremental frequency steps. Nevertheless, a computer-based, automatic monitoring system is a useful supplement to an attended system.

Observed TVI Effects Caused by Wind Turbines

For a receiver in the backward scattering zone of the wind turbine (Fig. 9-7) the delayed multipath signal leads to a *ghost image* on the TV screen that pulsates (*jitters*) horizontally at the blade passage frequency. If the ghost image is sufficiently strong, the resulting interference can be objectionable. As the interference increases, the entire picture (desired plus ghost) also shows a pulsed brightening. Still larger levels of interference can disrupt the vertical synchronization of the receiver, causing the picture to roll (*flip*) or even break up. This type of interference occurs as a result of scattering signals off the broad faces of the rotor blades, which is primarily a specular (*i.e.*, mirror-like) scattering.

As the angle ϕ_{RT} increases, the separation between the desired and ghost images decreases, and a somewhat stronger scattered signal is required to produce the same amount of video distortion. Video distortion in the backward scattering zone shows no significant dependence on either the ambient level of the signal (provided this is well above the noise level of the receiver) or on the specific receiver used.

In the forward scattering region (Fig. 9-7), where the wind turbine is almost on a line between the transmitter and the receiver, there is virtually no difference in the time of arrival of the direct and scattered signals. The ghost image is then superimposed on the desired picture, and any video distortion appears as a fluctuation in the brightness (*intensity*) of the picture in synchronism with the blade passage.

In both the backward- and forward-scattering cases, noticeable distortion occurs only when the modulation waveform is of a pulsed nature. Other things being equal, the modulation index and the resulting video distortion increase with increasing TV channel numbers (*i.e.*, increasing frequency and decreasing wave length) and with decreasing distance from the wind turbine. No audio distortion has been observed in any of the field testing to date.

Comparison of Observed and Idealized Signal Scatter Ratios

For the conditions of a given field test, Equations (9-26) can be used to calculate an idealized scatter ratio, while the observed scatter ratio is determined for that test using Equations (9-16). By comparing observed and idealized scatter ratios for a variety of wind turbines and field conditions, we can estimate the *probability* that signal interference will exceed the idealized value by a given amount. This has been done for the 75 field test cases listed in Table 9-1 [Spera and Sengupta 1994]. Scatter ratios are compared in Figure 9-13 for the backward-scatter zone and Figure 9-14 for the forward-scatter zone.

Table 9-1. TVI Cases Analyzed for Observed *vs.* Idealized Scatter Ratios
[Spera and Sengupta 1994]

Wind turbine	No. of units	Scatter zone	No. of cases	Wave lengths (m)	WT-Receiver distances (m)	Data Reference
Mod-1 HAWT	1	Backward	16	1.6 - 5.0	1041 - 2745	[Sengupta *et al.* 1981a]
"	"	Forward	5	1.5 - 3.7	"	"
Mod-2 HAWT	1	Backward	3	1.6 - 3.4	1603 - 6100	[Sengupta *et al.* 1983a]
"	"	Forward	1	0.6	1445	"
"	2	Backward	1	3.4	6254	"
"	2,3	Forward	4	0.5 - 1.4	1354 - 1717	"
17-m VAWT	1	Backward	33	0.4 - 4.2	32 - 133 27 - 31	[Sengupta *et al.* 1981b]
"	"	Forward	12	"	"	"

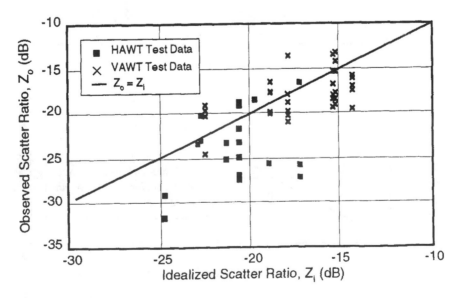

Figure 9-13. Comparison of observed and idealized signal scatter ratios for receivers in the backward-scatter zone. [Spera and Sengupta 1994]

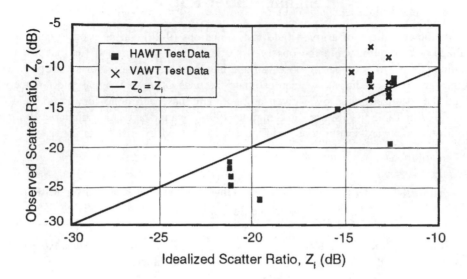

Figure 9-14. Comparison of observed and idealized signal scatter ratios for receivers in the forward-scatter zone. [Spera and Sengupta 1994]

The assumption that signal scatter ratios are linearly additive for a small cluster of wind turbines can be evaluated by examining Figure 9-15. Signals scattered by two and three large-scale Mod-2 HAWTs operating simultaneously were found to be as strong or stronger than the sum of the idealized scattered signals from each turbine.

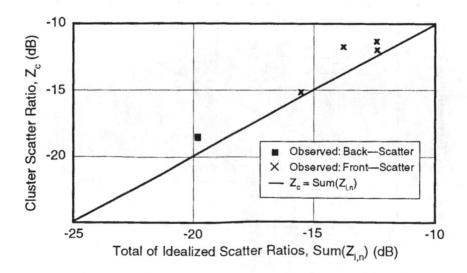

Figure 9-15. Comparison of observed and idealized signal scatter ratios for two and three Mod-2 HAWTs operating simultaneously. [Spera and Sengupta 1994]

Observed scatter ratios often deviate significantly from idealized ratios, which is to be expected since the simplified model treats several variable parameters as constants (*e.g.*, ground reflection). We can estimate the effect of non-ideal field conditions by a statistical *probability of exceedance* model derived from the results of the cases in Table 9-1. This useful parameter is the probability that the observed signal scatter ratio will exceed the idealized ratio by a given amount, or

$$Y(\Delta Z_O > \Delta Z) = Y_E(\Delta Z) \tag{9-27a}$$

$$\Delta_O = Z_O - Z_I \tag{9-27b}$$

where
$$Y(\ ) \ = \ \text{probability of }(\ ); 0 \leq Y \leq 1$$
$$Y_E(\ ) \ = \text{probability of exceeding }(\ )$$
$$\Delta Z_O, \Delta Z \ = \text{observed and selected deviations in the signal scatter ratio (dB)}$$

Figure 9-16 shows the probability of exceedance as a function of the deviations for the 75 cases in Table 9-1. A linear fit to the central portion of this distribution is

$$Y_E(\Delta Z) = 0.39 - 0.11\,\Delta Z \qquad -5.5 \leq \Delta Z \leq 3.5 \tag{9-28a}$$

from which

$$Z = Z_I + 3.5 - 9.0 Y_E \tag{9-28b}$$

where all quantities are in dB units. In ratio form,

$$Z = F_E Z_I \qquad \log F_E = 0.35 - 0.090 Y_E \tag{9-28c}$$

where F_E = empirical exceedance factor

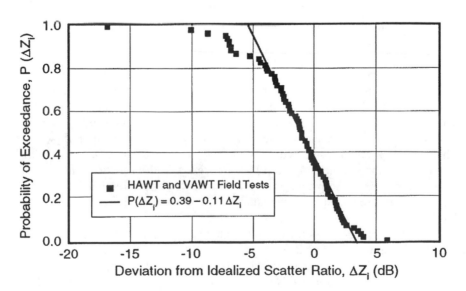

Figure 9-16. Probability analysis of deviations between observed and idealized signal scatter ratios. Data points are the cases listed in Table 9-1. [Spera and Sengupta 1994]

Laboratory Simulation of Television Interference

Full-scale field measurements have verified that a wind turbine can distort the video portion of a TV signal and have shown how this distortion depends on some of the parameters involved. Because of environmental conditions near the wind turbine and the lack of a full range of receiver sites, field test results are necessarily limited in scope. However, information gained from these measurements makes it possible to simulate the interference, and to examine distortion under the controlled conditions that only a laboratory can provide.

Threshold of Perception and Tolerance of TV Interference

Laboratory simulations of HAWT-induced TVI have been conducted to establish the following three levels of the *modulation perception index*, defined previously as m_P in Equation (9-13):

-- $m_{P,T}$ *Threshold of perception:* Interference is first perceived on a TV screen. This is defined as the smallest modulation which produces video distortion detectable by an observer viewing the picture from a distance of about 1.5 m.
-- $m_{P,L}$ *Long-term tolerance:* Interference is no longer acceptable for most prolonged viewing purposes. This modulation level usually dictates the maximum allowable extent of the interference region around a wind turbine.
-- $m_{P,S}$ *Short-term tolerance:* Interference at this level is no longer acceptable even for short viewing times. Disruptive distortion occurs at larger modulations, occasionally resulting in picture breakup caused by loss of vertical hold.

Although these criteria are obviously subjective, results are found to be reproducible when the tests are repeated at a later time with the same observers.

Laboratory Testing Procedure

Details of the test equipment and procedures that have been used for laboratory simulations of TVI are described in [Sengupta and Senior 1978 and 1979a]. Figure 9-19 shows a block diagram of this equipment. The TV signal is received with a commercial, roof-mounted *log-periodic antenna.* With the weaker UHF channels a *preamplifier* is desirable, but this can be bypassed for VHF channels with strong signals. The signal is taken through a set of *variable attenuators* to a *coaxial T-junction* where it is split into two branches: a *direct line* representing the primary signal from the transmitter to the receiver, and a *multipath line* representing the signal scattered by the wind turbine. The two branched signals are then combined and fed to a *receiver* and a *spectrum analyzer.* The latter is tuned to the channel's audio carrier frequency, and its vertical output is fed to a *strip chart recorder.* The strip chart is used to monitor the effective modulation level which can be set to a desired value with the attenuators and the DC bias at the terminals of the receiver's tuner.

Signal scattering can be simulated in the multipath line by a time delay followed by a repetitive pulse amplitude modulation, producing waveforms typical of those obtained during the full-scale measurements with HAWTs. In the studies cited, laboratory simulations of the scattered signal were all made with a pulse repetition period of 0.5 s, which is representative of two-bladed rotors about 30 m in diameter. Although the field tests were conducted on larger HAWTs with longer repetition periods, this change in frequency did not affect the nature of the interference phenomenon.

Backward-zone interference is simulated by introducing a time delay between the direct and multipath signals and suitably alternating the latter. For the forward zone, where the actual time delay is negligible, the direct line shown in Figure 9-17 is eliminated and only the modulated signal is applied to the receiver and the spectrum analyzer.

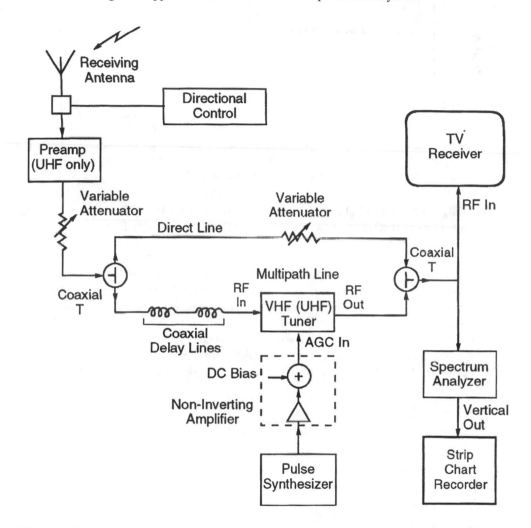

Figure 9-17. Block diagram showing an experimental arrangement for simulating TV interference by a wind turbine, with which to measure the perception of multi-path interference by test subjects. [Sengupta and Senior 1979a]

HAWT TVI Perception and Tolerance Levels

Figure 9-18 illustrates thresholds of perception of interference on Channels 7 and 50 in the form of the modulation index, $m_{P,T}$ as a function of the strength of the carrier signal. Two types of color receivers were used during the measurement of these responses: Receiver A is typical of older TV technology and Receiver B represents newer receivers. For the interference simulating the backward zone thresholds of perception on any given channel are reasonably independent of the ambient signal level for either receiver, but smaller modulations can be perceived with the newer receiver. Modulation thresholds are higher for forward-zone interference (*i.e.*, without time lag), for both receivers. With the newer receiver the threshold was quite large at strong ambient signal levels.

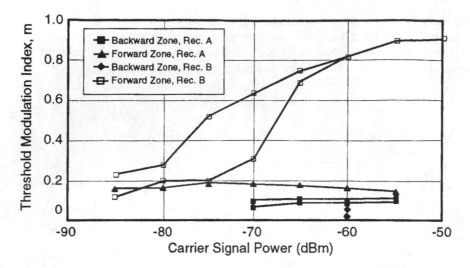

Figure 9-18. Thresholds of perception of simulated TVI by a HAWT on channels 7 and 50, for two types of receivers. [data from Sengupta and Senior 1978, 1979a].

Figure 9-19 shows the long- and short-term tolerances of test subjects to backward-zone TVI with Receiver A, again with signals on Channels 7 and 50. Simulation tests like these indicate that there is no single value of the modulation perception index that is applicable to all signal levels in both the backward and forward interference zones of a HAWT. However, the long-term tolerance index of the dominant backward-zone interference is almost constant at 0.15. Therefore, this is the value recommended for the boundary of the entire region about a HAWT, within which unacceptable video distortion may occur.

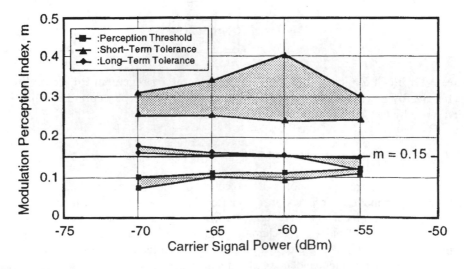

Figure 9-19.   Tolerance levels for simulated TVI in the backward zone of a HAWT. A long-term modulation tolerance of 0.15 is indicated. [Sengupta and Senior 1978, 1979a]

Darrieus VAWT TVI Perception and Tolerance Levels

Most of the above comments also apply to the interference caused by a VAWT like the Darrieus, but there are differences in detail. In particular, a laboratory study has shown that the modulation waveform introduced by a Darrieus blade contains broad pulses and narrow spikes, both repeating at the blade passage frequency [Sengupta and Senior 1979b]. The pulses produce video distortion with a modulation threshold of perception, $m_{P,T}$, equal to the relatively high value of 0.28. The spikes, however, did not produce observable distortion. During field tests of TV interference around a Darrieus VAWT [Sengupta *et al.* 1981b], video distortion levels were observed to be as follows:

$$m_{P,T} \approx 0.025$$
$$m_{P,L} \approx 0.14$$
$$m_{P,S} \approx 0.17$$

From these preliminary results, we can assume that the long-term tolerance for video distortion caused by a Darrieus VAWT is about the same as for a HAWT, *i.e.*, $m_{P,L} = 0.15$.

TVI Effects of Small Wind Turbines

With the growing interest in small-scale wind turbines (rated at about 10 kW or less) for supplying power to individual residences, there is the possibility that a small wind turbine could be in the vicinity of a TV receiving antenna. It is of interest to know under what conditions these two devices are compatible, but at the small distances involved, interference predictions based on data from large HAWTs are questionable. Laboratory simulations with scale models of wind turbine rotors are a convenient method for studying these size effects.

In one study [Sengupta *et al.* 1983b], TVI measurements were carried out using 1/8-scale models of small HAWTs in conjunction with a microwave TV system inside an anechoic chamber. Two and three-bladed rotors were tested, with metallized and wooden blades rectangular and trapezoidal in shape. A locally-broadcast TV signal was used as the RF source of the microwave TV system, after its center frequency was up-converted to 4.0 GHz. Simulated television interference results were then extrapolated to represent TVI effects of full-scale wind turbines at commercial TV channel frequencies.

Generally, it was found that the two- and three-bladed machines produce similar TVI effects, the metallized blades providing stronger interference than the wooden blades. Extrapolation of these results indicates that even at the highest TV frequency, video distortion produced by small-scale wind turbines with metal blades does not extend beyond a distance larger than about 2.5 times the rotor diameter. With wood blades the interference zone is much smaller. At lower TV frequencies the interference distance is correspondingly smaller. Within this distance, the observed effects are generally small-to-negligible, except possibly in the forward interference direction.

Theoretical Prediction of Interference Distances and Zones

The *interference zone* around a wind turbine or group of turbines is defined as the region where the modulation index m_R exceeds the long-term tolerance value $m_{P,\ L}$ when the maximum scattered signal is directed toward the receiver [Sengupta and Ferris 1983, Sengupta 1984]. It should be emphasized that the zone boundary is not sharply defined, since neither the modulation by the turbine nor the tolerance of the viewer is constant.

Some Practical Aspects of the TVI Problem

Theoretical predictions of interference are usually based on "worst case" assumptions, so it should be noted that the interference zone is merely the region where video distortion *could possibly* occur. To estimate the actual percentage of viewing time during which unacceptable TV reception *may* occur within the interference zone, one must take into account relevant statistical parameters such as wind speed and direction [Sengupta and Senior 1979b]. In practice, the interference problem could be much less severe than is suggested by the size of the basic interference zone obtained under the deterministic assumptions used here. However, it is still true that this zone defines the region where detailed interference calculations and measurements are justified.

Interference Zone Around a Single HAWT

With $B_E = B_{E,\ max}$ and $F_{A,\ T} = 1$, Equations (9-15), (9-26), and (9-28) give

$$\Lambda = \frac{\zeta}{D} \frac{|E_{R,D}|}{\eta_S |E_{WT,D}|} = \frac{F_E}{2m_R} \sqrt{F_{A,W}} \cos(k\phi_{RT}) \qquad (9\text{-}29a)$$

$$
\begin{aligned}
k &= 0.5 \quad for \quad -0.8\pi \leq \phi_{RT} \leq 0.8\pi \\
k &= 2.0 \quad for \quad 0.8\pi < \phi_{RT} < 1.2\pi
\end{aligned}
\qquad (9\text{-}29b)
$$

where Λ = normalized distance from a HAWT to its interference zone boundary

Figure 9-20 shows the size of the interference zone when $m_R = 0.15$ and the probability of exceedance is 0.01 [Spera and Sengupta 1994]. The antenna response factor, $F_{A,\ w}$, has been approximated by

$$F_{A,\ W} \approx 10^{F_{B/F}[1-\cos(\phi_S)]/20} \qquad (9\text{-}30)$$

where $F_{B/F}$ = *back-to-front ratio* of the antenna (see Figure 9-3) (dB)

Figure 9-20 clearly indicates the significant reduction of the backward-interference zone by the antenna's discrimination against signals coming from other than the transmitter. However, the narrow forward zone is unaffected by the antenna directivity. Since the blade-scattered signal will be received *via* the side or back lobes of the antenna throughout most of the interference zone, the desirability of an antenna with a high back-to-front ratio is now evident. In fact, were a directional antenna oriented incorrectly (*e.g.*, pointed at the wind turbine instead of the transmitter), interference could exist well beyond the outer zone in Figure 9-20.

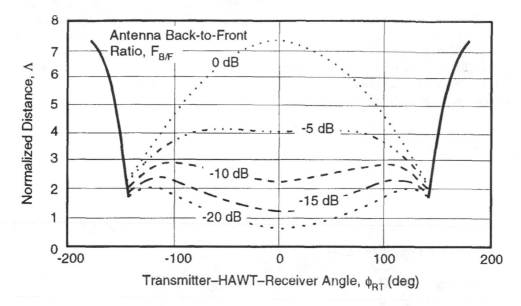

Figure 9-20. Theoretical interference zone around a HAWT, expressed in radial coordinates. The modulation index is 0.15, the receiver antenna is beamed at the transmitter, and the probability of exceedance is set at 1 percent. [Spera and Sengupta 1994]

Darrieus VAWT Interference Zones

Referring again to Equations (9-26), we see that Equations (9-29) can be modified to define a Darrieus VAWT interference zone by replacing η_S with $0.8\eta_M$ and D with the blade chord c. Thus, the interference zone around a Darrieus VAWT is much smaller than that around a HAWT of the same diameter, by a factor approximately equal to $4c/D$.

Interference Zones Around a HAWT Wind Power Station

To estimate the size of the interference zone around a wind power station composed of many turbines, we first divide the station into M clusters, each containing N turbines. The turbines within a cluster are assumed to be operating in synchronism and their scatter ratios are additive, in accordance with Equation (9-26h). Therefore, a reasonable value of N is ten or less. However, we will assume that the cluster scatter ratios add in a random or *root-sum-square* fashion, which leads to the following equations:

$$m_R = \frac{\eta_S N F_E}{2} \frac{|E_{PS,D}|}{|E_{R,D}|} \sqrt{\sum_{i-1}^{M} \left[\frac{F_{A,W}\cos^2(k\phi_{RT})}{(\zeta/D)^2} \right]_i} \tag{9-31a}$$

$$k = 0.5 \quad \text{for the Backward Zone} \tag{9-31b}$$
$$k = 2.0 \quad \text{for the Forward Zone}$$

$$|E_{PS,D}| = \frac{1}{M}\sum_{i-1}^{M} |E_{C,D}|_i \tag{9-31c}$$

where

$|E_{PS,D}|$ = average amplitude of the direct signal incident on the station (mV/m)

The application of Equations (9-31) is illustrated in Figure 9-21, which represents a hypothetical rectangular array of 60 HAWTs divided into 12 clusters of 5 turbines each [Spera and Sengupta 1994]. The modulation index at the zone boundaries is 0.15, and the probability of exceedance is 0.05. Other parameters are $\eta_S = 0.5$ and $|E_{PS, D}| = |E_{R, D}|$. Again we see that the back-to-front ratio of a properly-oriented receiving antenna only affects backward scatter.

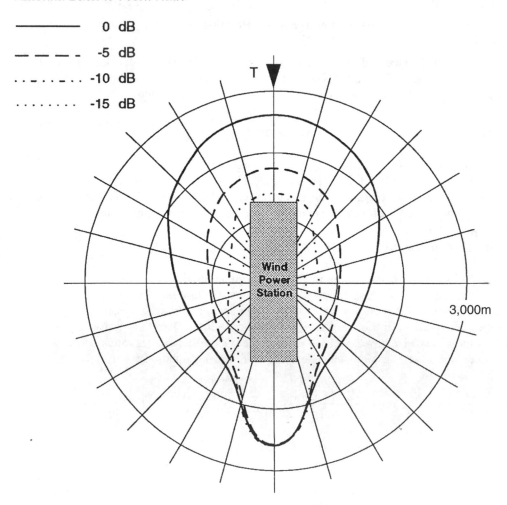

Figure 9-21. Theoretical TV interference zone around a hypothetical HAWT power station, for various back-to-front ratios of the receiving antenna. The station contains 60 HAWTs divided into 12 clusters. It is assumed that turbines within a cluster operate in synchronism, but cluster interferences add randomly. [Spera and Sengupta 1994]

Radio Interference Effects

Amplitude Modulated (AM) Systems

AM broadcast signal reception is susceptible to interference from various man-made and natural sources of background noise. Due to the fact that a rotating wind turbine blade predominantly modulates the amplitude of an electromagnetic signal in its vicinity, interference with AM radio reception can be anticipated. However, since AM broadcast frequencies are low and signal wave lengths are very long, any interference will be confined to the immediate vicinity of a wind turbine. For example, assuming a perfectly conducting flat earth, an omnidirectional receiving antenna, and equal signal strength at the turbine and the receiver, Equations (9-15) and (9-26b) give

$$\left(\frac{\zeta}{D}\right)_{max} = \frac{\eta_S B_E c}{2 m_R \lambda}$$

(9-32)

where c = mean blade chord (m)

Assuming $\eta_S = 0.8$, $B_E = 2$, $m_R = 0.1$, and $\eta = 300$ m, it is found that $(\zeta/D)_{max}$ is less than 0.03 c, which indicates that the maximum interference distance is only a few meters even for a large-scale wind turbine.

Frequency Modulated (FM) Systems

Laboratory simulation techniques with an experimental setup similar to that illustrated in Figure 9-17 have been used to investigate the interference of a wind turbine with FM broadcast reception [Sengupta and Senior 1978]. In this study, a modulation synthesizer was attached to the input terminals of an FM test receiver to amplitude-modulate the input signal. The test receiver was typical of the high-quality FM stereo receivers used in automobiles. Interference was assessed by evaluating the quality of the audio reproduction as a function of the ambient signal level and the applied modulation index.

The results of these laboratory simulations indicate that when the ambient level of the input signal is high (*i.e.*, a *signal-to-noise ratio* > 15 dB), no audio distortion is found until the modulation index, m_R, reaches 0.72, which is equivalent to a modulation range, Δ, of 16 dB. Even with a weaker signal, there is no significant distortion for m_R less than 0.36 (Δ < 7 dB). As m_R is increased beyond this level, however, there is an increasing amount of audio distortion in the form of a pulsed, high-frequency "hiss" superimposed on the desired sound.

These results are consistent with the fact that ordinary FM receivers are only susceptible to noise interference while operating in their threshold regions (*i.e.*, with signal-to-noise ratios less than 12 dB) [Peebles 1976]. They also imply that the effects of wind turbine interference on FM radio reception are negligible, except possibly within a few tens of meters of a wind turbine located in a region of low signal-to-noise ratio for a particular FM station.

Potential Interference with Microwave Communication

Terrestrial microwave communication links, which are widely used for transmitting and receiving a wide variety of information, operate within the 300 MHz to 300 GHz portion of the electromagnetic frequency spectrum. Telephone companies employ such links for long- and short-distance audio communication as well as television and data transmissions [Bell Laboratories 1971]. These link systems use *microwave repeater stations* at selected sites where received signals are detected and passed on to local customers or amplified and transmitted to the next repeater station. A typical link or *TD-system* operates at 4 GHZ and uses a highly-directional *pencil beam antenna* with a beam width of ± 0.6 deg, an effective gain (G_0) of 35 db, and side-lobe antenna factors (F_A) less than -30 dB.

A rigorous investigation of any potential interference with such communications by wind turbines requires detailed knowledge of each specific link, including its function, method of operations, radiation patterns of transmitting and receiving antennas, and propagation characteristics in the region around the wind turbines. Problems of interference in microwave link systems are usually treated statistically [Jakes 1974, Bennett *et al.* 1955], but in the present case it is more convenient to follow a deterministic approach [Sengupta and Senior 1978]. In a manner similar to that used previously for TV interference, a wind turbine located in the vicinity of a microwave repeater station is treated as a time-varying multipath source producing amplitude and phase modulation of the signal that is picked up by the microwave receiving antenna.

The effect of a rotating blade on the detected signal (and, hence, on the interference produced) has been qualitatively assessed, by examining the basic detection process in the TD receiver system [Sengupta and Senior 1978]. It has been found that the interference effect is a *frequency smearing* of the received base band signal energy, the maximum amount of which depends on the blade size, rotor orientation, and rotor speed. For the relatively low rotational speed of a HAWT, this maximum is much less than the total bandwidth of the signal. The degrading influence of the smear depends on the amplitude of the scattered signal at the receiver relative to the desired signal, m_R, and it is sufficient to investigate potential wind turbine interference on the basis of a threshold value of m_R.

The flat plate model of a HAWT blade developed for analyzing TV interference can also be used to estimate the amplitude and phase of the interfering microwave signal, relative to the desired carrier signal at the receiver. In general, the equations presented earlier for TVI can be used to evaluate potential interference with microwave signals. Microwave interference zones differ somewhat from TVI zones because of *lower thresholds of perception* and *higher antenna directionality*. These two subjects will be discussed briefly here.

Thresholds of Perception

The region about a microwave link receiver where a wind turbine could produce unacceptable interference with reception will be called the *forbidden zone*. Within this zone the amplitude of the blade scattered signal relative to the desired (direct) signal exceeds a predetermined threshold value, where this threshold depends on the characteristics of the link system [Curtis 1960, Campbell 1958]. Telephone companies customarily demand a threshold of -40 dB (*i.e.*, $m_R = 0.01$), but if a -40 dB *margin for fading* is required, the resulting threshold is -80 dB (*i.e.*, $m_R = 0.0001$).

It is often unnecessary to determine the complete forbidden zone for the purpose of estimating interference potential. The satisfactory performance of a microwave link system requires that there be adequate clearance between the *link path* (*i.e.*, the optical line-of-sight transmission path between two antennas) and any nearby scattering objects such as a wind turbine. It is often sufficient that all scattering objects lie outside the *first Fresnel zones* of the antennas in a microwave link [Bell Laboratories 1971, Griffiths 1987], as shown in Figure 9-22. For any distance ζ from antenna A, the lateral distance from the link path between antennas A and B to the boundary of the nth Fresnel zone is approximately

$$y_n \approx \sqrt{n\lambda\zeta_A(1 - \zeta_A/x_{AB})} \qquad (9\text{-}33a)$$

$$y_{min} = 3y_1 \qquad (9\text{-}33b)$$

where

$\quad y_n$ = width of the nth Fresnel zone (m)
$\quad n$ = Fresnel zone number
$\quad \zeta_A$ = distance from the wind turbine to antenna A (m)
$\quad x_{AB}$ = distance from antenna A to antenna B (m)
$\quad y_{min}$ = minimum clearance between the link path and the wind turbine (m)

A clearance between the wind turbine and the link path equal to three times the width of the first Fresnel zone is sufficient to ensure immunity from interference. In some cases it may also be necessary to evaluate m_R to make sure that it is less than the appropriate threshold value.

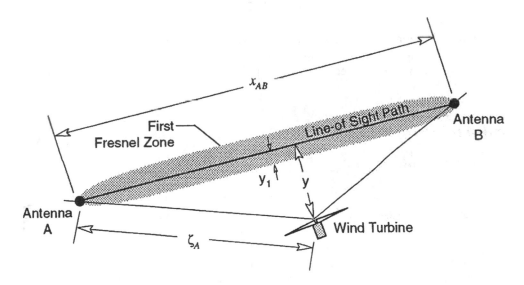

Figure 9-22. **Schematic diagram showing a plan view of a wind turbine located outside the first Fresnel zone of a link between two microwave antennas.**

Antenna Azimuthal Response Factors

The azimuthal response pattern for a typical microwave antenna is highly directional, as illustrated in Figure 9-23. This diagram represents the response function of a parabolic dish antenna of the horn-reflector type, 2.4 m in diameter and shows the differences between microwave and TV antenna patterns. It is obvious that to minimize interference, a wind turbine must be outside the main beam and the near side lobes of a microwave receiving antenna. Because of the very high back-to-front ratio of the dish, only forward-zone interference need be considered, with the wind turbine located between the microwave transmitter and the receiver.

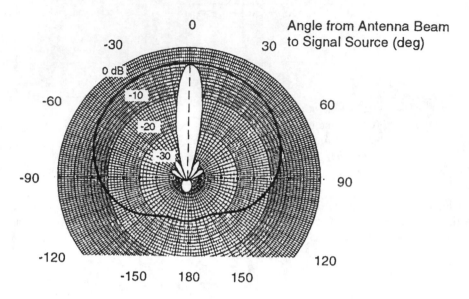

Figure 9-23. Comparison of typical azimuthal response factors for a microwave receiving antenna (white area) and a directional TV antenna (solid line). The microwave antenna dish is of the horn-reflector type, and the TV antenna response is the same as that in Figure 9-2.

Satellite Microwave Link

The potential impact of a wind turbine on the performance of an earth station receiving microwave signals in the 6 to 11 Ghz frequency range from a *geostationary satellite* can be estimated from the ratio of the scattered-to-desired signal amplitudes at the receiver. This ratio can be determined in a manner similar to that described for a terrestrial microwave link, with knowledge of the azimuthal and elevation response patterns of the receiving station's dish antennas. A modulation index threshold is not commonly specified for satellite microwave systems, but a value of -40 Db has been used for m_R to estimate the significance of interference signals at an earth station.

Potential interference with an earth station communicating with a geostationary satellite can also be assessed by using the Fresnel distance criterion given in Equations (9-33). In this case $x_{AB} = \infty$, and the clearance criterion reduces to

$$y_{\min} = 3\sqrt{\lambda \zeta_A} \qquad (9\text{-}34)$$

Interference with Navigation Communication Systems

VOR and DVOR Systems

The VOR (*Very High-Frequency Omnidirectional Range*) and DVOR (*Doppler VOR*) systems are used for short-range ground-to-air communication which provides navigation information to flying aircraft [Beck 1971, FAA 1968]. In an ideal situation, the VOR ground station effectively radiates a VHF carrier signal (in the 108-118 Mhz frequency range) which contains two synchronous 30 Hz modulation signals. One is known as the *reference signal* and is constant in phase and independent of the aircraft azimuth. The phase of the other signal, known as the *variable signal*, varies directly in accordance with the bearing of the aircraft from the VOR ground station. A phase-measuring device in airborne receivers enables pilots to find their bearings by determining the phase difference between these two modulation signals.

In a conventional VOR system, the variable signal is a 30-Hz amplitude-modulation (AM) on the carrier, while the reference signal is a 30-Hz frequency modulation (FM) of a 9.96 KHz sub-carrier. In the DVOR the roles of 30 Hz AM and FM signals are interchanged (*i.e.*, the variable signal is FM while the reference signal is AM). This makes the DVOR system relatively immune to interference from a potential multipath source in the vicinity, such as a wind turbine.

Scalloping errors in the bearing indications of a VOR (DVOR) system that might be produced by the rotating blades of a wind turbine near a ground station have been investigated theoretically with the following assumptions: The system is in free space, the airborne VOR receiver is ideal, the vertical-plane response characteristics of the transmitting antenna are constant, the antenna is located above a perfectly-conducting plane earth, and the scattering effects of the wind turbine tower are negligible [Sengupta and Senior 1978]. The rectangular metal plate model of a HAWT blade expressed in Equations (9-17) is used to obtain the strength of the scattered VOR signals at the airborne receiver for both the static and rotating blade conditions.

For a rotating blade, a Fourier-series analysis of the modulation function f_m obtained with Equations (9-17) indicates that it contains even harmonics of the rotor speed Ω, and that its spectrum may extend up to a frequency many times larger than Ω. In fact, the frequency spectrum of f_m may contain the VOR AM modulation frequency of 30 Hz and its harmonics. However, calculations indicate that the theoretical modulation index, m_R, obtained for a rotating HAWT blade is much smaller than that for a static blade. An analysis of DVOR performance also leads to the same conclusion [Sengupta and Senior 1978]. The reduced scalloping errors produced by rotating blades may be explained by the fact that the scattered energy is distributed over a band of frequencies in the time-varying case, whereas the VOR receiver is sensitive only to 30-Hz modulated signals. It is therefore concluded that potential wind turbine interference with VOR or DVOR communications may be adequately investigated by following the procedures developed by the Federal Aeronautics Administration for static interference sources [FAA 1968].

LORAN-C System

LORAN-C (*Long-Range Navigation System: Version C*) is a hyperbolic maritime navigation system that enables users to determine their positions very precisely anywhere within the designated coverage area, by measuring the arrival times of signals received from at least three stations [Beck 1971, Kayton and Fried 1969]. The coverage area encompasses about one-fourth of the earth's surface. LORAN-C is a pulse system with a common carrier frequency in the band from 90 to 110 Khz for all stations, which means that wave lengths are approximately 3,000 m. A chain of transmitting stations consists of one master station and two to four slave stations. Master and slave stations are separated by distances of 1,000 to 1,300 km to increase the accuracy of triangulation, and the operating range from a vessel to the stations is in excess of 1,600 km.

Since the objective of a LORAN-C system is to provide over-the-ocean coverage, its transmitting stations are often located on islands or in coastal regions where, with favorable wind conditions, wind turbines may also be sited. A wind turbine in the vicinity of a LORAN-C transmitting station could affect communications by interacting with the transmitting antenna or by acting as a secondary transmitter, radiating similar LORAN-C signals delayed in time. Both of these effects have been investigated theoretically [Sengupta and Senior 1979b]. Since the antenna interaction effect could be corrected at the transmitting station by using proper *impedance matching* of the antenna input terminals, it is unlikely to affect communication system performance. In particular, it has been found that such effects will be insignificant even for a large-scale HAWT (like the 2.5-MW Mod-2) located as close as $\lambda/12$ (approximately 250 m) to the transmitter.

The second effect, that of acting as a secondary signal source, is potentially more severe. Assuming that the transmitting antenna is linear and oriented vertically above an ideally conducting flat earth and again modelling a HAWT blade as a rectangular metal plate oriented so as to direct the maximum scattered signal to the receiver, the modulation index, m_R, assumes its largest value when the transmitter, the receiver and the wind turbine are collinear. Generally, forward-scattering is the more critical, with the wind turbine much closer to the transmitter than to the receiver. Under these conditions, the maximum modulation index can be estimated as follows [Sengupta and Senior 1979b]:

$$m_R \approx \frac{\pi R^3}{6\lambda^2 \zeta} \frac{\sqrt{(\zeta^2 + z_w^2)/\zeta^2 + z_T^2}}{\exp(0.184\pi R^2/\eta_S A_P)} \tag{9-35}$$

where

        ζ = distance from the wind turbine to the transmitter (m)
        z_T = elevation of the transmitter (m)

Since the rotor radius of even the largest HAWT is very small compared to the 3,000-m wave length of LORAN-C signals, the signals that could be scattered by a wind turbine are about 100 dB lower than the direct signal at the receiver, even when the wind turbine is located as close as 100 m to the transmitter or the receiver. Therefore, it is unlikely that LORAN-C communications would be degraded by such low levels of interfering signals. From these results, it appears that a wind turbine located at a distance greater than $\lambda/12$ (about 250 m) from any transmitter or receiver will not have a significant effect on the performance of a LORAN-C system.

Radar Systems

Radar is a system that uses electromagnetic waves to identify the range, altitude, direction, or speed of both moving and fixed objects such as aircraft, ships, motor vehicles, weather formations, and terrain. Wind turbine interference with radar signals is a concern to both the *Federal Aviation Administration* (FAA) and the *Department of Defense* (DOD), because of potential impacts on air traffic control and military missions. Experience to date shows that only a small percentage of wind turbine installations (less than 5 percent) produce interference with radar signals that is significant [Seifert 2005 and 2006, Meyers and Seifert

2008]. Radar interference assessments, the development of permitting processes, and techniques and technology for *mitigation* of any radar interference by wind turbines are all ongoing activities involving the FAA, DOD, manufacturers of radar systems, and developers of wind power stations.

A large-scale wind turbine tower has a *radar cross section* (RCS) equivalent to that of a building, a hill, or the tower of a high-voltage transmission line. The major components of a wind turbine contribute to its RCS approximately as follows [Seifert and Meyers 2008]:

- tower: 75 percent
- blades: 20 percent
- nacelle: 4 percent
- nose cone: 1 percent

There are two main types of potential interference with radar by wind turbines: *Direct interference* and *Doppler interference*. Direct interference can cause

- high reflectivity
- reduced sensitivity
- false images
- shadow areas

Doppler interference, caused by the rotating blades, alters signal frequencies and produces false targets. Doppler interference can impact both airborne and fixed radar. In general, the effects of Doppler interference with radar signals can be studied using the techniques described earlier for analyzing TV interference.

Seifert [2005] provides the following examples of potential impact and non-impact locations involving radar installations and wind turbines:

Potential Impact Locations

- airport approach and exit zones
- training routes used for testing radar systems
- routes used for training air crews
- low-altitude radar beam areas
- helicopter zones
- attack corridors
- specialized radar and telemetry areas
- test and evaluation ranges
- specialized military mission impact areas

Non-Impact Locations

- normal high altitude routes
- regions more than 15 miles from airports or transmitters
- wind power stations in areas served by radar systems with specialized filters (*e.g.* Palm Springs, California, and Boston, Massachusetts) more than 95 percent of wind power stations in the U.S.

References

Beck, G. E., 1971, *Navigation Systems - A Survey of Modern Electronic Aids*, London: Van Nostrand - Reinhold, Chapter 9.

Bell Laboratories, Inc., 1971, *Transmission Systems for Communications*, Revised Fourth Edition, Winston Salem, North Carolina: Western Electric Company, Inc., Technical Publications.

Bennett, R., H. E. Curtis, and S. O. Rice, 1955, "Interchannel Interference in FM and AM Systems under Noise Loading Conditions," *B. S. T. Journal*, Vol. 34: pp. 601- 636.

Campbell, R. D., 1958, "Radar Interference to Microwave Communication Services," *Proceedings of AIEE*, Vol. 77: pp. 717-722.

Curtis, H. E., 1960, "Radio Frequency Interference Considerations in the TD-2 Radio Relay Systems," *B. S. T. Journal*, Vol. 39: pp. 369-387.

FAA, 1968, *Handbook: VOR/VORTAC Siting Criteria*, Report 6700.11, Washington, DC: Federal Aviation Administration.

Griffiths, J., 1987, *Radio Wave Propagation and Antennas*, Englewood Cliffs, NJ: Prentice-Hall International, pp. 238-140.

IEA, 1986, *Recommended Practices for Wind Turbine Testing and Evaluation*, Vol. 5: Electromagnetic Interference, Preparatory Information, R. J. Chignell, ed., International Energy Agency, Lyngby, Denmark: Technical University of Denmark, Department of Fluid Mechanics.

Jakes, W. C, Jr., 1974, *Microwave Mobile Communications*, New York: John Wiley & Sons.

Kayton, M. and W. R. Fried, 1969, *Avionics Navigation Systems*, New York: John Wiley & Sons, Inc., pp. 192-197.

Kerr, D. E., 1964, *Propagation of Short Radio Waves*, Lexington, MA: Boston Technical Publisher, Inc., pp. 34-41.

LaHaie, I. J. and D. L. Sengupta, 1970, "Scattering of Electromagnetic Waves by a Slowly Rotating Rectangular Metal Plate," *IEEE Trans. Antennas Propagation*, Vol. AP-27, No. 1: pp. 40-46.

Peebles, P. Z., Jr., 1976, *Communication System Principles*, Reading, MA: Addison-Wesley Publishing Co., Inc., pp. 273-281.

Seifert, G., 2005, *Wind-Radar Interference*, NWCC meeting, INL presentation, Idaho Falls, Idaho: Idaho National Laboratory. Available at www.nationalwind.org/events/siting/presentations/seifert-radio_interference.pdf

Seifert, G., 2006, *Wind-Radar Interference*, Wind Powering America workshop, INL presentation, Idaho Falls, Idaho: Idaho National Laboratory. Available at www.windpoweringamerica.gov/pdfs/workshops/2006_summit/seifert.pdf

Seifert, G. and K. Meyers, 2008, *Wind Radar and FAA Issues*, Wind Powering America workshop, INL presentation, Idaho Falls, Idaho. Idaho National Laboratory. Available at www.windpoweringamerica.gov/pdfs/workshops/2008_weats/meyers.pdf

Sengupta, D. L., and T. B. A. Senior, 1977, *TV and FM Interference by Windmills*, Final Report No. DOE/TIC-11348, Washington, D.C.: U.S. Department of Energy.

Sengupta, D. L., and T. B. A. Senior, 1978, *Electromagnetic Interference by Wind Turbine Generators*, TID-28828, Contract No. EY-76-02-184, Washington, D.C.: U.S. Department of Energy.

Sengupta, D. L., and T. B. A. Senior, 1979a, "Electromagnetic Interference to Television Reception Caused by Horizontal Axis Windmills," *Proc. IEEE*, Vol. 67, No. 8: pp. 1133-1142.

Sengupta, D. L., and T. B. A. Senior, 1979b, *Wind Turbine Generator Interference to Electromagnetic Systems*, DOE/ET/20234-T1, Washington, D.C.: U.S. Department of Energy.

Sengupta, D. L., and T. B. A. Senior, 1980, *Wind Turbine Interference to Television Reception*, Radiation Laboratory Final Report No. 014438-4-F, Ann Arbor, Michigan: University of Michigan.

Sengupta, D. L., T. B. A. Senior, and J. E. Ferris, 1980, *Television Interference Tests on Block Island, Rhode Island*, Technical Report No. 3, Contract EY-76-S-02-2846, A004, Washington, D.C.: U.S. Department of Energy.

Sengupta, D. L., T. B. A. Senior, and J. E. Ferris, 1981a, *Measurements of Interference to Television Reception Caused by the Mod-1 Turbine at Boone, NC*, Technical Report No. 1, Contract DE-SERI-XH-0-9263-1, Golden, Colorado: National Renewable Energy Laboratory.

Sengupta, D. L., T. B. A. Senior, and J. E. Ferris, 1981b, *Measurements of Television Interference Caused by a Vertical Axis Wind Machine*, SERI/STR-215-1881, Golden, CO: National Renewable Energy Laboratory.

Sengupta, D. L. and J. E. Ferris, 1983, *Assessment of Electromagnetic Interference Effects of the Ellenville Windfarm*, Radiation Laboratory Final Report No. 388629-1-F, Ann Arbor, Michigan: University of Michigan.

Sengupta, D. L., T. B. A. Senior, and J. E. Ferris, 1983a, *Television Interference Measurements Near the Mod-2 WT Array at Goodnoe Hills, Washington*, SERI/STR-211-2086, Golden, Colorado: National Renewable Energy Laboratory.

Sengupta, D. L., T. B. A. Senior, and J. E. Ferris, 1983b, *Study of Television Interference by Small Wind Turbines*, SERI/STR-215-1880, Golden, Colorado: National Renewable Energy Laboratory.

Sengupta, D. L., 1984, *Preliminary Assessment of Electromagnetic Interference Effects of the Cape Blanco Windfarm*, Radiation Laboratory Final Report No. 388799- 1-F, Ann Arbor, Michigan: University of Michigan.

Senior, T. B. A., D. L. Sengupta, and J. E. Ferris, 1977, *TV and EM Interference by Windmills*, Radiation Laboratory Final Report No. 014438-1-F, Ann Arbor, Michigan: University of Michigan.

Senior, T. B. A., and D. L. Sengupta, 1983, *Large Wind Turbine Handbook: Television Interference Assessment*, SERI/STR-215-1879, Golden, Colorado: National Renewable Energy Laboratory.

Spera, D. A., and D. L. Sengupta, 1994, *Equations for Estimating the Strength of TV Signals Scattered by Wind Turbines*, NASA CR-194468, Cleveland, Ohio: NASA Lewis Research Center.

10

Structural Dynamic Considerations in Wind Turbine Design

Glidden S. Doman
System Design Manager
Wind Energy Systems Taranto S.p.A.
Taranto, Italy

Introduction

Experience has shown that the structural dynamic behavior of a wind turbine system, good or bad, must always be dealt with by the design and operations teams. Through the use of known analysis methods, the effects of physical changes, controls, and subsystem features on the dynamic behavior of a new or existing wind turbine can be understood. When this is accomplished, the structural dynamic behavior of any given system can be improved. Here the term "improve" means to increase fatigue life and reduce cost without sacrificing energy capture.

A necessary tool for accomplishing these improvements is a *computer simulation* of the dynamic system, with which to study and understand its dynamic behavior and determine how the system should be modified to change that behavior in favorable directions. Adequate simulation models are complete as to all the subsystem features and dimensional properties that affect system dynamics in any significant way. They can and must permit designers to study and improve the system behavior as though they were running the real turbine. Successful system designs evolve as they are modified in simulated form.

Wind Turbine Design Philosophies

There are certain design philosophies that transcend many of the other determinants of turbine configuration or "type", such as HAWT *vs.* VAWT, large *vs.* small, upwind *vs.* downwind, *etc.* Modern wind turbine configurations follow a *system architecture* that embodies one or more of these design philosophies, an architecture which seeks to combine efficient and durable static and dynamic structures with high-performance machinery and controls. Engineers are often more comfortable with requirements than philosophies, but nevertheless, the end product of their work is always a record of their design philosophy.

The emphasis in this chapter is on system design philosophy and, in particular, on system architectures based on *compliance* with the forces of nature, rather than *resistance* to these forces. For the moment, we shall depart from the more-objective approach followed elsewhere in this book and advocate the compliant architecture philosophy of wind turbine design. The opinions expressed are based on design and operating experience with several large-scale prototype HAWTs which have demonstrated the benefits of compliant architecture, as well as on observations of commercial wind turbines that do or do not embody this philosophy. Contradictory opinions can easily be found if one consults other experienced wind turbine designers. One only has to look at the wide variety of installed turbine configurations illustrated in Appendix E to see that design tradeoffs, to date, have usually been resolved in favor of resistance rather than compliance.

Two factors may explain the current emphasis on resistive system architectures. First, the great majority of installed wind turbines are medium-to-small in scale, and any weight and cost penalties associated with this type of architecture are manageable. Second, the compliant architecture is viewed as less reliable or less durable because of a lack of long-term, successful field experience with turbine configurations based upon it. However, as the world-wide trend toward larger and more powerful commercial turbines continues and as positive experience with compliant subsystems accumulates, we can expect more of a balance between compliant and resistive architectures in future wind turbines.

Stiff *vs.* Soft Design Philosophy

A prime issue in establishing a design philosophy is that of *stiff vs. soft design.* All structural dynamic systems can be simulated by a series of interconnected *masses, springs*, and *dampers.* In a stiff system concept, transient air loadings go into the structural springs as physical strains and (sooner or later) induce fatigue failure. In the soft system concept, transient air loadings are reacted primarily by subsystem masses with little strain energy involvement. Hence, there is a low proclivity for fatigue failure in a dynamically soft system.

Representative of wind turbines with stiff structural systems are the early 200-kW NASA/DOE *Mod-OA* and the 2.0-MW *General Electric Mod-1* HAWTs (Figs. 3-29 and -31) and typical 3-bladed Danish-type turbines currently in service worldwide (*e.g.,* Fig. 4-22). Softer structural systems are embodied in the *Boeing* 2.5-MW *Mod-2* and 3.2-MW *Mod-5B* HAWTs (Figs. 3-39 and 2-1), and, to an even greater extent, in the *Hamilton Standard/KKRV WTS-3* and *WTS-4* (3.0 and 4.0 MW, respectively; Figs. 3-34 and 4-24) and, most recently, in the 2.0-MW *WEST Gamma 60* shown in Figure 10-1.

In addition to the *load response* aspects of selecting a philosophy of system design, the choice of concept will also affect both the achievable *load control* and the ease or difficulty of an adequate simulation of the system.

Figure 10-1.   The 2.0-MW Gamma 60 HAWT embodies the compliant architecture philosophy in its flexible tower, teetered hub, low-modulus blade material, and broad-range variable-speed power train. Power is controlled by yawing the rotor, a maneuver made possible by the load alleviation resulting from soft-system design. (*Courtesy of Wind Energy Systems Taranto S.p.A.*)

Philosophy of Load Control

A second fundamental design philosophy is that of controlling or *attenuating* the aerodynamic loading transients at their point of application: on the blade surfaces. Softening of the supporting structure can provide some reduction of these *cyclic air loads* (through damping effects), but features such as the *rotor teeter hinge* are more effective in eliminating major components of the cyclic air loads. *Active pitch control* could also contribute to direct alleviation of cyclic air loads. All control methods of air load alleviation involve some form of feedback from the external disturbance to the airfoil *angle of attack*. Some, therefore, present possible *dynamic instability* problems. These must be understood through adequate models in the system simulation.

Aerodynamic Stall Philosophy

A third choice that has divided HAWT designers into two schools is that of encouraging aerodynamic *stall* as a means of control or of intentionally keeping the rotor out of stall. While control through stall permits mechanical simplification, it presents difficult additional problems in predicting and controlling periodic and transient loads arising from *stall hysteresis*. By selecting stall control the HAWT designer introduces not only a new source of fatigue loads, but also the need to account for almost unpredictable unsteady aerodynamic effects in the system simulation.

The VAWT designer doesn't have the option of avoiding periodic stall and is therefore confronted with the difficult problem of accounting for it in system simulations. This explains the greater interest among VAWT designers in stall-hysteresis effects and the development of special airfoils to reduce undesirable stall loads.

Rotor Geometry

The choice of the basic geometry of a wind turbine rotor has a major influence on the way the structural system responds to air and acceleration loadings and develops internal loads. The configuration elements that can be used to control these loads are *blade coning angle, blade-to-shaft hinges*, and *number of blades*.

Blade Coning Angle

The steady *thrust* component of blade air loading can be alleviated as a dominant source of blade bending stress by providing a suitable fixed *coning angle*, in which the blades are inclined downwind from the plane of revolution. Coning combined with centrifugal force acts to counter a selected portion of the downwind bending that would otherwise be created by the thrust air loads. Both the steady air loads and centrifugal loads can be simply and accurately calculated and simulated, so this aspect of rotor design is straightforward. It can and should be accomplished independently of other features that involve dynamic behavior of the system.

Blade-to-Shaft Hinges

A second fundamental influence of rotor type on loads encountered is the presence or absence of a *teetering hinge* that will permit suitable freedom for the blades to move upwind and downwind under the influence of unsteady air loadings. Individual *flapping hinges* accomplish the same result but are impractical (except perhaps on small rotors) because the centrifugal force will be insufficient to prevent excessive free coning and the loss of effective disk area. The rigid interconnection of two blades restrains coning while permitting the rotor to teeter freely with respect to the shaft. Blade pitch angles may or may not be coupled to the teetering motion. A passive means for accomplishing this coupling is the δ_3 *hinge*, in which the teeter axis is not perpendicular to the blade axis. The δ_3-angle effect is discussed in more detail later.

Unsteady air loads are attenuated as a result of the upwind-downwind motion that the teeter hinge permits. This motion causes passive changes in blade angles of attack to oppose and eliminate the major *once-per-revolution* (1P) component of these air loads. For example, a steady vertical *wind shear*, which would create large fluctuating air loads on blades rigidly attached to the shaft, will create only a small teetering motion which in turn creates a cyclic fluctuation of angle of attack and eliminates the fluctuating air loading. The fluctuating loads that would result from a side wind component or *yawed flow* on a rigid-hub rotor are similarly eliminated by a teeter hinge.

A teeter hinge also attenuates, but to a lesser degree, the air loads that result from *small-scale turbulence* (spatial turbulence on a scale smaller than the rotor diameter) and the wind speed deficiency in the *tower shadow*, which is important for rotors operating downwind of the tower. In summary, the advantage of the teeter hinge is that no major out-of-plane bending moments from the rotor can enter the rotor shaft, whatever their source. Some or all of this advantage can be lost if teetering is restrained by dampers, springs, *etc.*

Number of Blades

The third major influence of rotor type is the choice of the number of blades. If the rotor is not hinged there will be major blade bending loads caused by wind shear, small-scale turbulence, yawed flow, and possibly tower shadow. These loads will bend the blades cyclically once per revolution of the shaft. If the hingeless rotor has two blades, these cyclic blade loads will shake the nacelle heavily at two cycles per revolution (2P). If there are three or more blades, the nacelle will not shake as heavily, but each blade will still suffer the cyclic bending strains. Either a teeter hinge or individual flapping hinges will eliminate both the blade bending and the nacelle shaking. A three- or four-bladed rotor could be provided with the equivalent of the teeter hinge effect by a concept called *gimbaling*, but this would result in an excessively complex and costly hub construction.

Based upon the above fundamental aspects of rotor configuration influence, the two-bladed, teeter-hinged rotor emerges as a highly attractive choice except for one additional factor. The mechanics of the teeter motion introduce cyclic in-plane accelerations of the blades because of *Coriolis effects*. If the power train is torsionally stiff and the lateral retention of the hub bearing is stiff, there will be substantial cyclic torque in the shaft and cyclic force on the nacelle at a 2P frequency. However, low shaft stiffness and low lateral nacelle stiffness are usually adopted for other reasons, so these Coriolis effects appear as very small motions rather than significant cyclic loads. Therefore, the load-attenuating behavior of the teetered rotor on a compliant nacelle structure becomes nearly ideal.

A final rotor feature that influences cyclic loads and prospective fatigue life is the *modulus of elasticity* of the material used in blade construction. Low blade stiffness enhances the cyclic displacements and velocities that can occur at a given vibratory stress level. This in turn increases the *aerodynamic damping* and the *inertial impedance* that the blade experiences in response to transient loadings. Thus, the selection of rotor blade materials for high ratios of *fatigue stress allowables* to modulus of elasticity can reduce loads and increase fatigue life. These ratios are generally favorable in wood and some composites and generally unfavorable in metals.

Rotor Controls

The fundamental purpose of wind turbine rotor control elements is, of course, the adjustment of steady and cyclic air loadings that would otherwise occur in response to spatial and temporal variations of wind speed. One school of conceptual design, motivated by an urge for simplicity as discussed earlier, has chosen to operate the rotor near aerodynamic stall. In this *stall control or passive control* concept, load increases are limited by the *maximum lift coefficient* that the airfoil generates at the advent of stall. This has led to the design of new airfoils for wind turbines with lower maximum lift coefficients and a smoother transition to stalled behavior (called *soft stall*) than available aircraft airfoils (*e.g.*, see Table 6-1).

Another design-concept school chooses to operate blades at lower lift coefficients (*i.e.*, lower angles of attack), in order to keep away from stall and increase the rotor *coefficient of performance*, C_p, because of the higher *lift-to-drag ratios* in the low-lift regime. Control of aerodynamic loading must then be accomplished by control of the blade's angle of attack, which leads to *pitch control* of either the full length of the blade or at least the outboard highly-active portion of the blade span. Although pitch-controlled rotors are in service that have quite successfully achieved control of structural loads, they are generally criticized for their complexity. This complexity and the sophisticated design involved in getting the desired sharpness of load control led to the selection of stall control for the most of the wind turbines in service at the end of the 1980s. However, the majority of HAWTs installed in the '90s and later have variable-pitch rotors.

A third approach to aerodynamic load control of a HAWT is that of *yawing the rotor* out of alignment with the wind. This approach has been used for centuries without much attention to the cyclic load aspects. As explained by Drees [1977] blade bending flexibility in a hingeless rotor can provide substantial alleviation of the cyclic loads that would otherwise arise during yawed operation. Hohenemser [1981] demonstrated in the early-1980s that a teeter-hinged rotor can be forced by the control system to yaw at very high rates without creating load problems in either the rotor or the nacelle. The structural feasibility of high-rate yaw control of a teetered rotor was also demonstrated in the mid-1980s by field tests on the WTS-4 4-MW HAWT [Stoltze 1985]. These demonstrations led to the selection of yaw control and fixed blade pitch for the Gamma 60 HAWT, which started operating in 1992.

Figure 10-2 illustrates the relationships between wind speed, rotor speed, yaw angle, turbine torque, and power output that exist in a HAWT with broad-range variable speed and yaw control of peak power, using the Gamma 60 operating map as an example. Where *broad-range variable speed* is adopted as the basic operating mode of a HAWT, aero-dynamic torque control is required only to limit rotor speed in high winds and large gusts, and to control speed in the event that electrical load is lost [DiValentin *et al.* 1986]. Under all other operating conditions, the rotor speed follows the wind speed, while changes in rotor momentum store and then give back most of the energy associated with large-scale turbulence. Aerodynamic torque control can be accomplished either by blade pitch or by yawing the rotor. In either event, aerodynamic control is confined to managing overspeed and is not active over most of the operating envelope.

It is significant to note that while the yaw control concept has the advantage of a simple fixed-pitch rotor, it also applies true *stall avoidance* principles. Aerodynamic loads are adjusted by combining reduction of effective disk area with reduction in the mean effective angle of attack. Meanwhile, cyclic air loadings that would otherwise result from yawed operation are essentially eliminated by the teetering action of the rotor. Control of the air loading on the rotor is achieved from outside the rotor system, without the need of blade pitch change and without stalling the active airfoils.

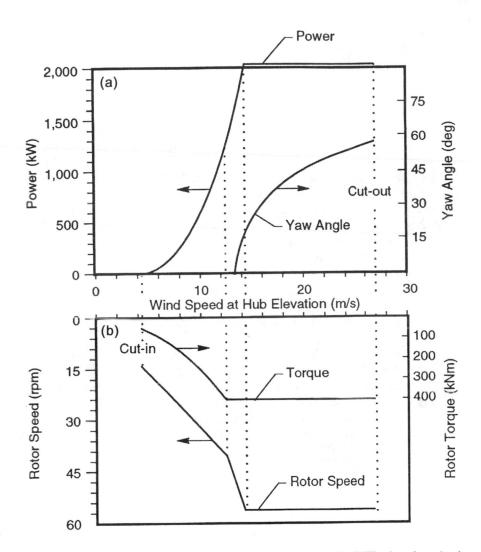

Figure 10-2. Operating map of the fixed-pitch Gamma 60 HAWT, showing the inter-action of broad-range variable speed with yaw control of peak power. (a) Power output and yaw angle *vs.* wind speed   (b) Rotor speed and turbine shaft torque *vs.* wind speed

In most of the wind turbine control concepts currently applied or under study, the prime objective is control of torque or power. The accompanying thrust load is usually not selec-ted as a parameter to be directly controlled. One reason for this is that the limiting of shaft torque can be accomplished with a much slower control rate than is required for limiting thrust. This is especially true if the power train is made torsionally soft, which is becoming a more common design practice. The significance of thrust loads in the design of bearings, towers, and foundations varies substantially between turbines that are stall-, pitch-, or yaw-controlled, and also between constant-speed and variable-speed power trains. It is foreseeable that in the future, the overall control scenario of a fully-optimized wind turbine will give more (and different) attention to the control of thrust loads.

Power Trains and Power Extraction

If the wind turbine rotor is stiffly coupled to an essentially constant-speed electrical generator, the control of fluctuations in torque and power is made difficult, requiring high rates of aerodynamic load control. Furthermore, the action of controlling torque transients will necessarily spill energy that might otherwise be captured. Reducing the stiffness of the power train, either mechanically or by adopting a *constant-torque control* on the generator, permits a great reduction in the required rate of aerodynamic load control. Constant torque control allows the rotor to operate at variable speed, so that the rotor inertia becomes both a source and a sink for energy transients, torque loads can be held within limits, and less energy is spilled.

If a constant electrical torque generator is adopted, the shaft stiffness felt by the turbine rotor is effectively reduced to zero, and the need to control transient aerodynamic torques is eliminated. Any method that reduces the *torsional impedance* of the power train also has the effect of decoupling the structural dynamic behavior of the rotor from that of the power train. This can simplify the simulation of the complete dynamic system and the analysis of internal rotor loads. However, to get these energy and load benefits, power train impedance must be reduced far below that obtained from the *slip* of an induction generator of reasonable efficiency. Therefore, merely substituting an induction generator for a synchronous generator leaves the system with essentially the same dynamic response.

A power train can be softened mechanically by introducing torsional springs with relatively low stiffness, as shown in Figure 10-3. Here the input shaft of the planetary gearbox of the WTS-4 HAWT is connected to the forward end of the turbine shaft and the gearbox case is connected to the bedplate through swing-links and springs. The latter are composed of stacks of *Belleville washers*. When the power train is softened by springs in this way, it becomes very beneficial to provide damping of the lowest-frequency torsional vibration mode wherein the turbine rotor oscillates like a torsional pendulum. The required amount of damping is not effectively available from pitch control when the rotor is operating near its peak of performance, since any change in pitch lowers torque even when an increase in torque is called for. Damping can be readily obtained, however, by adding damping devices to the torsional springs in the power train, as indicated in Figure 10-3.

If low power-train impedance is obtained electrically from generator torque control, analogous damping can also be obtained easily from the same electrical torque control. In that event, neither mechanical softening nor mechanical damping of the power train are necessary or beneficial.

As discussed above, control systems are very interactive with power train impedance characteristics. Design and development progress can only occur when the interaction of the controls with the system response dynamics is rigorously examined and adjusted. To repeat a previous point, there is a fortunate synergism when the soft-system structural design philosophy is applied to the power train. Not only are transient loads reduced and required control rates eased, but there is a decoupling of rotor torsional dynamics from that of the nacelle and its equipment in ways that permit simplification of analysis.

Tower Design

The advantages of the compliant structural design philosophy apply very clearly in tower design. There was an early traditional approach to HAWT tower design which sought to keep the lowest *system natural frequency*, in which the masses of the rotor and

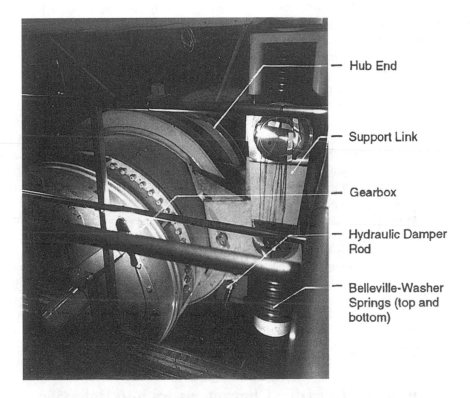

— Hub End

— Support Link

— Gearbox

— Hydraulic Damper Rod

— Belleville-Washer Springs (top and bottom)

Figure 10-3. Torsional springs and dash pots in the power train of the WTS-4 HAWT provide compliance and damping. (*Courtesy of Medicine Bow Energy, Inc.*)

the nacelle oscillate as a flexural or torsional pendulum on the structural spring of the tower, above the blade passage frequency. Arising from a fear of large resonant responses while crossing a tower natural frequency, this approach was supported by the fact that the first hingeless design of the experimental Mod-0 HAWT (Fig. 3-10) experienced resonant 2P torsional oscillations of the nacelle/yaw drive/tower structure, forced by the passage of the two rotor blades through the tower shadow. Blade passage frequency still serves to define the lower bound on the natural frequencies of a stiff structural system, with the exception of the power train.

In two-bladed HAWTs, adoption of the teeter hinge has eliminated the large 2P moment loads formerly transmitted from the rotor to the nacelle. This has enabled the lowest tower modal frequency to be reduced, not only below the blade passage frequency, but also below the rotor speed itself. The latter case is often called a *soft-soft system*. Designing a soft-soft tower becomes primarily a matter of supplying adequate bending strength, letting the stiffness fall as low as it will. In larger turbines this results in huge weight and cost savings. A very successful example of a soft-soft tower design is the WTS-4 HAWT, in which the first tower bending modal frequency is in the vicinity of 0.5P.

When a tower is stiff and a teeter hinge is not used, the analysis of blade bending loads in a two-bladed HAWT becomes extremely complex. Rotor blade models must be coupled to the nacelle and tower models in a manner that involves large cyclic variations in the spring and inertial coupling terms in the *equations of motion.* The necessary analysis can certainly be achieved, but it is of declining interest because the large cyclic loads in such a system prohibit economic success.

The difficulties of analyzing a two-bladed hingeless HAWT on any tower (stiff or soft) and the knowledge that adding a third blade eliminates both the tower shake and the need for a coupled rotor-tower analysis have led to the prevalence of three-bladed rotors. In three-bladed HAWTs much more freedom of tower frequency placement has been evident. Nothing done to adjust tower frequency has any influence on the blade bending and shaft bending loads that occur. The tower sees these loads as large steady loads that it can be designed to carry. Thus, if the fatigue loads on hingeless blades are acceptable and three or more blades are used, the way is opened for use of relatively soft, lightweight towers without the need for a sophisticated dynamic analysis of the coupled rotor and tower.

When the teeter hinge is adopted in the design of a two-bladed HAWT, the most troublesome dynamic coupling of the blades to the nacelle and tower is eliminated. This greatly simplifies the adequate simulation of the resulting system. It is then found that dynamic motions of the rotor are very little influenced by tower stiffness. However, loads are generally benefitted by reducing the stiffness of the tower, both in bending and torsion. It is of course true that some tower modal frequencies may respond excessively to whatever loads may remain at integer multiples of the rotor speed. It has been found that with soft or soft-soft tower design, tower vibration modes that involve substantial upwind-downwind motion will pick up good aerodynamic damping from the rotor blades. Those modes that involve only lateral or vertical motion of the rotor hub will have to be avoided, or suitable hardware provided that is designed to mechanically dampen a specific mode of vibration.

Modeling Dynamic Characteristics and Behavior

Natural Vibration Modes of HAWTs and VAWTs

It is of very great significance that the periodic loads in any subsystem of a wind turbine will depend upon the response of the system as a whole in its *natural* or *free-vibration modes.* These are the patterns of vibration that the system assumes after it has been set in motion but without any external applied loads. Each pattern has a distinctive shape and frequency, termed the *mode shape* and *modal* or *natural frequency*, respectively. It is only when the natural modes of elements such as the blades, power train, and support system have been fully coupled together to derive the system modes that structural dynamic analysis can be successful. Thus, the simulation model of the wind turbine must start with suitably detailed models of the subsystem elements and then combine them in a manner that enables *aeroelastic analysis* (*i.e.* analysis that permits elastic deflections to change angles of attack) of the system as a whole.

The complete wind turbine will, of course, have a very large number of natural modes, each at its own natural frequency. They may differ greatly from the uncoupled mode shapes and natural frequencies of the isolated subsystem elements. Also, the number and type of structurally significant system modes will depend upon how stiffly the subsystem elements are coupled together. As discussed earlier, there are design features that can uncouple subsystem elements, reducing the number of system modes that will present significant responses. There is little value or dependability in modal analysis except that of the fully-coupled system. In a wind turbine designed according to the soft-system philosophy, there can be so much aerodynamic damping and cross-modal damping that only a rigorous aeroelastic analysis of the complete system can be meaningful. This must include a full accounting for the action of the control system.

The accuracy with which a subsystem model need be constructed will depend on the degree to which the subsystem modes persist as significant components of the system modes or act as an influence upon dynamic stability. Often this will be best understood only after a preliminary version of the complete simulation is running and subjected to study. There are, however, some results of dynamic modeling experience that can be cited here.

Rotor Models

Whether a HAWT rotor is hinged or hingeless, a blade simulation model should be adequate if it consists of a beam analysis that is divided into 10 to 15 spanwise segments for both aerodynamic and inertial loads. There should be little need for finer segmentation, because a blade designed for fatigue life under cyclic gravity loading (at a frequency of 1*P*) will have a beam structure that is stiff in-plane, strongest at the hub, and essentially a continuous structure from blade-to-blade through the hub. Design for adequate fatigue life under gravity loading will also result in a high torsional stiffness of the blade and of the pitch control elements in series with the blade. Thus, a 10-segment model will be more than adequate for simulating the torsional deflections of the blade and the resulting changes in blade angle.

Because VAWT blades are not subjected to cyclic gravity loading and tend to have stiffness discontinuities at their attachment fittings, a VAWT blade model requires much greater detail and many more segments than a HAWT blade model. Further complicating the blade responses are the constant cycling of air loadings and the difficulty in supplying damping from the non-rotating part of the VAWT structure. Rotor hub modeling is first a matter of correctly modeling the inter-blade structure and the impedance with which the hub is restrained. Then, to enable correct aeroelastic results, the influence upon blade angle of attack of hub motions and blade-to-hub motions must be accounted for. This means that modeling of the rotor as an isolated subsystem will usually not provide a realistic simulation. This is discussed in detail in the next chapter.

Power Train Models

Constant-Speed Generators

If the rotor is stiffly coupled to a synchronous generator, the resonant modal frequencies will not be importantly affected by the generator model. Moreover, the coupled rotor/power train vibration modes will be poorly damped, making frequency placements relatively critical. If, however, the drive system is made torsionally soft, the lower rotor modal frequencies will be essentially uncoupled and unaffected by drive system modeling details. In such a system, elements that provide damping should be included in the model, since the analyst will want to explore cases in which beneficial damping will reach the lower-frequency rotor modes.

Attempts to provide attenuation of power train rotor loads by varying *excitation control* on a synchronous generator have often been made but have failed. Because the speed of the synchronous generator is very stiffly locked to an electrical grid frequency (60 Hz in the U.S. and 50 Hz in Europe), it acts as a clamped end for the rotor/power train torsional modes, with no significant response to excitation forces. Therefore, there is no useful purpose in including the detailed dynamic features of a synchronous generator in the power-train model for an on-line analysis.

An induction generator with its slip behavior can be readily modeled as a dynamic element. It should be noted again that the apparent softness of an induction generator cannot provide the degree of reduced impedance of the power train that is necessary to attenuate periodic rotor/power train coupled loads.

Variable-Speed Generators

When a variable-speed generator provides the opportunity to control generator *air-gap torque*, it becomes very important to include the mass, stiffness, and damping characteristics of the generator in the power train model. Except for minor inertial effects, such a drive train subsystem presents a zero torsional impedance to the wind turbine rotor. The lowest mode of rotor vibration in such a system will act as though the turbine rotor was totally uncoupled from the power train. However, the mass of the generator rotor becomes free to "ring" in torsion, with the wind turbine rotor acting as a clamped end in the mode shape. Generator air-gap torque damping can and should be included in the model, since this damping can be readily and usefully supplied by the actual generator.

Support Structures

The response dynamics of the coupled tower, yaw drive, nacelle, and drive shaft support structures must be included in the system simulation model. In a VAWT, simulation of the dynamic behavior of the support cables is critical. In a HAWT, it will be generally necessary to construct detailed *finite element models* (FEMs) of all of the structural elements between the rotor shaft and the ground, and to give close attention to nacelle mass distribution.

Aeroelastic Instability

Rotor Aeroelastic Instability

Aeroelastic instability can occur whenever any modal response creates an accompanying periodic aerodynamic force that feeds a buildup of that mode. The induced periodic aerodynamic force, which acts as a *negative damping mechanism*, is created by some form of feedback from the modal motion to the angle of attack of the airfoil. This instability can occur with airfoils unstalled (as with *classical flutter*) or as a result of stall (as with *stall flutter*).

Because the prime mechanism of aeroelastic instability is a cyclic disturbance of blade angles of attack, it is beneficial to retain high torsional stiffness in the blade and all the elements of the pitch control system. It is also beneficial to provide the favorable blade details of *chordwise mass balance* and *shear center location* that will reduce or eliminate torsion deflections from periodic twist loads. In typical HAWTs, however, the robust blade design needed to carry cyclic gravity loadings results in such a high torsional stiffness that the blade need not have full chordwise balance to preclude flutter. The same is true for bending modes within a rotor blade that might also be made unstable by pitch disturbances.

Deflection of control surfaces is a different matter. With either full- or partial-span pitch control (*i.e.*, control through an outboard movable portion of the blade) any loss or inadequacy of control stiffness can enable flutter. Therefore, attention to chordwise balance can be valuable protection against flutter in the event control stiffness deteriorates.

Coupling between teeter motion and blade pitch angle is introduced when the axis of the teeter hinge is skewed from being perpendicular to the blade axis by a so-called δ_3 *angle,* as illustrated in Figure 10-4. As a rotor blade teeters upwind and downwind of the plane of revolution, the δ_3 angle causes the blade to twist on its longitudinal axis and change pitch in proportion to the teeter angle. Thus, a δ_3 angle introduces *cyclic pitch* in a two-bladed rotor, which means that the pitch on one blade increases while the pitch on the second blade decreases. The purpose of having a δ_3 hinge in a wind turbine hub is to restrain the amplitude of teetering by means of an aerodynamic "spring." However, the resulting feedback from teetering to blade pitch creates negative aerodynamic damping of the first unsymmetrical flatwise bending mode. This important mode is superimposed on the rigid-body teeter motion. Under typical design conditions, a δ_3 angle can reduce the aerodynamic damping of this mode from a normal 3 percent to 1 percent of *critical.* If the rotor is to run at an unusually high rotational speed, it may be necessary to reduce or eliminate the δ_3 angle to keep this bending mode from becoming unstable.

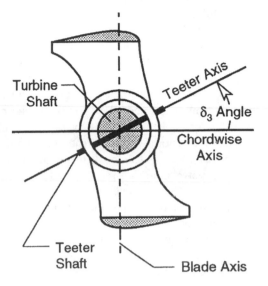

Figure 10-4. The so-called δ_3 angle introduced into a teetered hub to restrain the teeter motion. However, a δ_3 angle also reduces aerodynamic damping.

HAWT Nacelle/Tower Instability

In the absence of corrective control system action, there are aeroelastic instabilities that may be encountered as motions of the nacelle and tower of a HAWT. There are two modes of importance, each essentially driven by a disturbance of the direction the rotor thrust vector relative to the rotor shaft axis.

First, a *whirl mode instability* can occur in which the shaft moves through a conical locus, with the tower deflecting and the hub moving in a nearly circular path. This mode requires approximately equal stiffness of shaft restraint in the vertical and lateral directions before it will become unstable, which is not difficult to avoid in a practical design. Bending coupling between the rotor and the shaft, which is present in hingeless hubs, exacerbates the prospect of a whirl mode instability and warrants careful analytical investigation.

A second vibration mode involving yaw angle oscillations of the turbine shaft is inherently unstable for downwind rotors and stable for upwind rotors. When the center of gravity of the nacelle/rotor combination is not on the tower centerline and the yaw drive is soft or disengaged, there will be cyclic angular motion of the shaft centerline in a horizontal plane whenever there is lateral bending motion. The rotor thrust vector will lag behind the shaft angular motion because of teetering action, resulting in a periodic lateral thrust component that can feed the lateral tower bending. In the upwind-rotor configuration, the result is positive damping of this mode. However, in the downwind configuration the result is destabilizing. Fortunately, the negative damping involved can be of about the same magnitude as the positive damping available trom a typical yaw bearing. Depending on system geometry, however, it may be necessary to provide additional mechanical damping to the yaw drive system of a downwind-rotor HAWT.

Control System Interaction

While control system activity may not be perceived as a factor in aeroelastic stability, control actions can have the same effects as those induced by elastic deflections. Thus, controls can stabilize or destabilize any system dynamic mode which falls within the their frequency range. If the control rate is designed for a response to changes sensed in the rotor torque or speed without regard to thrust effects, it can create the following two adverse results: First, it may create large transient overloads in thrust, because torque and speed changes are very slow relative to the buildup of thrust and blade flatwise bending loads. A control that appears to be sufficiently stable and successful as viewed through torque time histories may be creating or permitting excessive periodic thrust loads.

Secondly, a control that is merely tasked to hold power constant in high winds will tend to negatively damp the first longitudinal mode of tower bending. The good aerodynamic damping that a fixed-pitch rotor would supply to this mode will tend to be overcome by pitch control action. Thus, the control system should be provided with the means to sense the offending tower motion and act to correct it.

Instability of the overall *pitch control loop* can also affect the system like an aeroelastic instability. Unexpectedly high *control gains* can reduce stability. This may arise from neglect of the time lag of induced flow changes behind sharp changes of rotor aerodynamic loads. A fast-acting pitch control should be considered suspect in any situation where sustained periodic thrust loads are observed.

The above is a general description of the instabilities explored in simulations during the development of large-scale HAWTs. No instabilities within the rotor blades or in shaft whirl have been encountered in practice, primarily because of high blade and control stiffness.

Typical Dynamic Responses

The following discussion briefly covers the expected responses of a wind turbine system to various sources of excitation and the parameters that influence the magnitude of these responses.

Aeroelastic Effects

Aeroelastic effects that are usually encountered in wind turbines will be favorable. In a teetered HAWT rotor, for example, both the rigid-body teeter motion and first flatwise bending mode of the blades are highly-damped aerodynamically. Thrust load excitations can create blade and tower deflections rather than stresses. Transient wind loadings involve upwind-downwind velocities of the structure and accompanying aerodynamic damping. Generally, when the control system of a HAWT is strong enough to handle gravity and flatwise bending loads, blade torsional stiffness will be high enough to prevent flutter and divergence.

As discussed previously, a δ_3 angle that is too large can destabilize the first unsymmetrical flatwise bending mode in a high-speed rotor, and a control system which only controls power can negatively damp the first tower bending mode in high winds.

Response to Wind Shear

Wind shear loadings, both vertical and horizontal, can impact the design of blades when the hub is hingeless. However, the use of a teetered hub reduces wind shear effects to an insignificant level. The teeter angle amplitudes caused by wind shear are usually only a few degrees, and present no design problems such as lack of clearance between the blade and the tower. On a stall-controlled rotor with a hingeless hub, wind shear in high winds can be an aggravation when one blade is stalled and the other is not. The seriousness of this problem is reduced considerably if the stall-controlled rotor is also teetered.

Response to a Tower Shadow

Tower shadow (*i.e.* the wake behind the tower structure in which wind speed is reduced and turbulence is increased) is a particularly severe excitation source for a downwind HAWT rotor. In a VAWT, the excitation of the downwind blade is much less severe because of the central column is usually much smaller in diameter than a HAWT shell tower and the column-to-blade distance at the rotor equator is a relatively large number of column diameters. Like wind shear, tower shadow responses are high in a hingeless HAWT rotor, but are not design drivers when the rotor is teetered. However, tower shadow loads may still design the outboard shell structure of the blade airfoil, even with teetering. As discussed in Chapter 7, the sudden change in air loading on a blade as it enters and leaves the tower shadow can create an acoustic noise problem that is severe enough to warrant abandonment of the downwind rotor location entirely.

Rotor blades operating upwind of a cylindrical tower experience a *bow wave* effect not unlike a tower shadow, but much less severe. *Potential flow theory* is adequate for simulating the wind speed disturbance ahead of the tower.

Response to Gusts

Once a turbine design has been optimized to the point of reducing or eliminating large dynamic responses such as those from tower shadow, gust responses become the design drivers. If a control system is designed to "clip" the effects of a gust on peak power, it will at best sacrifice energy capture and may worsen transient thrust loads. It is much more desirable to avoid a control response to a gust and substitute a rotor speed response instead. If this is done, the largest gust load responses will still occur, but there will be many fewer of them in the fatigue load spectrum. It should not be a significant burden to design for long fatigue life under the action of gusts, if all the opportunities for passive attenuation of the machine's response to air load transients are grasped.

In accordance with Equation (5-13), a wind turbine rotor slows the wind as it extracts power, by a fractional amount called the *axial induction factor*. At the turbine design point, the wind may be retarded by one-third of its free-stream velocity, and at higher wind speeds the induction factor may approach unity (see Fig. 5-11). Retardation by the rotor, however, takes time to develop and as such should be viewed as a *quasi-steady state phenomenon* that is appropriate for predicting power performance. For a transient control or load simulation, however, retardation is usually neglected on the theory that there is insufficient time for it to develop during the period of the gust. The blades are assumed to feel the full, *unretarded wind speed transient.* Failure to eliminate retardation effects on wind transients in a dynamic simulation can be a large factor in underestimating gust loads.

Response to Yawed Flow and Nacelle Yaw Motion

This response is the most distinguishing difference between a hingeless and a teetered HAWT rotor. In the hingeless system, yawed flow and yaw motion are sources of relatively large cyclic loads. On the other hand, they create no design-driving loads in a teetered system, even at yaw angles and yawing rates not usually believed feasible.

Response to Control Inputs

The rapid application of pitch angle control can be lethal to the fatigue life of a wind turbine. Control systems must be designed to assure limited blade pitching rates. Differences in lag times between the thrust and torque responses to a wind transient must be carefully considered in the control design, to avoid a damaging increase in thrust loads.

Response to Loss of Electrical Load

The initial response to a loss of electrical load is benign, without any significant dynamic loads. However, the rise in rotor speed will be rapid and require control actions to prevent overspeed. These actions can result in a thrust reversal that is a very damaging cycle in the fatigue load spectrum. Thus, the control system response to loss of electrical load must be carefully shaped to assure a safe shutdown without structural damage.

Response to Gravity

Gravity applies large cyclic *lead-lag* or *in-plane* bending moments to a rotating HAWT blade that are its primary design-driving loads. The strength required for adequate fatigue life under this 1P excitation results in a high blade stiffness and a high lead-lag (in-plane) natural frequency. As a result, the gravity bending response is not amplified significantly, nor can anything be done to attenuate it. A hand-calculation of the gravity-induced dynamic response is adequate for preliminary design. In a coned rotor, components of the cyclic gravity load can enter the control linkage, and other components can act on the rotor shaft as a 2P load unless the rotor is teetered.

Response to Rotor Speed Fluctuations

Fluctuation in rotor speed is desirable as both a source and a sink for the energy in wind gusts. While a torsionally-soft power train permits some fluctuation, a constant-torque generator is the design feature that takes full advantage of this energy benefit. Power quality, in terms of steadiness of the power output of the generator, can be better with rotor speed fluctuations than with pitch control and a constant rotor speed. Energy capture is much enhanced when rotor speed is free to follow the wind.

Response to Resonance Crossings

The dynamic response of the wind turbine system when a natural frequency coincides with an integer multiple of the rotor speed can vary from "critical" to "no problem," depending on the design philosophy employed, the mode in question, and the damping features of that mode. Potential resonances are displayed graphically by means of a *Campbell diagram*, which is discussed in detail in Chapter 11. In a variable-speed HAWT with a broad speed range there will be one or more poorly-damped "tower" modes *that are in resonance within the operating regime.* In such cases the control system can be programmed to "step over" the resonant rotor speeds [DiValentin 1985].

Prediction of Design Loads

It is most useful to determine and evaluate blade and hub loads as time histories, as they would be seen by an observer in a *rotating coordinate system* attached to a HAWT or VAWT rotor. For the nacelle, tower, and cable structures, time histories in a *fixed coordinate system* attached to the ground are easiest to comprehend and apply to design. For the power train, either a rotating or a fixed coordinate system (or both systems) can be suitable. A simulation which analyzes the system in the *time domain* will make the needed computations involved in coupling the rotating and fixed systems. By suitable post-processing of the simulation output, the loads transferred between the rotating and fixed systems (*e.g.*, at the apex of a coned HAWT rotor) can thus be obtained as time histories in both the rotating and fixed coordinates. Generally, the rotor designer will think and work in the rotating system, and others will think in the fixed systems as they perform analyses of internal loads, stresses, and fatigue lives.

The prime method of load computation should be to run steady-state or transient response cases on the system dynamic simulation and then adapt the time-history output to obtain *maximum, minimum, mean* (*i.e.* (maximum + minimum)/2), and *cyclic* (*i.e.* (maximum - minimum)/2) loads that occur. Included must be the essential aspects of control action and all the elements of aeroelastic behavior that could influence loads. If there are potential stability problems, these will show up as poor damping during transients, which can usually be corrected without resorting to the added complexity of a *frequency-domain analysis.* The latter is usually of little value in a loads analysis.

Mean Operating Loads

Simulation runs at steady wind speeds (with rotor-induced retardation effects) give ready access to the mean operating loads upon which cyclic loads can later by superimposed for fatigue life prediction. It should be noted that mean load analysis is required in the simulation model because of the complex counterbalancing of centrifugal and thrust loads.

Limit Loads

Establishing design limit loads is a matter of identifying sources of *one-time* or *infrequent events* that might do structural or mechanical damage to the system. Major gusts (without retardation effects), dropped electrical load events, and control failure events can be examined in the system simulation to identify which cause limit loads. Often, the resulting limit load will not cause damage in a structure that is designed for long fatigue life (*i.e.* fatigue, not limit loading, is the *design driver*). Moreover, if damage is indicated from a one-time event, steps can often be taken to reduce or control the limit load so that it does not remain the design driver.

Fatigue Loads

As discussed earlier, the sources of design-driving cyclic loads are longer-term changes in wind speed, large-scale temporal turbulence, small-scale spatial turbulence (within the swept area of the rotor), gravity, wind shear, and tower shadow. In theory, all of these can be readily modeled for input to the simulation. Determinate cyclic loads (like those from gravity, for example) will be readily and accurately obtained from the simulation. Also, the response to any discrete gust can be examined and steps can be taken to attenuate it. Unfortunately, the magnitude and number of turbulence cycles, whether large or small in scale, is not well known. Wind turbine designers are therefore forced to make very conservative (*i.e.* pessimistic) assumptions as to gust spectra, and then input these into the simulation in order to obtain cyclic loads for fatigue life prediction. Admittedly, this will result in an over-strength system, but the goal of achieving a design fatigue life that meets specifications will have been realized.

Fortunately, a compliant and well-damped system will respond very little to the higher-frequency components in a wind velocity spectrum [Stoltze 1986]. As a result, only those temporal gusts that envelop the whole rotor need be given major attention. A major use of the system simulator will be to evaluate the effectiveness of design features such as teeter hinges and low-stiffness elements in attenuating responses to these large-scale gusts.

Small-scale spatial turbulence is seen by a rotating blade as a quasi-steady, periodic series of gusts with frequencies at integer multiples of the rotor speed and amplitudes that are inversely proportional to frequency (see Eqns. 2-35). Placing the blade flatwise frequencies away from even multiples of the rotor speed in a teetered HAWT rotor (and all multiples in a hingeless rotor) can avoid a resonant response to these excitations. Without resonance, the amplitudes of small-scale spatial turbulence are too small to be of design significance. However, in actual practice, resonant frequencies can usually be crossed without a problem, because there is sufficient damping present.

Blade Internal and External Pressures

If a hollow blade is not suitably compartmented and vented, internal air pressure or suction caused by centrifugal force can exert design-driving loads on the shell structure. Typically, the *afterbody* of the airfoil (*i.e.*, the triangular section between the structural spar and the trailing edge) with its typically lightweight construction and small curvatures can be badly distorted with attendant material stresses that are excessive. For structural integrity, pressure-induced stresses must be calculated and, if necessary, reduced to acceptable levels by use of compartment walls and venting.

Optimum Turbine Size

As wind turbine design, development, and operation progress in machines of a wide range of diameters, it will become clearer whether or not there is a diameter that is in any way "optimum." Land utilization costs have favored larger size, while transportation and installation costs have favored smaller size. Meanwhile, design concepts and simulations of system dynamic behavior have not been pushed to the degree of perfection that would enable one to predict a rotor diameter at which economic performance may eventually become optimized. What is clear is that the trend in the sizes of wind turbines installed in wind power stations has been and continues to be steadily upward.

In searching for an optimum or desirable rotor diameter, it seems that one should explore the effects of size upon *prototype* turbine systems that are in other respects most promising. Such wind turbines would be those with high energy-capture performance and low fatigue loading. Under these criteria, it can be informative to look at size effects in a two-bladed HAWT with a teetered rotor, which embodies the soft-system design philosophy. This baseline design would be presumed to operate with fixed pitch over a broad range of speeds and with a yaw drive for limiting aerodynamic loads. As guidance in the exploration of the effects of size on the economic performance of this or any other prototype design, there are some basic principles of scaling which apply to wind turbines, and these are discussed in the following section.

Dimensional Scale Factors

Assuming that the rotor of the scaled wind turbine (scaled up or down) is to operate at the same mean lift coefficient as its known predecessor, it can be expected that its loads will follow all the basic aerodynamic and inertial scaling laws. Tip speeds will usually remain constant for performance and acoustic noise reasons. If all dimensions of the prototype are scaled in proportion to the diameter ratio D_S/D_P, where D_S is the scaled diameter and D_P is the prototype diameter, other structural parameters would scale as indicated in Table 10-1, to a first-order approximation.

Table 10-1. First-Order Scaling Relationships for HAWTs

Parameter	Proportionality[1]
Rotor speed	$\propto (D_S/D_P)^1$
Blade centrifugal stress	$\propto (D_S/D_P)^0$
Blade gravity-induced stress	$\propto (D_S/D_P)^1$
Blade centrifugal force	$\propto (D_S/D_P)^2$
Rotor power	$\propto (D_S/D_P)^2$
Rotor thrust	$\propto (D_S/D_P)^2$
Rotor torque	$\propto (D_S/D_P)^3$
System weight	$\propto (D_S/D_P)^3$

[1] D_S = scaled rotor diameter; D_P = prototype rotor diameter

Use of Scaled Load Data from a Prototype Wind Turbine

In actual design practice it is found that the scaling of the parameters listed above will vary somewhat from their first-order approximations. Weights do not grow as rapidly as the cube law, but are proportional to about $D^{2.4}$, because local departures from proportional dimensioning are adopted. For example, a minor increase in the relative size of a blade root will defeat the first-order rise of gravity-induced stress that would otherwise come with an increase in diameter. Power and thrust, on the other hand, will change more rapidly than indicated by the square law, because of the effects of tower height and wind shear. If due consideration is given to the influence of scale on the frequency and amplitude of gusts enveloping the rotor, it can also be expected that unsteady air loads will scale in proportion to the steady air loads.

When the structural *margins of safety* in the prototype wind turbine are known to be positive (*i.e.* prototype design stresses are less than allowable stresses), the machine can be scaled up or down moderately with little risk of miscalculation. This does not mean that mere geometric scaling will be sufficient to insure that all margins of safety in the scaled design will also be positive, but it does mean that the scaling influences upon loads and stresses can be readily accounted for. Local design changes can then be made where necessary without changing the overall prediction of the economic performance of the scaled wind turbine system.

References

DiValentin, E., 1985, *Variable Frequency Modal Response Testing of the WTS-4*, Hamilton Standard Report HSER 9577, Denver, Colorado: U.S. Bureau of Reclamation.

DiValentin, E., H. Healy, J. M. Kos, and C. L. Stoltze, 1986, *An Investigation of the Feasibility of Converting the WTS-4 to Fixed Pitch, Broad Range Variable Speed Operation*, Hamilton Standard Report HSER 10804, Denver, Colorado: U.S. Bureau of Reclamation.

Drees, J. M., Spring 1977, "Blade Twist, Droop Snoot, and Forward Spars," *Wind Technology Journal*, 1:1, pp. 10-16; see also "Speculations about the Origin of Sails for Horizontal Axis Turbines," *Wind Technology Journal*, 1984, 2:1/2, pp. 13-31.

Hohenemser, K. H., and A. H. P. Swift, 1981, "Atmospheric Testing of a Two-Bladed Furl-Controlled Wind Turbine with Passive Cyclic Pitch Variation," *Proceedings, Fifth Biennial Wind energy Conference and Workshop*, Vol. I, SERI/CP-635-1340, Golden, Colorado: National Renewable Energy Laboratory, pp. 457-467.

Stoltze, C. L., 1985, *Teeter Response of a Large-Scale, Two-Bladed HAWT Rotor in Rapid Yaw Maneuvers*, Hamilton Standard Report HSER 10224, Denver, Colorado: U.S. Bureau of Reclamation.

Stoltze, C. L., 1986, *An Evaluation of Turbulence-Induced Fatigue Loads Sustained by a Large-Scale, Compliant HAWT Rotor*, Hamilton Standard Report HSER 10805, Denver, Colorado: U.S. Bureau of Reclamation.

11

Structural Dynamic Behavior
of Wind Turbines

Robert W. Thresher, Ph.D.
National Renewable Energy Laboratory
Golden, Colorado

Louis P. Mirandy, Ph.D.
The General Electric Company
King of Prussia, Pennsylvania

Thomas G. Carne, Ph.D.   Donald W. Lobitz, Ph.D.
Sandia National Laboratories
Albuquerque, New Mexico

and

George H. James III, Ph.D.
NASA Johnson Space Center
Houston, Texas

Introduction

The structural dynamicist's areas of responsibility require interaction with most other members of the wind turbine project team. These responsibilities are to *predict structural loads and deflections* that will occur over the lifetime of the machine, *ensure favorable dynamic responses* through appropriate design and operational procedures, *evaluate potential design improvements* for their impact on dynamic loads and stability, and *correlate load and control test data* with design predictions. Load prediction has been a major concern in wind turbine designs to date, and it is perhaps the single most important task faced by the structural dynamics engineer. However, even if we were able to predict all loads perfectly, this in itself would not lead to an economic system. Reduction of dynamic loads, not merely a "design to loads" policy, is required to achieve a cost-effective design.

The two processes of load prediction and structural design are highly interactive: loads and deflections must be known before designers and stress analysts can perform structural sizing, which in turn influences the loads through changes in stiffness and mass. Structural design identifies "hot spots" (local areas of high stress) that would benefit most from dynamic load alleviation. Convergence of this cycle leads to a turbine structure that is neither under-designed (which may result in structural failure), nor over-designed (which will lead to excessive weight and cost).

This chapter introduces some of the physical principles and basic analytical tools needed for the structural dynamic analysis of a horizontal-axis or vertical-axis wind turbine (HAWT or VAWT). This is done through discussions of two subjects that are fundamental building blocks of our understanding of the structural-dynamic behavior of wind turbines:

-- **single-degree-of-freedom dynamic load model of a HAWT blade,** following the development of the *FLAP* computer code at the National Renewable Energy Laboratory [Thresher *et al.* 1986, Wright *et al.* 1988]
-- **theoretical and experimental analysis of the vibration modes of a VAWT system,** following methods developed at the Sandia National Laboratories [Carne *et al.* 1982]

These discussions are written for engineers familiar with structural mechanics, but no specific knowledge of wind turbine rotors is required. Other technical disciplines with which the wind turbine structural dynamicist should become familiar are *aerodynamics* (required to determine wind forces and *aeroelastic* effects), *rotor dynamics* (with its own special set of accelerations and responses), *statistics* (required in dealing with wind turbulence and fatigue life prediction), *mathematical modeling*, and *testing*.

Role of Structural Dynamics in the Overall System Design

The design cycle in a typical wind turbine project is usually divided into *concept definition* and *detail design* stages. During the concept definition period trade studies are conducted to determine the overall configuration. The structural dynamics engineer is concerned most with the relative load levels and the operational and technological risks associated with the various design alternatives. Initially, only the feasibility of each concept is judged, with little time spent on refinements. For example, aeroelastic stability might be addressed only for unconventional concepts where experience is lacking.

In the detail design stage the structural dynamicist must furnish the design loads and determine design and operational requirements that insure acceptable dynamic behavior. These requirements might include drive train damping and spring rates, balance tolerances, stiffness ranges for bearings, maximum operating wind speed, allowable yaw misalignment, yaw rate limits, *etc.*, all based on the results of analysis using various mathematical models.

The key objectives that the design team seeks to achieve by relying on the guidance and skill of the structural dynamicist include the following:

-- to select a configuration and design approach which alleviate dynamic loads
-- to minimize the sensitivity of dynamic responses to inevitable differences between the "as-designed" and "as-built" physical properties of the structure
-- to place natural frequencies favorably with respect to turbine operating speeds
-- to insure aeroelastic stability, without which safe operation is impossible

Accomplishment of the first of these objectives must be based on the team's collective understanding of the dynamic behavior of various wind turbine configurations. It is guided by past experience and frequently involves trade-offs with factors outside the realm of structural dynamics (*e.g.*, cost, manufacturing, and energy capture). The remaining key objectives are accomplished by straightforward (though sometimes complex) mathematical modelling of the selected configuration. Of course, the likelihood of a good detail design solution should be evident at the completion of the concept definition stage.

Wind Turbine Substructures and Subsystems

General background material on wind turbine nomenclature, configurations, and major substructures may be found in Chapters 2 and 4. It is its rotor which sets a wind turbine apart from other structural systems. Therefore, the thrust of this chapter is the understanding and analysis of wind turbine rotors. The structural dynamic behavior of the power train and support structures of a wind turbine can usually be analyzed by well-established methods, so those subsystems will not be discussed here.

Mathematical Models

While certain analyses can be accomplished with simple formulas, experience has shown that the analyst will need separate, specialized computer models to determine the following:

-- *system structural modes and natural frequencies* which incorporate rotational effects
-- *system loads*, both steady-state and transient, incorporating all important degrees of freedom; includes deflections, vibrations, and power output quality
-- *aeroelastic stability*, including the effects of structure/control system interaction

In theory, one comprehensive computer code could be used for all three of these tasks, but it might not be as efficient to develop or to run as separate specialized codes.

Modal Analysis Models

Modal analysis is the determination of the set of discrete patterns or *mode shapes*, each with at its own *modal frequency* and *modal damping*, which a vibrating structure describes. These patterns are also known as *natural modes*. Modal analysis of a wind turbine is generally based on well-developed *finite element* procedures (*e.g. NASTRAN*). Each substructure in the turbine system is modeled separately to determine its own mode shapes and frequencies. With special care, substructures can be coupled together to produce system modes. Modal models can be derived directly or spawned from more-detailed stress analysis models using standard *modal reduction* techniques. For rotors, centrifugal stiffening and gyroscopic effects can be incorporated directly into the finite element code, or they can be computed externally and applied later.

System Loads Model

The system loads code should be tailored to suit a specific wind turbine configuration, since a code that attempts to analyze all possible wind turbine types can become unwieldy. For example, the important dynamic loads usually result from motion in only a few dominant modes that are characteristic of a particular configuration, so the degrees of freedom used in the system loads model are generally a selected subset of the natural modes computed with the finite element models described above. While it is possible to use all of the finite-element degrees of freedom in the loads model, a modal approach is commonly used for analyzing system loads.

Direct numerical integration of the forced equations of motion in the time domain (*i.e.*, output is load *vs.* time) is the most straightforward and informative method of solution. This procedure, which is often referred to as *system simulation*, handles both steady-state and transient response. Since the governing equations of motion contain time-varying coefficients and are generally nonlinear, frequency domain techniques have limited value. Furthermore, seeking closed-form analytical solutions to these complex equations is usually unnecessary with the capabilities of today's digital computers. The engineer's time is best spent in understanding the results of a system simulation and then exploring methods for reducing dynamic loads.

Aeroelastic Stability Models

Aeroelastic stability analysis is often separate and distinct from the modal and loads analyses for two reasons: First, it is preferable to linearize the equations of motion and examine the resulting *eigenvalues* for stability, rather than to examine a myriad of time histories to determine if the response is stable. Of course, if the mechanism of instability is highly nonlinear, simulation in the time domain may be the only recourse. Most potential wind turbine instabilities can be adequately assessed using linearized models.

The second reason for a separate model is that the modes which are important for aeroelastic stability often differ from those needed for loads analysis. In a successful system, instabilities will occur outside the limits of planned operation. Hence, a blade torsional mode which incites instability at a high rotor speed may well be dormant during normal operation and, hence, be omitted in the loads analysis. It should be noted that aeroelastic stability analysis, like modal analysis, can be carried out with a finite-element model.

Dynamic Load Model of a HAWT Blade

A beam model of a wind turbine blade is generally suitable for structural-dynamic analysis. It will differ from the small-deflection theory beam models used in conventional analyses of non-rotating structures, however, because the axial loads on the blade significantly effect the lateral and torsional deflections. In this respect it is more like the beam-column representations used in elastic stability analysis. General three-dimensional theory is quite complicated and has provided fodder for more than one doctoral thesis. It is still subject to controversy as to which terms are important and which are not. Rather than beginning with the general case, we will develop the equations for uncoupled *flapwise* (*i.e.* out-of-plane or flatwise) motions to acquaint the reader with the physics involved. In many cases, these will be suitable for the analysis of HAWT rotor loads.

Elastic blade flapping equations appear in many sources and are in a sense, classical. A relatively brief derivation is given here to highlight the approximations and assumptions that are implicit in the equations which are commonly used. We begin by assuming that the strains everywhere in the blade are small (less than 10 percent) and below the elastic limit. Other assumptions will be introduced at appropriate points in the derivation.[1] Equations of motion for wind turbine components are generally derived using a "strength of materials" rather than a "theory of elasticity" approach.

---

[1] Note here that the assumptions of small strain and elasticity do not mean that the deflections must be small. Everyone is aware that a steel ruler can be bent into an arc with deflections much greater than its cross-sectional dimensions, and yet it will spring back elastically to its original shape.

Coordinate System Definitions

Figure 11-1 shows the orientation of the HAWT blade under analysis with all the intermediate coordinates required to represent the blade motion. The X-Y-Z coordinates are the fixed reference system. The mean wind velocity at the hub elevation, V_{hub}, and its fluctuating components -- δV_x, δV_y, δV_z -- are given in this system. The rotor axis may be tilted at a fixed angle χ. It may also have a prescribed, time-dependent yawing motion given by the yaw angle $\phi(t)$. The yaw axis is coincident with the Z-coordinate axis. The center of the rotor hub, H, is located a distance a from the yaw axis. The blade may be rigid from the axis outward to a distance h, which may also be interpreted as the outer radius of the hub. The blade airfoil shape may begin at h or at a station further out along the blade z axis. The blade may be coned at a prescribed angle β_0, as shown in the figure.

The x,y,z coordinates are located in the surface of revolution that a rigid blade would trace in space, with the y-axis normal to this surface. The x_p-y_p-z_p system are the blade's principal bending coordinates, where the z_p-axis is coincident with the elastic axis of the undeformed blade. Bending takes place about the x_p-axis. It is further assumed that the blade principal axes of area inertia remain parallel along the z_p-axis. Any influence of blade twist on bending displacements is neglected. As a reference for computations, the angle 0_p is set equal to the orientation of the principal bending plane relative to the cone of rotation (*i.e.*, the x-y-z system) at a selected station of primary interest along the blade span. The third coordinate system illustrated in Figure 11-1 is the η-ζ-ξ system, which is on the principal axes of the deformed blade along each point along the elastic axis.

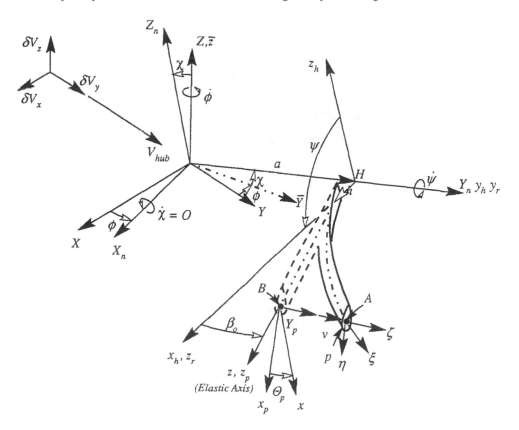

Figure 11-1.   Fixed, rotating, and deformed-blade coordinate systems.   Positive displacements and rotations are shown. [Wright *et al.* 1988]

The coordinate transformation that takes vector components from the fixed X-Y-Z system to the rotating x_p-y_p-z_p coordinates of the undeformed blade is as follows:

$$\begin{bmatrix} x_p \\ y_p \\ z_p \end{bmatrix} = \begin{bmatrix} a_{11} & a_{12} & a_{13} \\ a_{21} & a_{22} & a_{23} \\ a_{31} & a_{32} & a_{33} \end{bmatrix} = \begin{bmatrix} X \\ Y \\ Z \end{bmatrix} \tag{11-1}$$

where

$$a_{11} = c\theta_p\,c\psi + \beta_0 s\theta_p\,s\psi + \phi s\theta_p$$

$$a_{12} = c\theta_p\,c\psi\phi - s\theta_p + \chi c\theta_p\,s\psi$$

$$a_{13} = -\chi s\theta_p - c\theta_p s\psi + \beta_0 s\theta_p\,c\psi$$

$$a_{21} = s\theta_p c\psi - \beta_0 c\theta_p s\psi - \phi c\theta_p$$

$$a_{22} = s\theta_p c\psi\phi + c\theta_p + \chi s\theta_p s\psi$$

$$a_{23} = \chi c\theta_p - s\theta_p s\psi - \beta_0 c\theta_p c\psi$$

$$a_{31} = s\,\psi$$

$$a_{32} = \phi s\psi + \beta_0 - \chi c\psi$$

$$a_{31} = c\psi$$

$$c\theta_p = \cos\theta_p$$

$$c\psi = \cos\psi$$

$$s\theta_p = \sin\theta_p$$

$$s\psi = \sin\psi$$

This transformation has been linearized by assuming a small yaw angle ϕ, a small tilt angle χ, and a small coning angle β_0. The angle of the principal axis, θ_p, may be large. Of course, the blade azimuth angle, ψ, can vary from 0 to 2π radians. The inverse transformation taking the x_p-y_p-z_p components into the X-Y-Z system is given by the transpose of the 3x3 matrix in Equation (11-1). The blade position for $\psi = 0$ is straight up.

Assuming small deformations, the transformation for obtaining η-ζ-ξ components is

$$\begin{bmatrix} \eta \\ \zeta \\ \xi \end{bmatrix} = \begin{bmatrix} 1 & 0 & 0 \\ 0 & 1 & -v' \\ 0 & v' & 1 \end{bmatrix} \begin{bmatrix} x_p \\ y_p \\ z_p \end{bmatrix} \tag{11-2}$$

The necessary intermediate coordinate transformations for components in the x-y-z, x_r-y_r-z_r, and x_h-y_h-z_h coordinate systems shown in Figure 11-1 can be obtained as follows:

-- set $\theta_p = 0$ for transforming to the x-y-z system
-- set $\theta_p = \beta_0 = 0$ for transforming to the x_r-y_r-z_r system
-- set $\theta_p = \beta_0 = \psi = 0$ for transforming to the x_h-y_h-z_h system

Moment-Curvature Relationship

The HAWT blade is assumed to be a long, slender beam so that the normal strength-of-materials assumptions concerning the bending deformations are valid. Figure 11-2 shows an infinitesimal element of the deformed blade. It is assumed that the blade bends only about its weakest principal axis of inertia, which is about the x_p-axis in the figure. No other deformations are considered in this analysis. The strength-of-materials model for elastic bending assumes a one-dimensional form for Hooke's Law that neglects all stresses except the longitudinal (in this case *spanwise*) bending stress. It results in the following differential equation relating bending moment to curvature in our blade:

$$M_{x,p} = -EI_{x,p}\frac{d^2v}{dz_p^2} - EI_{x,p}v'' \tag{11-3}$$

where

$M_{x,p}$ = bending moment about the blade principal axis of inertia, x_p (N-m)
$[\ ]' = d[\ ]/dz_p$, etc.
E = modulus of elasticity (N/m²)
$I_{x,p}$ = area moment of inertia of the airfoil section about the x_p-axis (m⁴)
v = displacement in the y_p (flap) direction (m)

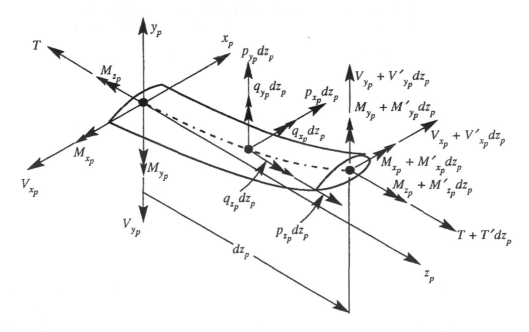

Figure 11-2. Deformed blade element showing forces and moments, all acting in a positive sense. [Wright *et al.* 1988]

Equilibrium Equations

The equations of equilibrium are derived by summing forces and moments in the three coordinate-axis directions and equating the sums to zero. Referring to Figure 11-2, summing of forces and moments gives the following six equations:

$$V'_{x,p} + p_{x,p} = 0 \tag{11-4a}$$

$$V'_{y,p} + p_{y,p} = 0 \tag{11-4b}$$

$$T' + p_{z,p} = 0 \tag{11-4c}$$

$$M'_{x,p} - V_{y,p} + Tv' + q_{x,p} = 0 \tag{11-5a}$$

$$M'_{y,p} + V_{x,p} + q_{y,p} = 0 \tag{11-5b}$$

$$M'_{z,p} - V_{x,p}\, v' + q_{z,p} = 0 \tag{11-5c}$$

where

$p_{x,p}, p_{y,p}, p_{z,p}$ = applied (external) force loading per unit length (N/m)
$q_{x,p}, q_{y,p}, q_{z,p}$ = applied (external) moment loading per unit length (N-m/m)
$M_{x,p}, M_{y,p}, M_{z,p}$ = internal bending moment loads (N-m)
$V_{x,p}, V_{y,p}, T$ = internal force loads (N)

The terms *load* and *loading* are often used interchangeably in the literature, so some clarification of their use in this chapter (and generally in this book) is in order. We shall use the term *loading* to describe an external action applied onto the structure. Thus, *p* and *q* are the sum of all aerodynamic pressures and inertial body forces and are here called *loadings*. The internal moment and force responses of the structure to these loadings -- *M*, *V*, and *T* -- are termed *loads*. It has become the general practice in the structural analysis of wind turbines that if the term "load" stands alone it refers to internal moment and force responses.

Differentiation twice of Equation (11-3) and once each of Equations (11-5a) and (11-5b) allows these three equations to be combined with each other and with Equations (11-4a) and (11-4b) to eliminate $M_{x,p}$, $V_{x,p}$, and $V_{y,p}$. In addition, Equations (11-5b) and (11-5c) can be combined to eliminate the shear force load $V_{x,p}$. These eliminations reduce the previous seven equations to the following four combined moment-curvature-equilibrium equations in the four unknowns v, $M_{x,p}$, $M_{y,p}$, and T:

$$\textit{Flapwise Bending:} \quad (-v''EI_{y,p})'' + (Tv')' + q'_{x,p} + p_{y,p} = 0 \tag{11-6a}$$

$$\textit{Edgewise Bending:} \quad M''_{y,p} + q'_{y,p} - p_{x,p} = 0 \tag{11-6b}$$

$$\textit{Pitchwise Torsion:} \quad M'_{z,p} + M'_{y,p}v' + q_{y,p}v' + q_{z,p} = 0 \tag{11-6c}$$

$$\textit{Spanwise Tension:} \quad T' + p_{z,p} = 0 \tag{11-6d}$$

Wind Speed Models

Wind Shear Gradient

The vertical gradient in steady wind speed with elevation above and below the elevation of the rotor axis is often modeled by a simple power law (see Chapters 2 and 8) as

$$U(z) = U_H(z/H)^m \tag{11-7}$$

where

$$
\begin{aligned}
U(z) &= \text{free-stream wind speed at elevation } z \text{ (m/s)} \\
U_H &= \text{free-stream wind speed at hub elevation (m/s)} \\
z &= \text{elevation above ground level (m)} \\
H &= \text{elevation of the rotor hub (m)} \\
m &= \text{empirical wind shear exponent (same as } \alpha \text{ in Eqs. (2-4 and 8-11))}
\end{aligned}
$$

The wind speed gradient may be described in polar coordinates centered at the hub elevation by a binomial series, as follows:

$$U(z) = U_H \left(\frac{r \cos \psi + H}{H} \right)^m = U_H[1 + W_S(r, \psi)] \tag{11-8}$$

$$W_S(r, \psi) \simeq m \left(\frac{r}{H} \right) \cos \psi + \frac{m(m-1)}{2} \left(\frac{r}{H} \right)^2 \cos^2 \psi \tag{11-9}$$

where

$$
\begin{aligned}
r &= \text{radial distance from the rotor axis (m)} \\
W_s &= \text{wind shear shape function}
\end{aligned}
$$

In this series expression, powers of r/H higher than 2 have been neglected.

Tower Shadow Deficit

The presence of the tower causes a deficit in the wind speed, with the downwind reduction often referred to as *tower shadow* and the much-smaller upwind reduction as a *bow wave* effect. Here we shall refer to both as tower shadow. Neglecting any effects of retardation by the rotor, we will assume that the deficit has a spatial distribution of the form

$$\delta V_T(z, \psi) = -W_T(\psi)U(z)$$

$$W_T(\psi) = \begin{bmatrix} t_0 + t_p \cos[p(\psi - \pi)] \\ \quad for \ \pi - \psi_0 \le \psi \le \pi + \psi_0 \\ \\ 0 \quad elsewhere \end{bmatrix} \tag{11-10}$$

$$p = k\pi / \psi_0 \tag{11-11}$$

where

δV_T = wind speed deficit caused by the tower (m/s)
W_T = tower shadow shape function
t_o, t_p = empirical scale constants
ψ_0 = half-angle of tower shadow sector (rad)
k = number of waves in the tangential profile of the shadow

The parameters t_o, t_p, ψ_0, and k are selected to give the desired approximation for the tangential profile of wind speed in a pie-shaped region with a central angle of $2\psi_0$, centered on a blade azimuth angle of π. For example, the shadow of a shell tower is often modeled as a $\sin^2$ function using the following parameter values:

$t_0 = t_p$ = one-half the maximum ratio of deficit to free-stream wind speed
$k = 1$

A truss tower with three legs produces a shadow with three peaks, and this is modeled by selecting $k = 3$.

Spatial Turbulence Model

As discussed in Chapter 8 and illustrated in Figures 8-16 and 8-29, small-scale turbulence (*i.e.* having significant spatial variations in wind speed within the swept area of the rotor) will cause a moving rotor blade to experience a wind power spectrum with peaks at multiples of the rotor speed. In effect, the blade will be subject to harmonic forcing functions with frequencies at integer multiples of its rotational speed, and it will respond dynamically to them. Estimating the amplitudes of these turbulence forcing functions is beyond the scope of this chapter and is, in fact, the subject of continuing research on wind characteristics. For example, see Equations (2-28) to (2-35) [Spera 1995]. Here we will make provision for wind turbulence excitations and emphasize that the harmonic content of these excitations plays an important role in determining the structural-dynamic response of wind turbine blades.

We shall assume that the distribution of turbulence-induced variations in wind speed across the rotor swept area changes slowly compared to the period of rotor rotation, so our model will be non-uniform in space but uniform in time. In the general case illustrated in Figure 11-1, the wind turbulence will have components in all three of the *X-Y-Z* directions, and each of these may vary with position in the rotor disk area. Thus

$$\delta U(r, \psi, t) = \begin{bmatrix} \delta u(r, \psi, 0) \\ \delta v(r, \psi, 0) \\ \delta w(r, \psi, 0) \end{bmatrix} \qquad (11\text{-}12)$$

where

$\delta U(r,\psi,t)$ = turbulence in the free-stream horizontal wind (m/s)
$\delta u(r,\psi,0)$ = spatial turbulence in the X direction (m/s)
$\delta v(r,\psi,0)$ = spatial turbulence in the Y direction (m/s)
$\delta w(r,\psi,0)$ = spatial turbulence in the Z direction (m/s)

Induced (Retardation) Wind Speeds

In an unducted rotor, extraction of power reduces the speed of the wind at the rotor blades compared to the free-stream wind speed, as discussed in Chapter 5. This reduction is termed the *induced* wind speed and its magnitude is usually given as a fraction of the free-stream wind speed by a parameter called the *axial induction factor* (see Figs. 2-14 and 5-11). Calculation of axial induction factors is routinely performed using aerodynamic performance models, balancing the rate of change of the momentum in the wind stream with the rotor thrust. Because thrust is not normally uniform across the swept area of the rotor, advanced performance codes determine radial and circumferential distributions of the induction factors.

Some dynamic-load computer models also include the calculation of induced wind speeds. Whatever their source (*i.e.*, external or internal to the load model), provision must be made for induced wind speeds, because they can have a significant effect on airfoil angles of attack and on aerodynamic loading. In our wind model this is done as follows:

$$V_i(r, \psi) = a(r, \psi)U_H \tag{11-13}$$

where

V_i = induced wind speed (m/s)

$a(r,\psi)$ = prescribed spatial distribution of axial induction factors

Combined Wind Effects

The net wind speed at a point in the plane of rotation is the sum of free-stream, wind shear, tower shadow, turbulence, and induction components. These are usually combined in a linear fashion, neglecting minor interaction terms. To help avoid ill-conditioned simultaneous equations later, wind speeds are then normalized by the tip speed of the rotor. The net wind velocity vector is specified in the *X-Y-Z* coordinate system and is composed of steady and turbulence terms, as follows:

$$V_w(X\text{-}Y\text{-}Z) = R\Omega \begin{bmatrix} 0 \\ V_r + \delta V_r \\ 0 \end{bmatrix} + R\Omega \begin{bmatrix} \delta u \\ \delta v \\ \delta w \end{bmatrix} \tag{11-14a}$$

$$V_r = \frac{V_H[1 - a(r, \psi)]}{R\Omega} \tag{11-14b}$$

$$\delta V_r = \frac{V_H[W_S(r, \psi) - W_T(r, \psi)]}{R\Omega} \tag{11-14c}$$

$$\delta u = \delta u(r,\psi)/R\Omega \tag{11-14d}$$

$$\delta v = \delta v(r,\psi)/R\Omega \tag{11-14e}$$

$$\delta w = \delta w(r, \psi)/R\Omega \tag{11-14f}$$

where

$Vw(r, \iota)$ = net wind speed at the plane of rotation (m/s)

R = rotor tip radius (m)

Ω = rotor rotational speed (rad/s)

Kinematics of the Blade Motion

Velocity Analysis

We will now derive equations for the velocity of a point on the deformed blade in the η-ζ-ξ deformed blade system, starting in the *X-Y-Z* ground-based coordinate system and then transforming the results first into x_p-y_p-z_p principal blade coordinates and finally into the η-ζ-ξ coordinates. Referring to Figure 11-1, we can illustrate the required velocity vector analysis as follows: Designate the *X-Y-Z* coordinates as the α reference frame. Call the x_p-y_p-z_p principal blade coordinates, located at a point B on the elastic axis of the blade, as the β reference frame. The velocity of the same point on the deformed blade in the α system, here designated as A, may then be written symbolically as

$$V_A(\alpha) = V_B(\alpha) + \frac{d}{dt}r_{A/B}(\beta) + \Omega x r_{A/B} \qquad (11\text{-}15)$$

where subscript *A/B* indicates a vector from point B to point A. Performing the computations indicated by this equation and transforming the result, the result using Equation (11-1) and then (11-2) gives

$$V_A(\eta - \zeta - \xi) = R\Omega \begin{bmatrix} rc\theta_p - \dot{\phi}[(d + \beta_0 r)c\theta_p + v]c\psi - \dot{\phi}(rs\theta_p)s\psi \\ rs\theta_p + \dot{v} - \dot{\phi}[(d + \beta_0 r)s\theta_p]c\psi + \dot{\phi}(rc\theta_p)s\psi \\ (v'r - v)s\theta_p - \dot{\phi}[d + (v - v'r)c\theta_p]s\psi \end{bmatrix} \qquad (11\text{-}16a)$$

$$r = \frac{r}{R} \qquad (11\text{-}16b)$$

$$\dot{\phi} = \frac{\dot{\phi}}{\Omega} \qquad (11\text{-}16c)$$

$$d = \frac{d}{R} \qquad (11\text{-}16d)$$

$$v = \frac{v}{R} \qquad (11\text{-}16e)$$

where ϕ = yaw angle of the rotor axis (rad)
             d = distance from the tower axis to the rotor hub (m)

and bold print designates a dimensionless variable. In addition, it has been assumed that the order of magnitude for the various terms are as follows:

$$Order\ 1\ variables: \quad r$$
$$Order\ \varepsilon_0^{1/2}\ variables: \quad \dot{\phi}$$
$$Order\ \varepsilon_0\ variables: \quad d, \beta_0, v, \dot{v}, v', H/R$$

Terms of order ε_0^2 and higher have been neglected in Equation (11-16a).

The relative velocity of the wind with respect to the moving rotor blade is computed by transforming the wind velocity components from Equation (11-14) into the deformed blade coordinates and then subtracting the blade velocity of Equation (11-16a). This gives

$$\frac{V_{rel}}{R\Omega} = \sqrt{\left(\frac{V_\eta}{R\Omega}\right)^2 + \left(\frac{V_\zeta}{R\Omega}\right)^2 + \left(\frac{V_\xi}{R\Omega}\right)^2} \qquad (11\text{-}17a)$$

$$\frac{V_\eta}{R\Omega} = -rc\theta_p - (V_r + \delta V_r + \delta V_y)s\theta_p + c\psi[(\delta V_x + \phi V_r)c\theta_p \\ + \dot{\phi}\{(d + \beta_0 r)c\theta_p + v\}] \qquad (11\text{-}17b)$$

$$\frac{V_\eta}{R\Omega} = -rs\theta_p - \dot{v} + (V_r + \delta V_r + \delta V_y)c\theta_p + c\psi[(\delta V_x + \phi V_r)s\theta_p \\ + \dot{\phi}\{(d + \beta_0 r)s\theta_p\}] \\ + s\psi\{-(\delta V_z - \chi V_r)s\theta_p - \dot{\phi}(rc\theta_p)\} \qquad (11\text{-}17c)$$

$$\frac{V_\zeta}{R\Omega} = -(v'r - v)s\theta_p + c\psi(\delta V_z - \chi V_r) + (\beta_0 + v'c\theta_p)V_r \\ + s\psi[\delta V_x + \phi V_r + \dot{\phi}\{d + (v - v'r)c\theta_p\}] \qquad (11\text{-}17d)$$

where

V_{rel} = relative wind speed at airfoil section in η-ζ-ξ coordinates (m/s)
V_η = relative chordwise wind speed at airfoil section (m/s)
V_ζ = relative normal wind speed at airfoil section (m/s)
V_ξ = relative spanwise wind speed at airfoil section (m/s)

Terms of order ε_0^2 have been discarded in these equations. The following assumptions have been made regarding the order of the velocity components:

$$Order\ 1\ velocities: \quad V_r$$
$$Order\ \varepsilon_0\ velocities: \quad \delta V_r, \delta V_x, \delta V_y, \delta V_z$$

Acceleration Analysis

Referring again to Figure 11-1, our acceleration analysis will recognize the fact that the mass of the blade is not all concentrated on the elastic axis but is distributed across the airfoil section. Acceleration equations will, therefore, be derived for an arbitrary point P at coordinates (η, ζ) within the cross-section. The point A in the velocity analysis is the same as $P(0,0)$. Using the same α-β reference frame designations as for the velocity vector analysis, the acceleration of P is given by the usual five-term acceleration equation as

$$a_P(\alpha) = a_B(\alpha) + \frac{d\Omega}{dt} x r_{P/B} + 2\Omega x \frac{d}{dt} r_{P/B}(\beta) + \Omega x(\Omega x r_{P/B}) + \frac{d^2}{dt} r_{P/B} \qquad (11\text{-}18)$$

The indicated operations of Equation (11-18) must be carried out, and the results must then be transformed to the β coordinate system and linearized. This tedious activity gives the expressions for the acceleration components in the x_p-y_p-z_p coordinates. For the small-deformation theory of this analysis, there is no difference between the x_p-y_p-z_p and η-ζ-ξ coordinate systems with respect to structural equations. For convenience, therefore, we will use the same η-ζ-ξ notation for the accelerations as for the relative velocities. This gives

$$a_P = \begin{bmatrix} a_{P,\eta} \\ a_{P,\zeta} \\ a_{P,\xi} \end{bmatrix} \qquad (11\text{-}19a)$$

$$a_{P,\eta} = -s\theta_p(r\Omega^2\beta_0 + 2\dot{\phi}\,\Omega rc\psi + \ddot{\phi}rs\psi + v\Omega^2c\theta_p) \qquad (11\text{-}19b)$$
$$- \zeta\Omega^2 c\,\theta_p s\theta_p - \eta\Omega^2 c\,\theta_p^2$$

$$a_{P,\zeta} = -c\theta_p(r\Omega^2\beta_0 + 2\,\dot{\phi}\Omega rc\psi + \ddot{\phi}rs\psi) - v\Omega^2 c\theta_p^2 \qquad (11\text{-}19c)$$
$$- \zeta\,\Omega^2 s\theta_p^2 - \eta\,\Omega^2 s\theta_p c\theta_p + \ddot{v}$$

$$a_{P,\xi} = -r\Omega^2 - 2\Omega\dot{v}s\theta_p \qquad (11\text{-}19d)$$

where

$a_{P,\eta}, a_{P,\zeta}, a_{P,\xi}$ = components of the acceleration vector at an arbitrary point P within the blade, in the normal, chordwise, and spanwise directions, respectively (m/s^2)

Aerodynamic Loading

Referring to Figure 6-1, the lift and drag forces on an airfoil section are given by

$$dL = \frac{1}{2}\rho\, C_L c\, V_{rel}^2 d\xi \qquad (11\text{-}20a)$$

$$dD = \frac{1}{2}\rho\, C_D c\, V_{rel}^2\, d\xi \qquad (11\text{-}20b)$$

where

dL = increment of lift force, normal to the relative wind (N)
dD = increment of drag force, parallel to the relative wind (N)
ρ = air density (kg/m^3)

C_L, C_D = lift and drag force coefficients, respectively; prescribed functions of α
α = angle of attack, from the relative wind to the chord line (rad)
c = chord length of the airfoil section (m)
$d\xi$ = increment of spanwise length (m)

We see from Figure 6-1 that the directions of the lift and drag forces are determined by the direction of the relative wind and not by the blade geometry. As a result, lift and drag directions will vary along the span of the blade and change with changing wind and operating conditions, which is not convenient for structural analysis.

The force components of interest for structural analysis are those related to the principal inertia axes of the entire blade. To locate these directions, we first define a *reference station* along the spanwise axis of the blade. The reference station is any location of special structural or aerodynamic interest. Referring to Figure 11-3, the angle between the cone of rotation and the chordline at this reference station is defined here as the *pitch angle*, θp.

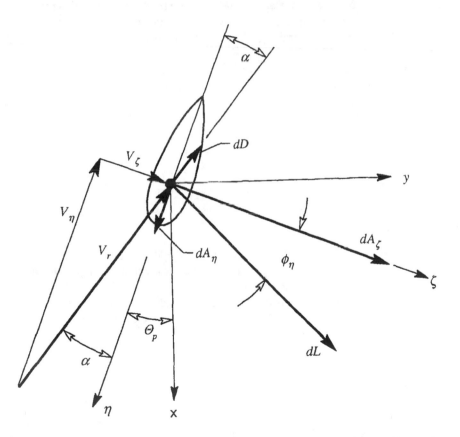

Figure 11-3. Velocity triangle at the blade reference station, and coordinate systems used for computing aerodynamic loading. The twist angle is zero. [Thresher *et al.* 1986]

The chord line at the reference station becomes the η-axis. Next, we define a *reference plane* as the surface generated by a η-axis line moving along the spanwise axis, remaining parallel to the chord line at the reference station. Finally, we define the *twist angle*, θ_t, as the angle between the local airfoil chord line and the reference plane. The force components of structural interest can now be expressed by

$$dA_\eta = \frac{1}{2}\rho c V_{rel}[C_L(\alpha)V_\zeta - C_D(\alpha)V_\eta]d\xi \qquad (11\text{-}21a)$$

$$dA_\zeta = \frac{1}{2}\rho c V_{rel}[C_L(\alpha)V_\eta + C_D(\alpha)V_\zeta]d\xi \qquad (11\text{-}21b)$$

$$\alpha = \theta_\eta - \theta_t \qquad (11\text{-}21c)$$

$$\theta_\eta = \arctan(V_\zeta/V_\eta) \qquad (11\text{-}21d)$$

where

dA_η, dA_ζ = increments of aerodynamic loading normal to and parallel to the blade reference plane (N/m)

θ_η = angle of the relative wind from the blade reference plane (rad)

θ_t = local twist angle relative to the blade reference plane; positive twist rotates the trailing edge downwind, as does positive pitch, θ_p (rad)

Most HAWT blades and their pitch control mechanism (if they have one) are relatively stiff in torsion about the spanwise axis, so aerodynamic pitching moments can be neglected. However, if a blade has pitch flexibility (*e.g.* one that is *self-twisting* for controlling power at high wind speeds) pitching moments must be included in the loading analysis, as follows:

$$q_{\xi,a} = \frac{1}{2}\rho c^2 V_{rel}^2 C_{M,a.c.}\,d\xi - e_{e.a.}\,cdA_\zeta \qquad (11\text{-}22a)$$

$$e_{e.a.} = \frac{x_{a.c.} - x_{e.a.}}{c} \qquad (11\text{-}22b)$$

where

dM_ζ = increment of aerodynamic pitch moment loading (N-m)

$C_{M,a.c.}$ = pitch moment coefficient about the aerodynamic center of the airfoil

e_{ea} = relative eccentricity of the elastic center of the airfoil; positive for elastic center between leading edge and aerodynamic center

x_{ac} = distance from leading edge to aerodynamic center; usually $c/4$ (m)

x_{ec} = distance from leading edge to elastic center (m)

Inertia Loading

The distributed inertia forces and moments acting along a unit length of the blade can be computed using Newton's laws as follows:

$$\boldsymbol{p}_I = \int_{\xi-12}^{\xi+1/2} -\boldsymbol{a}_p dm = \int\int_{Section} -\boldsymbol{a}_P \rho_b d\eta d\zeta \qquad (11\text{-}23)$$

$$\boldsymbol{q}_I = \int_{\xi-1/2}^{\xi+1/2} -\boldsymbol{r}_c \times \boldsymbol{a}_p dm = \int\int_{Section}(\boldsymbol{r}_c \times \boldsymbol{a}_P)\rho_b d\eta d\zeta \qquad (11\text{-}24a)$$

$$r_c = \begin{bmatrix} \eta \\ \zeta \\ 0 \end{bmatrix}$$

(11-24b)

where

dm = mass of an incremental length of blade (kg)
ρ_b = average mass density of the blade section (kg/m³)

If the force of gravity is considered as deriving from an acceleration in the -Z direction, it can be treated in a similar manner.

Distributed Loading on the Blade

The combined loading on a HAWT blade caused by aerodynamic forces, shaft rotation, yaw motion, and gravity are

$$p_\eta = dA_\eta + m[(r\Omega^2\beta_0 + 2\Omega\dot{\phi}rc\psi + \ddot{\phi}rs\psi)s\theta_p + v\Omega^2 s\theta_p c\theta_p + e_\eta\Omega^2 c\theta_p^2$$
$$\qquad + mg(\chi s\theta_p + c\theta_p s\psi - \beta_0 s\theta_p c\psi)$$

(11-25a)

$$p_\zeta = dA_\zeta + m[-(r\Omega^2\beta_0 + 2\Omega\dot{\phi}rc\psi + \ddot{\phi}rs\psi)c\theta_p + v\Omega^2 s\theta_p^2$$
$$\qquad + e_\eta\Omega^2 s\theta_p c\theta_p - \ddot{v}] + mg(-\chi c\theta_p + s\theta_p s\psi + \beta_0 c\theta_p c\psi)$$

(11-25b)

$$p_\xi = m(r\Omega^2 + 2\dot{v}\Omega s\theta_p) - mgc\psi$$
$$q_\eta = v'\Omega^2 s\theta_p^2 I_{\eta\eta}^m$$

(11-25c)

$$q_\zeta + - mr\Omega^2 e_\eta - v'\Omega^2 c\theta_p s\theta_p I_{\eta\eta}^m + mge_\eta c\psi$$

(11-25d)

$$q_\xi = q_{\xi,a} + \Omega^2 s\theta_p c\theta_p(I_{\zeta\zeta}^m - I_{\eta\eta}^m) + mge_\eta s\psi\theta_p$$

(11-25e)

$$m = \int\limits_{Section} \rho_b d\eta\, d\zeta$$

(11-25f)

$$e_\eta = \frac{1}{m} \int\limits_{Section} \int \rho_b \eta\, d\eta\, d\zeta$$

(11-25g)

$$I_{\eta\eta}^m = \int\limits_{Section} \int \rho_b \zeta^2 d\eta\, d\zeta$$

(11-25h)

$$I_{\eta\eta}^m = \int\limits_{Section} \int \rho_b \eta^2 d\eta\, d\zeta$$

(11-25i)

where

m = blade mass per unit length (kg/m)
e_η = η-coordinate of the section center of mass (m)
$I_{\zeta\zeta}{}^m$ = minimum mass moment of inertia of section, per unit length (kg-m)
$I_{\eta\eta}{}^m$ = maximum mass moment of inertia of section, per unit length (kg-m)

Governing Equation of Motion

For this single-degree-of-freedom system, with flap displacement v as the only dynamic motion, the standard equation of motion for an incremental length of the blade has the form

$$[M]\,\ddot{v} + [J]\,\dot{v} + [K]\,v = [L] \tag{11-26}$$

where

$[M]$ = mass function (kg/m)
$[J]$ = damping function (N-s/m²)
$[K]$ = stiffening function (N/m²)
$[L]$ = loading function (N/m)

To develop this equation governing flap motion, it is necessary to begin with Equations (11-6) defining the equilibrium-moment curvature relationships for a blade element. Equation (11-6a) for flapwise bending will be converted into an equation of motion.

Spanwise Tension

The spanwise tension, T, in Equation (11-6a) can be obtained by directly integrating Equation (11-6d) to give

$$T(\xi) = \int_{\xi}^{R} m(r\Omega^2 + 2\dot{v}\Omega s\theta_p - gc\psi)d\xi \tag{11-27}$$

For the integration limit, we neglect any difference between the tip radius R and the length of the blade.

Modal Flap Displacement Model

In order to reduce the flap motion equation to an ordinary differential equation required for a computerized solution, a *modal model* is used, in which the flap displacement is assumed to be of the form

$$v(\xi,t) = \sum_{k} s_k(t)\gamma_k(\xi) \qquad k = 1, 2, \ldots \tag{11-28}$$

where

k = number of *mode shapes* in the displacement model
s_k = time-dependent displacement scaling factor for the kth mode (m)
γ_k = spanwise shape function for the kth mode

The spanwise shape functions, which must satisfy the blade kinematic and natural (force) boundary conditions, are usually prescribed as the mode shapes for free vibration of the blade. The time-history of $s_k(t)$ is determined by numerical integration of the equation of motion during consecutive revolutions of the rotor, until each s_k and its first time derivative at $\psi = 2\pi$ are equal to the same parameters at $\psi = 0$, within a set tolerance. This is the so-called *trim solution*.

The flap mode shapes of importance to the structural-dynamic analysis of HAWT rotors can be divided into the following two classes: (1) *dependent* blade modes, in which the motions of a given blade are constrained or influenced by the motions of another blade in the rotor, and (2) *independent* blade modes, in which the motions of a given blade can be analyzed separately from the other blades. An example of a dependent mode is the rigid-body teetering motion of two-bladed HAWT rotor, in which the mode shapes of opposite blades must always be antisymmetric and without curvature at the hub. Dependent blade motions will not be discussed further here because of the additional complexity introduced by the constraints at the hub. The reader is referred to Wright *et al.* [1988] and [Wright and Butterfield 1991] for the equations required to perform a dynamic analysis of a teetered HAWT rotor and sample comparisons with test data.

The two classic examples of independent blade modes are the *cantilevered* shape, in which both v and v' are zero at the hub, but v'' may be non-zero; and the *flapping hinge* shape, in which both v and v'' are zero at the hub, but v' may be non-zero [Spera 1975]. In the remainder of this analysis, we will limit ourselves to one independent flapwise mode, for simplicity (*i.e.*, $k = 1$). Fortunately, experience has shown that one cantilever flapwise mode is usually sufficient for the calculation of blade loads in a rigid-hub rotor. Moreover, blade loads are much more sensitive to the scale of the wind input models (for wind shear, turbulence, and tower-shadow) than they are to higher blade modes. The exception to this general observation is the case in which the natural frequency of the next-higher mode is equal to an integer multiple of the rotor speed, and a resonance condition may be present.

Equation of Motion for One Independent Blade Mode

Substitution of the assumed form of the blade displacement from Equation (11-28) into Equation (11-6a) with k set equal to 1 gives

$$[-(s\gamma)''EI_\xi]'' + (Ts\gamma')' + q'_\xi + p_\zeta = 0 \qquad (11\text{-}29)$$

Now, use of a *Galerkin* approach [*e.g.* Dym and Shames 1973] gives us

$$\int_0^R \left\{ [-(s\gamma)''EI_\xi]'' + (Ts\gamma')' + q'_\xi + p_\zeta \right\} \gamma d\xi = 0 \qquad (11\text{-}30)$$

Integrating this equation by parts and using the boundary conditions required of $\gamma(\xi)$ lead to an ordinary differential equation in $s(t)$, which can then be integrated numerically in the time domain.

The numerical integration approach is best explained by rearranging Equation (11-29) so that the acceleration term remains on the left-hand side but all other terms are moved to the right-hand side. If we perform our time-domain integration in time steps of duration dt, we can now calculate the acceleration at time t in terms of the right-hand side of the equation evaluated with calculations made previously at time $t - dt$. We continue this forward-integration process in time until convergence to a trim solution is obtained. As discussed previously, trim is achieved when the flap displacement and velocity at $\psi = 2\pi$ are equal to the same parameters at $\psi = 0$, within a specified tolerance.

Carrying out the integration of Equation (11-30) by   parts   gives the following expressions for the flap accelerations of an independent blade:

$$\ddot{s}M = -sK_B - s\left(\Omega^2 K_\Omega + 2\Omega s\theta_p \dot{s} K_C - c\psi g K_g\right)$$
(**Bending**)   (**Tension Stiffening**)

$$-c\theta_p \left(\Omega^2 + 2\Omega\dot{\phi}c\psi + \ddot{\phi}s\psi\right) M_R$$
(**Rigid – Body Motion**)

$$+s\Omega^2 s\theta_p^2 K_q$$
(**Inertia Moment Stiffening**)

(11-31)

$$+\Omega^2 s\theta_p c\theta_p M_B \qquad\qquad + F_a$$
(**C.G. Imbalance**)   (**Aero Forace**)

$$+ s\Omega^2 s\theta_p^2 M$$
(**Inertia Force Stiffening**)

$$+ g(-\chi c\theta_p + s\theta_p s\psi + \beta_0 c\theta_p c\psi) M_g$$
(**Gravity**)

The various coefficients in this acceleration equation are given by the integrals in Equations (11-32) which are evaluated numerically. A *trim* solution of the equation of motion is then found by forward integration in time (or, equivalently, in azimuth ψ) until the following *convergence criteria* are met:

$$s[2(n + 1)\pi/\Omega] = s[2n\pi/\Omega] \pm \varepsilon_1$$
$$\dot{s}[2(n + 1)\pi/\Omega] = \dot{s}[2n\pi/\Omega] \pm \varepsilon_2$$

where

n = number of rotor rotations
$\varepsilon_1, \varepsilon_2$ = convergence tolerances (m, m/s)

$$M = \int_0^R m\gamma^2 d\xi$$

$$M_R = \int_0^R mr\gamma d\xi$$

$$M_B = \int_0^R e_\eta m\gamma d\xi$$

$$M_g = \int_0^R m\gamma d\xi$$

$$K_B = \int_0^R EI_\zeta(\gamma'')^2 d\xi$$

$$K_\Omega = \int_0^R T_\Omega(\xi)(\gamma')^2 d\xi$$

$$K_q = I_{\eta\eta}(R)\gamma'(R)\gamma(L) - \int_0^R I_{\eta\eta}(\gamma')^2 D\Xi$$

$$K_c = \int_0^r T_c(\xi)(\gamma')^2 d\xi \qquad\qquad (11\text{-}32)$$

$$F_a = \int_0^R dA_\zeta \gamma d\xi$$

$$T_\Omega(\xi) = \int_\xi^R mr d\xi$$

$$T_c(\xi) = \int_\xi^{RT} m\gamma d\xi$$

$$T_g(\xi) = \int_\xi^R md\xi)$$

After a trim solution has been found or after each revolution of a yaw-motion solution, the flap displacement, slope, and velocity can be computed from the following equations:

$$v = s(t)\gamma(\xi) \qquad\qquad (11\text{-}33\text{a})$$

$$v' = s(t)\gamma'(\xi) \qquad\qquad (11\text{-}33\text{b})$$

$$\dot{v} = \dot{s}(t)\gamma(\xi) \qquad\qquad (11\text{-}33\text{c})$$

Rotor Blade Loads

Blade loads distributed along the ξ-axis are calculated by *the force-integration* method, as follows: From Equations (11-27) and (11-32c), the spanwise tension force load is

$$T(\xi) = \Omega^2 T_\Omega(\xi) + 2\Omega s\theta_p \dot{s} T_c(\xi) - gc\psi T_g(\xi) \qquad (11\text{-}34)$$

Integrating Equation (11-4b) gives the blade flapwise shear force load as

$$
\begin{aligned}
V_\zeta(\xi) = \int_\xi^R dA_\zeta d\xi &- (\Omega^2 \beta_0 + 2\Omega\dot{\phi}c\psi + \ddot{\phi}s\psi)c\theta_p T_\Omega(\xi) \\
&+ (\Omega^2 s\theta_p^2 s - \ddot{s})T_c(\xi) + \Omega^2 s\theta_p c\theta_p M_e(\xi) \\
&+ g(-\chi c\theta_p + s\theta_p s\psi + \beta_0 c\theta_p c\psi)T_g(\xi)
\end{aligned}
\qquad (11\text{-}35)
$$

Integrating Equation (11-4a) gives the blade chordwise shear force load as

$$
\begin{aligned}
V_\eta(\xi) = \int_\xi^R dA_\eta d\xi &+ (\Omega^2 \beta_0 + 2\Omega\dot{\phi}c\psi + \ddot{\phi}s\psi)s\theta_p T_\Omega(\xi) \\
&+ \Omega^2 s\theta_p c\theta_p s T_c(\xi) + \Omega^2 c\theta_p^2 M_e(\xi) \\
&+ g(\chi s\theta_p + c\theta_p s\psi - \beta_0 s\theta_p c\psi)T_g(\xi)
\end{aligned}
\qquad (11\text{-}36)
$$

Integrating Equation (11-5c) gives the blade pitch moment load as

$$
\begin{aligned}
M_\xi(\xi) = -\int_\xi^R V_\eta(\xi)v'(\xi)d\xi &+ \int_\xi^R q_{\xi,a}(\xi)d\xi \\
&- \Omega^2 s\theta_p c\theta_p \Delta I(\xi) + gs\psi s\theta_p M_e(\xi)
\end{aligned}
\qquad (11\text{-}37)
$$

Integrating Equation (11-5b) gives the blade edgewise moment load as

$$
\begin{aligned}
M_\zeta(\xi) = \int_\xi^R V_\eta(\xi)d\xi &- \Omega^2 \int_\xi^R e_\eta m(\xi)rd\xi \\
&- \Omega^2 c\theta_p s\theta_p s \int_\xi^R I_{\eta\eta}(\xi)d\xi + gc\psi M_e(\xi)
\end{aligned}
\qquad (11\text{-}38)
$$

Finally, integrating Equation (11-5a) gives the blade flatwise moment load as

$$M_\eta(\xi) = - \int_\xi^R V_\zeta(\xi)d\xi + \int_\xi^R T(\xi)v'(\xi)d\xi$$

$$+ \Omega^2 s\theta_p^2 s \int_\xi^R I_{\eta\eta}(\xi)\gamma'(\xi)d\xi$$

(11-39)

In the above equations,

$$M_e(\xi) = \int_\xi^R e_\eta(\xi)m(\xi)d\xi$$

(11-40a)

$$\Delta I(\xi) = \int_\xi^R [I_{\zeta\zeta}(\xi) - I_{\eta\eta}(\xi)]d\xi$$

(11-40b)

Typical Computer Model for Blade Load Calculations

The calculation of wind turbine blade loads for even the simplest of cases, such as the one described here (*i.e.*, one degree of freedom and one blade mode), requires a computer model. A variety of special-purpose models for HAWT loads have been developed in the past, dating back to the 1970s when only mainframe computers had sufficient capability for the task [Spera 1977]. Today, HAWT load computer models for personal computers are available, such as the FLAP Code [Wright *et al.* 1988]. The following discussion is based on experience with the FLAP Code, which is a current analysis tool in the public domain.

Requirements

A computerized solution of the equations of motion and computation of displacements and loads requires a sophisticated, interactive model that is flexible, well-documented, and easily modified. The model must be capable of performing a variety of tasks, including

-- input of data and output of results;
-- matrix inversion;
-- time-domain analysis of differential equations;
-- computation of spatially-dependent blade properties and aerodynamic parameters.

The computational requirements of the equations of motion and other quantities associated with it determine which portions of the model can forego efficiency in favor of flexibility and readability, and which portions need to be as efficient as possible. This can be accommodated in a computer model composed of two modules, as illustrated by the flow chart in Figure 11-4.

Module 1 is a data pre-processor, in which a file of raw blade and turbine property is processed to produce a data file that can be used to solve the equation of motion. Flexibility and readability are more important than efficiency in this module. Module 1 also computes all the coefficient matrices since they are independent of most of the non-structural variables, such as wind speed and tower shadow. Module 2 performs the actual model run including solution of the equation of motion, computation of the loads and output of the results. Here, efficiency is a primary requirement.

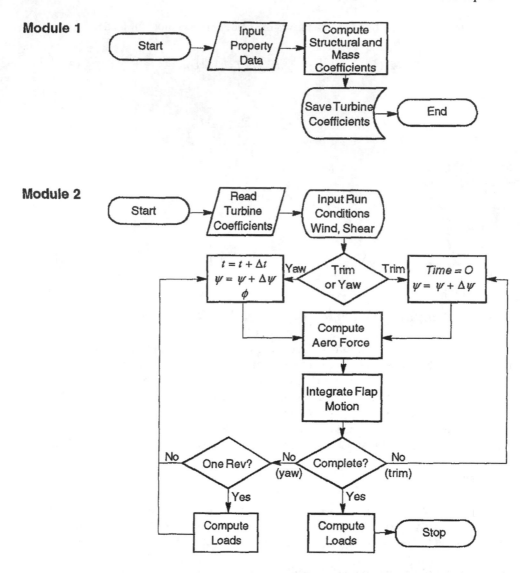

Figure 11-4. Typical flow chart for a computer model used to calculate HAWT blade loads. [Thresher *et al.* 1986]

Computational Procedure

The procedure followed in the FLAP Code for solving Equation (11-31), the fundamental equation of motion for a HAWT blade, can serve as an illustration of an approach that has been used successfully by the creators of such computer models. The general approach is as follows:

1. All coefficient matrices are multiplied by the inverse of the mass matrix. Thus the mass matrix associated with blade tip accelerations multiplied by its inverse gives the identity matrix. This results in a set of equations with the blade tip accelerations on one side of the equality and the computed generalized forces on the other. Multiplication by the inverse mass matrix is only done once, at the beginning of Module 2.

2. In this way, the accelerations associated with each flap coordinate function can be evaluated numerically be substituting the current values for the flap velocities and displacements into the force side of the equation.

3. The blade tip displacements are computed by solving the second-order differential equation relating acceleration, velocity, and displacement. The solution is performed *via* the modified Euler predictor-corrector method, which uses the current blade tip accelerations and the previous values of the displacements and velocities.

4. The blade loads are computed only at the completion of a trim solution.

5. The solution during time-dependent prescribed yawing motions is run at the completion of the trim solution. Loads are computed at the completion of each revolution during the yawing solution. Yawing motion is prescribed according to the simple equation

$$\phi(t) = \phi_0 + \phi_u \sin(\omega_\phi t) \qquad (11\text{-}41)$$

6. A sct of nine blade motion and blade load parameters is computed for each specified spanwise location on the blade axis (typically 11 stations) at each azimuthal position of the blade during a revolution (typically 36 azimuths). These parameters are the

- -- flap displacement, v;
- -- flapwise slope, v';
- -- flap velocity, dv/dt;
- -- edgewise shear load, S_η;
- -- flapwise shear load, S_ζ;
- -- spanwise tension load, T;
- -- flapwise moment load, M_η;
- -- edgewise moment load, M_ζ;
- -- pitching moment load, M_ξ.

Sample Blade Load Analysis

The turbine modeled in this example is the *Grumman* prototype HAWT (see Table 3-1), which has a three-bladed, rigid-hub rotor 10.1-m in diameter that runs downwind of a shell tower [Adler *et al.* 1980]. The lift and drag models are simplifications of the airfoil data presented by Sweeney and Nixon [1978]. Other input data are as follows:

$$
\begin{array}{ll}
U_H = 9.1 \text{ m/s} & m = 0.143 \\
t_0 = 0.25 & t_p = 0 \\
\psi_0 = 15 \text{ deg} & \phi_0 = 10 \text{ deg} \\
\Omega = 7.54 \text{ rad/s} & c = 0.46 \text{ m} \\
\beta_0 = 3.5 \text{ deg} & \theta_t = 0 \text{ deg}
\end{array}
$$

The natural frequency of the blade's first (*i.e.* lowest frequency) flapwise bending mode is approximately 3.95 Hz, or about 3.3 times the rotor rotational frequency.

Flap bending moments in the operating turbine were measured at a spanwise station 1.8 m from the rotor axis (20 percent of span) and averaged over 20 revolutions. The test data were azimuth-averaged in this manner to remove random wind effects that are not modeled in this analysis. After averaging, the mean value of the measured flap bending moment was subtracted, producing the variable flap moment history labeled "Test Data" in Figure 11-5. The calculated flap moment load at 20 percent of span, again with the average removed, is also plotted in Figure 11-5, permitting a comparison of the blade's flapping response to gravity, tower shadow, and average wind shear.

The test and calculated wave forms are very similar in Figure 11-5, indicating a validated computer model for this turbine. However, another HAWT would require a similar test/calculation comparison for validation of its structural, aerodynamic, and wind input data. These data are an integral part of the blade load analysis model, and as critical to validation as the basic equations of motion.

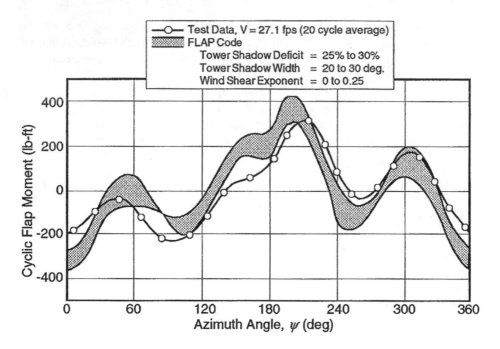

Figure 11-5. Sample comparison between calculated and measured flapwise moment loads. The test turbine is the 10.1-m diameter *Grumman* prototype HAWT. [Thresher *et al.* 1986]

Current and Future Developments in Structural-Dynamic Modeling

The demands placed on structural dynamics models are becoming more severe as turbine designs evolve. Models are being asked to account for more-complex phenomena, including the following:

-- non-steady air flow
-- blade-to-blade interactions
-- yaw and pitch motions of the nacelle and tower
-- blade pitch control
-- teetering motion
-- variable-speed
-- startup transient
-- shutdown and braking transients

The desire to account for all such effects within one or two comprehensive models is driving engineers to develop computer codes of ever-increasing complexity.

Improved structural dynamic models of wind turbines appear to be developing along two different paths. The first approach involves *special purpose* codes, each written to analyze the structural dynamic behavior of a specific configuration of wind turbine. Such codes generally have 12 to 15 *degrees of freedom* (DOF) tailored to the significant motions of the "target" structural system. The second approach seeks to adapt *general purpose* structural analysis codes to simulate wind turbine dynamics. These general purpose codes are powerful enough to handle almost any turbine configuration, but they are complex and expensive. Moreover, the structural dynamicist must have a significant level of expertise in their use, and each new configuration requires considerable time for modeling and verification efforts.

Typical Special-Purpose Model

The structural analysis program *PHATAS-II*, developed at the *Netherlands Energy Research Foundation* (ECN), is an example of a special-purpose code [Lindenburg and Snel 1993]. The PHATAS computer model can be used to calculate dynamic displacements and loads for the main structural subsystems of a HAWT. Wind inputs to this model include these:

-- steady wind, including vertical and horizontal gradients
-- wind direction as a function of elevation
-- rotationally-sampled wind turbulence
-- turbulent wind in time-series form

A number of structural degrees of freedom are available within the code, such as the following:

-- flapwise and chordwise elastic blade deformations
-- blade pitch motion
-- blade flapping and teetering hinges
-- torsional deflection of the power train
-- variable rotor speed
-- pitch and yaw motions of the nacelle
-- bending and torsional deflections of the tower

Typical General-Purpose Model

The *ADAMS* code, developed at *Mechanical Dynamics, Inc.*, is an example of a general-purpose, multi-body structural analysis program that has been adapted for wind turbine applications [Malcolm and Wright 1994]. This code has virtually unlimited degrees of freedom and places no limits on the type or magnitude of displacements. While the ADAMS code can be run on a high-end personal computer (PC), it is exceptionally slow when modeling systems with many degrees of freedom or when performing simulations involving wind turbulence inputs. For example, it can take several days of PC run time to simulate 10 minutes of real time operation of a HAWT subjected to turbulent winds.

The strengths of this type of computer model are that almost any structural system (wind turbine or otherwise) can be modeled with it once the required skills have been learned by the user, and a mixture of complex external loadings and operating conditions can be applied. As PC performance increases, this type of general-purpose computer model is expected to used more and more for the structural analysis of modern wind turbines.

Natural Vibration Modes in a Wind Turbine

Dynamic effects can be substantial in a wind turbine system because of the periodic nature of its aerodynamic loading. As a result, the design process must depend on accurate predictive tools and/or measurement techniques for determining the dynamic response characteristics of the rotating structure. An elastic structure responds to periodic *forcing functions* by vibrating in one or more discrete geometric patterns called *mode shapes*, each of which has a companion rate of vibration called a *modal frequency*. Mode shapes and frequencies are determined primarily by the distributions of mass and stiffness throughout the structure and by its *boundary conditions*. Rotation can alter the natural frequencies of certain mode shapes, when centrifugal and Coriolis forces change stiffnesses. Because mode shapes and frequencies are relatively independent of applied loadings, they are often referred to as *natural* or *free-vibration modes*. Both analytical and experimental procedures for determining mode shapes and frequencies are referred to as *modal analysis*.

Figure 11-6 shows the results of a typical modal analysis of a HAWT [Sullivan 1981], displayed in a frequency plot called a *Campbell diagram*. Natural frequencies are plotted *vs.* rotational speed, with one generally-horizontal line per mode. Rays from the origin are plots of integer multiples of the rotational frequency. In a rotating structure, excitation loadings normally occur at these integer-multiple frequencies, often abbreviated 1*P*, 2*P*, *etc.*

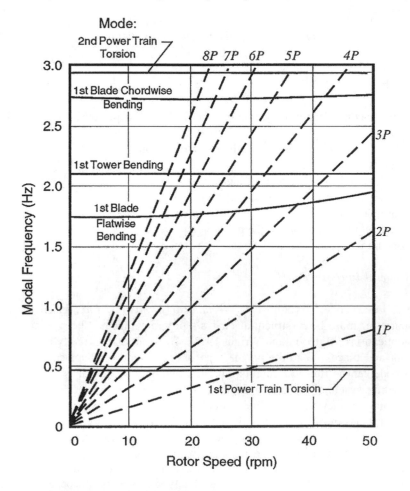

Figure 11-6.   Typical Campbell diagram displaying the results of a HAWT modal analysis. Radial lines indicate the frequencies of potential excitation forces [Sullivan 1981]

Defining modal characteristics is of paramount importance in that, with this information, potentially harmful *resonant behavior* can be avoided. Resonance is a phenomenon which occurs in a structure when the frequency of periodic loading or *excitation* equals or nearly equals one of the modal frequencies of the structure. Thus, in Figure 11-6, every intersection of a radial line and a modal-frequency line is a potential resonance. Resonance if often characterized by a large increase in displacements and internal loads. The relative size of this increase is measured by a *dynamic amplification factor*, which is the ratio of the amplitude of the dynamic displacement to the static displacement under the same forcing function.

Not all resonances cause large amplification factors, because rotation can sometimes eliminate the matching of a forcing function with a natural mode, even when both have the same frequency in a non-rotating frame of reference. A general discussion of amplified and non-amplified resonances is beyond the scope of this chapter, but background information on resonances significant in HAWT systems is given in [Sullivan 1981].

In rotating structures, the acquisition of dynamic response characteristics is complicated by their dependence on the structure's rotational speed. Therefore, special methods have to be employed for both the predictive and experimental modal analysis. Predictive techniques for wind turbines are usually based on *the finite-element method*, in order to take advantage of its superior versatility. As the turbine is modeled in a frame of reference which rotates at the turbine operating speed (assumed constant), rotating coordinate system effects must be taken into account.

The finite-element method has been used extensively to predict the modal characteristics of rotating structures. When this method is applied to HAWT blades and other structures which are comprised of rotating beams reminiscent of fan or helicopter blades, some of the rotating coordinate system effects are neglected. This is usually justified when motions perpendicular to the axis of rotation (*i.e.*, along the axis of a HAWT blade) are relatively small. This is not the case for a VAWT, however, and these effects cannot be omitted. Here we will use a method similar to the one described by Patel and Seltzer [1971] for analyzing the modes of the spinning "Skylab" satellite, in which all rotating coordinate system effects are included.

Other analytical techniques which have been used to predict the dynamic characteristics of rotating structures include various *Galerkin* procedures [Klahs and Ginsberg 1979, Warmbrodt and Friedmann 1980], the *modal method*, and the *component mode method* in combination with various eigensolution schemes [Vollan and Zwaan 1977, Ottens and Zwaan 1978a and 1978b]. A major experimental obstacle to obtaining the natural frequencies and mode shapes of a rotating structure is that mechanisms must be devised for exciting the structure while it is rotating. Sliprings or telemetry must also be employed for data transmission from instruments such as *accelerometers* on the structure.

Finite Element Modal Analysis of a VAWT

When performing a modal analysis of a rotating structure, the motion is most conveniently measured relative to a coordinate system that is rotating with the structure. This permits the displacements to be small by eliminating the large rigid-body rotation. The coordinate system employed in this analysis rotates at the turbine operating speed, Ω, with the origin fixed in space at the lower bearing of the rotor. The angular velocity vector is directed vertically upward along the positive z axis. The finite-element matrices are developed on the basis of the motions that remain small within this rotating frame. For an accurate representation of the dynamics of a VAWT relative to a coordinate system rotating at a constant speed, the following effects must be included:

> -- *tension stiffening*
> -- *centrifugal forces*
> -- *Coriolis accelerations*

The tension stiffening that affects primarily the blades is induced by steady centrifugal and gravity forces. The centrifugal and Coriolis terms are a direct result of using a rotating coordinate system.

Equations of Motion

The total acceleration at a point in the rotating structure, with respect to fixed coordinates, can be represented in terms of the acceleration in the rotating coordinate system by an equation similar to Equation (11-18), as follows:

$$a_t = \ddot{u} + 2\Omega x \dot{u} + \Omega x [\Omega x (r + u)] \tag{11-42}$$

where

a_t = total acceleration vector, excluding gravity (m/s^2)
r = original position vector of an element; origin at lower bearing (m)
u = displacement vector; observed in the rotating coordinate system (m)
Ω = constant angular velocity vector of the rotating coordinate system (rad/s)

Using this expression for the acceleration in the *equation of motion* for a structure, the usual damping and stiffness matrices are altered from those of a static structure. The resulting differential equations are represented by

$$M\ddot{u} + C\dot{u} - Su + Ku = F_c + F_g \tag{11-43}$$

where

M = normal (unaltered) mass matrix (kg)
u = displacement observed in the rotating coordinate system; dot signifies derivative with respect to time (m)
C = additional Coriolis matrix (N-s/m)
S = additional centrifugal softening matrix (N/m)
K = normal (unaltered) stiffness matrix (N/m)
F_c = additional static load vector representing steady centrifugal force (N)
F_g = gravitational forces; steady in a VAWT (N)

The Coriolis matrix, C, is *skew-symmetric* and results from the second term on the right side of Equation (11-42). Here we have ignored any structural (internal) damping because most VAWTs are lightly damped. Therefore, the C matrix is totally the result of Coriolis effects. The centrifugal softening matrix, S, comes from the variable portion (*i.e.*, dependent on u) of the last term on the right in Equation (11-42). F_c also comes from the last term in Equation (11-41). To obtain the mode shapes and frequencies of the turbine as observed in the rotating system, Equation (11-41) is reduced to the following form:

$$M\ddot{u} + C\dot{u} + (K + K_G - S)\,u = 0 \tag{11-44}$$

where K_G = geometric stiffness matrix resulting from steady centrifugal and gravitational loadings (N/m)

Thus, the solutions correspond to small motions about a prestressed state. *Aeroelastic effects* (*i.e.*, aerodynamic loading caused by structural motions) have not been included in this analysis. While strongly influencing the aerodynamic stability of a VAWT [*e.g.*, Popelka 1982], aeroelasticity seems to have little effect on the natural frequencies and modes of the turbine.

Eigenvalue Solution

Obtaining mode shapes and frequencies from Equation (11-44) involves solving what is known as a *characteristic-value problem*, which has the standard form

$$
\begin{bmatrix}
a_{11}x_1 + a_{12}x_2 + \cdots + a_{1n}x_n = \omega\, x_1 \\
a_{21}x_1 + a_{22}x_2 + \cdots + a_{2n}x_n = \omega\, x_2 \\
\cdot\ \cdot\ \cdot\ \cdot\ \cdot\ \cdot\ \cdot\ \cdot\ \cdot\ \cdot\ \cdot\ \cdot\ \cdot\ \cdot\ \cdot\ \cdot\ \cdot \\
a_{n1}x_1 + a_{n2}x_2 + \cdots + a_{nn}x_n = \omega\, x_n
\end{bmatrix}
\tag{11-45}
$$

where, for our modal analysis problem

n = number of degrees of freedom; number of finite elements times three displacement components per element

$X_j...x_n$ = *eigenvector* or *characteristic vector*; defines one mode shape

ω = *eigenvalue* or *characteristic value*; defines one modal frequency (rad/s)

Each non-trivial solution of this characteristic-value problem has an eigenvector and a corresponding eigenvalue, and the set of eigenvector/eigenvalue pairs comprises the results of a theoretical modal analysis of our structural system. Modes are usually presented in ascending order of frequency, from the lowest or *fundamental* mode to the highest frequency of interest. The latter seldom needs to exceed ten times the maximum rotational speed of the turbine (10*P*).

To find the eigenvalues of the system of equations represented by Equation (11-44), it is more convenient to use *state vectors* instead of the displacements *u*. Hence, we will introduce the state vector

$$
v = \begin{bmatrix} \dot{u} \\ -\,-\,- \\ u \end{bmatrix}
\tag{11-46a}
$$

and the matrices

$$
A = \begin{bmatrix} M & | & 0 \\ -\,- & | & -\,- \\ 0 & | & K' \end{bmatrix}
$$

$$
B = \begin{bmatrix} C & | & K' \\ -\,- & | & -\,- \\ -K' & | & 0 \end{bmatrix}
\tag{11-46b}
$$

in which

$$K' = K + K_G - S$$

Equation (11-43) can now be written as

$$A\dot{v} + Bv = 0 \tag{11-47}$$

Here A is *real symmetric*, and B is *real skew-symmetric*. Seeking a solution for Equation (11-46) in the form

$$v(t) = \phi e^{i\omega t} \tag{11-48a}$$

where i is the square root of -1, the resulting eigenvalue problem is

$$[B + i\omega A]\phi = 0 \tag{11-48b}$$

We now define an eigenvector **x** as

$$x = A^{1/2}\phi \tag{11-48c}$$

assuming that A is now positive definite. We can now transform Equation (11-47b) into the standard eigenvalue form of Equation (11-44), with all terms on the left side, as follows:

$$[G - \omega I]x = 0 \tag{11-49a}$$

$$G = iA^{-1/2}BA^{-1/2} \tag{11-49b}$$

Because of the skew-symmetry of the matrix B, the matrix G is *Hermetian*. Consequently, it can be shown that the eigenvectors are, in general, complex; but the eigenvalues are real [Franklin 1968].[2] If structural damping is included, the system loses its Hermetian character and the eigenvalues as well as the eigenvectors are complex. Again, in the present analysis, structural damping has been ignored because VAWTs are lightly damped. The Hermetian character of the eigenvalue system has important ramifications to the rotating modal test. Even if the modes of the non-rotating structure are perfectly real, the modes of the rotating structure will be complex.

---

[2] If A is not positive-definite, then G is no longer Hermetian and the eigenvalues are not necessarily real. This condition can lead to dynamic instability.

NASTRAN Finite Element Modeling

The general-purpose structural analysis code *NASTRAN* is the software commonly used for modeling a wind turbine in finite elements to determine its natural vibration modes. Before presenting an example of the modeling, a short description of the NASTRAN calculational procedure is in order, as it is applied to this specific problem. Two of the available NASTRAN *rigid format* options are used in the modeling effort. First, a static analysis with geometric stiffening effects is performed on the model under the action of centrifugal, gravitational, and boundary forces. The resulting modified stiffness matrix ($K + K_G$ in Eq. (11-43)) is then retained *via* a modest amount of *DMAP* programming for use in the subsequent complex eigenvalue analysis (the second rigid format).

NASTRAN users are provided with a choice of several complex eigensolvers. One employs the *upper-Hessenberg* method and is very efficient. A method that has been used successfully for the modal analysis of wind turbines is to first model the structure with the necessary number of degrees of freedom, then use a standard *Guyan reduction* procedure (also a NASTRAN option) to reduce the degrees of freedom to a manageable number (approximately 100) and then apply the upper-Hessenberg method.

Sample Theoretical Modal Analysis

In this sample modal analysis, taken from Carne *et al.* [1982], modal shapes and frequencies are determined for the *Sandia/DOE 2-m VAWT* shown in Figure 11-7. The configuration of the finite-element model of this VAWT is shown in Figure 11-8. The model is composed of two blades, a central column, and elastic supports. The curved blades are modeled with 20 beam elements each. The truss column, which adds a fair degree of complexity to the model, is represented by 150 beam elements. The masses of the hardware associated with the upper bearing and its support-cable connections are represented by a single concentrated mass placed at the top of the truss model. The support cables are modeled with horizontal, linear springs and a vertical downward force on the truss, as shown in Figure 11-8.

The masses of the test instrumentation installed on the rotor (which are not negligible compared to the mass of the blades) are included by adding appropriate concentrated masses. The base structure is modeled as a concentrated mass in combination with torsional and linear springs. The horizontal springs at the upper and lower bearings have equal stiffnesses in two orthogonal directions, producing the desired isotropic boundary conditions typical of VAWTs. The stiffness and mass properties of the support system must usually be estimated, at least to some degree. The initial estimates for this model are given in parenthesis in the figure. The final properties of the supports were obtained by "tuning" the model to match frequencies measured when the VAWT was not rotating, as will be discussed in the next section.

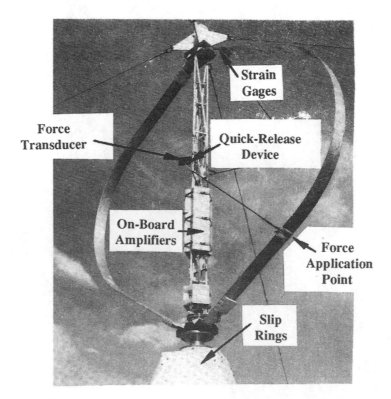

Figure 11-7. Sandia/DOE 2-m test VAWT with typical modal-survey instrumentation and excitation devices. (*Courtesy of Sandia National Laboratories*)

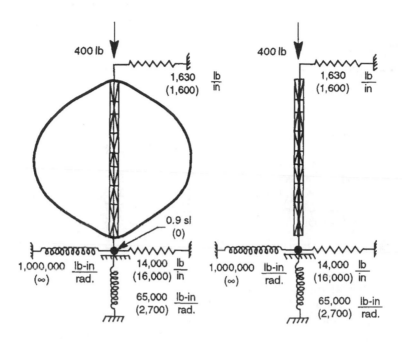

Figure 11-8. Finite-element model of the 2-m VAWT. [Carne *et al.* 1982]

Non-Rotating System Modes

Table 11-1 lists the frequencies of the first ten modes of vibration on the non-rotating (*i.e.*, parked) 2-m VAWT, with its rotor brake on. The first column gives the mode number, usually assigned in increasing order of frequency. Mode identification terminology is given in the second column. The corresponding mode shapes, in three views, are sketched in Carne *et al.* [1982]. The third column lists the frequencies measured with the rotor parked, and the fourth column gives the frequencies calculated using initial estimates of the support mass and stiffness properties. Again, these estimates are the data in parenthesis in Figure 11-8. In the fifth column of Table 11-1 are the "tuned" calculations of the modal frequencies for the non-operating condition. The last column in the table gives the relative changes in frequency resulting from tuning the model.

Tuning the Finite-Element Model

The differences between the initial and tuned frequencies, ranging from -20 to +8 percent, indicate the scale of the errors that can be present when (1) the support system and the resulting theoretical boundary conditions are complicated, and (2) some of the auxiliary components in the structural system are not modeled in detail. For practical reasons, the latter is usually the case. As a result, a finite-element model must be calibrated against a basic set of measured responses, on the same or a similar structure, to establish confidence in predictions made with it of natural frequencies under operating conditions. Modal analysis is clearly recognized to be a combination of theoretical and experimental methods.

Table 11-1.
Modal Frequencies of the Non-Rotating 2-m VAWT [Carne *et al.* 1982]

Mode Number	Mode name	Test (Hz)	Initial model (Hz)	Tuned model (Hz)	Tuning change (Hz)
1	1st Anti-Symmetric Flatwise	12.3	12.5	12.3	-1.6%
2	1st Symmetric Flatwise	12.5	12.6	12.4	-1.6%
3	1st Rotor Out-of-Plane	15.3	17.1	15.2	-11.1%
4	1st Rotor In-Plane	15.8	17.2	15.9	-7.6%
5	Dumbbell	24.4	22.6	24.4	+8.0%
6	2nd Rotor Out-of-Plane	26.2	30.5	26.2	-14.1%
7	2nd Rotor In-Plane	28.3	30.6	28.0	-8.5%
8	2nd Symmetric Flatwise	29.7	30.9	30.6	-1.0%
9	2nd Anti-Symmetric Flatwise	31.5	39.7	31.7	-20.2%
10	3rd Rotor Out-of-Plane	36.5	42.3	36.5	-13.7%

Calculated Rotating System Modes

The Campbell diagram in Figure 11-9 presents the modal frequencies of the 2-m VAWT at rotor speeds from 0 to 600 rpm. The curves in this figure are the calculated frequencies obtained using the tuned NASTRAN model, with mode numbers corresponding to those in Table 11-1. The variations of the frequencies with rotor speed are quite complex. While most increase monotonically with speed, some decrease monotonically. Others increase and then decrease and *vice versa*. These variations are similar to those associated with the classical whirling-shaft problem. In addition to complicated frequency variations, the mode shapes also change in character with rotor speed. Specifically, mode shapes which are completely *uncoupled* when the turbine is parked become coupled during rotation. This coupling is discussed in more detail in Carne *et al.* [1982]. In mathematical terms, the modes change from real to complex as the turbine rotates.

Sample Experimental Modal Analysis

An overview of the general techniques for modal testing of wind turbines will now be presented, again using the Sandia/DOE 2-m VAWT as an example. Portable data acquisition tools of the test engineer are based on *the fast Fourier transform*. The FFT technique, which involves exciting the structure with a force having a linear spectrum containing the frequency band of interest, is generally faster and more versatile than the classical *swept-sine* technique. The applied forces and responses are measured in the time domain and transformed to the frequency domain using the FFT. The frequency response functions are then computed from the cross-spectral and auto-spectral densities of the applied force and the responses.

Typically, several measurements are averaged to reduce the effects of uncorrelated noise. A more-complete description of FFT modal testing is contained in [Klosterman and Zimmerman 1975]. The greater versatility of this technique is a result of its more-relaxed requirements on the excitation force. For example, the FFT technique is equally applicable to shaker-driven excitations, instrumented-hammer impacts, and excitations by wind forces.

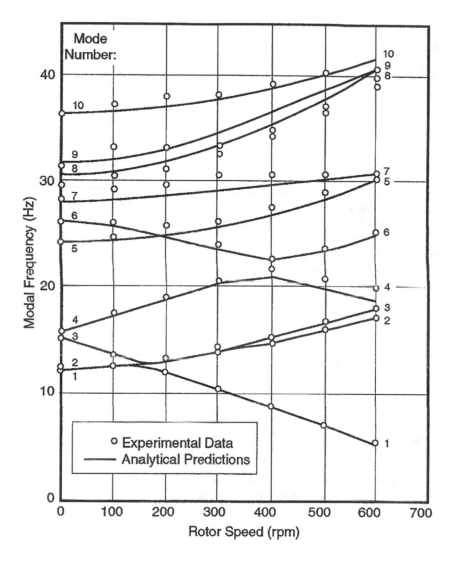

Figure 11-9. Campbell diagram of the Sandia/DOE 2-m VAWT, presenting both theoretical and experimental modal frequencies *vs*. rotor speed. [Carne *et al*. 1982]

Instrumentation

The general layout of instrumentation on the test VAWT is illustrated in Figure 11-7. For non-rotating tests, the turbine was instrumented with 18 piezoelectric accelerometers on the blades, 10 on the central column, 20 on the supporting pedestal, and 6 on the turbine shaft assembly. All input and response signals were low-passed at 100 Hz, amplified as necessary with signal-conditioning amplifiers, and recorded. In the rotating modal test the high, steady centrifugal accelerations prevented the use of accelerometers except on the central column. Strain gages are normally installed on newly-designed wind turbines undergoing development tests, and these gages are commonly used as response transducers for modal tests as well.

For this rotating test, strain gages were placed on the blades and central column, using a double-active gage in the bridge so that bending about a single axis would be measured. Gages were placed near the blade roots on leading and trailing edges (to measure chordwise bending), on the inside and outside surfaces at maximum blade thickness (to measure flat-wise bending), and on the central column (to measure in-plane and out-of-plane bending). On-board amplifiers boosted the strain signal to 2,000 psi/V for all gages, in order to give an adequate level for passing through the sliprings and the long wires that are typically required for modal tests on operating wind turbines. Because of expected difficulties in understanding the modes of the rotating central column, two piezoresistive accelerometers were attached to the upper column to record in- and out-of-plane motions.

Excitation

The scope of methods used to excite vibrations in wind turbines for modal analysis is broad. These methods fall into the four general categories of human, impact, step-relaxation, and natural-wind. As mentioned previously, the FFT-testing procedure allows flexibility in the excitation method, which is of significant benefit.

Human

Human excitation is the simplest approach to implement, but has the most limitations. The large displacement, low frequency modes of large wind turbines allow a test engineer to manually input a quasi-sinusoidal force into the structure to excite a single mode of the system. The frequency and damping parameters of the mode can be determined as the systems rings down after the excitation is removed. This approach was used to test over 20 commercial 17-m VAWTs in as little as 1¼ hours each. This study also provided valuable insight into the structural-dynamic variability of duplicate units in the field [Lauffer 1986].

Impact

Impact excitation methods range from using an instrumented 3-lb hammer to strike the 2-m VAWT at various locations on its blades and central tower to swinging a 1,000-lb ram from a crane to strike the tip of the NASA/DOE 2.5-MW Mod-2 HAWT [Boeing 1982]. In both cases, the impact force must be of sufficient amplitude and duration to excite the modes and frequencies of interest. This is accomplished by prior calibration tests.

Impact testing was also one of the techniques used to test the Sandia 34-m VAWT Test Bed in various stages of assembly. This allowed model correlation of substructure models before final model assembly. The substructures tested included two blade sections and a blade assembly. Step-relaxation and natural wind excitation were also used for other substructures, as appropriate [Carne *et al.* 1989].

Impact testing was also used to study the use of a Laser Doppler Velocimeter (LDV) for non-contact measurements of a HAWT in the field. This study used impact excitation to compare the LDV to traditional accelerometers as well as to study damage detection techniques. A comparison to natural wind excitation was also performed. The test object in this study, a 15 m diameter HAWT in its parked configuration [Rumsey *et al.* 1997].

Step-Relaxation

Step-relaxation relies of the quick release of strain energy in a pretensioned cable for the excitation of the structure. A small device applying a force of approximately 50 lb was used successfully on the 2-m VAWT [Carne *et al*. 1982]. A heated wire was used to burn through a nylon cord to affect the quick release of the excitation force. For a modal test of the 64-m diameter *Eolé* (Fig. 3-36) two separate force-application points were used, one on the central column and one on a blade [Carne *et al*. 1988]. Forces were applied through high-strength steel cables loaded with a diesel-powered winch and restrained by a nylon strap. Loads of 135 kN (30,000 lb) and 45 kN (10,000 lb) were applied to the column and blade, respectively.

One difficulty in using step relaxation is related to the FFT algorithm. A quick release basically applies a force which is a *Heaviside* function, for which there is no Fourier transform. However, if the Heaviside function is passed through a high-pass filter, it can be converted to a well-behaved function that is easily transformable with the FFT.

Natural Wind

Using the natural wind to excite the turbine has certain advantages, as discussed in [Carne *et al*. 1988]. Two of the most significant are the greatly reduced cost and complexity of the test and the ability to test during windy conditions. Wind-excitation methods were developed during test of the Eolé VAWT, and these produced modal frequencies in excellent agreement with those obtained by step relaxation. However, damping information is not as readily available from the power spectra obtained using wind excitation. Development of the *natural excitation technique* (NExT) rectified this situation and will be discussed below [James *et al*. 1995].

Measured Rotating System Modes of the 2-m VAWT

The results of the modal testing performed on the 2-m VAWT using step-relaxation excitation are shown by the data points in Figure 11-9. The correlation here is typical of wind turbine modal analysis. The average absolute deviation of the experimental frequencies from the theoretical (tuned-model) frequencies is 0.5 Hz, or 2.2 percent. Deviations generally increase with increasing rotor speed, which is a characteristic of finite-element models that do not have a sufficient number of degrees of freedom to accurately represent complex mode shapes.

Development of Natural Wind Excitation Testing

The structural dynamics of rotating VAWT machines are critical design conditions that require experimental techniques to match the fidelity of the advanced analyses described previously. The previous example using the 2-m VAWT showed the power of performing step-relaxation testing during rotation. However, this approach is technically intensive and would not be generally applicable for many operational scenarios. As mentioned previously, the use of natural wind excitation is generally a more applicable approach and allows testing in the actual operating environment. This section details the development of the natural excitation technology in the form of the *natural excitation technique* or NExT [James *et al.* 1995]. A more complete background on the historical development of the natural wind excitation testing for wind turbines is also available [Carne and James 2008].

Theoretical Development of NExT

A critical step in the development of NExT was to find a relationship between modal parameters of structural interest (such as frequency and damping) and a function that could be produced from measured response data. For NExT, the *cross-correlation function* between two response–only measurements was used. The derivation of the relationship of modal parameters and the cross-correlation function begins by assuming the standard matrix equations of motion:

$$[M]\{\ddot{x}(t)\} + [C]\{\dot{x}(t)\} + [K]\{x(t)\} = \{f(t)\} \tag{11-50}$$

where

$$[M] = \text{mass matrix (g)}$$
$$[C] = \text{damping matrix(N-s/m)}$$
$$[K] = \text{stiffness matrix (N/m)}$$
$$\{f\} = \text{vector of random forcing functions (N)}$$
$$\{x\} = \text{vector of random displacements (m)}$$
$$t = \text{elapsed time (s)}$$
$$\dot{x}, \ddot{x} = \text{first and second derivatives with respect to time (m/s, m/s}^2)$$

Equation (11-50) can be expressed in modal coordinates using a standard modal transformation after performing a standard matrix diagonalization (assuming proportional damping). A solution to the resulting scalar modal equations can be obtained by means of a Duhamel integral, assuming a general forcing function, $\{f\}$, with zero initial conditions (*i.e.* zero initial displacement and zero initial velocity). The solution equations can be converted back into physical coordinates and specialized for a single input force and a single output displacement using appropriate mode shape matrix entries. The following equations result:

$$x_{ik}(t) = \sum_{r=1}^{n} \psi_{ir}\psi_{kr} \cdot \int_{-\infty}^{t} f_k(\tau)g^r(t-\tau)d\tau \tag{11-51}$$

$$g^r(t) = 0, \ for \ t < 0$$

$$g^r(t) = \frac{1}{m^r\omega_d^r} \exp\left(-\zeta^r\omega_n^r t\right) \sin\left(\omega_n^r t\right), \ for \ t \geq 0;$$

where

$$\omega_d^r = \omega_n^r (1 - \zeta^{r2})^{1/2} \text{ is the damped modal frequency (rad/s)}$$
$$\omega_n^r = rth \text{ modal frequency (rad/s)}$$

$$\zeta^r = rth \text{ modal damping ratio}$$
$$m^r = rth \text{ modal mass (kg)}$$
$$n = \text{number of modes;}$$
$$\psi_{ir} = ith \text{ component of mode shape r}$$

The next step of the theoretical development is to form the cross-correlation function of two responses (x_{ik} and x_{jk}) to a white noise input at a particular input point k. The cross-correlation function $\Re_{ijk}(T)$ can be defined as the expected value of the product of two responses evaluated at a time separation of T:

$$\Re_{ijk}(T) = E[x_{ik}(t+T)\, x_{jk}(t)] \tag{11-52}$$

where E is the *expectation operator*.

Substituting Equation (11-51) into (11-52), recognizing that $f_k(t)$ is the only random variable, allows the expectation operator to depend only on the forcing function. Using the definition of the *autocorrelation function*, and assuming for simplicity that the forcing function is white noise, the expectation operation collapses to a scalar times a *Dirac delta function*. The Dirac delta function then collapses one of the Duhamel integrations embedded in the cross-correlation function. The resulting equation can be simplified via a change of variable of integration, setting $\lambda = t$-τ. Using the definition of g from Equation (11-51) and the trigonometric identity for the sine of a sum results in all the terms involving T being separated from those involving λ. This separation allows terms that depend on T to be factored out of the remaining integral and out of one of the modal summations, with the following results:

$$\Re_{ijk}(T) = \sum_{r=1}^{n} \left[\begin{array}{l} A_{ijk}^{r}\ \exp\left(-\zeta^r\omega_n^r\,T\right)\ \cos\left(\omega_d^r\,T\right) \\[2mm] +\ B_{ijk}^{r}\ \exp\left(-\zeta^r\omega_n^r\,T\right)\ \sin\left(\omega_d^r\,T\right) \end{array} \right] \tag{11-53}$$

where A_{ijk}^r and B_{ijk}^r are independent of T, are functions of only the modal parameters, and completely contain the remaining modal summation, as shown in the following equations:

$$\left\{ \begin{array}{c} A_{ijk}^{r} \\[4mm] B_{ijk}^{r} \end{array} \right\} = \sum_{s=1}^{n} \frac{\alpha_k\,\psi_{ir}\psi_{kr}\psi_{js}\psi_{ks}}{m^r\,\omega_d^r\,m^s\,\psi_d^s} \cdot \int_0^{\infty} \exp\left(\begin{array}{c} -\zeta^r\omega_n^r \\[2mm] -\zeta^s\omega_n^s \end{array} \right) \lambda \cdot \sin\left(\omega_d^s\,\lambda\right) \left\{ \begin{array}{c} \sin\left(\omega_d^r\,\lambda\right) \\[2mm] \cos\left(\omega_d^r\,\lambda\right) \end{array} \right\} d\lambda. \tag{11-54}$$

Equation (11-53) is the key result of this derivation. Examining Equation (11-53), the cross-correlation function is seen to be a sum of decaying sinusoids, with the same characteristics as the impulse response function of the original system. Thus, cross-correlation functions can be used as impulse response functions in time-domain modal parameter estimation schemes. This is more clearly seen after further simplification using a new constant multiplier (Gj^r), as follows:

$$R_{ij}(T) = \sum_{r=1}^{n} \frac{\psi_{ir}G_j^r}{m^r\,\omega_d^r} \exp\left(-\zeta^r\omega_n^r\,T\right) \sin\left(\omega_d^r\,T + \Theta r\right). \tag{11-55}$$

James *et al.* [1995] present more definition of the intermediate steps in this derivation and the results of analytical verification checks of the derivation.

Comparison of NExT Test Data from a Rotating VAWT to Simulated Data

An interesting comparison was performed between NExT data processing results and simulated data and using the known modal parameters. For this activity, a simulation code (VAWT-SDS), was used to compute the time domain response of the DOE/Sandia 34-m VAWT test bed (Figure 2-2) during rotation in turbulent wind [Dohrmann and Veers 1989]. The structural model (including rotational effects) used in VAWT-SDS was available to calculate the analytical modal frequencies and analytical modal damping. VAWT-SDS was used to generate analytical data, which were then input to NExT. The results were compared to the known frequencies and damping information to test the capabilities of NExT.

Simulated data were generated for the 34-m testbed using a 30 rpm rotation rate and 20 mph (8.9 m/s) turbulent winds with 15 percent turbulence intensity. Stiffness proportional damping, sufficient to produce a damping ratio of 0.2 percent at 1.4 Hz was added to the model. Time histories of 2,048 points for eight strain gauge outputs were generated using a step size of 0.04 s. Sensor noise was simulated by adding a white-noise signal to each simulated time history. The standard deviation of this additive signal was 2 percent of the standard deviation of each time history. The analytical modal frequencies and damping ratios were also calculated by extracting the complex eigenvalues from the structural matrices used in the VAWT-SDS code. VAWT-SDS used the Newmark-Beta numerical integration scheme, and the approximations inherent in this procedure produced period elongations. The frequency shifts created by numerical integration were calculated and a correction was added to the analytical values [James *et al.* 1996].

NExT was used to estimate modal frequencies and damping ratios from the simulated data so that the NExT results and the analytical values could be compared. Table 11-2 shows the results of the comparison. With NExT, it was possible to correctly extract the frequencies and damping from the data with the exception of the first two modes at 1.27 Hz and 1.35 Hz. These two modes are very closely spaced, making it difficult to obtain accurate

Table 11-2.
Comparison of NExT With Simulated Results [James *et al.* 1993]

Mode	Frequency (Hz)		Damping (%)	
	Simulated	**NExT**	**Simulated**	**NExT**
1st Flatwise Antisymmetric	1.27	1.31	0.2	0.4
1st Flatwise Symmetric	1.35	1.32	0.2	0.3
1st Blade Edgewise	1.59	1.59	0.3	0.3
1st Tower In-Plane	2.02	2.01	0.3	0.4
2nd Flatwise Symmetric	2.43	2.44	0.4	0.5
2nd Flatwise Antisymmetric	2.50	2.50	0.4	0.4
1st Tower Out-of-Plane	2.80	2.80	0.3	0.5
2nd Rotor Twist	3.39	3.39	0.5	0.6
2nd Tower In-Plane	3.46	3.45	0.5	0.4
3rd Flatwise Antisymmetric	3.65	3.63	0.5	0.4
3rd Flatwise Symmetric	3.73	3.73	0.6	0.4
2nd Blade Edgewise	3.88	3.87	0.5	0.3

frequency or damping results from NExT. It should also be noted that the higher modes (3.65 Hz, 3.73 Hz, and 3.88 Hz) have NExT-estimated damping ratios that are lower than the specified damping ratios. The amplitudes of these modes are low compared to the noise level, which affected the estimates. This finding suggested the need for longer time histories to improve the accuracy. Also, this comparison shows the validity of natural wind excitation as a tool for wind turbine testing, as well as pointing out the issues (such as period elongations) associated with numerical integration schemes used in analysis codes.

Comparison of NExT to Non-Rotating Step Relaxation Excitation

Another insightful comparison is between NExT processing using wind excitation and traditional analysis using step-relaxation excitation. For this comparison, a *FloWind Corporation* 19-m VAWT in Altamont Pass, CA, was tested using step-relaxation modal testing techniques during quiescent daytime winds. NExT was then used during periods of more substantial nighttime winds (above 7 m/s or 16 mph). The turbine was parked (nonrotating) during all testing. Accelerometers were used to measure the response at predetermined locations on the turbine. This allowed a comparison between modal parameters estimated by NExT and modal parameters estimated using conventional testing techniques (step-relaxation testing).

Table 11-3 compares the modal frequencies and modal damping ratios of the 19-m VAWT as determined from conventional testing and from NExT. The two methods produced estimates of the modal frequencies that are in good agreement, particularly in view of the temperature difference between day and night. The average difference for the ten modes is only 0.5 percent. Also, the modal damping ratios of all six of the tower modes (rotor twist, tower in-plane, tower out-of-plane) are very similar. However, the modal damping ratios of all four blade flatwise modes (flatwise symmetric and flatwise antisymmetric) are substantially higher from NExT estimates.

Table 11-3.
Comparison of NExT With Step-Relaxation Excitation [James *et al.* 1993]

Mode	Frequency (Hz)		Damping (%)	
	Step Relax	NExT	Step Relax	NExT
1st Rotor Twist	2.37	2.38	0.2	0.1
1st Flatwise Antisymmetric	2.48	2.49	0.2	1.3
1st Flatwise Symmetric	2.51	2.51	0.1	1.4
1st Tower Out-of-Plane	2.72	2.76	0.4	0.4
1st Tower In-Plane	3.11	3.15	0.4	0.4
2nd Tower Out-of-Plane	4.53	4.53	0.1	0.1
2nd Flatwise Antisymmetric	5.30	5.31	0.3	0.8
2nd Flatwise Symmetric	5.64	5.65	0.1	0.6
2nd Rotor Twist	6.59	6.62	0.1	0.1
2nd Tower In-Plane	6.64	6.71	0.3	0.6

Subsequent analysis suggested that these differences were due to a drag phenomenon experienced as the largest cross-section of the blade oscillated normal to the stronger night-time winds [James *et al.* 1995]. This exercise showed the utility of performing modal testing in an operating environment to allow the extraction of the total damping (structural and aeroelastic). This test series also pointed out the reduced infrastructure associated with the natural excitation test while clarifying the utility of traditional testing to separate out the different physical phenomena acting on wind turbine structures.

Application of NExT to a Rotating VAWT

The significant utility of wind excitation is the ability to extract structural dynamics information from a rotating system. NExT was used to extract modal frequency and damping using data from the DOE/Sandia 34-m testbed during rotation at 0, 10, 15, 20, 28, 34, and 38 rpm. The wind speed during these tests was also approximately 10 m/s (22 mph). Twelve strain gauges (accessed through slip rings) were used as the sensors in all of these analyses. Time history length, sample rate, correlation block size, and the time domain analysis parameters were all the same as provided in the previous section. Figure 11-10 is a plot of the rotation-rate dependent modal frequencies of this turbine from analysis and from experiment using NExT.

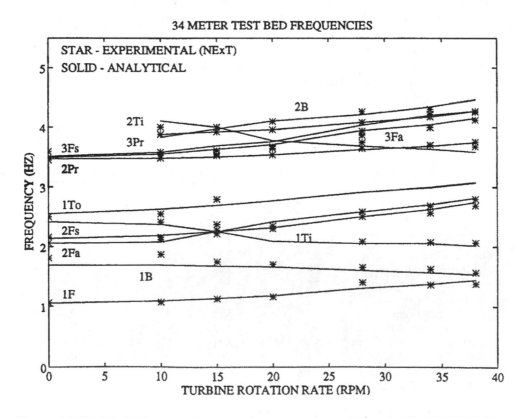

Figure 11-10. Modal frequencies versus rotor rotation rate for the Sandia/DOE 34-m VAWT. [James *et al.* 1993]

Figure 11-11 shows a plot of modal damping ratios, as calculated with NExT, versus turbine rotation rate for the blade flatwise modes. The damping ratios of these modes generally increase significantly with turbine rotation rate. For example, the first flatwise damping ratio increases from 2 percent to 7 percent. The notable exception is the second blade flatwise mode that drops between 15 rpm and 28 rpm. Such a drop in damping ratio is likely due to modal coupling to a more lightly damped mode, since modal coupling varies with rotation speed. This data validates the need to understand the variation in parameters in operating conditions including rotation rate effects, especially in variable-speed machines or during start-up/shut-down.

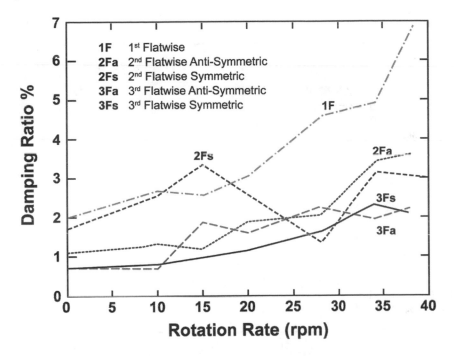

Figure 11-11. Modal damping versus rotor rotation rate for the Sandia/DOE 34-m VAWT. [James *et al.* 1993]

Application of NExT to HAWT Systems

Table 11-4 shows the results of comparison between impact excitation and natural excitation on a parked HAWT. In this case the machine was an *Atlantic Orient Corporation* (AOC) 15/50, which is a three-bladed turbine with a 15-m rotor mounted downwind of the tower. The results show a very close agreement between the measured frequencies and a typical agreement for the damping values [Rumsey *et al.* 1997].

A controlled yaw or upwind-rotor HAWT is the most direct application of NExT to a rotating turbine, as the kinematics of the system are primarily limited to rotation of the blades. This class of turbine shows the same dynamic characteristics as the rotating VAWT, including distinct natural frequencies, environment-dependent damping, and rotation-dependent forced harmonics. NExT has been applied to field results from a *Northern Power Systems* 100-kW HAWT, which has a two-bladed teetering rotor, 17.8 m in diameter [James 1994]. Figure 11-12 displays some representative data from this analysis. In this case the damping in the 5.34 Hz edgewise mode is plotted versus wind speed. The damping is seen to have a clear and increasing trend as the wind speed increases.

Table 11-4.
Comparison of Two Modes from a Parked AOC 15/50 HAWT
Using NExT and Impact Excitation [Rumsey *et al.* 1997]

Mode	Frequency (Hz)		Damping (%)	
	Impact	**NExT**	**Impact**	**NExT**
Teeter	3.19	3.18	1.18	1.58
Umbrella	3.72	3.73	1.11	1.27

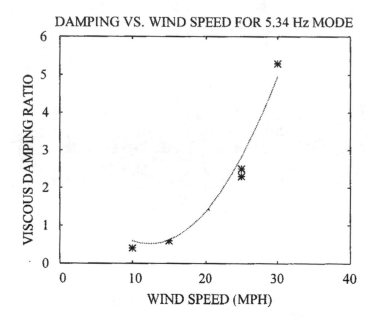

Figure 11-12. Damping versus wind speed for the 5.34 Hz edgewise mode of the Northern Power Systems 100-kW HAWT. [James 1994]

A more challenging dynamical system is a free-yaw HAWT with its rotor mounted downwind of the tower. This turbine configuration can be simpler to build, but the uncontrolled yaw degree-of-freedom couples to the other in-plane dynamic modes and creates entire families of frequency-dependent phenomena. Figure 11-13 shows an example from the free-yaw 26-m diameter *Advanced Wind Turbines* AWT-26, which is a downwind teetering-rotor machine [Malcolm and James 1995]. The expected per-rev harmonic forcing function frequencies are plotted as the dotted lines. The solid lines denote the analytically-predicted natural frequency families. These frequencies are only distinct in the non-rotating condition.

As the machine rotates, natural frequencies split into multiple rotation-rate dependent frequencies. NExT was used to process field data from this HAWT as it was rotating at 58 rpm. There is a slight mismatch between the analysis and the field test data. This mismatch illustrates the utility of extracting structural dynamics data from an operating machine for comparisons to analytical data.

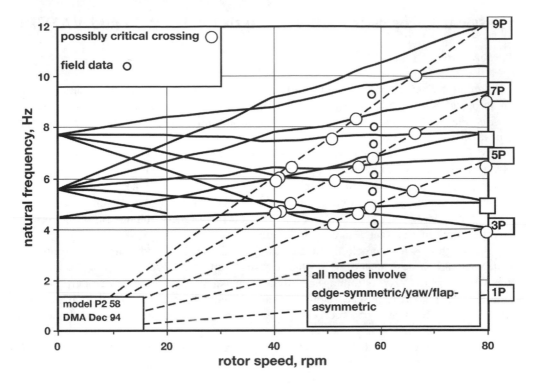

Figure 11-13. Periodic natural frequency families including NExT-derived field data *vs.* **rotor speed for the AWT-26 HAWT.** [Malcolm and James 1995]

Classical Aeroelastic Flutter Analysis for HAWTs and VAWTs

Background

   Flutter is a self-starting and potentially destructive vibration where aerodynamic forces on an object couple with a structure's *natural mode of vibration* to produce rapid *periodic motion*. Flutter can occur in any object within a strong fluid flow, under the conditions that a *positive feedback* occurs between the structure's natural vibration and the aerodynamic forces. That is, when the vibrational movement of an object like a flexible airfoil increases an aerodynamic load which in turn drives the object to move further. If the energy during the period of aerodynamic excitation is larger than the natural *damping* of the system, the level of vibration will increase. The vibration levels can thus build up and are only limited when the aerodynamic or mechanical damping of the object match the energy input. This often results in large amplitudes and can lead to rapid failure.

   Although classical aeroelastic flutter has generally not been a driving issue in utility-scale wind turbine design, one case in which classical flutter was observed involved a vertical-axis wind turbine (VAWT) turning in still air [Popelka 1982]. The rotor was purposefully motored at ever-increasing speeds until the flutter boundary was breached and dramatic classical flutter oscillations were observed. Flutter occurred at approximately twice the design operating speed of the rotor.

   For very large horizontal-axis wind turbines (HAWTs) fitted with relatively soft (flexible) blades, classical flutter becomes a more important design consideration. Innovative blade designs involving the use of *aeroelastic tailoring*, wherein the blade twists as it bends under the action of aerodynamic loads to shed loads resulting from wind turbulence, increase the blades proclivity for flutter.

Analytical Model of Flutter

   The analysis of classical flutter in wind turbines necessitates the use of unsteady aerodynamics. As pointed out by Leishman [2002], for horizontal axis wind turbines there are two interconnected sources of unsteady aerodynamics. The first is a result of the *trailing wake* of the rotor and is addressed by investigating the interactions between the rotor motion and the inflow. The second, which will be the focus here, is due to the *shed wake* of the individual blades and can be addressed using techniques developed for analyzing flutter in fixed-wing aircraft.

   To simplify the analysis, the rotor is assumed to be turning in still air on a hub fixed in space. Because there is no wind inflow, unsteady aerodynamics caused by the trailing wake can be neglected. Consequently, the aerodynamic behavior of a single blade is similar to that of a fixed wing with a free stream velocity that varies linearly from the root to the tip, assuming that the shed wake of the preceding blade dies out sufficiently fast so that the oncoming blade will encounter essentially still air. Focusing on aeroelastic stability associated with the shed wake from an individual HAWT blade, the technique developed by Theodorsen [Theodorsen 1935, Fung 1969, Dowell 1995] for fixed wing aircraft has been adapted for use with HAWT blade flutter [Lobitz 2004].

   The Theodorsen technique specifically addresses classical flutter in an infinite (*i.e.* two dimensional with no end effects) airfoil undergoing oscillatory pitching and plunging motion in an incompressible flow, as illustrated in Figure 11-14. The airfoil's pitching motion is represented by the angle α and its plunging motion by the vertical translation h. L represents the lift force vector positioned at the quarter-chord, M is the pitching moment about the *elastic axis* and U is the free-stream wind velocity. The origin of the coordinate system is positioned along the chordline at the section elastic axis. Note that a is defined as the fraction of b (the half-chord) that the elastic axis is aft of the mid-chord. Thus in Figure 11-14, a is negative since the elastic axis is ahead of the mid-chord.

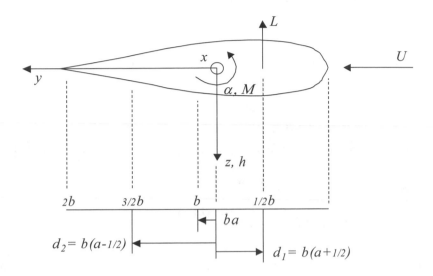

Figure 11-14. Schematic diagram of motions of a blade cross section assumed in the application of the Theodorsen technique for analyzing the flutter of a fixed wing and adapted for wind turbine blades.

The blade is assumed to be simultaneously pitching and plunging in an oscillatory fashion at a modal natural frequency ω, as follows:

$$h = h_0 e^{i\omega t} \quad \text{and} \quad \alpha = \alpha_0 e^{i\omega t} \tag{11-56}$$

where

h = vertical displacement (m)
h_0 = complex constant (m)
ω = modal natural frequency (rad/s)
t = elapsed time (s)
α = pitch angular displacement (rad)
α_0 = complex constant (rad)

According to the Theodorsen model, the lift force, L, and the pitching moment, M, are given by Equations (11-57), where a, b, d_1, and d_2 are as shown in Figure 11-14.

$$L = 2\pi\rho U^2 b \left\{ \begin{array}{l} \dfrac{i\omega C(k)}{U} h_0 + C(k)\alpha_0 + [1 + C(k)(1 - 2a)]\dfrac{i\omega b}{2U}\alpha_0 \\[2mm] -\dfrac{\omega^2 b}{2U^2} h_0 + \dfrac{\omega^2 b^2 a}{2U^2}\alpha_0 \end{array} \right\} \tag{11-57a}$$

$$M = 2\pi\rho U^2 b \left\{ \begin{array}{l} d_1 \left[\dfrac{i\omega C(k)}{U} h_0 + C(k)\alpha_0 + [1 + C(k)(1-2a)] \dfrac{i\omega b}{2U}\alpha_0 \right] + d_2 \dfrac{i\omega b}{2U}\alpha_0 \\[3mm] -\dfrac{\omega^2 ab^2}{2U^2} h_0 + \left(\dfrac{1}{8} + a^2 \right) \dfrac{\omega^2 b^3}{2U^2}\alpha_0 \end{array} \right\}$$

$$(11\text{-}57b)$$

where

L = lift force intensity; force per unit length of span (N/m)
M = pitch moment intensity, moment per unit length of span (N-m/m)
U = resultant airspeed (m/s)
$C(k)$ = complex-valued Theodorsen function
k = reduced frequency (rad)

The Theodorsen function, $C(k)$ and the reduced frequency, k, are defined as follows:

$$k = \omega b/U \qquad\qquad (11\text{-}58a)$$

$$C(k) = \frac{H_1^{(2)}(k)}{H_1^{(2)}(k) + iH_0^{(2)}(k)} \qquad\qquad (11\text{-}58b)$$

where H denotes the Hankel function. The real and imaginary parts of C are displayed graphically in Figure 11-15.

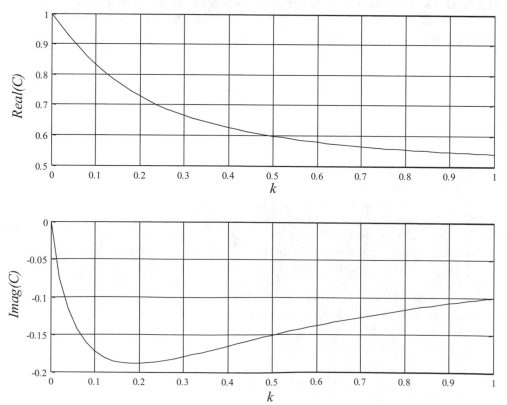

Figure 11-15. Real and imaginary parts of the Theodorsen function $C(k)$ where k is the reduced frequency.

In order to incorporate Equations (11-57) in a finite element procedure for subsequent *complex eigenvalue analysis*, they must be recast in a pseudo time domain form for developing contributions to the finite element mass, stiffness and damping matrices. This can be accomplished by leaving the Theodorsen function as is, but using the explicit ω's in the equations to construct time derivatives, producing the following equations:

$$L = 2\pi\rho U^2 b \left\{ \frac{C(k)}{U}\dot{h} + C(k)\alpha + [1 + C(k)(1 - 2a)]\frac{b}{2U}\dot{\alpha} + \frac{b}{2U^2}\ddot{h} - \frac{b^2 a}{2U^2}\ddot{\alpha} \right\} \quad (11\text{-}59a)$$

$$M = 2\pi\rho U^2 b \left\{ \begin{array}{l} d_1\left[\frac{C(k)}{U}\dot{h} + C(k)\alpha + [1 + C(k)(1 - 2a)]\frac{b}{2U}\dot{\alpha}\right] + d_2\frac{b}{2U}\dot{\alpha} \\[2mm] + \frac{ab^2}{2U^2}\ddot{h} - \left(\frac{1}{8} + a^2\right)\frac{b^3}{2U^2}\ddot{\alpha} \end{array} \right\} \quad (11\text{-}59b)$$

Now, using Equations (11-59) with the principle of virtual work, contributions to the finite element stiffness, mass and damping matrices can be developed that include the complex-valued Theodorsen function.

However, before that is done, it is noted that the HAWT blade does not conform to the configuration of an infinitely long uniform wing to which the above equations apply. Rather, such quantities as the semi-chord, b, and the resultant velocity, U, vary significantly along the span of the blade. Moreover, the lift curve slope, which in the above equations is assigned the theoretical value of 2π corresponding to a flat plate, varies with blade span. These variations can be approximated by assembling a conglomerate of uniform blades, wherein the above equations are assumed to be applicable incrementally. Specifically, the quantities mentioned above are represented by a linear variation over the length of the element and included within the integral over the element length associated with the principle of virtual work.

Significant detail can be incorporated in this analysis by refining some of the variable quantities mentioned above. For example, inboard stall for a twisted blade turning in still air might be accommodated by reducing the lift curve slope in that region of the blade. More complicated inflow variations can also be accommodated if they are known *a priori*. However, in comparison with other approximations that have been made in this analysis, these types of refinements are deemed excessive. The simple aeroelastic stability analysis presented here is meant to serve primarily as a common sense check for use during the blade design process.

In matrix notation, the finite element equation to be used for investigating the aeroelastic stability of the wind turbine blade is as follows:

$$[M + M_a(\Omega)]\{\ddot{u}\} + [C_C(\Omega) + C_a(\omega, \Omega)]\{\dot{u}\}$$
$$+ [K(u_0, \Omega) + K_{tc} + K_{cs}(\Omega) + K_a(\omega, \Omega)]\{u\} = 0 \quad (11\text{-}60)$$

where M is the conventional mass matrix and $K(u_0, \Omega)$ is the stiffness matrix, centrifugally stiffened commensurate with displacements, u_0, resulting from centrifugal loads corresponding to the rotor rotational speed, Ω. The displacement, u, and its time derivatives represent motion about this centrifugally loaded state. The α and h degrees of freedom of Equations (11-59) are included in u.

The blade is modeled with NASTRAN tapered beam (CBEAM) elements (developed by the MSC Software Corp.) that have six degrees of freedom at either end. These elements also have provisions for shear deformation and warping of the cross section, although these features were not used in this analysis. Additional terms are added to the various matrices to provide for rotating coordinate system effects. These are the Coriolis and centrifugal softening terms, $C_c(\Omega)$ and $K_{cs}(\Omega)$, respectively, that will be discussed in detail later. The aeroelastic matrices, $M_a(\Omega)$, $C_a(\omega,\Omega)$ and $K_a(\omega,\Omega)$, all depend on Ω, since the free-stream velocity at any blade radius is governed by it (it is assumed that the rotor is turning in still air). Linear shape functions are used in the development of these aeroelastic matrices.

As a demonstration of the method used to generate the aeroelastic matrices, a single beam element's contribution to K_a is presented in Equation (11-61). The element has a length, l, and is bounded by nodes 1 and 2. The stiffness contribution is due to the α terms of Equations (11-59) and can be developed with the principle of virtual work using only α virtual displacements. Thus, for α varying linearly along the element (i.e. $\alpha(x) = \alpha_1(1\text{-}x/l) + \alpha_2 x/l$) the contribution of this single element is as follows:

$$
\begin{Bmatrix} L_1 \\ L_2 \\ M_1 \\ M_2 \end{Bmatrix} = \rho \int_0^l a_0(x)U^2(x)b(x)C(k(x)) \begin{bmatrix} \left(1-\frac{x}{l}\right)^2 & \frac{x}{l}\left(1-\frac{x}{l}\right) \\ \frac{x}{l}\left(1-\frac{x}{l}\right) & \left(\frac{x}{l}\right)^2 \\ d_1(x)\left(1-\frac{x}{l}\right)^2 & d_1(x)\frac{x}{l}\left(1-\frac{x}{l}\right) \\ d_1(x)\frac{x}{l}\left(1-\frac{x}{l}\right) & d_1(x)\left(\frac{x}{l}\right)^2 \end{bmatrix} dx \begin{Bmatrix} \alpha_1 \\ \alpha_2 \end{Bmatrix}
$$

$$(11\text{-}61)$$

Here the quantity 2π of Equations (11-59), which represents the lift curve slope for a flat plate, is replaced by $a_0(x)$ representing the lift curve slope of an airfoil. This slope varies along the length of the element as does the free stream velocity, U, the semichord, b, the Theodorsen function C, and the distance from the elastic axis to the quarter-chord, d_1. The integral is evaluated numerically. Elemental contributions to M_a and C_a are developed in an entirely similar manner.

Commensurate with the use of the NASTRAN CBEAM element above, the NASTRAN commercial finite element software is used for this aeroelastic stability investigation. The contributions to the stiffness, mass, and damping matrices discussed above are supplied by means of a NASTRAN input option. NASTRAN can accommodate non-symmetric, complex-valued matrices as are required in this effort, and it provides a number of complex eigenvalue solvers for the stability analysis.

In addition to Ω, the aeroelastic matrices, $C_a(\omega,\Omega)$ and $K_a(\omega,\Omega)$, also depend on ω, the natural frequency of the mode shape of interest, which occurs in the argument of the Theodorsen function. Since this frequency is unknown at the onset of the computations an iterative process is required for obtaining accurate results. The iterative procedure developed for the aeroelastic stability analysis of the rotor blade is composed of the following steps:

1. Select a value for Ω .
2. In a quasi-static NASTRAN run, create $K(u_0,\Omega)$ for subsequent eigenvalue analysis.

3. Provide an initial guess for ω or update it from the prior calculation.
4. Using a NASTRAN complex eigenvalue solution procedure, compute modes, frequencies and damping coefficients.
5. Select a mode with a small or negative damping coefficient and return to step 3 with corresponding frequency update.
6. When the prior updated frequency is sufficiently close to the subsequently computed one, either suspend computations or modify Ω and return to step 1.

The end goal in this process is to identify the eigenmode that exhibits a negative damping coefficient for the lowest rotor rotational speed. This speed is designated as the *classical flutter speed* for the blade. A comprehensive approach would be to sequentially investigate the stability of each mode by choosing its frequency as the initial guess for ω, and tracking that frequency over a range of increasing Ω. The flutter speed would then correspond to the lowest value of Ω where the damping becomes negative over the set of modes investigated. In practice, however, the first torsional blade mode produces the lowest values of Ω, coupling with other blade modes (primarily flapwise modes) as the rotational speed increases. Of course it is always prudent to check other modes that appear suspicious during the course of the analysis.

Sample Flutter Calculation

After developing some confidence in the above computational procedure by comparing its results to those of Theodorsen, it was applied to the large relatively soft wind turbine blade shown in the wire-frame illustration of Figure 11-16. This blade, termed the baseline blade, was designed as part of the WindPACT Rotor Design Study [Malcolm and Hansen 2002]. Some characteristics of this blade and the associated rotor are listed in Table 11-5.

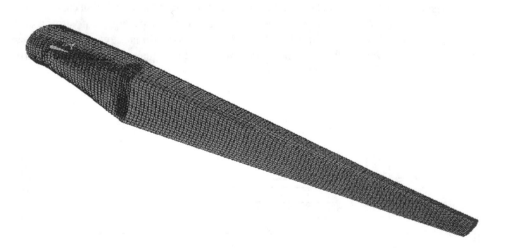

Figure 11-16. Wire-frame illustration of the 1.5 MW baseline WindPACT blade.
[Malcom and Hansen 2002]

Table 11-5.
Characteristics of the Baseline 1.5 MW WindPACT Blade and Rotor

Characteristic	Units	Value
Rated power	MW	1.5
Rotor diameter	m	70
Max rotor speed	rpm	20.5 (0.342 Hz)
Max tip speed	m/s	75
Blade coning	deg	0
Max blade chord	m	8% of radius
Radius to blade root	m	5% of radius
Rotor solidity		0.05
Blade mass	kg	4,230
Hub mass	kg	15,104
Total rotor mass	kg	32,016
1st flapwise frequency	Hz	1.233 (1.199)
1st edgewise frequency	Hz	1.861 (1.714)
2nd flapwise frequency	Hz	3.650 (3.596)
1st torsional frequency	Hz	9.289 (9.846)

Although not readily apparent from Figure 11-16, the blade has a modest amount of twist, from 11.1 deg. at the root to 0.0 deg. at the tip. The modal frequencies in parentheses in this table were computed with the software developed herein with Ω set to zero and the Theodorsen function, $C(k)$, set to 1.0. They differ from the companion frequencies taken from the WindPACT study, which were computed using the ADAMS software (developed by the MSC Software Corp.). The two models are inherently different, yet the frequency differences are small. Only minimal effort was made to minimize these differences.

In Figure 11-17 a planform view of the blade is presented showing the positions of the elastic axis and the axis representing the locus of the cross-sectional center of gravity. As the elastic axis is forward of the mid-chord, the quantity a in Equations (11-59) is negative, varying along the span from -0.354 to -0.284. Since in this implementation a constant value of a is required, a mid value of -0.32 was selected.

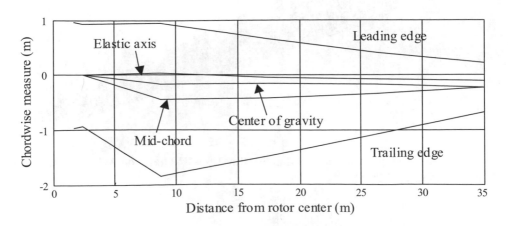

Figure 11-17. Diagram of the planform of the baseline 1.5 MW WindPACT blade showing the locations of the elastic axis and the cross-sectional centers of gravity.

Flutter Analysis Results

Aeroelastic stability predictions with the rotor turning in still air were made for this baseline model. Results indicate that the onset of flutter occurs at a rotational speed of 43.4 rpm (0.723 Hz) which is 2.14 times the maximum design operating speed of the rotor. As shown in Figure 11-18, the mode shape associated with the onset of flutter contains significant amounts of edgewise (motion in the direction of the chord), flapping, and pitching motion. Its frequency of oscillation is 6.234 Hz, which is significantly lower than the 1st torsional frequency of 9.289 Hz.

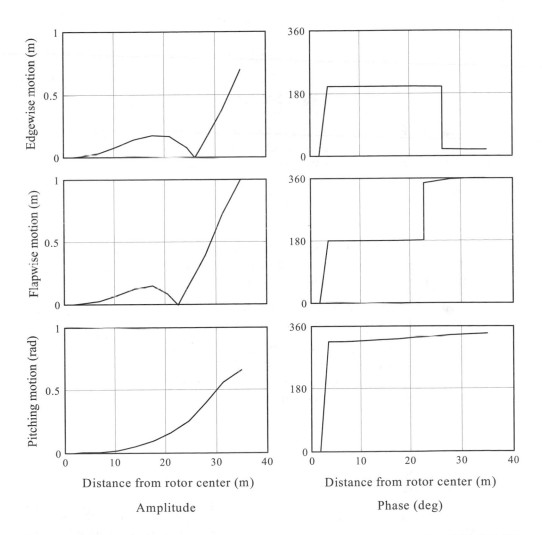

Figure 11-18. Mode shape at the onset of flutter for the 1.5 MW baseline WindPACT uncoupled blade.

In classical flutter, motion in the edgewise direction is generally minimal, and the large amount predicted here is thought to be due to the fact that the blade is twisted, and that the frequency of the second edgewise mode (6.153 Hz) is close to the flutter frequency of 6.234 Hz. The amount of twist in the blade appears to have an important influence on the flutter mode shape. In fact, when flutter predictions are made for an entirely similar blade that is not twisted, edgewise motion in the associated flutter mode is minimal.

According to this analysis, the 2nd flapwise mode combines with the 1st torsional mode to form the flutter mode. This combination of modes was also observed by Hansen [2004] in predicting classical flutter in HAWTs. In contrast, for fixed wing aircraft, it is the 1st flapwise mode that combines with the 1st torsional mode.

Experimental Verification

Unfortunately, no experimental data exist to corroborate these results. However, the flutter analysis technique described herein has been verified by comparison to the observed flutter in a VAWT [Lobitz and Ashwill 1985]. The experimental flutter frequency was measured at 745 rpm [Popelka 1982], reasonably close to the predicted frequency of 680 rpm. The inclusion of a small amount of structural damping in the model greatly improved the agreement. The flutter mode shape was also well predicted using this technique.

References

Adler, F. M., *et al.*, 1980, *Development of an 8 kW Wind Turbine Generator for Residential Type Applications, Phase I - Design and Analysis*, Technical report RFP-3007-2, Vol. II, Bethpage, New York: Grumman Energy Systems, Inc.

Boeing, 1982, *Mod-2 Wind Turbine System Development Final Report, Vol. II: Detailed Report*, NASA CR-168007, DOE/NASA 0002-82/2, Cleveland, Ohio: NASA Lewis Research Center, p. 3-9 to 3-10.

Carne, T. G., D. W. Lobitz, A. R. Nord, and R. A. Watson, 1982, *Finite Element Analysis and Modal Testing of a Rotating Wind Turbine*, Report No. SAND82-0345, Albuquerque, New Mexico: Sandia National Laboratories.

Carne, T. G., J. P. Lauffer, and A. J. Gomez, 1988, *Modal Testing of the Eolé*, SAND87-1506, Albuquerque, New Mexico: Sandia National Laboratories.

Carne, T. G., J. P. Lauffer, A. J. Gomez, and T. D. Ashwill, 1989, "Model Validation Testing of the Sandia 34-meter Test Bed Turbine using Substructure Modal Testing," *Proceedings, ASME Energy Sources Technology Conference and Exhibition and 7th Intersociety Cryogenic Symposium*, New York: American Society of Mechanical Engineers.

Carne, T. G., and G. H. James, 2008, "The Inception of OMA in the Development of Modal Testing Technology for Wind Turbines," Keynote Address at 2007 International Operational Modal Analysis Conference, to be published in *Mechanical Systems and Signal Processing*.

Dohrmann, C. R. and P. S. Veers, 1989, "Time Domain Response Calculations for Vertical Axis Wind Turbines", *Proceedings of the 8th ASME Wind Energy Symposium*, Houston, TX.

Dowell, E. E. (Editor), 1995, *A Modern Course in Aeroelasticity*, Kluwer Academic Publishers: Dordrecht, pp. 217-227.

Dym, C. L., and I. H. Shames, 1973, *Solid Mechanics: A Variational Approach*: McGraw Hill: pp. 164-176.

Franklin, J. N., 1968, *Matrix Theory*, Englewood Cliffs, New Jersey: Prentice-Hall Inc., p. 99.

Fung, Y. C., 1969, *An Introduction to the Theory of Aeroelasticity*, Dover Publications Inc., New York, pp. 210-216.

Hansen, M. H., 2004, "Stability Analysis of Three-Bladed Turbines Using an Eigenvalue Approach," *Proceedings of the 2004 ASME/AIAA Wind Energy Symposium*, Reno, pp. 192-202.

James, G. H., T. G. Carne, and J. P Lauffer, 1993, "The Natural Excitation Technique (NExT) for Modal Parameter Extraction from Operating Wind Turbines," SAND92-1666, Albuquerque, New Mexico: Sandia National Laboratories.

James, G. H., 1994, "Extraction of Modal Parameters from an Operating HAWT using the Natural Excitation Technique (NExT)," *Proceedings, 13th ASME Wind Energy Symposium*, New York: American Society of Mechanical Engineers.

James, G. H., T. G. Carne, and J. P. Lauffer, 1995, "The Natural Excitation Technique (NExT) for Modal Parameter Extraction from Operating Structures," SEM *International Journal of Analytical and Experimental Modal Analysis*, Vol. 10, No. 4.

James, G. II., T. G. Carne, and P. S. Veers, 1996, "Damping Measurements Using Operational Data," *ASME Journal of Solar Energy Engineering*, Vol. 18, No. 3, New York: American Society of Mechanical Engineers.

Klahs, J. W., Jr., and J. H. Ginsburg, 1979, "Resonant Excitation of a Spinning, Nutating Plate," *Journal of Applied Mechanics*, Vol. 46: pp. 132-138.

Klosterman, A. L., and R. Zimmerman, 1975, *Modal Survey activity Via Frequency Response Functions*, SAE Paper No. 751068: Society of Automotive Engineers.

Lauffer, J. P., 1986, "The Statistical Variation of Modal Parameters within Production Units," *Proceedings, 4th International Modal Analysis Conference*, Los Angeles, California.

Leishman, J. G., 2002, "Challenges in Modeling the Unsteady Aerodynamics of Wind Turbines," *Wind Energy*, 5, pp. 85-132.

Lindenburg, C., and H. Snel, "PHATAS-II: Program for Horizontal Axis Wind Turbine Analysis and Simulation, Version II," *Proceedings, Twelfth ASME Wind Energy Symposium*, SED-Vol. 14, New York, NY: American Society of Mechanical Engineers, pp. 133-138.

Lobitz, D. W., and T. D. Ashwill, 1985, "Aeroelastic Effects in the Structural Dynamic Analysis of Vertical Axis Wind Turbines," *Proceedings, Windpower '85*, San Francisco, CA, pp. 82-88.

Lobitz, D. W., 2004, "Aeroelastic Stability Predictions for a MW-Sized Blade," *Wind Energy*, Vol. 7, pp. 211-224.

Malcolm, D. J., and A. D. Wright, 1994, "The Use of ADAMS to Model the AWT-26 Prototype," *Proceedings, Wind Energy - 1994 Symposium*, SED-Vol. 15, W. Musial *et al.*, eds., New York: American Society of Mechanical Engineers, pp. 125-131.

Malcolm, D. J. and G. H. James, 1995, "Stability of a 26m Teetered, Free-Yaw Wind Turbine," *Proceedings, 10th ASCE Engineering Mechanics Conference*: American Society of Civil Engineers.

Malcolm, D. J., and A. C. Hansen, 2002, "WindPACT Rotor Turbine Design Study," NREL/SR-500-32495, Golden, Colorado: National Renewable Energy Laboratory.

Ottens, H. H., and R. J. Zwaan, 1978a, *Description of a Method to Calculate the Aeroelastic Stability of a Vertical Axis Wind Turbine*, NLR TR 78072 L, The Netherlands: National Aerospace Laboratory.

Ottens, H. H., and R. J. Zwaan, 1978b, "Investigation of the Aeroelastic Stability of a Vertical Axis Wind Turbine," *Proceedings, Second International Symposium on Wind Energy Systems*, Paper C3, The Netherlands.

Patel, J. S., and S. M. Seltzer, 1971, *Complex Eigenvalue solution to a Spinning Skylab Problem*, Vol. II, NASA TMX-2378, Houston, Texas: NASA Johnson Space Flight Center.

Popelka, D., 1982, *Aeroelastic Stability of a Darrieus Wind Turbine*, SAND82-0672, Albuquerque, New Mexico: Sandia National Laboratories.

Rumsey, M., J. Hurtado, B. Hansche, T. Simmermacher, T. Carne, and E. Gross, 1997, "In-Field Use of Laser Doppler Vibrometer on a Wind Turbine Blade," *Sound and Vibration*, Vol. 32, No. 2, p 14-19.

Sweeney, T. E., and W. B. Nixon, 1978, *Two-Dimensional Lift and Drag Characteristics of Grumman Windmill Blade Section*, (unpublished), Princeton, New Jersey: Princeton University.

Spera, D. A., 1975, *Structural Analysis of Wind Turbine Rotors for NSF-NASA Mod-0 Wind Power System*, NASA TM X-3198, Cleveland, Ohio: NASA Lewis Research Center.

Spera, D. A., 1977, "Comparison of Computer Codes for Calculating Dynamic Loads in Wind Turbines," NASA TM-73773, Cleveland, Ohio: NASA Lewis Research Center.

Spera, D. A., 1995, "A Model of Rotationally-Sampled Wind Turbulence for Predicting Fatigue Loads in Wind Turbines," *Collected Papers on Wind Turbine Technology*, NASA CR-195432, Cleveland, Ohio: NASA Lewis Research Center, pp. 17-26.

Sullivan, T. L., 1981, "A Review of Resonance Response in Large, Horizontal-Axis Wind Turbines," *Proceedings, Wind Turbine Dynamics Workshop*, R. W. Thresher, ed., NASA Conference Publication 2185, DOE Publication CONF-810226, Cleveland, OH: NASA Lewis Research Center, pp. 237-244.

Theodorsen, T., 1935, "General Theory of Aerodynamic Instability and the Mechanism of Flutter," NACA Report 496, Washington, DC: National Advisory Committee for Aeronautics.

Thresher, R. W., A. D. Wright, and E. L. Hershberg, 1986, "A Computer Analysis of Wind Turbine Blade Dynamic Loads," *Journal of Solar Energy Engineering*, Vol. 108: pp. 17-25.

Vollan, A. J., and R. J. Zwaan, 1977, *Contribution of NLR to the First Phase of the National Research Program on Wind Energy, Part II: Elastomechanical and Aeroelastic Stability*, NLR TR 77030 U Part II, The Netherlands: National Aerospace Laboratory.

Warmbrodt, W., and P. Friedmann, 1980, "Coupled Rotor/Tower Aeroelastic Analysis of Large Horizontal Axis Wind Turbines," *AIAA Journal*, Vol. 18, No. 9: pp. 1118-1124.

Wright, A. D., M. L. Buhl, and R. W. Thresher, 1988, *FLAP Code Development and Validation*, SERI/TR-217-3125, Golden, Colorado: National Renewable Energy Laboratory.

12

Fatigue Design of Wind Turbines

David A. Spera, Ph.D.
Formerly Chief Engineer of Wind Energy Projects
NASA Lewis (now Glenn) Research Center
Cleveland, Ohio

and

Wind Energy Consultant
DASCON Engineering, LLC
Bonita Springs, Florida

Introduction

Wind turbine structures present many difficult fatigue design problems because they are (1) relatively slender and flexible, (2) subject to vibration and resonance, (3) acted upon by loads which are often non-deterministic, (4) operated continuously in all types of weather with a minimum of maintenance, and (5) constantly competing with other energy sources on the basis of life-cycle costs. Progress in the development of modern wind turbines has been paced by progress in our understanding of fatigue loads, modeling of structural-dynamic responses to unsteady winds, and the designing of innovative structural elements to reduce these responses.

In this chapter we describe the *types of fatigue loads* common to wind turbines, *typical statistical distributions* of fatigue loads, *empirical equations* for estimating fatigue loads during preliminary design, and typical *fatigue design procedures*. The sequence of tasks to be performed in a typical fatigue life analysis is as follows:

1. Define the *system configuration*, from requirements and trade-off studies.
2. Define the *external environment*, including wind loadings, and operating conditions.
3. Determine *system dynamic loads*, both average and cyclic, at subsystem interfaces.
4. Calculate local *average and cyclic stresses* at critical locations such as joints.
5. Calculate local *fatigue lives* using material test data and a selected damage theory.

Structural Design Drivers

It is a common engineering practice to divide structural design requirements into three general classes, as follows:

-- *Limit strength* requirements, to resist the infrequent application of the highest loads expected during the useful lifetime of the structure.
-- *Fatigue life* requirements, to resist the repeated application of lesser loads for a specified lifetime, with or without repairs.
-- *Stiffness* requirements, to control deflections and place vibration frequencies within specified ranges to avoid resonances.

All structural and mechanical components are subject to these three requirements, at least to some degree. The requirement that governs the final selection of the materials, dimensions, manufacturing processes, and inspection procedures is referred to as the *design driver* of the component. The design-driving requirement is often followed closely in importance by a second type of requirement, in which case the design may be referred to as *balanced*. When one type of requirement strongly dominates the design of a major component, changes in the *system design* may be in order, as discussed in Chapter 9.

Figure 12-1 illustrates a typical pattern of design drivers for a horizontal-axis wind turbine (HAWT) [Boeing 1988]. In general, fatigue is a design driver for at least one-half of the primary structure of a HAWT above the foundation. This still leaves large portions of a wind turbine that are designed by limit loads and by stiffness requirements for proper placement of natural frequencies.

Fatigue Life Requirements in Wind Turbines

The reason that fatigue requirements often drive the design of the majority of the primary structure of a wind turbine is that wind turbines must achieve very long operating lives in order to be cost-effective. Fatigue life is usually expressed by the term *cycles to failure*, which is the number of repetitions of significant loads that can be sustained before cracks begin (the *initiation* phase) and grow to an allowable length (the *propagation* phase). As shown schematically in Figure 12-2 by a curve of allowable cyclic stress *vs.* the number of cycles to failure at that cyclic stress (called an *S-N* curve), fatigue life is very sensitive to the amplitude of stress variations. This is emphasized by plotting lifetime on a logarithmic scale. Design lifetimes of wind turbines contain more load cycles than those of airplanes, bridges, hydrofoils, and helicopters, so cyclic stresses in a wind turbine must be lower than those allowable in other structures.

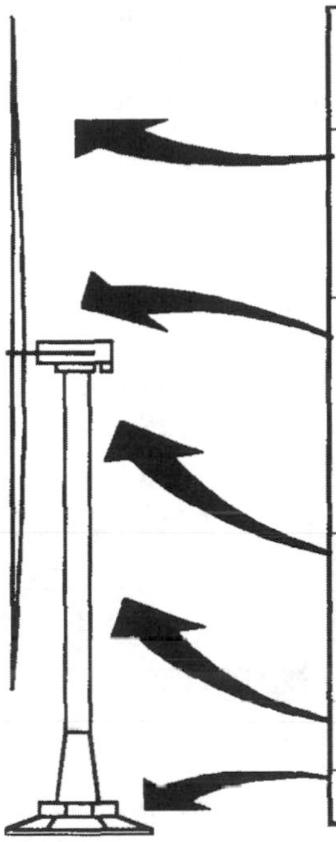

Component	Design Drivers		
	Fatigue	Limit	Stiffness
Rotor			
Upwind Side	✔	—	—
Teeter Stops	✔	✔	—
Downwind Side	—	✔	—
Drivetrain			
Rotor Cap	✔	—	—
Low-Speed Shaft	✔	—	✔
Gearbox	✔	—	—
Quill Shaft	—	✔	✔
High-Speed Shaft	—	✔	—
Nacelle			
Truss	✔	✔	✔
Yaw Drive	—	✔	—
Yaw Brakes	—	✔	—
Tower	✔	—	✔
Foundation	—	✔	—

Figure 12-1. Typical pattern of design drivers for the primary structural and mechanical components in a HAWT. [Boeing 1988]

    HAWT blades, hubs, and turbine shafts are subjected to full reversals of a large dead load once each rotor revolution. These large *gravity loading cycles* are seldom, if ever, sustained by other fabricated and jointed structures. They are normally carried only by one piece precisely-machined components, such as vehicle axles. We can quickly estimate the number of gravity cycles that a HAWT rotor must sustain in its lifetime by assuming a typical

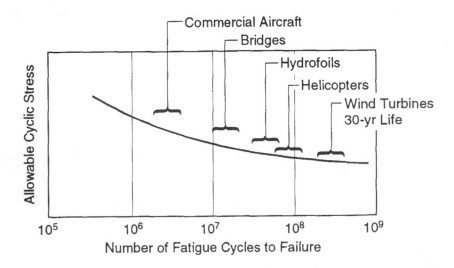

Figure 12-2. Schematic S-N curve illustrating the severity of wind turbine fatigue requirements compared to those of other structural systems.

tip speed of 90 m/s and an on-line time of 7,000 hours per year. On this basis, a rotor 60 m in diameter will experience over 350 million reversals of its dead weight in a 30-year lifetime. This number varies inversely with rotor diameter, for a constant tip speed.

Limit and Stiffness Design Requirements

Limit strength requirements are usually critical when a wind turbine is acted upon by the specified *extreme wind* (see Chapter 8). This is not an operating condition, so limit-strength design of a wind turbine is normally a conventional problem of wind loading on a static structure. Furthermore, conventional finite-element models with centrifugal stiffening are suitable for determining stiffness requirements, so that natural vibration frequencies will fall within allowable ranges. Because structural engineers are able to apply conventional design methods to meet both limit and stiffness requirements, only fatigue loads and fatigue design methods will be addressed here.

Fatigue Load and Stress Spectra

In a structure with highly-variable loading, such as a wind turbine, fatigue damage is determined by the amplitudes and frequencies of stress cycles or *stress spectra* at critical locations. These cycles are normally idealized into a time sequence of alternating minimum and maximum values connected by straight lines. In other words, the shape of the path between a given minimum and the next maximum is usually disregarded, and only the size and number of the minimum and maximum stresses are modeled.

Sample Stress Spectra

Figure 12-3 is an example of an idealized stress spectrum for a conventional non-rotating structure subject to fatigue loading. In this case the structure is a wing of a bomber aircraft, the *B-52G-H*, undergoing a full-scale, ground fatigue test in the 1960s. The spectrum represents a typical stress history at one location in the structure, with loads applied to simulate a four-hour flight. We may think of this graph as the idealization of the output of a strain gage mounted on the wing at this point. In defining patterns of stress or load, it is common to normalize the data, in this case by dividing all values by the maximum of all the stress- or load-maxima of the cycles in the spectrum. This is often referred to as the *maxmax* value of the spectrum. The *minmin* value is similarly defined.

During its simulated four-hour flight, our sample airplane wing experienced about 100 measurable and significant cycles of fatigue loading, which is an average rate of only 0.4 cpm. Thus, significant loading cycles are separated by periods of relatively steady stress. A partial reversal of gravity loads occurs only once per flight and, with its transient stresses, causes the so-called *ground-air-ground* (GAG) cycle. The GAG amplitude is equal to one-half the difference between the maxmax and minmin stresses. Amplitudes of intermediate load cycles are generally small fractions of the GAG amplitude.

Figure 12-4 illustrates a time-history of stress that is common to wind turbines, with the data again normalized by the maxmax value in the spectrum. The graph presents data for a six-hour run, in which the wind speed was low at startup, became high and gusty halfway through the run, and then dropped prior to shutdown. In contrast to Figure 12-3, there would typically be about 5,000 to 20,000 measurable fatigue stress cycles during this period, the number varying inversely with rotor diameter for a given blade tip speed. These estimates of wind turbine life requirements are based on cycling rates from 15 to 60 cpm, far larger than that of the sample bomber wing.

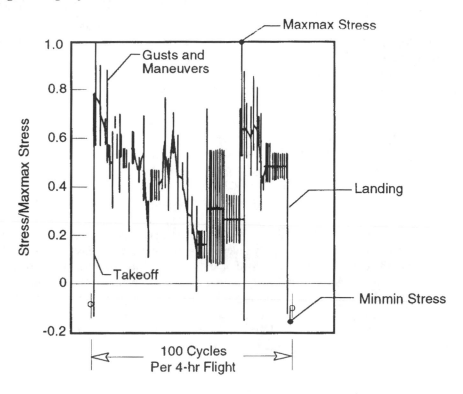

Figure 12-3. Sample of an idealized stress spectrum used for the full-scale fatigue testing of a wing of a typical bomber aircraft.

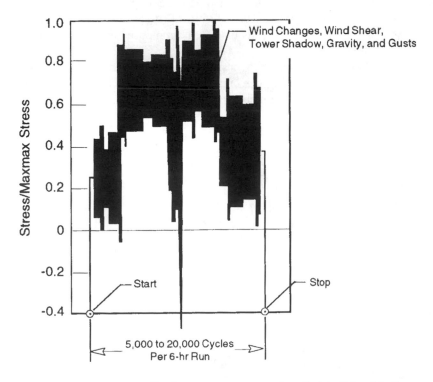

Figure 12-4. Sample of an idealized stress spectrum of a wind turbine blade.

Wind turbine components also experience the equivalent of a GAG cycle, in this case the "start-run-stop" cycle shown in Figure 12-4. In addition, there are other types of fatigue cycles which must be considered by the designer. One of these occurs at least once per rotor revolution and is caused by gravity, wind shear (*i.e.* vertical gradient of the wind speed near the ground), constant inflow distortion (*e.g.* tower blockage of the wind), and small-scale turbulence.

Significant fatigue cycles on certain components can occur more often than once per rotor revolution under two general conditions: (1) When the primary loading on the component is related to the blade passage (*e.g.*, the yaw drive on a two-bladed HAWT), in which case the number of fatigue cycles per revolution equals the number of blades, and (2) when a resonance exists, in which case there are ω/Ω fatigue cycles per revolution. Here ω is the natural frequency of the component and Ω is the rotor speed, and their ratio is typically three or larger.

Small-scale and large-scale wind turbulence (such as gusts that envelop the entire rotor) can cause fatigue cycles that are relatively independent of rotor revolutions.

Because of the complex nature of a load or stress time-history, a systematic procedure is required for identifying and counting the individual fatigue cycles in a spectrum [ASTM 1990]. Several of these are evaluated using wind turbine field data by Murtha-Smith [1985] and Bovarnick *et al.* [1985]. A counting method is usually selected for compatibility with the analytical model used for predicting fatigue life. From the point of view of fatigue life assessment, the cycle counting scheme cannot be separated from the fatigue life model.

Early Method for Counting Fatigue Cycles

A simple cycle-counting model will be described here that was used successfully in the design of the NASA/DOE Mod-1, Mod-2, and Mod-5B experimental HAWTs and in the analysis of field test data from these and other turbines [Finger 1985]. As illustrated in Figure 12-5, a time-history of load or stress is assumed to be composed of three types of cycles:

-- *Type I*:   Minimum to maximum during one or a part of one rotor revolution
-- *Type II*:   Minimum to maximum during one large-scale change in wind speed
-- *Type III*: Minimum to maximum during one run, from startup to shutdown

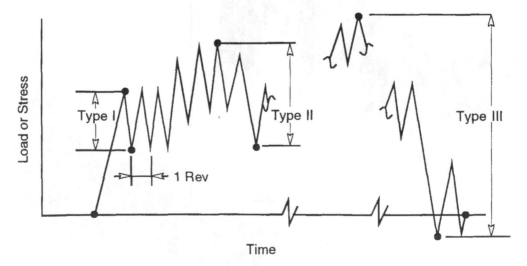

Figure 12-5. Idealized types of fatigue cycles. [Finger 1985]

As an example, Table 12-1 contains a listing of a sequence of loads calculated for a wind turbine blade, normalized by the maximum load expected in the sequence. This is one of several load spectra developed during the conceptual design of the 2.5-MW Mod-2 HAWT [Boeing 1979]. A run period is first divided into "layers" of various durations during which

Table 12-1.
9.0-Hour Blade Fatigue Design Spectrum for the DOE/NASA Mod-2 Wind Turbine in Time Order

Layer No.	Wind speed (m/s)	Cycle per layer	Flatwise Blade Load / Maxmax Load, L_F / L_{maxmax}				R-Ratio R_f
			Max	Min	Cyclic	Average	
Start	5.4	0	-0.400	-0.400	0.000	-0.400	1.00
1	"	908	0.433	0.061	0.186	0.247	0.14
2	"	333	0.496	-0.002	0.249	0.247	-0.00
3	"	605	0.383	0.111	0.136	0.247	0.29
4	"	1,059	0.463	0.030	0.217	0.247	0.06
5	"	91	0.525	-0.030	0.278	0.248	-0.06
6	"	30	0.550	-0.055	0.303	0.248	-0.10
7	9.4	613	0.880	0.446	0.217	0.663	0.51
8	"	56	0.923	0.405	0.259	0.664	0.44
9	"	1,951	0.855	0.473	0.191	0.664	0.55
10	"	1,115	0.787	0.540	0.124	0.664	0.69
11	"	1,673	0.828	0.498	0.165	0.663	0.60
12	"	167	0.903	0.423	0.240	0.663	0.47
13	15.6	237	0.645	0.192	0.227	0.419	0.30
14	"	415	0.798	0.038	0.380	0.418	0.05
15	"	356	0.740	0.098	0.321	0.419	0.13
16	"	36	0.918	-0.080	0.499	0.419	-0.09
17	"	12	0.968	-0.130	0.549	0.419	-0.13
18	"	130	0.863	-0.025	0.444	0.419	-0.03
19	18.3	3	0.838	-0.471	0.655	0.184	-0.56
20	"	37	0.710	-0.343	0.527	0.184	-0.48
21	"	67	0.447	-0.080	0.264	0.184	-0.18
22	"	118	0.632	-0.265	0.449	0.184	-0.42
23	"	101	0.561	-0.195	0.378	0.183	-0.35
24	"	10	0.778	**-0.411**	0.595	0.184	-0.53
25	13.0	1,305	0.890	0.520	0.185	0.705	0.58
26	"	130	0.976	0.433	0.272	0.705	0.44
27	"	870	0.841	0.568	0.137	0.705	0.68
28	"	1,522	0.918	0.491	0.214	0.705	0.54
29	"	44	**1.000**	0.410	0.295	0.705	0.41
30	"	478	0.950	0.460	0.245	0.705	0.48
31	7.2	927	0.527	0.206	0.161	0.367	0.39
32	"	139	0.702	0.031	0.336	0.367	0.04
33	"	1,622	0.626	0.108	0.259	0.367	0.17
34	"	1,390	0.588	0.145	0.222	0.367	0.25
35	"	46	0.732	0.002	0.365	0.367	0.10
36	"	510	0.666	0.068	0.299	0.367	0.10
Stop	5.4	0	-0.400	-0.400	0.000	-0.400	1.00
Spectrum:	**9.6**	**19,105**	**0.724**	**0.299**	**0.212**	**0.511**	**0.37**

wind speed, maximum load, and minimum load are relatively constant. A shortcoming of this simple counting scheme is that identifying Type II cycles requires some judgment.

Three parameters which appear in almost all models of the fatigue cycles are as follows:

$$L_{cyc} = 0.5\,(L_{max} - L_{min})$$ (12-1a)

$$L_{avg} = 0.5\,(L_{max} + L_{min})$$ (12-1b)

$$R_f = L_{min}/L_{max}$$ (12-1c)

where

L_{max}, L_{min} = maximum and minimum loads in one fatigue cycle (kN or kN-m)
L_{cyc} = cyclic, *alternating*, or *half-range* load; also load *amplitude* (kN or kN-m)
L_{avg} = average, *mean*, or *mid-range* load (kN or kN-m)
R_f = fatigue cycle shape parameter; also called *R-ratio*

Rainflow Model for Counting Fatigue Cycles

Other models have now been developed that automatically include the cyclic effects of transient wind conditions and control changes. A very useful one is the *rainflow model* [Matsuiski and Endo 1969, Downing and Socie 1982, Rychlik 1987, Sutherland and Schluter 1989 and 1990, Schluter 1991, Endo Memorial Volume 1991, ASTM 2005], which is often included in the software of data systems used to monitor loads and stresses because it can be applied reliably to a general spectrum. The rainflow model is currently the most common algorithm used for counting fatigue cycles when analyzing test data from wind turbines.

A major advantage of using the rainflow counting algorithm to define wind turbine fatigue spectra is that cycle counting is de-coupled from the rotational rate of the turbine. In this way, fatigue cycles caused by wind turbulence can be captured at whatever frequency they occur. Often, large numbers of smaller-amplitude load cycles are counted by rainflow algorithms that were not recognized in earlier, simpler counting methods.

There are at least two ways to diagram the rainflow cycle counting algorithm, which for convenience we may term the *drain* and *drip* methods. Figure 12-6 illustrates the drain method (Cosmos 2005). The time axis of a 6-cycle idealized time-series of fatigue loads (or stresses) is oriented horizontally, and the series is figuratively filled with water. Water is then assumed to be drained from the lowest point in the series, at an elapsed time of approximately 7 seconds. The resulting change in water level above this point defines the maximum and minimum loads in Cycle 1, which are 600 and – 200 kN-m in this example. Applying Equations (12-1), average and cyclic loads for Cycle 1 are calculated to be 200 and 400 kN-m, respectively, and its R-ratio is – 3.00. The same parameters for Cycles 2 through 6 are obtained by successively draining water from progressively higher minimum points. The cycle-counting results for this example are listed in Table 12-2.

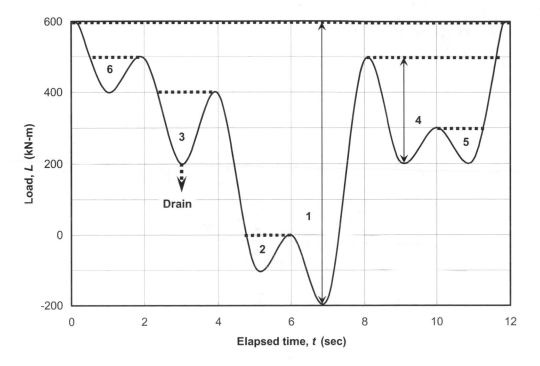

Figure 12-6. Schematic diagram of a fatigue load-time series in which cycles are counted by the rainflow algorithm, using the "drain" method. [Cosmos 2005]

Table 12-2.
Fatigue Load Parameters for the Time-Sequence in Figure 12-6,
by the Rainflow Counting Algorithm

Cycle	Minimum Load, L_{min} (kN-m)	Maximum Load, L_{max} (kN-m)	Average Load, L_{avg} (kN-m)	Cyclic Load, L_{cyc} (kN-m)	R-Ratio
1	-200	600	200	400	- 3.00
2	-100	0	-50	50	0
3	200	400	300	100	+ 2.00
4	200	500	350	150	+ 2.50
5	200	300	250	50	+ 1.50
6	400	500	450	50	+ 1.25

An alternative (and more common) visualization of the rainflow counting method is illustrated in Figure 12-7. Here the time axis is oriented vertically, and water is imagined to be flowing down off a series of "pagoda" roofs. The drip lines from the pagoda roof sections define the minimum and maximum loads in each cycle. The cycle-counting results are the same as those listed in Table 12-2.

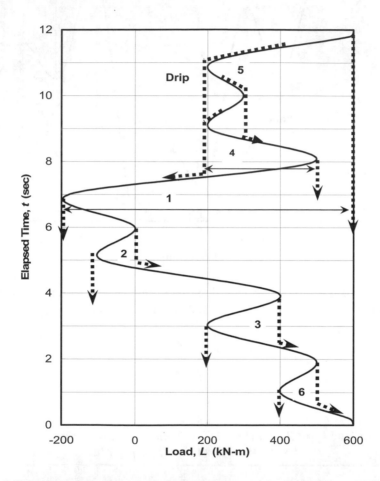

Figure 12-7. Schematic diagram of a fatigue load-time series in which cycles are counted by the rainflow algorithm, using the "drip" method.

Statistical Analysis of Examples of Fatigue Load Spectra

After the individual fatigue cycles in the spectrum have been identified and the parameters of each have been tabulated, a statistical analysis is usually required to convert the resulting time-history data into fatigue design information. Depending on the *damage mechanism* assumed in the fatigue model, fatigue life may be expressed as a function of maximum stresses, cyclic stresses, or a combination of cyclic and average stresses. In the examples that follow, cyclic stress has been selected as the key fatigue-damage parameter, and three different distributions of fatigue cycle rates with respect to loads are illustrated.

Example 1: Log-Normal Distribution of Blade Flatwise Loads

Normalized blade load design spectra for the Mod-2 2.5-MW wind turbine were listed in Table 12-1, in time-series order [Boeing 1979]. These same data are now converted to a dimensional format based on an assumed maxmax load of 1,800 kN-m. Cyclic loads are then sorted into *bins* each 50 kN-m in width. The bins and their populations are tabulated in the first four columns of Table 12-3. The 1-hr count data from Table 12-3 are then curve-fit as

Table 12-3. Calculation of Probability Density Function for Flatwise Blade Fatigue Loads on the Mod-2 2.5-MW Wind Turbine

	9.0-hr Design Spectrum Data Sorted into Load Bins 50-kN-m Wide			Log-Normal Curve-Fit of 1-hr Count	Probability Density Function
Bin No.	Cyclic Load $L_{F,mid}$ (kN-m)	9-hr Count n' (cyc/bin)	1-hr Count n (cyc/hr/bin)	n (cyc/hr/bin)	*PDF* (cyc/hr/kN-m)
1	1,275	1	0.11	0.11	0.0022
2	1,225			0.17	0.0034
3	1,175	3	0.33	0.27	0.0054
4	1,125			0.43	0.0087
5	1,075	10	1.11	0.70	0.0140
6	1,025			1.13	0.0225
7	975	12	1.33	1.83	0.0365
8	925	37	4.11	2.97	0.0594
9	875	36	4.00	4.85	0.0970
10	825	118	13.11	7.93	0.1586
11	775	130	14.44	12.97	0.2594
12	725			21.18	0.4237
13	675	562	62.44	34.45	0.6889
14	625	139	15.44	55.55	1.1110
15	575	356	39.56	88.38	1.7675
16	525	675	75.00	137.65	2.7531
17	475	1,875	208.33	207.63	4.1526
18	425	2,605	289.44	298.43	5.9686
19	375	3,194	354.89	398.79	7.9757
20	325	4,164	462.67	476.42	9.5284
21	275	2,600	288.89	476.61	9.5322
22	225	2,590	287.78	354.96	7.0992
Totals		**19,107**	**2,123**	**2,583 cycles/hr**	

illustrated in Figure 12-8. In this case, a *log-normal distribution* is the best fit to the design data, which indicates that there is a random distribution of the *logarithm* of the load parameter. The governing equations are as follows:

$$n = 490 \exp\left\{-0.5\left[\frac{\ln(L_F) - \mu}{\sigma}\right]^2\right\}$$ (12-2a)

$$PDF = n/50$$ (12-2b)

where

n = bin cycle count (cycles/hr/bin)
$\ln(\ )$ = natural logarithm of ( )
L_F = flatwise cyclic bending moment load (kN-m)
μ = 5.705, natural logarithm of median flatwise blade cyclic load L_F
σ = 0.35, natural logarithm of standard deviation of L_F
PDF = probability density function (cycles/hr/kN-m)

Using these log-normal curve-fit parameters, the median size of the flatwise cyclic load is exp (μ) = 300 kN-m, and the 98th percentile cyclic load is exp ($\mu + 2\sigma$) = 605 kN-m.

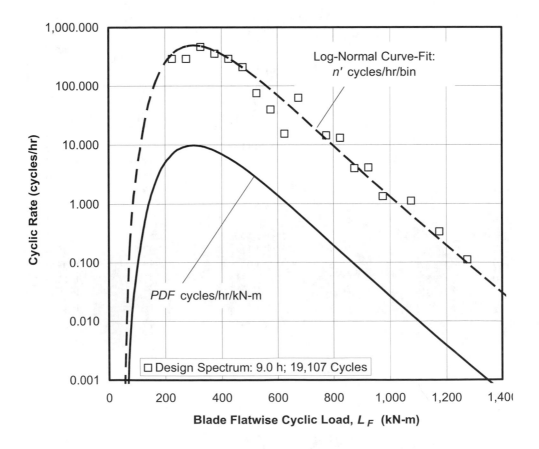

Figure 12-8. Curve-fitting of a Mod-2 blade flatwise cyclic load spectrum with a log-normal distribution, and calculation of the related probability density function.

As noted in Table 12-3, the curve-fit cyclic rate is 2,583 cycles per hour, slightly more than the design rate that does not include all bins or very small loads. The design rotational speed of the Mod-2 wind turbine is 17 rpm, so the overall cyclic rate is approximately 2.5 cycles per rotor revolution, or *2.5P*. This is a reasonable rate for a two-bladed turbine.

Example 2: Exponential Distribution of Blade Flatwise Loads

A heavily-instrumented Micon 65/13 115-kW wind turbine with SERI 8m blades was the primary test turbine in a program to collect long-term inflow and structural test data [called *LIST*, Sutherland 2002]. This turbine has a fixed-pitch, 3-bladed rotor located up-wind of a tubular tower. A 20-hour set of blade cyclic load data was collected during this program, and cycles in this set were counted by the rainflow method. The count was then sorted into bins 0.5-kN-m wide and converted to 1-hr rates. The flatwise cyclic load results are illustrated in Figure 12-9.

In this figure we note that the cycle count continued to increase as the blade loads decreased, unlike the Mod-2 blade design spectrum in Figure 12-8. This distribution with many cycles of small loads is best fit with an *exponential equation*, as follows:

$$n = 4,550 \exp[-0.3836 \, L_F] \qquad (12\text{-}3a)$$

$$PDF = n/0.5 \qquad (12\text{-}3b)$$

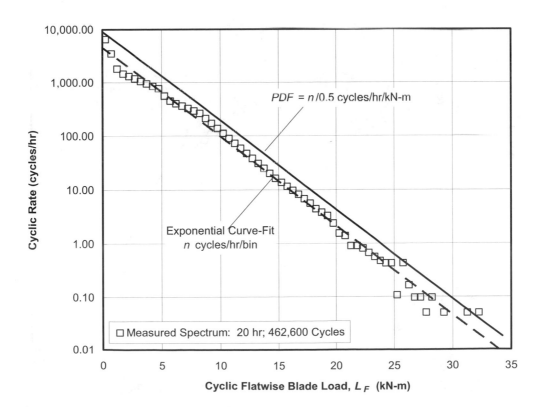

Figure 12-9. Curve-fitting of a Micon 65/13 wind turbine blade flatwise cyclic load spectrum with an exponential distribution, and calculation of the related probability density function.

The very large number of cycles at small loads lowers the median flatwise cyclic load to approximately 2.2 kN-m. A spectrum shape like that in Figure 12-9 can be attributed to the rainflow counting method that captures blade responses to all levels of wind turbulence. Because conventional fatigue damage decreases rapidly with decreasing stress, this large population of very low loads is generally of second-order importance in the calculation of fatigue life. The total cyclic population is calculated to be 23,700 cycles per hr. Because the rotor speed during data collection was 55 rpm, the average cyclic rate is calculated to be approximately 7 cycles per revolution, or 7P.

Example 3: Dual-Mode Log-Normal Distribution of Blade Edgewise Loads

As illustrated in Figure 12-10, the distribution of edgewise cyclic loads on the blades of the Micon 65/13 LIST wind turbine was found to be significantly different from the flatwise load distribution. Here we see that there is a large population of cycles at low loads (Mode 1) and a second population centered on a moderately high load (Mode 2). The Mode 1 loads can be attributed to the blade's random cyclic *torque response* to wind turbulence, analogous to the flatwise *flap response* shown in Figure 12-8. The Mode 2 loads are dominated by the blade's deterministic response to *gravity loading*, at a rate of one cycle per rotor revolution, or *1P*.

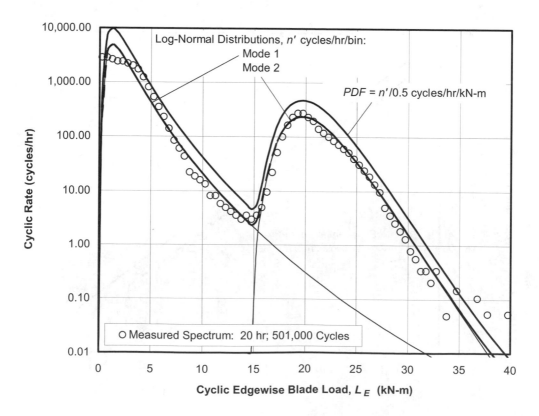

Figure 12-10. Curve-fitting of a Micon 65/13 wind turbine blade edgewise cyclic load spectrum with a dual-mode log-normal distribution, and calculation of the related probability density function.

The log-normal curve-fit equations for the two modes are as follows:

Mode 1:

$$n_1 = 4,876 \exp\left\{-0.5 \left[\frac{\ln(L_F) - \mu_1}{\sigma_1}\right]^2\right\}$$

(12-4a)

Mode 2:

$$n_2 = 235 \exp\left\{-0.5 \left[\frac{\ln(L_F - 12.3) - \mu_2}{\sigma_2}\right]^2\right\}$$

(12-4b)

$$n = n_1 + n_2$$

(12-4c)

$$PDF = n/0.5$$

(12-4d)

where

n_1, n_2 = bin cycle counts in Modes 1 and 2, respectively (cycles/hr/bin)
μ_1 = 0.20, natural logarithm of median edgewise blade cyclic load L_F in Mode 1
σ_1 = 0.90, natural logarithm of standard deviation of L_F in Mode 1
μ_2 = 2.00, natural logarithm of median edgewise blade cyclic load L_F in Mode 2
σ_1 = 0.39, natural logarithm of standard deviation of L_F in Mode 2

Using these log-normal curve-fit parameters, the median sizes of the edgewise cyclic loads are 1.2 kN-m for Mode 1 and 19.7 kN-m for Mode 2. The 98th percentile cyclic loads are 14.1 and 28.4 kN-m for Modes 1 and 2, respectively.

Fatigue Load Spectra from HAWT Tests

Deterministic and Probabilistic Models

Mathematical models of the structural-dynamic behavior of wind turbines are now developed to the point where median (*i.e.*, 50th percentile) loads can be predicted with sufficient accuracy for design purposes. Median loads result from normal wind conditions and normal aerodynamic responses that are predictable enough to be referred to as *deterministic*, and these have been incorporated successfully into computer models of wind turbine structures. Unfortunately, it is not these median loads that cause fatigue damage but the infrequent, higher-amplitude loads with probabilities of exceedance in the range of 10 percent or less. These fatigue-driving loads, often called *probabilistic* or *stochastic* loads, are much more difficult to predict. Their frequency of occurrence depends on random combinations of configuration, site, wind, and control characteristics that are not known *a priori*, if ever.

One solution to the difficult problem of predicting probabilistic loads is to modify a deterministic model on the basis of operating load data like that illustrated in Table 12-3. For example, the variability of the wind input in time and space can be made larger and larger until the output loads from the model equal higher and higher percentile loads measured at selected points in the structure. This produces a probabilistic wind model that can then be used as an input to new structural-dynamic models, and the resulting loads will have known probabilities of exceedance. Analytical tools of this type can be used to

-- modify deterministic mathematical models for the estimation of probabilistic loads;
-- guide the preliminary fatigue design of critical wind turbine components;
-- establish baseline loads for future field tests.

The semi-empirical approach just described has been found to be both practical and accurate when field test data on a prototype turbine are available, or when an operating turbine is being modified. In most other cases, however, there is usually a lack of operational test data on the specific configuration being designed. Instead, engineers must look for guidance from the probabilistic loads measured on similar wind turbines. In this section we will examine a set of fatigue load data from tests on a variety of medium- and large-scale HAWTs as well as empirical equations which correlate these data with basic wind turbine and site properties.

Representative Field Test Data

The fatigue loads data presented here were measured during various HAWT research and development projects at the NASA Lewis Research Center (Spera *et al.* 1985). The configurations of the rotors tested are described in Table 12-4, and tower and site data are given in Table 12-5. The parameters listed in these two tables are those that were found to correlate with cyclic load levels, and they appear in empirical equations to be discussed later. The 13 data cases in the set include 11 two-bladed rotor designs with diameters from 28 m to 91.4 m, power ratings from 50 kW to 4,000 kW, truss and shell towers, and eight coastal and inland sites in the U.S.

Defining the scope of the data set is important when using an empirical or semi-empirical method to predict fatigue loads, since the objective is to make the process one of interpolation rather than extrapolation. For example, only isolated HAWTs with two-bladed rotors are represented in this data set, so additional tests would be required to model the effect of

Table 12-4. HAWT Test Rotor Configuration Data

Case	Rotor diam. D (m)	Rated power P (kW)	Teeter axis angle δ_3 (deg)	Hub coning angle θ (deg)	Tip chord c_t (m)	Rotor speed N (rpm)	Blade 1st chordwise frequency ω_c (cpm)	Blade gravity moment M_g (kN-m)	Ref.
A	28.0	200	-90	7	1.11	40.0	174	71	(a)
B	38.1	200	-90	7	0.46	40.0	160	62	"
C	38.1	150	-90	7	0.46	31.0	160	62	"
D	38.1	200	-90	7	0.61	40.0	164	69	"
E	38.1	150	-90	7	0.61	31.0	164	57	"
F	60.9	2,000	-90	9	0.86	35.0	146	739	(b)
G	30.6	100	0	3	1.03	34.0	171	91	(c)
H	39.1	100	-20	3	0.67	31.0	117	130	"
I	39.0	104	0	3	0.67	20.0	163	104	"
J	39.0	50	0[1]	3	0.67	20.0	163	104	"
K	39.0	50	0[1]	3	0.67	20.0	163	82	"
L	79.0	4,000	-30	6	1.16	30.0	125	1,112	(d)
M	91.4	2,500	0	0	1.44	17.5	108	3,254	(e)

[1] Maximum allowable during test
(a) [Spera and Janetzke 1981, Birchenough 1981]
(b) [Spera and Janetzke 1981, Collins *et al.* 1981, Spera *et al.* 1979]
(c) [McDade and Pfanner 1985, Ensworth 1985]
(d) [Young and Hasbrouck 1983, Johnson and Young 1985]
(e) [Sullivan 1984, Boeing 1985]

the number of blades or the effect of turbulence generated by upwind turbines on fatigue load spectra.

The test results from these 13 data cases are graphed in Figures 12-11(a) to (m), with sketches of the individual blade planforms and control surfaces. This reference data base contains almost a million measurements each of (1) wind speed at hub elevation, (2) cyclic torque in the turbine shaft, calculated from the generator output power and rotor speed, (3) cyclic flatwise (or flapwise) blade moment near the root, and (4) cyclic chordwise (or edge-wise) blade moment at the same location. One sample of tower bending moment is also included. The load cycles are all of Type I. Taken as a whole, these data are representative of the fatigue load spectra that can be expected during long-term operation of HAWTs and (to a certain extent) VAWTs, and can be used as a guide for preliminary design and testing.

Empirical Equations for Fatigue Load Spectra

If a new HAWT design is very similar to one in Figures 12-11, the accompanying data are satisfactory of estimating fatigue load spectra. But how can these data be applied to different configurations? The standard engineering approach is to derive *empirical equations* with which to interpolate within the multi-dimensional space represented by the rotor, tower, and site data in Tables 12-4 and 12-5. When the same parameters are measured in a variety of test cases using relatively consistent procedures, empirical equations can usually be developed that will model the test results with acceptable accuracy, as functions of selected input parameters. The test results in Figures 12-11 form this type of a consistent data set, so we should be able to define empirical equations with which to calculate these same fatigue loads for new combinations of configurations, sites, and winds.

Table 12-5. HAWT Tower and Site Data

Case	Side of rotor; Type of structure	Tower blockage factor d	Hub elevation H (m)	Nearest town	Altitude ASL Z (m)	Surface roughness Length z_0 (m)	Surface roughness Exponent α_0
A	Upwind; Truss	2.5	30.5	Clayton, NM	1,540	0.024	0.30
B	" "	2.5	30.5	"	1,540	0.024	0.30
C	" "	2.5	30.5	Culebra, PR	80	0.0032	0.20
D	" "	2.5	30.5	Kahuku, HI	20	0.0098	0.25
E	" "	2.5	30.5	Block Island, RI	110	0.16	0.44
F	" "	2.5	42.7	Boone, NC	1,340	0.50	0.55
G	Upwind; Shell	1.0	31.4	Sandusky, OH	200	0.25	0.48
H	" "	1.0	31.4	"	200	0.25	0.48
I	" "	1.0	31.4	"	200	0.25	0.48
J	" "	1.0	31.4	"	200	0.25	0.48
K	" "	1.0	31.4	"	200	0.25	0.48
L	" "	1.0	80.0	Medicine Bow, WY	2,180	0.024	0.30
M	Downwind; Shell	0.2	60.9	Goldendale, WA	800	0.024	0.30

Deriving empirical equations to represent test data is a common task for engineers and scientists, and the procedure usually consists of

-- defining *factors* that represent key configuration parameters and test conditions in quantitative form;
-- adding *coefficients* and *exponents* to these factors and other input parameters;
-- adjusting the coefficients and exponents until both the mean and standard deviations between calculated and measured results are minimized.

This process has been applied to derive empirical equations which model the cyclic load spectra in Figures 12-11 [Spera 1993], and the results are presented in the following sections.

Two limitations on the scope of the underlying data set are (1) two-bladed HAWTs only, and (2) only natural turbulence (*i.e.*, no *generated* turbulence from upwind turbines).

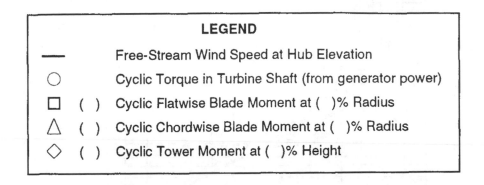

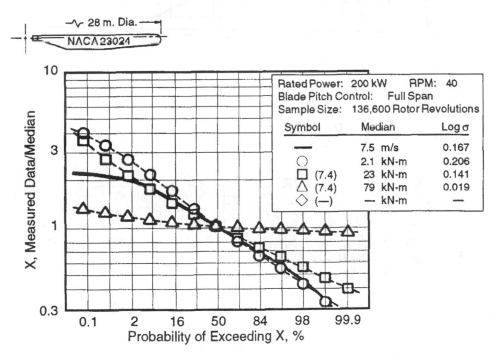

(a) Configuration A: Mod-0A Wind Turbine at Clayton, New Mexico

Figure 12-11. Representative fatigue load distributions measured on medium- and large-scale HAWTs. [Spera *et al.* 1985]

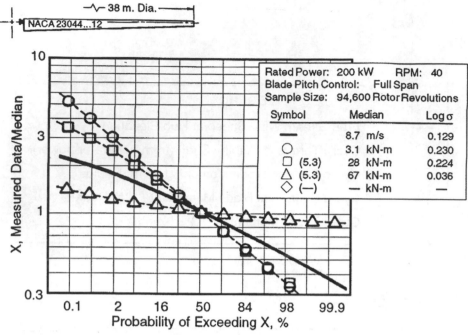

(b) Configuration B: Mod-0A Wind Turbine at Clayton, New Mexico

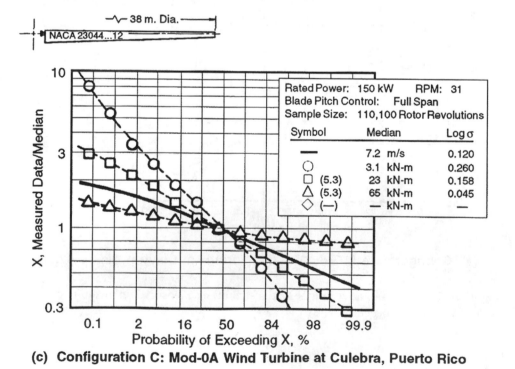

(c) Configuration C: Mod-0A Wind Turbine at Culebra, Puerto Rico

Figure 12-11 (Continued). Representative fatigue load distributions measured on medium- and large-scale HAWTs. [Spera *et al.* 1985]

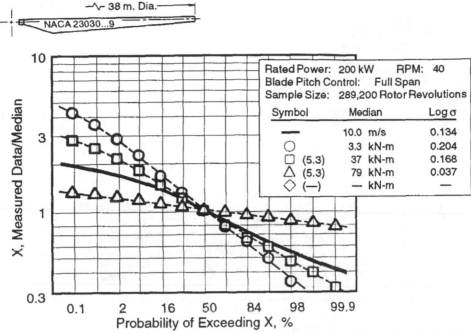

(d) Configuration D: Mod-0A Wind Turbine at Kahuku, Hawaiian Islands

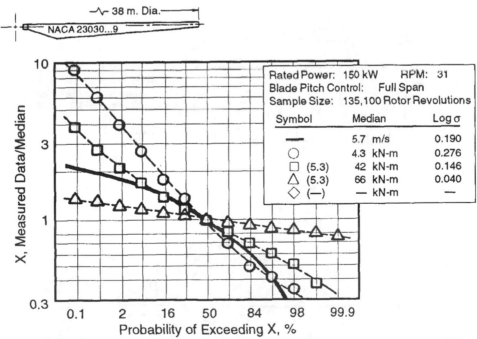

(e) Configuration E: Mod-0A Wind Turbine at Block Island, Rhode Island

Figure 12-11 (Continued). Representative fatigue load distributions measured on medium- and large-scale HAWTs. [Spera *et al.* 1985]

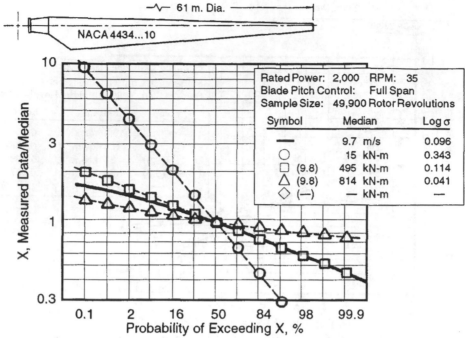

(f) Configuration F: Mod-1 Wind Turbine at Boone, North Carolina

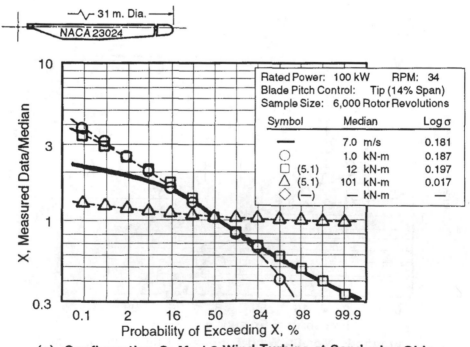

(g) Configuration G: Mod-0 Wind Turbine at Sandusky, Ohio

Figure 12-11 (Continued). Representative fatigue load distributions measured on medium- and large-scale HAWTs. [Spera *et al.* 1985]

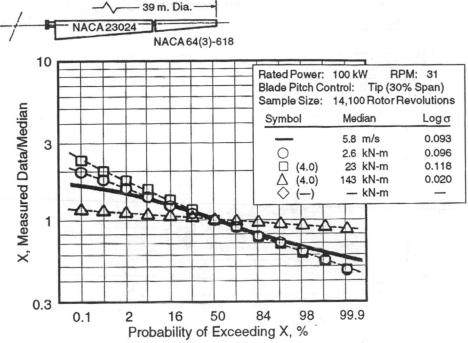

(h) Configuration H: Mod-0 Wind Turbine at Sandusky, Ohio

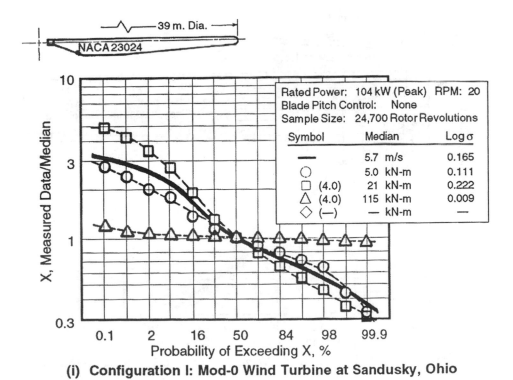

(i) Configuration I: Mod-0 Wind Turbine at Sandusky, Ohio

Figure 12-11 (Continued). Representative fatigue load distributions measured on medium-and large-scale HAWTs. [Spera *et al.* 1985]

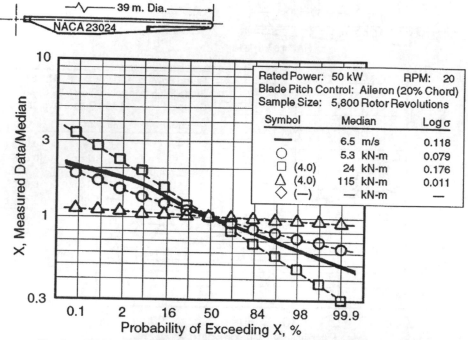

(j) Configuration J: Mod-0 Wind Turbine at Sandusky, Ohio

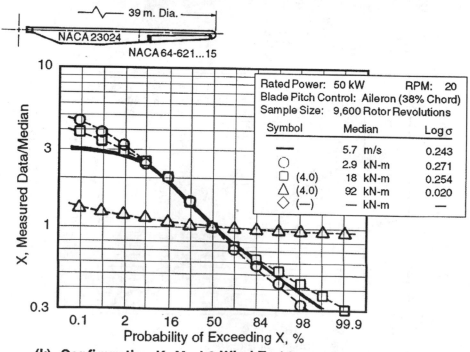

(k) Configuration K: Mod-0 Wind Turbine at Sandusky, Ohio

Figure 12-11 (Continued). Representative fatigue load distributions measured on medium- and large-scale HAWTs. [Spera *et al.* 1985]

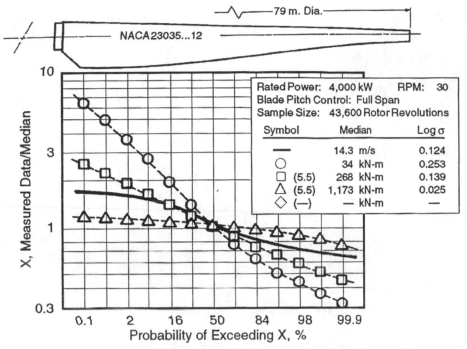

(l) Configuration L: WTS-4 Wind Turbine at Medicine Bow, Wyoming

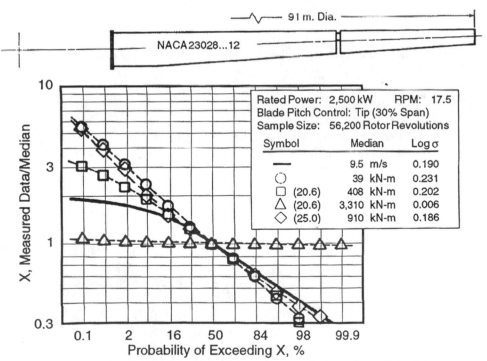

(m) Configuration M: Mod-2 Wind Turbine at Goldendale, Washington

Figure 12-11 (Concluded). Representative fatigue load distributions measured on medium- and large-scale HAWTs. [Spera *et al.* 1985]

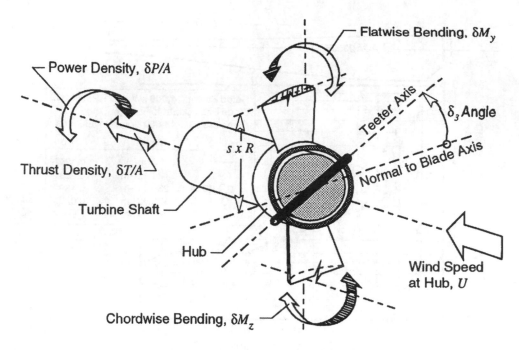

Figure 12-12. Hub model geometry and loads. [Spera 1994]

Figure 12-12 shows the geometry of the HAWT hub area and the nomenclature used in the empirical equations [Spera 1994].

Equation Factors

The derivation of equation factors relies heavily on knowledge of the physical phenomena present, plus intuition and trial-and-error. For this set of fatigue load data, the following six factors have been defined that combine basic rotor, tower, and site parameters from Tables 12-4 and 12-5:

$$\text{Hub rigidity factor:} \quad a = 0.5\,(1 - \cos 2\delta_3) \qquad (12\text{-}5a)$$

$$\text{Tower blockage factor:} \quad b = 2.5 \;\text{Upwind truss} \qquad (12\text{-}5b)$$

$$= 1.0 \;\; \text{Upwind shell}$$

$$= 0.2 \;\; \text{Downwind shell}$$

$$\text{Tip chord factor:} \quad c = 50\, c_t\,/D \qquad (12\text{-}5c)$$

$$\text{Air density factor:} \quad d = 1 - 0.00009\,(Z + H) \qquad (12\text{-}5d)$$

$$\text{Chordwise dynamic amplification factor:} \quad e = 1\,/[1 - (\,N/\omega_c\,)^2] \qquad (12\text{-}5e)$$

$$\text{Wind variability factor:} \quad g = \alpha\, D/H \qquad (12\text{-}5f)$$

$$\text{Wind shear power-law exponent:} \quad \alpha = \alpha_0 \frac{1 - 0.55 \log(U_0)}{1 - 0.55 \alpha_0 \log(H/10)} \quad \text{(12-6a)}$$

$$\text{Surface roughness exponent:} \quad \alpha_0 = (z_0/10)^{0.2} \quad \text{(12-6b)}$$

where

δ_3 = inclination of teeter axis from a normal to the blade axis (deg)

c_t = blade chord at tip (m)

D = rotor diameter (m)

Z = site altitude above sea level (m)

H = elevation of rotor hub above ground level (m)

N = rotor speed (rpm)

ω_c = blade first chordwise natural frequency (cpm)

z_0 = surface roughness length (m) (See Table 8-3)

U_0 = 50th percentile (median) free-stream wind speed at hub elevation (m/s)

The hub-rigidity factor, *a*, varies from zero for a "pure" teetered hub to 1.0 for a rigid hub and is a measure of the hub's out-of-plane resistance to once-per-revolution or *1P* loads. (Modeling a rigid hub with a δ_3 angle of 90 deg is done for mathematical convenience only; there is no teeter shaft in this configuration.) The tower-blockage factor, *b*, defines the severity of tower wake effects on blade air loads. The tip-chord factor, *c*, is a measure of outboard blade solidity (*i.e.,* ratio of blade *planform* area to swept area). The air-density factor, *d*, approximates the air density ratios in the U.S. Standard Atmosphere and altitude effects on the dynamic pressure of the wind.

The chordwise dynamic-amplification factor, *e*, is the theoretical undamped amplification of *1P* cyclic forcing by gravity in the plane of rotation. The comparable amplification factor for flatwise gravity loading of a coned rotor had no observable effect on the correlation of calculated and measured cyclic loads. The apparent reason is the high aerodynamic damping of the first flatwise bending mode. Finally, the wind-variability factor, *g*, is the ratio of the total wind shear across the rotor disk to the mean wind speed at hub elevation. Fatigue loading has been found to be very sensitive to this factor, because it is a measure of the *wind speed variance* from both large-scale and small-scale turbulence. For wind power station applications, this factor would be increased to account for generated turbulence from other turbines.

Empirical Equations

In a large data base of this type, specific data points are selected to which the empirical equations are fitted to obtain zero mean deviations and minimum standard deviations. In this case the selected points are the 50th and 98th percentile loads (*i.e.,* 50 percent and 2 percent probabilities of exceedance) for each of the 13 cases. One reason for selecting these two percentiles is that little fatigue damage occurs under loads smaller than the 50th percentile load, so it is only necessary to define the lines in Figures 12-11 for ordinates larger than unity. A second reason is that the 98th percentile load is near the upper limit of accuracy in most probability distributions of test data. Very long run times with large numbers of data points are needed to accurately define percentile levels at the logical next-higher choice, the 99.9th percentile or "+ 3σ" level.

When the factors in Equations (12-5) are combined with the rotor diameter, *D*, and the wind speed, *U*, and coefficients and exponents are adjusted for minimum overall deviations (both mean and standard) between calculated and measured cyclic loads at the 50th and 98th percentile levels for all 13 cases in Figures 12-11, the following empirical equations result:

$$\delta M_{y,n} = a\,M_g \sin\theta + 432\,(1 + 1.47a)\;cd\,(g + 0.012\;b)\;x$$
$$U_n\,(1 - s)\exp(0.134n)\,(D/100)^4 \tag{12-7a}$$

$$\delta M_{z,n} = e\,M_g + 46.8\;cd\,(g + 0.100\;b)\,U_n\,(1 - s)\exp(0.276\;n)(D/100)^3 \tag{12-7b}$$

$$\delta P_n/A = 6.4 + (0.014\,P_R/A + 4\,a)\exp\{0.807\,[1 + 0.0027\,b\,(D - 37)]n\} \tag{12-7c}$$

$$\delta T_n/A = 2.91(1 + 1.47\,a)\;cd\,(g + 0.012\;b)\,U_n \exp(0.140\;n)(D/100) \tag{12-7d}$$

where

δM_y	=	blade cyclic flatwise bending load (kN-m)
n	=	number of standard deviations, σ, from the mean in a *log-normal* probability distribution:
	=	0 for the 50th percentile load, = 1 for the 84th percentile load
	=	2 for the 98th percentile load, = 3 for the 99.9th percentile load
M_g	=	blade maximum static gravity moment (kN-m)
Θ	=	hub coning angle (deg)
s	=	blade station at which loads are measured, as a fraction of span
U_n	=	wind speed at hub elevation that is $n\sigma$ from the mean in a log-normal probability distribution (m/s)
δM_z	=	blade cyclic chordwise bending load (kN-m)
$\delta P/A$	=	generator cyclic power density (W/m²)
A	=	rotor swept area (m²)
P_R	=	generator rated power (W)
$\delta T/A$	=	rotor cyclic thrust density (N/m²)

For convenience, cyclic shaft torque and tower bending have been replaced by cyclic power and cyclic rotor thrust. The first terms in Equations (12-7a) and (12-7b) are the effects of gravity with undamped dynamic amplification in the chordwise direction. All other terms are wind effects. The exponential terms define the load probability distributions.

Correlation between Calculated and Measured Fatigue Loads

As listed in Table 12-6, *coefficients of variation (i.e.,* ratios of standard deviation to the mean value of the variable) of about 12 percent can be expected between cyclic loads in service and those calculated in accordance with Equations (12-7a) to (12-7d). These range from 5 percent for blade cyclic chordwise moment to 20 percent for generator cyclic power density. A typical correlation graph is shown in Figure 12-13, in which measured blade cyclic flatwise moments at the 50th and 98th percentile levels are plotted *vs.* calculated values. The standard deviation for this fatigue load is 10 percent, and calculations within one standard deviation of perfect correlation fall in the shaded band.

Deviations between calculated and measured fatigue loads are caused by several factors. First, the empirical equations are *idealized models* and as such do not contain all the parameters affecting cyclic loads. Two of these that are under the control of the design engineer are (a) airfoil properties such as maximum lift coefficient and stall behavior, and (b) flexible

Table 12-6. Correlation Results
(All mean deviations = 0)

Cyclic Load	Standard Deviation	Coefficient of Variation
Blade flatwise bending	10.0%	10.0%
Blade chordwise bending	4.5%	4.5%
Generator power density	4.4 W/m^2	19.9%
Rotor thrust density (est.)	0.8 N/m^2	12.6%
Representative:		**11.8%**

[Spera 1993]

elements in the power train and tower. Second, the equations are based on a *limited test time* without a systematic sampling of diurnal and seasonal effects. Third and most important, there is an *expected range of dynamic responses* between similar wind turbines, based on the accuracy with which the turbine has been modeled and the extent to which the modeling results have been translated into a design that alleviates fatigue loads.

An interpretation of Figure 12-13 that is potentially useful to both design and test engineers is as follows: Fatigue load responses within the shaded band can be viewed as typical of the *first-generation* HAWTs in the data set (*e.g.*, many have rigid hubs), and therefore reducing fatigue loads to this level should be readily achievable for future turbines. Responses

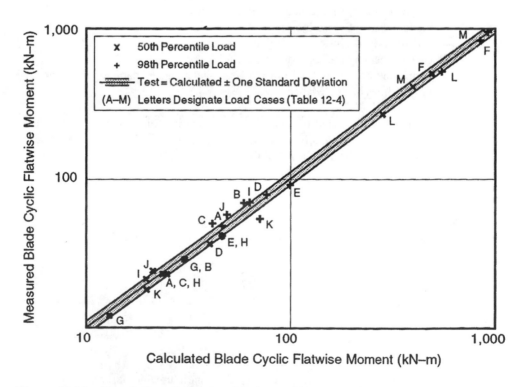

Figure 12-13. Correlation of calculated fatigue loads with test data. Letters designate test cases.

above the band indicate fatigue load problems that are probably cost-effective to solve by design changes. Responses below the band are a reasonable goal for improved HAWTs, since these would be expected to experience fatigue loads that are smaller than those in most of the test cases of the data set. It should be carefully noted that empirical equations of this type are not to be relied upon to form the sole basis for a complete fatigue load analysis.

A useful application of Equations (12-7a) to (12-7d) is in the verification of structural dynamic computer models, such as those discussed in Chapter 11. As mentioned before, computer codes of this type are becoming more and more reliable for the prediction of *median* cyclic loads. The above empirical equations are a convenient method for calibrating a computer model of the structure, to verify that predicted median fatigue loads are consistent with HAWT field test data. Once this is done, high-percentile load predictions from the computer model can be compared to predictions made with the empirical equations in order to "tune" the model and its wind turbulence inputs.

Design Load Trend Charts

One of the most important uses of empirical equations is to evaluate the effects of potential design changes through *parametric studies* during the conceptual design phase of a project. These are also referred to as *trade-off studies*, because technical benefits must be balanced or traded off against any impacts on system costs and schedules. Design trend charts are convenient tools for displaying the relative effects of configuration and operational changes, and the general procedure for preparing them is as follows:

-- Define a baseline configuration and calculate baseline fatigue loads.
-- Vary one parameter as an independent variable while holding the remaining parameters constant at their baseline values or scaling them in proportion to the independent variable.
-- Calculate loads for the changed configurations and normalize (*i.e.*, divide) each of them by the appropriate baseline load.
-- Graph the results as load ratios *vs.* the independent parameter, for desired probabilities of exceedance.

This process will now be illustrated by sample calculations in which Equations (12-7a) to (12-7d) are used to calculate the relative changes in both fatigue loads and stresses that are caused by increasing and decreasing the rotor diameter [Spera 1994].

Baseline Configuration and Loads

A baseline configuration that represents the mid-range of the rotor, tower, and site data in Tables 12-4 and 12-5 can be defined in terms of the following parameters and factors:

$$
\begin{aligned}
D &= 46.0 \text{ m} & U_0 &= 7.9 \text{ m/s} \\
P_R &= 410 \text{ kW} & U_2 &= 14.3 \text{ m/s} \\
\delta_3 &= 0 \text{ deg} & M_g &= 456 \text{ kN-m} \\
c_t &= 0.80 \text{ m} & \theta &= 5 \text{ deg} \\
Z &= 660 \text{ m} & s &= 0.066 \\
H &= 38.0 \text{ m} & \alpha_0 &= 0.40 \\
N &= 30.0 \text{ rpm} & \alpha &= 0.232 \\
\omega_c &= 152 \text{ cpm} & \text{Tower type} &= \text{shell, upwind of rotor}
\end{aligned}
$$

The factors in Equations (12-5) for the baseline case are then calculated to be

$$a = 0.00 \qquad\qquad d = 0.94$$
$$b = 1.00 \qquad\qquad e = 1.04$$
$$c = 0.87 \qquad\qquad g = 0.28$$

Figure 12-14 presents the probability distributions of the wind and fatigue loads as calculated for the baseline case. The format here is the same as in Figures 12-11, except that the ordinate scale is linear rather than logarithmic, which is more convenient for design purposes. With the exception of the blade cyclic chordwise moment, we see that the low-probability, high-percentile cyclic loads that are most damaging in fatigue can be 2.5 to 5 times the median cyclic loads on our baseline turbine. Because chordwise moments in larger HAWTs are dominated by the force of gravity, they typically have a *narrow-band distribution* about their mean levels.

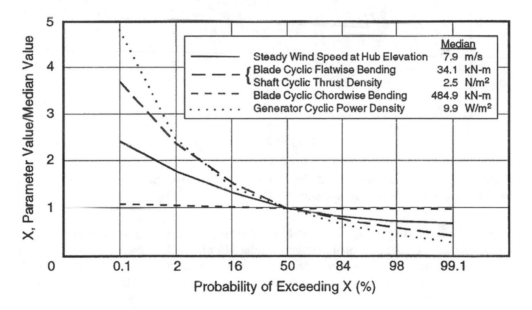

Figure 12-14. Wind and fatigue load probability distributions calculated for the sample baseline HAWT. Configuration: Diameter = 46 m, rated power = 410 kW, teetered hub, upwind shell tower.

Effect of Changing the Rotor Size

To preserve the turbine shape while varying its size, several parameters must be expressed in terms of the rotor diameter or converted to ratios, as follows:

$$c_t = 0.0174\,D \text{ m} \qquad\qquad U_0 = 7.9\,(D/46.0)^{0.232} \text{ m/s}$$
$$H = 0.826\,D \text{ m} \qquad\qquad U_2 = 1.81\,U_0 \text{ m/s}$$
$$N = 1{,}380/D \text{ m} \qquad\qquad P_R = 0.194\,D^2 \text{ kW}$$
$$N/\omega_c = 0.20 \qquad\qquad M_g = 456\,(D/46.0)^3(c_t/0.80) \text{ kN-m}$$

Re-calculating loads as the diameter is changed by ± 20 percent from its baseline size of 46 m, normalizing the results by the baseline loads to obtain load ratios, and plotting the results produces the design trend graph in Figure 12-15. These load ratios apply to all probabilities of exceedance. Blade cyclic loads are changed the most by scale changes, while the generator cyclic power density (which is already a normalized quantity) changes the least. The effects on loads of changing other parameters (hub δ_3 angle, steady wind speed, surface roughness, and tip chord) are given in [Spera 1994] for this same baseline case.

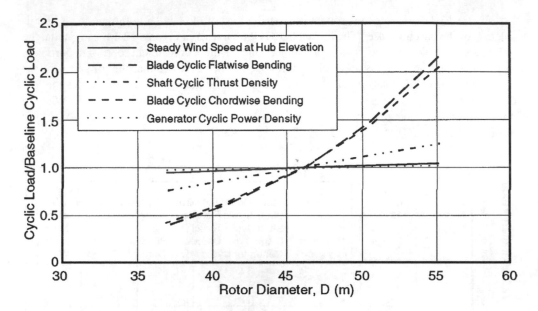

Figure 12-15. Typical design trend chart showing the relative changes in fatigue loads as the size of a HAWT is changed. [Spera 1994]

The load trends in Figure 12-15 can be used to estimate stress trends when the size of a turbine is changed but its geometry remains similar. In wind turbine components like blades and towers, bending stress is usually the design driver and *section modulus* is the strength parameter of interest. Because the section moduli of geometrically similar cross-sections are proportional to the cube of their relative dimensions, it follows that

$$\frac{Bending\ stress}{Baseline\ bending\ stress} = \left(\frac{Bending\ moment}{Baseline\ bending\ moment} \right) \frac{D_o^3}{D^3} \tag{12-8a}$$

where D_0 is the baseline diameter. Bending moments on the tower are directly proportional to the thrust density times the cube of the diameter (from rotor area x tower height), so the cubic ratios cancel out and

$$\frac{Tower\ bending\ stress}{Baseline\ tower\ bending\ stress} = \frac{Shaft\ thrust\ density}{Baseline\ shaft\ thrust\ density} \tag{12-8b}$$

Torque on the turbine shaft is proportional to the power density times the cube of the diameter (from rotor area/shaft speed) and so is torsional section modulus. Again, cubic ratios cancel out and

$$\frac{Shaft\ shear\ stress}{Baseline\ shaft\ shear\ stress} = \frac{Generator\ power\ density}{Baseline\ generator\ power\ density} \tag{12-8c}$$

Combining the data in Figure 12-15 with Equations 12-9 produces the stress trend graph in Figure 12-16. Here we see that blade and tower cyclic bending stresses in geometrically-similar HAWTs increase linearly with the scale factor, to a first approximation. This is similar to a well-known scaling property of cantilever beams loaded by their own weight. In other words, a baseline wind turbine can be geometrically-scaled *down* without increasing fatigue stresses in the blades and tower, but not *up*. Cyclic shear stresses in the turbine shaft, however, remain approximately constant with scale changes.

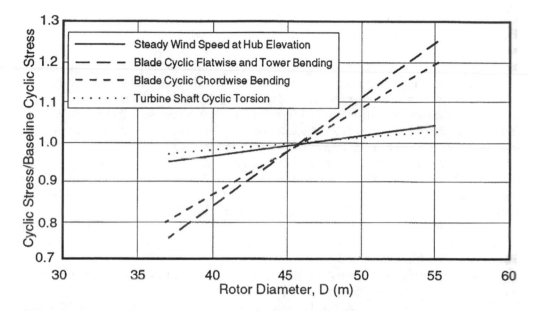

Figure 12-16. Fatigue stress trend chart derived from the load-trend chart in Figure 12-15. In geometrically-similar wind turbines, blade and tower fatigue stresses vary linearly with rotor diameter while shaft cyclic shear stresses remain relatively constant.

Fatigue Design Procedures

When fatigue load spectra have been defined for critical sections in a wind turbine (*e.g.,* welded and bolted joints, access holes, fillets, *etc.*) the problem of designing wind turbine components to resist these loads for the design lifetime becomes one for which accepted and validated procedures are available. Standard stress analysis procedures are used to convert load spectra to stress spectra, and allowable fatigue stresses are based on standard laboratory tests of the materials of construction [see, *e.g.*, Manson 1965 and 1966, Mitchell 1979, Fuchs and Stephens 1980, Dowling 1993, Mandel *et al.* 1993, 1994, 1997, 2002].

There is little that is unique about either fatigue theory or the fatigue design process as applied to wind turbines, once the load spectra are specified. Figure 12-2 illustrates one of the few differences between fatigue design of wind turbines and that of other structures. The magnitudes of the allowable stresses in critical wind turbine components are significantly lower because of their longer fatigue life requirements. Thus, structural integrity of wind turbines requires very conservative yet standard fatigue design procedures.

One of the most important elements in the fatigue design process is the definition of allowable fatigue stresses. This is a technical specialty of its own, in which engineers modify the results of laboratory fatigue tests on material and joint specimens to account for the expected spectrum of stress, size effects, cost-effective manufacturing and inspection procedures, environmental effects, and maintenance planned for the structural system during its lifetime. Thus, the specification of fatigue allowable stresses for various materials in a structure is an integral part of the manufacture and operation of that structure.

Verification of the *total* fatigue design process by field experience with prototype (or at least similar) structures is necessary for confidence in the long-term structural integrity of the design. As a result, different fatigue design procedures and different fatigue allowable stresses will be utilized by different manufacturers, because each will have its own field experience and will have modified its design procedures accordingly.

Discussion of all of the factors that enter into the specification of a fatigue allowable stress is well beyond the scope of this chapter. Instead, two general methods will be discussed that have been used successfully to account for the *stress spectrum effect*, which is one of the first steps needed to modify laboratory test data into fatigue allowable stresses. These will be referred to as the *S-N linear damage* and the *fracture-mechanics* methods.

S-N Linear Damage Method

Fatigue damage is both a physical process (*e.g.* the initiation and propagation of defects in the material) and a mathematical representation of that process. Here we shall deal only with the latter. In general, fatigue damage and fatigue lifetime are inversely proportional. One of the simplest models of the accumulation of fatigue damage during repeated cycles of stress is the *linear damage hypothesis*, proposed by Palmgren [1924] and Miner [1945]. According to this hypothesis, if the stress cycle remains constant throughout a fatigue lifetime equal to N, then the fraction of that lifetime consumed on every cycle is constant and equal to $1/N$. This fraction is also defined as the *damage per cycle*, and it follows that the total damage at failure is equal to unity. Furthermore, if the stress cycles change during the lifetime, the damage fractions per cycle are linearly additive, and fatigue failure still occurs when the accumulated damage reaches unity. If a stress spectrum is subdivided into groups of cycles or *layers* within which the stress cycles are relatively uniform, then

$$m\sum_{l}^{I}\frac{n_i}{N_i} = 1.0 \quad at\ fatigue\ failure \tag{12-9a}$$

$$m = \frac{N_f}{\sum\limits_{1}^{I} n_i} \qquad (12\text{-}9b)$$

where

i = index of layers in the spectrum

I = number of layers in one spectrum

m = number of repetitions of the spectrum required to cause fatigue failure

n_i = number of cycles applied at stress level S_i

N_i = fatigue lifetime at a constant stress level S_i (cycles)

S_i = stress parameter in the cycle upon which fatigue damage is primarily dependent (*e.g.*, maximum stress, cyclic stress, *etc.*) (kN/m²)

N_f = fatigue lifetime under spectrum loading (cycles)

Inherent in Equation (12-12) is the assumption that the order of applying the layers of stress to the material is not important. While this may not be true for a single spectrum with only a few layers, it has been found to be acceptable for design when high- and low-stress layers are randomly mixed and a large number of spectra is applied.

The dependence of N on repeated cycles at a constant level S, illustrated by the *S-N curve* in Figure 12-2, is typically expressed by a power-law equation [Basquin 1910, Manson 1965 and 1966], as

$$S = S_1 N^{\alpha} \qquad (12\text{-}10a)$$

$$S \geq S_e \qquad (12\text{-}10b)$$

where

S_1 = empirical stress coefficient (kN/m²)

α = empirical exponent

S_e = *endurance limit*, below which no fatigue damage occurs (kN/m²)

On log-log coordinates Equation (12-10) is plotted as a straight line with an intercept of S_1 at N equal to one cycle and a slope of α, down to a stress of S_e. Typically, α is less than about - 1/8. Solving Equation (12-10a) for N with $S = S_e$, we obtain the fatigue life at the so-called "knee" of the S-N curve. To the right of this point the curve becomes a horizontal line.

The next step in defining an allowable stress for the spectrum is to express the layer stresses as fractions of the largest stress in the spectrum, S_{maxmax}. This produces a set of normalized stresses analogous to the normalized loads in Table 12-1. In this way S_{maxmax} can be used as a scale factor to proportionately increase or decrease all stress levels in the spectrum until the required fatigue lifetime is achieved. Let

$$S_i = s_i S_{maxmax} \qquad (12\text{-}11)$$

where

s_i = stress ratio for the ith layer

S_{maxmax} = largest stress in the spectrum (kN/m²)

Combining Equations (12-9), (12-10), and (12-11) produces the following equation for calculating allowable stresses by the S-N linear damage method:

$$S_{maxmax} = S_1 \left(N_f \frac{\sum_1^I n_i \, s_i^{-1/\alpha}}{\sum_1^I n_i} \right)^{\alpha}$$

$(12\text{-}12)$

This value of S_{maxmax} becomes the allowable fatigue stress for the loading condition represented by L_{maxmax} in Table 12-1. The structural-dynamic analysis of loads has been converted into a static structural design problem by this method.

Laminated Wood Blade Fatigue

To illustrate the S-N linear damage method for calculating the effect of spectrum loading on fatigue strength, consider the following problem: A HAWT blade is to be manufactured from *laminated Douglas fir/epoxy* material, whose fatigue strength is known to depend strongly on the maximum stresses in the applied fatigue cycles. The design spectrum for the blade is composed of the layers and numbers of cycles in Table 12-1, and the stress ratios in each layer are the same as the load ratios listed in this table. Each spectrum represents nine hours of operation, and the blade is to be designed for a 30-yr lifetime running continuously.

The laboratory fatigue data for the laminated wood material are shown in Figure 12-17 [Spera *et al.* 1990]. The veneer grade is assumed to be *A+*, veneer-to-veneer joints are scarfed, and large-volume properties are to be used, to account for the *size effect* present in this material. The lowest S-N curve in Figure 12-17(a) is the appropriate one for the material requirements in this problem, and its empirical exponent α is equal to -0.0676. The empirical coefficient S_1 must be determined using the *Goodman diagram* in Figure 12-17(b), since the S-N curves shown are for an R-ratio of 0.1 whereas the average R-ratio in Table 12-1 is 0.37. The line along which combinations of the cyclic stress and the average stress produce a constant R-ratio of 0.37 is shown in the figure. Its slope is calculated as

$$\frac{S_{cyc}}{S_{avg}} = \frac{(1 - 0.37)/2}{(1 + 0.37)/2} = 0.46$$

$(12\text{-}13a)$

The stress cycle parameters at the intersection of this R-ratio line and the fatigue strength line for 10^7 cycles to failure of the specified material are

$$\begin{aligned}
S_{cyc} &= 1{,}650 \text{ psi} \ \ (11{,}380 \text{ kN/m}^2) \\
S_{avg} &= 3{,}600 \text{ psi} \ \ (24{,}820 \text{ kN/m}^2) \\
S_{max} &= 5{,}250 \text{ psi} \ \ (36{,}200 \text{ kN/m}^2)
\end{aligned}$$

The S-N curve for the average fatigue strength of laboratory specimens with an R-ratio of 0.37 then becomes

$$S_{max} = 5{,}250 \left(\frac{N}{10^7} \right)^{-0.0676} = 15{,}610 \, N^{-0.0676}$$

$(12\text{-}13b)$

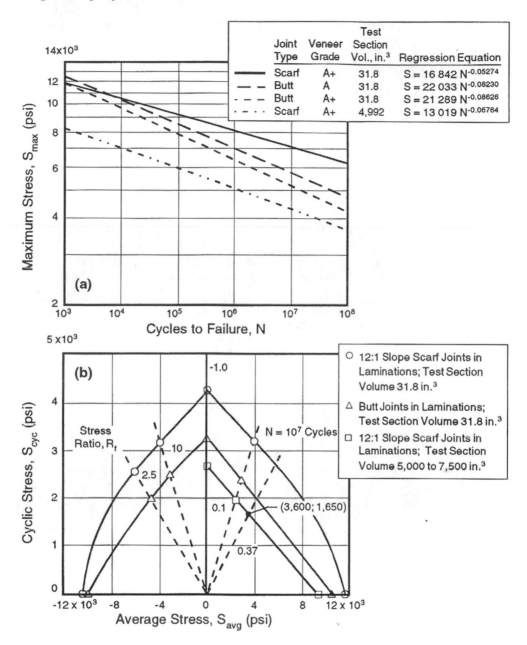

Figure 12-17. Reference fatigue properties of laminated Douglas fir/epoxy material, from laboratory tests. (a) Effects of joint type, veneer grade, and specimen size on tension-tension fatigue ($R_f = 0.1$) parallel to the grain (b) Goodman diagram of the interaction of cyclic and average stresses for lifetimes of 10 million cycles [Spera *et al.* 1990]

from which $S_l = 15,610$ psi $(107,600 \text{ kN/m}^2)$. For the required 30-year design life,

$$N_f = 30 \ yr \times 8,760 \ \frac{hr}{yr} \times \frac{19,105 \ cyc}{9.0 \ hr} = 5.6 \times 10^8 \ cyc \qquad (12\text{-}13c)$$

From Table 12-1,

$$\sum_{1}^{36} n_i s_i^{-1/(-0.0676)} = 1,618 \ cycles \quad \sum_{1}^{36} n_i = 19,105 \ cycles \qquad (12\text{-}13d)$$

Substituting the factors calculated in Equations (12-13) into Equation (12-12) then gives

$$S_{maxmax} = 4,230 \text{ psi} \ (32,590 \text{ kN/m}^2)$$

Again, this stress represents the average fatigue strength of laboratory specimens and therefore **must not be used directly for a fatigue allowable stress**. It must first be multiplied by one or more so-called *knock-down factors* (each less than 1.0) to account for (a) scatter in the laboratory test data and (b) several conditions that can reduce fatigue strength in full-scale structures below that of laboratory specimens.

Fiberglass Blade Fatigue

Fiberglass composites (also termed *glass fiber-reinforced-plastics* or *GRP*) are currently the most common materials used in the manufacture of modern wind turbine blades [Ancona and McVeigh 2001, Griffin 2001]. The fatigue properties of fiberglass have been studied extensively and databases of S-N curves have been compiled [Mandel and Samborsky 1997, Delft *et al.* 1997, Wahl 2001, Wahl *et al.* 2002].

Because fiberglass is a non-homogeneous material, like the laminated fir-epoxy material discussed previously, its fatigue resistance is sensitive not only to the selection of its constituent materials and the processing variables used to combine these materials, but also to the ratio of applied tensile and compressive stresses.

A *Goodman diagram* of the type shown in Figure 12-18 [Sutherland and Mandel 2005] is the conventional method for displaying the dependence of fatigue life on cyclic stress combined with various amounts of mean stress. The asymmetry evident in Figure 12-18 illustrates the significantly different effects that tensile and compressive mean stresses have on the fatigue resistance of this composite material. An extensive amount of fatigue testing of material samples is normally required to produce a complete Goodman diagram for fiberglass. To reduce the required amount of testing, Sutherland and Mandell [2005] verified that fatigue tests at the 5 R-ratios shown in Figure 12-18 are sufficient to capture the essential features of this material's fatigue resistance in a Goodman diagram.

Fracture-Mechanics Method

A fracture-mechanics model of the fatigue damage process is more complex than the S-N linear damage model, but it is considered by many fatigue specialists to be the most representative model of the physical processes leading to fatigue failure [*e.g.*, see Broek 1982, Tada *et al.* 1985, Suresh 1991]. Fracture-mechanics analysis is a standard, validated tool of structural engineering, and its application to wind turbines is straight-forward [Finger 1980, Finger 1985]. As with any fatigue design methodology, experience and verification by field testing are much more critical to success than the level of complexity.

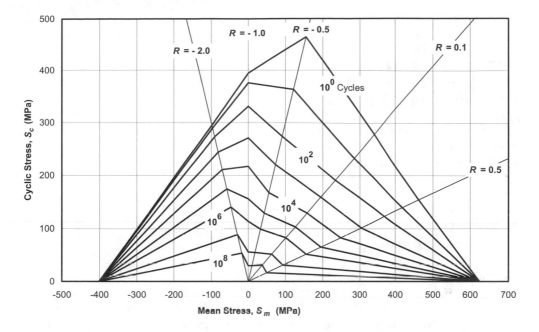

Figure 12-18. Sample Goodman diagram for fiberglass composite material based on fatigue tests at 5 R-ratios [Southerland and Mandell 2005].

A fundamental concept of the fracture-mechanics method is that all structures contain small flaws when they are first placed in service. The sizes of these *pre-existing* or *initial flaws* are related to the inspection methods used and the specified *acceptance criteria*. Flaws in critical areas can develop into cracks when fatigue loads are applied, and these cracks continue to grow in length or *propagate* during the service life of the structure. Crack length is the measure of fatigue damage, and the *crack propagation rate* (expressed in length units per cycle) is a highly non-linear function of crack length and elapsed time. At the start of service, crack lengths and crack propagation rates are very low. As the crack grows the propagation rate accelerates. Eventually a crack length is reached that is unstable, the propagation rate becomes infinite, and fracture occurs.

Application of the fracture-mechanics method to the fatigue analysis of wind turbine structures can be divided into the following steps:

1. Define a *load spectrum* for each major component, in terms of interface forces and moments and their probability of exceedance.
2. Estimate section sizes (*e.g.*, skin thicknesses) and calculate a *stress spectrum* for selected points around each critical section in the component, such as welds and joint flanges.
3. *Normalize stresses* in each spectrum by its largest stress, S_{maxmax}, which serves as a scale factor to be defined.
4. Specify a *design initial flaw size* for the selected material, fabrication method, and inspection capability, and a *design failure crack size*.
5. Derive and verify a *crack-propagation model* for the selected material, expressed in terms of the an incremental growth rate *vs.* stress parameters away from the crack and the instantaneous crack length or depth.
6. Estimate S_{maxmax} and calculate *crack size vs time* by forward integration of the propagation model over the design lifetime, at each selected point around each critical

section. Adjust S_{maxmax} at each selected point until the crack size at the end of the design lifetime equals the design failure crack size around each critical section in the component.

7. Specify a set of *fatigue allowable stresses* equal to the adjusted values of S_{maxmax} from Step 6 and an associated set of *fatigue design loads* equal to the maximum set of loads in the spectrum defined in Step 3.

Steps 1, 2, and 3 were discussed earlier and result in spectrum data like that in Table 12-1. The remaining steps will be discussed here, drawing upon methods used for the structural design of the 2.5-MW Mod-2 and the 3.2-MW Mod-5B HAWTs [Boeing 1979, Boeing 1988]. The major components in these two machines, including the skins and spars of their 91.4-m and 97.5-m rotors, were fabricated from welded-steel plates, forgings, or rolled sections. The fracture-mechanics method was used throughout both design efforts, in part because of the extensive experience at the *Boeing Company* and its subsidiaries with its application and verification.

Design Initial Flaw Size

The initial flaw sizes specified for fatigue design are based on the *flaw size acceptance criteria* specified for manufacturing the component in question. The latter specification includes not only technical considerations (like material properties) but also cost and schedule requirements. Thus, system-level decisions are required in order to set flaw size acceptance criteria. After this has been accomplished, design flaws are defined which are (1) located in the most fatigue-critical areas and in their highest-stress orientation, and (2) significantly larger than the acceptance criteria. Both of these procedures provide conservatism (*i.e.*, additional reliability) to the design.

Table 12-7 illustrates the relationship between the sizes of acceptable flaws and design flaws used for the Mod-5B HAWT. Here and in the equations and calculations to follow, customary U.S. units will be used for compatibility with the literature of fracture mechanics technology.

Table 12-7. Inspection and Design Flaw Sizes for ASTM A-6 Steel in the Mod-5B HAWT [Boeing 1988]

Defect planes perpendicular to the surface and all dimensions in inches

Inspection Method	Detection Capability	Acceptance Criteria
Visual	0.005 wide x 0.060 long	Linear indications: 0.060 long
Dye-Penetrant	0.005 wide x 0.030 long	Rounded indications: 0.125 long
Radiological	2% of thickness deep x 0.040 long	
Ultrasonic	0.030 deep x 0.090 long	

| Material | Design Initial Flaw Sizes | |
	Surface Flaws	Internal Flaws
Base Metal	0.022 deep x 0.110 long	0.044 deep x 0.110 long
"B" Weld [1]	0.050 deep x 0.250 long	0.100 deep x 0.250 long

[1] Full penetration; constant thickness in base and weld metal; grinding parallel to major stresses

Crack Propagation Model

In the fracture-mechanics method, the stress parameter that governs the rate at which fatigue cracks grow is the *stress intensity factor*, defined as follows for tensile loading:

$$K = S \sqrt{Q\, \pi a} \qquad (12\text{-}14)$$

where

K = stress intensity factor (psi-in$^{0.5}$)
S = tensile stress normal to the crack (psi)
Q = crack shape factor
a = crack radius for a circular crack or length of the semi-minor axis of an elliptical crack (in)

A stress cycle that is completely in compression (*i.e.*, $S_{max} < 0$) is usually assumed to cause neither crack growth nor retardation. Referring to Figure 12-1, this explains why fatigue is usually not the design driver on the downwind surfaces of HAWT rotor blades, where the steady thrust loads cause compressive stress. Values of the crack shape factor in Equation 12-14 for some common crack configurations are as follows:

-- Through-crack in an infinitely-wide plate $\quad Q = 1.00$
-- Internal circular or "penny-shaped" crack $\quad Q = 4/\pi^2 = 0.405$
-- Surface semi-circular crack $\quad Q \approx 4.8/\pi^2 = 0.486$
-- Internal elliptical crack $\quad Q \approx 1/\phi^2$
-- Surface semi-elliptical crack $\quad Q \approx 1.2/\phi^2$

$$\phi = \int_0^{\pi/2} \sqrt{1 - [1 - (a/c)^2]\sin^2\theta}\; d\theta \qquad (12\text{-}15a)$$

where $\quad c$ = length of the semi-major axis of the crack (in)

Evaluating φ for a range of ratios a/c from 0 (*i.e.*, a linear crack) to 1.0 (a circular crack), we can derive the following convenient empirical equation for a surface crack:

$$Q \approx 1.200 - 0.714(a/c) \quad 0 \le a/c \le 1.0 \qquad (12\text{-}15b)$$

Fatigue crack propagation models have been developed that are based on either the maximum stress intensity in a given layer in the spectrum, K_{max}, or the stress-intensity range, $\Delta K = K_{max} - K_{min}$, or both. The following three crack growth models, based on K_{max} and the ratio R_f of minimum-to-maximum stress, were considered during the development of structural design methods for the Mod-2 HAWT:

(a) A *retardation model*, which accounts for any beneficial effects of load excursions during and between load spectra. Excursions are known to cause local plastic "wakes" that close crack tips and slow crack growth. This model, which considers all cycles to be damaging, has the general mathematical form

$$da/dn \propto (1 - R_f)^k K_{max}^l \left(\frac{S_{max}}{R_{OL}\, S_{maxmax}} \right)^m \quad S_{max} \ge 0;\; R_{OL} \ge 1.0 \qquad (12\text{-}16a)$$

where

da/dn = crack growth rate (in/cyc)

k, l, m = empirical exponents

R_{OL} = overload ratio; ratio of peak stress between spectra to S_{maxmax}

(b) A *threshold model*, which contains a minimum size of stress intensity below which there is no crack growth (equivalent to the endurance limit, S_e, on an *S-N* curve):

$$da/dn = 0 \quad if\ K_{max} \leq K_{TH}$$

$$da/dn \propto (1 - R_f)^k K_{max}^l \quad if\ K_{max} > K_{TH} \qquad (12\text{-}16b)$$

where

K_{TH} = threshold stress intensity (psi-in$^{0.5}$)

and

(c) a combination of these two models.

When correlating model predictions with the results of laboratory fatigue crack propagation tests of pre-cracked specimens under spectrum loading, the retardation model was found to underestimate the lives of lower-stress specimens while the threshold model underestimated the lives of higher-stress specimens. Equations for a combined model that best corrects these deficiencies are as follows, with empirical constants for *ASTM A-6* steel, both base and weld metal:

$$da/dn = 0 \quad if\ K_{max} \leq K_{TH} \qquad (12\text{-}17a)$$

$$da/dn = 3\ x\ 10^{-10}(1 - R_f)^{2.4}(K_{max}/1,000)^3(S_{max}/S_{maxmax})^2(1/R_{OL}^2) \qquad (12\text{-}17b)$$

$$if\ K_{TH} < K_{max} < K_C$$

$$da/dn = \infty \quad if\ K_{max} = K_C \qquad (12\text{-}17c)$$

$$R_f \geq 0 \quad R_{OL} \geq 1.0 \qquad (12\text{-}17d)$$

$$K_{TH} = 5.00 + 52.0 \exp\ [-11.0(1 - R_f)]\ psi-in^{0.5} \qquad (12\text{-}17e)$$

$$K_C = 125,000\ psi-in^{0.5} \qquad (12\text{-}17f)$$

where

K_C = *fracture toughness* material property in tension (psi-in$^{0.5}$)

Fracture occurs upon a single application of a stress intensity equal to the fracture toughness, a material property that can be determined by standardized test procedures such as *ASTM E*-399 [ASTM 1993]. Figure 12-19 is a graph of Equations (12-17). The primary line defines crack-propagation rates for constant-amplitude cycles with $R_f = 0$, at maximum stress intensities between a threshold of 5,000 psi-in$^{0.5}$ and the fracture-toughness of 125,000

$$K_{TH} < K_{max} < K_c: \quad da/dn = 3 \times 10^{-10} (1-R_I)^{2.4}(K_{max})^3(K_{max}/K_{maxmax})^2$$

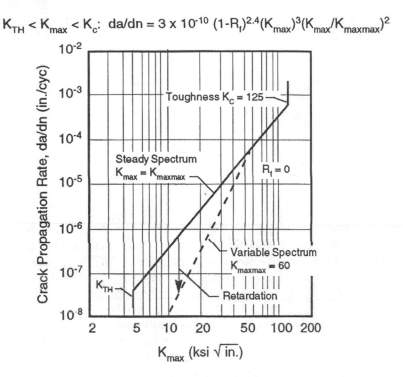

Figure 12-19. Sample crack-propagation rate model. [Boeing 1988]

psi-in$^{0.5}$, the latter property determined by test. The secondary line illustrates the effect of occasional overloads (*i.e.*, load excursions) to a stress intensity of 60,000 psi-in$^{0.5}$.

Sample Problem

Application of the fracture-mechanics method to the calculation of fatigue design allowable stresses in a wind turbine will be illustrated by a sample problem in which the following parameters are specified:

-- Material...Category "B" weld in ASTM A-6 steel
-- Normalized Stress Spectrum...............Table 12-1; 19,105 cycles
-- Design life time....................................30 years, 330 spectra per year; 8 x 10^8 cycles
-- Initial crack size...................................Table 12-6, semi-elliptical surface flaw
-- Final crack size.....................................10 x Initial crack depth
-- Crack propagation model....................Equations (12-17)
-- Overload stresses between spectra.......From 1.0 to 2.0 x S_{maxmax}

Fatigue design allowables are to be calculated from these parameters and expressed in terms of S_{maxmax} *vs.* $S_{overload}$.

From Table 12-7, the initial crack dimensions are $a = 0.050$ in. and $c = 0.125$ in. The crack is assumed to first propagate only in the depth direction, from its initial semi-elliptical shape to a semi-circular shape of radius $a = 0.125$ in. Propagation then continues as a semi-circular crack until $a = 0.500$ in., in accordance with the final crack size specification. Thus, the shape factor in Equation (12-15b) is

$$Q = 1.200 - 0.714 \, (a/c)^2 \qquad \text{if} \qquad 0.050 \le a \le 0.125 \text{ in.}$$

$$Q = 0.486 \qquad \text{if} \qquad 0.125 \le a \le 0.500 \text{ in.}$$

(12-18)

Defining a threshold stress as the stress producing the threshold stress intensity with the initial crack and an R-ratio of zero, we obtain from Equations (12-14), (12-17e), and (12-18)

$$Q = 1.200 - 0.714 (0.050/0.125)^2 = 1.086 \qquad \text{(12-19a)}$$

$$S_{TH} = 5,000 / \text{sqrt} [1.086 \, \pi \, x \, 0.050] = 12,100 \text{ psi} \qquad \text{(12-19b)}$$

where

S_{TH} = threshold stress (psi)

Crack depths *vs.* operating time can now be calculated by setting the maximum stress in the spectrum equal to values larger than the threshold stress since, in accordance with Equation (12-17a), no crack growth takes place if S_{maxmax} is less than or equal to S_{TH}.

Typical results of the forward-integration of Equations (12-17) for an overload ratio, R_{OL}, of unity are shown in Figure 12-20 for increasing values of S_{maxmax}. Here we see the sensitivity of crack growth to the amount by which S_{maxmax} exceeds S_{TH} when the number of fatigue cycles is very high, as it is in this case. When the maximum stress in the spectrum is only 4 percent larger than the threshold stress, the crack will grow to the specified final depth of 0.500 in. after 30 years of operating time. If this is increased to 8 percent, the same crack depth is reached in about 15 years. The effect of overloads between load spectra is shown in Figure 12-21, which is the solution to this sample problem. As the overload ratio increases from 1.0 to 2.0 the fatigue design allowable stress increases from 12,100 psi to 13,600 psi for the given design parameters.

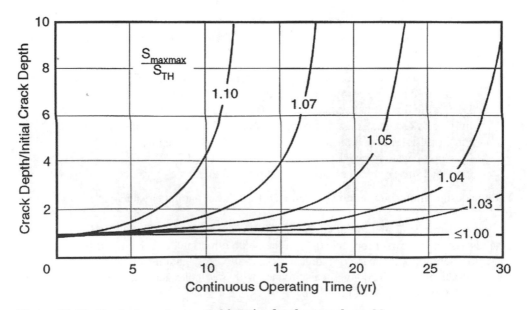

Figure 12-20. Typical crack growth histories for the sample problem.

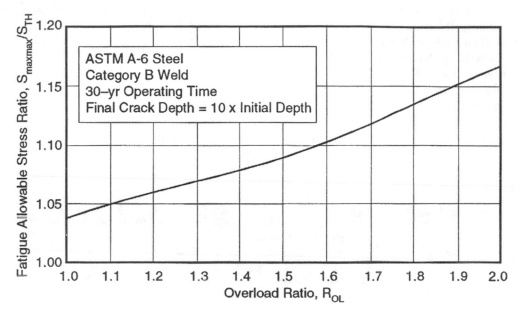

Figure 12-21. Fatigue allowable design stresses for the sample problem calculated according to the fracture-mechanics method, showing the beneficial effect of overloads.

Margin of Safety

Both the S-N linear damage method and the fracture-mechanics method have been used successfully in the fatigue design of wind turbines. The end product of the application of either method is a fatigue allowable stress specified for a given component and sometimes for a given location in that component. Paired with this fatigue allowable stress is a set of fatigue design loads, which in our sample cases are the L_{maxmax} loads in Table 12-1. Generally, a fatigue design is judged to be acceptable if, after the structural engineer adds required *factors of safety*, the *margin of safety* is positive at all critical points in a component. Margin of safety is calculated as follows:

$$MOS = \frac{Fatigue\ allowable\ stress}{(Design\ stress)\ x\ (FOS)} - 1.0 \geq 0 \tag{12-20}$$

where

MOS = margin of safety; figure of merit for structural integrity
$Design\ stress$ = calculated stress in the component under the specified loads (kN/m²)
FOS = factor(s) of safety

For a cost-effective design, margins of safety for design-driving loads should be close to zero.

Figure 12-22 illustrates an acceptable fatigue design of a cross-sectional weld in a steel rotor blade, in this case for an inboard blade section of the Mod-5B HAWT. The margin of safety is zero wherever the design stress line (multiplied by the applicable factors of safety) intersects the allowable stress lines. In this design most of the upwind-surface stresses are within the "B" weld allowable. However, in a small area near the leading edge (where the most critical combination of in-plane and out-of-plane loads often occurs) it was found to be more economical to locally increase the weld quality and with it the fatigue allowable stress than to increase the plate thickness. This so-called "B+" weld category has an initial crack size smaller than that of the "B" welds, which requires additional inspection. This trade-off between component size and the cost and time of additional inspection illustrates how the specification of a design allowable stress is a system-level decision.

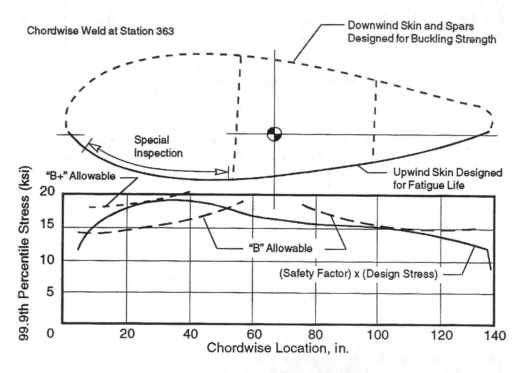

Figure 12-22. Example of fatigue design stresses in a HAWT blade cross-sectional weld compared with fatigue allowable stresses. [based on Boeing 1988]

Fatigue Testing of Wind Turbine Blades

Current methods for designing, conducting, and analyzing the results of fatigue tests of wind turbine blades are described by White [2004]. Much of the following discussion contains information from this reference. Readers are referred to White's comprehensive report for details not only on fatigue testing but also on the following supporting topics:

-- Historical review of blade fatigue testing
-- Wind turbine blade loads, deflections, natural frequencies, and phase angles
-- Fatigue damage analysis, and
-- Controller design of a multi-function fatigue testing machine

While dynamic computer models using programs such as *ADAMS* and *FAST* [Kelley 1997] have now been developed to simulate the real-time response of a wind turbine blade to wind conditions over its lifetime, there is still a need to physically test blades under controlled conditions. By applying a schedule of fatigue loads to a full-size blade it is possible to

-- Verify the design strength, stiffness, and natural frequencies as predicted using a *finite- element model* (FEM) of the blade
-- Identify potential manufacturing defects, and
-- Validate blade fatigue endurance in months instead of decades

Current Major Fatigue Test Facilities for Wind Turbine Blades

At this time there are four national laboratories worldwide that maintain major facilities for performing static and fatigue tests of both small and large wind turbine blades, as follows:

RISO National Laboratories (RISO) in Denmark

RISO operates under the Danish Ministry of Science, Technology, and Innovation. Static, fatigue, and modal tests of wind turbine blades are conducted at the *Sparkuer Blade Test Centre* in Roskilde.

Centre for Renewable Energy Sources (CRES) in Greece

CRES operates under the Greek Ministry of Development. Facilities located in Attiki are capable of testing wind turbine blades up to 25 m in length.

Wind Turbine Materials and Constructions Knowledge Centre (WMC) in the Netherlands

A modern facility opened in 2003, WMC is a joint project between *Delft University* and the *Energy Research Centre of the Netherlands* (ECN). It has the capability of testing blades up to 60 m in length.

National Renewable Energy Laboratory (NREL) in the United States

NREL, a laboratory operated for the U.S. Department of Energy, is responsible for wind turbine blade testing at its *National Wind Technology Center* (NWTC) in Golden, Colorado. More than 100 blades have been tested at the NWTC facility, which is unique in the U.S. Most of these tests have been conducted for commercial wind turbine manufacturers under various NREL development agreements. At the NWTC, engineers can perform static, fatigue, and modal testing of blades at least 45 m in length.

Methods of Blade Fatigue Testing

Each of the four facilities listed above has independently developed methods for performing blade tests. Over time, some original test methods have been replaced by newer techniques, and several of these test facilities now use similar equipment and procedures [White 2004]. There are three general methods used currently at national laboratories to test the fatigue endurance of full-scale wind turbine blades. These may be referred to as (1) the *dual-axis forced-displacement method*, (2) the *single-axis resonance method*, and (3) the newly-developed *dual-axis hybrid resonance method*. In all three methods, the root end of the test blade is attached to a very stiff, vertical load frame or *strongback*, and the blade is cantilevered outward.

Dual-Axis Forced-Displacement Method

The dual-axis forced-displacement loading system, currently used at NREL and other research facilities, employs a servo-hydraulic system with dual actuators to load the blade cyclically and simultaneously in both the flatwise (*flap*) and edgewise (*chordwise*, or *lead-lag*) directions [Hughes and Musial 1999, Larwood *et al.* 2001, Hughes and Musial 2002]. Cyclic loads are applied at frequencies well below the blade's flatwise first (*fundamental*) natural frequency. The dual-axis fatigue test system at NREL is shown in Figure 12-23. The

Figure 12-23. NREL's dual-axis forced-displacement test system [Courtesy of the National Renewable Energy Laboratory]

hydraulic actuator for applying flatwise loads is oriented vertically below the outboard end of the blade. Edgewise loads are applied by a horizontal bar loaded through a bell-crank mechanism by a vertical actuator, as seen at the far right in the figure.

The primary advantage of this system is that its biaxial loading can be independently adjusted in the flatwise and edgewise directions for accurate simulation of the actual operating loads on a blade. While the dual-axis forced-displacement technique is the most accurate method currently used to test wind turbine blades, it also has several drawbacks that become significant as capacity requirements increase to match the increasing size of modern utility-scale wind turbines:

-- Ever-larger forces and displacements are required from the hydraulic actuators.
-- New actuators must be designed and built each time a larger blade is to be tested.
-- Hydraulic pumping requirements increase as actuator size increases.
-- Substantial equipment costs are incurred with each capacity increase.
-- Increased hydraulic fluid flow requirements create significant operating energy costs.
-- Blade tip sections are normally removed, in order to avoid unwanted vibration modes and to fit larger blades into the existing facility, and therefore are not included in the test.

Flatwise actuators have the larger requirements for both displacement and force. Edgewise actuators have smaller displacement requirements because of the higher blade stiffness in the edgewise direction.

Single-Axis Resonance Method

In the single-axis resonance method, currently used at the RISO laboratory, an electric motor is mounted on the blade and spins an eccentric mass in order to excite blade resonance in either the flatwise or edgewise direction. Whereas the dual-axis forced-displacement method permits applying flatwise and edgewise loads simultaneously, the single-axis resonance method is restricted to applying these two cyclic load components in separate test periods, effectively doubling the test duration.

Figure 12-24 shows a single-axis resonance blade fatigue test in progress at the RISO facility. In this test, fatigue loads are being applied in the flatwise direction. By attaching additional masses to the test blade, it is possible to adjust the bending moment distribution along the blade length for increased accuracy in matching operating loads. These added masses lower the fundamental natural frequency of the blade, and therefore the rate of testing. In general, the added masses lower the natural frequency by 25 percent to 30 percent [White 2004]. However, in spite of testing separately in flatwise and edgewise directions and the lower cyclic frequency with auxiliary masses, the test cycle frequency remains substantially higher than that of the forced-displacement method. Thus, increased speed of testing is a major advantage of the single-axis resonance method.

By taking advantage of the displacement magnification that occurs naturally when cyclic loads are applied at frequencies near resonance, the force required to achieve the required bending moments is substantially lower for this method. This results in much lower energy consumption during testing. Because the testing is performed at the blade's natural frequency, the blade bends into a natural *mode shape*. This makes it possible to test the entire blade without introducing unwanted vibration modes, so the tip section does not need to be removed before testing.

Compared to force-displacement testing, resonance excitation presents several advantages for fatigue testing of larger wind turbine blades, including lower testing costs, faster results, and the ability to test the entire blade. However, there are limitations to this method. One of these limitations is the fact that the rotating eccentric mass also applies horizontal

Figure 12-24. RISO's single-axis resonance test system. The frame holding the rotating eccentric mass is attached at approximately 2/3-span. Additional masses attached nearer to the blade tip are designed to shape or "tune" the distribution of bending moments along the blade axis [Courtesy of the RISO National Laboratories, Denmark]

cyclic forces in the axial direction of the blade, forces that are not present in service. As the size of the test blade increases, these unwanted axial forces also increase and may ultimately interfere with the accuracy of the test results.

Dual-Axis Hybrid Resonance Method

The dual-axis hybrid resonance method is one of the latest developments in blade fatigue testing technology [White 2004]. This method is designed to combine the accuracy advantage of forced-displacement testing with the increased speed and reduced energy of resonance testing. As shown schematically in Figure 12-25, resonance excitation is provided by a vertically oscillating mass mounted in a frame attached to the blade. The frame also carries the hydraulic actuator that drives the mass at the blade's resonant frequency. Because the energy consumed by blade *damping* at resonance is relatively small, both the actuator force and stroke are much reduced, compared to force-displacement bending of the blade in the flatwise direction. Not shown in Figure 12-25 is the actuator and bell-crank mechanism that apply cyclic edgewise loading. This force-displacement part of the hybrid system is exactly the same as that shown in Figure 12-23.

In preliminary tests, a blade system damping ratio of 0.010 was measured, which was somewhat higher than a reference value of 0.006 for composite structures [White 2004]. Compared to dual forced-displacement, testing speed was increased from 0.30 Hz to 0.75 Hz and hydraulic flow was reduced by 50 percent, from 200 gallons per minute (gpm) to 100 gpm. These tests confirmed the expected advantages of the hybrid system, leading to shorter test times and reduced energy consumption during testing.

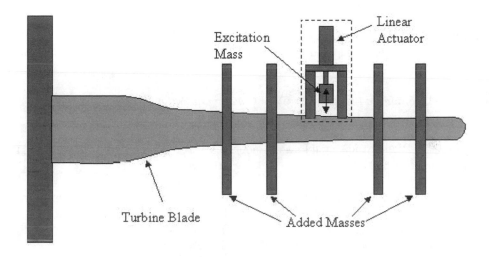

Figure 12-25. Schematic elevation view of a wind turbine blade in a dual-axis hybrid test facility, in which flatwise loads are applied by the resonance method and edgewise loads (not shown) are applied by the force-displacement method. An excitation mass mounted on the blade oscillates vertically at the blade's flatwise fundamental natural frequency, driven by a small hydraulic actuator. [White 2004]

References

Ancona, D., and J. McVeigh, 2001, *Wind Turbine – Materials and Manufacturing Fact Sheet*, Princeton Energy Resources International, prepared for Office of Industrial Technologies, US Department of Energy.

ASTM, 1990, "Standard Practice E 1049-85 (Re-Approved 1990) for Cycle Counting in Fatigue Analysis, "1993 *Annual Book of ASTM Standards*, Vol. 03.01, Philadelphia, Pennsylvania: American Society for Testing and Materials.

ASTM, 1993, *1993 Annual Book of ASTM Standards*, Section 3: Metals Test Methods and Analytical Procedures, Philadelphia, Pennsylvania: American Society for Testing and Materials.

ASTM, 2005, *Standard Practices for Cycle Counting in Fatigue Analysis*, E 1049-85 (Re-Approved 2005).

Basquin, O. H., "The Exponential Law of Endurance Tests," *Proceedings ASTM*, Vol. 11, 1910, pp. 625-630.

Birchenough, A. G., A. L. Saunders, T. W. Nyland, and R. K. Shaltens, 1981, "Operating Experience with Four 200-kW Mod-0A Wind Turbine Generators," *Proceedings, Conference on Large Horizontal-Axis Wind Turbines*, NASA CP-2230, DOE Publ. CONF-810752, Cleveland, Ohio: NASA Lewis Research Center, pp. 469-490.

Boeing Engineering and Construction, 1979, *Mod-2 Wind Turbine System Concept and Preliminary Design Report, Volume II Detailed Report*, NASA CR-159609, DOE/NASA 0002-80/2, Cleveland, Ohio: NASA Lewis Research Center.

Boeing Aerospace Company, 1985, *Goodnoe Hills Mod-2 Cluster Test Program, Vol. 2: Structural and Mechanical Loads Tests and Analysis*, EPRI AP-4060, Palo Alto, California: Electric Power Research Institute.

Boeing Aerospace Company, 1988, *Mod-5B Wind Turbine System Final Report, Volume II Detailed Report*, NASA CR 180897, DOE/NASA 0200-4, Cleveland, Ohio: NASA Lewis Research Center.

Broek, D., 1982, *Elementary Engineering Fracture Mechanics*, Boston, Massachusetts: Martinus Nijhoff Publishers.

Collins, J. L., R. K. Shaltens, R. H. Poor, and R. S. Barton, 1981, "Experience and Assessment of the DOE/NASA Mod-1 2000-kW Wind Turbine Generator at Boone, North Carolina," *Proceedings, Conference on Large Horizontal-Axis Wind Turbines*, NASA CP-2230, DOE Publ. CONF-810752, Cleveland, Ohio: NASA Lewis Research Center, pp. 491-574.

Cosmos Solidworks Corp., 2005, *Design to Prevent Fatigue*, White Paper, Santa Monica, California.

Delft, D.R.V., G.D. van de Winkel, and P.A. Joosse, 1997, "Fatigue Behaviour of Fiberglass Wind Turbine Blade Material Under Variable Amplitude Loading," *Wind Energy 1997*, AIAA-97-091, ASME/AIAA.

Endo Memorial Volume, 1991, *The Rainflow Method in Fatigue*, Y. Murakami, ed., Oxford, England: Butterworth-Heinemann Ltd.

Ensworth, C. B. F., III, 1985, "Comparison of Blade Loads for Aileron and Tip Control Rotors," *Proceedings, ASME 4th Wind Energy Symposium*, A. H. P. Swift, ed., pp. 115-123.

Dowling, N. E., 1993, *Mechanical Behavior of Materials*, Englewood Cliffs, New Jersey: Prentice Hall.

Downing, S. D., and D. F. Socie, 1982, "Simple Rainflow Counting Algorithms," *International Journal of Fatigue*, Vol. 4, No. 1: pp. 31-40.

Finger, R. W., 1980, "Prediction Model for Fatigue Crack Growth in Windmill Structures," *Effect of Load Spectrum Variables on Fatigue Crack Initiation and Propagation*, ASTM STP 714, Philadelphia, Pennsylvania: American Society for Testing and Materials, pp. 185-204.

Finger, R. W., 1985, "Methodology for Fatigue Analysis of Wind Turbines," *Proceedings, Windpower '85 Conference*, SERI/CP-217-2902, Washington, DC: American Wind Energy Association, pp. 52-56.

Fuchs, H. O., and R. I. Stephens, 1980, *Metal Fatigue in Engineering*, New York: John Wiley & Sons.

Griffin, D.A., 2001, *WindPACT Turbine Design Scaling Studies Technical Area 1 – Composite Blades for 80- to 120-Meter Rotor*, Global Energy Concepts, Kirkland, Washington, NREL/SR-500-29492, Golden, Colorado: National Renewable Energy Laboratory.

Hughes, S., and W. Musial, 1999, "Implementation of Two-Axis Servo-Hydraulic System for Full-Scale Fatigue Testing of Wind Turbine Blades," *Proceedings, WindPower '99*, Washington, DC: American Wind Energy Association.

Hughes, S., and W. Musial, 2002, *Wind Turbine Blade Fatigue Test Plan for Enron 37B Rotor Blade*, NWTC Document: NWTC-ST-EW-FAT-04-TP, Golden, Colorado: National Renewable Energy Laboratory.

Johnson, W. R., and W. R. Young, "Status of the DOI SVU Wind Turbine Project," *Proceedings, Windpower '85 Conference*, SERI/CP-217-2902, Washington, DC: American Wind Energy Association, pp. 604-609.

Larwood, S., W. Musial, G. Freebury, A. Beattie, 2001, *NedWind 25 Blade Testing at NREL for the European Standards Measurement and Testing Program*, NREL/TP-500-29103, Golden, Colorado: National Renewable Energy Laboratory.

Mandell, J. F., R. M. Reed, D. D. Samborsky, and Q. Pan, 1993, "Fatigue Performance of Wind Turbine Blade Composite Materials," *Proceedings, Wind Energy – 1993 Symposium*, SED-Vol. 14, S. Hock, ed., New York: American Society of Mechanical Engineers.

Mandell, J. F., R. F. Creed, Jr., and Q. Pan, 1994, "Fatigue of Fiberglass Generic Materials and Substructures," *Proceedings, Wind Energy - 1994 Symposium*, SED-Vol. 15, W. D. Musial, *et al.*, eds., New York: American Society of Mechanical Engineers, pp. 207-213.

Mandell, J. F., and D. D. Samborsky, 1997, *DOE/MSU Composite Material Fatigue Database Test Methods, Materials and Analysis*, Contractor Report SAND97-3002, Sandia National Laboratories, Albuquerque, New Mexico.

Mandell, J.F., D.D. Samborsky, and D.S. Cairns, 2002, *Fatigue of Composite Materials and Substructures for Wind Turbine Blade*, Contractor Report SAND2002-077, Albuquerque, NM: Sandia National Laboratories.

Manson, S. S., 1965, "Fatigue: A Complex Subject – Some Simple Approximations," *Journal of the Society of Experimental Stress Analysis*, Vol. 5, July 1965, pp. 193-226.

Manson, S. S., 1966, "Interfaces Between Fatigue, Creep, and Fracture," *International Journal of Fracture Mechanics*, Vol. 2, 1966, pp. 327-363.

Matsuiski, M., and T. Endo, 1969, *Fatigue of Metals Subjected to Varying Stress*, Japan Society of Mechanical Engineering.

McDade, J., and H. G. Pfanner, 1985, "The Mod-0 Large Horizontal-Axis Wind Turbine Research Facility," *Proceedings, Windpower '85 Conference*, SERI/CP-217-2902, Washington, DC: American Wind Energy Association, pp. 546-551.

Miner, M. A., "Cumulative Damage in Fatigue," *Journal of Applied Mechanics*, Vol. 12, 1945, pp. A-159-164.

Palmgren, A., "Die Lebendauer von Kugellagern," *Zeitschrift von deutche Ingenieurring*, Vol. 68, No. 14, 1924, pp. 339-341.

Rychlik, I., 1987, "A New Definition of the Rainflow Cycle Counting Method," *International Journal of Fatigue*, Vol. 9:2, pp. 119-121.

Schluter, L., 1991, *Programmer's Guide for LIFE2's Rainflow Counting Algorithm*, SAND90-2260, Albuquerque, New Mexico: Sandia National Laboratories.

Spera, D. A., L. A. Viterna, T. R. Richards, and H. E. Neustadter, 1979, *Preliminary Analysis of Performance and Loads Data from the 2-Megawatt Mod-1 Wind Turbine Generator*, NASA TM-81408, DOE/NASA/1010-79/5, Cleveland, Ohio: NASA Lewis Research Center.

Spera, D. A., and D. C. Janetzke, 1981, "Performance and Load Data from Mod-0A and Mod-1 Wind Turbine Generators," *Proceedings, Conference on Large Horizontal-Axis*

Wind Turbines, NASA CP-2230, DOE Publ. CONF-810752, Cleveland, Ohio: NASA Lewis Research Center, pp. 447-467.

Spera, D. A., C. B. Ensworth III, and D. C. Janetzke, 1985, "Dynamic Loads in Large Horizontal-Axis Wind Turbines - Part I: Field Test Data," *Proceedings, Windpower '85Conference*, SERI/CP-217-2902, Washington, DC: American Wind Energy Association, pp. 557-462.

Spera, D. A., 1993, "Dynamic Loads in Horizontal-Axis Wind Turbines - Part II: Empirical Equations," *Proceedings, Windpower '93 Conference*, Washington, DC: American Wind Energy Association, pp. 282-289.

Spera, D. A., 1994, "Dynamic Loads in Horizontal-Axis Wind Turbines - Part III: Design Trend Charts," *Proceedings, Wind Energy - 1994 Symposium*, SED-Vol. 15, W. D. Musial *et al.*, eds., New York: American Society of Mechanical Engineers.

Suresh, S., 1991, *Fatigue of Materials*, New York: Cambridge University Press.

Sullivan, T. L., 1984, *Effect of Vortex Generators on the Power Conversion Performance and Structural Dynamic Loads of the Mod-2 Wind Turbine*, NASA TM-83680, DOE/NASA/20320-59, Cleveland, Ohio: NASA Lewis Research Center.

Sutherland, H. J., and L. L. Schluter, 1989-1990, "Fatigue Analysis of WECS Components Using a Rainflow Counting Algorithm," *Proceedings, Windpower '90 Conference*, Washington, DC: American Wind Energy Association.

Sutherland, H.J., and J. F. Mandell, 2005, *Optimized Goodman Diagram for the Analysis of Fiberglass Composites Used in Wind Turbine Blades*, Paper AIAA-2005-0196, Albuquerque, New Mexico: Sandia National Laboratories.

Tada, H., P. C. Paris, and G. R. Irwin, 1985, *The Stress Analysis of Cracks Handbook*, Saint Louis, Missouri: Paris Productions, Inc.

Wahl, N.K., 2001, *Spectrum Fatigue Lifetime and Residual Strength for Fiberglass Laminates*, Ph.D. Thesis, Department of Mechanical Engineering, Montana State University, Bozeman, Montana.

Wahl, N.K., J.F. Mandell, and D.D. Samborsky, 2002, *Spectrum Fatigue Lifetime and Residual Strength for Fiberglass Laminates*, Report SAND2002-0546, Albuquerque, New Mexico: Sandia National Laboratories.

White, D., 2004, *New Method for Dual-Axis Fatigue Testing of Large Wind Turbine Blades Using Resonance Excitation and Spectral Loading*, NREL/TP-500-35268, Golden, Colorado: National Renewable Energy Laboratory.

Young, P., and T. M. Hasbrouck, 1983, "Installation, Checkout, and Acceptance Testing of the Hamilton Standard 4-MW Wind Turbine System (WTS-4)," *Proceedings, Sixth Biennial Wind Conference*, American Solar Energy Society, pp. 121-132.

Bibliography of Additional Publications
Relating to the Fatigue Analysis of Wind Turbines

Ashwill, T. D., H. J. Sutherland, and P. S. Veers, 1990, "Fatigue Analysis of the Sandia 34-Meter Vertical Axis Wind Turbine," *Proceedings, Ninth ASME Wind Energy Symposium*, SED-Vol. 9, D. E. Berg, ed., New York: American Society of Mechanical Engineers, pp. 145-151.

Jackson, K. L., 1992, "Deriving Fatigue Design Loads from Field Test Data," *Proceedings, Windpower '92 Conference*, Washington, DC: American Wind Energy Association, pp. 313-320.

Jackson, K. L., 1994, "Scaling Wind Turbine Fatigue Design Loads," *Proceedings, Wind Energy - 1993 Symposium*, SED-Vol. 14, S. Hock, ed., New York: American Society of Mechanical Engineers, pp. 189-196.

Hacker, C. L., 1993, "Wood Composites - Properties and Performance in Fatigue," *Proceedings, Windpower '93 Conference*, Washington, DC: American Wind Energy Association, pp. 358-365.

Malcolm, D. J., 1990, "Prediction of Peak Fatigue Stresses in a Darrieus Rotor Wind Turbine under Turbulent Winds," *Proceedings, Ninth ASME Wind Energy Symposium*, SED-Vol. 9, D. E. Berg, ed., New York: American Society of Mechanical Engineers, pp. 125-136.

Schluter, L. L., and H. J. Sutherland, 1990, "Rainflow Counting Algorithm for the LIFE2 Fatigue Analysis Code," *Proceedings, Ninth ASME Wind Energy Symposium*, SED-Vol. 9, D. E. Berg, ed., New York: American Society of Mechanical Engineers, pp. 121-123.

Sutherland, H. J., and L. L. Schluter, 1989, "Crack Propagation Analysis of WECS Components Using the LIFE2 Computer Code," *Proceedings, Eighth ASME Wind Energy Symposium*, SED-Vol. 7, D. E. Berg and P. C. Klimas, eds., New York: American Society of Mechanical Engineers, pp. 141-145.

Sutherland, H. J., 1989, *Analytical Framework for the LIFE2 Computer Code*, SAND89-1397, Albuquerque, New Mexico: Sandia National Laboratories.

Sutherland, H. J., 1993, "Effect of the Flap and Edgewise Bending Moment Phase Relationships on the Fatigue Loads of a Typical HAWT," *Proceedings, Wind Energy 1993 Symposium*, SED-Vol. 14, S. Hock, ed., New York: American Society of Mechanical Engineers, pp. 181-187.

Sutherland, H. J., P. S. Veers, and T. D. Ashwill, 1994, "Fatigue Life Prediction of Wind Turbines: A Case Study on Loading Spectra and Parameter Sensitivity," *Proceedings, Symposium on Case Studies for Fatigue*, Philadelphia, Pennsylvania: American Society for Testing and Materials.

VanDenAvyle, J. A., and H. J. Sutherland, 1989, "Fatigue Characterization of a VAWT Blade Material," *Proceedings, Eighth ASME Wind Energy Symposium*, SED-Vol. 7, D. E. Berg and P. C. Klimas, eds., New York: American Society of Mechanical Engineers, pp. 125-129.

Veers, P. S., 1983, *A. General Method for Fatigue Analysis of Vertical Axis Wind Turbine Blades*, SAND83-2543, Albuquerque, New Mexico: Sandia National Laboratories.

Veers, P. S., 1989, "Simplified Fatigue Damage and Crack Growth Calculations for Wind Turbines," *Proceedings, Eighth ASME Wind Energy Symposium*, SED-Vol. 7, D. E. Berg and P. C. Klimas, eds., New York: American Society of Mechanical Engineers.

Veers, P. S., H. J. Sutherland, and T. D. Ashwill, 1992, "Fatigue Life Variability and Reliability Analysis of a Wind Turbine Blade," *Probabilistic Mechanics and Structural and Geotechnical Reliability*, Y. K. Lin, ed.: American Society of Civil Engineers, pp. 424-427.

Veers, P. S., C. H. Lange, and S. R. Winterstein, 1993, "FAROW: A Tool for Fatigue and Reliability of Wind Turbines," *Proceedings, Windpower '93 Conference*, Washington, DC: American Wind Energy Association, pp. 342-349.

13

A Utility Perspective of Wind Energy

Daniel F. Ancona III

Vice-President for Renewable Energy
Princeton Energy Resources International
Rockville, Maryland

and

Carl J. Weinberg

Manager of Research and Development (Ret.)
Pacific Gas and Electric Company
and Weinberg Associates
Walnut Creek, California

Introduction

The wind is rapidly becoming a practical source of energy for electric utilities around the world. In the U.S., development of commercial wind power plants (also referred to as *wind power stations*) has been carried out largely by independent firms operating under the 1978 *Public Regulatory and Policy Act* (PURPA) either as *independent power producers* (IPP) or as *qualifying facilities* (QF). In the post-2000 period utilities have been investing directly in wind power plants and operating them like conventional generation. In the early 1980s utilities were concerned about many technical and economic issues regarding the integration of wind power plants into existing *grids*, which are highly-interactive networks of generating stations, transmission lines, and distribution wires. They feared possible adverse impacts from wind generation on critical performance and economic factors such as power quality, grid stability, automatic generation control, spinning reserve requirements, and capacity credits. As operating experience with wind power plants accumulates, many of these early fears are being dispelled.

Since 1980 the installed capacity of wind power plants in the United States has grown dramatically surpassing 25,000 MW in 2008. This rise was kindled by financial incentives from federal and state governments and by rising prices for fossil fuels. More of this rapid growth is expected in the future, as other climate change drives the low carbon and other emission reductions. A similar process began in Europe in the early 1990s, and installed wind capacity there grew faster than the United States, driven by clean energy concerns and the lack of indigenous fossil fuel resources. By the end of 2007, installed capacity in Europe had exceeded 57,100 MW, 61 percent of the global total. World-wide, wind turbines totaling almost 100,000 MW of capacity are now generating 184 TW-hr of bulk power annually.

A variety of economic projections show that future growth in wind energy production could be even more dramatic. Figure 13-1 summarizes the findings of a 1990 study conducted jointly by five National Laboratories of the U.S. *Department of Energy* (DOE). Wind-generated power totaling 4,000 to 8,000 MW was projected to be on-line by the year 2000 [DOE 1990].

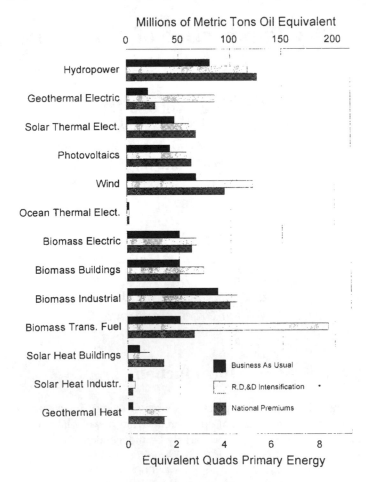

* Added R.D&D funding is variable, depending on opportunities and technology needs

Figure 13-1. Projected contributions of renewable energy sources in the U.S. by the year 2030. Wind power estimates range from 120,000 MW to 200,000 MW, depending on the scenario assumed. [DOE 1990]

That capacity was exceeded in 2005. This growth was projected to spread by 2010 to other regions, including New England and the Pacific Northwest. The study concluded that by 2030 the wind will be a leading renewable-energy source, comparable to hydropower.

As shown by the bars in Figure 13-1, three different scenarios were considered. The first is the "business as usual" scenario which assumes that (1) there are no major changes in the market prices of conventional fossil fuels, and (2) there is only modest support for improvements in renewable-energy technologies. Next is the "research and development intensification" scenario, which also assumes no major changes in the energy market but projects a more-rapid development of renewable-energy technologies in the near term. The third set of conditions considered make up the "national premiums" scenario. Here it is assumed that heightened national awareness of energy issues results in a premium value being placed on electricity from renewable sources. For example, a *generation credit* of $0.015 per kilowatt-hour, granted in the *Energy Policy Act* of 1992, might be extended or increased in order to stimulate the near-term market for renewable energy. These projections have proven to be quite accurate.

20 Percent Wind-Powered Electricity Generation by 2030

A new DOE study completed in 2008 describes a scenario in which by 2030 20 percent of U.S. electricity could come from wind plants. This study projects the addition of 241 GW of land-based and 54 GW of turbines located offshore, based on a *National Renewable Energy Laboratory* (NREL) model analyzing wind resources, turbine technology trends, manufacturing capacity, plant siting, environmental effects and assumes major expanding transmission systems to bring wind power to load centers [DOE 20 percent]. Offshore installations on the east coast and in the Great Lakes could be located near electrical load centers, reducing distribution line lengths and much of the needed transmission expansion.

A central aspect of the DOE study was development of *supply curves* showing future wind generation potential as a function of levelized cost. The national supply curve for bus-bar energy costs—for the wind plant alone, excluding transmission costs—is shown in Figure 13-2. The figure illustrates that more than 8,000 GW of wind energy is available in the United

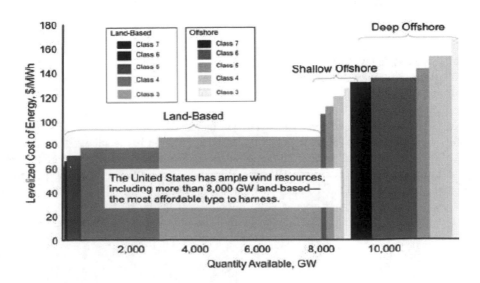

Figure 13-2. Supply curve for wind energy – current bus-bar energy cost (in $2007).

States at $85/MWh or less. This amount of capacity is equivalent to roughly 8 times the existing net summer generating capacity in the country, which is estimated at 986 GW [EIA 2008]. This wind power price, however, excludes the cost of transmission or integration.

The supply curve in Figure 13-2 uses 2008 wind technology cost and performance figures, which are projected to improve with future technology development. The current busbar energy costs, based on the wind plant only (excluding transmission and integration costs and Production Tax Credits and other incentives), vary by type of site and the wind power class. Onshore sites are less costly than offshore, and higher wind classes offer more productivity that reduces the *cost of energy* (COE). Depending on the scenario assumed, from 120,000 MW to 300,000 MW of installed wind power is projected to be in place by the year 2030, according to DOE studies. A similar overseas study shows that as much as one-third of the electricity in Europe could technically be produced from wind energy [Van Wijk 1993].

Changing Energy Mix

In recent years, interest in wind as an energy source has fluctuated, responding mainly to the availability and price of fossil fuels and environmental regulations. Initial growth in the wind industry occurred in response to the 1973 oil embargo when oil prices doubled, as illustrated in Figure 13-3. Oil prices dramatically increased again prior to the Iran/Iraq war. During that period utilities were generating about 16 percent of U.S. electricity from heavy oil and 2 percent from natural gas. By 2000, fossil fueled electric power generation had shifted to 18 percent natural gas and 2 percent oil, due to oil supply shortages and price fluctuations combined with new atmospheric emission regulations.

In the future, a combination of restrictions on atmospheric carbon and nitrogen emissions from coal-fired plants, carbon sequestration costs, price increases resulting from the necessary use of imported liquefied natural gas, and unresolved nuclear waste disposal issues are likely to cause more power price increases and restrictions. These fossil-fuel economic and supply issues, combined with wind energy technology improvements have now made wind the fastest growing power source for electricity worldwide for the fourth straight year. In the United States, wind power contributed 35 percent of all new generating capacity added during 2007, an amount that is up from 19 percent in 2006 and 12 percent in 2005, compared to less than 4 percent in 2000-2004.

An important driver of introduction of wind energy into the changing energy mix was the establishment of *Renewable Portfolio Standards* (RPS) beginning in about 2000. Texas led the way by passing a law with the goal of adding 2,000 MW of renewable generating capacity by 2009, through a renewable energy credit trading system that included a significant penalty for any utility that did not meet its quota. The penalty was the lesser of $50 per MWh or 200 percent of the average market value of credits for that compliance period. The Texas program exceeded initial expectations, so its goal was increased several times. By the beginning of 2009 more than 7,000 MW of wind power had been installed in the state.

By 2008, 25 states and the District of Columbia have set RPS goals, which are typically 25 percent renewable energy in 5 to 10 years. Wind has been fulfilling about 60 percent of these goals, because it is often the least-cost renewable source. Some states have set separate allocations for solar and other renewable technologies. If the current RPS policies are fulfilled it is projected that 60 GW of new renewable capacity will be added by 2025, much of it be wind power. Legislation establishing a national RPS has been proposed.

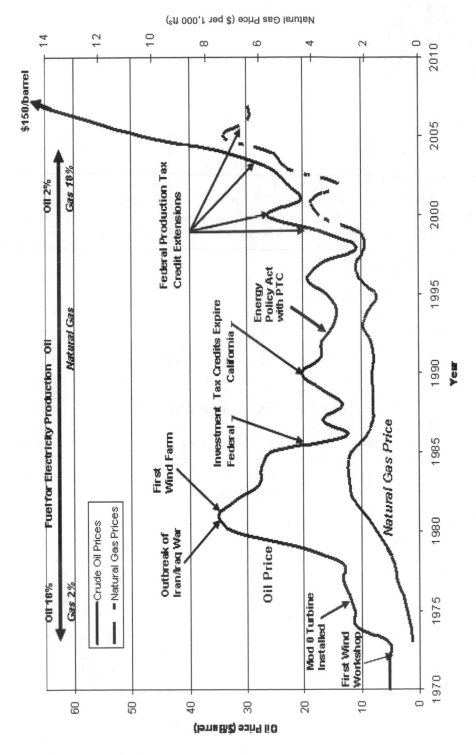

Figure 13-3. Historic crude oil and natural gas prices and milestones in wind energy development in the United States.

Introduction of New Technology into Utilities

Utilities in the United States and around the world are inherently conservative and cautious about adopting new technologies such as wind power. Consumers of electricity demand power that is high in quality, available on-demand, and reasonable in cost. Consequently, utilities are allowed by their regulatory commissions to pass on to the consumer the high cost of equipment needed to maximize reliability. New technologies represent uncertainty to both the utilities and regulators, although the level of that uncertainty is rapidly being reduced for wind power.

It is interesting to note that there was a time when wind was the established, reliable energy source and steam represented a new and uncertain technology. The first steam power in Holland was a Newcomen engine and pump installed in 1776 by Steven Hoogendijk to drain the inner city of Rotterdam [Lintsen 1991]. However, 100 years later wind machines still performed about 50 percent of the polder draining in the country. Historians in The Netherlands have recently been studying why Dutch industry was extremely slow to adopt steam power, compared to the rest of Europe. Cautiousness was probably one of the main factors delaying the introduction of this new technology. Concern on the part of the engineers of that time and place is understandable when one considers the possible effects on the population if the steam-driven pumps failed to work.

We see from this example that caution can be a force that is strong enough to influence the energy policy of a nation or of an industry. Utilities need tangible assurance that wind power plants are reliable and cost-effective. To meet this need, U.S. utilities work with the wind power industry and DOE in wind turbine development, and field testing of new turbines is done in partnership with a variety of utilities in different parts of the country. In some cases, system demonstration programs are also performed on a cost-sharing basis.

Utility Cost Estimation Models

A utility's perspective of wind energy is often determined by its view of the costs associated with this form of generation. It is therefore important for wind power developers and engineers alike to understand how a utility determines what is and is not cost effective. Electric utilities and the wind industry commonly use two basic economic models to estimate the COE produced by a wind power system. These are (1) the *Electric Power Research Institute, Technical Analysis Guide* or "EPRI TAG" method [EPRI 1979], and (2) what will be called the *cash flow* method. Each of these models will be discussed briefly in the following sections. COE methodologies can be used to either create an index with which to rank competing power supplies, or to create an actual price structure for a given power plant.

EPRI TAG Method

The EPRI TAG method produces a *levelized* COE. A levelized COE is the constant-annual COE stream over time whose present value is equal to the present value of a variable-annual COE stream over the same period of time. In its simplest form, COE is estimated by the EPRI TAG method as follows:

$$COE = FCR \times \frac{Capacity\ Cost}{8,760 \times Capacity\ Factor} + \frac{Fuel\ Cost}{Heat\ Rate} + OM \qquad (13\text{-}1)$$

where

COE = cost of energy ($/kWh)

FCR = fixed charge rate; present value factor which includes utility debt and equity costs, state and federal income taxes, insurance, and property taxes

Capacity Cost = total cost of constructing the facility normalized by rated power ($/kW)

Capacity Factor = ratio of annual average power output to rated power

Fuel Cost = unit market cost for heat-producing fuel; zero for wind energy ($/MBTU)

MBTU = million British thermal units

Heat Rate = factor measuring plant thermal energy conversion efficiency (kWh/MBTU)

OM = direct cost of operations and maintenance per unit of energy ($/kWh)

Equation (13-1) does not produce an actual estimate of the cost of power generation, but rather it is an approximation of the levelized unit cost [Karas 1992]. In other words, EPRI TAG "dollars" will not actually pay the annual costs of generating electricity. However, if this method is applied consistently to a variety of technologies, it does produce a useful and equitable index for comparison purposes. Some limitations of the EPRI TAG method are

-- It assumes a debt term period equal to the life of the power plant, which may not always be the case; and
-- It does not easily allow for variable equity return, variable debt repayment, or variable costs.

As a consequence of the second limitation, the EPRI TAG model is not particularly suitable for estimating the cost or revenue of a project owned by an independent power producer or of a qualifying facility. These two terms are commonly used by utilities in the process of purchasing power, and often include wind power projects.

Cash Flow Method

The *Cash Flow* method is based on an accounting spreadsheet that is a year-by-year listing of estimated income and expenses over the life of the project. COE is calculated by performing the following operations:

-- identifying each cost component of the plant and its operation, including the costs of land, permits, plant construction, insurance, power transmission, O&M, and administration
-- projecting the amount of each cost component in each year of the plant's anticipated service life
-- projecting the depreciation, debt service, equity returns, and taxes in each year of the plant's anticipated service life
-- discounting the resulting cash flows to present value using the utility's cost of capital as established by the state regulatory agency
-- levelizing the cost by determining the needed payment stream (*i.e.*, the annuity) which has a present value equal to the results of the previous calculation, at the utility's cost of capital

The spreadsheet produced by the Cash Flow method provides annual COE estimates which are the actual revenue requirements of the project. This method allows for the real variations that must be expected in cost, operational, and economic data, such as price increases, inflation, and changing interest rates. Like any model, the accuracy and validity of the COE estimates depend on the quality of the input data. In this case the economic projections for the out-years of the project are the most vulnerable to error.

COE Case Study

A large utility in California, the *Pacific Gas and Electric Company* (PG&E), applied the EPRI TAG model in a case study reported by Smith and Ilyin [1991]. COE estimates were made for the period from 1981 to 1990 for a wind power station located in the Altamont Pass area of California.

Wind Turbine Costs

During the 10-yr period from 1981 to 1990, the price of the wind turbines alone, installed on a prepared site, was estimated to decrease from $2,200/kW to $900/kW. These estimates are for medium-scale wind turbines, with power ratings from 100 to 500 kW at a rated wind speed of about 13 m/s.

Balance of Station Costs

The capacity cost in Equation (13-1) is for a fully-operational or *turnkey* power plant, and therefore it includes not only the cost of the generating equipment but also the *balance of station* costs for land, roads, substations, transformers, and transmission lines within the plant area. For a wind power plant these add from $200/kW to $300/kW onto the installed cost of the wind turbines themselves.

Operations and Maintenance Costs

Industry-wide, O&M costs decreased from about $0.04/kWh in 1981 to less than $0.01/kWh by 1993. For this study O&M costs in the Altamont Pass area were estimated to range downward from $0.03/kWh in 1981 to $0.009/kWh in 1990.

Capital Recovery Factor

This case study used a capital recovery factor of 0.065, based on a real annual interest rate (*i.e.*, one that factors out inflation) of 5 percent and a plant service life of 20 years.

COE Estimates

With these input data, the EPRI TAG method produced COE estimates that decreased from $0.52/kWh in 1981 to $0.06/kWh in 1990, with both estimates expressed in 1990 dollars. Future COEs can potentially decrease to under $0.05/kWh at sites that are rated as *Class 5* or higher (see Table 8-1). Such sites have an annual average wind speed of at least 7.5 m/s at an elevation of 50 m, and an annual average wind power density of at least 500 W/m^2. This reduction in COE would require reductions in installed capacity cost down to the $700/kW to $900/kW range and O&M costs of about $0.007/kWh. This amount of cost reduction is considered possible with the innovative, efficient, and lightweight turbines that are being developed.

Direct Value of Wind-Generated Energy to a Utility

The COE of a wind power station has meaning only when it is compared to the value which the utility places on the electricity generated. Wind power can be evaluated on two primary bases and on other secondary but possibly equally important values. Primarily wind energy is a *fuel saver* and that it is a means of *deferring construction* of other types of power plants or of transmission and distribution lines. When the wind is blowing, fossil-fueled or hydroelectric plants can be turned down or off, thus saving their fuel or water.

Wind power plants can also have some *capacity value*, which will allow a utility to defer the building of additional conventional plants. In addition, the *modularity* and *disbursed generation* of wind turbines is a valuable asset in obtaining construction capital, because the utility needs to install only the capacity required meeting near-term, not long-term, demand and the wind plants can be located near load centers in areas where additional energy is needed. The later is true if wind projects are developed by utilities rather than IPP's who have no incentive for that site selection.

Fuel Savings

The value of wind-generated electricity based on the cost of fuel saved appears to be relatively straight-forward, but utilities have some limitations on the amount of fuel that actually can be saved. As discussed in Chapter 4, the diesel-electric generators that power many small utilities operate inefficiently at partial load, so there is a limit on how much their output can be reduced without actually increasing their O&M costs more than the savings on fuel. In a hydroelectric plant, the water "fuel" may have to be released anyway, in accordance with agriculture, fish-maintenance and navigation requirements, even if power generation requirements are offset by wind plants [EPRI 1993]. In all utilities, generating unit sizes and requirements for *spinning reserve* also limit the rate and the amount by which conventional generation can be turned down or off.

Capacity Value

A measure of capacity value used by utilities is the *load carrying capability* or LCC of a power source. This is the power of an equivalent absolutely-reliable source (an ideal not achieved by any real source) that would give the same contribution to the reliability of the utility system during peak demand periods. If a utility were certain that a wind power plant would be producing its full rated power during peak demand hours, than the LCC of this plant would equal its power rating. At the other extreme, if a utility were certain that the wind would never be in the operating range during peak hours, then the LCC of the wind power plant would be zero.

In reality, the LCC of commercial wind power stations is somewhere between zero and the station rating, depending on the specific "fit" of the wind power source to the utility's seasonal and diurnal (daily) demand cycles. In the Pacific Gas and Electric (PG&E) service area the greatest demand for electricity occurs during hot summer afternoons, when air conditioner use and irrigation are at their highest rates. The output of the wind power stations in the Altamont Pass is a good seasonal fit to this peak load, but their maximum daily output occurs later than the maximum daily PG&E demand.

To obtain a rough approximation of the wind energy LCC for a specific location, the actual or projected wind power outputs during the hours of peak utility loads need to be studied in detail. As an example, two wind power stations on sites in northern California were examined for their fit to the PG&E load cycle during a period of five years. These are an actual station in the Altamont Pass and a projected station in Solano County [Smith and Ilyin 1991]. During the summer the PG&E load usually peaks at 1600 Pacific Daylight Time.

Figure 13-4 illustrates the results of this study for 1987 and 1988. The rising wind speed in the morning in Solano County usually occurs about two hours in advance of that in the Altamont Pass. As a result, the average wind speed at this site during the peak load hour equaled or exceeded the rated wind speed during four of the five years studied. The average wind speeds at the Altamont site during the peak hour, however, were usually only between *cut-in* (the lowest speed in the operating range) and half of the rated speed. Thus, a significantly higher LCC is indicated for the projected Solano County wind power plant.

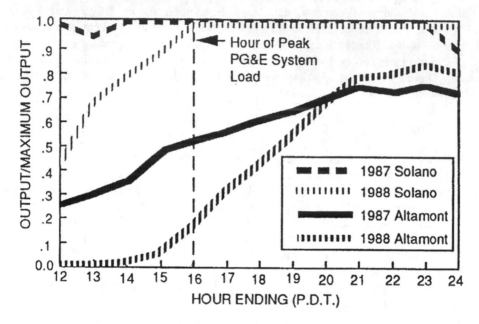

Figure 13-4. Typical hourly output data required for calculating the load carrying capability (LCC) of a wind power plant, with the peak load time of the utility system noted. Hourly data are averages over one year. Altamont Pass site data are measured and Solano County site data are projected. [Smith and Ilyin 1991]

To obtain a more-accurate LCC, the output at all hours of the year needs to be examined and compared to the system load, hour by hour. Another factor that is needed for the determination of the LCC of a power source is its relative *loss of load probability*, or LOLP. The LOLP is the chance that the utility cannot meet its load demand, because of unexpectedly high loads, equipment failures, or other reasons that the necessary plant capacity would not be available. An LOLP needs to be computed for each hour of the year using *Garver's formula* [Garver 1966]. The projected hourly outputs of a power source are then multiplied by their associated LOLPs before calculating the LCC of the source. An LCC of about 80 percent of the rated power times the projected *on-line availability* was calculated for the Solano County site in this study. This relatively high value is the result of the excellent fit of the Solano wind resource to the PG&E load cycle.

Indirect Value

Wind energy has additional indirect value to the utility and the community. These benefits are typically not counted in a utilities generation planning models or in the *regional transmission operator* (RTO) economic dispatch decision making. However, the trend is for regulators to factor these considerations into *Renewable Portfolio Standards* (RPS), *Renewable Energy Credits and Carbon Cap and Trade* mechanisms that are discussed in more detail in Chapter 4. Examples of indirect benefits of wind generation are as follows:

-- reduced use of large quantities of cooling water in fossil fuel plants
-- reduced water pollution from feed water additives
-- reduced atmospheric emissions of carbon, mercury, and other pollutants
-- reduced removal of mountain tops and other impacts of coal and uranium mining
-- reduced water contamination from coal washing
-- reduced security vulnerability through disbursed generation

Selecting an Energy Portfolio

Electricity can be produced from renewable resources in many different ways. This diversity of options means that most parts of the world have at least one renewable resource that can be used to generate electricity at competitive prices. But diversity frustrates a simple assessment of value. Techniques used to evaluate traditional utility investments do not work well when applied to renewable equipment whose value depends heavily on details of local conditions, such as

-- local wind or sunlight resources
-- hourly patterns of electric demand in the region
-- characteristics of other equipment operating in the local utility system, such as fuel costs and atmospheric emissions.

Existing evaluation techniques, however, are also inadequate for evaluating many other areas open for utility investment. In addition to investments in renewable electric generation equipment, future utility portfolios are also expected to include interacting investments in

-- thermal generating equipment with characteristics substantially different from those in place today
-- electric storage systems
-- advanced transmission and distribution systems
-- advanced control systems

In an attempt to better understand these interacting investments, planners at PG&E analyzed a variety of investment portfolios, concentrating on one simple measure of portfolio value: the annual cost of meeting the demand for electric service in the utility's region if the utility receives a *fixed rate of return* on its capital investments. This measure forces attention to the complex interactions between different kinds of utility equipment. For simplicity, it was assumed in this analysis that all cost-effective measures for improving the energy efficiency of each type of generating equipment had already been taken. Some of the results of this study for a *summer-peaking* utility like PG&E are as follows [Weinberg and Kelley 1993]:

1. Utility portfolios that rely on intermittent renewable sources for 30 percent of their electricity can serve loads at lower costs than utilities using typical new equipment and at costs about 5 percent higher than those achievable using advanced coal and gas production equipment now in development.

2. The cost of the renewable portfolios just described would be unchanged if *biomass* (com-bustible refuse and agricultural products) is co-fired with coal. The utility system would then rely on renewables for over 90 percent of its energy.
3. A utility meeting 50 percent of its loads from intermittent renewable resources would cost 10 percent more to operate than a system using intermittent sources for 30 percent of its loads. At lower renewable energy penetration levels the actual increases in operating costs are much less, typically below $5/MWh (See Table 4-10).
4. The three conclusions just described are based on the assumption of a 6 percent real dis-count rate (which is typical of regulated utilities in industrialized nations) and equipment costs that appear to be reasonable for the first quarter of the next century. If a 12 percent real rate of return is used (or if all capital costs are increased by two-thirds) the annual cost of operating the utility would be approximately 20 percent higher.
5. Generating equipment that is relatively small and can be located close to demand centers can reduce transmission and distribution costs. This can reduce the annual operating cost of a utility with 30 percent of its power from intermittent sources to well within 5 percent of one using advanced fossil-fuel generators. Again, the discount rate is 6 percent.
6. Advanced gas turbines using natural gas as a fuel provide an excellent complement to a utility portfolio that contains a significant percentage of renewable power sources. These advanced turbines operate efficiently while following load variations and they can be quickly added to and dropped from the line. Their relatively low cost makes them an excellent way to satisfy near-term demand while a utility determines its best investment strategy for the future.
7. Hydroelectric sites with large reservoirs also provide an excellent match for intermittent re-newable technologies. The output of a hydroelectric system can be adjusted (within limits) to fill in the production gaps caused by the intermittent nature of the renewable source.
8. Electric storage equipment is not needed to achieve the high levels of penetration by intermittent renewable sources just described (*i.e.*, 30 percent to 50 percent). In fact, the value of storage to a utility is decreased if wind, photovoltaic, or solar-thermal equip-ment is added to a utility system.

It is important to recognize that these evaluations were made under the assumption that utili-ties are free to optimize their investments to minimize costs based on new investments. No attempt was made to develop a detailed schedule of plant additions and retirements.

Grid Interconnection

Wind power stations are dispatched and operated like other electricity generating plants. As such, transmission grid connection, line capacity, access priority, and operations are util-ity issues that must be addressed. When the wind blows, the station feeds electricity to the grid, so other generating plants can be curtailed or throttled back. Improved forecasting of the output from a wind power plant and distributing wind turbines over a large geographic area help to minimize the potential impact on the utility system. Grid integration studies have concluded that large-scale wind deployment does not have negative impacts at penetration levels up to 25 percent energy and 35 percent capacity [IEEE 2007]. The *New York State Energy Research and Development Authority* (NYSERDA) supported analysis of potential im-pacts on the existing grid system and concluded, "New York Bulk Power System can reliably accommodate at least 10 percent penetration, 3,300 MW of wind generation with only minor adjustments to its existing planning, operational and reliability practices" [Piwko 2005].

The cost of connecting wind plants (and any other renewable power source) to the grid is an issue that can be solved in several ways. In the United States, some states have established *Clean Renewable Energy Zones (CREZ)* with additional transmission capacity built as needed, so individual wind projects do not have to bear the cost of new transmission lines. In Germany, power companies are required to bring the power lines to new projects, again at no cost to the project developer. Of course, in both cases the consumer or tax payers cover the transmission line and substation cost, which is similar to the process used when major hydro electric plants were built.

Grid Access Rules

In addition to transmission lines, *grid access priority* is an increasingly important issue as wind penetration grows above 15 percent. In regions with strict "least-cost" and "first-in last-out" power generation dispatch policies, this becomes the primary impediment to realizing the full potential of wind power. In February 2007, the Federal Energy Regulatory Commission made a ruling titled *Preventing Undue Discrimination and Preference in Transmission Service* to allow greater access to transmission lines for power generators of all generation types, including renewable energy projects [FERC 2007]. The new rule exempts intermittent power generators, such as wind power plants, from excessive "imbalance" charges when the amount of energy they deliver is different from the amount of energy they were scheduled to deliver. To help accommodate less predictable forms of renewable power generation, the rule creates a "conditional firm" service to deliver power from a generator to a customer, allowing the power supplier to provide firm service for most, but not all, hours in the requested time period.

A key aspect of the new rule is that it eliminates the broad discretion that transmission providers currently possess in calculating the unused, available capacity on their transmission lines. Instead, the new rule requires public utilities to work with the *North American Reliability Corporation* to develop consistent methods of calculating the available capacity and to publish those calculations to increase transparency.

In the future it is expected that grid access priority policy will shift from "least cost" to "least carbon" in determining priority access to the grid. "Least carbon" may in fact be least cost, if health affects, climate change costs, and security costs are considered along with the price of electricity.

Power line capacity rating systems may also unduly restrict wind plants. Transmission line capacity rating is typically based on the high-temperature thermal capacity under worst-case conditions. This occurs on the hottest summer days with no wind for convective cooling of the wires. Of course wind plants would not be operating under those conditions.

Technical Issues

As discussed in Chapter 4, integrating wind power into existing utility power systems has been successfully accomplished in a variety of geographical locations, using many different turbine configurations. Utilities are still concerned, however, that as the size and numbers of wind power plants increase, there could be adverse effects on power system operating strategies, system protection, and personnel safety. The expanding experience of utilities working with independent power producers in operating large-scale wind power plants has reduced these concerns. By 2008, wind generation capacity has surpassed 7.5 percent in Minnesota and Iowa with no significant impact on grid system operations, and wind power has more than 4 percent penetration in Colorado, South Dakota, Oregon and New Mexico. Texas has over 7,000 MW of wind plants and 6 percent wind penetration.

An excellent review of the integration issues faced by wind power producers and utilities [Milborrow 1993] is the source of much of the following information. From the utility viewpoint, integration issues can be categorized as follows:

   -- *macro-scale* issues involving controls, interconnection, and operation with high
      penetration (*i.e.*, fraction of the system demand filled by wind power)
   -- *micro-scale* issues relating to connection and control of individual wind turbines
      or small clusters of wind turbines

It is important to note that the cost of resolving either macro- or micro-scale operating problems must be weighed against the value of the wind-generated electricity to the utility.

Macro-Scale Integration Issues

Macro-scale concerns include problems with the control of a wind power plant that extends over a very large area and with grid electrical stability, which may result from high penetration. An early concern was that arrays of wind turbines might lack sufficient controls to prevent them from injecting power into the utility system that could cause operational problems, such as significant deviations in frequency and voltage. When the wind-generated electricity is purchased by the utility from an independent power producer, the burden of correcting any frequency, voltage, and line loading deviations caused by the wind plant will fall on the utility's conventional generating system. As a result, the operating procedures of a utility need to be examined to determine the best means to manage the penetration of a grid by wind power.

For moderate penetrations by wind power sources, operating performance of the utility can be managed by modifying the utility's sequence and rate of committing generating units as demand increases, through the scheduling of additional conventional *regulating* units. This increases the ramp rate of control capability (*i.e.*, the control authority), but regulating units are more expensive to operate than the intermediate units they would displace. If operating problems become more acute, the development of more sophisticated grid control systems could be required, but this has not been the case, even in Denmark where wind penetration has exceeded 50 percent in some regions.

Micro-Scale Integration Issues

Utility technical concerns at the micro-scale level involve determining any impact of wind turbines as an intermittent generating source. When penetration is low, such impacts have been inconsequential. However, any potential impact on a substation or other local equipment will increase with penetration level. Portions of the sub-transmission and distribution networks may be subject to overloading and *counter flows*, which can give the utility problems with reliability and voltage regulation. *Reactive power* or VAR support for the utility line could be an additional requirement, although newer electronically controlled variable-speed asynchronous wind turbine generators allow full control of power factor and can provide VAR support to the grid and have virtually eliminated this concern.

Transmission and Distribution System Effects

At the local level, there are three factors that need to be considered in regard to this issue. First, *power swings* caused by wind speed variations might cause voltage fluctuations that are large enough to be detectable as flicker in fluorescent lights. Utilities can avoid this potential problem through integrated resource planning or by connecting a wind power plant into the network at a voltage sufficiently high to ensure that any fluctuations are easily absorbed by the "pool" into which they are fed. Secondly, *line loading* could become a problem in rural areas. Individual wind turbines could be connected to single phase lines potentially causing phase imbalances. Larger wind plants avoid this problem by being connected at substations.

Thirdly, there is the safety issue of controlling small turbines and disconnecting them in the event that grid connection is lost. Turbine controls are needed to prevent "islanding" where disbursed wind turbines could continue to operate after a fault caused disconnection to the grid. Several turbines could possibly self excite without regulating voltage and frequency control from the grid. These issues are avoided by safety shut-down systems built into the turbine controls. In addition, utilities can require a mandatory lock out of isolated wind turbines to prevent accidental re-energization of distribution lines that may be under repair.

Any negative effects on the transmission system become more pronounced on a so-called *weak grid*, where long lower-voltage distribution lines are used to connect into the higher-voltage system (33 kV or higher). It is important to carefully consider large concentrations of wind power in areas where the grid is weak. To avoid this problem, wind turbines are connected to a higher-voltage line. As a very rough guide, it has been suggested that wind power capacity in megawatts should not exceed the network voltage in kilovolts, but on a weak grid only a tenth of this capacity may be allowable.

Wind turbines can potentially to provide power to the grid during "black outs" or other emergency conditions. In the event of major grid disruptions wind turbines working with other generation, with voltage and frequency regulating capabilities, can help to provide emergency power during so called "black starts" when central control and full generating capacity may not be available. Interest in distributed generation, grid security and optional operating configurations has increased since recent terrorist attacks.

Electric Power Quality

Electric power quality has been precisely defined by the *International Electrotechnical Commission* (IEC) and the *North American Reliability Council* (NARC). These organizations aim to ensure that frequency and voltage remain within specifications. A key aspect of this effort is to accommodate the load changes through automatic generation control. This produces several benefits in power quality. One of these is to insure that any light flicker caused by voltage changes does not cause undue annoyance to the utility customers. Wind power plants are able to meet these requirements, with proper interface design.

The IEC and NARC define how often a voltage change of a given magnitude can occur, as shown by the chart in Figure 13-5 [Ewart *et al.* 1978]. This graph is read by following the bold "normal operations" line until it intersects with the desired diagonal line titled " percent Excursion," interpolating as necessary. The abscissa of this intersection, read on the bottom scale, gives the minimum allowable time for the excursion to take place. For example, a 1 percent excursion is marked by the point *A* and must not occur more often than every 0.03 minutes. An excursion of 3 percent -- which would produce a flicker that is quite noticeable -- is marked *B* and must not occur more often than once every minute. Although a 1 percent excursion would not be detectable in lighting, this standard is also designed to ensure that fluctuations do not upset electronic equipment.

In addition to industry standards, most utilities have local reliability standards. These specify limits for disconnection times that must not be exceeded. For example, a 99.9 percent security standard means that a consumer must not be cut off from the electrical supply for more than 0.1 percent x 8,760 or 8.6 hours per year, on average. To meet these standards, utilities consider system interaction issues at both regional and local levels. In general, most

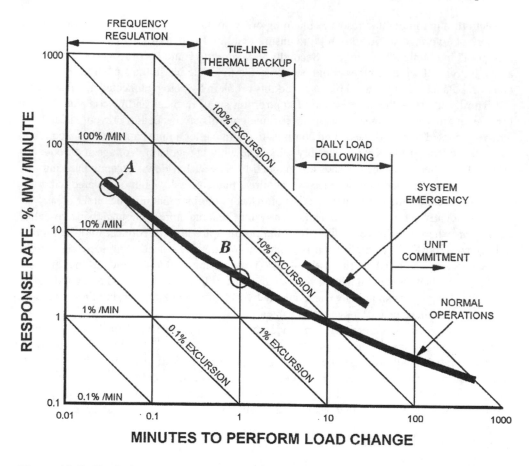

Figure 13-5. Typical standard relating voltage excursions in utility power systems to the minimum allowable time for the change to take place. [Ewart *et al.* 1978]

cutoffs of service occur at the local level, except under unusual circumstances like those that caused the "Northeast Blackout" in 1968 and subsequent emergency events.

System Dynamic Stability

Relatively small amounts of wind-generated power will simply not be noticed on a utility network, because normal consumer load fluctuations will dominate over the typical variations in the output of a wind power plant. However, as the capacity of a wind power plant increases there may come a point at which the loss of the wind power is noticeable at the system level. Normally, a high wind speed cut-out is the most severe loss-of-output test. If a significant amount of the plant (conventional or wind) has been operating at full power and then suddenly stops operating, two things happen: First, all the remaining generators slow down slightly, causing the frequency of the power to drop. Secondly, a small voltage drop occurs throughout the utility system.

Extra generators need to be brought up to load immediately to replace the dropped power. For this reason all utilities maintain generators on partial load or as spinning reserve (*i.e.*, turning without load), so that the power deficiency can be corrected along with the system voltage and frequency. The converse occurs if a large consumer load is switched off without warning or if a storm front moves across a wind power station and causes an upsurge in its output. Weather and wind forecasting becomes more important as the level of penetration by wind power plants increases. Although weather has always been an important generation planning consideration since it drives power demand for space heating and cooling, and forecasting models have been improved to the point that this is not considered problematic.

Generation Dynamic Response

Variability within the grid is not new to a utility, as illustrated by the display of time scales in Figure 13-6. The longest of these is the *daily load following* scale. All utilities need to cope with the continual fluctuations in demand caused by variations in industrial, commercial, and residential consumption. The left end of this bar, at a time scale of approximately 250 seconds, represents the local level where there is a continual load "ripple" caused by the many and varied demands on the utility system. The next-shorter time scale is that for *tie-line regulation*. Fluctuations in transmission line demand are at the right-hand end, with perturbations that tend to occur on time scales approaching 1,000 seconds. At the left-hand end of this bar, voltage fluctuations on distribution lines are minimized by ensuring that consumer loads are connected at a voltage level which ensures they can be absorbed by the power-carrying capacity of the local network. *Long-term dynamic* behavior occurs on a similar time scale, but extends down to about 0.3 second.

The next time scale is the one of greatest concern to utilities dealing with current or future wind power generation, that of *transient and dynamic stability* phenomena. The interaction between wind plants and the grid are important design issues that are well understood. We see that if significant fluctuations in the output of a wind power plant have periods of about 15 seconds or more -- by virtue of their multiple units, extended land area, active controls, or a combination of all these factors -- utilities need have no concern about its impact on system dynamic stability.

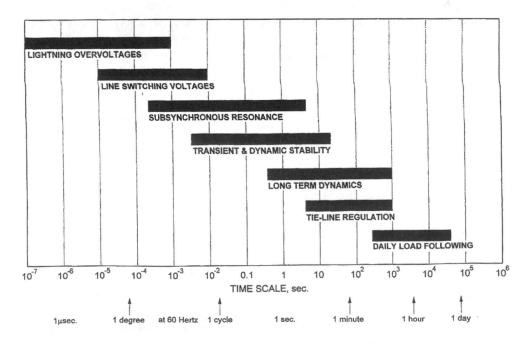

Figure 13-6. Time scales of utility system dynamic phenomena.

Wind Energy Utility Business Outlook and Trends

Since the passage of PURPA in 1978, the United States electric utility industry has seen increasing competition from independent energy producers. By the late 1980s, nearly 10 percent of California's electricity was generated by non-utility sources, and much of the state's new generation was built by independent energy producers. That trend has continued and in 2007 independent power producers owned 83 percent of the wind market [Wiser 2008].

Structural changes in the electric utility industry also played a major role in enabling independent energy producers to gain and maintain a foothold. Major financial burdens from large power plants during a period of high interest rates brought several utilities to the brink of bankruptcy during the 1980s. High prices slowed growth in electricity consumption, and regulatory agencies were under pressure to protect rate payers from the cost of expensive and potentially risky new facilities. Independent energy producers offered both the utilities and their regulators lower-cost, lower-risk alternatives to building these large new plants.

Changes in the Regulatory Environment

Because some electric utilities are monopolies, regulatory changes affecting them also have wide-reaching consequences for the wind industry. If, for example, regulatory agencies dictate that utilities must hold an auction for the next increment of generation, and the lowest bid must be used to build the capacity, wind generation will often (if not always) lose to *cogeneration* because cogeneration receives *firm capacity credits*.

Unregulated Subsidiaries

Under PURPA, utilities were initially prohibited from owning a controlling interest (more than 50 percent) in a so-called *qualifying facility (QF)*, such as a wind power station, but that prohibition was later eliminated. Under the Economic Stimulus Act of 2008 utility investors could also qualify for QF status and Production Tax Credits. Originally this prohibition was designed to encourage competition and eliminate self-dealing (*e.g.*, offering more favorable treatment to a utility subsidiary than to independent producers). For several years after its passage, this provision caused little concern to the utilities. But by the mid-1980s, as the independent producers became increasingly successful, this prohibition frustrated utilities who sought to diversify into unregulated subsidiaries that would build, own, and operate renewable energy plants.

Participation as an unregulated subsidiary permits the utility to act much like an independent energy producer, raising capital, building plants, and selling energy in an unregulated environment. Several U.S. utilities have established unregulated subsidiaries that finance and invest in cogeneration and other small power plants.

Least-Cost Planning and Fixed-Price Contracts

Another change that proved to be detrimental to the wind energy industry is the regulatory emphasis on *least-cost planning*. Under this type of regulation, a new generation source of electricity can be added only if it has been found to be the least costly, regardless of who the developer is. Bidding by independent energy producers to sell least-cost electricity pits wind generation against cogeneration and other fossil-fuel plants, but the true costs of environmental and health effects are not considered. Bidding, however, works fairly well when long-term costs are used. In the future it expected that regional transmission organizations will transition to *least-carbon dispatch* even if the price is higher.

Carbon Markets

Carbon offsets are expected to create a market for wind power. Wind energy can be sold to offset CO_2 and other greenhouse gas (GHG) emissions from power plants. A carbon market was created initially by the European Union in 2005. The first auction of CO_2 allowances in the United States was in September 2008. Six Northeastern states offered allowances that sold at a clearance price of $3.07 per allowance. Companies purchasing the allowances can use them to offset future GHG emissions. The money raised will be distributed to the states for activities that reduce emissions. This could include financial support for wind power.

As climate change becomes more visible, regulatory efforts to mitigate climate change are expected to expand. The World Bank estimated that the emerging carbon market in 2007 is valued at US$64 billion (€47 billion) worldwide [Capoor]. This, in turn, has stimulated wind energy development as a means of accomplishing carbon abatement, and has motivated individuals, communities, companies and governments to cooperate to reduce emissions.

Concluding Remarks

Most utilities consider the issues discussed in this chapter to be part of the challenges that can and are being overcome by joint efforts between the wind power plant developers and the electric utility industry. Load swings, problems in balancing and controlling the flow of electricity throughout the grid, and complex financial decisions are routine in the electric power industry. The simplicity and modularity of wind power plants makes them adaptable to the solutions required to overcome these challenges. Modern wind turbine technology has made wind energy an attractive option for electric power generation world-wide.

References

Capoor, K., and P. Ambrosi, 2008, *State and Trends of the Carbon Market*, Washington, D.C.: World Bank Institute.

DOE, 1990, "The Potential of Renewable Energy: An Interlaboratory White Paper," SERI/TP-260-3674, Golden, Colorado: National Renewable Energy Laboratory.

DOE, 2008, "20% Wind Energy by 2030 – Increasing Wind Energy's Contribution to U.S. Electricity Supply," DOE/GO-102008-2578: U.S. Department of Energy.

Energy Information Administration (EIA), 2007, Electric Power Annual, Report Released October 22, 2007.

EPRI, 1979, "Electric Power Research Institute, Technical Assessment Guide," Special Report EPRI-PS-1201-SR, Palo Alto, California: Electric Power Research Institute.

EPRI, 1993, "Wind as a Utility-Grade Supply Resource: A Planning Framework for the Pacific Northwest," EPRI TR-102094, Palo Alto, California: Electric Power Research Institute.

Ewart, D. N., *et al.*, "Power Response Requirements for Electric Utility Generating Units," *Proceedings 1978 American Power Conference*, Vol. 40.

Federal Energy Regulatory Commission (FERC), 2007, "Preventing Undue Discrimination and Preference in Transmission Service," *Fed. Reg.,* vol. 71, no. 108, pp. 32685-32734.

Garver, L. L., 1966, "Effective Load Carrying Capacities of Generating Units," *IEEE Transactions Power Apparatus and Systems.*

IEEE, 2007, "Wind Integration–Driving Technology, Policy, and Economics," *Power and Energy Magazine*, Vol. 5, No. 6.

Karas, K. C., 1992, "Wind Energy: What Does It Really Cost?" *Proceedings, Windpower '92 Conference*, Washington, DC: American Wind Energy Association.

Lintsen, H. W., 1991, "From Windmill to Steam Engine Waterpumping," *Proceedings, European Wind Energy Conference, EWEC '91*, Part II, Amsterdam: Elsevier, pp. 7-12.

Milborrow, D., 1993, "Understanding Integration," *Windpower Monthly News Magazine*, Vol. 9, No. 9, Redding, California: pp. 27-33.

Piwko, R., 2005, *The Effects of Integrating Wind Power on Transmission Planning, Reliability and Operations*, for New York State Energy Research and Development Agency (NYSERDA): GE Energy.

Smith, J. C., and B. Parsons, 2007, "What does 20% look like? Developments in Wind Technology and Systems," *IEEE Power and Energy Magazine,* Volume 5, Number 6, November/December 2007.

Smith, D. R., and M. A. Ilyin, 1991, "Wind and Solar Energy, Costs and Value," *Proceedings, Tenth ASME Wind Energy Symposium*, 1991, SED-Vol. 11, D. E. Berg and P. S. Veers, eds., New York: American Society of Mechanical Engineers, pp. 29-32.

Van Wijk, A. J. M., 1993, *Wind Power Potential in OECD Countries*, Utrecht, Holland: University of Utrecht.

Weinberg, C. J., and H. Kelley, 1993, *Renewable Energy: Sources for Fuels and Electricity*, T. B. Johnnsson, et al., eds., Washington, DC: Island Press, Chapter 23.

Wiser, R., and M. Bolinger, 2008, *Annual Report on United States Wind Power Installation, Cost, and Performance Trends: 2007*, NREL Report No. DOE/GO-102008-2590, Golden, Colorado: National Renewable Energy Laboratory.

14

Wind Turbine Control Systems

Alan D. Wright, Ph.D.
National Renewable Energy Laboratory
Golden, Colorado

Introduction

Wind turbines are complex, nonlinear, dynamic systems forced by gravity, stochastic wind disturbances, and gravitational, centrifugal, and gyroscopic loads. The aerodynamic behavior of wind turbines is nonlinear, unsteady, and complex. Turbine rotors are subjected to a complicated three-dimensional turbulent wind inflow field that drives fatigue loading. Wind turbine modeling is also complex and challenging. Accurate models must contain many degrees of freedom (DOF) to capture the most important dynamic effects. The rotation of the rotor adds complexity to the dynamics modeling. Designs of control algorithms for wind turbines must account for these complexities. Algorithms must capture the most important turbine dynamics without being too complex and unwieldy. Off-the-shelf commercial software is seldom adequate for wind turbine dynamics modeling. Instead, specialized dynamic simulation codes are usually required to model all the important nonlinear effects.

As illustrated in Figure 14-1, a wind turbine control system consists of sensors, actuators and a system that ties these elements together. A hardware or software system processes input signals from the sensors and generates output signals for actuators. The main goal of the controller is to modify the operating states of the turbine to maintain safe turbine operation, maximize power, mitigate damaging fatigue loads, and detect fault conditions. A supervisory control system starts and stops the machine, yaws the turbine when there is a significant yaw misalignment, detects fault conditions, and performs emergency shut-downs. Other parts of the controller are intended to maximize power and reduce loads during normal turbine operation.

- Control Actuators

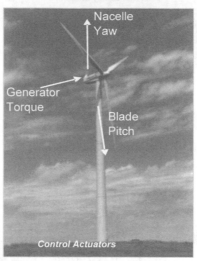

- Speed sensors

- Accelerometers

- Anemometers and wind vanes

- Electrical power sensors

- Strain gages

Figure 14-1. Wind turbine control actuators and typical sensors.

Figure 14-2 shows different regions of operation that are typical for utility-scale wind turbines. In Region 2, when wind speeds are in the operating range but lower than the *rated wind speed*, the goal of the controller is to maximize turbine power. In Region 3, when winds exceed the rated wind speed, the goal of the controller is to maintain turbine power at a constant *rated power*, in order to limit turbine blade loads and generator torques. Other regions of operation include *startup* (Region 1) and machine *shutdown*.

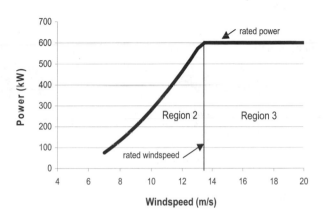

Plot of Turbine Power Versus Windspeed

Figure 14-2. Typical regions of operation for utility-scale wind turbines. Not shown are the startup (Region 1) and shutdown regions.

In the past, designers of wind turbines have employed different control strategies to achieve these different goals. Extensive research on a variety of control systems was performed during the evolution of modern wind turbines [see **Prior References** at the end of this chapter]. In Regions 2 and 3, generator speed was often held constant. Some turbines achieved control in Region 3 with blades designed so that power was limited passively through *aerodynamic stall*. Power output was not constant, but no pitch mechanism was required for over-power control. Typically, active control of these machines involved only starting and stopping the turbine.

Rotor blades with adjustable pitch have often been used in constant-speed machines to provide better control of turbine power than is possible with blade stall. Blade pitch can be regulated to provide constant power in Region 3. The pitch mechanisms in these machines must be fast, to provide good power regulation in the presence of gusts and turbulence. However, operating the turbine at constant turbine rotational speed in Region 2 (through the use of synchronous or induction generators) reduces the power output of the machine. To maximize power output in Region 2, the rotational speed of the turbine must vary with wind speed to maintain an optimum, relatively constant *tip-speed ratio* (see Figure 2-9). Most large commercial wind turbines are now variable-speed pitch-regulated machines. This allows the turbine to operate at near optimum tip-speed ratio over a range of wind speeds and generate maximum power in Region 2. Blade pitch control is used in Region 1 for startup, in Region 3 to maintain rated power, and for shutdown.

Classical Design of Wind Turbine Controllers

Regions 2 and 3 controls for current variable-speed pitch-controlled wind turbines are typically designed using simple, classical control design techniques such as proportional-integral-derivative (PID) control [Bossanyi 2000; Burton *et al.* 2001]. A typical control diagram for such a machine is shown in Figure 14-3. Generator torque is controlled in accordance with the following equation in Region 2, as shown in the upper control loop in Figure 14-3:

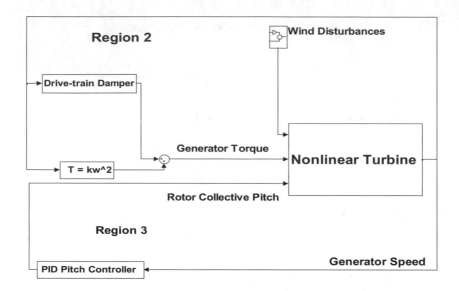

Figure 14-3. Typical control diagram of a commercial wind turbine.

$$Q_{gen} = k\Omega^2 \tag{14-1a}$$

$$k = \frac{1}{2}\, \rho\, \pi R^5\, \frac{C_{P,\max}}{(\lambda_{opt})^3} \tag{14-1b}$$

where

Q_{gen} = generator torque (N-m)
Ω = rotor rotational speed (rad/s)
k = proportionality constant for optimum rotor power (N-m-s^2)
ρ = air density (kg/m^3)
R = tip radius of rotor (m)
$C_{P,\max}$ = maximum of rotor power coefficient; Eq. (2-14)
λ_{opt} = optimum of tip-speed ratio corresponding to $C_{P,\max}$ at a particular blade pitch angle; Eq. (2-20)

Refinements to the Region 2 generator torque control can lead to enhanced energy capture. Fingersh and Johnson [2004] reported improved energy capture using a variation of the Region 2 baseline controller approach named the *optimally tracking rotor control*.

For Region 3, classical PID control design techniques are typically used to design the blade pitch controller [Wright and Fingersh 2008]. Referring to the lower loop in Figure

14-3, we see that generator or rotor speed is measured and passed to the pitch controller. The goal is to use PID pitch control to regulate turbine speed in the presence of wind-speed disturbances. The expression for the blade pitch command is:

$$\Delta\theta(t) = K_p\Delta\Omega(t) + K_I \int \Delta\Omega(t)dt + K_D\Delta\dot{\Omega}(t) \qquad (14\text{-}2)$$

where

$\Delta\theta$ = commanded blade pitch change (rad)
$\Delta\Omega$ = generator or rotor rotational speed error relative to set-point (rad)
K_P = proportional feedback gain (s)
K_I = integral feedback gain
K_D = derivative feedback gain (s^2)

The goal in Region 3 control design is to determine values for the three gains that give the desired tracking of rotor speed to the desired set-point and maintain closed-loop stability. Methods for choosing these gains are further given by Hansen *et al* [2005] and Wright and Fingersh [2008].

To verify satisfactory control performance with these gains, the control designer usually simulates the closed-loop turbine response before implementing and testing such a controller in the field. Some refinements to Equation (14-2) to improve performance include gain scheduling, including anti-windup, and introducing filters to prevent excitation of high frequency modes [Wright and Fingersh 2008].

An additional control design goal is to mitigate structural dynamic loads. One way is to design controls that actively dampen the motions of turbine components. In commercial turbines, an additional generator torque control loop in Region 2 is often used to actively damp the drive train torsion mode of the turbine, as illustrated in Figure 14-3. In Region 3, classical control design methods have been used to design controllers to add damping to the tower's first fore-aft mode with blade pitch changes [Bossanyi 2000]. The pitch control to actively damp this tower motion is usually implemented by adding a single-input single-output (SISO) control loop to the basic Region 3 speed control loop just discussed.

Other studies with classical control include the use of independent pitch control to mitigate asymmetric wind variations across the swept area of the rotor [Bossanyi 2003]. In this approach, two separate SISO control loops were used to mitigate the tilt- and yaw- oriented loads in the fixed frame of reference with independent pitch control of each blade. This work was extended with alternative sensors to measure the asymmetric loading on the rotor [Bossanyi 2004]. Good results were obtained when suitable sensors were used.

In these various applications, the use of classical controls to address more than one control objective is not straightforward. Often, multiple control loops are used, which adds complexity to the control design and the dynamic behavior of the wind turbine system. If these complex controls are not designed with great care, the control loops interfere with each other and destabilize the turbine. The potential for instability increases as turbines become larger and more flexible, and the degree of coupling between turbine components increases.

Design of Advanced Wind Turbine Controllers

As already mentioned, wind turbines are flexible systems acted on by stochastic wind disturbances and time-varying gravitational, centrifugal, and gyroscopic loads. While the simple classical controllers discussed previously can give good speed regulation performance in Region 3, other control objectives may not be met. These include accounting for stochastic wind disturbances, accounting for turbine flexibility, and mitigating loads and deflections. When additional control objectives must be met, additional SISO control loops must be added to the classical controllers. These additional loops can destabilize the machine if they are not very carefully designed.

Modern control designs using state-space methods more adequately address these issues, because the controller uses a model to determine system states. Controllers can be designed not only to maximize power and regulate speed but also to add damping to important flexible modes. Integrating all the available turbine actuators in a single control loop to maximize load-alleviating potential is advantageous. One of the most prevalent advanced control methods that have been applied to wind turbines is called *full state feedback* [Kwakernaak and Sivan 1972]. This method allows multiple control actions to be performed in a single loop, including speed regulation, adding active damping to low-damped turbine dynamic modes, and mitigating loads caused by stochastic wind disturbances.

Some Advanced Control Design Methods

Most advanced controls are based on linear control design methods and linear time-invariant models [Kwakernaak and Sivan 1972]. These linear models can be represented in the following form:

$$\Delta \dot{x} = A \underline{\Delta x} + B \underline{\Delta u} + B_d \underline{\Delta u_d} \tag{14-3a}$$

$$\Delta y = C \underline{\Delta x} + D \underline{\Delta u} + D_d \underline{\Delta u_d} \tag{14-3b}$$

where $\underline{\Delta x}$ is the state vector, $\underline{\Delta u}$ is the control input vector, $\underline{\Delta u_d}$ is the disturbance input vector, $\underline{\Delta y}$ is the measured output, A represents the state matrix, B the control input gain matrix, B_d the disturbance input gain matrix, C relates the measured output $\underline{\Delta y}$ to the turbine states, D relates the measured output to the control input, and D_d relates the measured output to the disturbance states; $\underline{\Delta \dot{x}}$ represents the time derivative of $\underline{\Delta x}$; Δx, $\Delta \dot{x}$, Δy, Δu, and Δu_d (perturbed values) represent small perturbations from the calculated operating point values $\underline{x_{op}}$, $\underline{\dot{x}_{op}}$, $\underline{y_{op}}$, $\underline{u_{op}}$, and $u_{d_{op}}$.

Values in the state vector $\underline{\Delta x}$ might represent generalized coordinates describing the flexible modes of the turbine, such as blade flapwise and edgewise motions, tower-top fore-aft and side-to-side motions, and rotor or generator speeds. They may also include states to describe the control actuator dynamics [Wright and Fingersh 2008]. The values in $\underline{\Delta y}$ are the measurements from the turbine, such as generator or rotor speed, tower-top acceleration, and blade-root flap bending moments obtained from strain gages.

Values in the vector $\underline{\Delta u}$ represent the control inputs. An example is blade pitch for control of turbine speed in Region 3. For Region 3 speed regulation, each blade is pitched identically, so that only one pitch value is commanded. This is called rotor *collective pitch control*. In other control actions, the pitch of each blade may be commanded separately, which is called *individual* or *independent pitch control*. Individual pitch control becomes important when the controller is trying to mitigate the effects of asymmetric variations of wind speed across the rotor. In such applications, the pitch of one blade may be different than the others because of differences between the local wind speeds at each blade.

In full state feedback, the control law formulates Δu as a linear combination of the turbine states, as follows:

$$\Delta \underline{u}(t) = G \Delta \underline{x}(t) \tag{14-4}$$

where G is the "gain" matrix. If the system consisting of (A, B) in Equation (14-3a) is "controllable", this feedback law can be used to place the poles of this system arbitrarily in the complex plane]. This is what allows active damping to be added to normally low-damped turbine components and thereby significantly reduce fatigue loads. Either pole placement or linear quadratic regulation (LQR) can be used to calculate the gain matrix G [Kwakernaak and Sivan 1972].

With LQR, a unique linear feedback control signal is found that will minimize the following quadratic cost function:

$$ J = \int_0^\infty \left(\Delta \underline{x}(t)^T Q \, \Delta \underline{x}(t) + \Delta \underline{u}(t)^T R \, \Delta \underline{u}(t) \right) dt \qquad (5) $$

where, $\Delta \underline{x}(t)$ represents the system states, $\Delta \underline{u}(t)$ represents the control inputs, Q contains weightings for the states, and R contains weightings for actuator usage. Fast state regulation and low actuator use are competing objectives; therefore, the Q and R weightings allow a trade-off of performance objectives with actuator use. Kwakernaak and Sivan [1972] provide more details regarding LQR.

To use full state feedback as the final control design, one would have to measure every state contained in the linear model described by Equation (14-3). Most current commercial wind turbines are not instrumented to the extent needed to measure all these states, especially as the order of the model increases. Usually only a limited number of measured signals are available for control, such as generator or rotor azimuth angle, rotor speed, tower-top acceleration, and blade-root bending moments. However, according to Kwakernaak and Sivan, *observability* allows us to estimate the states contained in the linear model based on just a few turbine measurements.

Progress in Advanced Controls for Wind Turbines

Some very early work in developing advanced state-space controls for wind turbines was performed by Mattson [1984] who used a state estimator in combination with LQR. He described regulation of power for a fixed-speed machine using blade pitch, reporting good results except in frequency intervals close to the natural frequency of the first drive train torsional mode. This was due to the amplified effects of measurement noise caused by the controller attempting to compensate for phase lag at this natural frequency. Liebst [1985] developed an individual pitch control system for a wind turbine using LQR design to alleviate blade loads caused by wind shear, gravity and tower deflection.

Important progress has been made in simplifying dynamic models as expressed in Equation (14-3) for use in control design. Large multi-body dynamics codes [*e.g.* MSC. ADAMS®] divide a complex structure into numerous rigid-body masses and then connect these parts with springs and dampers. This approach leads to nonlinear dynamic models with hundreds or thousands of *degrees of freedom* (DOF). The order of these models must be greatly reduced to make them practical for control design. In addition, these nonlinear models need to be linearized in order to apply linear control theory.

An example of tools for simulation and linear model generation is a nonlinear wind turbine simulation tool called DUWECS that was developed at the Delft University of Technology [Bongers 1994]. This code models the effects of blade and tower flexibility, nacelle yaw, and rotor teeter motion, as well as drive train torsional flexibility. Blade and tower flexibility are modeled using a rigid blade/tower hinge approach. This approach leads to models with fewer DOF than those developed with either multi-body dynamics codes or finite element methods. DUWECS tools not only simulate the entire nonlinear turbine but also generate a linear model of a turbine for control design.

In the U.S., another code named SymDyn was developed for wind turbine simulation and control design using a similar modeling approach [Stol and Bir 2003]. Blade and tower flexibility were also modeled using the rigid blade/tower hinge approach. The code generated linear models for control design and simulated the system behavior once the control system was active. The development of such codes as SymDyn and DUWECS for control design and simulation paved the way for additional progress in advanced wind turbine controls.

Another issue in wind turbine control design is *periodicity* of the dynamic system. For a real wind turbine, the state matrices (A, B, B_d, etc.) in Equation (14-3) are not constant but vary as the rotor rotates. Stol and Balas [2003] developed *periodic control gains* using time-varying LQR techniques for a two-bladed, teetering-hub turbine operating in Region 3. They concluded that when blade load reduction is the primary goal, periodic control gives the best results when full state feedback is used. They also concluded that if speed regulation is the only objective, then periodic control is not the most appropriate method. In these studies, SymDyn was used to provide the proper linear models with periodic matrices.

In another modeling approach, the *assumed modes method* is used to discretize the wind turbine structure. This method usually models components with higher fidelity than the rigid blade/tower hinge approach. With the assumed modes method, the most important turbine dynamics can still be modeled with just a few degrees of freedom. An example is the BLADED code developed by Garrad Hassan and Partners, Ltd [Bossanyi 2003a, 2003b] which can also generate linear models suitable for control design. The nonlinear turbine can also be simulated with a controller in the loop [Stol and Bir 2003; Stol and Balas 2003]. Another assumed modes code is the FAST dynamics code [Jonkman and Buhl 2005] (developed at Oregon State University by Robert Wilson and modified at NREL), which also models and simulates an entire nonlinear turbine. The FAST code has been modified recently to produce linear state-space models of turbine systems and has been extensively tested and validated.

Disturbance Accommodating Controls

Wind turbines must be able to operate in a highly turbulent wind environment. Turbulent winds cause fluctuations in the blade aerodynamic forces, and thus they influence the power, torque, and cyclic loading of the machine. What is needed is a control approach that counteracts or accommodates these disturbances and permits full-state feedback and state estimation. *Disturbance accommodating control* (DAC) is a way to reduce or counteract persistent disturbances [Johnson 1976]. The basic concept of DAC is to augment the usual state estimator-based controller to re-create disturbance states via an assumed-waveform model. Johnson used these disturbance states as part of the feedback control to reduce (*i.e.* accommodate) or counteract any persistent disturbance effects.

During the late 1990s, Balas *et al.* [1998] made significant progress in applying DAC to wind turbine control. Later, Stol and Balas [2003] studied the use of state-space methods to design *disturbance accommodating controls* (DACs). Using SymDyn, they developed a linear model of a turbine containing only the rotor rotation degree of freedom. This simple DAC was shown to adequately control a turbine as modeled in their nonlinear simulator with just the rotor rotation DOF.

Reducing Cyclic Loads by Advanced Controls

Further work in advanced controls, using DAC, has been performed by Wright [2004]. In this work, a multiple-input multiple-output (MIMO) controller was used to perform multiple control objectives and accommodate wind disturbances. The MIMO was designed and simulated for the 2-bladed *Controls Advanced Research Turbine* (CART) at NREL's National Wind Technology Center (NWTC). That study used independent blade pitch control for attenuating loads caused by asymmetric wind variations across the rotor, demonstrating the advantages of performing multiple control objectives with a single MIMO control loop. This controller performs speed regulation, damping of the tower's fore-aft and side-to-side modes, damping of the first drive train torsion mode and mitigation of asymmetric wind variations across the rotor.

Compared to simulated cyclic loads in the CART system with a standard PID controller, this state-space controller achieved the following load reductions:

- Low-speed shaft torque ....28 percent reduction
- Tower side-side bending moment ....76 percent reduction
- Tower fore-aft bending moment ....20 percent reduction
- Blade flap bending moment ....16 percent reduction

Another control design for attenuating tower motions is presented by Malcom and Hansen [2002].

Control Methods to Accommodate Nonlinearity Effects

The nonlinear behavior of a wind turbine can make control design difficult. For example, rotor aerodynamic behavior is highly nonlinear. In pitch control, the control input gains are usually the partial derivative of the rotor aerodynamic torque with respect to blade pitch angle. These input gains are not constant but vary with wind speed, rotor speed, and pitch angle. Figure 14-4 shows the variation of rotor aerodynamic torque with blade pitch angle for the CART. A linear controller designed for a turbine at one operating point (such as the *control design point* shown in this figure) may give poor results for other operating points.

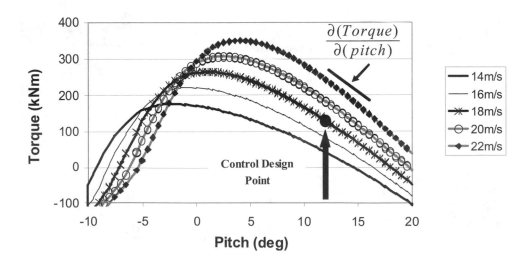

Figure 14-4. Variation of control input gains with pitch angle and wind speed. Gains are the slopes of the torque *vs.* pitch angle curves.

To address this nonlinearity issue, multiple controllers can be optimized for different turbine control design points. Then, as the turbine's operating point varies from one region to another, the controllers can be switched. This usually requires a *scheduling* or *switching parameter*, such as wind speed or pitch. Some work has been performed to study switching between controllers for wind turbines [Kraan 1992]. Often, switching between different controllers can be problematic, in that if one simply switches between one controller and the next, undesirable *switching transients* can occur. Kraan and Bongers [1993] describe the use of *controller pre-conditioning* in which the next controller predicted to be activated is prepared in advance for this task, so that switching transients are minimized.

Another modern control technique to account for changing operating points and therefore changing optimum gains is *adaptive control*. Bossanyi [1987] researched an adaptive scheme that consisted of a time-varying state estimator that applied optimal control to account for varying gains. The author reported satisfactory simulation results for schemes based on a combination of measurements of power output and shaft speed. An adaptive control method for a three-bladed turbine has been studied by Freeman and Balas [1999]. Other more recent work in adaptive controls has been performed by Johnson [2004] and Johnson *et al.* [2004], who reported using adaptive control to improve energy capture in Region 2. Additional work has been done by Frost *et al.* [2008] in the use of nonlinear adaptive controls for speed regulation in Region 3.

Tools for Control Design and Evaluation

Developing advanced controls for wind turbines is a complex process, as illustrated in Figure 14-5. This process consists of 1) determining control objectives, 2) developing simplified dynamic models from which to design the controller, 3) applying control design tools, and 4) performing dynamic simulations to test closed-loop performance. These steps may need to be repeated in order to design a controller with satisfactory performance in simula-

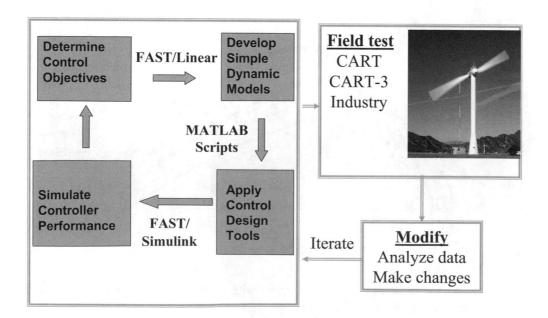

Figure 14-5. Illustration of a typical control design, simulation, and field testing process. The complete process is iterative until the desired control performance is achieved.

tion. The next steps include implementing and field-testing the controller in the prototype turbine. The controller must be tested in the field for a variety of wind and operating conditions. The whole process from start to finish may need to be repeated in order to optimize the controller for the final wind turbine design.

Simulation Tools

NREL researchers have documented the use of control design tools based on the FAST code [Wright and Fingersh 2008], providing demonstrations of available control design and simulation tools. Various advanced control design methods are also discussed in this report, based on full state feedback, state estimation, DAC, and LQR.

Testing the controller through simulation can be a challenging step. The control design must be tested under a broad range of simulated wind and turbine operating conditions. This necessitates the generation of *turbulent wind input cases* that represent the unsteady winds at a selected site, using specialized codes such as TurbSim [Jonkman and Buhl 2007]. Stochastic inflow codes of this type simulate full-field flows that contain coherent turbulence structures, with the space and time relationships seen in instabilities associated, for example, with nocturnal boundary layer flows

Another software tool that was developed to test turbine controls is illustrated in Figure 14-6. This tool is an interface developed between NREL's FAST wind turbine simulation code [Jonkman and Buhl 2005] and SIMULINK [Grace *et al.* 1992], a popular simulation tool for controls design. This combination introduces tremendous flexibility in wind turbine controls implementation during simulation. Generator torque control, nacelle yaw control, and blade pitch control modules can all be designed in the SIMULINK environment and simulated while making use of the complete nonlinear aeroelastic wind turbine equations of motion available in FAST.

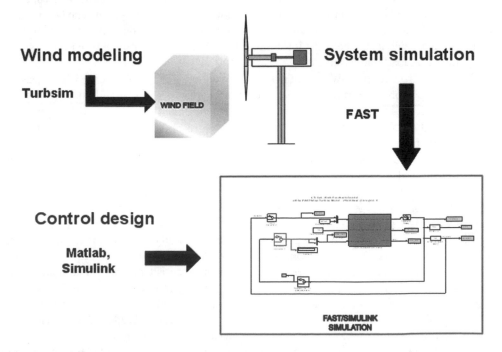

Figure 14-6. Schematic diagram of the application of NREL's advanced software tool, FAST/SIMULINK Simulation, combining control design and system simulation, including modeling of wind inputs with TURBSIM.

Field Testing

After testing control performance through simulation, the next critical step is field implementation and testing. Wright and Fingersh [2008] documented the steps involved in field-implementing and testing advanced state-space controllers. NREL has configured the 2-bladed CART system, with its upwind teetering-hub rotor, to field-test advanced controls such as multivariable designs [Stol and Fingersh 2008]. State-space controls for speed regulation and drivetrain damping were implemented and tested [Wright *et al.* 2005].

Direct, quantitative comparison of the benefits of different control designs is possible with the CART system configured as a controls test bed. For example, researchers compared the tower motions present when generator torque and blade pitch were used together in a single MIMO (multiple input, multiple output) control loop against those experienced by the turbine when a standard Region 2 torque controller was used. Back-to-back preliminary field tests showed that active damping added by the MIMO controller reduced tower fore-aft cyclic motion by 75 percent and side-side cyclic motion by almost 70 percent, during turbine operations in Region 2 [Wright *et al.* 2007].

Future Advanced Control Approaches

Currently, most control algorithms depend on measured turbine signals in the control feedback loop for load mitigation. Often these turbine measurements are unreliable or too slow. As a result, turbine controls must react to complex atmospheric disturbances after their effects have already been "felt" by the turbine. There is a certain lag between the time that the measured signal is detected by the controller and the time that the control actuator acts to mitigate the loads caused by these effects. A major improvement in load-mitigating capability could be attained by measuring atmospheric phenomena upwind of the turbine before they are felt by the turbine rotor. The needed control actuation signals could then be prepared in advance and applied as the detected inflow structure enters the rotor area, providing significantly increased load mitigation.

Future controls research will explore the use of anticipatory (look-ahead) wind measuring sensors for improved turbine control. Such remote sensors as Lidar and Sodar [Kelley *et al.* 2007] are being investigated for use in advanced controls. Wind characteristics measured ahead of the turbine can then be used in a *feed forward* approach, as sketched in Figure 14-7. Initial studies have documented some of the advantages of using Lidar to sense the wind shear upwind of the rotor for use in algorithms providing feed-forward independent blade pitch control [Harris *et al.* 2006].

The wind inflow displays a complex variation of speed and character across the swept area of the rotor. As rotor sizes increase, these localized inflow structures can cause blade loads to vary dramatically and rapidly along the rotor blade and from blade to blade. Pitching the entire rotor blade may not be the most optimum method of controlling the loads from these localized effects. Moreover, controlling the pitch of the entire blade may be too slow to effectively mitigate local loads, because of limits on pitch actuator rates and blade mass acceleration. In addition, pitching the entire blade can only mitigate the average integrated effects of inflow wind variations along the blade. For rapid and localized wind-speed variations across the rotor, local and faster acting blade actuators may be needed. New sensors are also needed to measure the flow at different span-wise positions along the blade.

Some of the advanced control technologies under investigation include such devices as *trailing edge flaps, micro-tabs*, and *adaptive trailing edge devices*. New sensors being investigated include localized flow-measuring devices (such as pitot tubes) and *imbedded*

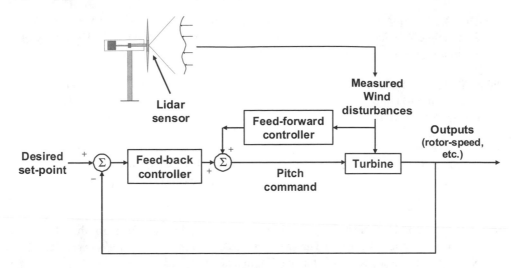

Figure 14-7. Control diagram illustrating a potential application of feed-forward controls. LIDAR measurements of disturbances in the wind field upwind of the turbine are factored into blade pitch commands.

fiber optic sensors. The goal is to develop "smart" rotor blades with imbedded sensors and actuators that are reliable, maintenance free, do not add significant weight and cost, and are effective for blade fatigue load mitigation. Recent advances in the technology of wind turbine controls are described in [International Energy Agency (IEA) 2008]. Future research will include further development of actuators and sensors, and proof of their effectiveness through wind tunnel and full-scale tests.

Current References

Burton, T., D. Sharpe, N. Jenkins, and E. Bossanyi, 2001, *Wind Energy Handbook*, John Wiley & Sons, Ltd, (New York), pp. 488–489.

Bossanyi, E.A, 1987, "Adaptive Pitch Control for a 250 kW Wind Turbine," *Proceedings, 9th BWEA Wind Energy Conference,* Edinburgh, Scotland: British Wind Energy Association, pp. 85-92.

Bossanyi, E. A., 2000, "The design of closed loop controllers for wind turbines." *Wind Energy* 2000; 3: 149–163.

Bossanyi, E. A., 2003, "Individual Blade Pitch Control for Load Reduction," *Wind Energy*, 6: pp. 119-128.

Bossanyi, E. A., 2003a, *GH Bladed Theory Manual*, Issue No. 12, 282/BR/009, Bristol, UK: Garrad Hassan and Partners Limited.

Bossanyi, E. A., 2003b, *GH Bladed Version 3.6 User Manual*, 282/BR/010, Bristol, UK: Garrad Hassan and Partners Limited.

Bossanyi, E. A., 2004, "Developments in Individual Blade Pitch Control," EWEA Conference *The Science of Making Torque from Wind,* DUWIND, Delft University of Technology, The Netherlands.

Balas, M. J., Y. J. Lee, and L. Kendall, 1998, "Disturbance Tracking Control Theory with Application to Horizontal Axis Wind Turbines," *Proceedings, 1998 ASME Wind Energy Symposium*: American Society of Mechanical Engineers, pp. 95-99.

Bongers, P.M., 1994, *Modeling and Identification of Flexible Wind Turbines and a Factorizational Approach to Robust Control*, Ph.D. Thesis, Department of Mechanical Engineering and Marine Technology, Delft University of Technology, The Netherlands.

Fingersh, L. J., and K. Johnson, 2004, "Baseline Results and Future Plans for the NREL Controls Advanced Research Turbine," *23rd ASME Wind Energy Conference*, Reno, NV: American Society of Mechanical Engineers, pp. 87-93.

Freeman, J. B., and M. J. Balas, 1999, "An Investigation of Variable-Speed Horizontal-Axis Wind Turbines Using Direct Model-Reference Adaptive Control," *Proceedings, 37th AIAA Aerospace Sciences Meeting and Exhibit*, Reno, Nevada, pp. 66-76.

Frost, S. A., M. J. Balas, and A. D. Wright, 2008, "Direct Adaptive Control of a Utility-Scale Wind Turbine for Speed Regulation," *International Journal of Robust Nonlinear Control*.

Grace, A., A. J. Laub, J. N. Little, and C. M. Thompson, 1992, *Control System TOOLBOX for Use with MATLAB*, Natick, MA: The MATHWORKS, Inc.

Hansen, M. H., A. Hansen, T. J. Larsen, S. Oye, P. Sorensen, P. Fuglsang, 2005, *Control Design for a Pitch-Regulated Variable-Speed Wind Turbine*, Riso-R-1500(EN), Roskilde, Denmark: Riso National Laboratory.

Harris, M., M. Hand, and A. Wright, 2006, *Lidar for Turbine Control*, NREL/TP-500-39154, Golden, Colorado: National Renewable Energy Laboratory.

International Energy Agency (IEA), 2008, "The Application of Smart Structures for Large Wind Turbine Rotor Blades," *Proceedings, 56th IEA Topical Expert Meeting (Preliminary)*, Albuquerque, New Mexico: Sandia National Laboratories.

Johnson, C.D., 1976, "Theory of Disturbance Accommodating Controllers," in *Advances in Control and Dynamic Systems*, **12**, ed. C. T. Leondes.

Johnson, K. E., 2004, *Adaptive Torque Control of Variable Speed Wind Turbines*, NREL/TP-500-36265, Golden, Colorado: National Renewable Energy Laboratory.

Johnson, K. E., L. Fingersh, L., M. Balas, and Y. L. Pao, 2004, "Methods for Increasing Region 2 Power Capture on a Variable-Speed HAWT," *ASME J. Solar Energy Engineering*, Vol. 126, No. 4: American Society of Mechanical Engineers, pp. 1092–1100.

Jonkman, J. M. and M. L. Buhl, 2005, *FAST User's Guide*, NREL/EL-500-38230, Golden, CO: National Renewable Energy Laboratory.

Jonkman, B. J., and M. L. Buhl, Jr., 2007, *TurbSim User's Guide*, NREL TP-500-41136, Golden, Colorado: National Renewable Energy Laboratory.

Kelley, N. D., B. J. Jonkman, G. N. Scott, and Y. L. Pichugina, 2007, *Comparing Pulsed Doppler LIDAR with SODAR and Direct Measurements for Wind Assessment*, NREL/CP-500-41792, Golden, Colorado: National Renewable Energy Laboratory.

Kraan, I., 1992, *Control Design for a Flexible Wind Turbine* (in Dutch), TUD-WBMR-A-613: Delft University of Technology, The Netherlands.

Krann, I, and P. M. Bongers, 1993, "Control of a Wind Turbine Using Several Linear Robust Controllers," *Proceedings, 32nd Conference on Decision and Control*, San Antonio, Texas.

Kwakernaak, H. and R. Sivan, 1972, *Linear Optimal Control Systems*, Wiley Interscience, (New York).

Liebst, B.S., 1985, "A Pitch Control System for the KaMeWa Wind Turbine," *Journal Dynamic Systems and Control*, 107, 47.

Mattson, S.E., 1984, *Modeling and Control of Large Horizontal Axis Wind Power Plants*, Ph.D. Thesis, Department of Automatic Control, Lund Institute of Technology, Lund Sweden.

MSC.ADAMS® (Automatic Dynamic Analysis of Mechanical Systems)

Stol, K., and M. Balas, 2003, "Periodic Disturbance Accommodating Control for Blade Load Mitigation in Wind Turbines," *ASME Journal Solar Energy Engineering*, Vol. 125, No. 4: American Society of Mechanical Engineers, pp. 379-385.

Stol, K. A. and G. S. Bir, 2003, *SymDyn User's Guide*, NREL/EL-500-33845, Golden, CO: National Renewable Energy Laboratory.

Stol, K., and L. Fingersh, 2004, *Wind Turbine Field Testing of State-Space Control Designs*, NREL/SR-500-35061, Golden, CO: National Renewable Energy Laboratory.

Wright, A. D., 2004, *Modern Control Design for Flexible Wind Turbines*, NREL TP-500-35816, Golden, CO: National Renewable Energy Laboratory.

Wright, A., K. Stol, and L. Fingersh, 2005, "Progress in Implementing and Testing State-Space Controls for the Controls Advanced Research Turbine," *Proceedings, 24th ASME Wind Energy Conference*: American Society of Mechanical Engineers, pp. 88-100.

Wright, A. D., L. J. Fingersh, and K. A. Stol, 2007, *Design and Testing Controls to Mitigate Tower Dynamic Loads in the Controls Advanced Research Turbine*, NREL/CP-500-40932, Golden, CO: National Renewable Energy Laboratory.

Wright, A. D., and L. J. Fingersh, 2008, *Advanced Control Design for Wind Turbines, Part I: Control Design, Implementation, and Initial Tests*, NREL/TP-500-42437, Golden, Colorado: National Renewable Energy Laboratory.

Prior References

Anderson, T. S., 1983, "Recent Advances in Variable-Speed Generator Technology for Large Wind Turbines," *Proceedings, Wind Energy Expo '83 and National Conference,* American Wind Energy Association, Washington, DC, pp. 27-33.

Anonymous, 1985, *Control Dynamics Testing of the WTS-4*, HSER-9836, Hamilton Standard, East Hartford, Connecticut.

Barton, R. S., T. J. Hosp, and G. P. Schanzenbach, 1995, "Control System Design for the Mod-5A 7.3-MW Wind Turbine Generator," *Collected Papers on Wind Turbine Technology,* D. A. Spera, ed., NASA-CR-195432, Cleveland, Ohio: NASA Lewis Research Center, pp. 157-164.

Bissell, W. A., 1995, "Use of Blade Pitch Control to Provide Power Train Damping for the Mod-2 2.5-MW Wind Turbine," *Collected Papers on Wind Turbine Technology,* D. A. Spera, ed., NASA-CR-195432, Cleveland, Ohio: NASA Lewis Research Center, pp. 175-184.

Bottrell, G. W., 1981, "Passive Cyclic Pitch Control for Horizontal Axis Wind Turbines," *Proceedings, Wind Turbine Dynamics Conference,* NASA-CP-2185, DOE CONF-810226, Cleveland, Ohio: NASA Lewis Research Center, pp. 271-275.

Brady, F. J., 1985, *Description and Test Results of a Variable Speed, Constant Frequency Generating System*, NASA-TM-87181, DOE/NASA/20320-67, Cleveland, Ohio: NASA Lewis Research Center.

Cao, H. V., W. H. Wentz, Jr., and M. H. Snyder, 1987, *Summary of Control Effectiveness of Vented Deflector-Ailerons*, WER-32A, Wichita State University, Wichita, Kansas.

Corrigan, R. D., C. B. F. Ensworth, Jr., and D. R. Miller, 1987, *Performance and Power Regulation Characteristics of Two Aileron-Controlled Rotors and a Pitchable Tip-Controlled Rotor on the Mod-0 Turbine*, NASA-TM-100136, DOE/NASA/20320-73, Cleveland, Ohio: NASA Lewis Research Center.

Currin, H., 1981, "North Wind 4 kW 'Passive' Control System Design," *Proceedings, Wind Turbine Dynamics Conference,* NASA-CP-2185, DOE CONF-810226, Cleveland, Ohio: NASA Lewis Research Center, pp. 265-270.

Emmitt, G. D., 1984, *NASA/MSFC Ground-Based Doppler Lidar Nocturnal Boundary Layer Experiment (NOBLEX)*, NASA-CR-3778, Huntsville, Alabama: NASA Marshall Space Flight Center.

Gnecco, A. J. and G. T. Whitehead, 1978, *Microprocessor Control of a Wind Turbine Generator*, NASA-TM-79021, DOE/NASA/1028-78/20, Cleveland, Ohio: NASA Lewis Research Center.

Hinrichsen, E. N., 1995, "Analysis Methods for Wind Turbine Control and Electrical System Dynamics," *Collected Papers on Wind Turbine Technology,* D. A. Spera, ed., NASA-CR-195432, Cleveland, Ohio: NASA Lewis Research Center, pp. 153-156.

Hinrichsen, E. N., 1983, *Control of Large Wind Turbine Generators Connected to Utility Networks*, DOE/NASA/0252-1, Cleveland, Ohio: NASA Lewis Research Center.

Hinrichsen, E. N. and P. J. Nolan, 1981, "Dynamics and Stability of Wind Turbine Generators," *Proceedings, Wind Turbine Dynamics Conference,* NASA-CP-2185, DOE CONF-810226, Cleveland, Ohio: NASA Lewis Research Center, pp. 315-323.

Hirschbein, M. S. and M. I. Young, 1980, *Stability of Large Horizontal-Axis Axisymmetric Wind Turbines*, NASA-TM-81623, Cleveland, Ohio: NASA Lewis Research Center.

Hoffman, J. A., R. Gluck, and S. Sridhar, 1995, "Design of a Real-Time Wind Turbine Simulator Using a Custom Parallel Architecture," *Collected Papers on Wind Turbine Technology,* D. A. Spera, ed., NASA-CR-195432, Cleveland, Ohio: NASA Lewis Research Center, pp. 157-164.

Holley, W. E. and R. W. Thresher, 1981, "Atmospheric Turbulence Parameters for Modeling Wind Turbine Dynamics," *Proceedings, Wind Turbine Dynamics Conference,* NASA-CP-2185, DOE CONF-810226, Cleveland, Ohio: NASA Lewis Research Center, pp. 101-112.

Malcolm, D. J., and A. C. Hansen, 2002, *WindPACT Turbine Rotor Design Study – June 2000 to June 2002*, NREL/SR-500-32495, Appendix E, Golden, Colorado: National Renewable Energy Laboratory.

Nyland, T. W. and A. G. Birchenough, 1982, *Microprocessor Control System for 200-kilowatt Mod-0A Wind Turbines*, NASA-TM-82711, DOE/NASA/20370-22, Cleveland, Ohio: NASA Lewis Research Center.

Perley, R., 1981, "Kaman 40-kW Wind Turbine Generator – Control System Dynamics," *Proceedings, Wind Turbine Dynamics Conference,* NASA-CP-2185, DOE CONF-810226, Cleveland, Ohio: NASA Lewis Research Center, pp. 3235-332.

Scott, G. W., V. F. Wilreker, and R. K. Shaltens, 1983, *Wind Turbine Generator Interaction with Diesel Generators on an Isolated Power System*, Paper, Summer Meeting, Institute of Electrical and Electronics Engineers: 5 pp.

Seidel, R. C. and A. G. Birchenough, 1981, *Variable Gain for a Wind Turbine Pitch Control*, NASA-TM-82751, DOE/NASA/20320-34, Cleveland, Ohio: NASA Lewis Research Center.

Appendix A

General References on Modern Wind Turbines

Ackermann, T., ed., 2005, *Wind Power in Power Systems*, Hoboken, NJ: Wiley.

Blaabjerg, F. and Z. Chen, 2006, *Power Electronics for Modern Wind Turbines* (Synthesis Lectures on Power Electronics), San Rafael, California: Morgan & Claypool Publishers.

Boyle, G., ed., 2007, *Renewable Electricity and the Grid: The Challenge of Variability*, Sterling, Va.: Earthscan.

Burton, Tony, et al., 2001, *Wind Energy Handbook*, New York: Wiley.

Casale, C., and M. Giuseppe, 1993, "Wind Energy - Present Situation and Future Prospects," *Proceedings, World Solar Summit Conference*, Paris: UNESCO, 58 p.

Dodge, D. M., and R. W. Thresher, 1989, "Wind Technology Today," *Advances in Solar Energy: An Annual Review of Research and Development*, Vol. 5, K. W. Böer, ed., New York: Plenum Press, pp. 306-359.

Eldridge, F. R., 1975, *Wind Machines,* NSF RA-N-75-051, MTR-6971, Washington, DC: U.S. Department of Energy.

Eggleston, D. M., and F. S. Stoddard, 1987, *Wind Turbine Engineering Design*, New York: Van Nostrand Reinhold Company.

Fox, B., 2007, *Wind Power Integration: Connection and System Operational Aspects*, London: Institution of Engineering and Technology.

Foreman, K. M., 1986, "The Fluid Mechanics of Wind Turbines," *Encyclopedia of Fluid Mechanics,* Chapter 27, Houston, Texas: Gulf Publishing Company, pp. 1113-1128.

Gasch, R. and J. Twele, eds., 2002, *Wind Power Plants: Fundamentals, Design, Construction, and Operation*, London: James & James.

Gipe, P., 1993, *Wind Power for Home & Business*, Post Mills, Vermont: Chelsea Green Publishing Company.

Gipe, P., 1990, *Wind Energy Comes of Age,* Tehachapi, California: Paul Gipe & Associates.

Gipe, P., 2004, *Wind Power: Renewable Energy for Home, Farm, and Business*, Junction, VT: Chelsea Green Publishing Company.

Golding, E. W., 1977, *The Generation of Electricity by Wind Power*, London: E. & F.N. Spon Ltd.

Hansen, M. O. L., 2008, *Aerodynamics of Wind Turbines* (2nd ed.), Sterling Va.: Earthscan.

Hau, E., 2006, *Wind Turbines: Fundamentals, Technologies, Application, Economics* (Horst von Renouard, trans.), New York: Springer.

Heier, S., 2006, *Grid Integration of Wind Energy Conversion Systems*, Hoboken, NJ: Wiley.

Manwell, J. F., J. G. McGowan, and A. L. Rogers, 2002, *Wind Energy Explained: Theory, Design, and Application*, New York: Wiley.

Mathew, S., 2006, *Wind Energy: Fundamentals, Resource Analysis and Economics*, Berlin: Springer.

Mertens, S., 2006, *Wind Energy in the Built Environment: Concentrator Effects of Buildings*, Essex, UK: Multi-science Publishing.

National Research Council (Committee on Environmental Impacts of Wind Energy Projects), 2007, *Environmental Impacts of Wind Energy Projects*, Washington, D.C.: National Academies Press.

Putnam, P. C., 1948, *Power from the Wind*, New York: Van Nostrand Reinhold Company.

Scheffler, R. L., and M. C. Wehrey, 1985, *Wind Power, Handbook of Energy Systems Engineering,* L. C. Wilbur, ed., New York: John Wiley & Sons, pp. 1223-1251.

Spera, D. A., 1995, *Bibliography of NASA-Related Publications on Wind Turbine Technology 1973 – 1995*, DOE/NASA/5776-3, NASA CR-195462, Cleveland, Ohio: National Aeronautics and Space Administration.

Touryan, K. J., J. H. Strickland, and D. E. Berg, 1987, "Electric Power from Vertical-Axis Wind Turbines," *Journal of Propulsion*, Vol. 3, No. 6, pp. 481-493.

Vosburgh, P. N., 1983, *Commercial Applications of Wind Power,* New York: Van Nostrand Reinhold Company.

Wizelius, T., 2007, *Developing Wind Power Projects: Theory and Practice* (R. Washington, trans.), Sterling, Va.: Earthscan.

Additional books on wind turbine technology available at www.amazon.com

Appendix B

Bibliography of Wind Energy Conference Proceedings[1]

Compiled by

Tami Sandberg
National Renewable Energy Laboratory
Golden, Colorado

National Science Foundation and U.S. Department of Energy

Proceedings, Workshop on Wind Energy Conversion Systems, 1973, Savino, J. M., ed., NSF/RA/W-73-006, Washington, DC: National Science Foundation; also Cleveland, OH: NASA Lewis Research Center.

Proceedings, Second Workshop on Wind Energy Conversion Systems, 1975, F. R. Eldridge, ed., NSF-RA-75-050, Washington, DC: National Science Foundation.

Proceedings, Third Biennial Conference and Workshop on Wind Energy Conversion *Systems*, Vols. I and II, 1977, T. R. Kornreich, ed., DOE Conference Publication CONF-770921, Washington, DC: JBF Scientific Corporation.

Proceedings of the Fourth Biennial Conference and Workshop on Wind Energy Conversion Systems, 1979, Richard J. Kottler, Jr., ed., CONF-791097, Washington D.C.: Department of Energy.

Third Wind Energy Workshop: Proceedings of the Third Biennial Conference and Workshop on Wind Energy Conversion Systems, 2 volumes, 1978, CONF-770921/1, Washington, D.C.: Department of Energy.

Proceedings, European Wind Energy Association Conference, Vols. I & II, 1986, Palz, W., and E. Sesto, eds., Rome: A. Raguzzi Bookshop for Scientific Publications.

---

[1] While extensive, this listing of primarily U.S.-based wind energy conferences is not necessarily comprehensive. *Ed.*

National Aeronautics and Space Administration

Proceedings, Workshop on Wind Turbine Structural Dynamics, 1977, D. R. Miller, ed., NASA Conference Publication 2034, DOE Publication CONF-771148, Cleveland, Ohio: NASA Lewis Research Center.

Proceedings, Workshop on Large Wind Turbine Design Characteristics and R&D Requirements, 1979, S. Lieblein, ed., NASA CP 2106, DOE, Publication CONF-7904111, Cleveland, Ohio: NASA Lewis Research Center.

Proceedings, Workshop on Wind Turbine Dynamics, 1981, R. W. Thresher, ed., NASA CP-2034, DOE CONF-771148, Cleveland, Ohio: NASA Lewis Research Center.

Proceedings, Workshop on Large Horizontal-Axis Wind Turbines, 1981, R. W. Thresher, ed., NASA CP 2230, DOE Publication CONF-810752, Cleveland, Ohio: NASA Lewis Research Center.

Rocky Flats and Solar Energy Research Institute

Proceedings, Small Wind Turbines Systems: A Workshop on R&D Requirements and Utility Interface/Institutional Issues, 2 volumes, 1979, D. M. Dodge and J. V. Stafford, eds., RFP/3014/3533/79-8, Golden, Colorado: Rockwell International, Energy Systems Group, Rocky Flats Plant, Wind Systems Program.

Wind Energy Innovative Systems Conference Proceedings, 1979, I. E. Vas, ed., SERI/TP-245-184, Golden, Colorado: Solar Energy Research Institute.

Proceedings: Panel on Information Dissemination for Wind Energy, August 2nd and 3rd, 1979, P. Weis, ed., SERI/TP-732-343, Golden, Colorado: Solar Energy Research Institute.

SERI Second Wind Energy Innovative Systems Conference, 1980, SERI/CP-635-938, Golden, Colorado: Solar Energy Research Institute.

Proceedings, Second Wind Energy Innovative Systems Conference, Vols. I and II, 1980, SERI/CP-635-1061, Golden, Colorado: National Renewable Energy Laboratory.

Proceedings, Small Wind Turbine Systems, 1981, SERI/CP-635-1212, Golden, Colorado: Solar Energy Research Institute.

Wind Turbine Dynamics: Proceedings of a Workshop, 1981, R. W. Thresher, ed., SERI/CP-635-1238, Golden, Colorado: Solar Energy Research Institute.

Fifth Biennial Wind Energy Conference & Workshop (WWV), 3 volumes, 1981, I. E. Vas, ed., SERI/CP-635-1340, Golden, Colorado: Solar Energy Research Institute.

Rural Electric Wind Energy Workshop, 1982, SERI/CP-254-1689, Golden, Colorado: Solar Energy Research Institute.

Proceedings, Sixth Biennial Wind Energy Conference and Workshop, 1983, B. H. Glenn, ed., Boulder, CO: American Solar Energy Society, Inc.

Papers Presented at Windpower '87, 1987, SERI/SP-217-3296, Golden, Colorado: Solar Energy Research Institute.

National Renewable Energy Laboratory
See http://nrelpubs.nrel.gov/Webtop/ws/nich/www/public/SearchForm
or http://nrelpubs.nrel.gov/wind/publications.html

IMTS: 14th Annual International Meeting of Wind Turbine Test Stations, 1994, NREL/CP-442-6877, Golden, Colorado: National Renewable Energy Laboratory.

Proceedings of National Avian-Wind Power Planning Meeting, 1995, NREL/SP-441-7814, Golden, Colorado: National Renewable Energy Laboratory.

Proceedings of Village Power '98: Scaling Up Electricity Access for Sustainable Rural Development, 1999, NREL/CP-500-26264, Golden, Colorado: National Renewable Energy Laboratory.

Report on a Workshop Concerning Code Validation, 1996, NREL/CP-440-21976, Golden, Colorado: National Renewable Energy Laboratory.

Proceedings of Village Power '97, 1997, NREL/CP-440-23409, Golden, Colorado: National Renewable Energy Laboratory.

Proceedings of National Avian-Wind Power Planning Meeting II, 1998, NREL/CP-500-23821, Golden, Colorado: National Renewable Energy Laboratory.

Village Power 2000: Empowering People and Transforming Markets [CD-ROM], 2000, NREL/EL-500-30718, Golden, Colorado: National Renewable Energy Laboratory.

Pacific Northwest Laboratory

Proceedings, Conference and Workshop on Wind Energy Characteristics and Wind Energy Siting 1979, 1979, PNL-3214, Richland, Washington: Pacific Northwest Laboratory.

Sandia National Laboratories
See http://www.sandia.gov/wind

Proceedings, Vertical Axis Wind Turbine Design Technology Seminar for Industry, 1980, S. Johnston, ed., SAND80-0984, Albuquerque, New Mexico: Sandia National Laboratories.

Proceedings, 1983 Wind and Solar Energy Technology Conference, 1983, Albuquerque, New Mexico: Sandia National Laboratories.

American Society of Mechanical Engineers and American Institute of Aeronautics and Astronautics

See http://asme.org/publications

International Wind Energy Symposium: presented at 5th Annual Energy-Sources Technology Conference and Exhibition; New Orleans, Louisiana March 7-10, 1982, H. Finlayson, ed., New York: American Society of Mechanical Engineers.

Second ASME Wind Energy Symposium: presented at 6th Annual Energy-Sources Technology Conference and Exhibition; Houston, Texas, January 30-February 3, 1983, New York: American Society of Mechanical Engineers.

Third ASME Wind Energy Symposium: presented at 7th Annual Energy-Sources Technology Conference and Exhibition; New Orleans, Louisiana, February 12-16, 1984, New York: American Society of Mechanical Engineers.

Fourth ASME Wind Energy Symposium: presented at the Eighth Annual Energy-Sources Technology Conference and Exhibition; Dallas, Texas, February 17-21, 1985, A. H. P. Swift, ed., New York: American Society of Mechanical Engineers.

Fifth ASME Wind Energy Symposium: presented at the Ninth Annual Energy-Sources Technology Conference and Exhibition; New Orleans, Louisiana, February 23-27, 1986, SED-Vol. 2, A.H.P. Swift, ed., New York: American Society of Mechanical Engineers.

Sixth ASME Wind Energy Symposium: presented at the Tenth Annual Energy-Sources Technology Conference and Exhibition; Dallas, Texas, February 15-18, 1987, SED-Vol. 3, R. W. Thresher, ed., New York: American Society of Mechanical Engineers.

Seventh ASME Wind Energy Symposium: presented at the Eleventh Annual Energy-Sources Technology Conference and Exhibition; New Orleans, Louisiana, January 10-13, 1988, SED-Vol. 5, R. W. Thresher, ed., New York: American Society of Mechanical Engineers.

Eighth ASME Wind Energy Symposium: presented at the Twelfth Annual Energy-Sources Technology Conference and Exhibition; Houston, Texas, January 22-25, 1989, SED-Vol. 7, D. E. Berg and P. C. Klimas, eds., New York: American Society of Mechanical Engineers.

Ninth ASME Wind Energy Symposium: presented at the Thirteenth Annual Energy-Sources Technology Conference and Exhibition; New Orleans, Louisiana, January 14-18, 1990, SED-Vol. 9, D.E. Berg, ed., New York: American Society of Mechanical Engineers.

Tenth ASME Wind Energy Symposium: presented at the Fourteenth Annual Energy-Sources Technology Conference and Exhibition; Houston, Texas, January 20-23, 1991, SED-Vol. 11, D. E. Berg and P. S. Veers, eds., New York: American Society of Mechanical Engineers.

Eleventh ASME Wind Energy Symposium: presented at the Energy-Sources Technology Conference and Exhibition; Houston, Texas, January 26-30, 1992, SED-Vol. 12, P. S. Veers and S. M. Hock, eds., New York: American Society of Mechanical Engineers.

Wind Energy-1993: presented at the 16ᵗʰ Annual Energy-Sources Technology Conference and Exhibition; Houston, Texas, January 31-February 4, 1993, SED-Vol. 14, S. M. Hock, ed., New York: American Society of Mechanical Engineers.

Wind Energy-1994: presented at the Energy-Sources Technology Conference; New Orleans, Louisiana, January 23-26, 1994, SED-Vol. 15, Walter D. Musial, S. M. Hock, and D. E. Berg, eds., New York: American Society of Mechanical Engineers.

Wind Energy-1995: presented at the Energy and Environmental Expo '95-the Energy-Sources Technology Conference and Exhibition; Houston, Texas, January 29-February 1, 1995, SED-Vol. 16, W. D. Musial, S. M. Hock, and D. E. Berg, eds., New York: American Society of Mechanical Engineers.

Wind Energy: Energy Week Conference & Exhibition, Incorporating ETCE, Book VIII, Volume I; G. R. Brown Convention Center-Houston, Texas, January 29-February 2, 1996, Texas: Penn Well Conferences & Exhibitions.

A Collection of the 1997 ASME Wind Energy Symposium Technical Papers: presented at the 35ᵗʰ AIAA Aerospace Sciences Meeting and Exhibit, Reno NV, January 6-9, 1997, W. Musial, ed., Reston, VA: American Institute of Aeronautics and Astronautics.

A Collection of the 1998 ASME Wind Energy Symposium Technical Papers: presented at the 36ᵗʰ AIAA Aerospace Sciences Meeting and Exhibit, Reno, NV, January 12-15, 1998, W. Musial and D. Berg, eds., New York: American Institute of Aeronautics and Astronautics and American Society of Mechanical Engineers.

A Collection of the 1999 ASME Wind Energy Symposium Technical Papers: presented at the 37ᵗʰ AIAA Aerospace Sciences Meeting and Exhibit, Reno, NV, 11-14 January, 1999, T. Ashwill, ed., New York: American Institute of Aeronautics and Astronautics and American Society of Mechanical Engineers.

A Collection of the 2000 ASME Wind Energy Symposium Technical Papers: presented at the 38ᵗʰ AIAA Aerospace Sciences Meeting and Exhibit, Reno, NV, 10-13 January, 2000, New York: American Institute of Aeronautics and Astronautics and American Society of Mechanical Engineers.

A Collection of the 2001 ASME Wind Energy Symposium Technical Papers: presented at the 39ᵗʰ AIAA Aerospace Sciences Meeting and Exhibit, Reno, NV, 11-14 January, 2001, New York: American Institute of Aeronautics and Astronautics and American Society of Mechanical Engineers.

A Collection of the 2002 ASME Wind Energy Symposium Technical Papers: presented at the 40ᵗʰ AIAA Aerospace Sciences Meeting and Exhibit, Reno, NV, 14-17 January, 2002, Reston, Va.: American Institute of Aeronautics and Astronautics and New York: American Society of Mechanical Engineers.

A Collection of the 2003 ASME Wind Energy Symposium Technical Papers: presented at the 41ˢᵗ AIAA Aerospace Sciences Meeting and Exhibit, Reno, NV, 6-9 January, 2003, Reston, Va.: American Institute of Aeronautics and Astronautics and New York: American Society of Mechanical Engineers.

A Collection of the 2004 ASME Wind Energy Symposium Technical Papers: presented at the 42nd AIAA Aerospace Sciences Meeting and Exhibit, Reno, NV, 5-8 January, Reston, Va.: American Institute of Aeronautics and Astronautics and New York: American Society of Mechanical Engineers.

A Collection of the 2005 ASME Wind Energy Symposium Technical Papers: presented at the 43rd Aerospace Sciences Meeting and Exhibit, Reno, NV, 10-13 January, 2005, Reston, Va.: American Institute of Aeronautics and Astronautics and New York: American Society of Mechanical Engineers.

44th AIAA Aerospace Sciences Meeting and Exhibit [CD-ROM], Reno, Nevada, January 9-12, 2006, Reston, Va.: American Institute of Aeronautics and Astronautics.

45th AIAA Aerospace Sciences Meeting and Exhibit [CD-ROM], Reno, Nevada, January 8-11, 2007, Reston, Va.: American Institute of Aeronautics and Astronautics.

46th AIAA Aerospace Sciences Meeting and Exhibit [DVD-ROM], Reno, Nevada, January 7-10, 2008, Reston, Va.: American Institute of Aeronautics and Astronautics.

American Wind Energy Association
See http://www.awea.org/publications

Proceedings of the National Conference, Spring 1978, V. Nelson, ed., Washington, D.C.: American Wind Energy Association.

Proceedings of the National Conference Fall 1978, V. Nelson, ed., Washington, D.C.: American Wind Energy Association.

Proceedings of the National Conference, Spring 1979, V. Nelson and M. Mooring, eds., Washington, D.C.: American Wind Energy Association.

Proceedings, American Wind Energy Association National Conference, 1979, V. Nelson, ed., Washington, DC: American Wind Energy Association.

Proceedings of the National Conference, Summer 1980, V. Nelson and M. Mooring, eds., Washington, D.C.: American Wind Energy Association.

Proceedings, Wind Energy Expo '82 and National Conference American Wind Energy Association, 1982, V. Nelson, ed., Washington, D.C.: American Wind Energy Association.

Proceedings, Wind Energy Expo '83 and National Conference American Wind Energy Association, 1983, Alexandria, Va.: American Wind Energy Association.

Proceedings, Wind Energy Expo '84 and National Conference American Wind Energy Association, 1984, Alexandria, Va.: American Wind Energy Association.

Proceedings, Windpower '85 Conference, 1985, SERI/CP-217-2902, Washington, DC: American Wind Energy Association.

Proceedings, Wind Energy Expo '86 and National Conference American Wind Energy Association, 1986, Alexandria, Va.: American Wind Energy Association.

Windpower '85, 1985, SERI/CP-217-2902, Alexandria, Va.: American Wind Energy Association.

Windpower '87, 1987, SERI/CP-217-3315, Alexandria, Va.: American Wind Energy Association.

Windpower '88, 1988, Alexandria, Va.: American Wind Energy Association.

Windpower '89, 1989, SERI/TP-257-3628, Arlington, Va.: American Wind Energy Association.

Windpower '90, 1990, Washington, D.C.: American Wind Energy Association.

Windpower '91 Proceedings, 1991, Washington, D.C.: American Wind Energy Association.

Windpower '92 Proceedings, 1992, Washington, D.C.: American Wind Energy Association.

Windpower '93, 1993, Washington, D.C.: American Wind Energy Association.

Windpower '94, Proceedings American Wind Energy Association, 1994, Washington, D.C.: American Wind Energy Association.

Windpower '95, Proceedings American Wind Energy Association, 1995, Washington, D.C.: American Wind Energy Association.

Windpower '96, Proceedings American Wind Energy Association, 1996, Washington, D.C.: American Wind Energy Association.

Windpower '97, Proceedings American Wind Energy Association, 1997, Washington, D.C.: American Wind Energy Association.

Windpower '98, Proceedings American Wind Energy Association, 1998, Washington, D.C.: American Wind Energy Association.

Windpower '99, Conference Proceedings [CD-ROM], 1999, Washington, D.C.: American Wind Energy Association.

Windpower 2000, Conference Proceedings [CD-ROM], 2000, Washington, D.C.: American Wind Energy Association.

Windpower 2001, Conference Proceedings [CD-ROM], 2001, Washington, D.C.: American Wind Energy Association.

Windpower 2002, Conference Proceedings [CD-ROM], 2002, Washington, D.C.: American Wind Energy Association.

Windpower 2003, Conference Proceedings [CD-ROM], 2003, Washington, D.C.: American Wind Energy Association.

Global Windpower, 2004 Conference Proceedings [CD-ROM], 2004, Washington, D.C.: American Wind Energy Association.

Windpower 2005, Conference and Exhibition [CD-ROM], 2005, Washington, D.C.: American Wind Energy Association.

Windpower 2006, Conference & Exhibition [CD-ROM], 2006, Washington, D.C.: American Wind Energy Association.

Windpower 2007, Conference & Exhibition [CD-ROM], 2007, Washington, D.C.: American Wind Energy Association.

Windpower 2008, Conference & Exhibition [CD-ROM], 2008, Washington, D.C.: American Wind Energy Association.

British Wind Energy Association

Proceedings, First British Wind Energy Association Wind Energy Workshop, 1979, London: Multi-Science Publishing Company.

Proceedings of the First BWEA Wind Energy Workshop, 1979, London: Multi-Science Publishing Company Ltd. (Edited version of above citation)

Proceedings of the Second BWEA Wind Energy Workshop, 1980, London: Multi-Science Publishing Company Ltd.

Proceedings of the Third BWEA Wind Energy Conference, 1981, P.J. Musgrove, ed., Cranfield, U.K.: BHRA Fluid Engineering.

Proceedings of the Fourth BWEA Wind Energy Conference, 1982, P.J. Musgrove, ed., Cranfield, U.K.: BHRA Fluid Engineering.

Wind Energy Conversion: Proceedings of the 1983 Fifth BWEA Wind Energy Conference, 1983, Peter Musgrove, ed., London: Cambridge University Press.

Wind Energy Conversion: Proceedings of the 1984 Sixth BWEA Wind Energy Conference, 1984, Peter Musgrove, ed., London: Cambridge University Press.

Wind Energy Conversion: Proceedings of the 1985 Seventh BWEA Wind Energy Conference, 1985, A.D. Garrad, ed., London: Mechanical Engineering Publications Ltd.

Wind Energy Conversion: Proceedings of the 1986 Eighth BWEA Wind Energy Conference, 1986, M.B. Anderson and S.J.R. Powles, eds., London: Mechanical Engineering Publications Ltd.

Wind Energy Conversion: Proceedings of the 1987 Ninth BWEA Wind Energy Conference, 1987, J.M. Galt, ed., London: Mechanical Engineering Publications Ltd.

Wind Energy Conversion: Proceedings of the 1988 Tenth BWEA Wind Energy Conference, 1988, D.J. Milborrow, ed., London: Mechanical Engineering Publications Ltd.

Wind Energy Conversion: Proceedings of the 1990 Twelfth BWEA Wind Energy Conference, 1990, T.D. Davies, J.A. Halliday, and J.P. Palutikof, eds., London: Mechanical Engineering Publications Ltd.

Wind Energy Conversion: Proceedings of the 1991 Thirteenth BWEA Wind Energy Conference, 1991, D.C. Quarton and V.C. Fenton, eds., London: Mechanical Engineering Publications Ltd.

Wind Energy Conversion: Proceedings of the 1992 Fourteenth BWEA Wind Energy Conference, 1992, B.R. Clayton, ed., London: Mechanical Engineering Publications Ltd.

Wind Energy Conversion: Proceedings of the 1993 Fifteenth BWEA Wind Energy Conference, 1993, K.F. Pitcher, ed., London: Mechanical Engineering Publications Ltd.

Wind Energy Conversion: Proceedings of the 1994 Sixteenth BWEA Wind Energy Conference, 1994, G. Elliot, ed., London: Mechanical Engineering Publications Ltd.

Wind Energy Conversion: Proceedings of the 1995 Seventeenth BWEA Wind Energy Conference, 1995, J. Halliday, ed., London: Mechanical Engineering Publications Ltd.

Wind Energy Conversion: Proceedings of the 1996 Eighteenth BWEA Wind Energy Conference, 1996, Mike Anderson, ed., London: Mechanical Engineering Publications Ltd.

Wind Energy Conversion: From Theory to Practice, Proceedings of the 1997 Nineteenth BWEA Wind Energy Conference, 1997, Ray Hunter, ed., London: Mechanical Engineering Publications Ltd.

Wind Energy: Switch to Wind Power, Proceedings of the 1998 Twentieth BWEA Wind Energy Conference, 1998, Simon Powles, ed., London: Professional Engineering Publishing.

Wind Energy: Wind Power Comes of Age, Proceedings of the 1999 Twenty-First BWEA Wind Energy Conference, 1999, Peter Hinson, ed., London: Professional Engineering Publishing.

Wind Energy: Building the 10%, Proceedings of the 2000 Twenty-Second BWEA Wind Energy Conference, 2000, David Still, ed., London: Professional Engineering Publishing.

Turning Things Around: Presentation Given BWEA23 [CD-ROM], 2001, London: British Wind Energy Association.

Deep Green Power: Presentations Given at BWEA 24 [CD-ROM], 2002, London: British Wind Energy Association.

UK Offshore Wind 2002: Presentations Given at BWEA's Special Topic Conference [CD-ROM], 2002, London: British Wind Energy Association.

UK Offshore Wind 2003: Presentations Given at BWEA's Special Topic Conference [CD-ROM], 2003, London: British Wind Energy Association.

Wind Connections: Presentations Given at BWEA25 [CD-ROM], 2003, London: British Wind Energy Association.

BWEA26: Build and Deliver, Conference Proceedings [CD-ROM], 2004, London: British Wind Energy Association.

UK Offshore Wind 2004: Presentations Given at BWEA's Special Topic Conference [CD-ROM], 2004, London: British Wind Energy Association.

BWEA27: Embrace the Revolution, Conference Proceedings [CD-ROM], 2005, London: British Wind Energy Association.

UK Offshore Wind 2005: BWEA's Fourth Dedicated Offshore Wind Conference [CD-ROM], 2005, London: British Wind Energy Association.

BWEA28: Securing Our Future, 2006, London: British Wind Energy Association. Available online: http://bwea.com/28/proceedings/index.html.

UK Offshore Wind 2006, 2006, London: British Wind Energy Association. Available online: http://www.bwea.com/offshore/conference2006.html.

BWEA29: Conference Proceedings, 2007, London: British Wind Energy Association. Available online: http://www.bwea.com/29/proceedings.html.

BWES30: BWEA's 30th Anniversary Conference & Exhibition, 2008, London: British Wind Energy Association. Available online: http://www.bwea30.com/conference.html.

Other proceedings and publications available on the British Wind Energy Association website: http://www.bwea.com/index.html.

Commission of the European Communities and European Wind Energy Association

European Wind Energy Conference: Proceedings of an International Conference Held at Hamburg, F.R. Germany, 1984, W. Palz, ed., Bedford, England: H.S. Stephens and Associates.

EWEC '86: European Wind Energy Association, Conference and Exhibition, 2 volumes, 1986, W. Palz and E. Sesto, eds., Rome: A. Raguzzi, Bookshop for Scientific Publications, for the Italian Section of the International Solar Energy Society.

European Community Wind Energy Conference: Proceedings of an International Conference Held at Herning Congress Centre, Denmark, 1988, W. Palz, ed., Bedford, England: H.S. Stephens and Associates.

EWEC '89: European Wind Energy Conference and Exhibition, 2 volumes, 1989, London: Peter Peregrins Ltd.

European Community Wind Energy Conference: Proceedings of an International Conference Held at Madrid Spain, 1990, W. Palz, ed., Bedford, England: H.S. Stephens and Associates.

Wind Energy Technology and Implementation: Proceedings of Amsterdam EWEC '91, 2 volumes, 1991, F.J.L. Van Hulle, P.T. Smulders, and J.B. Dragt, eds., London: Elsevier.

European Community Wind Energy Conference: Proceedings of the International Conference Held at Lübeck-Travemunde, Germany, 1993, A.D. Garrad, W. Palz, and S. Scheller, eds., Felmersham, Bedford, UK: H.S. Stephens and Associates.

EWEC '94: 5th European Wind Energy Association Conference and Exhibition, 3 volumes, 1994, J.L. Tsipouridis, ed.

European Union Wind Energy Conference: Proceedings of an International Conference Held at Götebord, Sweden, 1996, A. Zervos, H. Ehmann, and P. Helm, eds., Felmersham, Bedford, UK: H.S. Stephens and Associates.

EWEC '97: European Wind Energy Conference, Proceedings of the International Conference Held at Dublin Castle, Ireland, 1997, Rick Watson, ed., Slane, County Meath, Ireland: Irish Wind Energy Association.

Wind Energy for the Next Millennium: Proceedings of the European Wind Energy Conference, Nice, France, 1999, E.L. Petersen et al., eds., London: James & James.

Wind Energy for the New Millennium: Proceedings of the European Wind Energy Conference, Copenhagen, Denmark, 2001, Peter Helm and Arthouros Zervos, eds., Munich, Germany: WIP-Renewable Energies and Florence, Italy: ETA.

2002 Global Windpower Conference & Exhibition, [CD-ROM], 2002, Brussels, Belgium: European Wind Energy Association.

2003 European Wind Energy Conference & Exhibition [CD-ROM], 2003, Brussels, Belgium: European Wind Energy Association.

2004 European Wind Energy Conference & Exhibition [CD-ROM], 2004, Brussels, Belgium: European Wind Energy Association.

2006 European Wind Energy Conference & Exhibition, 2006, Brussels, Belgium: European Wind Energy Association. Available online: http://www.ewec2006proceedings.info/.

2007 European Wind Energy Conference & Exhibition, 2007, Brussels, Belgium: European Wind Energy Association. Available online: http://www. ewec2007proceedings.info/index.php.

Other proceedings and publications available on the European Wind Energy Association website: http://www.ewea.org/index.php?id=40.

Other European Conferences

Proceedings, Workshop on Advanced Wind Energy Systems, Vols. 1 and 2, 1974 (published 1976), O. Ljungström, ed., Stockholm: National Swedish Board for Technical Development and Swedish State Power Board.

DEWI-German Wind Energy Institute, publishers of various conferences: http://www.dewi. de/dewi/index.php.

International Workshop on Large-Scale Integration of Wind Power into Power Systems as well as on Transmission Networks for Offshore Wind Farms. Annual Conference Proceedings can be purchased online: http://www.windintegrationworkshop.org/.

Offshore Wind and Other Marine Renewable Energy in Mediterranean and European Sites. Conference information available online: http://www.owemes.org/.

Wind Power Production in Cold Climate: BOREAS Proceedings, various years, Finnish Meteorological Institute: http://www.fmi.fi/research_meteorology/ meteorology_13.html.

Appendix C

Wind Turbines in Commercial Service in the U.S. in the Mid-1980s

To illustrate the wide variety of small-, medium-, and large-scale turbines that have been operated commercially, sketches of most of the wind turbines installed in major U.S. wind power stations in the mid-1980s are shown in Figures C-1 to C-13, together with pertinent technical data. The number of units installed of a given configuration is the total for all regions during this time period.

References

Johnson, W. R., and W. R. Young, 1985, "Status of the DOI SVU Wind Turbine Project," *Proceedings, Wind Power '85 Conference*, SERI/CP-217-2902, Washington, DC: American Wind Energy Association, pp. 604-609.

PG&E and U. S. Windpower, Inc., 1986 and 1988, *Altamont Pass Windfarms*, unpublished brochure, L. Pettibone, designer, San Francisco: Pacific Gas and Electric Company.

SCE and U. S. Windpower, Inc., 1986a, *San Gorgonio Wind Parks*, unpublished brochure, L. Pettibone, designer, Rosemead, California: Southern California Edison Company.

SCE and U. S. Windpower, Inc., 1986b, *Tehachapi Wind Parks*, unpublished brochure, L. Pettibone, designer, Rosemead, California: Southern California Edison Company.

Suehiro, D., and M. Miller, 1988, "HEI's Wind Farm Control Experience," *Proceedings, Windpower '88 Conference*, Washington, DC: American Wind Energy Association, pp. 260-269.

Figure C-1.

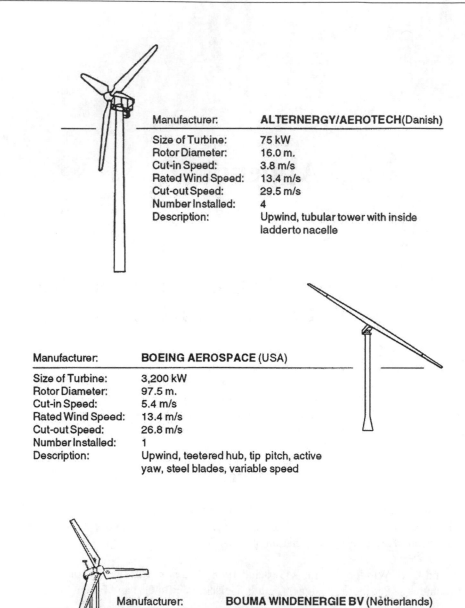

Manufacturer: **ALTERNERGY/AEROTECH**(Danish)

Size of Turbine:	75 kW
Rotor Diameter:	16.0 m.
Cut-in Speed:	3.8 m/s
Rated Wind Speed:	13.4 m/s
Cut-out Speed:	29.5 m/s
Number Installed:	4
Description:	Upwind, tubular tower with inside ladder to nacelle

Manufacturer: **BOEING AEROSPACE** (USA)

Size of Turbine:	3,200 kW
Rotor Diameter:	97.5 m.
Cut-in Speed:	5.4 m/s
Rated Wind Speed:	13.4 m/s
Cut-out Speed:	26.8 m/s
Number Installed:	1
Description:	Upwind, teetered hub, tip pitch, active yaw, steel blades, variable speed

Manufacturer: **BOUMA WINDENERGIE BV** (Netherlands)

Size of Turbine:	100 kW	200 kW
Rotor Diameter:	16.0 m.	20.4 m.
Cut-in Speed:	4.0 m/s	4.0 m/s
Rated Wind Speed:	24.1 m/s	28.2 m/s
Cut-out Speed:	30.0 m/s	30.0 m/s
Number Installed:	6	35
Description:	Upwind, fixed pitch, active yaw, fiberglass blades	

Figure C-2.

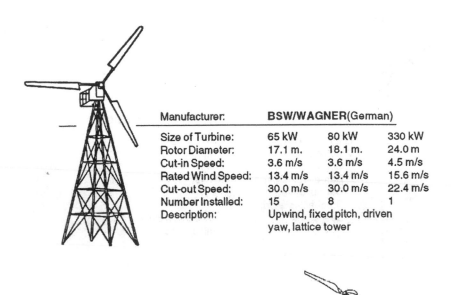

Manufacturer:	**BSW/WAGNER**(German)		
Size of Turbine:	65 kW	80 kW	330 kW
Rotor Diameter:	17.1 m.	18.1 m.	24.0 m
Cut-in Speed:	3.6 m/s	3.6 m/s	4.5 m/s
Rated Wind Speed:	13.4 m/s	13.4 m/s	15.6 m/s
Cut-out Speed:	30.0 m/s	30.0 m/s	22.4 m/s
Number Installed:	15	8	1
Description:	Upwind, fixed pitch, driven yaw, lattice tower		

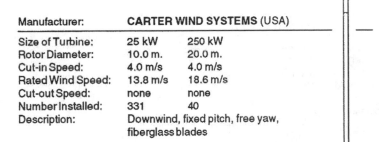

Manufacturer:	**CARTER WIND SYSTEMS** (USA)	
Size of Turbine:	25 kW	250 kW
Rotor Diameter:	10.0 m.	20.0 m.
Cut-in Speed:	4.0 m/s	4.0 m/s
Rated Wind Speed:	13.8 m/s	18.6 m/s
Cut-out Speed:	none	none
Number Installed:	331	40
Description:	Downwind, fixed pitch, free yaw, fiberglass blades	

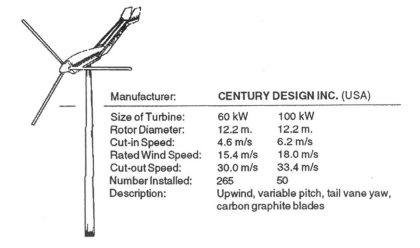

Manufacturer:	**CENTURY DESIGN INC.** (USA)	
Size of Turbine:	60 kW	100 kW
Rotor Diameter:	12.2 m.	12.2 m.
Cut-in Speed:	4.6 m/s	6.2 m/s
Rated Wind Speed:	15.4 m/s	18.0 m/s
Cut-out Speed:	30.0 m/s	33.4 m/s
Number Installed:	265	50
Description:	Upwind, variable pitch, tail vane yaw, carbon graphite blades	

Figure C-3.

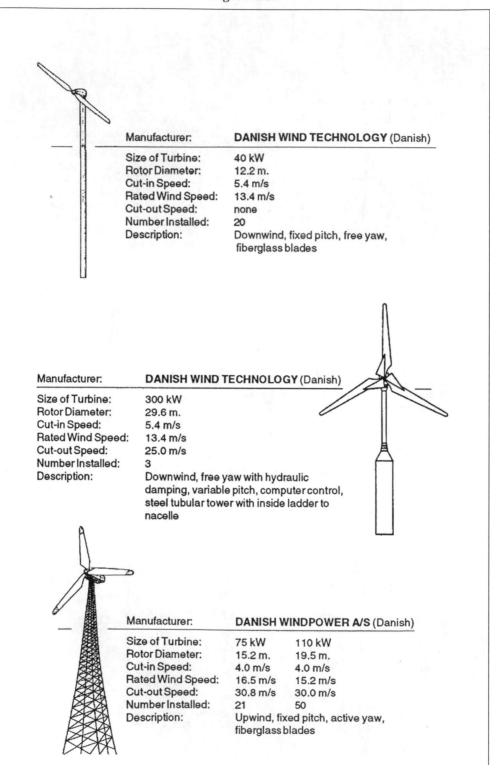

Manufacturer: **DANISH WIND TECHNOLOGY** (Danish)

Size of Turbine: 40 kW
Rotor Diameter: 12.2 m.
Cut-in Speed: 5.4 m/s
Rated Wind Speed: 13.4 m/s
Cut-out Speed: none
Number Installed: 20
Description: Downwind, fixed pitch, free yaw,
                       fiberglass blades

Manufacturer: **DANISH WIND TECHNOLOGY** (Danish)

Size of Turbine: 300 kW
Rotor Diameter: 29.6 m.
Cut-in Speed: 5.4 m/s
Rated Wind Speed: 13.4 m/s
Cut-out Speed: 25.0 m/s
Number Installed: 3
Description: Downwind, free yaw with hydraulic
                       damping, variable pitch, computer control,
                       steel tubular tower with inside ladder to
                       nacelle

Manufacturer: **DANISH WINDPOWER A/S** (Danish)

Size of Turbine: 75 kW   110 kW
Rotor Diameter: 15.2 m.   19.5 m.
Cut-in Speed: 4.0 m/s   4.0 m/s
Rated Wind Speed: 16.5 m/s   15.2 m/s
Cut-out Speed: 30.8 m/s   30.0 m/s
Number Installed: 21   50
Description: Upwind, fixed pitch, active yaw,
                       fiberglass blades

Figure C-4.

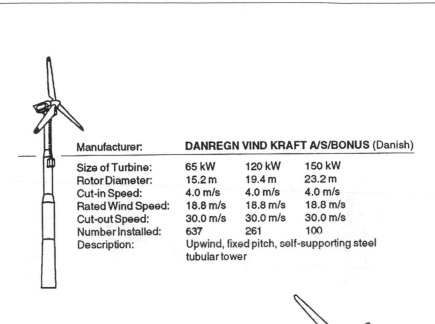

Manufacturer:	**DANREGN VIND KRAFT A/S/BONUS** (Danish)		
Size of Turbine:	65 kW	120 kW	150 kW
Rotor Diameter:	15.2 m	19.4 m	23.2 m
Cut-in Speed:	4.0 m/s	4.0 m/s	4.0 m/s
Rated Wind Speed:	18.8 m/s	18.8 m/s	18.8 m/s
Cut-out Speed:	30.0 m/s	30.0 m/s	30.0 m/s
Number Installed:	637	261	100
Description:	Upwind, fixed pitch, self-supporting steel tubular tower		

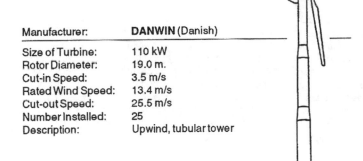

Manufacturer:	**DANWIN** (Danish)
Size of Turbine:	110 kW
Rotor Diameter:	19.0 m.
Cut-in Speed:	3.5 m/s
Rated Wind Speed:	13.4 m/s
Cut-out Speed:	25.5 m/s
Number Installed:	25
Description:	Upwind, tubular tower

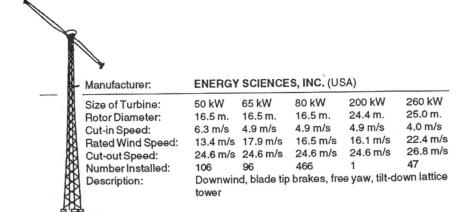

Manufacturer:	**ENERGY SCIENCES, INC.** (USA)				
Size of Turbine:	50 kW	65 kW	80 kW	200 kW	260 kW
Rotor Diameter:	16.5 m.	16.5 m.	16.5 m.	24.4 m.	25.0 m.
Cut-in Speed:	6.3 m/s	4.9 m/s	4.9 m/s	4.9 m/s	4.0 m/s
Rated Wind Speed:	13.4 m/s	17.9 m/s	16.5 m/s	16.1 m/s	22.4 m/s
Cut-out Speed:	24.6 m/s	24.6 m/s	24.6 m/s	24.6 m/s	26.8 m/s
Number Installed:	106	96	466	1	47
Description:	Downwind, blade tip brakes, free yaw, tilt-down lattice tower				

Figure C-5.

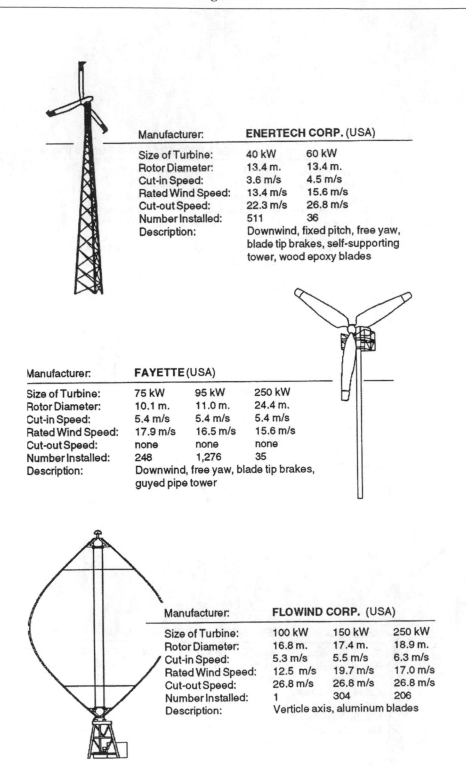

Manufacturer:	**ENERTECH CORP.** (USA)	
Size of Turbine:	40 kW	60 kW
Rotor Diameter:	13.4 m.	13.4 m.
Cut-in Speed:	3.6 m/s	4.5 m/s
Rated Wind Speed:	13.4 m/s	15.6 m/s
Cut-out Speed:	22.3 m/s	26.8 m/s
Number Installed:	511	36
Description:	Downwind, fixed pitch, free yaw, blade tip brakes, self-supporting tower, wood epoxy blades	

Manufacturer:	**FAYETTE** (USA)		
Size of Turbine:	75 kW	95 kW	250 kW
Rotor Diameter:	10.1 m.	11.0 m.	24.4 m.
Cut-in Speed:	5.4 m/s	5.4 m/s	5.4 m/s
Rated Wind Speed:	17.9 m/s	16.5 m/s	15.6 m/s
Cut-out Speed:	none	none	none
Number Installed:	248	1,276	35
Description:	Downwind, free yaw, blade tip brakes, guyed pipe tower		

Manufacturer:	**FLOWIND CORP.** (USA)		
Size of Turbine:	100 kW	150 kW	250 kW
Rotor Diameter:	16.8 m.	17.4 m.	18.9 m.
Cut-in Speed:	5.3 m/s	5.5 m/s	6.3 m/s
Rated Wind Speed:	12.5 m/s	19.7 m/s	17.0 m/s
Cut-out Speed:	26.8 m/s	26.8 m/s	26.8 m/s
Number Installed:	1	304	206
Description:	Verticle axis, aluminum blades		

Figure C-6.

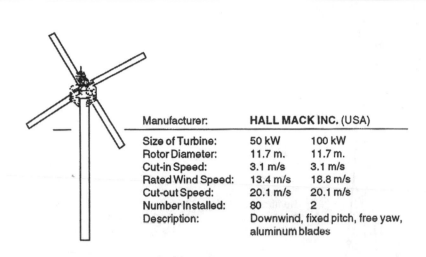

Manufacturer:	**HALL MACK INC.** (USA)	
Size of Turbine:	50 kW	100 kW
Rotor Diameter:	11.7 m.	11.7 m.
Cut-in Speed:	3.1 m/s	3.1 m/s
Rated Wind Speed:	13.4 m/s	18.8 m/s
Cut-out Speed:	20.1 m/s	20.1 m/s
Number Installed:	80	2
Description:	Downwind, fixed pitch, free yaw, aluminum blades	

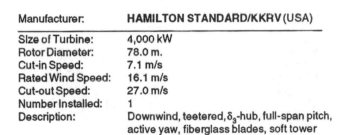

Manufacturer:	**HAMILTON STANDARD/KKRV** (USA)
Size of Turbine:	4,000 kW
Rotor Diameter:	78.0 m.
Cut-in Speed:	7.1 m/s
Rated Wind Speed:	16.1 m/s
Cut-out Speed:	27.0 m/s
Number Installed:	1
Description:	Downwind, teetered, δ_3-hub, full-span pitch, active yaw, fiberglass blades, soft tower

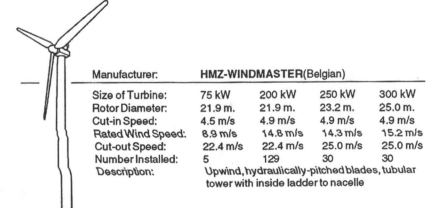

Manufacturer:	**HMZ-WINDMASTER** (Belgian)			
Size of Turbine:	75 kW	200 kW	250 kW	300 kW
Rotor Diameter:	21.9 m.	21.9 m.	23.2 m.	25.0 m.
Cut-in Speed:	4.5 m/s	4.9 m/s	4.9 m/s	4.9 m/s
Rated Wind Speed:	8.9 m/s	14.8 m/s	14.3 m/s	15.2 m/s
Cut-out Speed:	22.4 m/s	22.4 m/s	25.0 m/s	25.0 m/s
Number Installed:	5	129	30	30
Description:	Upwind, hydraulically-pitched blades, tubular tower with inside ladder to nacelle			

Figure C-7.

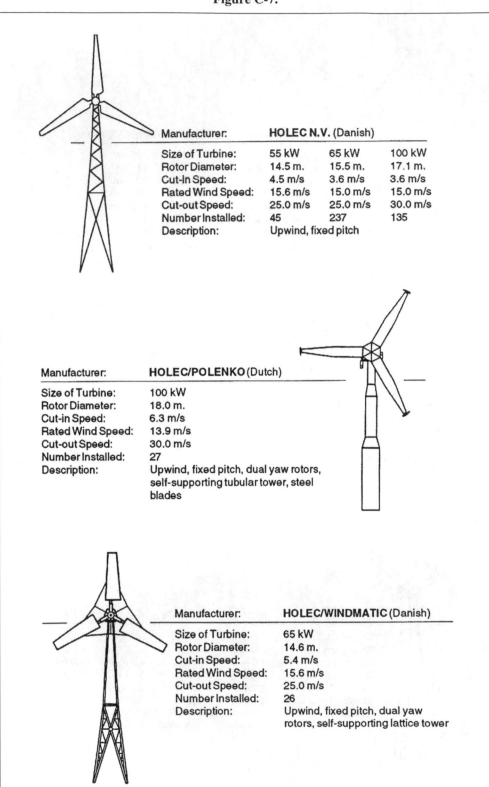

Manufacturer: **HOLEC N.V.** (Danish)

Size of Turbine:	55 kW	65 kW	100 kW
Rotor Diameter:	14.5 m.	15.5 m.	17.1 m.
Cut-In Speed:	4.5 m/s	3.6 m/s	3.6 m/s
Rated Wind Speed:	15.6 m/s	15.0 m/s	15.0 m/s
Cut-out Speed:	25.0 m/s	25.0 m/s	30.0 m/s
Number Installed:	45	237	135
Description:	Upwind, fixed pitch		

Manufacturer: **HOLEC/POLENKO** (Dutch)

Size of Turbine:	100 kW
Rotor Diameter:	18.0 m.
Cut-in Speed:	6.3 m/s
Rated Wind Speed:	13.9 m/s
Cut-out Speed:	30.0 m/s
Number Installed:	27
Description:	Upwind, fixed pitch, dual yaw rotors, self-supporting tubular tower, steel blades

Manufacturer: **HOLEC/WINDMATIC** (Danish)

Size of Turbine:	65 kW
Rotor Diameter:	14.6 m.
Cut-in Speed:	5.4 m/s
Rated Wind Speed:	15.6 m/s
Cut-out Speed:	25.0 m/s
Number Installed:	26
Description:	Upwind, fixed pitch, dual yaw rotors, self-supporting lattice tower

Figure C-8.

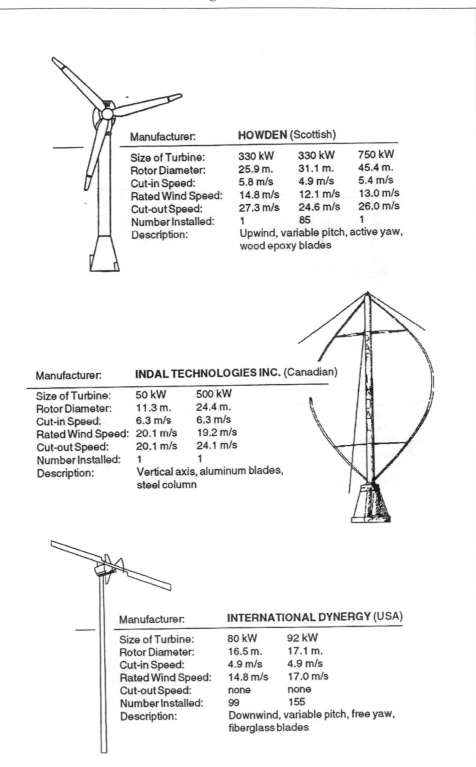

Manufacturer:	HOWDEN (Scottish)		
Size of Turbine:	330 kW	330 kW	750 kW
Rotor Diameter:	25.9 m.	31.1 m.	45.4 m.
Cut-in Speed:	5.8 m/s	4.9 m/s	5.4 m/s
Rated Wind Speed:	14.8 m/s	12.1 m/s	13.0 m/s
Cut-out Speed:	27.3 m/s	24.6 m/s	26.0 m/s
Number Installed:	1	85	1
Description:	Upwind, variable pitch, active yaw, wood epoxy blades		

Manufacturer:	INDAL TECHNOLOGIES INC. (Canadian)	
Size of Turbine:	50 kW	500 kW
Rotor Diameter:	11.3 m.	24.4 m.
Cut-in Speed:	6.3 m/s	6.3 m/s
Rated Wind Speed:	20.1 m/s	19.2 m/s
Cut-out Speed:	20.1 m/s	24.1 m/s
Number Installed:	1	1
Description:	Vertical axis, aluminum blades, steel column	

Manufacturer:	INTERNATIONAL DYNERGY (USA)	
Size of Turbine:	80 kW	92 kW
Rotor Diameter:	16.5 m.	17.1 m.
Cut-in Speed:	4.9 m/s	4.9 m/s
Rated Wind Speed:	14.8 m/s	17.0 m/s
Cut-out Speed:	none	none
Number Installed:	99	155
Description:	Downwind, variable pitch, free yaw, fiberglass blades	

Figure C-9.

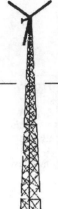

Manufacturer:	**JACOBS WIND ELECTRIC CO.** (USA)	
Size of Turbine:	17.5 kW	20 kW
Rotor Diameter:	7.9 m.	8.8 m.
Cut-in Speed:	3.6 m/s	3.6 m/s
Rated Wind Speed:	12.1 m/s	12.1 m/s
Cut-out Speed:	33.5 m/s	33.5 m/s
Number Installed:	346	274
Description:	Upwind, variable pitch, free yaw, wood blades	

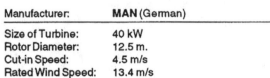

Manufacturer:	**MAN** (German)
Size of Turbine:	40 kW
Rotor Diameter:	12.5 m.
Cut-in Speed:	4.5 m/s
Rated Wind Speed:	13.4 m/s
Cut-out Speed:	None
Number Installed:	298
Description:	Upwind, variable pitch, fan-driven yaw, fiberglass blades

Manufacturer:	**MICON ENERGY SYSTEMS** (Danish)			
Size of Turbine:	60 kW	65 kW	108 kW	108 kW
Rotor Diameter:	15.8 m.	16.0 m.	18.9 m.	19.3 m.
Cut-in Speed:	4.0 m/s	4.0 m/s	4.0 m/s	3.6 m/s
Rated Wind Speed:	15.2 m/s	14.8 m/s	16.5 m/s	13.4 m/s
Cut-out Speed:	None	None	None	None
Number Installed:	221	401	586	23
Description:	Upwind, fixed pitch, active yaw, fiberglass blades, tubular tower with inside ladder to nacelle			

Figure C-10.

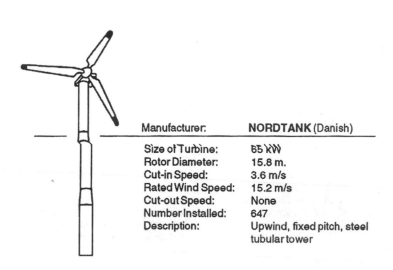

Manufacturer:	**NORDTANK** (Danish)
Size of Turbine:	65 kW
Rotor Diameter:	15.8 m.
Cut-in Speed:	3.6 m/s
Rated Wind Speed:	15.2 m/s
Cut-out Speed:	None
Number Installed:	647
Description:	Upwind, fixed pitch, steel tubular tower

Manufacturer:	**RIISAGER-VINDMOLLER** (Danish)
Size of Turbine:	90 kW
Rotor Diameter:	15.5 m.
Cut-in Speed:	6.0 m/s
Rated Wind Speed:	11.2 m/s
Cut-out Speed:	none
Number Installed:	13
Description:	Upwind, variable pitch, fan-driven yaw, wood blades

Manufacturer:	**U.S. WINDPOWER** (USA)	
Size of Turbine:	50 kW	100 kW
Rotor Diameter:	17.1 m.	17.1 m.
Cut-in Speed:	5.4 m/s	5.4 m/s
Rated Wind Speed:	9.8 m/s	13.0 m/s
Cut-out Speed:	19.7 m/s	19.7 m/s
Number Installed:	407	3,500
Description:	Downwind, free yaw, variable pitch blades, remote computer control, tripod tower	

Figure C-11.

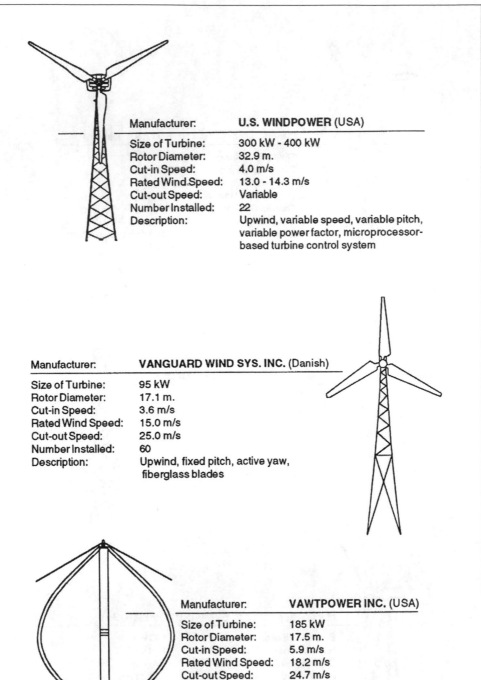

Manufacturer: **U.S. WINDPOWER** (USA)

Size of Turbine: 300 kW - 400 kW
Rotor Diameter: 32.9 m.
Cut-in Speed: 4.0 m/s
Rated Wind Speed: 13.0 - 14.3 m/s
Cut-out Speed: Variable
Number Installed: 22
Description: Upwind, variable speed, variable pitch,
                        variable power factor, microprocessor-
                        based turbine control system

Manufacturer: **VANGUARD WIND SYS. INC.** (Danish)

Size of Turbine: 95 kW
Rotor Diameter: 17.1 m.
Cut-in Speed: 3.6 m/s
Rated Wind Speed: 15.0 m/s
Cut-out Speed: 25.0 m/s
Number Installed: 60
Description: Upwind, fixed pitch, active yaw,
                         fiberglass blades

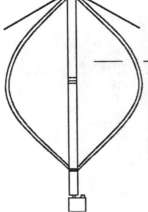

Manufacturer: **VAWTPOWER INC.** (USA)

Size of Turbine: 185 kW
Rotor Diameter: 17.5 m.
Cut-in Speed: 5.9 m/s
Rated Wind Speed: 18.2 m/s
Cut-out Speed: 24.7 m/s
Number Installed: 40
Description: Vertical axis, aluminum
                        blades

Figure C-12.

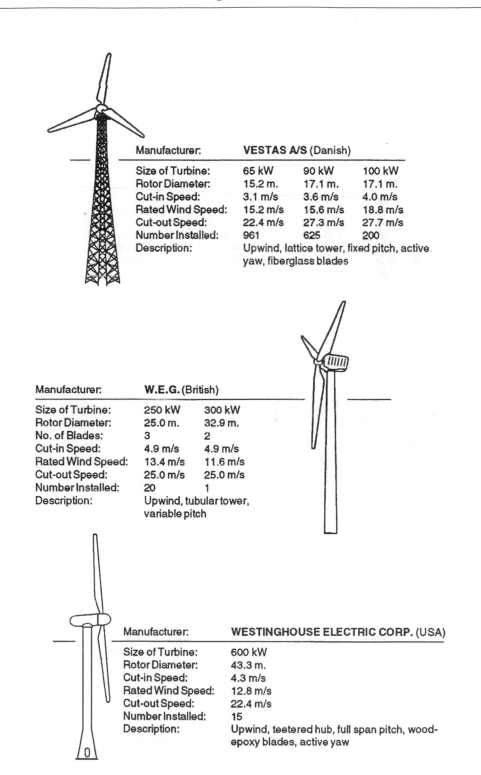

Manufacturer: **VESTAS A/S** (Danish)

Size of Turbine:	65 kW	90 kW	100 kW
Rotor Diameter:	15.2 m.	17.1 m.	17.1 m.
Cut-in Speed:	3.1 m/s	3.6 m/s	4.0 m/s
Rated Wind Speed:	15.2 m/s	15.6 m/s	18.8 m/s
Cut-out Speed:	22.4 m/s	27.3 m/s	27.7 m/s
Number Installed:	961	625	200
Description:	Upwind, lattice tower, fixed pitch, active yaw, fiberglass blades		

Manufacturer: **W.E.G.** (British)

Size of Turbine:	250 kW	300 kW
Rotor Diameter:	25.0 m.	32.9 m.
No. of Blades:	3	2
Cut-in Speed:	4.9 m/s	4.9 m/s
Rated Wind Speed:	13.4 m/s	11.6 m/s
Cut-out Speed:	25.0 m/s	25.0 m/s
Number Installed:	20	1
Description:	Upwind, tubular tower, variable pitch	

Manufacturer: **WESTINGHOUSE ELECTRIC CORP.** (USA)

Size of Turbine:	600 kW
Rotor Diameter:	43.3 m.
Cut-in Speed:	4.3 m/s
Rated Wind Speed:	12.8 m/s
Cut-out Speed:	22.4 m/s
Number Installed:	15
Description:	Upwind, teetered hub, full span pitch, wood-epoxy blades, active yaw

Figure C-13.

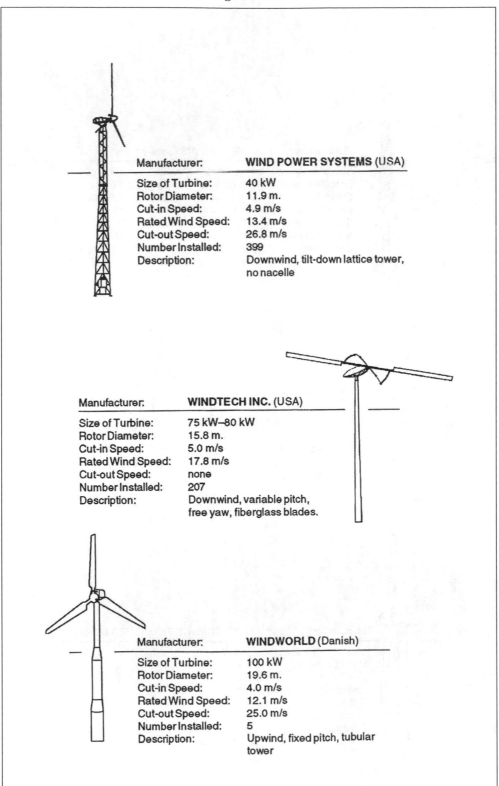

Manufacturer:	**WIND POWER SYSTEMS** (USA)
Size of Turbine:	40 kW
Rotor Diameter:	11.9 m.
Cut-in Speed:	4.9 m/s
Rated Wind Speed:	13.4 m/s
Cut-out Speed:	26.8 m/s
Number Installed:	399
Description:	Downwind, tilt-down lattice tower, no nacelle

Manufacturer:	**WINDTECH INC.** (USA)
Size of Turbine:	75 kW–80 kW
Rotor Diameter:	15.8 m.
Cut-in Speed:	5.0 m/s
Rated Wind Speed:	17.8 m/s
Cut-out Speed:	none
Number Installed:	207
Description:	Downwind, variable pitch, free yaw, fiberglass blades.

Manufacturer:	**WINDWORLD** (Danish)
Size of Turbine:	100 kW
Rotor Diameter:	19.6 m.
Cut-in Speed:	4.0 m/s
Rated Wind Speed:	12.1 m/s
Cut-out Speed:	25.0 m/s
Number Installed:	5
Description:	Upwind, fixed pitch, tubular tower

Appendix D

Airfoils Designed for Horizontal-Axis Wind Turbines

Introduction[1]

The development of special-purpose airfoils for horizontal-axis wind turbines (HAWTs) began in 1984 as a joint effort between the National Renewable Energy Laboratory (NREL), formerly the Solar Energy Research Institute (SERI), and *Airfoils, Incorporated*, State College, Pennsylvania. From 1984 to 2002, eleven airfoil families were designed for various size rotors, using the *Eppler Airfoil Design and Analysis Code*. A general performance requirement of the new airfoil families is that they exhibit a maximum lift coefficient ($C_{L,max}$) which is relatively insensitive to surface roughness effects. Together, these airfoil families address the needs of stall-regulated, variable-pitch, and variable-speed wind turbines.

For stall-regulated rotors, better peak-power control is achieved through the design of mid-span and outboard airfoils that restrain the maximum lift coefficient. Restrained maximum lift coefficients allow the use of more rotor swept area for a given generator size. Also, for stall-regulated rotors, thicker airfoil tips help accommodate devices for overspeed control.

For variable-pitch and variable-speed rotors, mid-span and outboard airfoils having high maximum lift coefficients lend themselves to lower blade solidity. Airfoils having greater thickness result in greater blade stiffness and tower clearance. Airfoils of low thickness result in less drag and less roughness sensitivity. These lower-thickness airfoils are better suited for downwind rotors.

Annual energy improvements from the NREL airfoil families are projected to be 23 percent to 35 percent for stall-regulated turbines, 8 to 20 percent for variable-pitch turbines, and 8 percent to 10 percent for variable-rpm turbines. The improvement for stall-regulated turbines has been verified in field tests.

Table D-1 lists the eleven families of NREL airfoils, with recommended blade lengths and generator sizes for each family. Also listed in this table are the specific airfoils designed to be located at blade sections from root to tip. Figures D-1 to D-11 illustrate the wide variety of section contours exhibited by these airfoils, together with their design specifications.

---

[1] Excerpts from Tangler, J. L. and D. M. Somers, 1995, *NREL Airfoil Families for HAWTs*, NREL/TP-442-7109, Golden, Colorado; also in *Proceedings, Windpower '95 Conference,* American Wind Energy Association: National Renewable Energy Laboratory.

Table D-1

Blade Length (m)	Generator Size (kW)	Thickness category	Airfoil Family			Tip Max Lift Coef. $C_{L,max}$
			Root Section	Intermediate Section	Tip Section	
1-5	2 – 20	Thick	S823		S822	Low
1-5 (a)	2 – 20	Thick	S835	S833	S834	Low
5-10	20 – 150	Thin	S804	S801	S803	High
5-10	20 – 150	Thin	S808, S807	S805A	S806A	Low
5-10	20 – 150	Thick	S821	S819	S820	Low
10-15	150 – 400	Thick	S815, S814	S809	S810	Lower
10-15	150 – 400	Thick	S815, S814	S812	S813	Low
10-15	150 – 400	Thick	S815, S814	S825	S826	High
15-30	400 – 2,000	Thick	S818	S816	S817	Low
15-30	400 – 2,000	Thick	S818	S827	S828	Lower
15-30 (a)	400 – 2,000	Thick	S818	S830, S831	S832	High

(a) Low noise trailing edge

Permission Requirements

Please note that some or all of these airfoil designs are covered by patents, and permission for their commercial use must be obtained from the National Renewable Energy Laboratory in Golden, Colorado.

References

Tangler, J., *et al.*, 1990, "Atmospheric Performance of the Special-Purpose SERI Thin-Airfoil Family: Final Results," European Wind Energy Conference, Madrid, Spain.

Tangler, J., *et al.*, 1991, "SERI Advanced Wind Turbine Blades," International Solar Energy Society Conference, Denver, Colorado.

Tangler, J., *et al.*, 1994, "Measured Structural Loads for the Micon 65/13," ASME/ETCE Wind Energy Symposium, New Orleans, Louisiana.

Eppler, R., 1990, "Airfoil Design and Data," Springer-Verlag, Berlin.

Eppler, R., 1993, "Airfoil Program System. User's Guide." (unpublished).

Somers, D., 1986, "Design and Experimental Results for the S805 Airfoil," Airfoils, Inc., State College, Pennsylvania.

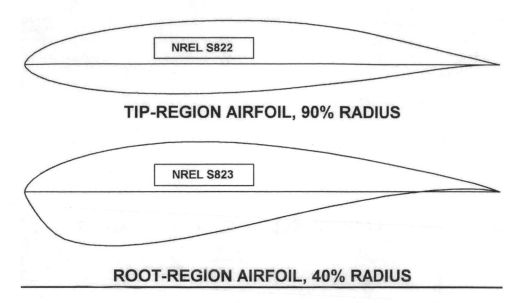

TIP-REGION AIRFOIL, 90% RADIUS

ROOT-REGION AIRFOIL, 40% RADIUS

Design Specifications

Airfoil	r/R	Re. No. (x10^6)	t/c	c_{lmax}	c_{dmin}	c_{mo}
S822	0.90	0.6	0.160	1.0	0.010	-0.07
S823	0.40	0.4	0.210	1.2	0.018	-0.15

Figure D-1. Thick-airfoil family for small-scale HAWT blades (low tip $C_{L,max}$).

Somers, D., 1987, "Design and Experimental Results for the S809 Airfoil," Airfoils, Inc., State College, Pennsylvania.

Somers, D., 1994, "Design and Experimental Results for the S814 Airfoil," Airfoils, Inc., State College, Pennsylvania.

Somers, D., 1999, "Design and Experimental Results for the S825 Airfoil," Airfoils, Inc., Port Matilda, Pennsylvania.

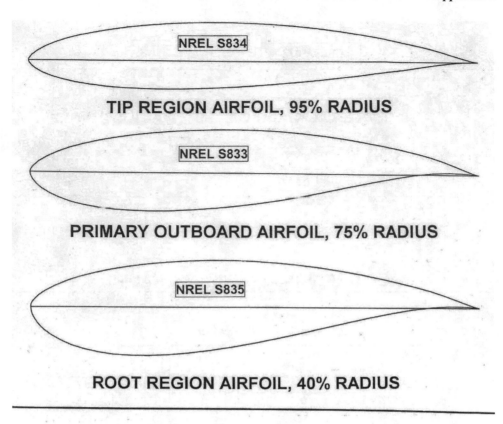

Airfoil	r/R	Re. No. (x10[6])	t/c	C_{lmax}	C_{dmin}	C_{mo}
S834	0.95	0.40	0.015	1.0	low	-0.15
S833	0.75	0.40	0.018	1.1	low	-0.15
S835	0.40	0.25	0.021	1.2	low	-0.15

Figure D-2. Low-noise thick-airfoil family for small-scale HAWT blades (low tip $C_{L,max}$).

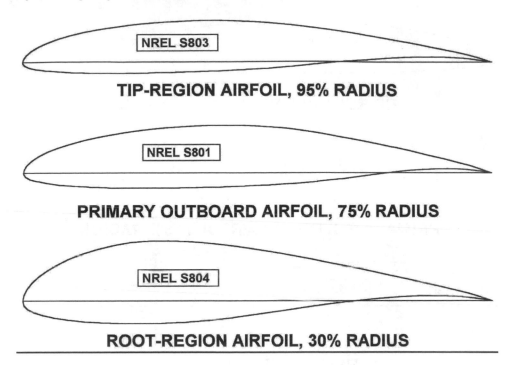

TIP-REGION AIRFOIL, 95% RADIUS

PRIMARY OUTBOARD AIRFOIL, 75% RADIUS

ROOT-REGION AIRFOIL, 30% RADIUS

Design Specifications

Airfoil	r/R	Re. No. (x10^6)	t/c	c_{lmax}	c_{dmin}	c_{mo}
S803	0.95	2.6	0.115	1.5	0.006	-0.15
S801	0.75	2.0	0.135	1.5	0.007	-0.15
S804	0.30	0.8	0.180	1.5	0.012	-0.15

Figure D-3. Thin-airfoil family for medium-scale HAWT blades (high tip $C_{L,max}$).

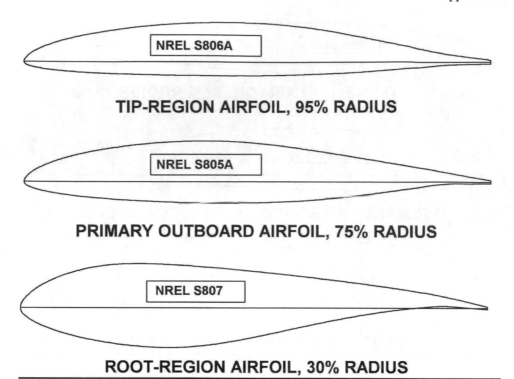

TIP-REGION AIRFOIL, 95% RADIUS

PRIMARY OUTBOARD AIRFOIL, 75% RADIUS

ROOT-REGION AIRFOIL, 30% RADIUS

Design Specifications

Airfoil	r/R	Re. No. (x10^6)	t/c	c_{lmax}	c_{dmin}	c_{mo}
S806A	0.95	1.3	0.115	1.1	0.004	-0.05
S805A	0.75	1.0	0.135	1.2	0.005	-0.05
S807	0.30	0.8	0.180	1.4	0.010	-0.10
S808	0.20	0.4	0.210	1.2	0.012	-0.12

Figure D-4. Thin-airfoil family for medium-scale HAWT blades (low tip $C_{L,max}$).

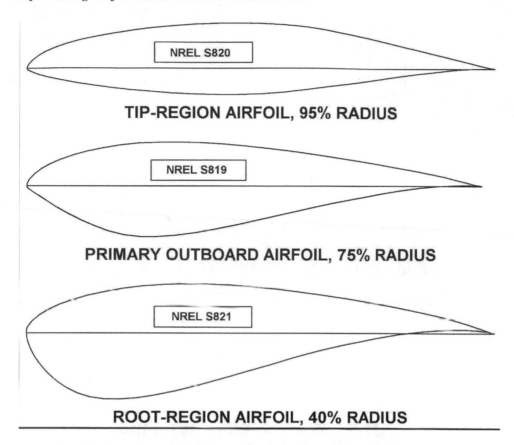

TIP-REGION AIRFOIL, 95% RADIUS

PRIMARY OUTBOARD AIRFOIL, 75% RADIUS

ROOT-REGION AIRFOIL, 40% RADIUS

Design Specifications

Airfoil	r/R	Re. No. (x10⁶)	t/c	c_{lmax}	c_{dmin}	c_{mo}
S820	0.95	1.3	0.160	1.1	0.007	-0.07
S819	0.75	1.0	0.210	1.2	0.008	-0.07
S821	0.40	0.8	0.240	1.4	0.014	-0.15

Figure D-5. Thick-airfoil family for medium-scale HAWT blades (low tip $C_{L,max}$).

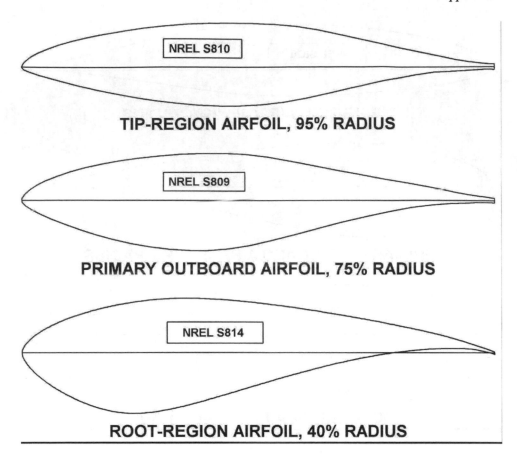

TIP-REGION AIRFOIL, 95% RADIUS

PRIMARY OUTBOARD AIRFOIL, 75% RADIUS

ROOT-REGION AIRFOIL, 40% RADIUS

Design Specifications

Airfoil	r/R	Re. No. (x10⁶)	t/c	c_{lmax}	c_{dmin}	c_{mo}
S810	0.95	2.0	0.180	0.9	0.006	-0.05
S809	0.75	2.0	0.210	1.0	0.007	-0.05
S814	0.40	1.5	0.240	1.3	0.012	-0.15
S815	0.30	1.2	0.260	1.1	0.014	-0.15

Figure D-6. Thick-airfoil family for large-scale HAWT blades (lower tip $C_{L,max}$).

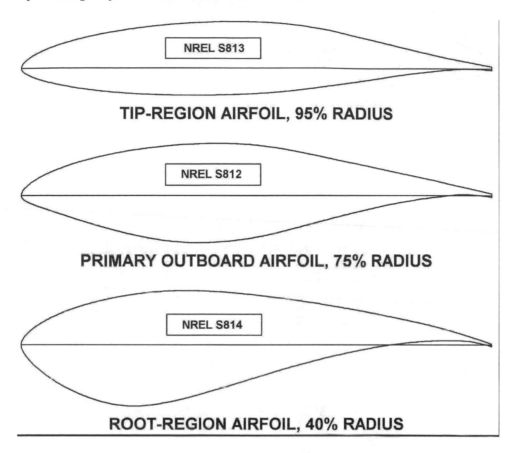

Design Specifications

Airfoil	r/R	Re. No. (x10^6)	t/c	c_{lmax}	c_{dmin}	c_{mo}
S813	0.95	2.0	0.160	1.1	0.007	-0.07
S812	0.75	2.0	0.210	1.2	0.008	-0.07
S814	0.40	1.5	0.240	1.3	0.012	-0.15
S815	0.30	1.2	0.260	1.1	0.014	-0.15

Figure D-7. Thick-airfoil family for large-scale blades (low tip $C_{L,max}$).

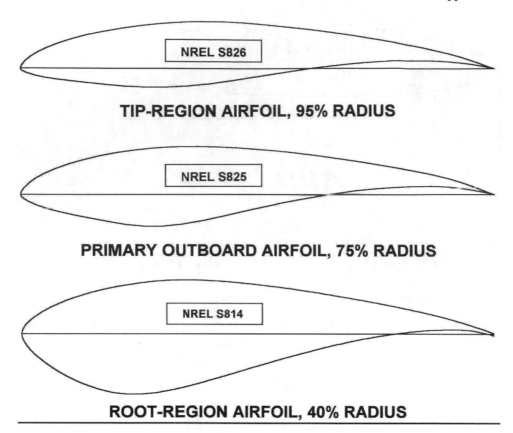

TIP-REGION AIRFOIL, 95% RADIUS

PRIMARY OUTBOARD AIRFOIL, 75% RADIUS

ROOT-REGION AIRFOIL, 40% RADIUS

Design Specifications

Airfoil	r/R	Re. No. (x10^6)	t/c	c_{lmax}	c_{dmin}	c_{mo}
S826	0.95	1.5	0.140	1.60	0.006	-0.14
S825	0.75	2.0	0.170	1.60	0.008	-0.14
S814	0.40	1.5	0.240	1.30	0.012	-0.15
S815	0.30	1.2	0.260	1.10	0.014	-0.15

Figure D-8. Thick-airfoil family for large-scale HAWT blades (high tip $C_{L,max}$).

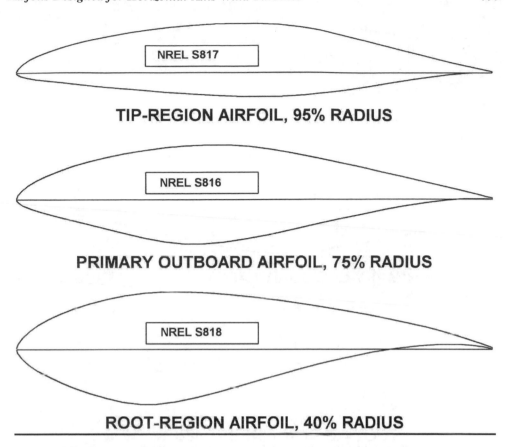

TIP-REGION AIRFOIL, 95% RADIUS

PRIMARY OUTBOARD AIRFOIL, 75% RADIUS

ROOT-REGION AIRFOIL, 40% RADIUS

Design Specifications

Airfoil	r/R	Re. No. (x10^6)	t/c	c_{lmax}	c_{dmin}	c_{mo}
S817	0.95	3.0	0.160	1.1	0.007	-0.07
S816	0.75	4.0	0.210	1.2	0.008	-0.07
S818	0.40	2.5	0.240	1.3	0.012	-0.15

Figure D-9. Thick-airfoil family for larger-scale HAWT blades (low tip $C_{L,max}$).

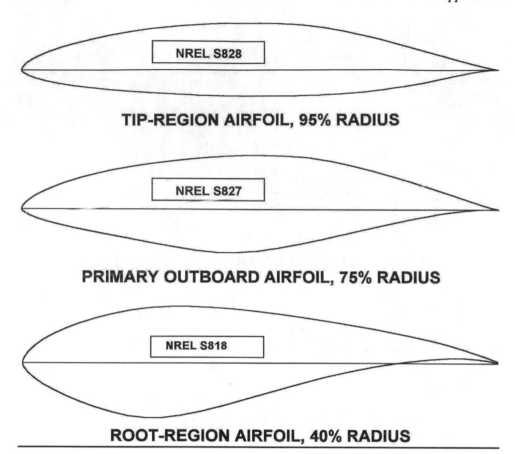

TIP-REGION AIRFOIL, 95% RADIUS

PRIMARY OUTBOARD AIRFOIL, 75% RADIUS

ROOT-REGION AIRFOIL, 40% RADIUS

Design Specifications

Airfoil	r/R	Re. No. (x10^6)	t/c	c_{lmax}	c_{dmin}	c_{mo}
S828	0.95	3.0	0.160	0.90	0.007	-0.07
S827	0.75	4.0	0.210	1.00	0.008	-0.07
S818	0.40	2.5	0.240	1.30	0.012	-0.15

Figure D-10. Thick-airfoil family for larger-scale HAWT blades (lower tip $C_{L,max}$).

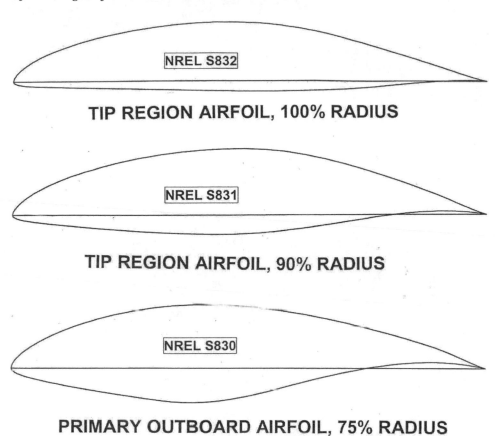

TIP REGION AIRFOIL, 100% RADIUS

TIP REGION AIRFOIL, 90% RADIUS

PRIMARY OUTBOARD AIRFOIL, 75% RADIUS

Design Specifications

Airfoil	r/R	Re. No. (x10^6)	t/c	C_{lmax}	C_{dmin}	C_{mo}
S832	1.00	2.5	0.015	1.4	low	-0.15
S831	0.90	3.5	0.018	1.5	low	-0.15
S830	0.75	4.0	0.021	1.6	low	-0.15
S818	0.40	2.5	0.240	1.3	0.012	-0.15

Figure D-11. Low-noise thick-airfoil family for larger-scale HAWT blades (high tip $C_{L,max}$).

Appendix E

Gallery of Representative
Commercial Wind Turbines

Figure E-1. (*Courtesy of Aerostar, Inc.*)

Designation: Aerostar 6 Meter Wind Turbine

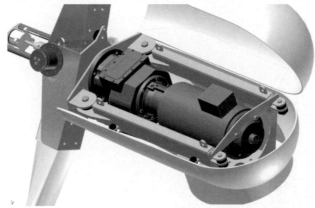

Rated Power: 10 kW   Rotor Speed: 170 rpm
Rotor Dia.: 6.7 m   Hub Height: 30 m - 43 m
SLS Wind Speeds: Cut-In .....3.6 m/s   Rated.... 13 m/s   Cut-Out ......23.5 m/s

Figure E-2. (*Courtesy of Atlantic Orient Canada, Inc.*)

Designation: Atlantic Orient AOC 15/50 Wind Turbine

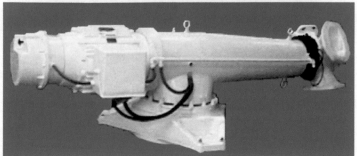

Rated Power: 50 kW   Rotor Speed: 62 rpm at rated wind speed
Rotor Dia.: 15m   Hub Height: 25m (Typical)
SLS Wind Speeds: Cut-In ......4.6 m/s   Rated .... 16.0 m/s   Cut-Out...... 22.4 m/s

Figure E-3. (*Courtesy of Bergey WindPower, Inc.*)

Designation: Bergey BWC Excel Wind Turbine

Rated Power: 10 kW   Rotor Speed: 310 rpm at Rated Power
Rotor Dia.: 6.7 m;   Hub Height: 18 – 43 m
SLS Wind Speeds: Cut-In ..... 3.1 m/s   Rated .... 13.8 m/s   Furling ..... 15.6 m/s

Figure E-4. (*Courtesy of Clipper Windpower, Inc.*)

Designation: Clipper Liberty 2.5 MW Wind Turbine

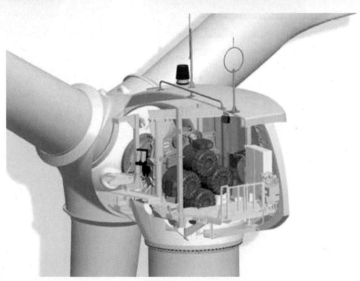

Rated Power: 2.5 MW   Rotor Speed: 9.6 – 15.5 rpm
Rotor Dia.: 89, 93, 96, and 99 m   Hub Height: 80 m; Other Options Available
SLS Wind Speeds: Cut-In ..... 4 m/s   Rated .....13 m/s   Cut-Out......25 m/s

Figure E-5. (*Courtesy of Composite Technology Corporation/DeWind Div.*)

Designation: DeWind D8.2 Wind Turbine

Rated Power: 2.0 MW   Rotor Speed: 11.1 – 20.7 rpm
Rotor Dia.: 80 m   Hub Height: 80 and 100 m
SLS Wind Speeds: Cut-In ....3 m/s; Rated . ... 13.5 m/s; Cut-Out .... 25 m/s

Figure E-6. (*Courtesy of Enertech*)

Designation: Enertech E48 Wind Turbine

Rated Power: 600 kW
Rotor Speed: 15/23 rpm
Rotor Dia.: 48 m
Hub Height: 50/60 m
SLS Wind Speeds:
   Cut-In ....3 m/s
   Rated...................12.5 m/s
   Cut-Out..................25 m/s

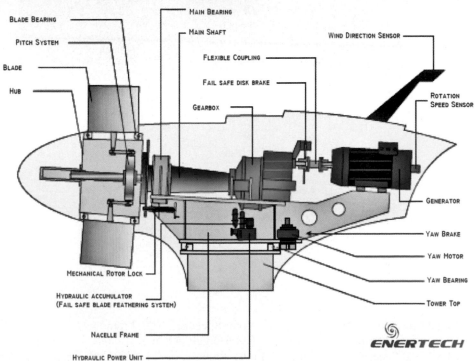

Figure E-7. (*Courtesy of GE Energy, Inc.*)

Designation: GE 1.5s Wind Turbine

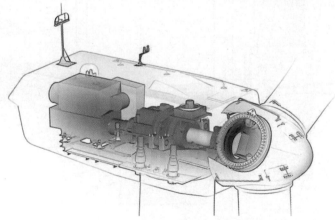

Rated Power: 1.5 MW   Rotor Speed: 12 – 22.2 rpm
Rotor Dia.: 70.5m   Hub Height: 64.7m
SLS Wind Speeds: Cut-In ..... 4 m/s; Rated .... 13 m/s; Cut-Out..........25 m/s

Figure E-8. (*Courtesy of Green Energy Technologies, Inc.*)

Designation: Green Energy Technologies WindCube Wind Turbine

Rated Power: 53 kW   Rotor Speed: 100 – 290 rpm
Rotor Dia.: 4.6 m   Hub Height: Customized for Site
SLS Wind Speeds: Cut-In ...... 2.1 m/s; Rated ......12 m/s; Cut-Out......... 15 m/s

Figure E-9. (*Courtesy of Wind Turbine Industries Corporation*)

Designation: Jacobs Model 31-20 Wind Turbine

Rated Power: 20 kW   Rotor Speed: 54 – 175 rpm
Rotor Dia.: 9.4 m   Hub Height: 27 to 39 m
SLS Wind Speeds: Cut-In ....3.6 m/s   Rated .....11.6 m/s   Furling....18 – 20 m/s

Figure E-10. (*Courtesy of Multibrid GmbH*)

Designation: Multibrid M5000 Wind Turbine

Rated Power: 5.5 MW   Rotor Speed: 5.9 – 14.8 rpm
Rotor Dia.: 116 m   Hub Height: 102 m (Onshore Prototype)
SLS Wind Speeds: Cut-In ....4 m/s; Rated 12 m/s; Cut-Out..........25 m/s

Figure E-11. (*Courtesy of Nordic Windpower, Ltd.*)

Designation: Nordic N1000 Wind Turbine

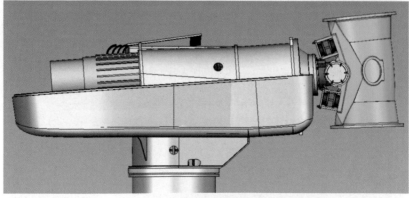

Rated Power: 1.0 MW   Rotor Speed: 25 rpm, 15 rpm
Rotor Dia.: 54 m, 59 m   Hub Height: 70 m (Standard)
SLS Wind Speeds: Cut-In ..... 4.0 m/s Rated .... 16.0 m/s   Cut-Out.... 25 m/s, 22 m/s

Figure E-12. (*Courtesy of REpower Systems AG*)

Designation: REpower 5M Wind Turbine

Rated Power: 5 MW   Rotor Speed: 6.9 – 12.1 rpm
Rotor Dia.: 126 m   Hub Height: 100 -120 m Onshore; 90 –100 m Offshore
SLS Wind Speeds: Cut-In ..... 3.5 m/s   Rated . 13 m/s   Cut-Out........ 25 - 30 m/s

Figure E-13. (*Courtesy of Siemens Energy, Wind Power Business Unit*)

Designation: Siemens SWT-2.3-93 Wind Turbine

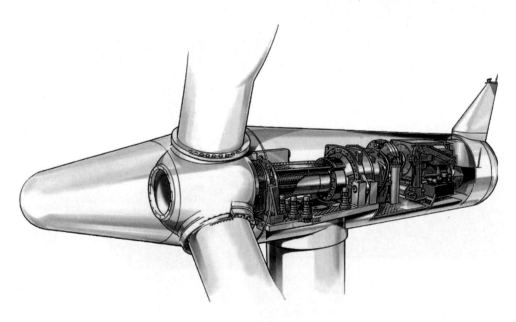

Rated Power: 2.3 MW   Rotor Speed: 6 - 16 rpm
Rotor Dia.: 93 m   Hub Height: 70 m, 80 m or site-specific
SLS Wind Speeds: Cut-In ... 4 m/s   Rated .....13-14 m/s   Cut-Out ..... 25 m/s

Figure E-14. (*Courtesy of Southwest Windpower, Inc.*)

Designation: Southwest Skystream 3.7 Wind Turbine

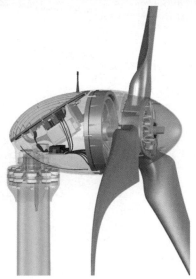

Rated Power: 2.4 kW   Rotor Speed: 50 - 330 rpm
Rotor Dia.: 3.7 m   Hub Height: 10 – 30 m
SLS Wind Speeds: Cut-In .... 3.5 m/s; Rated .... 13.0 m/s; Cut-Out .... 25.0 m/s

Figure E-15. (*Courtesy of Vergnet SA*)

Designation: Vergnet GEV HP 1 MW Wind Turbine

Rated Power: 1.0 MW   Rotor Speed: 12 – 23 rpm
Rotor Dia.: 62 m   Hub Height: 70 m
SLS Wind Speeds: Cut-In .... 3 m/s; Rated ....15 m/s; Cut-Out......... 25 m/s

Figure E-16. (*Courtesy of Vestas Wind Systems A/S*)

Designation: Vestas V90 Wind Turbine

Rated Power: 3.0 MW
Rotor Dia.: 90 m

Figure E-17. (*Courtesy of Wind Turbine Company*)

Designation: Wind Turbine Company WTC 250 Wind Turbine

Rated Power: 250 kW   Rotor Speed: 35 rpm
Rotor Dia.: 32 m   Hub Height: 40 - 65 m
SLS Wind Speeds: Cut-In .... 4 m/s; Rated . ... 11m/s; Cut-Out .... 24 m/s

Figure E-18. (*Courtesy of Windflow Technology Ltd*)

Designation: Windflow 500 Wind Turbine

IEC class 1A turbine

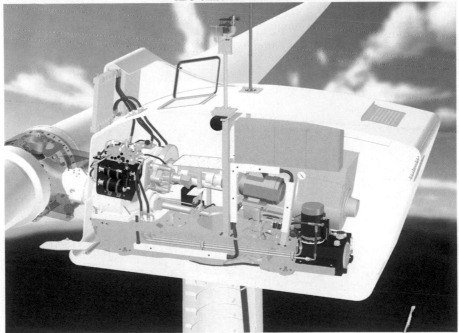

Rated Power: 500 kW   Rotor Speed: 48 – 50 rpm
Rotor Dia.: 33.2 m   Hub Height: 30 m
SLS Wind Speeds: Cut-In ..... 5.5 m/s   Rated .... 13.7 m/s;   Cut-Out ..... 30 m/s

Figure E-19. (*Courtesy of WinWinD Oy.*)

Designation: WinWinD WWD-3 Wind Turbine

Rated Power: 3.0 MW
Rotor Speed: 5 – 16 rpm
Rotor Dia.: 90 and 100 m
Hub Height: 80 – 100 m
SLS Wind Speeds:
     Cut-In .... 4 m/s
     Rated ....12.5 - 13 m/s
     Cut-Out..... 20-25 m/s

**Companion Product:
WWD-1 Wind Turbine**
Rated Power: 1.0 MW
Rotor Speed: 7.7 – 25.6 rpm
Rotor Dia.: 56, 60, and 64 m
Hub Height: 56, 66 and 70 m
SLS Wind Speeds:
     Cut-In................ 4 m/s
     Rated............ 12.5 m/s
     Cut-Out......20-25 m/s

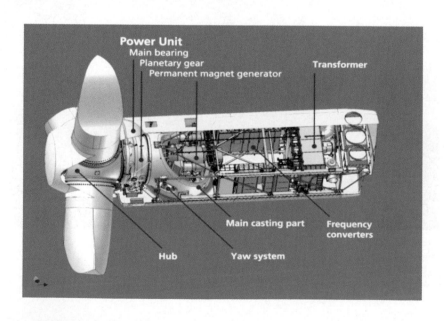

Index